ARSENIC IN THE ENVIRONMENT: BRIDGING SCIENCE TO PRACTICE FOR SUSTAINABLE DEVELOPMENT

The Congress and Exhibition Series "Arsenic in the Environment" offers an international, multi- and interdisciplinary discussion platform for research and innovation aimed towards a holistic solution to the challenges posed by the environmental toxin arsenic, with global societal impact. The Congress has focused on cutting edge and breakthrough research in physical, chemical, toxicological, medical, agricultural and other specific issues on arsenic across a broader environmental realm. The Biennial Congress and Exhibition "Arsenic in the Environment" was first organized in Mexico City (As2006) followed by As2008 in Valencia (Spain), As2010 in Tainan (Taiwan, Republic of China), As2012 in Cairns (Australia), As2014 in Buenos Aires (Argentina), As2016 in Stockholm (Sweden) and As2018 in Beijing (P.R. China).

The 8th International Congress As2021 was held June 7–9, 2021 (first time digitally owing to the global COVID-19 pandemic, in Wageningen, The Netherlands) and with a title *Arsenic in the Environment – Bridging Science to Practice for Sustainable Development.*

The Congress addressed the broader context of arsenic research aligned on the following themes:

- Theme 1: Arsenic in Natural Soil and Water Systems
- Theme 2: Arsenic in Agriculture and Food Production
- Theme 3: Health Impacts of Arsenic
- Theme 4: Technologies for Arsenic Removal from Water
- Theme 5: Sustainable Mitigation and Management for Sustainable Development

Arsenic in drinking water and food is a major health issue, affecting millions of people in many parts of the world. In recent years serious cases of arsenic exposure through different environmental matrices have been reported from, for example, Argentina, Bangladesh, Chile, China, Taiwan, Turkey, India, Mexico, UK, USA, Pakistan, Vietnam as well as other regions in the world. Arsenic can cause a number of carcinogenic and non-carcinogenic adverse effects on human health and therefore human exposure to arsenic should be avoided. Notably, The Netherlands has been in the forefront of research on arsenic removal technology and developed a cutting edge innovation to remove arsenic to levels below the WHO drinking water guideline to as low as less than 1 μg/L. This has created an enabling environment to discuss on policy issues for defining the new drinking water guideline. The Congress has attracted professionals involved in different segments of interdisciplinary research on arsenic in an open forum, and strengthened relations between academia, research institutions, government and non-governmental agencies, industries, and civil society organizations to share an optimal ambience for exchange of knowledge.

Arsenic in the Environment – Proceedings

Series Editors

Jochen Bundschuh
UNSECO Chair on Groundwater Arsenic within the 2030 Agenda for Sustainable Development & Faculty of Health, Engineering and Sciences, The University of Southern Queensland, Toowoomba, Queensland, Australia

Prosun Bhattacharya
KTH-International Groundwater Arsenic Research Group, Department of Sustainable Development, Environmental Science and Engineering, KTH Royal Institute of Technology, Stockholm, Sweden
KWR Water Research Institute, Nieuwegein, The Netherlands

ISSN: 2154-6568

Cover photo

The cover photo shows the dune area along the western coast of Schouwen-Duiveland, which is one of the islands of the province of Zeeland. These dune are used for infiltration of surface water for the production of drinking water. The surface water resides in the dunes for at least 60 days and during this time the water becomes fully anaerobic and disinfected. During transport in the dunes from the infiltration ponds to the extraction wells, the arsenic level in the water increases up to concentrations of several micrograms per liter due to leaching of arsenic from the dune sediments. After oxidation, the arsenic gets removed in the water treatment plant so that the arsenic level in the drinking water remains below 1 microgram per liter.

Photo taken by Jeroen Willemsen and owned by Evides Water Company.

First published 2023
by CRC Press/Balkema
4 Park Square, Milton Park, Abingdon, Oxon, OX14 4RN
and by CRC Press/Balkema
2385 NW Executive Center Drive, Suite 320, Boca Raton FL 33431

CRC Press/Balkema is an imprint of the Taylor & Francis Group, an informa business

Library of Congress Cataloging-in-Publication Data
A catalog record has been requested for this book

ISBN: 978-1-032-32928-4 (hbk)
ISBN: 978-1-032-32930-7 (pbk)
ISBN: 978-1-003-31739-5 (ebk)

DOI: 10.1201/9781003317395

Typeset in Times New Roman
by MPS Limited, Chennai, India

8TH INTERNATIONAL CONGRESS AND EXHIBITION: ARSENIC IN THE ENVIRONMENT, WAGENINGEN, THE NETHERLANDS, JUNE 7–9, 2021

Arsenic in the Environment: Bridging Science to Practice for Sustainable Development

Edited by

Albert van der Wal

Department of Environmental Technology, Agrotechnology and Food Sciences Group, Wageningen University and Research (WUR), Wageningen, The Netherlands
Evides Water Company, Research and Development, Water technology, Rotterdam, The Netherlands

Arslan Ahmad

KTH-International Groundwater Arsenic Research Group, Department of Sustainable Development, Environmental Sciences and Engineering, KTH Royal Institute of Technology, Stockholm, Sweden
SIBELCO Ankerpoort NV, Maastricht, The Netherlands (formerly at KWR Watercycle Research Institute, Nieuwegein, The Netherlands)

Branislav Petrusevski

IHE Delft Institute for Water Education, Delft, The Netherlands

Jan Weijma

Department of Environmental Technology, Agrotechnology and Food Sciences Group, Wageningen University and Research (WUR), Wageningen, The Netherlands

Dragan Savic

KWR Water Research Institute, Nieuwegein, The Netherlands

Patrick van der Wens

Brabant Water NV, Breda, The Netherlands

Erwin Beerendonk

KWR Water Research Institute, Nieuwegein, The Netherlands

Prosun Bhattacharya

KTH-International Groundwater Arsenic Research Group, Department of Sustainable Development, Environmental Sciences and Engineering, KTH Royal Institute of Technology, Stockholm, Sweden
KWR Water Research Institute, Nieuwegein, The Netherlands

Jochen Bundschuh

Former UNSECO Chair on Groundwater Arsenic within the 2030 Agenda for Sustainable Development & Faculty of Health, Engineering and Sciences, The University of Southern Queensland, Toowoomba, Australia

Ravi Naidu

Global Centre for Environmental Remediation (GCER), Faculty of Science & Information Technology, The University of Newcastle, Callaghan, NSW, Australia
Research Centre for Contamination Assessment and Remediation of the Environment (CRC CARE), University of Newcastle, Newcastle, New South Wales, Australia

CRC Press
Taylor & Francis Group
Boca Raton London New York Leiden

CRC Press is an imprint of the Taylor & Francis Group, an informa business
A BALKEMA BOOK

Table of Contents

1.4 Arsenic mobility and fate in soils, sediments and mining wastes

1.3 Arsenic and other trace elements in global groundwaters

Section 2: Arsenic in agriculture and food production

2.1 Processes and pathways of arsenic in agricultural ecosystems

2.2 Dynamics of arsenic in rhizosphere

2.3 Microbial ecology and molecular mechanisms of arsenic biotransformation in soils and plant uptake

1.5 Advances and challenges in arsenic analysis in solid and aqueous media

3.2 Genetic predisposition of chronic arsenic poisoning

3.3 Reliable biomarkers for arsenic exposure

3.4 Neurophysiological and IQ impacts of arsenic

3.5 Risk assessment of chronic ingestion

4.2 Nanomaterials related novel applications in arsenic treatment

4.3 *Biochar and Chitosan-based novel technologies for removal of arsenic*

4.4 *Membrane technologies and applications*

4.5 *Biological processes in arsenic removal*

4.6 *Pilot studies*

Section 5: Sustainable mitigation and management
5.1 Approaches for system strengthening and scaling up drinking water safety in Bangladesh

5.2 Arsenic in drinking water sources: citizen science and community based interventions

5.3 Arsenic in drinking water: technology and management challenges

 ISBN 978-1-032-32928-4

About the book series

Although arsenic has been known as a 'silent toxin' since ancient times, and the contamination of drinking water resources by geogenic arsenic was described in different locations around the world long ago—e.g. in Argentina in 1914—it was only three decades ago that it received overwhelming worldwide public attention. As a consequence of the biggest arsenic calamity in the world, which was detected more than thirty years back in West Bengal, India and other parts of Southeast Asia, there has been an exponential rise in scientific interest that has triggered high quality research. Since then, arsenic contamination (predominantly of geogenic origin) of drinking water resources, soils, plants and air, the propagation of arsenic in the food chain, the chronic effects of arsenic ingestion by humans, and their toxicological and related public health consequences, have been described in many parts of the world, and every year, even more new countries or regions are discovered to have elevated levels of arsenic in environmental matrices.

Arsenic is found as a drinking water contaminant, in many regions all around the world, in both developing as well as industrialized countries. However, addressing the problem requires different approaches which take into account the differential economic and social conditions in both country groups. It has been estimated that hundreds of millions of people worldwide are at risk from drinking water containing high concentrations of As, well above the recommended maximum level of 10 μg/L, by the World Health Organization.

The book series "Arsenic in the Environment – Proceedings" is an inter- and multidisciplinary source of information, making an effort to link the occurrence of geogenic arsenic in different environments and the potential contamination of ground- and surface water, soil and air and their effect on the human society. The series fulfills the growing interest in the worldwide arsenic issue, which is being accompanied by stronger regulations on the permissible Maximum Contaminant Levels (MCL) of arsenic in drinking water and food, which are being adopted not only by the industrialized countries, but increasingly also by developing countries.

Consequently, we see the book series *Arsenic in the Environment-Proceedings* with the outcomes of the International Congress Series – Arsenic in the Environment, which are organized biennially in different parts of the world, as a regular update on the latest developments of arsenic research. It is further a platform to present the results from international or regional congresses or other scientific events. This Proceedings series acts as an ideal complement to the books of the series *Arsenic in the Environment*, which includes authored or edited books from world-leading scientists on their specific field of arsenic research, giving a comprehensive information base. Supported by a strong multi-disciplinary editorial board, book proposals and manuscripts are peer reviewed and evaluated. Both of the two series will be open for any person, scientific association, society or scientific network, for the submission of new book projects.

We have an ambition to establish an international, multi- and interdisciplinary source of knowledge and a platform for arsenic research oriented to the direct solution of problems with considerable social impact and relevance rather than simply focusing on cutting edge and breakthrough research in physical, chemical, toxicological and medical sciences. It shall form a consolidated source of information on the worldwide occurrences of arsenic, which otherwise is dispersed and often hard to access. It will also play a role in increasing the awareness and knowledge of the arsenic problem among administrators, policy makers and company executives and improving international and bilateral cooperation on arsenic contamination and its effects.

Both of the book series cover all fields of research concerning arsenic in the environment and aims to present an integrated approach from its occurrence in rocks and mobilization into the ground- and surface water, soil and air, its transport therein, and the pathways of arsenic introduction into the food chain including uptake by humans. Human arsenic exposure, arsenic bioavailability, metabolism and toxicology are treated together with related public health effects and risk assessments in order to better manage the contaminated land and aquatic environments and to reduce human arsenic exposure. Arsenic removal technologies and other methodologies to mitigate the arsenic problem are addressed not only from the technological perspective, but also from an economic and social point of view. Only such inter- and multidisciplinary approaches will allow a case-specific selection of optimal mitigation measures for each specific arsenic problem and provide the local population with arsenic-safe drinking water, food, and air

Jochen Bundschuh
Prosun Bhattacharya
(Series Editors)

 ISBN 978-1-032-32928-4

Dedication

Gunnar Jacks, Ph.D.
Groundwater chemist with an exceptional dedication to work for the underpreviledged in the global south
Professor Emiritus, Department of Sustainable Development Environmental Science and Engineering, KTH Royal Institute of Technology, Stockholm, Sweden
* 13 April 1937 † 18 September 2019

We dedicate the 8th International Congress on Arsenic in the Environment (As 2021) and the Volume of Proceedings of the International Congress of Arsenic in the Environment – *Arsenic in the Environment: Bridging Science to Practice for Sustainable Development (As 2021)* to the memory of Professor Dr. Gunnar Jacks (popularly known as Gunnar), who passed away on the 18th of September 2019 at the age of 82.

Dr. Gunnar Jacks accomplished his basic education with a M.Sc. in mining engineering from KTH Royal Institute of Technology Stockholm in 1963. Soon after his graduation, he engaged himself in humanitarian work in Turkey between 1965–1967. He returned back to KTH Royal Institute of Technology where in 1973 he defended his PhD thesis entitled "Chemistry of Groundwater in Hard Rocks". During the period 1976–1978, he worked for Central Ground Water Board of India on a full-time basis stationed in Coimbatore, in the southern state of Tamil Nadu, India. After the completion of this overseas assignment, he joined KTH in 1978 as Associate Professor, but continued his affiliation with the Central Groundwater Board until 1984. In 1986, he was appointed as a full Professor of Groundwater Chemistry at the Department of Land and Water Resources at KTH Royal Institute of Technology. In 1994, he was appointed as the head of the Groundwater Chemistry Division within the Department of Civil and Environmental Engineering. Soon after his retirement, Gunnar Jacks continued his academic involvement as professor emeritus of groundwater chemistry at KTH's Department of Sustainable Development, Environmental Science and Engineering. Since 1996, he had strong ties with the Åbo Akademi University, Finland and later in 2002–2006 he was appointed as a guest professor at the Institute of Geology and Mineralogy. Through his versatile knowledge he contributed there to the research and education within environmental geology, and groundwater chemistry. While groundwater chemistry had been his principal field of expertise, he worked on diverse other contemporary environmental challenges in fields such as, acidification of soil and water; environmental effects of mining, geogenic contaminants in groundwater such as fluoride and arsenic, groundwater recharge in semi-arid areas and also on agricultural ecosystems such as arsenic in rice, and zinc deficiencies in soils, crops and human food intake.

Throughout his illustious career, he published more than 300 scientific papers, peer-reviewed conferences papers and book chapters. Based on his deep knowledge and understanding in the fields of hydrogeology, soil sciences and global water supply and his experience in a several water supply projects in developing countries, he had been unanimously elected as the President of the International Society of Ground water for Sustainable Development in 2006. In 2016, an honorary doctorate degree was conferred on Gunnar Jacks by the Åbo Akademi University, Finland in recognition of his outstanding contributions to research and education.

Gunnar Jacks's work on arsenic started during early 1990's, with the studies on the dynamics of arsenic in the forested soil profiles of Sweden. His research interest on arsenic in soil and groundwater grew successively in collaboration with the University of Kalyani in West Bengal in India with the conceptuatization of the mechanisms of *reductive dissolution of iron oxyhydroxides* for arsenic mobilization in groundwater in the Bengal Delta Plains and other similar geological settings. This work was published in the International Journal of Water Resources Development in 1997, and formed a landmark publication in the field of groundwater arsenic in groundwater of the sedimentary aquifers with more than 837 citations till date. Since then, along with his team he had not only been engaged in research on groundwater arsenic in the Ganga-Meghna-Brahmaputra (GMB) Plain, but also worked on fluoride one of the most widespread geogenic contaminant in groundwaters of India. He and his co-workers have played a pivotal role in documenting the magnitude of the arsenic calamity in the Bengal delta, both in India and Bangladesh. To the end of his life, though he continued to fight for clean water he became increasingly tolerant to failures of drinking water supply mitigation schemes across both countries.

He was member of several national and international affiliate of research funding organizations notably the Swedish Research Councils, VR and the FORMAS. The scholastic achievements of his publications are demonstrated through more than 5300 citations till date with a h-index of 39. Together with his colleagues of the KTH-International Groundwater Arsenic Group, he has been associated with the organization of the several international workshops and conferences on the groundwater arsenic problem including the well known International Congress on Arsenic in the Environment in Stockholm held in June 2016.

He was married to Birgitta Jacks, a pharmacist and nutritionist, who remained extremely supportive to his scientific work and achievements. He is survived by two sons Andreas and Johannes Jacks, who are both medical doctors by profession. He has also left behind his legacy through a number of his students who also made significant footprints in the field of arsenic research in Sweden and abroad. He had a sound life style including routine physical activities, healthy diet and daily cycling during his long working carreer.

The community of the interdisciplinary arsenic researchers across the world deeply mourn the death of Professor Gunnar Jacks. We have lost a beloved friend and colleague, with a kind and generous soul, who devoted his entire life to the support the people needing help. The arsenic community across the globe will always remember Gunnar Jacks his contributions in the field of arsenic research as well as for his simplicity, humble personality and his extreme passion for and devotion to working for the population from developing nations across the world. We will always miss his supportive and optimistic company.

Prosun Bhattacharya
Kazi Matin Ahmed
Debashis Chatterjee
Arslan Ahmad

"Ja mår han leva i vårt minne i hundrade år och inspirerar oss för att skapa en bättre vårld"

Arsenic in the Environment: Bridging Science to Practice for Sustainable Development – van der Wal, Ahmad, Petrusevski, Weijma, Savic, van der Wens, Beerendonk, Bhattacharya, Bundschuh & Naidu (Eds)
© 2023 the Editor(s), ISBN 978-1-032-32928-4

Organizers

ORGANIZERS OF BIENNIAL CONGRESS AND EXHIBITION SERIES: ARSENIC IN THE ENVIRONMENT

Jochen Bundschuh
University of Southern Queensland (USQ), Toowomba, QLD, Australia
International Society of Groundwater for Sustainable Development (ISGSD)
Stockholm, Sweden

Prosun Bhattacharya
KTH-International Groundwater Arsenic Research Group
Department of Sustainable Development, Environmental Sciences and Engineering, KTH Royal Institute of Technology, Stockholm, Sweden
SKWR Watercycle Research Institute, Nieuwegein, The Netherlands
International Society of Groundwater for Sustainable Development (ISGSD)
Stockholm, Sweden

Ravi Naidu
Global Centre for Environmental Remediation,
The University of Newcastle, Callaghan, NSW, Australia
CRC CARE, University of Newcastle, Callaghan, NSW
Australia

Local Organizing Committee

Albert van der Wal
Department of Environmental Technology, Agrotechnology and Food Sciences Group, Wageningen University and Research (WUR), Wageningen, The Netherlands

Evides Water Company, Research and Development, Water Technology, Rotterdam, The Netherlands

Arslan Ahmad
KTH-International Groundwater Arsenic Research Group, Department of Sustainable Development, Environmental Sciences and Engineering, KTH Royal Institute of Technology, Stockholm, Sweden

SIBELCO Ankerpoort NV, Maastricht, The Netherlands (formerly at KWR Water Research Institute, Nieuwegein, The Netherlands)

Branislav Petrusevski
IHE Delft Institute for Water Education, Delft, The Netherlands

Jan Weijma
Department of Environmental Technology, Agrotechnology and Food Sciences Group, Wageningen University and Research (WUR), Wageningen, The Netherlands

Dragan Savic
KWR Water Research Institute, Nieuwegein, The Netherlands

Patrick van der Wens
Brabant Water NV, Breda, The Netherlands

Erwin Beerendonk
KWR Water Research Institute, Nieuwegein, The Netherlands

Arsenic in the Environment: Bridging Science to Practice for Sustainable Development –
van der Wal, Ahmad, Petrusevski, Weijma, Savic, van der Wens, Beerendonk, Bhattacharya, Bundschuh & Naidu (Eds)
© 2023 the Editor(s), ISBN 978-1-032-32928-4

Sponsors and Contributors

Arsenic in the Environment: Bridging Science to Practice for Sustainable Development –
van der Wal, Ahmad, Petrusevski, Weijma, Savic, van der Wens, Beerendonk, Bhattacharya, Bundschuh & Naidu (Eds)
© 2023 the Editor(s), ISBN 978-1-032-32928-4

Scientific Committee

A. Fernández Cirelli: *University of Buenos Aires, Buenos Aires, Argentina*
A. Figoli: *Institute on Membrane Technology, ITM-CNR c/o University of Calabria, Rende (CS), Italy*
B. Figueiredo: *UNICAMP, Campinas, SP, Brazil*
R.B. Finkelman: *Department of Geosciences, University of Texas at Dallas, Richardson, Texas, USA*
A. Fiúza: *University of Porto, Porto, Portugal*
A.E. Fryar: *Department of Earth and Environmental Sciences, University of Kentucky, Lexington, KY, USA*
S.E. Garrido Hoyos: *Mexican Institute of Water Technology, Jiutepec, Mor., Mexico*
M. Gasparon: *The University of Queensland, Australia*
A. van Geen: *Lamont-Doherty Earth Observatory, Columbia University, New York, USA*
A. Ghosh: *Bihar State Pollution Control Board, Patna, Bihar, India*
A.K. Giri: *CSIR-Indian Institute of Chemical Biology, Kolkata, India*
W. Goessler: *University of Graz, Austria*
D.N. Guha Mazumder: *Institute of Post Graduate Medical Education & Research, Kolkata, India*
L.R. Guimaraes Guilherme: *Federal University of Lavras, Lavras, M.G., Brazil*
H.M. Guo: *State Key Laboratory of Biogeology and Environmental Geology & School of Water Resources and Environment, China University of Geosciences, Beijing, PR China*
X. Guo: *Department of Occupational & Environmental Health Sciences, Peking University SPH, Peking, P.R. China*
J.P. Gustafsson: *Swedish University of Agricultural Sciences, Uppsala, Sweden*
J. Hoinkis: *Karlsruhe University of Applied Sciences, Karlsruhe, Germany*
C. Hopenhayn: *University of Kentucky, Lexington, KY, USA*
M.F. Hughes: *Environmental Protection Agency, Research Triangle Park, NC, USA*
J. Ijumulana: *DAFWAT Research Group, Department of Water Resources Engineering, College of Engineering and Technology, University of Dar es Salaam, Dar es Salaam, Tanzania*
A.M. Ingallinella: *Centro de Ingeniería Sanitaria (CIS), Facultad de Ciencias Exactas, Ingeniería y Agrimensura, Universidad Nacional de Rosario, Rosario, Prov. de Santa Fe, Argentina*
M.T. Islam: *KTH-International Groundwater Arsenic Research Group, KTH Royal Institute of Technology, Stockholm, Sweden; Wateraid Bangladesh, Dhaka, Bangladesh*
J. Jarsjö: *Department of Physical Geography, Stockholm University, Stockholm, Sweden*
J.-S. Jean: *National Cheng Kung University, Tainan, RO China*
D. Johnston: *UNICEF, Dhaka, Bangladesh*
R. Johnston: *Department of Public Health, Social and Environmental Determinants of Health, World Health Organization, Geneva, Switzerland*
N. Kabay: *Chemical Engineering Department, Engineering Faculty, Ege University, Izmir, Turkey*
I.B. Karadjova: *Faculty of Chemistry, University of Sofia, Sofia, Bulgaria*
A. Karczewska: *Institute of Soil Sciences and Environmental Protection, Wroclaw University of Environmental and Life Sciences, Poland*
G. Kassenga: *Ardhi University, Dar es Salaam, Tanzania*
D.B. Kent: *US Geological Survey, Menlo Park, CA, USA*
N.I. Khan: *The Australian National University, Canberra, Australia*
K.-W. Kim: *Department of Environmental Science and Engineering, Gwangju Institute of Science and Technology, Gwangju, South Korea*
W. Klimecki: *Department of Pharmacology and Toxicology, University of Arizona, Tucson, Arizona, USA*
M. Kumar: *Discipline of Earth Sciences, Indian Institute of Technology Gandhinagar, Gujarat, India; Sustainability Cluster, School of Engineering, University of Petroleum & Energy Studies, Dehradun, Uttarakhand, India & Escuela de Ingeniería y Ciencias, Tecnologico de Monterrey, Campus Monterey, Monterrey, Nuevo Leon, Mexico*
J. Kumpiene: *Department of Civil, Environmental and Natural Resources Engineering, Luleå University of Technology, Luleå, Sweden*
M.I. Litter: *Comisión Nacional de Energía Atómica, and Universidad de Gral. San Martín, San Martín, Argentina*
D.L. López: *Ohio University, Athens, Ohio, USA*
L.Q. Ma: *Research Center for Soil Contamination and Environment Remediation, Southwest Forestry University, Kunming, PR China*
M. Mallavarapu: *Faculty of Science and Information Technology, The University of Newcastle Callaghan, NSW, Australia*
D. Mondal: *School of Science, Engineering and Environment, University of Salford, Manchester, UK*
N. Mañay: *De la República University, Montevideo, Uruguay*
R. Mato: *Ardhi University, Dar es Salaam, Tanzania*

J. Matschullat: *Interdisciplinary Environmental Research Centre (IÖZ), TU Bergakademie Freiberg, Freiberg, Germany*
A. Meharg: *Queen's University of Belfast, Belfast, UK*
F. Mtalo: *Department of Water Resources Engineering, University of Dar E Salaam, Dar es Salaam, Tanzania*
M. Mörth: *Stockholm University, Sweden*
A. Mukherjee: *Department of Geology and Geophysics, Indian Institute of Technology–Kharagpur, Kharagpur, West Bengal, India; School of Environmental Science and Engineering, Indian Institute of Technology–Kharagpur, Kharagpur, West Bengal, India*
R. Naidu: *Global Centre for Environmental Remediation (GCER), Faculty of Science, The University of Newcastle, Callaghan, NSW, Australia, Cooperative Research Centre for Contamination Assessment and Remediation of the Environment (CRC CARE), The University of Newcastle, Callaghan, NSW, Australia*
J.C. Ng: *The University of Queensland, Queensland Alliance for Environmental Health Sciences, QLD Australia*
B. Noller: *The University of Queensland, Australia*
D.K. Nordstrom: *U.S. Geological Survey, Menlo Park, CA, USA*
M. Ormachea Muñoz: *Instituto de Investigaciones Químicas, Universidad Mayor de San Andrés, La Paz, Bolivia*
K.M. Paknikar: *Agharkar Research Institute, India*
M.F. Parvez: *Colombia University, New York, USA*
P. Pastén González: *Pontificia Universidad Católica de Chile, Chile*
C.A. Pérez: *NLS-Brazilian Synchrotron Light Source Laboratory, Campinas, SP, Brazil*
A.L. Pérez-Carrera: *University of Buenos Aires, Buenos Aires, Argentina*
C. Pérez Coll: *National San Martín University, San Martín, Argentina*
B. Petrusevski: *UNESCO-IHE, Institute for Water Education, Delft, The Netherlands*
B. Planer-Friedrich: *University Bayreuth, Bayreuth, Germany*
D.A. Polya: *School of Earth and Environmental Sciences and Williamson Research Centre for Molecular Environmental Science, University of Manchester, Manchester, UK*
T. Pradeep: *Indian Institute of Technology Madras, Chennai, India*
I. Queralt: *Institute of Earth Sciences Jaume Almera – CSIC, Spain*
J. Quintanilla: *Institute of Chemical Research, Universidad Mayor de San Andrés, La Paz, Bolivia*
M.M. Rahman: *Global Centre for Environmental Remediation (GCER), Faculty of Science, The University of Newcastle, Callaghan, NSW, Australia*
AL. Ramanathan: *School of Environmental Science, Jawaharlal Nehru University, New Delhi, India*
O.E. Ramos Ramos: *Instituto de Investigaciones Químicas, Universidad Mayor de San Andrés, La Paz, Bolivia*
B.P. Rosen: *Department of Cellular Biology and Pharmacology, Florida International University, Herbert Wertheim College of Medicine, Miami, USA*
J. Routh: *Department of Thematic Studies – Environmental Change, Linköping University, Linköping, Sweden*
T. Roychowdhury: *School of Environmental Studies, Faculty of Interdisciplinary Studies (Law & Management), Jadavpur University, Kolkata, India*
D. Saha: *Central Groundwater Board, New Delhi, India*
A.M. Sancha: *Department of Civil Engineering, University of Chile, Santiago de Chile, Chile*
D. Savic: *KWR Water Research Institute, Nieuwegein, The Netherlands*
M. Schreiber: *Department of Geosciences, Virginia Polytechnic Institute and State University, VA, USA*
C. Schulz: *La Pampa National University, Argentina*
A. Sen Gupta: *Lehigh University, Bethlehem, PA, USA*
V.K. Sharma: *Florida Institute of Technology, Florida, USA*
A. Shraim: *The University of Queensland, Brisbane, Australia*
M. Sillanpää: *Lappeenranta University of Technology, Lappeenranta, Finland*
S. Singh: *Innervoice Foundation, Sarnath, Varanasi, Utar Pradesh, India*
P.L. Smedley: *British Geological Survey, Keyworth, UK*
A.H. Smith: *University of California, Berkeley, CA, USA*
M. Styblo: *University of North Carolina, Chapel Hill, USA*
J. Tapia: *Universidad Católica del Norte, Facultad de Ingeniería y Ciencias Geológicas, Departamento de Ciencias Geológicas, Antofagasta, Chile*
C. Tsakiroglou: *Foundation for Research and Technology, Hellas, Greece*
M. Vahter: *Karolinska Institutet Stockholm, Sweden*

C.M. van Genuchten: *Geological Survey of Denmark and Greenland (GEUS), Copenhagen, Denmark*
D. van Halem: *Sanitary Engineering Section, Faculty of Civil Engineering and Geosciences, Delft University of Technology, Delft, The Netherlands*
J. W. Vargas de Mello: *Federal University of Viçosa, Viçosa, MG, Brazil*
D. Velez: *Institute of Agrochemistry and Food Technology, Valencia, Spain*
M. Vithanage: *Ecosphere Resilience Research Center, Faculty of Applied Science, University of Sri Jayewardenepura, Nugegoda, Sri Lanka*
A. Voegelin: *Eawag, Swiss Federal Institute of Aquatic Science and Technology, Dübendorf, Switzerland*
A. van der Wal: *Wageningen University and Research (WUR), Wageningen, The Netherlands*
J. Weijma: *Wageningen University and Research (WUR), Wageningen, The Netherlands*
Y. Zheng: *School of Environmental Science and Engineering, Southern University of Science and Technology, Shenzhen, P R China*
Y.G. Zhu: *Chinese Academy of Sciences, Research Centre for Eco-environmental Sciences, Beijing, P R China*

Arsenic in the Environment: Bridging Science to Practice for Sustainable Development –
van der Wal, Ahmad, Petrusevski, Weijma, Savic, van der Wens, Beerendonk, Bhattacharya, Bundschuh & Naidu (Eds)
© 2023 the Editor(s), ISBN 978-1-032-32928-4

Foreword (President, KTH Royal Institute of Technology)

Arsenic is a natural or anthropogenic contaminant in many areas around the globe, where human subsistence is at risk. It is considered as a class 1 carcinogen, and its presence in groundwater has emerged as a major environmental calamity in several parts of the world. The biennial International Congress Series on Arsenic in the Environment is providing a common platform for sharing knowledge and experience on multidisciplinary issues on arsenic occurrences in groundwater and other environmental compartments on a worldwide scale to identify, assess, develop and promote approaches for management of arsenic in the environment and health effects.

KTH Royal Institute of Technology holds an apex position in the global arsenic research over the past three decades on the interdisciplinary research on Arsenic in the Environment and I deeply appreciate the continued efforts of the International Organizers from KTH-International Groundwater Arsenic Research Group, Department of Sustainable Development, Environmental Science and Engineering, School of Architecture and Built Environment KTH Royal Institute of Technology. I feel proud to write this foreword to this Volume of Arsenic in the Environment-Proceedings Series, containing the extended abstracts of the presentations made during the 8th International Congress & Exhibition on Arsenic in the Environment – As 2021. The present volume "Arsenic in the Environment – Bridging Science to Practice for Sustainable Development" being published as a new volume of the book series "Arsenic in the Environment-Proceedings" under the auspices of the International Society of Groundwater for Sustainable Development (ISGSD) based at the KTH Royal Institute of Technology, will be an important updated contribution, comprising a large number of over 190 extended abstracts submitted by various researchers, health workers, technologists, students, legislators, and decision makers around the world that would be discussed during the conference. Apart from exchanging ideas, and discovering common interests, the scientific community involved in this this specialized field needs to carry out researches, which not only address academic interests but also contribute to the societal needs through prevention or reduction of exposure to arsenic and its toxic effects in millions of exposed people throughout the world.

I acknowledge the efforts of the Local Organizers from Knowledge Group Water at the KWR Water Research Institute (KWR), the Wageningen University and Research (WUR), IHE Delft Institute for Water Education (IHE), Brabant Water, Sibelco Group and the entire editorial team for their untiring work for accomplishing this volume. I hope that the book will reflect the update on the current state-of-the-art knowledge on the interdisciplinary facets of arsenic in the environment documented for the period between 2018 through 2021 required for the management of arsenic in the environment for human health protection and societal development.

Dr. Anders Söderholm
President, KTH Royal Institute of Technology
Stockholm, Sweden
April, 2023

 ISBN 978-1-032-32928-4

Foreword (Former UNESCO Chair on Groundwater Arsenic within the 2030 Agenda for Sustainable Development, University of Southern Queensland)

UNESCO CHAIR ON GROUNDWATER ARSENIC WITHIN THE 2030 AGENDA FOR SUSTAINABLE DEVELOPMENT

An independent leadership think-tank for researchers, developers, industries, and policy makers for innovative ideas and best practices addressing global water challenges within the 2030 Agenda for Sustainable Development

UNIVERSITY OF SOUTHERN QUEENSLAND AUSTRALIA

The UNESCO Chair on Groundwater Arsenic within the 2030 Agenda for Sustainable Development based at the University of Southern Queensland (USQ) in Australia, opened up a platform for all from academia to industry including agriculture and mining, in collaboration with various institutions across the world. All sectors are concerned with searching for solutions to problems associated with arsenic of geogenic origin, in both small communities with decentralized water supply to large water industries and food sectors fulfilling the 2030 Agenda for Sustainable Development. We have the great pleasure in co-organising the 8th International Congress & Exhibition on Arsenic in the Environment (As2021) themed 'Environmental Arsenic in a Changing World' held digitally during June 2021 in Wageningen, The Netherlands.

Arsenic originating from geogenic sources is a global issue as over 200 million people, so far known from over 80 countries, is at risk due to ingestion of arsenic-contaminated food and drinking water. In food, arsenic is particularly accumulated as a result of irrigation with arsenic-rich water – the staple food rice is thereby especially affected. Despite the fact that the problem occurs equally in developing and industrialized countries, the problem is most severe in the first country group where the poor are those who are at the highest risk and suffer most. Hence, arsenic pollution is an increasing global problem that will require a global approach and worldwide solutions. Thereby, transdisciplinary research into the occurrence, mobility and bioavailability of arsenic in different environments including aquifers, soils, sediments as well as the food chain, will all become increasingly important.

It gives me pleasure to congratulate the organisers for their success in organizing the prestigious International Congress on Arsenic in the Environment in Netherlands digitally despite the challenges due to the COVID-19 pandemic. I would also acknowledge the collaborative and cooperative efforts of the KTH Royal Institute of Technology, KWR Water Research Institute, Wageningen University and Research (WUR), IHE Delft Institute for Water Education (IHE) and Sibelco Group. I hope that these proceedings will serve as a lasting record in co-organising this international Congress.

Professor Jochen Bundschuh
Former UNESCO Chair on Groundwater Arsenic
within the 2030 Agenda for Sustainable Development
The University of Southern Queensland
Toowoomba, QLD, Australia
April 2023

Arsenic in the Environment: Bridging Science to Practice for Sustainable Development –
van der Wal, Ahmad, Petrusevski, Weijma, Savic, van der Wens, Beerendonk, Bhattacharya, Bundschuh & Naidu (Eds)
© 2023 the Editor(s), ISBN 978-1-032-32928-4

Foreword (Chair, Environmental Technology Group, Wageningen University, The Netherlands)

WAGENINGEN
UNIVERSITY & RESEARCH

With great success the 8th International Congress and Exhibition on Arsenic in the Environment (As2021) was held digitally in Wageningen, The Netherlands, June 7–9, 2021. The special focus was on 'Bridging Science to Practice for Sustainable Development' and that offered great inspirations on how to improve (drinking)water quality worldwide. I am delighted to present here the foreword for the proceedings of that great conference.

Arsenic is a naturally occurring element of the earth's crust, nevertheless, exposure can result in adverse health effects to humans. The WHO standard for arsenic in drinking water is 10 micrograms per liter (μg/L). However, long term exposure to drinking water with arsenic concentration below 10 μg/L can still lead to an increased risk of cancer.

Inorganic arsenic is naturally present in the groundwater of many countries in the world, including the Netherlands, although concentrations here tend to be rather low. Worldwide, the greatest threat to public health from arsenic originates from contaminated water used for drinking, food preparation and irrigation of food crops. The most important action is therefore to reduce exposure to arsenic by the provision of a safe water supply for drinking water, food preparation and irrigation of food crops. As a result, drinking water companies in the Netherlands have updated their policy on arsenic levels in drinking water to stay below 1 μg/L.

Recognizing the global significance of arsenic for safe water supply, Wageningen University has gladly invested in realizing the As 2021 congress via participation in the organizing committee and the scientific board, via our Environmental Technology group, specifically by providing the chair of the conference, Prof. Dr. Ir. Bert van der Wal as well as the contribution of our former colleague Dr. Jan Weijma. This was all done in collaboration with KTH Royal Institute of Technology and KWR Water Research Institute.

I am grateful to all authors, reviewers and editors for making the As2021 congress a great success and I hope that these proceedings will help to increase our understanding of arsenic in the environment and at the same time can help policy makers and water professionals around the globe to ensure that all human beings will have access to safe and healthy (drinking) water.

Prof. Dr. Ir. Huub Rijnaarts
Chair holder of the Environmental Technology Group
Wageningen University and Research
Wageningen, The Netherlands

Foreword (Chief Executive Officer, KWR Water Research Institute)

It is with great pleasure and expectations that I write this Foreword to the Proceedings of the 8th International Congress and Exhibition on Arsenic in the Environment (As2021), themed 'Bridging Science to Practice for Sustainable Development' held digitally in Wageningen, The Netherlands, June 7–9, 2021.

The International Congress on Arsenic in the Environment has been previously held seven times: Mexico 2006, Spain 2008, Taiwan, Republic of China 2010, Australia, 2012, Argentina 2014, Sweden 2016 and the People's Republic of China 2018. Due to the Corona pandemic, the As2020 Congress in Wageningen was postponed, but finally organised during June 2021 as a digital Congress renamed to As2021. The Congress series has evolved into a highly reputed platform for sharing and accessing global knowledge on diverse aspects of arsenic research. Arsenic in drinking water is a global problem affecting populations on all five continents. Despite historical recognition of arsenic toxicity, more than 200 million people around the world are still exposed to above-acceptable arsenic levels. This situation is alarming and an urgent public health problem. Arsenic contamination of drinking water can be caused both by natural and anthropogenic processes. For example, in Poland and Brazil, arsenic in groundwater caused by anthropogenic mining activities has been reported. On the other hand, in some parts of Turkey elevated arsenic levels in groundwater are attributed to natural geothermal factors, and in Bangladesh geogenic processes are the major cause of large-scale arsenic contamination. Whatever the origin may be, once detected in drinking water sources, suitable arsenic remediation measures should be taken to ensure supply of safe drinking water – as this is the fundamental right of every human being.

In the Netherlands, drinking water companies have recently updated their policy on arsenic and they presented their rationale at As2018. KWR is collaborating with Dutch water companies in various fundamental and applied research projects to support the realisation of this policy. Moreover, KWR is involved in international projects on remediation of arsenic-contaminated water. In Argentina, a country facing wide-scale arsenic contamination of groundwater, KWR is collaborating with the drinking water company of Buenos Aires (AySA) to remove arsenic from water.

Recognising the global significance of arsenic for achieving safe water supply for all and making the event a success, KWR gladly invested their efforts in bringing As2021 to life via participation in the organising committee and the scientific board of As2021. This was achieved through a collaborative effort of our Knowledge Group Water, the Wageningen University and Research (WUR), IHE Delft Institute for Water Education (IHE), Sibelco and the KTH Royal Institute of Technology.

I congratulate all the authors, reviewers and editors for providing excellent content and structure to this book. I hope that these proceedings will serve as a deep-rooted record of the state-of-the-arsenic-related science developed since the past Congress between 2018 and 2021 and serve as a reference base for future research. Furthermore, it can support water suppliers and policymakers all over the world in addressing the arsenic problem efficiently and effectively.

Prof. Dragan Savic, FREng
Chief Executive Officer
KWR Water Research Institute
Nieuwegein, The Netherlands, April 2023

Arsenic in the Environment: Bridging Science to Practice for Sustainable Development –
van der Wal, Ahmad, Petrusevski, Weijma, Savic, van der Wens, Beerendonk, Bhattacharya, Bundschuh & Naidu (Eds)
© 2023 the Editor(s), ISBN 978-1-032-32928-4

Foreword (CEO and Managing Director crcCARE, The University of Newcastle)

It is with deep satisfaction that I write this foreword to the Proceedings of the 8th International Congress & Exhibition on Arsenic in the Environment (As 2021) with the theme 'Bridging Science to Practice for Sustainable Development' held digitally in Wageningen, The Netherlands, June 7–9, 2021. The crc for Contamination Assessment and Remediation of the Environment (crcCARE) is very proud to be part of this International Congress series as co-organizer as the first arsenic workshop (Arsenic in the Asia-Pacific Region) was organized by me in Adelaide, South Australia in 2001 where the extent, severity and potential risks arising from exposure to arsenic, as well as the fate of arsenic in water, soil and food was discussed. The continuation of this as a global congress – "Arsenic in the Environment" – was then held in Mexico in 2006. Since then, the international congress series has been held every two years at various locations around the globe. Thus, this arsenic congress has received enormous attention and is a platform where scientists, government officials, policy makers and regulators share their knowledge on the recent developments in arsenic research.

Arsenic is a toxic element and is categorized as a Class I carcinogen, which is ubiquitous in the environment. Arsenic is present in our environment as a naturally occurring substance and because of anthropogenic activities. It is generally found in waters (both surface and sub-surface), soil, food and the air and can occur in both organic and inorganic forms. Arsenic occurrence in water in the Australian landscape is generally low but major pollution can occur due to mining activities, the use of arsenic based pesticides and herbicides, as well as from CCA treated wood. Arsenic concentrations in cattle dip and sheep dip soils and railway corridor soils in Australia are also at levels two to five times above the health screening levels. These are the major causes of arsenic contaminated sites in Australia.

crcCARE is Australia's leading research centre, which contributes to both Australian and international social, economic, cultural and environmental well-being through its innovative research activities that supports research in identified areas of strength, to address national and international challenges. Having been established in 2005, crcCARE is currently supported by end-users who invest significant funding to ensure innovative research on environmental contamination.

crcCARE supported researchers have been working in Bangladesh, India and other SE Asian countries over many years and have made substantial contributions to various aspects of arsenic research including arsenic chemistry, toxicity and bioavailability, human health effects, and food quality and safety. By combining laboratory-based studies with field surveys, they have contributed significantly to the generation of new knowledge in this important research field. We have also made major contributions by developing new and novel analytical techniques for arsenic speciation in various environmental matrices, which has helped to understand the toxicity, bioavailability, and accurate estimation of human health risks. Our research on arsenic continues from management in the field to minimize uptake into food crops and hence human exposure.

As with the past Congress proceedings, we are confident that this proceeding will be accepted as another outstanding collection of papers on recent arsenic research. We would like to thank the contributors and conference delegates for their active participation. We would also like to express our whole-hearted appreciation to all co-organizers and others involved in the congress series and who have made this congress a great success.

Laureate Professor Ravi Naidu
CEO and Managing Director
crcCARE
The University of Newcastle (UON)
Newcastle, Australia
April 2023

Arsenic in the Environment: Bridging Science to Practice for Sustainable Development –
van der Wal, Ahmad, Petrusevski, Weijma, Savic, van der Wens, Beerendonk, Bhattacharya, Bundschuh & Naidu (Eds)
© 2023 the Editor(s), ISBN 978-1-032-32928-4

Editors' Foreword

Occurrence of elevated arsenic concentrations in groundwater and associated health risks were reported during the international conference on environmental arsenic, held in Fort Lauderdale, USA, almost 45 years ago; October, 1976. Over the past years arsenic in drinking water, and more recently, in plant based foods, especially rice, has been recognized as a major public health concern in many parts of the world. Latest surveys estimate that currently more than 200 million people around the world are exposed to unacceptably high arsenic levels. Recent research suggests that even 10 µg/L might not be a safe guideline for maximum arsenic in drinking water, therefore many water utilities are beginning to explore new arsenic removal methods that can remove arsenic to extremely low levels. The geological, geomorphological and geochemical reasons for high arsenic concentrations in groundwater vary from place to place and require different mitigation policies and practices.

Although the high income countries may invest in research and development of suitable remediation techniques, arsenic in private water sources is not always tested. On the other hand, low to lower-middle income countries, such as many areas in South-East Asia, Africa and South America, where millions of people still use arsenic-contaminated drinking water, are still coping with stagnated mitigation efforts and slow progress towards safe drinking water. It is disturbing to enter almost any village of the Bengal basin today and find that groundwater drawn from untested shallow wells continues to be used routinely for drinking and cooking, given that the arsenic problem was already recognized in the mid-1980s in West Bengal and the mid-1990s in Bangladesh. Equally problematic is the fact that hundreds of millions of wells worldwide are not yet tested for arsenic. Moreover, many low and lower-middle income countries have yet not been able to revise their standards for arsenic in drinking water to 10 µg/L, the guideline value of the World Health Organization. We sincerely believe that sharing knowledge and experience on arsenic related science and practices on a worldwide scale and across varied disciplines can serve as an effective strategy to support global arsenic management and mitigation efforts.

The biennial International Congress Series on Arsenic in the Environment aims at providing a common platform for sharing knowledge and experience on multidisciplinary issues on arsenic occurrences in groundwater and other environmental compartments on a worldwide scale for identifying and promoting optimal approaches for the assessment and management of arsenic in the environment. The International Congress on Arsenic in the Environment has previously been held seven times; Mexico 2006, Spain 2008, R.O. China (Taiwan) 2010, Australia 2012, Argentina 2014, Sweden 2016 and in P.R. China in 2018. The 8th International Congress on Arsenic in the Environment (As2021) was organized in the Netherlands and also digitally, with the theme "Bridging Science to Practice for Sustainable Development.

We envision As2021 as a global interdisciplinary platform to exchange and disseminate research results to improve our understanding of the occurrence, mobility, bioavailability, toxicity and dose-response relationship with various health effects of environmental arsenic in the current epoch of a changing world.

We have received a large number of extended abstracts, which were submitted mainly from researchers, but also health workers, technologists, students, legislators, and government officials. The topics to be covered during the Congress As 2021 have been grouped under the five inter-disciplinary generic thematic areas:

- Theme 1: Arsenic in Natural Soil and Water System
- Theme 2: Arsenic in Agriculture and Food Production
- Theme 3: Health Impacts of Arsenic
- Theme 4: Technologies for Arsenic Removal from Water
- Theme 5: Sustainable Mitigation and Management

During the Congress, we collected a few take home messages, important findings or conclusions:

- Success is when the poorest child has access to clean water
- Improve technologies with low costs and energy
- People are the key, not the technology. The system approach is the way forward
- Research is not only for journals, but also for policy making
- Biologic conversion of arsenic may be part of the sustainable water treatment solution
- Recover resources from the water sector, or use waste streams from other sectors, for removing arsenic from drinking water

We thank the international scientific committee members for their efforts on reviewing the extended abstracts. Further, we thank the sponsors of the Congress from around the world: KWR Watercycle Research Institute (The Netherlands), Wageningen University and Research (The Netherlands), IHE-Delft (The Netherlands), Brabant Water (The Netherlands), Evides Water Company (The Netherlands), KTH Royal Institute of Technology (Sweden), University of Southern Queensland (Australia), The University of Newcastle (Australia) and the CRC-CARE, at the University of Newcastle (Australia) for their generous support – Thank you all sponsors for your support that contributed to the success of the congress As2021.

Lastly, the editors thank Dr. Rakesh Kumar at the Department of Biosystems Engineering, Auburn University, United States for his support with the critical reading this manuscript for the final technical edits. We are also thankful to Janjaap Blom, Jahnavi Vaid and Lukas Goosen of the CRC Press/Taylor and Francis (A.A. Balkema) Publishers, The Netherlands for their patience, support, and skill for the production of this volume.

Albert van der Wal
Arslan Ahmad
Branislav Petrusevski
Jan Weijma
Dragan Savic
Patrick van der Wens
Erwin Beerendonk
Prosun Bhattacharya
Jochen Bundschuh
Ravi Naidu
(Editors)

 ISBN 978-1-032-32928-4

List of Contributors

Abbasi, S.: *Global Centre for Environmental Remediation (GCER), The University of Newcastle, Callaghan, NSW, Australia*
Abdoel Gafour, S.: *Evides Water Company, Department of Water Technology and Source Protection, Rotterdam, The Netherlands*
Abdulmutalimova, T.O.: *Institute of Geothermal and Renewable Energy, Joint Institute of High Temperature, Russian Academy of Sciences, Makhachkala, Russia*
Abelenda, E.: *Dirección Nacional de Minería y Geología, Ministerio de Industria, Energía y Minería, Montevideo, Uruguay*
Abiye, T.A.: *School of Geosciences, University of the Witwatersrand, Johannesburg, South Africa*
Addison, S.T.: *Department of Earth and Environmental Sciences and Williamson Research Centre for Molecular Environmental Science, University of Manchester, Manchester, UK*
Agbaba, J.: *Environmental Protection Research Group, Department of Chemistry, Biochemistry and Environmental Protection, University of Novi Sad, Faculty of Sciences, Novi Sad, Serbia*
Aguayo, A.: *Instituto de Geofísica, Universidad Nacional Autónoma de México, México City, Mexico*
Agui, L.: *Departamento de Química Analítica, Facultad de CC. Químicas, Universidad Complutense de Madrid, Madrid, Spain*
Ahmad, A.: *SIBELCO Ankerpoort NV, Maastricht, The Netherlands; SCR-Sibelco N.V., Antwerp, Belgium; Department of Environmental Technology, Wageningen University and Research (WUR), Wageningen, The Netherlands; KWR Water Cycle Research Institute, Nieuwegein, The Netherlands & KTH-International Groundwater Arsenic Research Group, Department of Sustainable Development, Environmental Science and Engineering, KTH Royal Institute of Technology, Stockholm, Sweden*
Ahmad, M.M.: *Regional and Rural Development Planning, Department of Development and Sustainability, School of Environment, Resources and Development, Asian Institute of Technology, Thailand*
Ahmed, F.: *Department of Civil, Environmental, and Construction Engineering, Texas Tech University, Texas, USA*
Ahmed, K.M.: *Department of Geology, University of Dhaka, Dhaka, Bangladesh*
Ahmed, S.: *Department of Chemistry, Sripat Singh College, Jiaganj, Murshidabad, India*
Ahmed, S.: *Environment and Population Research Centre, New DOHS, Mohakhali, Dhaka, Bangladesh*
Aitken, J.: *School of Chemistry, The University of Sydney, Sydney, NSW, Australia*
Akif, M.M.A.: *Department of Geology, University of Dhaka, Dhaka, Bangladesh*
Akter, N.: *Wash Section, UNICEF Bangladesh, Dhaka, Bangladesh*
Alam, M.A.: *Departamento de Geología, Facultad de Ingeniería, Universidad de Atacama, Copiapó, Región de Atacama, Chile*
Alam, M.J.: *Department of Geology, University of Dhaka, Dhaka, Bangladesh*
Alauddin, M.: *Department of Chemistry, Wagner College, Staten Island, New York, USA*
Ali, M.: *Mahavir Cancer Sansthan & Research Centre, Patna, Bihar, India*
Ali, N.: *Department of Environmental Sciences, Quaid I Azam University, Pakistan*
Alleoni, L.R.F.: *ESALQ/USP Soil Science Department, College of Agriculture, University of São Paulo, São Paulo, Brasil*

Alvareda, E.: *Water Department, Centro Universitario Regional Litoral Norte, Universidad de la República, Salto, Uruguay*
Alvarenga, I.F.S.: *Department of Soil Science, Federal University of Lavras, Lavras, Minas Gerais, Brazil*
Álvarez Cruz, J.L.: *Posgraduate Program in Environmental Engineering UNAM-IMTA, Jiutepec, Morelos, Mexico*
Alvarez Gonçalvez, C.V.: *Universidad de Buenos Aires, Facultad de Ciencias Veterinarias, Centro de Estudios Transdisciplinarios del Agua (CETA), Buenos Aires, Argentina & Universidad de Buenos Aires-CONICET, Instituto de Investigaciones en Producción Animal (INPA), Buenos Aires, Argentina*
Álvarez Vargas, A.: *University of Guanajuato, Division of Natural and Exact Sciences, Guanajuato, Mexico*
Amézaga-Campos, B.S.: *Posgrado en Ciencias de la Tierra, Universidad Nacional Autónoma de México, Del. Coyoacán, Ciudad de México., México*
Annaduzzaman, M.: *Water Management Department, Faculty of Civil Engineering and Geosciences, Delft University of Technology, Delft, The Netherlands*
Anwar, Z.: *Department of Civil, Environmental, and Construction Engineering, Texas Tech University, Texas, USA*
Arancibia, V.: *Pontificia Universidad Católica de Chile, Chemistry Faculty, Santiago, Chile*
Araújo, L.S.S.: *Environmental Engineering Graduating Program-ProAmb, Federal University of Ouro Preto, Ouro Preto, Brazil*
Araya, A.: *School of Construction, Instituto Tecnológico de Costa Rica, Cartago, Costa Rica*
Arellano, F.E.: *Universidad de Buenos Aires, Facultad de Ciencias Veterinarias, Centro de Estudios Transdisciplinarios del Agua (CETA), Buenos Aires, Argentina & Universidad de Buenos Aires-CONICET, Instituto de Investigaciones en Producción Animal (INPA), Buenos Aires, Argentina*
Armienta-Hernández, M.A.: *Instituto de Geofísica, Universidad Nacional Autónoma de México, Ciudad de México, México*
Armstrog-Altrin, J.S.: *Institute of Marine Sciences and Limnology, National Autonomous University of Mexico, Mexico City, Mexico*
Arya, P.C.: *Biogeochemistry Research Group, Centre for Earth Sciences, Indian Institute of Science, Bangalore, India*
Ashraf, R.: *Environmental Geochemistry Laboratory, Department of Environmental Sciences, Faculty of Biological Sciences, Quaid-i-Azam University, Islamabad, Pakistan*
Aslam, Z.: *Department of Agronomy, University of Agriculture Faisalabad, Faisalabad, Pakistan*
Baba, A.: *İzmir Institute of Technology, Department of International Water Resources, Urla, Izmir, Turkey*
Bahr, C.: *GEH Wasserchemie GmbH & Co. KG, Osnabrück, Germany*
Banerjee, N.: *Molecular Genetics Division, CSIR-Indian Institute of Chemical Biology, Kolkata, India & Biomedical Research and Diagnostic Centre, Manovikas Kendra Rehabilitation and Research Institute for the Handicapped, Kolkata, India*
Banerjee, P.: *Department of Earth and Environmental Sciences, National Chung Cheng University, Chiayi, Taiwan, R.O. China*
Barbiero, L.: *Institut de Recherche pour le Développement (IRD), GET, Toulouse, France*
Barbosa, M.V.: *Department of Soil Science, Federal University of Lavras, Lavras, Minas Gerais, Brazil*
Bari, A.S.M.F.: *Global Centre for Environmental Remediation (GCER), The University of Newcastle, Callaghan, NSW, Australia*
Barla, A.: *Department of Earth Sciences, Indian Institute of Science Education and Research (IISER) Kolkata, Mohanpur, West Bengal, India*
Barnwal, P.: *Department of Economics, Michigan State University, East Lansing, Michigan, USA*

Barua, S.: *Department of Geology, Kansas State University, Manhattan, KS, USA & School of Earth, Atmosphere and Environment, Monash University, Clayton, VIC, Australia*
Basak, S.: *ExcelDots AB, Bromma, Sweden*
Bashir, S.: *Sub-campus Depalpur, University of Agriculture Faisalabad, Okara, Pakistan*
Bassil, J.: *L2GE, Lebanese University, Faculty of Sciences 2, Fanar, Matn, Lebanon*
Bassil, N.M.: *School of Earth and Environmental Sciences, The University of Manchester, UK*
Basu, A.: *Sripat Singh College, Jiaganj, Murshidabad, India*
Battaglia-Brunet, F.: *BRGM, Water, Environment and Ecotechnology Division, Orléans, France & BRGM, Water, Environment, Process and Analyses Division, Orléans, France*
Batool, S.: *Department of Environmental Sciences, Quaid I Azam University, Pakistan*
Bea, P.A.: *Instituto de Hidrologa de Llanuras "Dr. Eduardo J. Usunoff" (IHLLA), Azul, Argentina*
Belkasmi, A.I.: *Institute for Water Resources and Water Supply, Hamburg University of Technology, Am Schwarzenberg-Campus 3, Hamburg, German*
Berg, M.: *Eawag, Swiss Federal Institute of Aquatic Science and Technology, Dübendorf, Switzerland & UNESCO Chair on Groundwater Arsenic within the 2030 Agenda for Sustainable Development, School of Civil Engineering and Surveying, University of Southern Queensland, QLD, Australia*
Bhattacharjee, S.: *Department of Civil, Environmental, and Construction Engineering, Texas Tech University, Texas, USA*
Bhattacharya, P.: *KTH International Groundwater Arsenic Research Group, Department of Sustainable Development, Environmental Science and Engineering, KTH Royal Institute of Technology, Stockholm, Sweden & KWR Water Research Institute, Nieuwegein, The Netherlands*
Bhattacharya, Piyal: *Department of Environmental Science, Kanchrapara College, Kanchrapara, West Bengal, India*
Bhowmick, S.: *Kolkata Zonal Center, CSIR – National Environmental Engineering Research Institute (NEERI), Kolkata, West Bengal, India*
Bibi, I.: *Institute of Soil and Environmental Sciences, University of Agriculture Faisalabad, Faisalabad, Pakistan*
Bieger, K.: *Departamento de Química y Biología, Facultad de Ciencias Naturales, Universidad de Atacama, Copiapó, Región de Atacama, Chile*
Bishal, M.M.: *Department of Geology, University of Dhaka, Dhaka, Bangladesh*
Biswas, M.: *Sripat Singh College, Jiaganj, Murshidabad, India*
Boberg, M.: *KTH-International Groundwater Arsenic Research Group, Department of Sustainable Development, Environmental Science and Engineering, KTH Royal Institute of Technology, Stockholm, Sweden*
Bose, S.: *Department of Earth Sciences, Indian Institute of Science Education and Research (IISER) Kolkata, Mohanpur, West Bengal, India*
Bostick, B.C.: *Lamont-Doherty Earth Observatory, Columbia University, New York, NY, USA*
Braeuer, S.: *Institute of Chemistry, University of Graz, Graz, Austria*
Brennwald, M.S.: *Eawag, Swiss Federal Institute of Aquatic Science and Technology, Dübendorf, Switzerland*
Broberg, K.: *Institute of Environmental Medicine, Karolinska Institutet, Stockholm, Sweden*
Brombach, C.-C.: *Department of Geosciences, Geochemistry & Hydrogeology, University of Bremen, Bremen, Germany*
Bueno, M.: *CNRS/IPREM, Université Pau & Pays de l'Adour, Pau, France*
Bundschuh, J.: *School of Civil Engineering and Surveying, University of Southern Queensland, Toowoomba, QLD, Australia & UNESCO Chair on Groundwater Arsenic within the 2030 Agenda for Sustainable Development & Faculty of Health, Engineering and Sciences, University of Southern Queensland, Toowoomba, QLD, Australia*

Bühl, V.: *Analytical Chemistry, DEC, Faculty of Chemistry, Universidad de la República, Montevideo, Uruguay*

Caballero-Chavarria, A.: *Environmental Protection Research Center (CIPA), School of Chemistry, Instituto Tecnológico de Costa Rica, Cartago, Costa Rica*

Caldeira, C.L.: *Department of Metallurgical and Materials Engineering, Universidade Federal de Minas Gerais – UFMG, Minas Gerais, Brazil & National Institute of Science and Technology on Mineral Resources, Water and Biodiversity, INCT-Acqua, Brazil*

Campolina, D.: *Campolina Toxicologia Comp. Nucl. e Audio ME, Belo Horizonte, Brazil*

Caner, L.: *CNRS/IC2MP, Université de Poitiers, Poitiers, France*

Cano Canchola, C.: *University of Guanajuato, Division of Natural and Exact Sciences, Guanajuato, Mexico*

Cañas Kurz, E.E.: *Center of Applied Research, Karlsruhe University of Applied Sciences, Karlsruhe, Germany*

Cardona-Benavides, A.: *Earth Sciences, Autonomous University of San Luis Potosi, San Luis Potosí, México*

Carneiro, M.A.C.: *Department of Soil Science, Federal University of Lavras, Lavras, Minas Gerais, Brazil*

Casiot, C.: *HydroSciences Montpellier, Univ. Montpellier-CNRS-IRD, Montpellier, France*

Castro-Larragoitia, J.: *Earth Sciences, Autonomous University of San Luis Potosi, San Luis Potosí, México*

Cauich-Kau, D.: *Institute of Hydrogeology, RWTH Aachen University, Germany & Earth Sciences, Autonomous University of San Luis Potosi, San Luis Potosí, México*

Cechova, A.: *GLAZURA, s.r.o., Dobřín, Czech Republic*

Centeno, J.A.: *Founding Member and Past President, International Medical Geology Association*

Cerveny, S.: *Centro de Física de Materiales (CSIC, UPV/EHU)-Materials Physics Center (MPC), San Sebastián, Spain*

Chacón-Madrid, K.: *Universidade Estadual de Campinas, Instituto de Química, Campinas, São Paulo, Brazil*

Chakraborti, B.: *National Institute of Hydrology, Patna, India*

Chakraborti, S.: *School of Environmental Studies, Jadavpur University, Kolkata, India*

Chakraborty, M.: *Department of Geology and Geophysics, Indian IIT Kharagpur, Kharagpur, West Bengal, India*

Chandrajith, R.: *Department of Geology, Faculty of Science, University of Peradeniya, Peradeniya, Sri Lanka*

Charron, M.: *BRGM, Water, Environment, Process Development and Analysis Division, Orléans, France*

Chatterjee, D.: *Department of Chemistry, University of Kalyani, Kalyani, West Bengal, India*

Chatterjee, S.: *Department of Geological Sciences, Jadavpur University, Kolkata, India*

Chaudhuri, S.: *Department of Civil Engineering, Indian Institute of Technology IIT-G, Guwahati, India*

Chen, C.Y.: *Department of Earth and Environmental Sciences, National Chung Cheng University, Chiayi, Taiwan, Republic of China*

Chen, J.: *Department of Cellular Biology and Pharmacology, Florida International University, Herbert Wertheim College of Medicine, Miami, USA*

Chen, Z.: *Department of Health and Environmental Sciences, Xi'an Jiaotong-Liverpool University, Jiangsu, PR China*

Chicken, A.: *Faculty of Chemistry, National Autonomous University of Mexico, Mexico City, Mexico*

Choudhury, I.: *Deparment of Geology, University of Dhaka, Dhaka, Bangladesh*

Cimadoro, J.: *Universidad de Buenos Aires. Facultad de Ciencias Exactas y Naturales, Departamento de Física, Laboratorio de Polímeros y Materiales Compuestos (LP&MC), Instituto de Física de Buenos Aires (IFIBA-CONICET), Buenos Aires, Argentina*
Ciminelli, V.S.T.: *Department of Metallurgical and Materials Engineering, Universidade Federal de Minas Gerais – UFMG, Minas Gerais, Brazil & National Institute of Science and Technology on Mineral Resources, Water and Biodiversity, Brazil & INCT-Aqua, Universidade Federal de Minas Gerais, Belo Horizonte, Brazil*
Cirpka, O.: *Geomicrobiology, Microbial Ecology and Hydrology, University of Tübingen, Tübingen, Germany*
Clemens, S.: *Department of Plant Physiology, University of Bayreuth, Bayreuth, Germany*
Cocciolo, S.: *World Bank, USA*
Collazo, P.: *Instituto de Ciencias Geológicas, Facultad de Ciencias – UDELAR, Montevideo, Uruguay*
Costa, J.S.: *Center for Nuclear Energy in Agriculture (CENA), University of Sao Paulo (USP), Piracicaba, SP, Brazil*
Cruz, O.: *Instituto de Geofísica, Universidad Nacional Autónoma de México, México City, Mexico*
Cruz Jiménez, G.: *University of Guanajuato, Division of Natural and Exact Sciences, Guanajuato, Mexico*
Cui, J.: *Guangdong Key Laboratory of Integrated Agro-environmental Pollution Control and Management, Guangdong Institute of Eco-environmental Science & Technology, Guangzhou, PR China*
Cui, Y.S.: *College of Resources and Environment, University of Chinese Academy of Sciences, Beijing, PR China*
Dako, F.: *Osijek Water Supply Company, Osijek, Croatia*
Dalmacija, B.: *Environmental Protection Research Group, Department for Chemistry, Biochemistry and Environmental Protection, University of Novi Sad Faculty of Sciences, Novi Sad, Republic of Serbia*
Dalton, C.: *FREDsense Technologies Corp., Calgary, Alberta, Canada*
da Silva, E.B.: *UF/ IFAS Soil and Water Sciences Department, University of Florida, Gainesville, USA*
Das, A.: *School of Environmental Studies, Jadavpur University, Kolkata, India*
Datta, S.: *Department of Geology, Kansas State University, Manhattan, KS, USA & Department of Geological Sciences, University of Texas at San Antonio, San Antonio, TX, USA*
Davis, J.A.: *U.S. Environmental Protection Agency, Center for Public Health and Environmental Assessments, Research Triangle, Park, USA*
De, A.: *School of Environmental Studies, Jadavpur University, Kolkata, India*
Deb, U.: *Department of Aquaculture and Fisheries, University of Arkansas, Pine Bluff, AR, USA*
De Loma, J.: *Institute of Environmental Medicine, Karolinska Institutet, Stockholm, Sweden*
de Ridder, D.J.: *Evides Water Company, Department of Water Technology, Rotterdam, The Netherlands*
Deng, Y.: *Agro-Environmental Protection Institute, Ministry of Agriculture and Rural Affairs, Tianjin, PR China*
Deonarine, A.: *Department of Civil, Environmental, and Construction Engineering, Texas Tech University, Texas, USA*
Devau, N.: *BRGM, Water, Environment, Process Development and Analysis Division, Orléans, France*
Devansh, A.: *School of Environmental Studies, Jadavpur University, Kolkata, India*
de Waal, L.: *KWR Water Research Institute, Nieuwegein, The Netherlands*
Dey, B.C.: *Department of Public Health Engineering, Dhaka, Bangladesh*

Dheeman, D.S.: *Department of Cellular Biology and Pharmacology, Florida International University, Herbert Wertheim College of Medicine, Miami, USA*
Diacomanolis, V.: *The University of Queensland, Centre For Mined Land Rehabilitation, St Lucia, Australia & The University of Queensland, Queensland Alliance for Environmental Health Sciences, QLD, Australia*
Dietrich, S.: *IHLLA – Instituto de Hidrologa de Llanuras "Dr. Eduardo J. Usunoff", Azul, Argentina*
Dissanayake, C.B.: *Department of Geology, Faculty of Science, University of Peradeniya, Peradeniya, Sri Lanka*
Dixit, A.: *Department of Civil Engineering, Indian Institute of Technology IIT-G, Guwahati, India*
Diyabalanage, S.: *Instrument Centre, Office of the Dean, Faculty of Applied Sciences, University of Sri Jayewardenepura, Sri Lanka*
Dold, B.: *Division of Geosciences and Environmental Engineering, Luleå University of Technology, Luleå, Sweden*
dos Santos, F.H.: *ESALQ/USP Soil Science Department, College of Agriculture, University of São Paulo, São Paulo, Brasil*
Dousova, B.: *University of Chemistry and Technology Prague, Prague, Czech Republic*
Druwe, I.: *U.S. Environmental Protection Agency, Center for Public Health and Environmental Assessments, Research Triangle, Park, USA*
Du, H.: *College of Resource and Environment, Hunan Agricultural University, Changsha, PR China*
Duan, G.L.: *State Key Lab of Urban and Regional Ecology, Research Center for Eco-Environmental Sciences, Chinese Academy of Sciences, PR China & University of Chinese Academy of Sciences, PR China*
Dunnican, L.: *Department of Civil, Environmental, and Construction Engineering, Texas Tech University, Texas, USA*
Dutta, S.: *School of Water Resources, Indian Institute of Technology Kharagpur, Kharagpur, India*
Duyen, V.T.: *CETASD, Vietnam National University, Hanoi, Vietnam*
Eiche, E.: *Institute of Applied Geosciences, KIT, Karlsruhe, Germany*
Ellis, T.: *Lamont-Doherty Earth Observatory of Columbia University, Palisades, NY, USA*
Ernst, M.: *Institute for Water Resources and Water Supply, Hamburg University of Technology, Am Schwarzenberg-Campus 3, Hamburg, Germany*
Espinoza-González, R.: *LabMAM, Departamento de Ingeniería Química Biotecnología y Materiales, FCFM, Universidad de Chile, Santiago, Chile*
Fang, L.: *Guangdong Key Laboratory of Integrated Agro-environmental Pollution Control and Management, Guang-dong Institute of Eco-Environmental Science & Technology, Guangzhou, PR China & National-regional joint Engineering Research Center for Soil Pollution Control and Remediation in South China, Guangzhou, PR China*
Farooqi, A.: *Department of Environmental Sciences, Quaid-i-Azam University, Pakistan*
Faulmino, C.: *Institute of Civil Engineering, College of Engineering, University of the Philippines, Diliman, Quezon City, Philippines*
Feitosa, M.M.: *Department of Soil Science, Federal University of Lavras, Lavras, Minas Gerais, Brazil*
Felipe-Sotelo, M.: *ICP-MS Facility, Chemistry Department, University of Surrey, Guildford, Surrey, UK*
Fernández-Cirelli, A.: *Universidad de Buenos Aires, Facultad de Ciencias Veterinarias, Centro de Estudios Transdisciplinarios del Agua (CETA), Buenos Aires, Argentina & Universidad de Buenos Aires-CONICET, Instituto de Investigaciones en Producción Animal (INPA), Buenos Aires, Argentina*
Fiore, M.: *Center for Nuclear Energy in Agriculture (CENA), University of Sao Paulo (USP), Piracicaba, SP, Brazil*

Flores, M.: *Laboratorio de Superficies, Depto. de Física, FCFM, Universidad de Chile, Santiago, Chile*

Flores-Ocampo, I.Z.: *Earth Sciences Postgraduate Program, Universidad Nacional Autónoma de México,Ciudad Universitaria No. 3000, Col. Copilco Universidad, Delegación Coyoacán, México*

Fonseka, S.: *Department of Pharmacology, Faculty of Medicine, University of Peradeniya, Peradeniya, Sri Lanka*

Fontaine, C.: *CNRS/IC2MP, Université de Poitiers, Poitiers, France*

Foster, S.: *Omic Research Group, Laboratory of Analytical Chemistry, Faculty of Science and Technology, University of Canberra, Australia*

Fostier, A.H.: *Universidade Estadual de Campinas, Instituto de Química, Campinas, São Paulo, Brazil*

Fröllje, H.: *University of Bremen, Department of Geosciences, Geochemistry & Hydrogeology, Bremen, Germany*

Fuentes-Hernandez, H.A.: *Instituto de Ecología A.C. (INECOL), Red Ambiente y Sustentabilidad, Chihuahua, México*

Galván, A.E.: *Department of Cellular Biology and Pharmacology, Florida International University, Herbert Wertheim College of Medicine, Miami, USA*

Gamazo, P.: *Water Department, Centro Universitario Regional Litoral Norte, Universidad de la República, Salto, Uruguay*

Gao, Y.: *Institute of Environment Remediation and Health, Southwest Forestry University, Kunming, Yunnan, PR China & Institute of Ecology and Environment, Southwest Forestry University, Kunming, Yunnan, PR China*

Gao, Z.P.: *State Key Laboratory of Biogeology and Environmental Geology, China University of Geosciences, Beijing, PR China & School of Water Resources and Environment, China University of Geosciences (Beijing), Beijing, PR China*

Garbinski, L.D.: *Department of Cellular Biology and Pharmacology, Herbert Wertheim College of Medicine, Florida International University, Miami, USA*

Gardon, J.: *Hydrosciences Montpellier, Institut de Recherche pour le Développement, CNRS, University of Montpellier, France*

Garrido Hoyos, S.E.: *Mexican Institute of Water Technology (IMTA), Postgraduate Subcoordination, Jiutepec, Morelos, Mexico*

Gaurav, A.: *Mahavir Cancer Institute & Research Center, Phulwari Sharif, Patna, India*

Gayantha, K.: *MaxPlank Institute of Biogeochemistry, Jena, Germany*

Ge, Y.: *Jiangsu Provincial Key Laboratory of Marine Biology, College of Resources and Environmental Sciences, Nanjing Agricultural University, Nanjing, PR China*

Ghasemi, S.: *Department of Civil, Environmental, and Construction Engineering, Texas Tech University, Texas, USA*

Ghisolfi, S.: *Institute for International Economic Studies, Stockholm University, Sweden*

Ghosh, A.K.: *Mahavir Cancer Institute & Research Centre, Patna, Bihar, India*

Ghosh, D.: *Biogeochemistry Research Group, Centre for Earth Sciences, Indian Institute of Science, Bangalore, India*

Ghosh, P.: *Department of Chemistry, University of Kalyani, Kalyani, Nadia, West Bengal, India*

Ghosh, U.C.: *Department of Chemistry, Presidency University, Kolkata, India*

Gift, J.S.: *U.S. Environmental Protection Agency, Center for Public Health and Environmental Assessments, Research Triangle, Park, USA*

Giri, A.K.: *Molecular Genetics Division, CSIR-Indian Institute of Chemical Biology, Kolkata, India*

Glodowska, M.: *Geomicrobiology, Microbial Ecology and Hydrology, University of Tübingen, Tübingen, Germany*

Goessler, W.: *Institute of Chemistry, University of Graz, Graz, Austria*

González, J.: *National Autonomous University of Mexico, Faculty of Chemistry, Mexico City, Mexico*
Gooddy, D.C.: *Groundwater Processes, British Geological Survey, Wallingford, UK*
Gómez-Hernández, A.: *Instituto Potosino de Investigación Científica y Tecnológica, División de Geociencias Aplicadas, San Luis Potosí, México*
González-Rodríguez, R.: *Environmental Protection Research Center (CIPA), School of Chemistry, Instituto Tecnológico de Costa Rica, Cartago, Costa Rica*
Goswami, R.: *Department of Environmental Science, The Assam Royal Global University, Guwahati, India*
Gourain, A.C.G.: *School of Ocean Sciences, University of Liverpool, UK*
Goyanes, S.: *Universidad de Buenos Aires. Facultad de Ciencias Exactas y Naturales, Departamento de Física, Laboratorio de Polímeros y Materiales Compuestos (LP&MC), Instituto de Física de Buenos Aires (IFIBA-CONICET), Buenos Aires, Argentina & Consejo Nacional de Investigaciones Científicas y Técnicas (CONICET), Buenos Aires, Argentina*
Guerra, M.B.B.: *Department of Chemistry, Federal University of Lavras, Lavras, Minas Gerais, Brazil*
Guilherme, L.R.G.: *Department of Soil Science, Federal University of Lavras, Lavras, Minas Gerais, Brazil*
Gutiérrez-Aviña, X.: *Programa de Maestría en Ingeniería Ambiental, Universidad Nacional Autónoma de México (UNAM), Mexico City, Mexico*
Gunn, A.: *Omic Research Group, Laboratory of Analytical Chemistry, Faculty of Science and Technology, University of Canberra, Australia & National Centre for Forensic Studies, Faculty of Science and Technology, University of Canberra, Australia*
Guo, H.M.: *State Key Laboratory of Biogeology and Environmental Geology, China University of Geosciences, Beijing, PR China & School of Water Resources and Environment, China University of Geosciences (Beijing), Beijing, PR China*
Gupta, A.K.: *School of Water Resources, Indian Institute of Technology Kharagpur, Kharagpur, India*
Gupta, K.: *Department of Chemistry, Presidency University, Kolkata, India*
Gupta, S.: *Department of Geology and Geophysics, Indian Institute of Technology Kharagpur, India*
Gustave, W.: *Department of Environmental Science, University of Liverpool, Brownlow Hill, Liverpool, UK & Department of Health and Environmental Sciences, Xi'an Jiaotong-Liverpool University, Jiangsu, PR China*
Habib, A.: *KTH-International Groundwater Arsenic Research Group, Department of Sustainable Development, Environmental Science and Engineering, KTH Royal Institute of Technology, Stockholm, Sweden & GO Forum for Public Health, Dhaka, Bangladesh*
Hansen, B.: *Geological Survey of Denmark and Greenland (GEUS), Aarhus, Denmark*
Harris, H.H.: *Department of Chemistry, University of Adelaide, SA, Australia*
Hasibuzzaman, M.M.: *Department of Biochemistry and Molecular Biology, University of Rajshahi, Rajshahi, Bangladesh*
Haugwitz, C.: *Ramboll Denmark A/S, Copenhagen, Denmark*
Hechavarría-Hernández, A.: *Instituto de Química, Campinas, Universidade Estadual de Campinas, São Paulo, Brazil*
Hellal, J.: *BRGM, Water, Environment, Process Development and Analysis Division, Orléans, France*
Hellriegel, U.: *Center of Applied Research, Karlsruhe University of Applied Sciences, Karlsruhe, Germany*
Herath, I.: *Faculty of Health, Engineering and Sciences, University of Southern Queensland, Toowoomba, Queensland, Australia*

Hernández-Martínez, J.L.: *Instituto Potosino de Investigación Científica y Tecnológica, División de Geociencias Aplicadas, San Luis Potosí, México*
Hery, M.: *HydroSciences Montpellier, Univ. Montpellier-CNRS-IRD, Montpellier, France*
Hicks, E.: *FREDsense Technologies Corp., Calgary, Alberta, Canada*
Himeno, S.: *Laboratory of Molecular Nutrition and Toxicology, Faculty of Pharmaceutical Sciences, Tokushima Bunri University, Tokushima, Japan*
Hofman-Caris, C.H.M.: *KWR Water Research Institute, Nieuwegein, The Netherlands*
Hofs, B.: *Evides Water Company, Department of Water Technology, Rotterdam, The Netherlands*
Hogendoorn, A.: *Evides Water Company, Department of Water Technology, Rotterdam, The Netherlands*
Hoinkis, J.: *Center of Applied Research, Karlsruhe University of Applied Sciences, Karlsruhe, Germany*
Hong, L.: *Guangdong Key Laboratory of Integrated Agro-environmental Pollution Control and Management, Guang-dong Institute of Eco-Environmental Science & Technology, Guangzhou, PR China & National-regional joint Engineering Research Center for Soil Pollution Control and Remediation in South China, Guangzhou, PR China*
Honma, T.: *Niigata Agricultural Research Institute, Nagaoka, Japan*
Hoque, B.A.: *Environment and Population Research Centre, New DOHS, Mohakhali, Dhaka, Bangladesh*
Hosen, F.: *Department of Biochemistry and Molecular Biology, University of Rajshahi, Bangladesh*
Hosen, Z.: *Department of Applied Nutrition and Food Technology, Islamic University, Bangladesh*
Hossain, E.: *Department of Biochemistry and Molecular Biology, University of Rajshahi, Bangladesh*
Hossain, K.: *Department of Biochemistry and Molecular Biology, University of Rajshahi, Rajshahi, Bangladesh*
Hossain, S.: *Department of Biochemistry and Molecular Biology, University of Rajshahi, Bangladesh*
Howlader, A.H.: *Department of Chemistry and Biochemistry, Florida International University, Miami, USA*
Huang, Y.H.: *Department of Earth and Environmental Sciences, National Chung Cheng University, Chiayi, Taiwan, Republic of China*
Hug, S.J.: *Eawag, Swiss Federal Institute of Aquatic Science and Technology, Dübendorf, Switzerland*
Hussain, M.M.: *Institute of Soil and Environmental Sciences, University of Agriculture Faisalabad, Faisalabad, Pakistan*
Huque, M.S.: *Environment and Population Research Centre, New DOHS, Mohakhali, Dhaka, Bangladesh*
Huynh, T.: *Centre For Mined Land Rehabilitation, The University of Queensland, St Lucia, QLD, Australia*
Huysman, K.: *PIDPA, Antwerp, Belgium*
Ijumulana, J.: *DAFWAT Research Group, Department of Water Resources Engineering, College of Engineering and Technology, University of Dar es Salaam, Dar es Salaam, Tanzania & KTH-International Groundwater Arsenic Research Group, Department of Sustainable Development, Environmental Science and Engineering, KTH Royal Institute of Technology, Stockholm, Sweden*
Irunde, R.: *DAFWAT Research Group, Department of Water Resources Engineering, College of Engineering and Technology, University of Dar es Salaam, Dar es Salaam, Tanzania; KTH-International Groundwater Arsenic Research Group, Department of Sustainable Development, Environmental Science and Engineering, KTH Royal Institute of Technology, Stockholm,*

Sweden & Department of Chemistry, College of Natural and Applied Sciences, University of Dar es Salaam, Tanzania
Ishikawa, S.: *Division of Hazardous Chemicals, National Institute for Agro-Environmental Sciences, NARO, Tsukuba, Ibaraki, Japan*
Islam, Md. N.: *Queensland Alliance for Environmental Health Sciences, The University of Queensland, Brisbane, QLD, Australia*
Islam, M.R.: *Global Centre for Environmental Remediation (GCER), Faculty of Science, The University of Newcastle, Callaghan, NSW, Australia & Cooperative Research Centre for Contamination Assessment and Remediation of the Environment (CRC CARE), The University of Newcastle, Callaghan, NSW, Australia*
Islam, M.S.: *Department of Applied Nutrition and Food Technology, Islamic University, Kushtia, Bangladesh*
Islam, M.T.: *KTH-International Groundwater Arsenic Research Group, Department of Sustainable Development, Environmental Sciences and Engineering, KTH Royal Institute of Technology, Stockholm, Sweden & WaterAid-Bangladesh, Dhaka, Bangladesh*
Islam, S.: *School of Earth and Environmental Sciences, University of Manchester, Manchester, UK & Department of Soil Science, Bangladesh Agricultural University, Mymensingh, Bangladesh*
Islam, S.: *Global Centre for Environmental Remediation (GCER), Faculty of Science, The University of Newcastle, Callaghan, NSW, Australia, Cooperative Research Centre for Contamination Assessment and Remediation of the Environment (CRC CARE), The University of Newcastle, Callaghan, NSW, Australia & Department of Soil Science, Bangladesh Agricultural University, Mymensingh, Bangladesh, Terengganu, Kuala Terengganu*
Jacob, J.: *BRGM, Water, Environment, Process and Analyses Division, Orléans, France*
Jain, N.: *Polymer & Process Engineering Department, IIT Roorkee Saharanpur Campus, Saharanpur, Uttar Pradesh, India*
Jaafar, M.: *ICP-MS Facility, Chemistry Department, University of Surrey, Guildford, Surrey, UK & Universiti Malaysia Jaiswal, M.K. Department of Earth Sciences, Indian Institute of Science Education and Research (IISER) Kolkata, Mohanpur, West Bengal, India*
Jakariya, M.: *Department of Environmental Science and Management, North South University, Dhaka, Bangladesh*
Jamil, N.B.: *Department of Earth and Environmental Studies, Montclair State University, Montclair, New Jersey, USA*
Javed, A.: *Environmental Geochemistry Laboratory, Department of Environmental Sciences, Faculty of Biological Sciences, Quaid-i-Azam University, Islamabad, Pakistan & Department of Earth and Environmental Sciences, Bahria University, Islamabad, Pakistan*
Jeelani, A.H.: *Centre of Social Medicine and Community Health, School of Social Sciences, Jawaharlal Nehru University, New Delhi, India*
Jeworrek, A.: *Evides Water Company, Department of Water Technology, Rotterdam, The Netherlands*
Jiang, Z.Q.: *Jiangsu Provincial Key Laboratory of Marine Biology, College of Resources and Environmental Sciences, Nanjing Agricultural University, Nanjing, PR China*
Joardar, M.: *School of Environmental Studies, Jadavpur University, Kolkata, India*
Johnston, D.: *UNICEF Bangladesh, Dhaka, Bangladesh*
José Triviño, J.: *Pontificia Universidad Católica de Chile, Facultad de Ciencias Biológicas, Santiago, Chile*
Joshi, H.: *Department of Hydrology, Indian Institute of Technology, Roorkee, India*
Joulian, C.: *BRGM, Water, Environment, Process Development and Analysis Division, Orléans, France*
Juárez-Aparicio, F.: *Earth Sciences Postgraduate Program, Universidad Nacional Autónoma de México, Ciudad Universitaria No. 3000, Col. Copilco Universidad, Delegación Coyoacán, México*

Jurji, Z.: *UNICEF Bangladesh, Dhaka, Bangladesh*
Kaegi, R.: *Eawag, Swiss Federal Institute of Aquatic Science and Technology, Dübendorf, Switzerland*
Kandavelu, P.: *SER-CAT and Department of Biochemistry and Molecular Biology, University of Georgia, Athens, USA*
Kappler, A.: *Geomicrobiology, Microbial Ecology and Hydrology, University of Tübingen, Tübingen, Germany*
Kar, K.K.: *Advanced Nano-engineering Materials Laboratory (ANML), Materials Science Programme (MSP) and Department of Mechanical Engineering, Indian Institute of Technology Kanpur, Kanpur, India*
Karim, Y.: *Department of Biochemistry and Molecular Biology, University of Rajshahi, Bangladesh*
Kaur, N.: *Department of Geology, Panjab University, Chandigarh, India*
Ke, T.: *School of Water Resources and Environment, China University of Geosciences (Beijing), Beijing, PR China*
Kelly, T.: *Omic Research Group, Laboratory of Analytical Chemistry, Faculty of Science and Technology, University of Canberra, Australia & National Centre for Forensic Studies, Faculty of Science and Technology, University of Canberra, Australia*
Kerl, C.F.: *Environmental Geochemistry, Bayreuth Center for Ecology and Environmental Research (BayCEER), University of Bayreuth, Bayreuth, Germany*
Khan, E.R.: *Department of Public Health Engineering (DPHE), Dhaka, Bangladesh*
Khan, M.A.I.: *Environment and Population Research Centre, New DOHS, Mohakhali, Dhaka, Bangladesh*
Khan, M.R.: *Department of Geology, University of Dhaka, Dhaka, Bangladesh*
Khan, R.: *Faculty of Civil Engineering, Institute of Technology, Shri Ramswaroop Memorial University, Lucknow, India*
Khan, T.M.S.: *Department of Public Health Engineering (DPHE), Dhaka, Bangladesh*
Khanam, S.: *Environment and Population Research Centre (EPRC), Dhaka, Bangladesh*
Kibria, M.G.: *Department of Geology, Kansas State University, Manhattan, KS, USA & Department of Geological Sciences, University of Texas at San Antonio, San Antonio, TX, USA*
Kimambo, V.: *DAFWAT Research Group, Department of Water Resources Engineering, College of Engineering and Technology, University of Dar es Salaam, Dar es Salaam, Tanzania & KTH-International Groundwater Arsenic Research Group, Department of Sustainable Development, Environmental Science and Engineering, KTH Royal Institute of Technology, Stockholm, Sweden*
Kipfer, R.: *Eawag, Swiss Federal Institute of Aquatic Science and Technology, Dübendorf, Switzerland & Institute for Geochemistry and Petrology, ETH Zürich, Zürich, Switzerland*
Kirschbaum, A.: *Instituto de Bio y Geo Ciencias del NOA (IBIGEO), CONICET-Universidad Nacional de Salta, Rosario de Lerma, Salta, Argentina*
Kleindienst, S.: *Geomicrobiology, Microbial Ecology and Hydrology, University of Tübingen, Tübingen, Germany*
Kolousek, D.: *University of Chemistry and Technology Prague, Prague, Czech Republic*
Kontny, A.: *Institute of Applied Geosciences, KIT, Karlsruhe, Germany*
Krause, S.: *School of Geography, Earth and Environmental Sciences, University of Birmingham, UK*
Kremenic, K.: *University of Chemistry and Technology Prague, Prague, Czech Republic*
Krishan, G.: *Groundwater Hydrology Division, National Institute of Hydrology, Roorkee, Uttarakhand, India*
Kristiansen, S.M.: *Department of Geoscience, Aarhus University, Denmark*
Kruse, E.: *CEIDE-Facultad de Ciencias Naturales y Museo, Universidad Nacional de La Plata, La Plata, Argentina*

Kshetrimayum, K.S.: *Department of Geology, Central University of Punjab, Ghudda, Bathinda, India*
Kulkarni, H.V.: *Department of Geology, Kansas State University, Manhattan, KS, USA & Department of Geological Sciences, University of Texas at San Antonio, San Antonio, TX, USA*
Kumar, A.: *Mahavir Cancer Institute & Research Centre, Patna, Bihar, India*
Kumar, D.: *School of Environmental Studies, Jadavpur University, Kolkata, India*
Kumar, M.: *Discipline of Earth Sciences, Indian Institute of Technology Gandhinagar, Gujarat, India; Sustainability Cluster, School of Engineering, University of Petroleum & Energy Studies, Dehradun, Uttarakhand, India & Escuela de Ingeniería y Ciencias, Tecnologico de Monterrey, Campus Monterey, Monterrey, Nuevo Leon, Mexico*
Kumar, M.: *Department of Environmental Science, School of Earth, Environment & Space Studies, Central University of Haryana, Jant Pali, Mahendergarh, India; School of Environmental Sciences, Jawaharlal Nehru University, New Delhi, India*
Kumar, R.: *Mahavir Cancer Institute & Research Centre, Patna, Bihar, India*
Kumar, S.: *Department of Civil Engineering, Indian Institute of Technology IIT-G, Guwahati, India*
Kurajica, L.: *Department for Water Safety and Water Supply, Croatian Institute of Public Health, Zagreb, Croatia*
Kuramata, M.: *Division of Hazardous Chemicals, National Institute for Agro-Environmental Sciences, NARO, Tsukuba, Ibaraki, Japan*
Lama, U.: *School of Environmental Studies, Jadavpur University, Kolkata, India*
Lamb, D.: *Global Centre for Environmental Remediation (GCER), The University of Newcastle, Callaghan, NSW, Australia*
Landa-Arreguín, J.F.A.: *Posgrado en Ciencias de la Tierra, Universidad Nacional Autónoma de México, Del. Coyoacán, Ciudad de México., México*
Lazo, A.: *Applied Research, Research & Development Department, Costa Rican Institute for Drinking Water and Sanitation (AyA), San José, Costa Rica*
Lee, J.S.: *U.S. Environmental Protection Agency, Center for Public Health and Environmental Assessments, Research Triangle, Park, USA*
Lehman, T.: *Department of Geoscience, Texas Tech University, Texas, USA*
Lei, M.: *College of Resource & Environment, Hunan Agricultural University, Changsha, PR China*
Leister, J.P.: *Institute of Chemistry, University of Campinas (UNICAMP), Campinas, São Paulo, Brazil*
Levi, M.: *Institute of Environmental Medicine, Karolinska Institutet, Stockholm, Sweden*
Lhotka, M.: *University of Chemistry and Technology Prague, Prague, Czech Republic*
Li, B.: *College of Resource & Environment, Hunan Agricultural University, Changsha, PR China*
Li, C.H.: *Jiangsu Provincial Key Laboratory of Marine Biology, College of Resources and Environmental Sciences, Nanjing Agricultural University, Nanjing, PR China*
Li, F.: *Guangdong Bioengineering Institute (Guangzhou Sugarcane Industry Research Institute), Guangdong Key Laboratory of Sugarcane Improvement and Biorefinery, Guangzhou, PR China; Guangdong Key Laboratory of Integrated Agro-environmental Pollution Control and Management, Guangdong Institute of Eco-Environmental Science & Technology, Guangzhou, PR China & National-regional joint Engineering Research Center for Soil Pollution Control and Remediation in South China, Guangzhou, PR China*
Li, H.F.: *Beijing Key Laboratory of Farmland Soil Pollution Prevention and Remediation, College of Resources and Environmental Sciences, China Agricultural University, Beijing, PR China*
Li, Q.: *Department of Earth and Environmental Sciences, School of Natural Sciences and Williamson Research Center for Molecular Environmental Science, University of Manchester, Manchester, UK*

Li, S.: *Key Laboratory of Integrated Agro-environmental Pollution Control and Management, Guangdong Institute of Eco-Environmental Science & Technology, Guangzhou, PR China*

Li, X.: *SCNU Environmental Research Institute, Guangdong Provincial Key Laboratory of Chemical Pollution and Environmental Safety, School of Environment, South China Normal University, Guangzhou, PR China*

Ligate, F.J.: *DAFWAT Research Group, Department of Water Resources Engineering, College of Engineering and Technology, University of Dar es Salaam, Dar es Salaam, Tanzania & KTH-International Groundwater Arsenic Research Group, Department of Sustainable Development, Environmental Science and Engineering, KTH Royal Institute of Technology, Stockholm, Sweden*

Lightfoot, A.: *Eawag, Swiss Federal Institute of Aquatic Science and Technology, Dübendorf, Switzerland*

Litter, M.I.: *Universidad de San Martín, San Martín, Prov. de Buenos Aires, Argentina*

Liu, C.: *Guangdong Key Laboratory of Integrated Agro-environmental Pollution Control and Management,Guangdong Institute of Eco-Environmental Science & Technology, Guangzhou, PR China*

Liu, X.: *Institute of Environment Remediation and Health, Southwest Forestry University, Kunming, Yunnan, PR China & Institute of Ecology and Environment, Southwest Forestry University, Kunming, Yunnan, PR China*

Lloyd, J.: *Williamson Research Centre for Molecular Environmental Science, School of Earth and Environmental Sciences, the University of Manchester, Manchester, UK*

Londonio, A.: *Comisión Nacional de Energía Atómica (CNEA), Gerencia Química, Buenos Aires, Argentina & Universidad de San Martín (UNSAM), 3iA, Argentina*

Lou, C.-Y.: *Department of Earth Sciences, National Cheng-Kung University, Tainan, Taiwan, Republic of China*

Lugo-Dorantes, A.E.: *Earth Sciences Student, México, Universidad Nacional Autónoma de México,Ciudad Universitaria No. 3000, Col. Copilco Universidad, Delegación Coyoacán, Mexico*

Luong, T.V.: *Lappeenranta University of Technology, Lappeenranta, Finland*

Luong, V.T.: *Department of Mechatronics and Sensor Systems Technology, Vietnamese-German University, Vietnam*

Ma, L.Q.: *Institute of Environment Remediation and Health, Southwest Forestry University, Kunming, Yunnan, PR China & Institute of Ecology and Environment, Southwest Forestry University, Kunming, Yunnan, PR China*

Macasieb, R.: *Institute of Civil Engineering, College of Engineering, University of the Philippines, Diliman, Quezon City, Philippines*

Machado, I.: *Analytical Chemistry, DEC, Faculty of Chemistry, Universidad de la República, Montevideo, Uruguay*

Mahanta, C.: *Department of Civil Engineering, Indian Institute of Technology, IIT-G, Guwahati, Assam, India*

Mahmud, M.N.: *UNICEF Bangladesh, Dhaka, Bangladesh*

Mai, T.T.: *CETASD, Vietnam National University, Hanoi, Vietnam*

Mailloux, B.J.: *Environmental Sciences Department, Barnard College, Columbia University, New York, USA*

Maiti, A.: *Polymer & Process Engineering Department, IIT Roorkee Saharanpur Campus, Saharanpur, Uttar Pradesh, India*

Maity, J.P.: *Department of Earth and Environmental Sciences, National Chung Cheng University, Min-Hsiung, Chiayi County, Taiwan & Department of Chemistry, School of Applied Sciences, KIIT Deemed to be University, Bhubaneswar, Odisha, India*

Majhi, K.: *School of Environmental Studies, Jadavpur University, Kolkata, India*

Majumdar, A.: *Department of Earth Sciences, Indian Institute of Science Education and Research Kolkata (IISER-K), Mohanpur, West Bengal, India*
Mallick, A.: *ENVIS RP on Environmental Biotechnology, Department of Environmental Science, University of Kalyani, West Bengal, India*
Mamani, J.: *Genotoxicity Unit, Genetic Institute, Medicine School, Universidad Mayor de San Andrés, La Paz, Bolivia*
Mandal, D.: *Sripat Singh College, Jiaganj, Murshidabad, India*
Mandal, U.: *Department of Chemistry, University of Kalyani, Kalyani, Nadia, West Bengal, India*
Mañay, N.: *Toxicology, DEC, Faculty of Chemistry, Universidad de la República, Montevideo, Uruguay*
Martínez-Jardines, L.G.: *Universidad Nacional Autónoma de México, Departamento de Recursos Naturales, Instituto de Geofísica, Ciudad Universitaria No. 3000, Col. Copilco Universidad, Delegación Coyoacán, México*
Martínez-Villegas, N.: *Instituto Potosino de Investigación Científica y Tecnológica, División de Geociencias Aplicadas, San Luis Potosí, México*
Mayall, R.M.: *FREDsense Technologies Corp., Calgary, Alberta, Canada*
Mazumder, M.: *Department of Chemistry, University of Kalyani, Kalyani, Nadia, West Bengal, India*
McDonald, M.: *FREDsense Technologies Corp., Calgary, Alberta, Canada*
Mendes, H.L.: *Anglo Gold Ashanti, Minas Gerais, Brazil*
Meza-Figueroa, D.: *Departamento de Geología, Universidad de Sonora, Sonora, Mexico*
Mhanna, R.: *CNRS/IC2MP, Université de Poitiers, Poitiers, France*
Michael, H.: *Department of Geological Sciences, University of Delaware, Delaware, USA*
Millerick, K.: *Department of Civil, Environmental, and Construction Engineering, Texas Tech University, Texas, USA*
Mishra, A.K.: *Department of Geology and Geophysics, Indian Institute of Technology–Kharagpur, Kharagpur, West Bengal, India*
Mohan, D.: *School of Environmental Sciences, Jawaharlal Nehru University, New Delhi, India*
Mondal, D.: *School of Science, Engineering & Environment, University of Salford, Salford, UK*
Mondal, P.: *Ceramic Membrane Division, CSIR-Central Glass and Ceramic Research Institute, Kolkata, India*
Mondal, V.: *Department of Biochemistry and Molecular Biology, University of Rajshahi, Bangladesh*
Monroy-Licht, A.: *Department of Chemistry and Biology, Universidad del Norte, Barranquilla, Colombia & Environmental and Computational Chemistry Group, School of Pharmaceutical Sciences, Universidad de Cartagena, Cartagena, Colombia*
Morales-Arredondo, J.I.: *Instituto de Geofísica, Universidad Nacional Autónoma de México, Ciudad de México, México*
Mozumder, M.R.H.: *Lamont-Doherty Earth Observatory, Columbia University, Palisades, New York, USA*
Mridha, D.: *School of Environmental Studies, Jadavpur University, Kolkata, India*
Mtalo, F.: *DAFWAT Research Group, Department of Water Resources Engineering, College of Engineering and Technology, University of Dar es Salaam, Dar es Salaam, Tanzania*
Mtamba, J.: *DAFWAT Research Group, Department of Water Resources Engineering, College of Engineering and Technology, University of Dar es Salaam, Dar es Salaam, Tanzania*
Mukherjee, A.: *Department of Geology and Geophysics, Indian Institute of Technology–Kharagpur, Kharagpur, West Bengal, India & School of Environmental Science and Engineering, Indian Institute of Technology–Kharagpur, Kharagpur, West Bengal, India & Applied Policy Advisory to Hydrogeosciences Group, Indian Institute of Technology–Kharagpur, Kharagpur, West Bengal, India*

Murray, J.: *Instituto de Bio y Geo Ciencias del NOA (IBIGEO), CONICET-Universidad Nacional de Salta, Rosario de Lerma, Salta, Argentina*
Nadar, V.S.: *Department of Cellular Biology and Pharmacology, Florida International University, Herbert Wertheim College of Medicine, Miami, USA*
Naidu, R.: *Global Centre for Environmental Remediation (GCER), Faculty of Science, The University of Newcastle, Callaghan, NSW, Australia & Cooperative Research Centre for Contamination Assessment and Remediation of the Environment (CRC CARE), The University of Newcastle, Callaghan, NSW, Australia*
Navarro-González, I.: *Instituto de Ingeniería, Universidad Nacional Autónoma de México (UNAM), Mexico City, Mexico*
Naveau, A.: *CNRS/IC2MP, Université de Poitiers, Poitiers, France*
Naveed, S.: *Jiangsu Provincial Key Laboratory of Marine Biology, College of Resources and Environmental Sciences, Nanjing Agricultural University, Nanjing, PR China*
Nawaz, M.F.: *Department of Forestry & Range Management, University of Agriculture Faisalabad, Faisalabad, Pakistan*
Neumann, T.: *Institute of Applied Geosciences, KIT, Karlsruhe, Germany*
Ng, J.C.: *The University of Queensland, Queensland Alliance for Environmental Health Sciences, Brisbane, QLD, Australia*
Niang, M.: *UNICEF Bangladesh, Dhaka, Bangladesh*
Niazi, N.K.: *Institute of Soil and Environmental Sciences, University of Agriculture Faisalabad, Faisalabad, Pakistan & School of Civil Engineering and Surveying, University of Southern Queensland, Toowoomba, Queensland, Australia*
Nikić, J.: *Environmental Protection Research Group, Department of Chemistry, Biochemistry and Environmental Protection, University of Novi Sad, Faculty of Sciences, Novi Sad, Serbia*
Nixon, S.: *Williamson Research Centre for Molecular Environmental Science, School of Earth and Environmental Sciences, The University of Manchester, Manchester, UK*
Nkiriti, N.S.: *IHE Delft Institute for Water Education, Delft, The Netherlands*
Noller, B.N.: *The University of Queensland, Centre For Mined Land Rehabilitation, St Lucia, Australia*
Nordstrom, D.K.: *United States Geological Survey, Boulder, CO, USA*
Núñez, C.: *Facultad de Química y de Farmacia, Pontificia Universidad Católica de Chile, Chile*
Ochoa-Rivero, J.M.: *Instituto Nacional de Investigaciones Forestales, Agrícolas y Pecuarias (INIFAP), Campo Experimental La Campana, Chihuahua, México*
Ogata, R.: *Japan International Cooperation Agency (JICA), Amman, Jordan & The University of Tokyo, Tokyo, Japan*
Oliveira, C.: *Department of Soil Science, Federal University of Lavras, Lavras, Minas Gerais, Brazil*
Ormachea Muñoz, M.: *Laboratorio de Hidroquímica, Instituto de Investigaciones Químicas, Universidad Mayor de San Andrés, La Paz, Bolivia*
Ortega-Gutiérrez, J.E.: *Posgrado en Ciencias de la Tierra, Universidad Nacional Autónoma de México, Del. Coyoacán, Ciudad de México., México*
Otomo, J.: *Environmental Chemical Energy Engineering, Department of Environment Systems, GSFS, The University of Tokyo, Japan*
Padhi, S.: *Geosphere Environment Systems, Department of Environment Systems, GSFS, The University of Tokyo, Japan*
Paikaray, S.: *Department of Geology, Panjab University, Chandigarh, India*
Palatnik, J.: *Comisión Nacional de Energía Atómica (CNEA), Gerencia Química, Buenos Aires, Argentina*
Pamoukaghlián, K.: *Instituto de Ciencias Geológicas, Facultad de Ciencias – UDELAR, Montevideo, Uruguay*

Pan, D.: *SCNU Environmental Research Institute, Guangdong Provincial Key Laboratory of Chemical Pollution and Environmental Safety, School of Environment, South China Normal University, Guangzhou, PR China & Guangdong Key Laboratory of Integrated Agro-Environmental Pollution Control and Management, Guangdong Institute of Eco-Environmental Science & Technology, Guangzhou, PR China*

Patel, A.K.: *Department of Environmental Sciences, Jawaharlal University, New Delhi, India*

Patzner, M.: *Geomicrobiology, Microbial Ecology and Hydrology, University of Tübingen, Tübingen, Germany*

Paul, D.: *National Centre For Microbial Resource, NCCS, Pune, India*

Penido, E.S.: *Department of Chemistry, Federal University of Lavras, Lavras, Minas Gerais, Brazil*

Penke, Y.K.: *Advanced Nano-engineering Materials Laboratory (ANML), Materials Science Programme (MSP) and Department of Mechanical Engineering, Indian Institute of Technology Kanpur, Kanpur, India*

Pérez-Carrera, A.L.: *Universidad de Buenos Aires, Facultad de Ciencias Veterinarias, Centro de Estudios Transdisciplinarios del Agua (CETA, UBA), e Instituto de Investigaciones en Producción Animal (INPA, UBA-CONICET), Buenos Aires, Argentina*

Petersen, M.M.: *Department of Geoscience, Aarhus University, Denmark & Brabant Water, Breda, The Netherlands*

Petrusevski, B.: *IHE Delft Institute for Water Education, Delft, The Netherlands*

Pichler, T.: *Department of Geosciences, Geochemistry & Hydrogeology, University of Bremen, Bremen, Germany*

Pincetti-Zúñiga, G.P.: *School of Earth, Atmospheric and Environmental Sciences and Williamson Research Centre for Molecular Environmental Science, University of Manchester, Manchester, UK*

Pizzaro, I.: *Analytical Chemistry Laboratory, Universidad de Antofagasta, Chile*

Pizzorno, P.: *Toxicology, Faculty of Chemistry, DEC, Universidad de la República (UdelaR), Montevideo, Uruguay*

Planer-Friedrich, B.: *Environmental Geochemistry, Bayreuth Center for Ecology and Environmental Research (BayCEER), University of Bayreuth, Bayreuth, Germany*

Podgorski, J.: *Eawag, Swiss Federal Institute of Aquatic Science and Technology, Dübendorf, Switzerland & University of Manchester, Department of Earth and Environmental Sciences, Manchester, UK*

Pokhrel, G.R.: *Fujian Provincial Key Laboratory of Agroecological Processing and Safety Monitoring, College of Life Sciences, Fujian Agriculture and Forestry University, Fuzhou, PR China*

Polya, D.A.: *School of Earth and Environmental Sciences and Williamson Research Centre for Molecular Environmental Science, University of Manchester, Manchester, UK*

Premaratne, K.M.: *Department of Geology, Faculty of Science, University of Peradeniya, Sri Lanka*

Priyadarshni, P.: *School of Environmental Studies, Jadavpur University, Kolkata, India*

Prommer, H.: *University of Western Australia and CSIRO Land and Water, Perth, Australia*

Qiao, J.: *Guangdong Bioengineering Institute (Guangzhou Sugarcane Industry Research Institute), Guangdong Key Laboratory of Sugarcane Improvement and Biorefinery, Guangzhou, PR China*

Qiao, W.: *School of Water Resources and Environment, China University of Geosciences (Beijing), Beijing, PR China*

Quino Lima, I.: *KTH-International Groundwater Arsenic Research Group, Department of Sustainable Development, Environmental Science and Engineering, KTH Royal Institute of Technology, Stockholm, Sweden & Laboratorio de Hidroquímica, Instituto de Investigaciones Químicas, Universidad Mayor de San Andrés, La Paz, Bolivia*

Quintanilla, J.: *Laboratorio de Hidroquímica, Instituto de Investigaciones Químicas, Universidad Mayor de San Andrés, La Paz, Bolivia*
Radhakrishnan, M.: *Department of Cellular Biology and Pharmacology, Florida International University, Herbert Wertheim College of Medicine, Miami, USA*
Rafferty, C.: *Department of Plant Physiology, University of Bayreuth, Bayreuth, Germany*
Rahman, I.: *Department of Geology, University of Dhaka, Dhaka, Bangladesh*
Rahman, M.: *Pure Earth, Dhaka, Bangladesh*
Rahman, M.A.: *Global Centre for Environmental Remediation (GCER), The University of Newcastle, Callaghan, NSW, Australia & Department of Public Health Engineering (DPHE), Zonal Laboratory, Khulna, Bangladesh*
Rahman, M.M.: *Global Centre for Environmental Remediation (GCER), Faculty of Science, The University of Newcastle, Callaghan, NSW, Australia, Cooperative Research Centre for Contamination Assessment and Remediation of the Environment (CRC CARE), The University of Newcastle, Callaghan, NSW, Australia*
Rahman, M.S.: *A.N. College (Magadh University, Bodh-Gaya) Bihar, India*
Rahman, M.S.: *Department of Public Health Engineering, Dhaka, Bangladesh*
Ramanathan, A.L.: *School of Environmental Sciences, Jawaharlal Nehru University, New Delhi, India*
Ramanayaka, S.: *Ecosphere Resilience Research Center, Faculty of Applied Sciences, University of Sri Jayewardenepura, Nugegoda, Sri Lanka*
Ramkumar, J.: *Advanced Nano-engineering Materials Laboratory (ANML), Materials Science Programme (MSP) and Department of Mechanical Engineering, Indian Institute of Technology Kanpur, Kanpur, India*
Ramos, J.: *Water Department, Centro Universitario Regional Litoral Norte, Universidad de la República, Salto, Uruguay*
Ramos Ramos, O.E.: *Laboratorio de Hidroquímica, Instituto de Investigaciones Químicas, Universidad Mayor de San Andrés, La Paz, Bolivia*
Ramsay, L.: *VIA University College, Horsens, Denmark*
Rainwater, K.: *Department of Civil, Environmental, and Construction Engineering, Texas Tech University, Texas, USA*
Rashid, S.M.A.: *NGO Forum for Public Health, Dhaka, Bangladesh*
Ratnayake, N.: *Ocean University of Sri Lanka & Department of Earth Resources Engineering, Faculty of Engineering, University of Moratuwa, Sri Lanka*
Rathi, B.: *Geomicrobiology, Microbial Ecology and Hydrology, University of Tübingen, Tübingen, Germany*
Ren, C.: *School of Water Resources and Environment, China University of Geosciences (Beijing), Beijing, PR China*
Resurreccion, A.: *Institute of Civil Engineering, College of Engineering, University of the Philippines, Diliman, Quezon City, Philippines*
Reyes-Gomez, V.M.: *Instituto de Ecología A.C. (INECOL), Red Ambiente y Sustentabilidad, Chihuahua, México*
Rezende-Filho, A.T.: *Federal University of Mato Grosso do Sul (UFMT), Campo Grande, Brazil*
Richards, L.A.: *Department of Earth and Environmental Sciences and Williamson Research Centre for Molecular Environmental Science, The University of Manchester, Manchester, UK*
Rietveld, L.C.: *Sanitary Engineering, Department of Civil Engineering and Geosciences, Delft University of Technology, Delft, The Netherlands & Water Management Department, Faculty of Civil Engineering and Geosciences, Delft University of Technology, Delft, The Netherlands*
Rocha-Gutierrez, B.A.: *Facultad de Ciencias Químicas, Universidad Autónoma de Chihuahua (UACH), Chihuahua, México*
Rodríguez Castrejón, U.E.: *Postgraduate Degree in Water Sciences, University of Guanajuato, Guanajuato, Mexico & Division of Engineering, University of Guanajuato, Guanajuato, Mexico*

Rojas-Carrillo, O.: *School of Chemistry, Universidad Nacional de Costa Rica, Heredia, Costa Rica*
Rola, A.: *Institute of Governance and Rural Development, College of Public Affairs and Development, University of the Philippines, Los Baños, College, Laguna, Philippines*
Román, D.: *Analytical Chemistry Laboratory, Universidad de Antofagasta, Chile*
Romero-Esquivel, L.G.: *Environmental Protection Research Center (CIPA), School of Chemistry, Instituto Tecnológico de Costa Rica, Cartago, Costa Rica*
Romić, Ž.: *Osijek Water Supply Company, Osijek, Croatia*
Rosen, B.P.: *Department of Cellular Biology and Pharmacology, Florida International University, Herbert Wertheim College of Medicine, Miami, USA*
Roy, M.: *Sanitary Engineering, Department of Civil Engineering and Geosciences, Delft University of Technology, Delft, The Netherlands*
Roychowdhury, T.: *School of Environmental Studies, Jadavpur University, Kolkata, India*
Roy Chowdhury, N.: *School of Environmental Studies, Jadavpur University, Kolkata, India*
Rüde, T.: *Institute of Hydrogeology, RWTH Aachen University, Germany*
Rutten, S.B.: *Department of Environmental Technology, Wageningen University and Research, Wageningen, The Netherlands & KWR Water Research Institute, Nieuwegein, The Netherlands*
Sacari, E.: *Laboratorio de Nanotecnología, Facultad de Ingeniería, Universidad Nacional Jorge Basadre Grohmann, Tacna, Perú*
Saha, D.: *Partners in Progress, India*
Saha, I.: *Sripat Singh College, Murshidabad, India*
Saha, P.: *Department of Chemical Engineering, Indian Institute of Technology Guwahati, Assam, India*
Sakamoto, M.: *Department of International Studies, GSFS, The University of Tokyo, Japan*
Salaun, P.: *School of Ocean Sciences, University of Liverpool, UK*
Samal, A.C.: *Department of Environmental Science, University of Kalyani, Nadia, West Bengal, India*
Sanchez-Palacios, T.: *Omic Research Group, Laboratory of Analytical Chemistry, Faculty of Science and Technology, University of Canberra, Australia*
Sanderson, P.: *Global Centre for Environmental Remediation (GCER), The University of Newcastle, Callaghan, NSW, Australia*
Sankaran, B.: *Berkeley Center for Structural Biology, Lawrence Berkeley Laboratory, Berkeley, USA*
Santo, V.: *Department for Water Safety and Water Supply, Institute of Public Health of Osijek-Baranja County, Osijek, Croatia*
Santos, A.: *Graduating Program in Environmental Engineering (PROAMB), Pharmacy Department, School of Pharmacy, Federal University of Ouro Preto (UFOP), Minas Gerais, Brazil*
Santos, J.V.: *Department of Soil Science, Federal University of Lavras, Lavras, Minas Gerais, Brazil*
Santra, S.C.: *Department of Environmental Science, University of Kalyani, Nadia, West Bengal, India*
Sarath, P.K.: *Biogeochemistry Research Group, Centre for Earth Sciences, Indian Institute of Science, Bangalore, India & LOCEAN, Sorbonne University, Paris, France*
Sarkar, S.: *Department of Chemical Engineering, Indian Institute of Technology Guwahati, Assam, India*
Sarker, M.M.R.: *Department of Agricultural Statistics, Sher-e-Bangla Agricultural University, Dhaka, Bangladesh*
Sathe, S.S.: *Department of Civil Engineering, Rajarambapu Institute of Technology Sakharale, Maharashtra, India & Department of Civil Engineering, Indian Institute of Technology IIT-G, Guwahati, India*

Saxena, A.: *Faculty of Civil Engineering, Institute of Technology, Shri Ramswaroop Memorial University, Lucknow, India*
Schaaf, B.: *Evides Water Company, Department of Water Technology and Source Protection, Rotterdam, The Netherlands*
Schneider, M.: *Institute of Applied Geosciences, KIT, Karlsruhe, Germany*
Schullehner, J.: *Department of Public Health, Aarhus University, Denmark & Geological Survey of Denmark and Greenland (GEUS), Aarhus, Denmark*
Selander, M.: *KTH-International Groundwater Arsenic Research Group, Department of Sustainable Development, Environmental Science and Engineering, KTH Royal Institute of Technology, Stockholm, Sweden*
Selim Reza, A.H.M.: *Department of Geology and Mining, University of Rajshahi, Rajshahi, Bangladesh*
Senaratne, A.: *Department of Geology, University of Peradeniya, Sri Lanka*
SenGupta, B.: *School of Energy, Geoscience, Infrastructure & Society, Heriot-Watt University. Water Academy, Scotland, UK*
Senn, A.-C.: *Eawag, Swiss Federal Institute of Aquatic Science and Technology, Dübendorf, Switzerland*
Serafín, V.: *Departamento de Química Analítica, Facultad de CC. Químicas, Universidad Complutense de Madrid, Madrid, Spain*
Serafín Muñoz, A.: *Postgraduate Degree in Water Sciences, University of Guanajuato, Guanajuato, Mexico & Division of Engineering, University of Guanajuato, Guanajuato, Mexico*
Shah, B.A.: *Department of Geological Sciences, Jadavpur University, Kolkata, India*
Shaha, S.K.: *Bangladesh Institute of Research and Rehabilitation in Diabetes, Endocrine and Metabolic Disorders (BIRDEM), Shahbag, Dhaka, Bangladesh*
Shahid, M.: *Department of Environmental Sciences, COMSATS University Islamabad, Vehari Campus, Vehari, Pakistan*
Shankar, P.: *Mahavir Cancer Institute & Research Centre, Patna, Bihar, India*
Sharma, S.: *Exceldot AB, Bromma, Sweden*
Shouche, Y.: *National Centre For Microbial Resource, NCCS, Pune, India*
Shrivastava, A.: *Department of Earth Sciences, Indian Institute of Science Education and Research Kolkata, India & Amity University Uttar Pradesh, Noida, India*
Shukla, S.: *Faculty of Civil Engineering, Institute of Technology, Shri Ramswaroop Memorial University, Lucknow, India*
Siddique, A.B.: *Global Centre for Environmental Remediation (GCER), Faculty of Science, The University of Newcastle, Callaghan, NSW, Australia & Cooperative Research Centre for Contamination Assessment and Remediation of the Environment (CRC CARE), The University of Newcastle, Callaghan, NSW, Australia*
Siddique, A.E.: *Department of Biochemistry and Molecular Biology, University of Rajshahi, Bangladesh*
Si-Liang, L.: *Institute of Surface Earth System Science, Tianjin University, PR China*
Silva, A.O.: *Department of Soil Science, Federal University of Lavras, Lavras, Minas Gerais, Brazil*
Silver, D.: *FREDsense Technologies Corp., Calgary, Alberta, Canada*
Singh, A.: *Discipline of Civil Engineering, Indian Institute of Technology Gandhinagar, Gujarat, India*
Singh, P.: *School of Environmental Sciences, Jawaharlal Nehru University, New Delhi, India & Shaheed Rajguru College of Applied Sciences for Women, Vasundhara Enclave, Delhi, India*
Singh, P.: *Kolkata Zonal Center, CSIR – National Environmental Engineering Research Institute (NEERI), Kolkata, West Bengal, India*
Singh, S.K.: *Department of Environment and Water Management, A.N. College, Patna, India*
Sinha, D.: *School of Environmental Studies, Jadavpur University, Kolkata, India*

Siregar, T.: *Omic Research Group, Laboratory of Analytical Chemistry, Faculty of Science and Technology, University of Canberra, Australia & Research Center for Marine and Fisheries Post-Harvest, Ministry of Marine and Fisheries, Indonesia*

Slater, G.: *School of Geography & Earth Sciences, McMaster University, Hamilton Ontario, Canada*

Slokar, Y.M.: *IHE Delft Institute for Water Education, Delft, The Netherlands*

Smichowski, P.: *Comisión Nacional de Energía Atómica (CNEA), Gerencia Química, Buenos Aires, Argentina & Consejo Nacional de Investigaciones Científicas y Técnicas (CONICET), Buenos Aires, Argentina*

Sohel, N.: *Department of Health Research Methods, Evidence, and Impact, McMaster University, Hamilton, Ontario, Canada*

Šolić, M.: *Environmental Protection Research Group, Department of Chemistry, Biochemistry and Environmental Protection, University of Novi Sad, Faculty of Sciences, Novi Sad, Serbia*

Solis, K.: *Institute of Civil Engineering, College of Engineering, University of the Philippines, Diliman, Quezon City, Philippines*

Sosa, A.: *Posgrado en Ciencias de la Tierra, Universidad Nacional Autónoma de México, México City, Mexico*

Srivastava, A.K.: *Nuclear Agriculture and Biotechnology Division, Bhabha Atomic Research Center, Mumbai, Maharashtra, India*

Srivastava, S.: *Institute of Environment & Sustainable Development, Banaras Hindu University, Varanasi, Uttar Pradesh, India*

Srivastava, S.K.: *Department of Chemistry, Indian Institute of Technology Kharagpur, Kharagpur, India*

Štiglić, J.: *Department for Water Safety and Water Supply, Croatian Institute of Public Health, Zagreb, Croatia*

Stopelli, E.: *Eawag, Swiss Federal Institute of Aquatic Science and Technology, Dübendorf, Switzerland*

Sufia, K.: *Environment and Population Research Centre, New DOHS, Mohakhali, Dhaka, Bangladesh*

Sullivan, L.A.: *University of Canberra, Australia*

Suman, S.: *Mahavir Cancer Institute and Research Center, Patna, India*

Suprasanna, P.: *Nuclear Agriculture and Biotechnology Division, Bhabha Atomic Research Center, Mumbai, Maharashtra, India*

Surya, G.: *Department of Environmental Health, Faculty of Public Health, Universitas Indonesia, Depok, Indonesia*

Suzol, S.H.: *Department of Chemistry and Biochemistry, Florida International University, Miami, USA*

Tamayo, R.: *LabMAM, Departamento de Ingeniería Química Biotecnología y Materiales, FCFM, Universidad de Chile, Santiago, Chile & Departamento de Ingeniería de Materiales, Facultad de Ingeniería de Procesos, Universidad Nacional San Agustín de Arequipa, Arequipa, Perú*

Tao, L.: *Guangdong Key Laboratory of Integrated Agro-environmental Pollution Control and Management, Guangdong Institute of Eco-environmental Science & Technology, Guangzhou, PR China*

Tapia, J.: *Universidad Católica del Norte, Facultad de Ingeniería y Ciencias Geológicas, Departamento de Ciencias Geológicas, Antofagasta, Chile*

Teixeira, M.C.: *Graduating Program in Environmental Engineering (PROAMB), Pharmacy Department, School of Pharmacy, Federal University of Ouro Preto (UFOP), Minas Gerais, Brazil & Pharmacy Department, School of Pharmacy, Federal University of Ouro Preto (UFOP), Minas Gerais, Brazil*

Thakur, B.K.: *Department of Economics, FLAME University, Pune, India*

Thayer, K.: *U.S. Environmental Protection Agency, Center for Public Health and Environmental Assessments, Research Triangle, Park, USA*
Thouin, H.: *BRGM, Water, Environment and Ecotechnology Division, Orléans, France & BRGM, Water, Environment, Process Development and Analysis Division, Orléans, France*
Tie, B.: *College of Resource and Environment, Hunan Agricultural University, Changsha, PR China*
Tirado, N.: *Genotoxicity Unit, Genetic Institut, Medicine School, Universidad Mayor de San Andrés, La Paz, Bolivia*
Thokchom, L.: *Department of Earth Science, Assam University, Silchar, India*
Todd-Supuy, S.: *School of Chemical Engineering, Faculty of Engineering, University of Costa Rica, San José, Costa Rica*
Tokunaga, T.: *Geosphere Environment Systems, Department of Environment Systems, GSFS, The University of Tokyo, Japan*
Tompsett, A.: *Institute for International Economic Studies, Stockholm University, Sweden & Beijer Institute for Ecological Economics, Royal Academy of Sciences, Sweden*
Toppler, N.: *Environmental Analysis Laboratory, Southern Cross University, Lismore, NSW, Australia*
Torasso, N.: *Universidad de Buenos Aires. Facultad de Ciencias Exactas y Naturales, Departamento de Física, Laboratorio de Polímeros y Materiales Compuestos (LP&MC), Instituto de Física de Buenos Aires (IFIBA-CONICET), Buenos Aires, Argentina*
Trang, P.T.K.: *CETASD, Vietnam National University, Hanoi, Vietnam*
Trembath-Reichert, E.: *Deparment of Geology, Arizona State University, Tempe, Arizona, USA*
Tris, H.: *BRGM, Water, Environment, Process and Analyses Division, Orléans, France*
Tubić, A.: *Environmental Protection Research Group, Department of Chemistry, Biochemistry and Environmental Protection, Faculty of Sciences, University of Novi Sad, Novi Sad, Serbia*
Ulapane, P.A.K.: *Department of Geology, University of Peradeniya, Sri Lanka*
Ujević Bošnjak, M.: *Department for Water Safety and Water Supply, Croatian Institute of Public Health, Zagreb, Croatia*
Upadhyay, M.K.: *Institute of Environment & Sustainable Development, Banaras Hindu University, Varanasi, Uttar Pradesh, India*
Usman, M.: *Institute for Water Resources and Water Supply, Hamburg University of Technology, Am Schwarzenberg-Campus 3, Hamburg, Germany*
Utturkar, S.M.: *Purdue University Center for Cancer Research, Purdue University, West Lafayette, USA*
Uzelli, T.: *İzmir Institute of Technology, Geothermal Energy Research and Application Center, Urla, İzmir, Turkey*
van der Voorn, T.: *University of Osnabrück, Institute of Environmental Systems Research, Osnabrück, Germany*
van der Wal, A.: *Evides Water Company, Department of Water Technology, Rotterdam, The Netherlands & Wageningen University, Department of Environmental Technology, Wageningen, The Netherlands*
van der Wens, P.: *Brabant Water, Breda, The Netherlands*
van Geen, A.: *Lamont Doherty Earth Observatory, Columbia University, Palisades, New York, USA*
van Genuchten, C.M.: *Department of Geochemistry, Geological Survey of Denmark and Greenland, Copenhagen, Denmark*
van Halem, D.: *Water Management Department, Faculty of Civil Engineering and Geosciences, Delft University of Technology, The Netherlands*
Vergara-Rubio, A.: *Universidad de San Martín (UNSAM), 3iA, Argentina*
Viet, P.H.: *CETASD, Vietnam National University, Hanoi, Vietnam*

Villanueva-Estrada, R.E.: *Instituto de Geofísica, Universidad Nacional Autónoma de México, Morelia, Mich., México*

Vithana, C.L.: *Department of Geology, University of Peradeniya, Sri Lanka*

Vithanage, M.: *Ecosphere Resilience Research Center, Faculty of Applied Science, University of Sri Jayewardenepura, Nugegoda, Sri Lanka*

Voegelin, A.: *Eawag, Swiss Federal Institute of Aquatic Science and Technology, Dübendorf, Switzerland*

Vollaard, P.: *Evides Water Company, Department of Water Technology, Rotterdam, The Netherlands*

von Brömssen, M.: *Ramböll Sweden AB, Stockholm, Sweden*

Voutchkova, D.: *Geological Survey of Denmark and Greenland (GEUS), Aarhus, Denmark*

Wan, Y.N.: *Beijing Key Laboratory of Farmland Soil Pollution Prevention and Remediation, College of Resources and Environmental Sciences, China Agricultural University, Beijing, PR China & Institute of Agricultural Resources and Regional Planning, Chinese Academy of Agricultural Sciences, Beijing, PR China*

Wang, H.T.: *KEY Laboratory of Urban Environment and Health, Institute of Urban Environment, Chinese Academy of Sciences, Xiamen, PR China*

Wang, K.: *Beijing Key Laboratory of Farmland Soil Pollution Prevention and Remediation, College of Resources and Environmental Sciences, China Agricultural University, Beijing, PR China*

Wang, K.T.: *Fujian Provincial Key Laboratory of Agroecological Processing and Safety Monitoring, College of Life Sciences, Fujian Agriculture and Forestry University, Fuzhou, PR China*

Ward, N.I.: *ICP-MS Facility, Chemistry Department, University of Surrey, Guildford, Surrey, UK*

Ward, N.J.: *Southern Cross Geoscience, Southern Cross University, Lismore, NSW, Australia*

Watson, M.: *Environmental Protection Research Group, Department for Chemistry, Biochemistry and Environmental Protection, University of Novi Sad Faculty of Sciences, Novi Sad, Republic of Serbia*

Weinzettel, P.: *IHLLA – Instituto de Hidrologa de Llanuras "Dr. Eduardo J. Usunoff", Azul, Argentina*

Weng, L.: *Agro-Environmental Protection Institute, Ministry of Agriculture and Rural Affairs, Tianjin, PR China & Department of Soil Quality, Wageningen University, Wageningen, The Netherlands*

Widyastuti, G.: *Department of Environmental Health, Faculty of Public Health, Universitas Indonesia, Depok, Indonesia*

Winkel, L.H.E.: *Eawag, Swiss Federal Institute of Aquatic Science and Technology, Dübendorf, Switzerland & Institute of Biogeochemistry and Pollutant Dynamics, ETH Zürich, Zürich, Switzerland*

Winkelnkemper, T.: *Winkelnkemper GmbH – Fermanox Wasseraufbereitung, Wadersloh, Germany*

Wnuk, S.F.: *Department of Chemistry and Biochemistry, Florida International University, Miami, USA*

Woobaidllah, A.S.M.: *Department of Geology, University of Dhaka, Dhaka, Bangladesh*

Wu, J.X.: *College of Resources and Environmental, South-Central University for Nationalities, Wuhan, PR China*

Wu, R.: *Department of Earth and Environmental Sciences, & Williamson Research Centre for Molecular Environmental Science, University of Manchester, Manchester, UK*

Xiu, W.: *State Key Laboratory of Biogeology and Environmental Geology, China University of Geosciences, Beijing, P.R. China & School of Water Resources and Environment, China University of Geosciences (Beijing), Beijing, PR China*

Xu, L.: *Department of Earth and Environmental Sciences, School of Natural Sciences and Williamson Research Center for Molecular Environmental Science, University of Manchester, Manchester, UK*

Xue, X.M.: *KEY Laboratory of Urban Environment and Health, Institute of Urban Environment, Chinese Academy of Sciences, Xiamen, PR China*

Yadav, M.K.: *School of Environmental Science and Engineering, Indian Institute of Technology Kharagpur, Kharagpur, India*

Yadav, S.K.: *School of Environmental Sciences, Jawaharlal Nehru University, New Delhi, India*

Yáñez-Sedeño, P.: *Departamento de Química Analítica, Facultad de CC. Químicas, Universidad Complutense de Madrid, Madrid, Spain*

Yang, G.D.: *Fujian Provincial Key Laboratory of Agroecological Processing and Safety Monitoring, College of Life Sciences, Fujian Agriculture and Forestry University, Fuzhou, PR China*

Yang, H.-J.: *Department of Earth Sciences, National Cheng-Kung University, Tainan, Taiwan, Republic of China*

Yang, Y.P.: *State Key Lab of Urban and Regional Ecology, Research Center for Eco-Environmental Sciences, Chinese Academy of Sciences, PR China & University of Chinese Academy of Sciences, PR China*

Ye, H.P.: *College of Resources and Environmental, South-Central University for Nationalities, Wuhan, PR China*

Yi, X.Y.: *State Key Lab of Urban and Regional Ecology, Research Center for Eco-Environmental Sciences, Chinese Academy of Sciences, PR China & University of Chinese Academy of Sciences, PR China*

Yin, N.Y.: *College of Resources and Environment, University of Chinese Academy of Sciences, Beijing, PR China*

Yoshinaga, M.: *Department of Cellular Biology and Pharmacology, Florida International University, Herbert Wertheim College of Medicine, Miami, USA*

Yoshinaga-Sakurai, K.: *Department of Cellular Biology and Pharmacology, Florida International University, Herbert Wertheim College of Medicine, Miami, USA*

Yu, H.: *Guangdong Key Laboratory of Integrated Agro-environmental Pollution Control and Management, Guangdong Institute of Eco-Environmental Science & Technology, Guangzhou, PR China*

Yu, N.Y.: *Key Laboratory of Sustainable Resources Processing and Advanced Materials of Hunan Province, National & Local Joint Engineering Laboratory for New Petrochemical Materials and Fine Utilization of Resources, Hunan Normal University, Changsha, PR China*

Yu, Q.: *College of Resources and Environmental, South-Central University for Nationalities, Wuhan, PR China*

Yu, Q.N.: *Jiangsu Provincial Key Laboratory of Marine Biology, College of Resources and Environmental Sciences, Nanjing Agricultural University, Nanjing, PR China*

Yuan, Z.-F.: *Department of Environmental Science, University of Liverpool, Brownlow Hill, Liverpool, UK & Department of Health and Environmental Sciences, Xi'an Jiaotong-Liverpool University, Jiangsu, PR China*

Yunus, M.: *icddr b, Dhaka Bangladesh*

Zahid, M.A.: *Environment and Population Research Centre (EPRC), New DOHS, Mohakhali, Dhaka, Bangladesh*

Zhang, C.: *School of Water Resources and Environment, China University of Geosciences (Beijing), Beijing, PR China*

Zhang, C.H.: *Laboratory Centre of Life Science, Nanjing Agricultural University, PR China*

Zhang, J.Y.: *Jiangsu Provincial Key Laboratory of Marine Biology, College of Resources and Environmental Sciences, Nanjing Agricultural University, Nanjing, PR China*

Zhang, Y.: *College of Resources and Environmental, South-Central University for Nationalities, Wuhan, PR China*
Zheng, J.: *The University of Queensland, Centre For Mined Land Rehabilitation, St Lucia, Australia*
Zhou, G.W.: *KEY Laboratory of Urban Environment and Health, Institute of Urban Environment, Chinese Academy of Sciences, Xiamen, PR China & STATE Key Laboratory of Urban and Regional Ecology, Research Center for Eco-Environmental Sciences, Chinese Academy of Sciences, Beijing, PR China*
Zhu, Y.G.: *KEY Laboratory of Urban Environment and Health, Institute of Urban Environment, Chinese Academy of Sciences, Xiamen, PR China; STATE Key Laboratory of Urban and Regional Ecology, Research Center for Eco-Environmental Sciences, Chinese Academy of Sciences, Beijing, PR China & University of Chinese Academy of Sciences, PR China*
Zhuang, H.M.: *Fujian Provincial Key Laboratory of Agroecological Processing and Safety Monitoring, College of Life Sciences, Fujian Agriculture and Forestry University, Fuzhou, PR China*
Zillany, A.N.M.K.: *CARE Bangladesh, Dhaka, Bangladesh*
Zima, J.: *Osijek Water Supply Company, Osijek, Croatia*

Plenary presentations

Arsenic in the Environment: Bridging Science to Practice for Sustainable Development –
van der Wal, Ahmad, Petrusevski, Weijma, Savic, van der Wens, Beerendonk, Bhattacharya, Bundschuh & Naidu (Eds)
 ISBN 978-1-032-32928-4

Second-generation global risk map of groundwater arsenic contamination

J. Podgorski[1,2] & M. Berg[1,3]
[1]*Department Water Resources and Drinking Water, Eawag, Swiss Federal Institute of Aquatic Science and Technology, Dübendorf, Switzerland*
[2]*Department of Earth and Environmental Sciences, University of Manchester, Manchester, UK*
[3]*UNESCO Chair on Groundwater Arsenic within the 2030 Agenda for Sustainable Development, School of Civil Engineering and Surveying, University of Southern Queensland, Queensland, Australia*

ABSTRACT: Naturally occurring arsenic in groundwater affects millions of people around the world. In order to assess the extent of this problem, we have used machine learning to create a global prediction map of groundwater arsenic greater than 10 µg/L. The resulting model includes already known arsenic-affected areas as well as new potentially contaminated areas. Combining the global arsenic prediction model with household groundwater-usage statistics, we estimate that 94–220 million people are potentially exposed to high arsenic concentrations in groundwater. As groundwater is increasingly utilized to support a growing population and buffer against increasing water scarcity due to a changing climate, this model will help raise awareness, identify suitable areas for safe wells and guide where testing for arsenic should be prioritized.

1 INTRODUCTION

The natural occurrence of arsenic in groundwater is a global problem with wide-ranging health effects, including disorders of the skin and vascular and nervous systems as well as cancer. The major source of inorganic arsenic in the diet is through arsenic-contaminated water, whether consumed directly by drinking or used for cooking or irrigating crops. As a consequence, the World Health Organization (WHO) has set a guideline concentration of 10 µg/L in drinking water.

In most large-scale cases of geogenic groundwater arsenic contamination, arsenic accumulates in aquifers composed of recently deposited alluvial sediments. Under anoxic conditions, arsenic is released from arsenic-bearing iron(III) minerals in the aquifer sediments (Berg *et al.*, 2008; Nickson *et al.*, 1998). Also under oxidizing high-pH conditions, arsenic can desorb from iron and aluminum hydroxides (Litter *et al.*, 2019). Other processes responsible for arsenic release into groundwater include oxidation of arsenic-bearing sulfide minerals as well as release from arsenic-enriched geothermal deposits.

In order to help identify areas likely to contain high concentrations of arsenic in groundwater, statistical learning methods can be employed to create arsenic risk maps that predict regions of safe and unsafe groundwater. The predictor variables used in these studies generally include various climate and soil parameters, geology and topography.

Here we use the random forest machine learning method (Breiman, 2001) to model an extensive compilation of groundwater arsenic concentrations with the latest global datasets of relevant environmental parameters. The result is a global risk map of groundwater arsenic that represents a large improvement over a first-generation global As-prediction map created over ten years ago (Amini *et al.*, 2008).

2 MATERIALS AND METHODS

2.1 *Preparation of arsenic data*

Measurements of arsenic concentration in groundwater (n = 208,097) were compiled from a wide variety of sources, excluding measurements known to have originated from a depth greater than 100 m. These were aggregated into 58,555 data points by taking the geometric mean of concentrations falling within 1-km square pixels and were then converted into binary form by setting all arsenic concentrations meeting the WHO guideline of ≤ 10 µg/L to zero and all concentrations >10 µg/L to one (Figure 1). The dataset of 58,555 arsenic points was then randomly divided into training (80%) and test (20%) datasets.

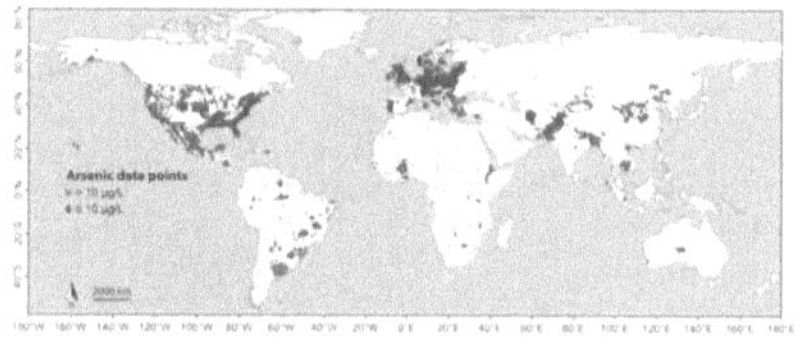

Figure 1. Arsenic concentration measurements excluding those known to originate from a depth greater than 100 m, averaged to 1-km square pixels (n = 58,555).

2.2 *Random forest modeling*

A collection of 52 spatially continuous predictor variables with global coverage representing various climatic, geologic, soil and other parameters known

DOI: 10.1201/9781003317395-1

or hypothesized to be related to the dissolution and accumulation of arsenic in groundwater was assembled. A subset of 11 variables was then identified for use in the final model by evaluating the variables' effect on model accuracy in an iterative random forest modeling scheme.

A random forest was grown with 10,001 trees using the training dataset and associated 11 independent variables. The model was then applied to the test dataset with its performance being evaluated by the Area Under the ROC (receiver operating characteristic (ROC)) Curve (AUC), which generally ranges from 0.5 (random model) to 1 (100% accurate). The variables' importance was measured as the mean decrease in accuracy and Gini node impurity when the values of each variable were randomly resorted.

2.3 *Calculation of population affected*

The random forest model was further employed to estimate the number of people potentially exposed to high levels of geogenic arsenic in drinking water. A probability cutoff at which the model equally well predicts high and low values was then used to delineate areas with high arsenic hazard and thereby calculate the number of people potentially affected, also accounting for rates of urban and rural household untreated groundwater use.

3 RESULTS AND DISCUSSION

3.1 *Random forest model*

The performance of the random forest model on the test dataset (20% of the data) is summarized in the confusion matrix in Table 1. Despite a prevalence of high values (>10 µg/L) of only 0.22 in the dataset, the model performs well in predicting both high values (sensitivity: 0.79) and low values (specificity: 0.85) at a

Table 1. Confusion matrix summarizing the results of applying the random forest model to the test dataset at a probability cut-off of 0.50. The receiver operating characteristic curve (AUC = 0.89) is also displayed.

	Measured	
	As ≤ 10 µg/L	As > 10 µg/L
Mod. As ≤ 10 µg/L	7710	555
Mod. As > 10 µg/L	1394	2037

Sensitivity:	0.79
Specificity:	0.85
Positive Pred. Value:	0.59
Negative Pred. Value:	0.93
Prevalence:	0.22
Balanced Accuracy:	0.82

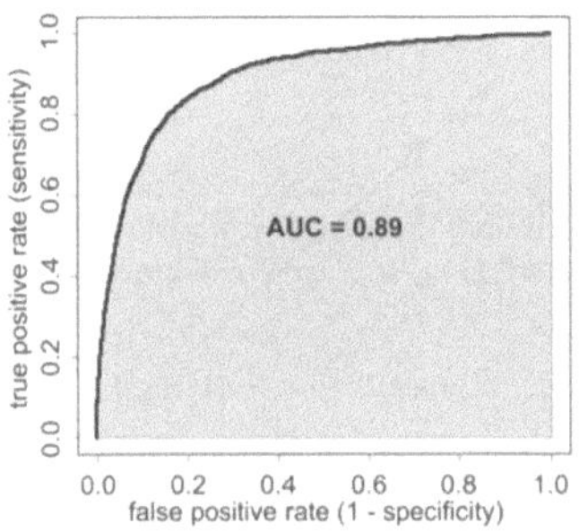

probability cutoff of 0.50. The average of these two figures, known as balanced accuracy, is correspondingly high at 0.82. Likewise, the model's Area Under the Curve (AUC), which considers the full range of possible cutoffs, has a very high value of 0.89 with the test dataset (Table 1).

3.2 *Regions and populations at risk*

Areas predicted to have high arsenic concentrations in groundwater exist on all continents. Known areas of groundwater arsenic contamination are generally well captured, and some new areas of potential geogenic arsenic contamination include large sections of Central Asia, the Sahel region and the Okavango Delta in Africa, and parts of the Arctic. The resulting global arsenic risk assessment indicates that approximately 94–220 million people around the world are potentially exposed to high concentrations of arsenic in groundwater from their domestic water supply.

4 CONCLUSIONS AND RECOMMENDATIONS

The arsenic prediction model should be used to guide further groundwater arsenic testing. This model highlights areas at risk and will provide a basis for targeted surveys and mitigation measures. If existing wells are not thoroughly tested, the already large number of people potentially affected can be expected to increase as groundwater use expands with a growing population and increasing irrigation, especially with warmer and drier conditions related to climate change.

ACKNOWLEDGEMENTS

We thank the Swiss Agency for Development and Cooperation (7F-09963.01.01) for their long-term support and co-funding of this study. We also thank the many providers of data, which were an essential component of this work.

REFERENCES

Amini M., Abbaspour K.C., Berg M., Winkel L., Hug S.J., Hoehn E., Yang H. & Johnson C.A. 2008. Statistical modeling of global geogenic arsenic contamination in groundwater. *Environ. Sci. Technol.* 42(10): 3669–3675.

Berg M., Trang P.T.K., Stengel C., Buschmann J., Viet P. H., Van Dan N., Giger W. & Stüben D. 2008. Hydrological and sedimentary controls leading to arsenic contamination of groundwater in the Hanoi area, Vietnam. *Chem. Geol.* 249(1–2): 91–112.

Breiman L. 2001. Random forests. *Mach. Learn.* 45(1): 5–32.

Litter M.I., Ingallinella A.M., Olmos V., Savio M., Difeo G., Botto L., FarfánTorres E.M., Taylor S., Frangie S., Herkovits J., Schalamuk I., González M.J., Berardozzi E., García Einschlag F.S., Bhattacharya P. & Ahmad A. 2019. Arsenic in Argentina: occurrence, human health, legislation and determination. *Sci. Total Environ.* 676: 756–766.

Nickson R., McArthur J., Burgess W., Ahmed K. M., Ravenscroft P. & Rahman M. 1998. Arsenic poisoning of Bangladesh groundwater. *Nature* 395(6700): 338–338.

Arsenic in the Environment: Bridging Science to Practice for Sustainable Development –
van der Wal, Ahmad, Petrusevski, Weijma, Savic, van der Wens, Beerendonk, Bhattacharya, Bundschuh & Naidu (Eds)
© 2023 the Author(s), ISBN 978-1-032-32928-4

Arsenic occurrence and research in Brazil

M.B.B. Guerra[1], C. Oliveira[2], A.O. Silva[2], I.F.S. Alvarenga[2], M.V. Barbosa[2], M.M. Feitosa[2], E.S. Penido[1], J.V. Santos[2], M.A.C. Carneiro[2], J. Bundschuh[3] & L.R.G. Guilherme[2]
[1]*Department of Chemistry, Federal University of Lavras, Lavras, Minas Gerais, Brazil*
[2]*Department of Soil Science, Federal University of Lavras, Lavras, Minas Gerais, Brazil*
[3]*School of Civil Engineering and Surveying, University of Southern Queensland, Toowoomba, QLD, Australia*

ABSTRACT: At least three Brazilian areas are well recognized by their anomalous arsenic levels: the Iron Quadrangle region and the Ribeira valley, both in the Southeast portion of the country and the Santana area, in the Amazon territory. Mining activities can contribute to the release of arsenic in the environment, especially gold exploration, since it is well documented that arsenic minerals are naturally found in gold deposits. Notwithstanding, mining is not the only anthropic activity that contributes to arsenic contamination in Brazil. Industrial, urban, and agricultural activities exert strong influence as well. Brazil is a continental country with high variability on geology and climate attributes as well as regarding the intensity of anthropic activities. Several studies focused on the evaluation of ecological risks and regulations for As in soils, sediments, water, air, and food. Nearly 70% of the studies are concentrated in the Southeast region of Brazil, 8.5% in the South, 12.5% in the Northeast region, 4% in the North region, and 4% in the Midwest region. Nevertheless, for most of the Brazilian States, there is still scarce data on As bioaccessibility.

1 INTRODUCTION

It is estimated that over 4 million people are exposed to high levels of arsenic (As) in Latin America countries (Esparza 2009). Both natural phenomena and anthropogenic activities are contributors to As enrichment in these areas. On a first review on As exposure in 14 Latin American countries published by Bundschuh *et al.* (2012) the authors acknowledge that As exposure can be minimized by following some feasible measures, such as: i) creating and updating an online database about As; ii) using international standards on maximum allowable As levels; iii) strengthening the international research collaboration. Recently, Bundschuh *et al.* (2021) expanded the literature survey (from 2010 to 2020) to all Latin American countries highlighting some concerns about As contamination and stressed the still limited availability of comprehensive data that hampers a holistic evaluation of the real scenario. Specific concerns were raised related to the pressing need for performing extensive risk assessment studies that evaluate the main routes of As exposure in the Latin America region.

In Brazil, there are at least three main areas with well documented anomalous As levels: the Iron Quadrangle and the Ribeira Valley, both in the Southeast region, and the Santana area, in the Amazon. The high exposure to this contaminant especially by low-income people living in rural communities in Brazil is a topic of particular concern (Figueiredo *et al.*, 2007).

The objective of this study is to provide an overview of natural and anthropic activities that promotes significant As contamination in Brazilian ecosystems from reviewing studies published in the last 20 years. A special emphasis is given to the studies involving environmental impact assessment derived from mining activities. In this context, it is important to mention the collapse of two large mining tailings dams in the State of Minas Gerais, in Mariana (in November 2015) and Brumadinho (in January 2019), both of them located in the Iron Quadrangle region.

2 ARSENIC CONTAMINATION IN BRAZIL

Mining activities and agriculture are the main pillars of the Brazilian economy. Iron and gold account for the largest share in the volume of sales, in which production is mainly concentrated in the States of Minas Gerais (Southeast) and Pará (North) (National Mining Agency 2020a,b). It is well-known that As minerals are naturally found in gold deposits (Borba *et al.*, 2003; Bossy *et al.*, 2010). Unfortunately, few studies evaluated As contamination in gold mining areas. Even more

DOI: 10.1201/9781003317395-2

worrying, is the widespread occurrence, all over the Brazilian territory, of gold mining tailings dams built with the less safe construction method, *i.e.*, the upstream lifting method (Figure 1), the same used in the Brumadinho and Mariana iron ore waste dams.

Figure 1. Distribution of gold mining tailings dams according to construction method. Tailings dams included in the Brazilian government's National Dam Safety Policy (National Mining Agency 2020b).

The use of fertilizers and pesticides is a common practice in the modern agriculture to reach high productivity. However, the long-term application of these products can contaminate the soil with potentially toxic elements (PTEs), As included. (Corguinha *et al.*, 2015). Consequently, agricultural soil becomes an important route of human exposure to PTEs (Wei & Yang 2010).

Brazil is a continental country with high variability on geology and climate attributes as well as regarding the intensity of anthropic activities. Several studies focused on the evaluation of ecological risks and regulations for As in soils, sediments, water, air, and food. Nearly 70% of the studies are concentrated in the Southeast region of Brazil, 8.5% in the South, 12.5% in the Northeast region, 4% in the North region, and 4% in the Midwest region. Nevertheless, for most of the Brazilian States, there is still scarce data on As bioaccessibility.

3 CONCLUSIONS

In spite of the fact that gold exploration in the Brazilian territory started in the 17th century, for most Brazilian States, there is still a large gap to be filled on human health risk assessments regarding As, particularly for the States in the Northern and Northeast regions.

REFERENCES

Borba R.P., Figueiredo B.R. & Matschullat J. 2003. Geochemical distribution of arsenic in waters, sediments and weathered gold mineralized rocks from Iron Quadrangle, Brazil. *Environ. Geol.* 44: 39–52.

Bossy A., Grosbois C., Beauchemin S., Courtin-Nomade A., Henershot W. & Bril H. 2010. Alteration of As bearing phases in a small watershed located on a high grade arsenic-geochemical anomaly (French Massif Central). *Appl. Geochem.* 25(12): 1889–1901.

Bundschuh J., Armienta M.A., Morales-Simfors N., Alam M.A., López D.L., Quezada V.D., Dietrich S., Schneider J., Tapia J., Sracek O., Castillo E., Parra L-M.M., Espinoza M.A., Guilherme L.R.G., Sosa N. N., Niazi N.K., Tomaszewska B., Allende K.L.A., Bieger K., Alonso D.L., Brandão P.F., Bhattacharya P., Litter M.I. & Ahmad A. 2021. Arsenic in Latin America: new findings on source, mobilization and mobility in human environments in 20 countries based on decadal research 2010–2020. *Crit. Rev. Env. Sci. Technol.* 5(16): 1727–1865.

Bundschuh J., Litter M.I., Parvez F., Román-Ross G., Nicolli H.B., Jiin-Shuh J., Chen-Wuing L., López D., Armienta M.A., Guilherme L.R.G., Cuevas A.G., Cornejo L., Cumbal L. & Toujaguez R. 2012. One century of arsenic exposure in Latin America: a review of history and occurrence from 14 countries, *Sci. Total Environ.* 429: 2–35.

Corguinha A.P.B., Souza G.A., Gonçalves V.C., Carvalho C. A., Lima W.E.A., Martins F.A.D., Yamanaka C.H., Francisco E.A.B. & Guilherme L.R. G. 2015. Assessing arsenic, cadmium, and lead contents in major crops in brazil for food safety purposes. *J. Food Compos Anal.* 37: 143–150.

Esparza M.L.C. 2009. *The Presence of Arsenic in Drinking Water in Latin America and its Effect on Public Health.* In: Bundschuh J., Armienta M.A., Birkle P., Bhattacharya P., Matschullat J., Mukherjee A.B., editors. *Natural Arsenic in Groundwater of Latin America.* Leiden, The Netherlands: CRC Press/ Balkema. p. 17–29.

Figueiredo B.R., Borba R.P. & Angélica R.S. 2007. Arsenic occurrence in brazil and human exposure. *Environ. Geochem. Health* 29: 109–118.

National Mining Agency (Agência Nacional de Mineração). 2020a. *Anuário Mineral Brasileiro: Principais Substâncias Metálicas – Brasília*: 35 p.

National Mining Agency (Agência Nacional de Mineração). 2020b. *Classificação Nacional de Barragens de Mineração.* https://app.anm.gov.br/SIGBM/Publico/ClassificacaoNacionalDaBarragem.

Wei B. & Yang L. 2010. A review of heavy metal contaminations in urban soils, urban road dusts and agricultural soils from China. *Microchem. J.* 94(1): 99–107.

Arsenic in the Environment: Bridging Science to Practice for Sustainable Development –
van der Wal, Ahmad, Petrusevski, Weijma, Savic, van der Wens, Beerendonk, Bhattacharya, Bundschuh & Naidu (Eds)
© 2023 the Author(s), ISBN 978-1-032-32928-4

Arsenic biotransformation in the gut of soil fauna

Y.G. Zhu[1,2], H.T. Wang[1], G.W. Zhou[1,2] & X.M. Xue[1]
[1]*KEY Laboratory of Urban Environment and Health, Institute of Urban Environment, Chinese Academy of Sciences, Xiamen, P.R. China*
[2]*STATE Key Laboratory of Urban and Regional Ecology, Research Center for Eco-Environmental Sciences, Chinese Academy of Sciences, Beijing, P.R. China*

ABSTRACT: Arsenic (As) is widely distributed in the environment and can be bio-accumulated through the food web. Most previous studies concentrated on As bioconcentration and toxicity to plants, animals, environment microbiota and humans. Few studies focus on As biotransformation in the gut of soil fauna and the effect of the transformation on the gut microbiota and the host. Recently, we systematically studied As biotransformation mediated by gut microbiota of earthworm and the effect of As on the microbiota. In addition, As reduction and methylation mediated by the gut microbiota exerted a critical influence on the reproduction of *Caenorhabditis elegans*. We found that gut microbiota of soil fauna is a hidden but critical hotspot of As biotransformation, and our studies provided new insights into the relationship between As toxicity and gut microbiome of soil fauna. Future works will focus on how the gut microbiota is involved in the degradation of organoarsenicals.

1 INTRODUCTION

Soil is a crucial component of planetary health and is related to human well-being and the long-term sustainable development (Zhu *et al.*, 2019). Soil fauna, accounting for about 23% of all described biodiversity, are essential components of the soil ecosystems, and involved in most of surface/subsurface ecological processes (Lavellea *et al.*, 2006), including the decomposition of organic matter, maintenance of biodiversity and the nutrient cycling of nitrogen and carbon, and degradation of pollutants. Previous studies indicated that the unique anoxic micro-environment of the gut from different environments and the gut microbiome of soil fauna make a significant contribution to these processes. In this presentation, we review recent studies on the effect of arsenic (As) on the gut microbiota of soil fauna and As metabolism mediated by the gut microbiota.

2 THE EFFECT OF ARSENIC ON THE GUT MICROBIOTA OF SOIL FAUNA

The gut microbiota of soil fauna plays a critical role in its host metabolism and health. The shifts in the communities of gut microbiota, known as dysbiosis, can trigger an imbalance of the host immune systems, and further affect the survival and growth of the host. In addition, the gut microbiota is very sensitive to environmental contamination. In our recent study (Wang *et al.*, 2019a), As species, As biotransformation genes (ABGs), and the composition of gut microbial community were characterized after the earthworm *Metaphire sieboldi* was cultured in soils spiked with different As concentrations after 28 days. The result indicated that ABGs encoding As(V) reductases and As efflux transporters in the gut were much more abundant than other genes. It is likely the main reason why most As(V) from soil was reduced into arsenite [As(III)] in the gut. Increasing As(III) concentrations in the gut shifted the community structure of the gut microbiota, and it was confirmed by the PD whole tree showing that the diversity of the gut microbiome first decreased and then increased with elevating As concentrations. The gut microbiota was more sensitive than soil fauna to low As concentration in soils. Thus, the gut microbiota of soil animals can serve as an efficient indicator of As contamination in soils.

We further demonstrated that co-pollution of microplastics can alleviate the effect of As on the gut microbiota likely via binding/adsorbing As(V) and lowering As bioavailability (Wang *et al.*, 2019b). However, the combined exposure to both of sulfamethoxazole and As(V) significantly decreased the microbial diversity of the earthworm *M. sieboldi* gut compared to the control ($P < 0.01$), likely due to the fact that microbiota resisting both As and sulfamethoxazole alone can survive (Wang *et al.*, 2019c). Taken together, As in the gut from soil or the transformed compounds will affect the microbiota community structure of the gut due to their toxicity. We envisage that some microbial taxa might be very sensitive to As contamination, and can be used as an indicator of As pollution in soil ecosystems.

DOI: 10.1201/9781003317395-3

3 ARSENIC TRANSFORMATION MEDIATED BY THE GUT MICROBIOTA OF SOIL FAUNA

Microorganisms play a critical role in As biogeochemical cycles (Zhu *et al.*, 2014, 2017), and involved in various pathways of As transformation (Figure 1). The gut of soil fauna is colonized by abundant microbiota which are significantly different from the surrounding soil. The gut microbiota makes a significant contribution to As speciation in the gut. Our previous studies have revealed that most As(V) was reduced to As(III), small amount of that was methylated in the earthworm gut (Wang *et al.*, 2019a, 2019b). Button *et al.* (2009) found lower proportions of monomethylarsonate [MAs(V)], dimethylarsinate [DMAs(V)], arsenobetaine (AB) and three arsenosugars in the earthworms and cast material from 23 locations at a former As mine in Devon, UK, suggesting that inorganic As can be transformed to organoarsenicals in the gut of earthworm. The definitive biotransformation pathway for As in the gut of earthworms remains to be elucidated.

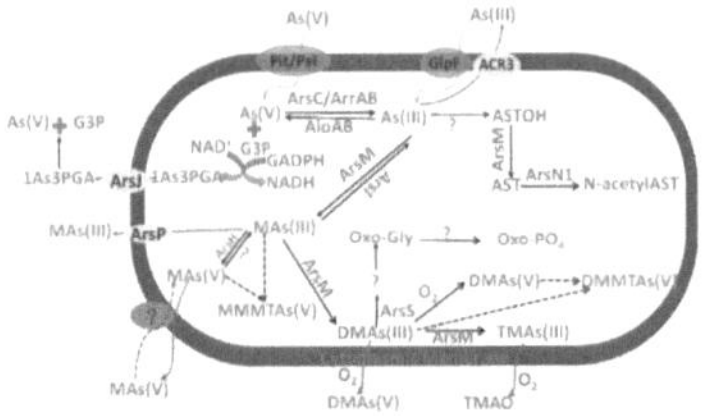

Figure 1. Proposed pathway of As biotransformation mediated by microbiota. ASTOH, 2-amino-4-(dihydroxyarsonoyl) butanoic acid; AST, 2-amino-4-(hydro-xymethylarsinoyl) butanoic acid (arsinothricin); N-acetyAST, arsinothricin by acetylation of the α-amino group; Oxo-Gly, Oxo-arsenosugar–glycerol; Oxo-PO4, Oxo-arsenosugar–phosphate; G3P, D-glceraldehylde 3-phosphate; GADPH, glyceraldehyde-3-phosphate dehydrogenase; 1As3PGA, the unstable organoarsenical 1-arseno-3-phosphoglycerate; MMMTAs(V), mono methylmonothioarsinic acid; DMMTAs(V), dimethyl monothioarsinic.

4 ARSENIC TRANSFORMATION MEDIATED BY GUT MICROBIOTA AFFECTS THE FECUNDITY OF *CAENORHABDITIS ELEGANS*

The gut microbiota of soil invertebrates mediates As biotransformation. However, little is known about the impact of the biotransformation on the fecundity of invertebrates. Using *Caenorhabditis elegans* as a model soil fauna, we found that As(III) and DMAs(V) produced by gut microbiota increased the levels of vitellogenin, glutathione S-transferases and superoxide dismutase, and decreased the reproduction of *C. elegans* (Zhou *et al.*, 2020). The results suggested that As metabolism mediated by the gut microbiota changed the bioavailability and toxicity of ingested As, and further affected the host growth and development.

5 CONCLUSIONS AND RECOMMENDATIONS

The environment, host and gut microbiota interact with each other. The microbial composition of the gut depends heavily on the host identity, diets and their living environment. The gut microbiota is from surrounding environment and is different significantly from the environment due to the unique niche in the gut. The host provides living space and nutrients for the microbiota, which affects the bio-availability and toxicity of various metal(loid)s, and further exerts fluence on host health. As far as As transformation is concerned, the gut microbiota should be involved in the degradation of complex organoarsenicals. Future efforts will solve the below questions: (1) the effect of various As species on the gut microbiota, (2) which species of gut microbes are involved in the degradation of organoarsenicals, (3) the biological and ecological function of organoarsenicals, (4) the function of the gut microbiota of soil fauna in degrading organoarsenicals in soil systems.

ACKNOWLEDGEMENTS

We acknowledge the National Natural Science Foundation of China (41877422 and 42077289) for supporting the program.

REFERENCES

Button M., Jenkin G.R.T., Harrington C.F. & Watts M.J. 2009. Arsenic biotransformation in earthworms from contaminated soils. *J. Environ. Monit.* 11(8): 1484–1491.

Lavellea P., Decaëns T., Aubert M., Barot S., Blouin M., Bureau F., Margerie P., Mora P. & Rossi J.P. 2006. Soil invertebrates and ecosystem services. *Eur. J. Soil Biol.* 42: S3–S15.

Wang H.T., Zhu D., Li G., Zheng F., Ding J., O'Connor P.J., Zhu Y.G. & Xue X.M. 2019a. Effects of arsenic on gut microbiota and its biotransformation genes in earthworm *Metaphire sieboldi. Environ. Sci. Technol.* 53: 3841–3849.

Wang H.T., Ding J., Xiong C., Zhu D., Li G., Jia X.Y., Zhu Y.G. & Xue X.M. 2019b. Exposure to microplastics lowers arsenic accumulation and alters gut bacterial communities of earthworm *Metaphire californica. Environ. Poll.* 251: 110–116.

Wang H.T., Chi Q.Q., Zhu D., Li G., Ding J., An X.L., Zheng F., Zhu Y.G. & Xue X.M. 2019c. Arsenic and sulfamethoxazole increase the incidence of antibiotic resistance genes in the gut of earthworm. *Environ. Sci. Technol.* 53(17): 10445–10453.

Zhu Y.G., Yoshinaga M., Zhao F.J. & Rosen B.P. 2014. Earth abides arsenic biotransformations. *Annu. Rev. Earth Planet. Sci.* 42(1): 443–467.

Zhu Y.G., Zhao Y., Zhu D., Gillings M., Penuelas J., Ok Y. S., Capon A. & Banwart S. 2019. Soil biota, antimicrobial resistance and planetary health. *Environ. Int.* 131: 105059.

Zhu Y.G., Xue X.M., Kappler A., Rosen B.P. & Meharg A. A., 2017. Linking genes to microbial biogeochemical cycling: lessons from arsenic. *Environ. Sci. Technol.* 51(13): 7326–7339.

Zhou G.W., Yang X.R., Zheng F., Zhang Z.X., Zheng B.X., Zhu Y.G. & Xue X.M. 2020. Arsenic Transformation mediated by gut microbiota affects the fecundity of *caenorhabditis elegans. Environ. Poll.* 260: 113991.

*Arsenic in the Environment: Bridging Science to Practice for Sustainable Development –
van der Wal, Ahmad, Petrusevski, Weijma, Savic, van der Wens, Beerendonk, Bhattacharya, Bundschuh & Naidu (Eds)
© 2023 the Author(s), ISBN 978-1-032-32928-4*

Arsenic in livestock – biotransfer to food

A.L. Pérez-Carrera
Facultad de Ciencias Veterinarias, Centro de Estudios Transdisciplinarios del Agua (CETA, UBA), e Instituto de Investigaciones en Producción Animal (INPA, UBA-CONICET), Universidad de Buenos Aires, Buenos Aires, Argentina

ABSTRACT: Arsenic (As) is an ubiquitous element found in the atmosphere, soils and rocks, natural waters and organisms. In groundwater, As has been detected in several regions of the world, with concentration levels exceeding the WHO drinking water guideline value of 10 μg/l, as well as the national regulatory standards; its presence in drinking water is one of the most important health problems in the world. Arsenic pollution affects several countries around the world. One of the most affected countries is Argentina, where As is naturally present in groundwater and soils of the main agriculture production areas. Arsenic can be taken up and bioaccumulated in plants and livestock, which means a potential risk to human health through agri-food. Arsenic contaminated groundwater is often used in the Argentinean Pampean Plain for livestock drinking water and to irrigate crops for human and animal consumption. This fact could potentially lead to As entering in the human food chain. In livestock, total As consumption depends on its concentration in drinking water and in the different foods and the percentage it represents in the diet. Biotransfer factors are used to relate either estimated daily exposure dosage, or chemical content in feed, to concentrations occurring in milk or dairy products. Arsenic transfer to milk is a complex process, the fact that a biotransfer factor (BTF) may be estimated through As water contribution reinforces the importance of dairy cattle drinking water quality not only from a productive point of view but also because of its incidence in the agricultural food chain.

1 INTRODUCTION

Arsenic (As) is an ubiquitous element found in the atmosphere, soils and rocks, natural waters and organisms. It is mobilized in the environment through a combination of natural processes (i.e. weathering reactions, biological activity and volcanic emissions), as well as through a range of anthropogenic activities (Bhattacharya *et al.*, 2007). In groundwater, As has been detected in several regions of the world, with concentration levels exceeding the WHO drinking water guideline value of 10 μg/L (WHO 2001), as well as the national regulatory standards; its presence in drinking water is one of the most important health problems in the world.

Arsenic pollution affects several countries around the world. One of the most affected countries is Argentina, where As is naturally present in groundwater and soils of the main agriculture production areas (Bundschuh *et al.*, 2012). Arsenic can be taken up and bioaccumulated in plants and livestock, which means a potential risk to human health through agri-food. The main results of our studies conducted to assess the impact of As in water on food quality is presented.

2 ARSENIC CONTAMINATED GROUNDWATER IN THE ARGENTINEAN PAMPEAN PLAIN

Arsenic contaminated groundwater is often used in the Argentinean Pampean Plain for livestock drinking water and to irrigate crops for human and animal consumption. This fact could potentially lead to As entering in the human food chain. In man and animals, the inorganic arsenic can be methylated to form the organic arsenic species such as monomethyl arsonic acid (MMA) and dimethylarsinic acid (DMA). This methylation process reduces the acute toxicity of the compound and facilitates its excretion (Tseng 2009).

As a secretion of the mammary gland, milk can carry numerous xenobiotic substances, which constitute a technological risk factor for dairy products and above all a health risk for the consumer (Licata *et al.*, 2004). Nevertheless,

DOI: 10.1201/9781003317395-4

information related to the incidence of high As livestock drinking water, its effect on livestock health, and its transfer to milk is scarce.

3 ARSENIC CONTAMINATED GROUNDWATER IN ARGENTINEAN PAMPEAN PLAIN FOR LIVESTOCK

In one of the first studies carried out in our laboratory, we analyzed the relation between As uptake through water and its transfer to milk, that being the first report of As milk content in Argentina. The study was carried out in five farms in the southeast of the province of Cordoba, which is one of the most affected areas in the Chaco Pampean region. This region is also an important milk production zone in Argentina, where dairy product consumption is up to 192 equivalent milk L/inhabitant/year. A BTF for milk was calculated considering drinking water as the only As exposure for cattle. (Pérez-Carrera & Fernández Cirelli 2005). In livestock, total As consumption depends on its concentration in drinking water and in the different foods and the percentage it represents in the diet. Biotransfer factors are used to relate either estimated daily exposure dosage, or chemical content in feed, to concentrations occurring in milk or dairy products.

The aim of that pioneer study was the validation of this transference factor as a tool for the estimation of As milk concentration from drinking water concentration. Arsenic content in livestock drinking water, soils, forages and milk, from thirty dairy farms located in an area of high As groundwater, was determined to analyze the relation between As uptake through water and food, and its transfer to milk.

In the last years, studies on the As content in water, soil and crops have increased in Latin American countries. On the other hand, there is scarce information about the As content in animal based food, and studies are mainly focused on bovine milk and tissues. Water is the principal source of As for livestock production. Our studies have been focused in the biotransference of this element from water to food and have been conducted mainly in the pampean plain of Argentina. The first studies were performed in the southeast of Córdoba province, which is one of the main dairy regions in our country.

4 CONCLUSIONS

Arsenic transfer to milk is a complex process, the fact that a biotransfer factor (BTF) may be estimated through As water contribution reinforces the importance of dairy cattle drinking water quality not only from a productive point of view but also because of its incidence in the agricultural food chain. On the other hand, in some regions, the presence of As in food can be a major source of exposure for humans, even greater than drinking water.

Since information about transference of As from environmental matrices to milk is very scarce, further studies on this subject are necessary, including speciation in milk and dairy products, to guarantee food safety.

REFERENCES

Bhattacharya P., Welch A.H., Stollenwerk K.G., McLaughlin M.J., Bundschuh J. & Panaullah G. 2007. Arsenic in the environment: Biology and chemistry. *Sci. Total Environ.* 379: 109–265.

Bundschuh J., Litter M.I., Parvez F., Román-Ross G., Nicolli H.B., Jean J.S., Liu C.-W., López D., Armienta M.A., Guilherme L.R.G., Cuevas A.G., Cornejo L., Cumbal L. & Toujaguez R. 2012. one century of arsenic exposure in Latin America: A review of history and occurrence from 14 countries. *Sci. Total Environ.* 429: 2–35.

Licata P., Trombetta D., Cristiani M., Giofre F., Martino D. Calo M. & Naccari F. 2004. Levels of "toxic" and "essential" metals in samples of bovine milk from various dairy farms in Calabria, Italy. *Environ. Int.* 30: 1–6.

Pérez-Carrera A. & Fernández-Cirelli A. 2005. Arsenic concentration in water and bovine milk in Cordoba, Argentina. Preliminary results. *J. Dairy Res.* 72: 122–124.

Pérez-Carrera A., Moscuzza C. & Fernández-Cirelli A. 2008. Transfer of arsenic from contaminated dairy cattle drinking water to milk (Córdoba, Argentina). In: J. Bundschuh, M.A. Armienta, P. Bhattacharya, J. Matschullat & A.B. Mukherjee (eds.), *Natural Arsenic in Groundwaters of Latin America*. Taylor & Francis/ Balkema, Leiden, The Netherlands. pp. 419–425.

Pérez-Carrera A., Alvarez-Gonçalvez C.V. & Fernández-Cirelli A. 2016a. Transference factors as a tool for the estimation of arsenic milk concentration. *Environ. Sci. Pollut. Res. Int.* 23(16): 16329–16335.

Pérez Carrera A., Arellano F. & Fernández Cirelli A. 2016b. Concentration of trace elements in raw milk from cows in the southeast of Córdoba province, Argentina. *Dairy Sci. Technol.* 96: 591–602.

Schoof R.A., Yost L.J., Eickhoff J., Crecelius E.A., Cragin D.W., Meachere D.M. & Menzele D.B. 1999. A market basket survey of inorganic arsenic in food. *Food Chem. Toxicol.* 37(8): 839–846.

Tseng C. 2009. A review on environmental factors regulating arsenic methylation in humans. *Toxicol. Appl. Pharmacol.* 235(3): 338–50.

WHO (World Health Organization) 2001. Arsenic and arsenic compounds. *Environmental Health Criteria* 224. Inter-Organization Programme for the Sound Management of Chemicals. Geneva.

Yost L.J., Schoof R.A. & Aucoin R. 1998. *Intake of Inorganic Arsenic in the North American Diet*. Human and Ecological Risk Assessment 4 137–152. Survey 2019, Key Findings. Bangladesh Bureau of Statistics Dhaka, Bangladesh.

Arsenic in the Environment: Bridging Science to Practice for Sustainable Development – van der Wal, Ahmad, Petrusevski, Weijma, Savic, van der Wens, Beerendonk, Bhattacharya, Bundschuh & Naidu (Eds)
© 2023 the Author(s), ISBN 978-1-032-32928-4

Arsinothricin, an arsenic-containing non-proteinogenic amino acid analog of glutamate, is a potent broad-spectrum antibiotic

V.S. Nadar[1], A.E. Galván[1], S.H. Suzol[2], A.H. Howlader[2], J. Chen[1], D.S. Dheeman[1], K. Yoshinaga-Sakurai[1], M. Radhakrishnan[1], P. Kandavelu[3], B. Sankaran[4], M. Kuramata[5], S. Ishikawa[5], S.M. Utturkar[6], S.F. Wnuk[2], B.P. Rosen[1] & M. Yoshinaga[1]
[1]*Department of Cellular Biology and Pharmacology, Florida International University, Herbert Wertheim College of Medicine, Miami, USA*
[2]*Department of Chemistry and Biochemistry, Florida International University, Miami, USA*
[3]*SER-CAT and Department of Biochemistry and Molecular Biology, University of Georgia, Athens, USA*
[4]*Berkeley Center for Structural Biology, Lawrence Berkeley Laboratory, Berkeley, USA*
[5]*Division of Hazardous Chemicals, National Institute for Agro-Environmental Sciences, NARO, Tsukuba, Ibaraki, Japan*
[6]*Purdue University Center for Cancer Research, Purdue University, West Lafayette, USA*

ABSTRACT: There is an urgent need for new antibiotics to combat the emergence and spread of antibiotic resistance. Bacteria evolved the ability to utilize the pervasive environmental toxic metalloid arsenic to produce the potent broad-spectrum antibiotic, arsinothricin (AST). AST is a non-proteinogenic analog of glutamate that inhibits glutamine synthetase. It is an effective broad-spectrum antibiotic against both Gram-positive and Gram-negative bacteria including major resistant pathogens, demonstrating the potential to address the global threat of drug resistance. New chemical synthesis methods have been devised for AST that are necessary for long process of drug tests and development. AST is biosynthesized via two steps involving sequential formation of two C-As bonds catalyzed by ArsL, the novel radical S-adenosylmethionine (SAM) protein, and ArsM, As(III) SAM methyltransferase. The arsN1 gene confers resistance to AST by acetylation of the α-amino group. Crystal structures of ArsN1 N-acetyltransferase, with or without AST, shed light on the mechanism of its substrate selectivity. Using this information inhibitors of AST have been identified. One inhibitor, 3-phosphonoalanine, reverses AST resistance. These findings have the potential for development of a new class of organoarsenical antimicrobials and ArsN1 inhibitors.

Arsenic is the most pervasive environmental toxic element. Here we describe how bacteria harnessed arsenic to create a potent broad-spectrum antibiotic. New antibiotics are urgently needed because the emergence of resistance has rendered nearly every clinically used antibiotic ineffectual. Human tuberculosis (TB), the top global infectious disease killer, which is caused by *Mycobacterium tuberculosis*, has become even more difficult to treat due to the drug resistance. The World Health Organization (WHO) declares multidrug-resistant tuberculosis a global public health crisis, calling for a pressing need for development of new and innovative antibiotics. In addition to *M. tuberculosis*, the WHO recently issued a global priority pathogen list of antibiotic resistant bacteria that pose the greatest threat to human health to guide and promote research and development of new antibiotics.

Here we demonstrate that the arsenic-containing natural product, arsinothricin 2-amino- 4-(hydroxymethylarsinoyl) butanoate, AST), produced by the rice rhizosphere microbe *Burkholderia gladioli* GSRB05, is a broad-spectrum antibiotic (Figure 1) (Nadar *et al.*, 2019). It is effective against pathogens such as *M. bovis*, the causative agent of bovine and zoonotic human TB. The WHO has declared the "last resort antibiotic" carbapenem-resistant *Enterobacter* (CRE) a global threat, and AST is very effective against carbapenem-resistant *E. cloacae.* AST is a non-proteinogenic amino acid analog of L-glutamate that inhibits of bacterial glutamine synthetase (GS). Most toxic arsenicals contain trivalent As(III). AST is unusual in being a highly toxic pentavalent organoarsenical.

We predict that AST and related arsenic-containing compounds may be the progenitors of a new class of antibiotics. While modest amounts of AST can be generated by the source organism, drug development requires a reliable source of the compound. For that reason, we embarked on a synthetic approach to produce AST. We developed a semisynthetic procedure that involves chemical synthesis of the precursor of AST, 2-amino-4-(dihydroxyarsonoyl)butanoic

DOI: 10.1201/9781003317395-5

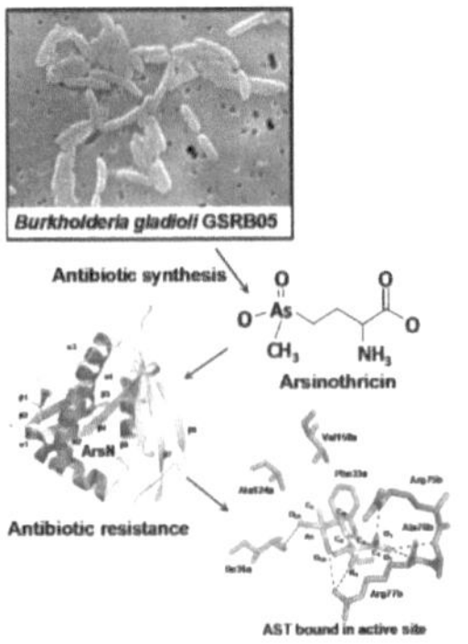

Figure 1. Arsinothricin (AST) is a novel arsenic-containing antibiotic. The soil isolate *B.* gradioli GRSB05 synthesizes the unnatural amino acid AST, an inhibitor of glutamine synthetase, from environmental Arsenite As(III). ArsN1 is an *N*-acetyltransferase that inactivates AST by acetylation.

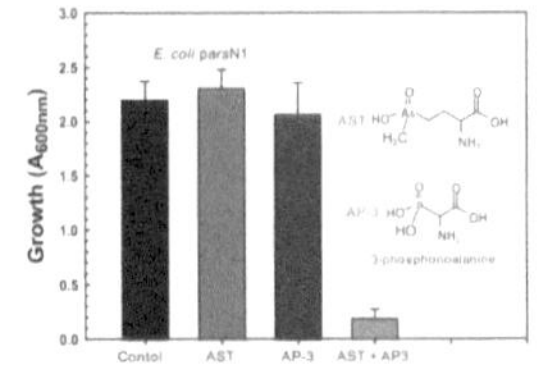

Figure 3. 3-phosphonoalanine (AP-3) reverses ArsN1 resistance to AST. Cells of ArsN1-expressing P. putida are resistant to AST (red). AP-3 does not inhibit growth of those cells (green). Addition of AP-3 reverses resistance and makes cells sensitive to AST again (yellow).

acid [hydroxyarsinothricin, or AST-OH], which is then enzymatically methylated to AST using the robust thermostable enzyme CmArsM the As(III) *S*-adenosylmethionine (SAM) methyltransferase from the acidothermophilic eukaryotic alga *Cyanidioschyzon* sp. 5508 (Figure 2) (Suzol *et al.*, 2020). Complete chemical synthetic methods for AST are under development.

The biosynthetic pathway of AST was unknown. We sequenced the genome of the producer *B. gladioli* GSRB05 and found a seven-gene cluster containing *arsM*. Heterologous expression of the gene cluster confers AST production in *Escherichia coli*. Mutagenesis analysis suggests that only two genes, *arsM* and an adjacent gene annotated to encode a radical SAM protein, named *arsL*, are necessary for AST biosynthesis (Figure 2). Our results demonstrate that the biosynthetic pathway of AST is much simpler in contrast to that of phosphinothricin (PT), the phosphorus mimetic of AST, which requires more than a dozen of genes and steps to be synthesized.

With every new antibiotic, resistance inevitably arises. Many arsenic resistance (*ars*) operons have an *arsN1* gene that confers resistance to AST. ArsN1 is an *N*-acetyl transferase that acetylates the α-amino group of AST, preventing it from binding to the active site of GS. We crystallized ArsN1 and solved the apo and substrate-bound structures (Figure 1). Using those structures we used *in silico* screening of millions of potential inhibitors and identified a number that could be predicted to bind to the active site of ArsN1. Several were purchased and shown to be noncompetitive inhibitors of ArsN1. In addition, 3-phosphonoalanine (AP-3), a phosphonate analog of AST, binds to the active site of ArsN1 and inhibits its enzymatic activity. Addition of 3-AP to cultures of *arsN1*-expressing *Pseudomonas putida* KT2440 reversed resistance to AST. This suggests that co-administration of AST and an *arsN1* inhibitor can extend the lifetime of this novel antibiotic.

AST is the first and only known arsenic-containing antibiotic, and there is still a huge knowledge gap to cover about this unique antibiotic. Our results elucidate the reactions and molecular mechanisms for AST biosynthesis and resistance, which can be utilized to design new and novel arsenic-containing drugs to combat WHO priority pathogens such as MTB and CRE. AST is chemically unrelated to other organoarsenicals and has the potential to be the progenitor of a new class of organoarsenical antibiotics.

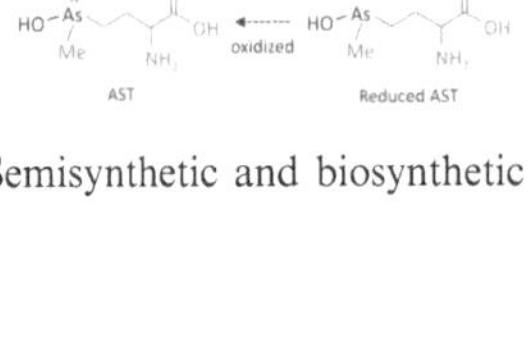

Figure 2. Semisynthetic and biosynthetic pathways of AST.

ACKNOWLEDGEMENTS

This work was supported by NSF BIO/MCB grant 1817962 to M.Y. and R35 GM136211, R01 GM55425 and R01 ES023779 to B.P.R.

REFERENCES

Nadar V.S. Chen J., Dheeman D.S., Galván A.E., Yoshinaga-Sakurai K., Kandavelu P., Sankaran B., Kuramata M., Ishikawa S., Rosen, B.P. & Yoshinaga M. 2019. Arsinothricin, an arsenic-containing non-proteinogenic amino acid analog of Glutamate, is a broad-spectrum antibiotic. *Commun. Biol.* 2: 131.

Suzol S.H., Hasan Howlader A., Galván A.E., Radhakrishnan M., Wnuk S.F., Rosen B.P. & Yoshinaga M. 2020. Semi-Synthesis of the Organoarsenical antibiotic arsinothricin. *J. Nat. Prod.* 83: 2809–2813.

Arsenic in the Environment: Bridging Science to Practice for Sustainable Development –
van der Wal, Ahmad, Petrusevski, Weijma, Savic, van der Wens, Beerendonk, Bhattacharya, Bundschuh & Naidu (Eds)
© 2023 the Author(s), ISBN 978-1-032-32928-4

Evaluation of the influence of main groundwater ions on arsenic removal by limestones at Zimapán, México

M.A. Armienta[1], A. Sosa[2], A. Aguayo[1] & O. Cruz[1]
[1]*Instituto de Geofísica, Universidad Nacional Autónoma de México, México City, Mexico*
[2]*Posgrado en Ciencias de la Tierra, Universidad Nacional de México, México City, Mexico*

ABSTRACT: Development of sustainable and technically accessible treatment alternatives of As-rich groundwater suitable for low-income populations is a relevant research field, mostly in locations relying on wells as drinking water source. Use of limestones for arsenic removal is specially promising in groundwater contaminated areas with limestone outcrops. This study was carried out with the aim to determine the influence of groundwater major anions on As removal by limestones through column experiments. Treatment experiments were carried out during 18 weeks with water containing As, fluoride and each of the main anions in concentrations corresponding to the most polluted well (As = 1.2 mg/L) at Zimapán, México. Results indicated that chloride and fluoride do not influence As removal that decreases more than 80%. Meanwhile, although As diminished to values below the Mexican drinking water standard (0.025 mg/L) in the solutions containing sulfate and bicarbonate since the 2nd week, it started to increase since the 12th week and 7th week respectively. This behavior may be due to a decrease of sorption sites due to their retention on limestones. Results of the study may be used to design a home filter to treat the contaminated water, and further exported to other locations.

1 INTRODUCTION

To remove arsenic from contaminated water and avoid the population exposure through drinking water various methods have been developed. Common treatment procedures include coagulation/filtration, oxidation, and membrane techniques. Other methods like adsorption, biological procedures, electrochemical technologies, ion exchange resins, zero-valent iron and photochemical methods have also been used (Baig *et al.*, 2014). Each method has advantages, draw-backs and limitations depending on local conditions. Native limestones from central Mexico have proven to remove arsenic from polluted water (Romero *et al.*, 2004). The efficiency and geochemical processes involved in the retention by the rocks have been previously studied, mainly through batch experiments with As synthetic solutions, and considered as an option to remove As from groundwater at Zimapán, México. However, to propose the use of limestones as a treatment option, the influence of common major ions on As retention must be assessed. To that aim, column experiments were performed to treat solutions containing arsenic and sulfate, chloride, bicarbonate or fluoride at concentrations corresponding to those of the most polluted well at Zimapán, México.

2 MATERIALS AND METHODS

2.1 *Study area*

The Zimapán Municipality is located in a valley at about 150 km north of Mexico City and 1800 m a.s.l. The stratigraphic sequence comprises the volcanic-sedimentary Jurassic Trancas Formation composed mainly by calcareous lutite, limonite, and micritic limestone, discordantly covered by sedimentary lower and upper Cretaceous Formations. These Formations correspond to the lower cretaceous El Doctor or El Abra limestones, underlying the upper Cretaceous pelitic-calcareous Soyatal Formation, and the Oligocene El Morro Fanglomerate, that concordantly underlies the volcanic Las Espinas Formation from late Miocene. Finally, the top of the sequence corresponds to a Quaternary alluvial and colluvial platform of sedimentary and volcanic origin. Zimapán has been a mining district since the 16th Century for the exploitation of silver, zinc and lead. In this area, natural processes due to the mineralization rich in As-bearing minerals, have released high As contents to deep groundwater wells (Armienta *et al.*, 2001). Although various alternatives have been stablished to avoid the population exposure such as small treatment plants, and an aqueduct that brings water from 30 km distance to the town, the problem has not completely been solved mainly due to energy costs, and difficulties on the chemicals supply, operation, and maintenance. Besides, small localities in the Municipality are not connected to the pipelines delivering the treated water.

2.2 *Experimental set-up*

Limestones from the Soyatal formation that is widely distributed at Zimapán were collected near the town, crushed, ground, and sieved. Transparent acrylic columns (50 cm height, 9 cm inner diameter, with 5 sampling ports located each 10 cm along the column length) were packed with the rock particles (0.5–1.41 mm) and fed with the solutions delivered by peristaltic pumps. The concentrations were chosen to mimic the arsenic, fluoride and other anions measured at the most polluted well in the Zimapán municipality that has also one of the

DOI: 10.1201/9781003317395-6

most As-rich wells of Mexico. Experimental solutions were prepared by dissolving analytical quality grade salts. Conductivity, pH and Eh were determined weekly at each sampling port along 4 months. Simultaneously, arsenic and each anion concentration were measured by HG- AAS (As), turbidimetry (SO_4^{2-}), potentiometry with selective electrodes (Cl^- and F^-) and titration (HCO_3^-), in samples collected from the first and fifth sampling ports. After the end of the experiments the rocks were taken from the top of the columns and analyzed by XRF, XRD, and SEM-WDS.

3 RESULTS AND DISCUSSION

3.1 *Arsenic speciation and retention*

Calcite and quartz were the main minerals identified in the Soyatal rock samples used in the experiments. Eh vs pH diagrams in the input solutions showed arsenic occurrence mainly as $HAsO_4^{2-}$ and $H_2AsO_4^-$; its predominance as As(V) was also assessed by specific analyses. This speciation indicates that As may be sorbed on the limestones since the pHpzc of Soyatal rocks is about 9.5, and at the pH of the solutions (from 6.9 to 9.4) the surface will mostly be positively charged.

3.2 *Effect of the anions on as removal*

Arsenic decreased from 1.2 mg/L (in the influent) to values below the Mexican drinking water standard of 0.025 mg/L in the fifth sampling port of the solutions containing fluoride (2.7 mg/L) and chloride (10 mg/L). This retention was observed since the first week until the end of the experiment (Figure 1), except for an increase from the 5th to the 8th week (up to 0.12 mg/L) probably due to heterogeneities of the packing and changes in the flow-paths. This behavior indicates that chloride and fluoride do not interfere with As removal. On the other hand, although As also strongly decreased in the solution containing sulfate since the first week, it started to increase slightly since the 12th week. In addition, sulfate concentrations decreased with time from 80 mg/L in the influent to 70 mg/L in the fifth port outflow at the end of the experiment. Sulfate seems to be retained on the limestones and competes with arsenic for the sorption sites as reflected by its decrease in the output solutions. Arsenic also decreased in the solution containing bicarbonate (300 mg/L) to concentrations below 0.025 mg/L from the 2nd to the 6th week, increasing afterward until reaching the input concentration in the 15th week. These results reflect the behavior observed in other experiments like that performed by Anawar *et al.* (2004) reporting As mobilization by bicarbonate water in Bangladesh. To get insight in the possible processes involved on As retention, geochemical modeling and mineralogical analyses of the rocks were performed. Saturation index calculations with Visual MINTEQ 3.1 program and DRX analyses did not evidence the formation of arsenic minerals. Meanwhile, elemental map analyses obtained by SEM-WDS showed the presence of arsenic scattered in low concentrations on the rocks' surface after the treatments. These results indicate As sorption on limestone as a possible retention process. Besides, XRF determinations revealed the increase of As concentration in the waste rocks with respect to that in the raw material before the treatments. The highest concentrations of As were measured in the rocks used for treating the solutions containing chloride and fluoride.

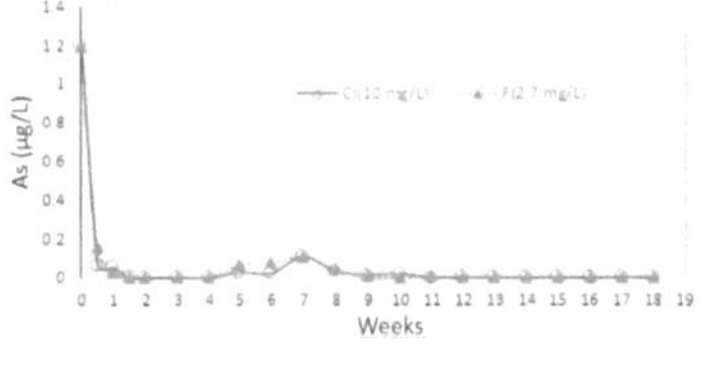

Figure 1. Arsenic concentration (mg/L) in the 5th sampling port along the experimental period in chloride and fluoride solutions.

4 CONCLUSIONS AND RECOMMENDATIONS

Native limestones of the Soyatal Formation outcropping at Zimapán are an option for treating As polluted water in a continuous mode. Experiments also indicated that among the main groundwater ions at the concentrations of the well water, sulfate and bicarbonate interfere with As removal. Although adsorption seems to be the retention mechanism, the actual links of As with calcite and the interfering geochemical processes should be studied with high-resolution techniques of surface analyses.

ACKNOWLEDGEMENTS

DGAPA (Dirección General de Asuntos del Personal Académico) UNAM Project PAPIIT IN106918 is acknowledged for funding. We thank Patricia Girón and Carlos Linares for XRF and SEM analyses.

REFERENCES

Anawar H.M., Akai J. & Sakugaua H. 2004. Mobilization of arsenic from subsurface sediments by effect of bicarbonate ions in groundwater. *Chemosphere* 54: 753–762.

Armienta M.A., Villaseñor G., Rodríguez R., Ongley L. K. & Mango H. 2001. The role of arsenic-bearing rocks in groundwater pollution at Zimapán Valley, México. *Environ. Geol.* 40: 571–581.

Baig S.A., Sheng T., Hu Y., Shu J. & Xu X. 2014. Arsenic removal from natural water using low cost granulated adsorbents: a review. *CLEAN-Soil Air Water* 43(1): 13–26.

Romero F.M., Armienta M.A. & Carrillo-Chavez A. 2004. Arsenic sorption by carbonate rich aquifer material, a control on arsenic mobility at Zimapán, México. *Arch. Environ. Contam. Toxicol.* 47: 1–13.

Arsenic in the Environment: Bridging Science to Practice for Sustainable Development –
van der Wal, Ahmad, Petrusevski, Weijma, Savic, van der Wens, Beerendonk, Bhattacharya, Bundschuh & Naidu (Eds)
© 2023 the Author(s), ISBN 978-1-032-32928-4

Arsenic in drinking water: Navigating towards positive water futures

D. van Halem
Water Management Department, Faculty of Civil Engineering and Geosciences, Delft University of Technology, Delft, The Netherlands

ABSTRACT: In this keynote address, I will investigate how arsenic (As) removal relates to the removal of other commonly found groundwater contaminants, like iron, ammonium and manganese. These groundwater contaminants are typically found in concentrations 10–100 fold that of As, making them key drivers for technology operation and design, and as such also their sustainable adoptation. I conclude that past research has enhanced the understanding of As behavior during conventional groundwater treatment, particularly in that of aerated rapid sand filters. This has led to a wide range of new, engineered solutions for improved As removal, either with operational adjustments (e.g., supernatant water level, filter media) or more advanced technology integration (e.g., Fe(0)-electrocoagulation). These solutions are robust and attractive for (decentralized) sustainable application, because they do not require chemical pre-oxidation or post-treatment (e.g., adsorbents) and are able to co-remove other groundwater contaminants.

1 INTRODUCTION

It has been 15 years since the first "Arsenic in the Environment" conference in 2006, yet, when it comes to mitigating this widespread problem, progress seems to be slow. For example, the Arsenic Primer (WHO/UNICEF 2018) reported that despite all efforts, As removal devices have had limited impact on overall mitigation efforts in Bangladesh. This calls for reflection. In this keynote address I will investigate the role of other commonly found groundwater contaminants, like iron, ammonium and manganese, in the high failure rates of As removal technologies. These groundwater contaminants are typically found in concentrations 10–100 fold that of As, causing clogging and biofilm growth, making them key drivers for technology operation and design, and as such also their sustainable adoptation.

The objective was therefore to investigate how As removal can be achieved alongside other groundwater contaminants, instead of adding pre-treatment (e.g., oxidant dosing) or post-treatment (e.g. adsorbent), see Figure 1. This knowledge will aid in development of strategies for chemical-free As removal, either for decentralized treatment or integration in current groundwater treatment trains.

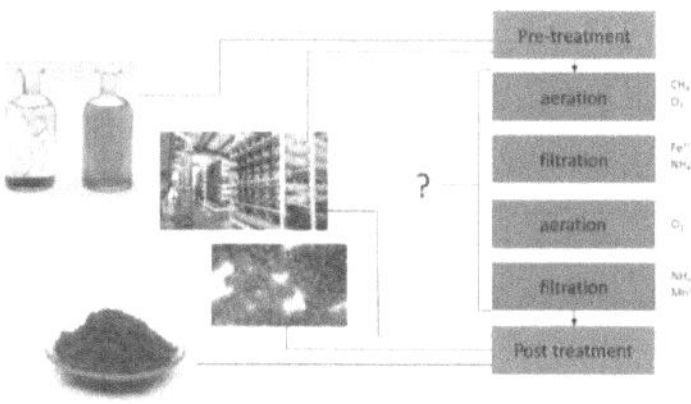

Figure 1. Arsenic removal is typically achieved as pre- or post-treatment of a conventional treatment train, but what if we could achieve integrative removal?

2 METHODS

The studies comprised of synthetic and natural groundwater experiments in the Netherlands (As «50 μg/L) and Bangladesh (As »50 μg/L). The specific aim of the research in the Netherlands was to reach the newly adopted target of 1 μg/L (VEWIN, 2016). In Bangladesh, work was executed in collaboration with governmental DPHE in Rajshahi. A wide range of batch, column and pilot experiments have been conducted over the past decade to unravel As behaviour during aeration-filtration, as well as the impact of groundwater matrix (e.g., pH, Fe concentration) and operational parameters like, supernatant water level, filtration rate or filter media. Also, innovative engineering solutions were studied to enhance As removal during the aeration-filtration sequence: triple-bed filtration, anoxic pre-storage and bio-Fe(0) electrocoagulation.

3 RESULTS AND DISCUSSION

3.1 *Rapid sand filters*

In existing aeration-filtration systems it was found that there are two main mechanisms that drive As immobilization, namely, biological oxidation of As (III) to As(V) in the top layer of filter beds and deep bed infiltration of co-occurring Fe (Gude 2018a). The growth of an NH_4^+ oxidizing biofilm (AOB) was not found to hamper As(III) oxidizing bacteria (AsOB) to grow. Also, accumulating MnO_2 could potentially aid in As(III) oxidation, but this abiotic process was found to be inhibited by Fe^{2+} and Mn^{2+}. The oxidation of As(III) to As(V) is a critical step towards effective adsorption onto hydrous ferric oxides (HFO) flocs, therefore, the infiltration of these flocs beyond the AsOB active layer was found to be

DOI: 10.1201/9781003317395-7

critical. Factors influencing the deep bed infiltration were found to be pH, supernatant water level, filter media and filtration rate (Gude *et al.*, 2018b, 2018c).

3.2 *Triple bed filtration*

This understanding of As mobility in filters has set the stage for follow-up studies to engineer these systems for enhanced removal. For example, a study was performed with Dutch non-chlorinated tap water, spiked with 2 mg Fe^{2+}/L and 12 µgAs/L, to investigate triple bed filtration (anthracite, sand, garnet). In such a system HFO flocs penetrate deeper into the filter, compared to a single bed filter (Figure 2, left). Consequently, the adsorption of As take place over the full height of the filter bed (Figure 2, right), resulting in better utilization of the same groundwater-native Fe^{2+}.

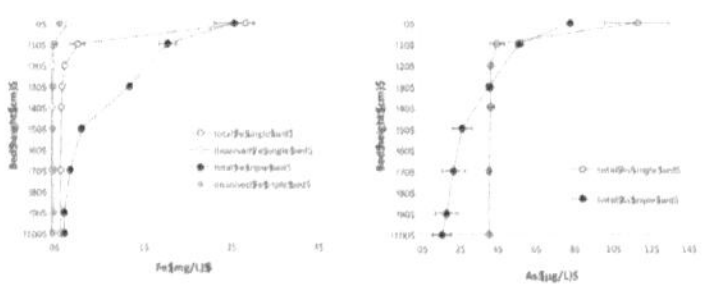

Figure 2. Aerated, biological triple bed filters utilize full bed height for As removal from groundwater, reaching As <1 µg/L.

3.3 *Anoxic storage*

A more innovative approach towards enhancing As removal with co-occurring Fe^{2+} was taken in Rajshahi, Bangladesh. Here, we investigated the sequence of anoxic storage prior to aeration-filtration. The idea was that by partially delaying Fe^{2+} oxidation, more HFO would be available for adsorption of oxidized As deeper in the filter bed. Figure 3 shows a summary of the results of this study (Annaduzzaman *et al.*, 2021), which illustrates that with the same groundwater Fe^{2+} concentration (2.3 mg/L) more As was captured, i.e., 329 µgAs/L was reduced to 128 µgAs/L and 28 µgAs/L for oxic and anoxic storage, respectively.

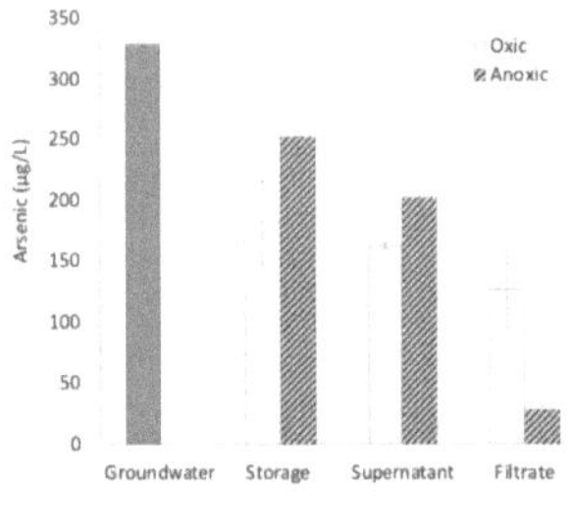

Figure 3. Anoxic storage prior to decentralized aeration-filtration increases As uptake by groundwater-native iron in Rajshashi, Bangladesh.

3.4 *Bio-Fe(0)electrocoagulation*

Not all groundwaters may contain sufficient naturally present Fe^{2+} to capture the high As concentrations. Therefore, we also investigated the integration of electrochemically dosing Fe^{2+} to biofilters. (Roy *et al.*, 2021). It was found in As-containing water (150 µg/L) that positioning Fe(0)-EC in sequence to the bio-active AsOB layer in filters reduced Fe dosing by 10-fold (Figure 4) This resulted in considerable sludge reduction and longer runtimes, making this technology more appealing for practice.

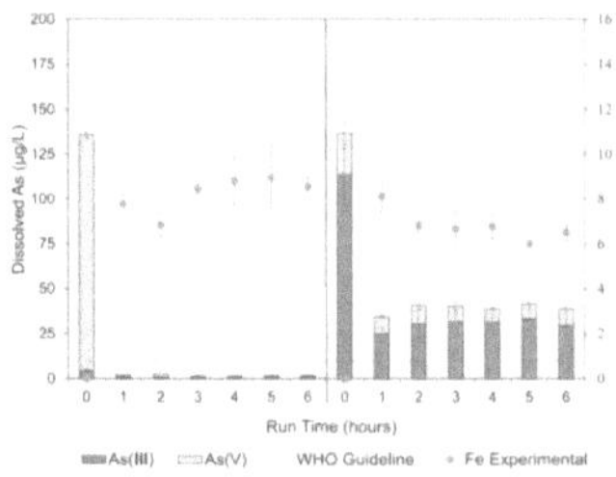

Figure 4. Fe(0)-electrocoagulation positioned after the bio-active filter layer (left) requires 10x lower Fe dosing than regular Fe(0)-EC.

4 CONCLUSIONS

In conclusion, enhanced understanding of As behavior in aerated rapid sand filters has led to new, engineered solutions to improve co-removal with Fe, either with elegant operational adjustments (e.g., supernatant water level, filter media) or more advanced technology integration (e.g., Fe(0)-electrocoagulation). These solutions are robust and attractive for (decentralized) sustainable application, because they do not require chemical pre-oxidation or post-treatment (e.g., adsorbents). Also, the integrated aeration-filtration step will ensure co-removal of other groundwater contaminants Fe^{2+}, NH_4^+, Mn^{2+} and CH_4^+, which will aid in sustainable application.

REFERENCES

Annaduzzaman M., Rietveld L.C., Hoque B.A., Bari M.N. & van Halem D. 2021. Arsenic removal from iron-containing groundwater by delayed aeration in dual-media sand filters *J. Hazard. Mater.* 41: 124823.

Gude J.C.J. 2018a. *Arsenic Removal in Rapid Sand Filters.* PhD Dissertation, Delft University of Technology.

Gude J.C.J., Rietveld L.C. & van Halem D. 2018b. As(III) removal in rapid filters: effect of pH, Fe(II)/Fe(III), filtration velocity and media size. *Wat. Res.* 147: 342–349

Gude J.C.J., Joris K., Huysman K., Rietveld L.C. & van Halem D. 2018c. Effect of supernatant water level on as removal in biological rapid sand filters. *Wat. Res. X* 1: 100013.

Roy M., van Genuchten C.M., Rietveld L.C. & van Halem D. 2021. Integrating biological As(III) oxidation with Fe(0) electrocoagulation for arsenic removal from groundwater. *Wat. Res.* 188: 116531.

VEWIN (2016). *Arsenic at Low Concentrations in Dutch Drinking Water: Assessment of Removal Costs and Health Benefits*

WHO/UNICEF 2018. *Arsenic Primer: Guidance on the Investigation & Mitigation of Arsenic Contamination.*

Arsenic in the Environment: Bridging Science to Practice for Sustainable Development – van der Wal, Ahmad, Petrusevski, Weijma, Savic, van der Wens, Beerendonk, Bhattacharya, Bundschuh & Naidu (Eds)
© 2023 the Author(s), ISBN 978-1-032-32928-4

Ending toxic arsenic exposure from well water in Bangladesh

A. van Geen[1], P. Barnwal[2], N.B. Jamil[3] & K.M. Ahmed[4]
[1]*Lamont-Doherty Earth Observatory, Columbia University, Palisades, New York, USA*
[2]*Department of Economics, Michigan State University, East Lansing, Michigan, USA*
[3]*Department of Earth and Environmental Studies, Montclair State University, Montclair, New Jersey, USA*
[4]*Deparment of Geology, University of Dhaka, Dhaka, Bangladesh*

ABSTRACT: Chronic exposure to arsenic from drinking well-water in Bangladesh has taken an immense toll. After an initial campaign to blanket test all wells in the country, mitigation efforts have faltered over the past decade. The government's new resolve to address the As problem is a unique opportunity for impactful mitigation by combining smartphone technology with the approaches that have been most effective in the past: blanket testing and using this information for the targeting of low As aquifers for private and truly public well installations.

1 BACKGROUND

In the past 40 years, millions of wells were installed privately by households in South Asia to avoid water-borne diseases, such as cholera, and to increase access to safe drinking water. This resulted in what has been described as the largest mass poisoning of a population in history due to high arsenic levels in a significant fraction of those wells. Bangladesh has been affected the most, with 40 million people exposed to arsenic levels 10 to 100 times higher than the World Health Organization guideline. Arsenic is an invisible killer that causes spontaneous abortions, still births, and infant deaths; reduces intellectual and motor function in children; and accounts for 1 of 18 adult deaths, primarily from cardiovascular disease (Flanagan *et al.*, 2012; Quansah *et al.*, 2015; Wasserman *et al.*, 2004). In addition, economists have demonstrated health effects of arsenic on wage earners that reduce the income of exposed households by an average of US$100 per year (Pitt *et al.*, 2018).

International organizations and the government recognized the magnitude of the well-water arsenic problem in the late 1990s. This led to a vast testing campaign of 5 million wells conducted in 2000–2005 and the most significant decline in exposure to arsenic to date. This is because most rural households live within walking distance of an existing safe well or within drilling distance to a safe aquifer that could be reached by local well drillers. Many hopes were initially placed in engineering solutions such as filters to remove arsenic that, despite extensive media coverage, turned out to be unsustainable. The government and the World Bank also supported rural piped-water supply systems but have since curtailed such efforts because of insufficient progress. To date, a much larger number of exposed people have been provided safe water through deep hand-pumped tube wells (Ravenscroft *et al.*, 2015). The past decade has been one of stagnation and even resignation because of failure of these technological solutions and a reduction funds allocated to arsenic mitigation. In 2018, however, the government of Bangladesh demonstrated its renewed interest in the arsenic issue with about US$240M in new funding because the well-documented health and economic impacts of arsenic can no longer be ignored. This presentation outlines a plan to effectively allocate this level of funding based on 20 years of research conducted in Bangladesh.

2 THE WAY FORWARD

The patterns of arsenic in groundwater are largely stable over time but so complex spatially that every well needs to be tested. Households reinstall their well on average once per decade and no testing of significance has taken place since 2005. The main reason for continued exposure to arsenic is therefore simply that most rural Bangladeshis don't know the status of their new well. This also means that a one-time blanket testing of the current estimate of over 10 million domestic hand tubewells in the country is not sufficient; about 1 million new wells need to be tested each year. One major change since the first blanket testing campaign is that smartphones are now ubiquitous in Bangladesh and can be used to both collect well-test data and process this information to guide both households installing wells that target low arsenic aquifers and the government installing

DOI: 10.1201/9781003317395-8

public sources of low arsenic water, which are also needed.

There are a number of large and experienced non-governmental organizations that could be entrusted with large-scale well testing using field kits and a smartphone app. The rationale for involving several NGOs, who are bound to manage implementation at scale somewhat differently, would be to monitor their effectiveness over time and install some level of competition for periodic contracts issued by the government. Processed data delivered by a smartphone app along with GPS coordinates can also rank those villages that are most in need of a public source of low-arsenic water and, within a village, identify a set of locations that would bring within walking distance the largest number of households with an unsafe well. Government funding to NGOs conducting these installations should again be renewed based on objective criteria of the impact of past installations. The task of installing public sources could also be transferred directly to the government as long as the same criteria are consistently used to measure impact. The impact of past allocations of low-arsenic deep wells has been limited to considerable extent by elite capture, whereby locally prominent households essentially privatize and restrict access to a source of safe water that should have been public, and this should be corrected (van Geen *et al.*, 2015).

3 COST ESTIMATES

What would be the cost of such mitigation program? Testing a well with a kit costs about $3, including the kit materials, labor, supervision, and durable placard with the test result to post on a well. Installing a deep public well costs about $1,000. For a quarter of the US$240M allocated to arsenic mitigation by the Bangladesh government, all 10 million existing wells in the country could be tested once and the 10 million replacement wells installed in the following decade could be tested as well. The remaining $180M could be used to install about 180,000 deep community wells across the tens of thousands of affected villages in the country. Given the price of a typical private household well of $100 and the 10-year turnover of the stock of wells, even this large public allocation will rapidly be dwarfed by private expenditures on the order of US$100M per year by households installing wells to access drinking water that, by and large, is free of microbial pathogens (Jamil *et al.*, 2019). By the best estimates that are available, public and private investment in lowering arsenic exposure of that order of magnitude would eliminate over 100,000 annual deaths and raise annual rural income by US$800 million.

ACKNOWLEDGEMENTS

Columbia University and the University of Dhaka's research in Araihazar has been supported for 20 years by the US National Institute of Environmental Health Sciences Superfund Research Program and several grants from the US National Science Foundation.

REFERENCES

Flanagan S.V., Johnston R.B. & Zheng Y. 2012. Arsenic in Tube Well Water in Bangladesh: Health and economic impacts and implications for arsenic mitigation. *Bull. World Health Organ.* 90: 839–846.

Jamil N.B., Feng H., Ahmed K.M., Choudhury I., Barnwal P. & van Geen A. 2019. Effectiveness of different approaches to arsenic mitigation over 18 years in Araihazar, Bangladesh: implications for national policy. *Environ. Sci. Technol.* 53: 5596–5604.

Quansah R., Armah F.A, Essumang D.K., Luginaah I., Clarke E., Marfoh K., Cobbina S.J., Nketiah-Amponsah E., Namujju P.B., Obiri S. & Dzodzomenyo M. 2015. Association of arsenic with adverse pregnancy outcomes/infant mortality: a systematic review and meta-analysis. *Environ. Health Perspect.* 123: 412–421.

Ravenscroft P., Kabir A., Hakim S.A.I., Ibrahim A.K. M., Ghosh S.K., Rahman M.S., Akhter F. & Sattar M.A. 2014. Effectiveness of public rural waterpoints in Bangladesh with special reference to arsenic mitigation. *J. Water Sanit. Hyg. Dev.* 4: 545–562.

Rosenzweig M., Pitt M. & Hassan N. 2018. Identifying the costs of a public health success: arsenic well water contamination and productivity in Bangladesh. *Review of Economic Studies*, forthcoming. (http://ibread.org/bread/system/files/bread_wpapers/546.pdf).

van Geen A., Ahmed K.M., Ahmed E.B., Choudhury I., Mozumder M.R., Bostick B.C. & Mailloux B.J. 2016. Inequitable allocation of deep community wells for reducing arsenic exposure in Bangladesh. *J. Water Sanit. Hyg. Dev.* 6: 142–150.

Wasserman G.A., Liu X., Parvez F., Ahsan H., Factor-Litvak P., van Geen A., Cheng Z., Slavkovich V., Hussain I., Momotaj H. & Graziano J.H. 2004. Water arsenic exposure and children's intellectual function in Araihazar, Bangladesh. *Environ. Health Perspect.* 112: 1329–1333, 2004.

Section 1: Arsenic in natural soil and water systems

1.1 Sources, transport and fate of arsenic in groundwater systems

Interplay of As (im)mobilisation processes in groundwater: Learning from hydrochemical investigations

E. Stopelli[1], V.T. Duyen[2], T.T. Mai[2], P.T.K. Trang[2], P.H. Viet[2], A. Lightfoot[1], R. Kipfer[1], M. Schneider[3], E. Eiche[3], A. Kontny[3], T. Neumann[4], M. Glodowska[5], M. Patzner[5], A. Kappler[5], S. Kleindienst[5], B. Rathi[5], O. Cirpka[5], B.C. Bostick[6], H. Prommer[7], L.H.E. Winkel[1,8] & M. Berg[1]
[1]*Eawag, Swiss Federal Institute of Aquatic Science and Technology, Dübendorf, Switzerland*
[2]*CETASD, Vietnam National University, Hanoi, Vietnam*
[3]*Institute of Applied Geosciences, Karlsruhe Institute of Technology, Karlsruhe, Germany*
[4]*Institute for Applied Geosciences, Technical University Berlin, Berlin, Germany*
[5]*Geomicrobiology, Microbial Ecology and Hydrology, University of Tübingen, Tübingen, Germany*
[6]*Lamont-Doherty Earth Observatory, Columbia University, New York, USA*
[7]*University of Western Australia and CSIRO Land and Water, Perth, Australia*
[8]*Institute of Biogeochemistry and Pollutant Dynamics, ETH Zürich, Zürich, Switzerland*

ABSTRACT: Millions of people worldwide are exposed to arsenic (As) contaminated groundwater. Despite decades of research and evidence of As mobilisation in anoxic aquifers being caused by reductive dissolution of iron minerals, the mechanisms behind the local scale variability of dissolved As remains unclear. Therefore, the trans-disciplinary AdvectAs project investigates the environmental behaviour and spatial heterogeneity of dissolved As in groundwater in the Red River delta, Vietnam. Here we present the results from hydrochemical and water isotope investigations. In particular, we will show how the large As variability (0.1–510 µg/L) is related to consecutive As (im)mobilisation steps, depending on site hydrology, geology and the interplay of Fe, Mn, S and organic matter cycles. Such complexity of (im)mobilisation processes can be simplified in 5 major hydro(geo) chemical zones, providing a conceptual tool with potential for application at other sites in Asia affected by geogenic As contamination of groundwater.

1 INTRODUCTION

Holocene sediments in river and river delta regions in Asia are a hotspot of geogenic contamination of groundwater resources. Large-scale groundwater pumping can exacerbate arsenic (As) contamination by causing horizontal and vertical advection from shallow aquifers to previously safe aquifers.

Despite decades of research, many processes concerning As release, transport and (im)mobilisation remain largely unknown, as well as the factors causing the heterogeneity of dissolved As over short distances and the involvement of further biogeochemical cycles in addition to iron and carbon.

2 MATERIALS AND METHODS

2.1 *Study area*

To investigate the biogeochemistry of As (im)mobilisation, we chose the location of Van Phuc (Red River delta, Vietnam), some 10 km southeast of the city of Hanoi (Figure 1, Eiche *et al.*, 2008; van Geen *et al.*, 2013). The site is characterized by a Holocene aquifer with Fe reducing conditions and generally high dissolved As concentrations, adjacent to a low-As Pleistocene aquifer. Groundwater abstraction from regional Pleistocene aquifers has tremendously increased as consequence of the expansion of the city of Hanoi. This has caused groundwater drawdown extending for several kilometres and stretching out to our study site. Therefore, the natural groundwater flow in Van Phuc has been reversed, now flowing from the Holocene high-As into the Pleistocene low-As aquifer. Furthermore, Van Phuc is characterised by large heterogeneity and patchiness of dissolved As concentrations in Holocene wells. These features make the site ideal for trans-disciplinary investigations of the factors responsible for As (im)mobilisation under advective conditions in Asia.

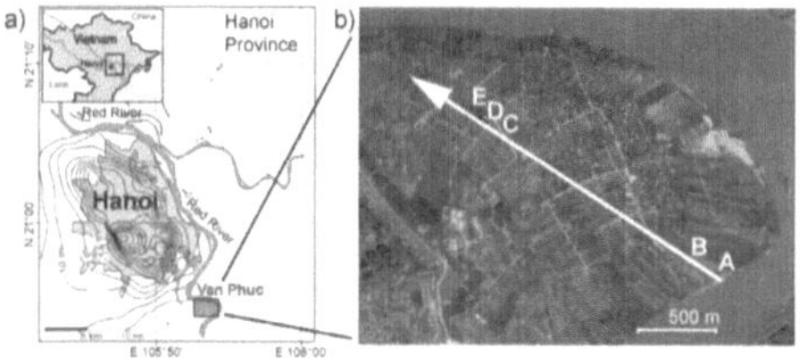

Figure 1. (a) Regional setting of the study site Van Phuc (Hanoi, Vietnam) and (b) current groundwater flow direction, from Stopelli *et al.* (2020).

2.2 *Water sampling and laboratory analysis*

Groundwater samples were collected between 2017 and 2019 and compared with previous datasets available

DOI: 10.1201/9781003317395-9

since 2010 for the same wells. Redox potential, pH and alkalinity were measured in the field; major and trace elements by ICP-MS; anions by IC; ortho-phosphate and ammonium by photometry; $\delta^{18}O$ and $\delta^{2}H$ by cavity ring-down spectrometer.

3 RESULTS AND DISCUSSION

3.1 *Hydro(geo)chemical zones involved in As (im)mobilisation*

Dissolved As concentrations ranged from 0.1 μg/L up to more than 500 μg/L. Several processes involved in As (im)mobilisation were identified, which can be associated with 5 hydro(geo)chemical zones (Figure 2): i) net As mobilisation occurs in freshly deposited riverbank sediments (A) with river water infiltration as main source of groundwater recharge (Wallis *et al.*, 2020); ii) arsenic is further transported (B) into the Holocene aquifer, where the co-occurrence of Fe(III) and SO_4^{2-} reduction causes no net change in dissolved As concentrations; iii) within the Holocene aquifer (C), an additional input of organic matter (OM) promotes further Fe(III) reduction and the onset of methanogenic conditions; iv) water isotopes indicate leaching of pore water from the aquitard into the aquifer, suggesting potential percolation of dissolved organic matter and As; v) at the redox transition zone (RTZ, D) of the Holocene and the Pleistocene aquifer sediments, As concentrations decrease as a consequence of sorption and incorporation on Fe(II)/Fe(III) minerals under shifting reducing conditions and onto Fe(III) (oxyhydr) oxides present in the Pleistocene sediments (E).

Manganese (Mn) and sulphur (S) cycles appeared to be crucial for the local variability of As concentrations, together with Fe and C redox biogeochemistry. The interlinked biogeochemical cycles at the RTZ seem to be stabilising the RTZ over time (at least between 2010–2019), thereby keeping As below 10 μg/L in the Pleistocene aquifer.

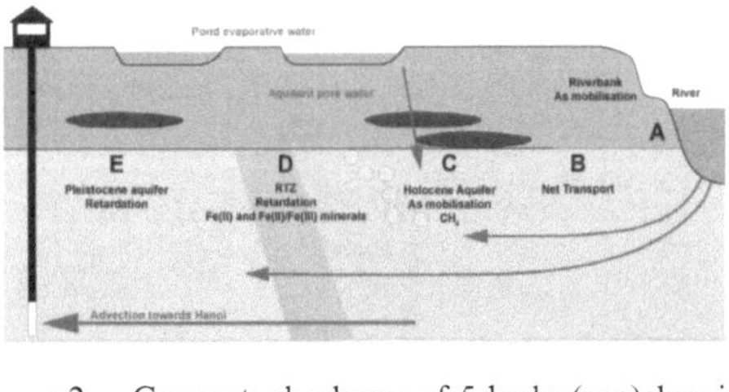

Figure 2. Conceptual scheme of 5 hydro(geo)chemical zones associated with biogeochemical As (im) mobilisation processes, adapted from Stopelli *et al.* (2020).

3.2 *Organic matter and As (im)mobilisation*

Given the association of dissolved organic matter with As contamination (Glodowska *et al.*, 2020), we are studying the impact of organic matter sources and cycling on As (im)mobilisation. For this purpose, $\delta^{13}C$ analyses of groundwater DIC, DOC and dissolved methane are currently being carried out and assessed together with isotopic signatures of sedimentary TOC. The latest results will be presented at the conference.

4 CONCLUSIONS AND OUTLOOKS

Large variability of As concentrations in groundwater depends on site hydrology, geology and on Fe, Mn, S, and OM biogeochemical cycles. This complex interplay of processes can be simplified in 5 major hydro (geo)chemical zones. The presented scheme provides a useful guideline to interpret the variability of dissolved As concentrations in other contaminated regions.

Further research is focusing on the link between the quality of organic matter and net As (im)mobilisation in the aquifer(s).

ACKNOWLEDGEMENTS

We thank C. Stengel, R. Britt, N. Pfenninger, T. Rüttimann for technical support; V. M. Lan, D. V. Nga and for support in field and H. Neidhardt for previous sampling campaigns; P. Kathriner, S. Robert and C. Schubert for support with ^{13}C-CH_4 analyses at Eawag; T. Maddox and SIEL Lab Uni Georgia for ^{13}C-DIC/DOC analyses; DFG and SNF for co-funding AdvectAs project with DACH grant #200021E-167821.

REFERENCES

Eiche E., Neumann T., Berg M., Weinman B., van Geen A., Norra S., Berner Z., Trang P.T.K., Viet P.H. & Stüben D. 2008: Geochemical processes underlying a sharp contrast in groundwater arsenic concentrations in a village on the Red River Delta, Vietnam. *Appl. Geochem.* 23: 3143–3154.

Glodowska M., Stopelli E., Schneider M., Lightfoot A., Rathi B., Straub D., Patzner M., Duyen V.T., AdvectAs Team Members, Berg M., Kleindienst S. & Kappler A. 2020. Role of in Situ Natural Organic Matter in Mobilizing As during microbial reduction of Fe^{III}-mineral-bearing aquifer sediments from Hanoi (Vietnam). *Environ. Sci. Technol.* 54(7): 4149–4159.

Stopelli E., Duyen V.T., Mai T.T., Trang P.T.K., Viet P.H., Lightfoot A., Kipfer R., Schneider M., Eiche E., Kontny A., Neumann T., Glodowska M., Patzner M., Kappler A., Kleindienst S., Rathi B., Cirpka O., Bostick B., Prommer. H., Winkel L.H.E. & Berg M. 2020: Spatial and temporal evolution of groundwater arsenic contamination in the Red River Delta, Vietnam: Interplay of mobilisation and retardation processes. *Sci. Total Environ.* 717:137143.

van Geen A., Bostick B.C., Trang P.T.K., Lan V.M., Mai N.N., Manh P.D., Viet P.H., Radloff K., Aziz Z., Mey J. L., Stahl M.O., Harvey C.F., Oates P., Weinman B., Stengel C., Frei F., Kipfer R. & Berg M. 2013: Retardation of arsenic transport through a pleistocene aquifer. *Nature* 501: 204–207.

Wallis I., Prommer H., Berg M., Siade A.J., Sun J. & Kipfer R. 2020. The river–groundwater interface as a hotspot for arsenic release. *Nat. Geosci.* 13: 288–295.

The role of (noble)gases in an As contaminated aquifer

A. Lightfoot[1], M.S. Brennwald[1], E. Stopelli[1], Advect As project members[2] & R. Kipfer[3]
[1]*Eawag, Swiss Federal Institute of Aquatic Science and Technology, Dübendorf, Switzerland*
[2]*Advect As project members * – see Acknowledgements*
[3]*Institute for Geochemistry and Petrology, ETH Zürich, Switzerland*

ABSTRACT: Arsenic (As) contamination of groundwater remains a problem for many of the river deltaic areas in South-East Asia; where concentrations regularly exceed the 10 µg/L currently recommended by the Word Health Organisation. The focus of this study is to determine inert and reactive gas concentrations in groundwaters at a location where As mobilization is active, to constrain the sites hydrogeology in a highly reducing environment. The small village of Van Phuc, Vietnam, presents an ideal opportunity for such research as is it well studied and accessible, however the As dynamics here are still not well understood. Gas concentrations in 21 wells at varying depths and locations were analysed in Van Phuc, Vietnam, with the miniRUEDI – a portable mass spectrometer (miniRUEDI, Gasometix GmbH, Switzerland) capable of measuring noble gases: He, Ar, Kr, and reactive gases: CO_2, CH_4, N_2 and O_2 in water. First results show As concentrations are highest where CH_4 production is observed. Furthermore, evidence of a gaseous CH_4 phase being produced within the aquifer is shown through excessive depletion of N_2, Ar and Kr elements. Such in-situ formation of gas bubbles can alter the permeability of the aquifer and modify groundwater flow. Simultaneous measurement of both inert and reactive gas species can therefore give an insight to groundwater dynamics of As contaminated aquifers.

1 INTRODUCTION

The Noble gases He, Ne, Ar, Kr, and Xe are commonly used as tracers in groundwaters. They are particularly useful for understanding the hydraulics of (groundwater) aquifer systems because they are biogeochemically inert. Combining noble gas measurements, with other dominant gas species in groundwaters, e.g. methane, can provide useful details for the hydrogeology of specific environmental systems where active and complex biogeochemical processes are also taking place. Whilst methane is frequently monitored for its (greenhouse) environmental impacts, it additionally plays an important role in aquifer systems with high Arsenic concentrations; where the two species are often positively correlated (Berg *et al.*, 2007; Nickson *et al.*, 2000).

To help predict As movement in contaminated aquifers, knowledge on groundwater residence times (i.e. groundwater dating) is usually of key importance to determine groundwater flow velocity. Yet, groundwater dating under such high CH_4 producing conditions is notoriously difficult, as many dating techniques which depend on trace gases (e.g. ^3H-^3He, SF6 etc.) are affected by the high CH_4 production via degassing (removing physically measurable tracer data from the system). Thus, new methods imparting information about the hydrogeology of such aquifers, can be extremely useful. Thus, by combining both noble and reactive gas measurements, we here try to determine information on groundwater flow, at a place where dating remains challenging.

2 MATERIALS AND METHODS

2.1 *Study area and geological setting*

Van Phuc is a small village situated around 10 km southeast of Hanoi, Vietnam. The village is located inside a meander of the Red River deltaic region (Figure 1). Such deltaic areas in Asia constitute of more recent sediment deposition commonly associated with high As concentrations.

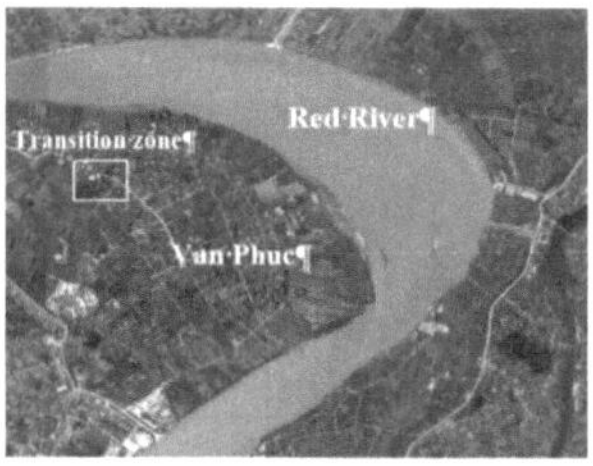

Figure 1. Study site: Van Phuc village, Vietnam, well locations sampled. Red, Orange and blue markers indicate high, Medium, and low As concentrations.

The south-east of the village comprises of young, highly reducing sediments, which lie both adjacent to, and above of, older less reducing sediments. At the horizontal interface between these two depositional ages, lies a redox front, also referred to here as the 'transition zone' (see Figure 2). As, is highly abundant in the younger sediments, whilst in the neighboring older sediments, remains practically non-existent. Furthermore, both adjoining aquifers are overlain by a clay aquitard, which varies in thickness up to around 23 m. Studies have shown that the consistency of this overlying clay layer is questionable (Stopelli *et al.*, 2020) and could potentially serve as a source of organic carbon, which in this case could play a role for As release into the aquifer. The general consensus for As release is microbial reduction of Iron – oxyhydroxides (FeOOH) though a sufficient carbon source is still not recognized to explain the abundance and patchy distribution of As in this region.

DOI: 10.1201/9781003317395-10

2.2 *Analysis of dissolved gas concentrations*

The concentrations of the aforementioned inert and reactive gas species dissolved in the groundwater were measured on-site using the miniRUEDI mass spectrometer (Brennwald *et al.*, 2016, Gasometrix GmbH, Switzerland). The groundwater was continuously pumped through a membrane contractor. The membrane allows for separation of the gas components in the (ground)water into a surrounding pre-evacuated head-space, where the total dissolved gas pressure (TDGP) is measured prior to separation and measurement of the individual gas species.

3 RESULTS AND DISCUSSION

First results indicate high As (Stopelli *et al.*, 2020) correlate to high total gas pressures (i.e. gas production) within the aquifer groundwater, in which CH_4 is the dominating species. This correlation supports the implication that microbial reduction of FeOOH (Glodowska *et al.*, 2020) is a major contributor to As release in the transition zone of this aquifer.Otherwise, we see a progressive depletion of the dissolved atmospheric gases (Ar, Kr and N_2) with increasing CH_4 concentrations (Figure 2). This would suggest that CH_4 tends to intermittently oversaturate and foster in-situ degassing of N_2, Ar and Kr; i.e. these gasses are depleted as a result of their partitioning into the free CH_4 gas bubbles. Interestingly, Helium, which would be expected to follow the same pattern, shows the opposite behavior such that it increases in concentration as CH_4 approaches in-situ saturation within the groundwater. Helium is the only noble gas whose concentrations are additionally affected by radioactive decay of U/Th in the aquifer sediments. Such an observation suggests Helium is accumulating via U/Th decay, in the groundwater where we also have CH_4 production.

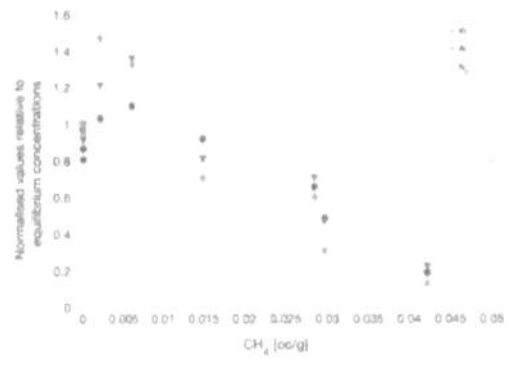

Figure 2. Concentrations of Ar, Kr and N_2 (normalized to equilibrium values) decrease with increasing CH_4.

The conceptual picture this combination of results indicate, is that the production of CH_4 bubbles reduces the hydraulic conductivity in the aquifer; allowing enough time for He to accumulate, whilst simultaneously depleting N_2, Ar and Kr in the groundwater as a result of their partitioning into the free CH_4 gas bubbles (Figure 3).

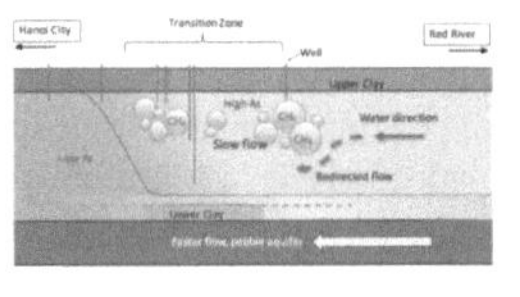

Figure 3. Conceptual picture of groundwater flow around CH_4 bubbles.

4 CONCLUSIONS

We simultaneously measured noble and reactive gases, to produce a conceptual picture (Figure 3) of groundwater flow; linking As, CH_4, inert and reactive gas species. In this case, CH_4 intermittently oversaturates and fosters in-situ degassing of N_2, Ar and Kr. Since tracer gases used for dating are also affected by the degassing, this makes the groundwater especially difficult to date. Thus, such measurements can help build an alternate picture of groundwater flow, in cases where groundwater dating is itself problematic.

ACKNOWLEDGEMENTS

We thank R. Britt, M.Mai, D. V. Nga and V. M. Lan for in field support. We acknowledge DFG and SNF for co-funding AdvectAs project with DACH grant #200021E-167821.

*Supporting AdvectAs project members: V.T. Duyen[I], M. Glodowska[II], M. Schneider[III], M. Patzner[II], S. Rudischer[IV], B. Rathi[V], E. Eiche[III], P. T. K. Trang[I], H. Neidhardt[IV], P. H. Viet[I], A. Kontny[III], T. Neumann[VI], A. Kappler[II], O. Cirpka[V], H. Prommer[VII], L. H. E. Winkel[VIII,1], M. Berg[I].

[I]CETASD, Vietnam National University, Hanoi, Vietnam; [II]Geomicrobiology, Microbial Ecology and Hydrology, University of Tübingen, Tübingen, Germany; [III]Institute of Applied Geosciences, KIT, Karlsruhe, German; [IV]Geoecology group, University of Tübingen, Germany; [V]Hydrogeology, Center for Applied Geosciences, University of Tübingen, Germany; [VI]Institute for Applied Geosciences, Technical University Berlin, Berlin, Germany; [VII]University of Western Australia and CSIRO Land and Water, Perth, Australia; [VIII]Institute of Biogeochemistry and Pollutant Dynamics, ETH Zürich, Zürich, Switzerland.

REFERENCES

Berg M. *et al.* 2007. Magnitude of arsenic pollution in the Mekong and Red River Deltas – Cambodia and Vietnam. *Sci. Total Environ.* 372(2–3): 413–425.

Brennwald M.S. *et al.* 2016. A portable and autonomous mass spectromic system for on-site environmental gas analysis. *Environ. Sci. Technol.* 50: 13455–13463.

Glodowska M. *et al.* 2020. CH4 oxidation coupled to Fe(III) reduction, unpublished data.

Nickson R.T. *et al.* 2000. Mechanism of arsenic release to groundwater, Bangladesh and West Bengal, *Appl. Geochem.* 15(4): 403–413.

Stopelli E. *et al.* 2020. Spatial and temporal evolution of groundwater arsenic contamination in the Red River Delta, Vietnam: interplay of mobilisation and retardation processes, *Environ. Sci. Technol.* 717.

van Geen A. *et al.* 2013. Retardation of arsenic transport through a pleistocene aquifer. *Nature* 501: 204–207.

Arsenic in the Environment: Bridging Science to Practice for Sustainable Development –
van der Wal, Ahmad, Petrusevski, Weijma, Savic, van der Wens, Beerendonk, Bhattacharya, Bundschuh & Naidu (Eds)
© 2023 the Author(s), ISBN 978-1-032-32928-4

Towards developing a logical framework on predominant occurrences of groundwater arsenic in large, tectonic-sourced basins across the globe

A. Mukherjee[1], P. Bhattacharya[2] & S. Gupta[3]
[1]*Department of Geology and Geophysics, Indian Institute of Technology Kharagpur, India*
[2]*School of Environmental Science and Engineering, Indian Institute of Technology Kharagpur, India*
[3]*KTH-International Groundwater Arsenic Research Group, Department of Sustainable Development, Environmental Science and Engineering, KTH Royal Institute of Technology, Stockholm, Sweden*
KWR Watercycle Research Institute, Nieuwegein, The Netherlands

ABSTRACT: More than 100 million people around the world today are endangered by toxic concentrations of geogenic arsenic (As) in groundwater, residing in high yielding aquifers hosted in large sedimentary basins. Nevertheless, not all sedimentary aquifers are enriched in groundwater As. The affected basins are preferentially located in orogenic forelands of present or ancient convergent tectonic margins. This correlation suggests that the worldwide distribution of As-enriched groundwater aquifers can be attributed to global-scale geodynamic processes. Elaborating on this hypothesis, we have advocated a model which involves the mobilization of As from its primary source in deep continental crustal materials, its extrusion at the surface in arc magmas, and the subsequent transport of As to foreland sedimentary basins that eventually act as the groundwater As-enriched aquifers.

1 INTRODUCTION

Arsenic (As) enrichment in groundwater, which potentially affects the health of more than 100 million inhabitants worldwide, has been one of the most well researched environmental concerns for about the last three decades. Consequently, a large scientific knowledge-base has already developed on the 3-dimensional distribution of As on local to global-scale, based mostly on geochemical surveys and/or statistical analyses. However, knowledge of the original source of the As has remained elusive. We try to establish "why?" and "how?", the primary source of groundwater As, globally, is potentially linked with arc-derived volcanic materials created by processes taking place at subduction zones, and more specifically, active continental margins (Mukherjee *et al.*, 2014, 2019).

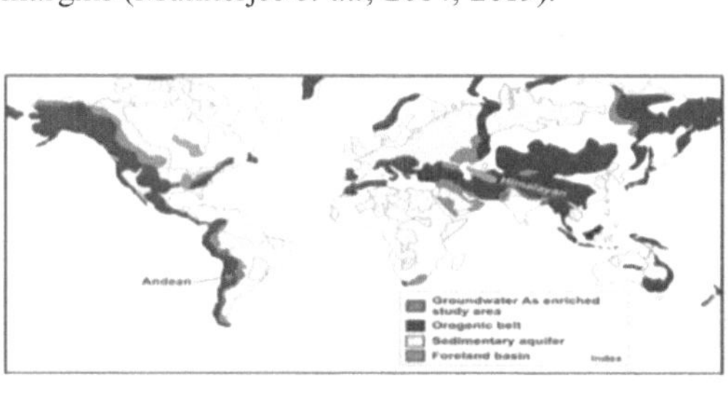

Figure 1. Major sedimentary aquifers across the world and their tectonic settings (Redrawn from Mukherjee *et al.* 2019) indicating the location of the arsenic contaminated aquifers.

2 RESULTS AND DISCUSSION

A close look at the global distribution of known groundwater As enriched areas shows a systematic pattern, where most of the major As polluted aquifers are hosted within extensive clastic sedimentary basins in or around tectonically controlled, active or ancient foreland basins (Figure 1). These foreland basins are located adjacent to major magmatic arc volcanics or orogenic belts at convergent plate boundaries, and are developed because of lithospheric flexure during orogenesis (Coomar *et al.*, 2019; Raychowdhury *et al.*, 2014). Prominent examples are the As-polluted clastic sedimentary foreland basin aquifers are the Indus-Ganges-Brahmaputra (IGB) basin (Bhattacharya *et al.*, 1997), including the Bengal basin in the Himalayan foreland, the western North American (American and Pacific Cordillera) foreland of western USA and Canada, the Appalachian foreland in eastern USA, and parts of Argentina, Bolivia, Chile and Peru in the forelands of the Andean orogenic belt of South America. Since the allocthonous clastics in these sedimentary aquifers are transported from upstream provenances, the enriched primary sources of the As must be ex-situ from the presently polluted aquifers. The sediments in these As-polluted aquifers therefore act as secondary and tertiary sources of the As in the adjacent groundwater. Thus, high groundwater As concentrations in aquifers may be ultimately related to enhanced primary solid-phase concentrations in the provenance, i.e. major magmatic arcs and orogenic belts at convergent plate boundaries.

DOI: 10.1201/9781003317395-11

Our proposed model is based on the following premises: (I) that As is primarily concentrated within continental crust, and therefore, in any tectonic regime, the continental crust must represent the ultimate natural reservoir from where As may be derived; (II) mobilization of As from continental crust must be effected by hydrothermal action; and (III) Arsenic mobilization would be enhanced in any situation that facilitates prolonged interaction between continental crust and hot hydrous fluids or magmas.

Arsenic mobilization from the subducting plate to the fluid will therefore be most effective in continental domains with elevated geothermal gradients, such as in the proximity of active volcanic centres. We argue that hot hydrous fluids that effect As mobilization are derived from the subducting slab. These fluids, that may also be associated with slab-derived melts, subsequently metasomatize and initiate the production of arc magmas. These hydrous arc magmas leads to the generation of the high alumina basalts/gabbros, andesites/diorites and dacites/granodiorites located in active or ancient magmatic arcs and are known to have high to very high As (up to 1500 mg/kg) in solid phase (Raychowdhury *et al.*, 2014). The ascending magma in the continental active margins would subsequently interact with the thick continental crust as it rises.

The arc magmas at the continental active margins erupt to form volcanic arcs. Volcanism in the arc leads to the deposition of As-laden volcanic ash and detritus layers that may become part of adjoining, downstream basins. Rhyolitic glass, volcanic ash, along with associated silicate rocks in the adjacent sediments and metamorphic deposits undergo hydrolytic dissolution and produce an influx of major (e.g. Na, K, Si, HCO_3^- etc.) and minor solutes, mostly oxy-anions (e.g. As, V, Mo etc.) to the groundwater and leads to precipitation of clays (Bhattacharya *et al.*, 2006). Some of the circulating As-laden hydrothermal fluids may also seep out of surficial hot springs (e.g. areas in and around Yellowstone). Subsequently, with the initiation of orogeny at these continental active margins, the As-laden magmatic arcs coalesce with the sediments and rocks of the rising orogen, and an accommodation space is created in the adjoining foreland basins. Sediments are derived from the rapidly rising orogenic belt by extensive weathering of the exhumed As-laden magmatic arc rocks, which are deposited in the adjacent foreland basins.

The earlier forelands basins, coalesced with the main orogenic belt, and can thus also act as interim repositories of As-bearing sediments eroded from the main orogenic belts, and in turn can act as secondary and even tertiary sources for newer generations of foreland basins. As erosional agents, such as rivers or glaciers, cut through the orogenic belt, they erode through the foreland sediments and any further exhumed arc magmatics, thus depositing the weathered As-rich sediments in the most recent foreland basins e.g. Himalayan IGB basin in the Indian subcontinent or Andean Chaco Pampean plains of South America (Mukherjee *et al.*, 2021).

3 CONCLUSION

We conclude that major groundwater occurrences with known As pollution are located in aquifers formed within the foreland basins, in the vicinity of active or ancient orogenic belts, specifically magmatic arcs that formed at continental convergence zones. The primary provenance of groundwater As in these in continental margins that have elevated solid-phase As either by effective scavenging of deep continental crust by rising arc magmas, or the involvement of subducted terrigenous sediments in magma generation. Subsequent As enrichment of aquifer matrix and groundwater is related to surficial extrusion and subsequent dispersal of particulate or dissolved As by hot, metasomatic fluids by extensive water-rock interactions.

REFERENCES

Bhattacharya P., Chatterjee D. & Jacks G. 1997. Occurrence of arsenic contaminated groundwater in Alluvial Aquifers from Delta Plains, Eastern India: Options for safe drinking water supply. *Int. J. Wat. Res.Dev.* 13:79–92.

Bhattacharya P., Claesson M., Bundschuh J., Sracek O. *et al.* 2006. Distribution and mobility of arsenic in the Río Dulce alluvial aquifers in Santiago del Estero province, Argentina. *Sci. Total Environ.* 358(1–3): 97–120

Coomar P., Mukherjee A., Bhattacharya P., Bundschuh J., Verma S., Fryar A.E. *et al.* 2019. Contrasting controls on hydrogeochemistry of arsenic-enriched groundwater in the homologous tectonic settings of Andean and Himalayan basin aquifers, Latin America and South Asia. *Sci. Total Environ.* 689: 1370–1387.

Mukherjee A., Gupta S., Coomar P., Fryar A.E., Guillot S., Verma S. *et al.* 2019. Plate tectonics influence on geogenic arsenic cycle: from primary source to global groundwater enrichment. *Sci. Total Environ.* 683: 793–807.

Mukherjee A., Sarkar S., Chakraborty M., Duttagupta S., Bhattacharya A., Saha D., Bhattacharya P. *et al.* 2021. Occurrence, predictors and hazards of elevated groundwater arsenic across India through field observations and regional-scale AI-based modeling. *Sci. Total Environ.* 759: 143511.

Mukherjee A., Verma S., Gupta S., Henke K.R. & Bhattacharya P. 2014. Influence of tectonics, sedimentation and aqueous flow cycles on the origin of global groundwater arsenic: paradigms from three continents. *J. Hydrol.* 518: 284–299.

Raychowdhury N., Mukherjee A., Bhattacharya P., Johannesson K. *et al.* 2014. Provenance and fate of arsenic and other solutes in the Chaco-Pampean Plain of the Andean foreland, Argentina: From perspectives of hydrogeochemical modeling and regional tectonic setup. *J. Hydrol.* 518: 300–316.

Arsenic in the Environment: Bridging Science to Practice for Sustainable Development – van der Wal, Ahmad, Petrusevski, Weijma, Savic, van der Wens, Beerendonk, Bhattacharya, Bundschuh & Naidu (Eds)
© 2023 the Author(s), ISBN 978-1-032-32928-4

Groundwater arsenic in the Manipur Valley of Northeastern India: controls of geomorphology and geology

K.S. Kshetrimayum[1] & L. Thokchom[2]
[1]*Department of Geology, Central University of Punjab, Ghudda, Bathinda, India*
[2]*Department of Earth Science, Assam University, Silchar, India*

ABSTRACT: Recent studies on arsenic (As) contamination in surface water and groundwater in the Manipur valley have been conducted to examine spatial distribution and role of geomorphology and geology. It was observed that As concentration in lakes and rivers are within the WHO and BIS limit, but springs and shallow groundwater exceed the recommended limits (76 μg/L). Newer alluvial plain is more prone to high As concentrations as compared to the springs located around piedmont zone and older alluvial plain owing to the presence of Holocene sediment deposits. The main source of As could be derived from geogenic and mobilized due to weathering and erosion of rocks and minerals like arsenopyrite and ferrous arsenic sulphite from Disang shale, Barail sandstone and Manipur Ophiolite rocks. Further, detail studies in terms of identification of the source of elevated As in groundwater needs to carried out in this region.

1 INTRODUCTION

Recent revelation on arsenic (As) contamination in groundwater from large parts of north-eastern region of India has put millions of people in alert and apprehension. Chronic As exposure is detrimental to human health being associated with cancer of the skin, lung, liver, urinary bladder, and kidney (Bunnell *et al.*, 2007; Bundschuh *et al.*, 2017; Kapaj *et al.*, 2006) and other diseases, including cardiovascular and peripheral vascular diseases, diabetes, peripheral neuropathies, portal fibrosis, and adverse birth outcomes (Xia *et al.*, 2009).

One of these regions is the Manipur Valley located at in the extreme eastern part of northeastern India and occupies ~1900 km^2 with about 2 million population. Study carried out on 628 tubewells in this valley revealed that 63.3% contained >10 μg/L of As, 23.2% contained between 10 and 50 μg/L, and 40% belongs to >50 μg/L (Chakraborti *et al.*, 2012). Another study found that As concentration exceed the Indian standard in 23.3% and the WHO limit in 73.3% of the groundwater sources with only 26.7% in the acceptable range (Singh *et al.*, 2013). Further study suggested that concentration of As is highest during post-monsoon (1–200 μg/L) as compared to pre-monsoon (1–108 μg/L) and monsoon (2–99 μg/L) where it has strong positive correlation with PO_4 and negative correlation with SO_4^{2-} suggesting a partial influence of anthropogenic sources (Oinam *et al.*, 2011). Arsenic concentration in groundwater used for irrigating the rice crops varied from 10 to 475 μg/L and for rice roots it varied from 32 to 52 mg/L (Chandrashekhar *et al.*, 2016). Bhattacharya *et al.* (2014) have studied insights into the occurrence of arsenic in groundwater of sedimentary aquifers. The co-occurrence of elevated fluoride and arsenic in groundwater and its health-related in Dharmanagar of northern Tripura, India has been assessed recently (Bhattacharya *et al.*, 2020). Keeping aforesaid studies in the background, this study explores the distribution and spatial variability of As in the Manipur valley (Figure 1).

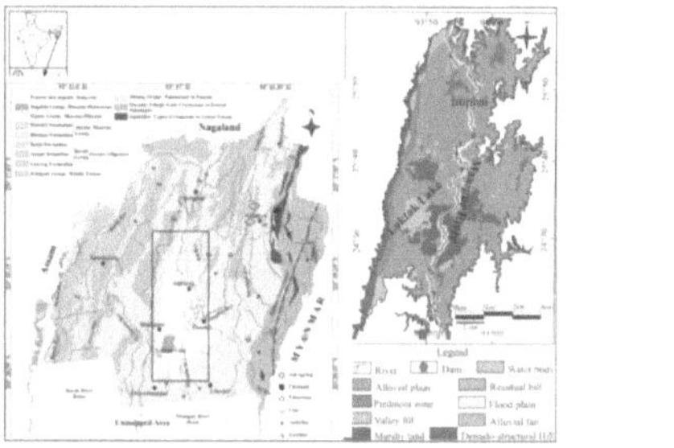

Figure 1. Map of the Imphal valley and regional geological set-up (modified after MARSEC, GSI).

2 METHODS

2.1 *Study area*

The Manipur valley is occupied by Quaternary alluvial deposits consisting of both lacustrine and fluvial types. The topography which varies from ~850 to 600 m with an average of ~700 m of this basin has been divided into various geomorphic landscapes such as the Loktak surface (late Holocene), Lilongsurface (late Holocene), Lamsang surface (early Holocene), Sekmai surface (late Pleistocene), Kanglatombi surface (middle Pleistocene) and Motbung surface (early Pleistocene) (GSI 2011).

2.2 *Water sampling and laboratory analysis*

A campaign of sampling of 46 samples were collected, out of which 7 were from rivers, 3 from lake/pond, 3

DOI: 10.1201/9781003317395-12

Table 1. Statistical summary of As concentration in surface water and shallow groundwater in the Manipur Valley of NE India.

Types of water	Average As (μg/L)	Minimum As (μg/L)	Maxmum As (μg/L)	Standard Deviation	Median
River	0.8	0.59	1.43	0.29	0.68
Lake	1.8	0.84	2.84	1.01	1.71
Spring	19.19	10.36	36.24	14.76	10.97
Shallow groundwater	25.04	0.1	76.56	27.14	11.87

from spring and 33 from shallow aquifers (Thokchom & Kshetrimayum 2020). The physio-chemical parameters such as, pH, temperature (T), Total dissolve solids (TDS), electrical conductivity (EC), redox potential (Eh) and elevation (H) were measured in the field using potable kits. Arsenic concentration was analyzed at Pondicherry University and SAIF, Indian Institute of Technology, Powai, India using ICP-AES.

3 RESULTS AND DISCUSSION

3.1 *Spatial distribution of As*

The statistical summary of As concentration in various types of surface water and shallow groundwater is given in Table 1. As evident form the Table, mean As concentrations in spring and groundwater (19.19 and 25.04 μg/L) exceed the WHO limit. The highest concentration in spring is found in a saline spring at Chandrakhong Thumkhong (36.24 μg/L) in piedmont zone while in groundwater the highest is found in alluvial plain at Kakching Wairi Sabal Pareng (76.56 μg/L). Among the surface water types, spring has highest concentration followed by lake (1.8 μg/L) and river (0.8 μg/L) while newer alluvial plain has generally higher elevation followed by older alluvial plain and piedmont zone. The spatial distribution pattern suggests that shallow aquifers located in and around newer alluvium are more prone to As contamination as the region has Holocene sediment deposits derived from surrounding Disang shales, Barail sandstone and rocks of the Manipur Ophiolite Belt.

3.2 *Probable occurrence and source of As concentration*

The groundwater in the Manipur Valley generally occurs in a reducing state as presence of relatively high concentration of sedimentary organic matter. Generally, arsenic is being released when arsenic – iron bearing minerals like arsenopyrite, ferrous arsenic sulphite, arsenic disulphide and arsenic trisulphide in the sediments are reduced by oxygen deficient groundwater. However, detail study regarding the source of As needs to be carried out in the study area.

4 CONCLUSIONS

Though surface water bodies like rivers and lakes have not yet been affected by As contamination, but shallow groundwater and some springs of Manipur valley has been effected. Geomorphologically, newer alluvial plain is more prone to this contamination as recently derived sediments belonging to Holocene age are being deposited. The source of As in the region could be geogenic through weathering and erosion of rocks having arsenopyrite minerals by the Imphal river and its tributaries. Therefore, it is recommended to thoroughly test the groundwater before it is used for different purposes like for domestic and drinking.

REFERENCES

Bhattacharya P., Adhikari S., Samal A.C., Das R., Dey D., Deb A., Ahmad S., Hussein J., De A., Das A., Joardar M., Panigrahi A.K., Roychowdhury T. & Santra S.C. 2020. Health risk assessment of co-occurrence of toxic fluoride and arsenic in groundwater of Dharmanagar region, North Tripura (India). *Groundw. Sustain. Dev.* 11: 100430.

Bhattacharya P., Welch A.H., Ahmad K.M., Jacks G. & Naidu R. 2014. Arsenic in groundwater of sedimentary aquifers. *Appl. Geochem.* 19(2): 163–260.

Bundschuh J., Maity J.P., Mushtaq S., Vithanage M., Seneweera S., Schneider J., Bhattacharya P., Khan N.I., Hamawand I., Guilherme L.R.G. Reardon-Smith K., Parvez F., Morales-Simfors N., Ghaze S., Pudmenzky C., Kouadio L. & Chen C.-Y. 2017. Medical geology in the framework of the sustainable development goals. *Sci. Total Environ.* 581–582: 87–104.

Bunnell J.E., Finkelman R. B., Centeno J.A. & Selinus O. 2007. Medical geology: a globally emerging discipline. *Geologica Acta.* 5(3): 273–281.

Chakraborti D., Singh J.E., Das S. & Singh N.R. 2012. Groundwater arsenic contamination in Nanipur, one of the seven north-eastern hill states of India: a future danger. *Environ. Geol.* 56(2): 381–390.

GSI 2011. Geology and Mineral Resources of Manipur, Mizoram, Nagaland and Tripura, *Geological Society of India, Miscelleaneous Publication No. 30 Part IV, Vol 1(Part-2).*

Kapaj S., Peterson H., Liber K. & Bhattacharya P. 2006. Human health effects from chronic arsenic poisoning – a review. *J. Environ. Sci. Health, Part A.*, 41(10): 2399–2428.

Oinam J.D., Ramanathan A.L., Linda A. & Singh G. 2011. A study of arsenic, iron and other dissolved ion variations in the groundwater of Bishnupur district, Manipur, India. *Environ. Earth Sci.* 62: 1183–1195.

Singh E.J., Gupta A. & Singh N.R. 2013. Groundwater quality in Imphal west district, Manipur, India, with multivariate statistical analysis of data. *Environ. Sci. Pollut. Res. Int.* 20(4): 2421–2134.

Thokchom L. & Kshetrimayum K.S. 2020. Delineation of sources of elevated trace elements in surface water and groundwater in quaternary aquifers of southeastern Imphal Valley, northeast India. *Sustain. Water Res. Manag.* 6: 33.

Xia Y., Wade T.J., Wu K., Li Y., Ning Z., Le X.C. *et al.* 2009. Well water arsenic exposure, arsenic induced skin-lesions and self-reported morbidity in Inner Mongolia. *Int. J. Environ. Res. Public Health* 6(3): 1010–1025.

Arsenic in the Environment: Bridging Science to Practice for Sustainable Development –
van der Wal, Ahmad, Petrusevski, Weijma, Savic, van der Wens, Beerendonk, Bhattacharya, Bundschuh & Naidu (Eds)
© 2023 the Author(s), ISBN 978-1-032-32928-4

Regional arsenic contamination transport model for safe drinking water aquifer delineation

S.S. Sathe[1,2] & C. Mahanta[2]
[1]*Department of Civil Engineering, Rajarambapu Institute of Technology Sakharale, Maharashtra, India*
[2]*Department of Civil Engineering, Indian Institute of Technology Guwahati, Guwahati, India*

ABSTRACT: Inadvertent daily use of arsenic (As) rich shallow aquifer groundwater has put many people on the edge of chronic As diseases. Management and delineation of As safe aquifers at regional scale is an urgent need in major flood plains. To manage and mitigate shallow aquifer groundwater As problem in alluvial aquifer for Brahmaputra flood plain, a conceptual model approach was used to build a numerical contaminant transport model. Hydrogeological and meteorological data, empirical formulae and soil characteristic values were used for development of a 3D transient state predictive MODFLOW and MT3DMS model. The model results showed that the computed water heads were perfectly matched with the observed hydrogeological data. This transient state MT3DMS results suggest that it can be more technically supportive and reliable method for management of sustainable groundwater source by considering different exploitation parameters to meet the increasing drinking water demands in the study area.

1 INTRODUCTION

Numerical contaminant transport groundwater modelling is a prerequisite to have dependable pre-information for avoiding major contamination catastrophe and management of safe groundwater sources. Numerous studies have used professional groundwater modelling software such as Visual MODFLOW, FEFLOW, SEAWAT, COMSOL etc. to solve various groundwater contamination problems in the last decade (Zhou & Li 2011). However, groundwater modelling software (GMS) with geographic information system (GIS) interface application has gathered more attention by hydrogeologist to solve various groundwater problems (Brindha & Elango 2016). In present study, GMS module, modular three-dimensional finite-difference ground water flow model (MODFLOW) and modular transport 3-dimensional multi species model (MT3DMS) were used to understand groundwater scenario and arsenic (As) contaminant gradient in a complex hydrological setting. Aim of this study was to delineate shallow As safe aquifer in the study area.

2 MATERIALS AND METHODS

2.1 *Study area*

The study area (of approximately ~218 km^2) is located on the southwestern region of Darrang district (latitude 26°33'N and 26°15'N; Longitude 91°45'E and 92°10'E), close to the northern banks of River Brahmaputra in Assam, India. This study region is one of the densely populated regions of the district (population ~ 1857 per sq. km) with the availability of limited pipe water supply schemes (PWSS), most people use readily available and economical sources of shallow depth groundwater for daily usage. Rivers namely Nanoi and Mangaldoi flow from Bhutan Himalayas through the study area and join the river Brahmaputra on the northern bank (Sathe *et al.*, 2019).

2.2 *Conceptual model development*

Stratigraphic model, fence diagram and finite difference grid for the As contaminant transport modelling were formed using lithology data obtained from drilling wells viz. M1, M2, M3, and M4 whereas lithology data for the wells M5 and M6 was obtained from Public Health Engineering Department, Darrang district, Assam, India. The collected sediment cores obtained from the successive 5ft depth were precisely classified into clay, silt, sand (fine, medium, and coarse) and gravel (mixed with sand and boulders). According to stratigraphic model, the finite difference grids were equally distributed into 150 cells in each X and Y direction whereas, in Z direction it divided into 17 numbers considering broad layers and covering complex heterogeneity of the study area domain.

2.3 *Boundary, initial conditions, and calibration*

The study area was delineated by georeferencing google earth 2017 image in such a way to cover all the sediment sampling and observation well sites in the model. Empirical equations were used to calculate potential recharge, river conductance and evapotranspiration using regional meteorological center data (from 1996 to 2016) for the study area. Empirical values for materials property (Table 1), especially, porosity, specific storage and yield were used (Todd & Mays 2005). The wells (n = 67)

DOI: 10.1201/9781003317395-13

pumping rate were assigned as 1.3 m^3/day considering average 10 people in one family. Hydraulic conductivity values for clay to medium sand samples were determined by falling head permeability test, whereas for remaining soil materials, empirical values mentioned by Todd & Mays (2005) were used. Model calibration was started from 1st January 1996 till 31st December 2016, whereas numbers of time steps 4 were assigned for each 23-stress period (till 2056 year). Predictive transient state groundwater simulation model was validated with several combinations of parameter values to attempt a successfully reproducible water Table results which were recorded at the observation wells by Central Ground Water Board of India, in the study area. Results of transient state MODFLOW simulation were used for prediction of As [i.e., As(III)] concentrations to delineate the safe aquifer in the study area.

Table 1. Hydrogeological properties of soil materials used for MODFLOW simulation.

Material \ Properties	Kx (m/d)	Ky (m/d)	S.S (1/m)	S.Y (%)	e (%)
Clay	6e-3	6e-4	0.005	3	42
Silt	0.08	0.008	0.001	8	46
Fine sand	0.16	0.32	0.001	23	43
Medium sand	0.33	0.65	0.001	28	39
Coarse sand	2.39	4.79	0.001	27	39
Sandy gravel	4.79	9.58	0.001	23	28
Gravel	150	300	0.001	23	28
River conductance [(m^2/d)/m]			8.55		
Tributary conductance [(m^2/d)/m]			1.83		

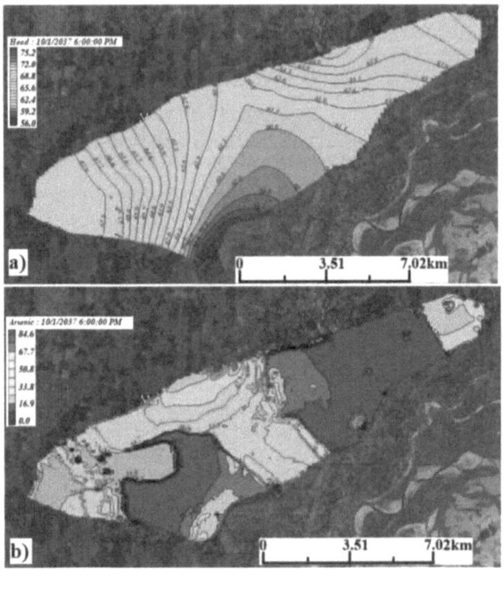

Figure 1. Groundwater contour map a) MODFLOW and b) MT3DMS simulation results.

3 RESULTS AND DISCUSSION

3.1 *Calibration, sensitivity, and validation*

The three-dimensional transient state groundwater flow and contaminant transport model results were well in agreement with observation wells data (Figure 1a,b). In the calibration model, groundwater Table fluctuation was observed as maximum of 5 m. Groundwater contour showed gradient in an average of 0.0006 m along the topography of the study area. It implied that the rates of aquifer flushing near the M1, M2 and M4 wells was slower than those for the aquifers at the M3, M5 and M6. The MT3DMS results showed that the aquifers closer to tributaries are acting as sink, while adjacent aquifers had low hydraulic conductivity and connectivity.

4 CONCLUSIONS

The results of the MODFLOW simulation indicated that hydraulic conductivity, river, and stream conductance had a major impact on the groundwater Table. Aquifers with high hydraulic conductivity and connectivity suggest that these aquifers could provide As safe groundwater and can be delineated as sustainable groundwater source. The results suggested that increasing anthropogenic activities such as groundwater extraction from deeper and shallow aquifers for irrigation and water consumption by the community wells may alter the groundwater table in future. It may influence on recharge rates and subsequent attenuation or magnification of As derived from pesticides used in agricultural fields. This further suggests that some shallow and deeper aquifers of the study area are vulnerable for contamination.

ACKNOWLEDGEMENTS

Authors thank sincerely the Department of Civil Engineering, IIT Guwahati for providing software facility. Authors would also like to thank Public Health Engineering Department, Divisional Officers, Mangaldai, Assam and AWRMI for their assistance and for sharing required data for this study.

REFERENCES

Brindha K. & Elango L. 2016. Application of groundwater modelling for sustainable management: few case studies from India. *Indian J. Mar. Sci.* 45(1): 111–116.

Sathe S.S. & Mahanta C. 2019. Groundwater flow and arsenic contamination transport modeling for a multi aquifer terrain: assessment and mitigation strategies. *J. Environ. Manage.* 231: 166–181.

Todd D.K. & Mays L.W. 2005. *Groundwater Hydrology*. Wiley.

Zhou Y. & Li W. 2011. A review of regional groundwater flow modeling. *Geosci. Front.* 2(2): 205–214.

Arsenic in the Environment: Bridging Science to Practice for Sustainable Development –
van der Wal, Ahmad, Petrusevski, Weijma, Savic, van der Wens, Beerendonk, Bhattacharya, Bundschuh & Naidu (Eds)
© 2023 the Author(s), ISBN 978-1-032-32928-4

Controls of lithology and groundwater pumping on arsenic contamination of deep groundwater in Bangladesh

M.R. Khan[1,2], A. van Geen[3], K.M. Ahmed[1] & H. Michael[2]
[1]*Department of Geology, University of Dhaka, Dhaka, Bangladesh*
[2]*Department of Geological Sciences, University of Delaware, Delaware, USA*
[3]*Lamont Doherty Earth Observatory, Columbia University, New York, USA*

ABSTRACT: Naturally elevated arsenic concentrations in shallow groundwater across the Bengal Basin threaten the health of millions of people. Deep (>150 m) groundwater low in arsenic is currently the most effective mitigation option in Bangladesh. However, there are both natural and anthropogenic processes in the Bengal basin limiting the use of deep groundwater as the primary arsenic mitigation option in Bangladesh. We have found that in areas where interbedded muds are scarce naturally deep and fast flow system develops resulting in the occurrence of young (<4.0 kyr) groundwater in aquifers as deep as 250 m. Additionally, groundwater model simulation suggests that large-scale pumping for the municipal supply of Dhaka are creating fast flow paths in a highly heterogeneous deltaic aquifer that could contaminate the aquifer within a century. Our findings put emphasis on the nationwide testing of deep groundwater for arsenic as well as setting up systematic monitoring networks around large pumping centers.

1 INTRODUCTION

Deep (>150 m) groundwater is currently one of the most popular arsenic (As) mitigation option in Bangladesh because of its reliability, cost effectiveness, and ease of use (Ravenscroft *et al.*, 2014). Therefore, the primary challenge in As mitigation is to understand the sustainability of this indispensable deep resource in the Bengal delta. Deep pumping could induce downward migration of As and young labile DOC from the overlying Holocene aquifers and contaminate the deep groundwater. However, the rate and spatial variability of groundwater flow in the vertical direction and the evolution of groundwater flowpaths depend on pumping intensity and on geological heterogeneity (Fogg 1986). In this abstract, groundwater flow model derived travel time for downward migration of As contaminated shallow groundwater to depths (>150 m) have been calculated and compared for two areas: i) highly heterogeneous aquifer affected by extensive groundwater abstraction, and ii) homogeneous aquifer with very little groundwater abstraction.

2 MATERIALS AND METHODS

2.1 *Study area*

Both of the study areas are located in the Bengal Basin. The area with mostly homogeneous lithology is located in the SW part of the Bengal basin across Bangladesh-India border (Figure 1). The other area with highly heterogeneous lithology and extensive pumping is located in greater Dhaka (Figure 1).

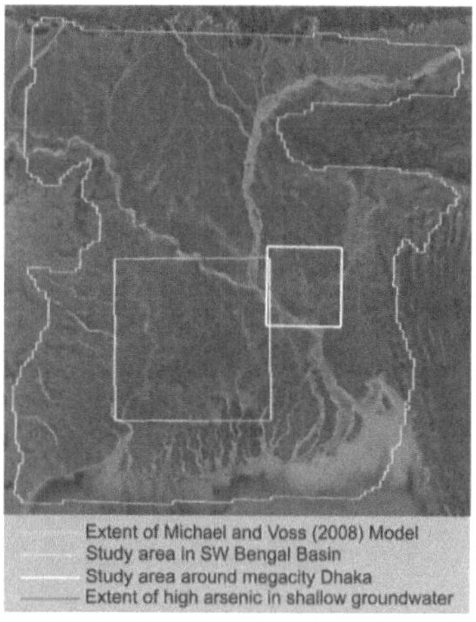

Figure 1. Map showing the study areas and the extent of the Bengal Basin Model of Michael & Voss (2008).

2.2 *Materials and methods*

The existing MODFLOW model of Michael and Voss (2008) was used for simulation of groundwater flow and USGS particle tracking code MODPATH was used for the simulation of advective travel time of groundwater. The model was locally refined in both the study areas to include the variation in hydraulic properties resulted from different lithology. Aquifer heterogeneity was simulated using the Sequential Indicator Kriging method based on hundreds of lithological logs (Figure 2). Equivalent horizontal and vertical hydraulic conductivities for both areas were estimated and used in the model.

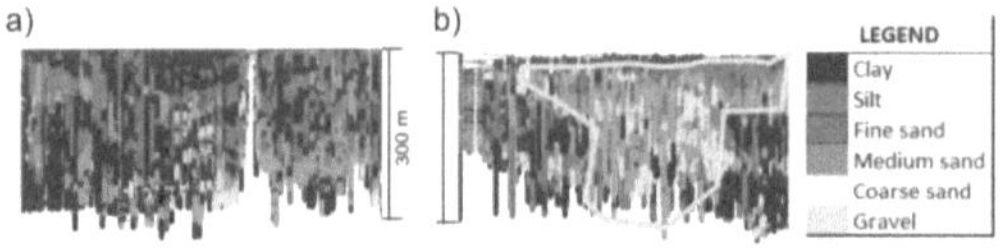

Figure 2. Comparison of lithologies encountered in boreholes. a) Typical of Dhaka region, b) SW Bengal Basin region, highlighted area shows a sandy almost homogeneous aquifer extending to as deep as 300 m.

3 RESULTS AND DISCUSSIONS

Our model simulation suggests that the median groundwater travel time in the SW Bengal basin area, dominated by sandy lithology, are on the order of a few thousands of years (Figure 3a). In contrast, the median groundwater travel time in the heterogeneous part of the delta is on the order of tens of thousands of years under unperturbed conditions (Figure 3b). However, large scale pumping can lower the median travel time to hundreds of years (Figure 3c).

The simulated median groundwater travel time, which can be approximated as groundwater ages, shows good conformity with radiocarbon ages of DIC in groundwater reported in Khan *et al.* (2019) for the SW Bengal basin and in Mihajlov *et al.* (2016) for greater Dhaka region. Since the large scale groundwater pumping in Dhaka city has started only a few decades ago, the radiocarbon ages are similar to that of the model simulated ages without pumping scenario (Figure 3b). If the current pumping continues, it is likely that the deep groundwater surrounding Dhaka city will be contaminated within a century or even before that through preferential flow paths as proposed by Khan *et al.* (2016).

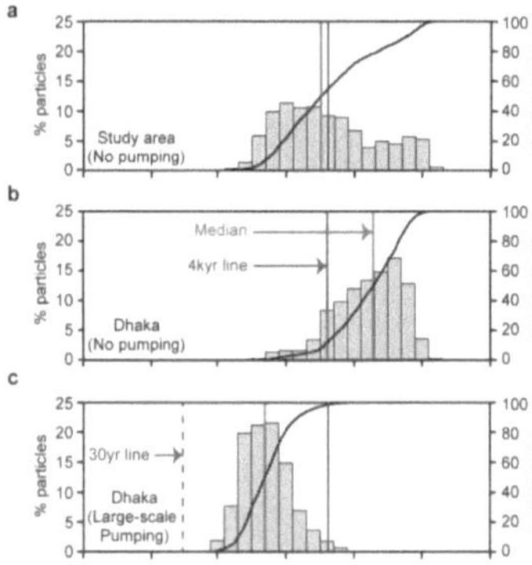

Figure 3. Distribution of particle travel time or model simulated groundwater ages for a) SW Bengal Basin with homogeneous lithology and no pumping, b) Dhaka city area with heterogeneous lithology and no pumping, and c) Dhaka city area with heterogeneous lithology and extensive pumping.

4 CONCLUSIONS AND RECOMMENDATION

Deep groundwater in SW Bengal basin naturally contains elevated As due to deep flow system aided by lack of interbedded muds. In highly heterogeneous parts of the basin deep groundwater are thousands of years old and naturally protected, however, extensive pumping can draw contaminated water at depths within a short period of time. Our findings put emphasis on the nationwide testing of deep groundwater for arsenic as well as setting up strong monitoring networks around large pumping centers.

ACKNOWLEDGEMENT

This work was funded in part by the National Science Foundation (ICER1414131) and by a Delaware Environmental Institute (DENIN) Fellowship to MRK.

REFERENCES

Fogg. G.E. 1986. Groundwater flow and sand body interconnectedness in a thick, multiple-aquifer system. *Water Resour. Res.* 22(5): 679–694.

Khan. M.R., Koneshloo. M., Knappett. P.S.K., Ahmed. K.M., Bostick. B.C., Mailloux. B.J. *et al.* 2016. Megacity pumping and preferential flow threaten groundwater quality. *Nat. Commun.* 7(1): 12833.

Khan. M.R., Michael. H.A., Nath. B., Huhmann. B.L., Harvey. C. F., Mukherjee. A. *et al.* 2019. High-arsenic groundwater in the southwestern Bengal basin caused by a lithologically controlled deep flow system. *Geophys. Res. Lett.* 46(22): 13062–13071.

Michael. H.A. & Voss. C.I. 2008. Evaluation of the sustainability of deep groundwater as an arsenic-safe resource in the Bengal basin. *Proc. Natl. Acad. Sci. U. S.A.* 105(25): 8531–8536.

Mihajlov. I., Stute. M., Schlosser. P., Mailloux. B.J., Zheng. Y., Choudhury. I. *et al.* 2016. Recharge of low-arsenic aquifers tapped by community wells in Araihazar, Bangladesh, inferred from environmental isotopes. *Water Resour. Res.* 52(5): 3324–3349.

Ravenscroft. P., Kabir. A., Hakim. S.A., Ibrahim. A.K. M., Ghosh. S.K., Rahman. S. & Sattar. F.A. 2014. Effectiveness of public rural water points in Bangladesh with special reference to arsenic mitigation. *J. Water Sanit. Hyg. Dev.* 4(4): 545–562.

Arsenic in the Environment: Bridging Science to Practice for Sustainable Development – van der Wal, Ahmad, Petrusevski, Weijma, Savic, van der Wens, Beerendonk, Bhattacharya, Bundschuh & Naidu (Eds)
© 2023 the Author(s), ISBN 978-1-032-32928-4

Investigating the sensitivity of fluvial processes on arsenic enrichment in the Mid-Gangetic floodplains of India

A. Singh[1], A.K. Patel[2] & M. Kumar[3,4,5]
[1]*Discipline of Civil Engineering, Indian Institute of Technology Gandhinagar, Gujarat, India*
[2]*Department of Environmental Sciences, Jawaharlal University, New Delhi, India*
[3]*Discipline of Earth Sciences, Indian Institute of Technology Gandhinagar, Gujarat, India*
[4]*Sustainability Cluster, School of Engineering, University of Petroleum & Energy Studies, Dehradun, Uttarakhand, India*
[5]*Escuela de Ingeniería y Ciencias, Tecnologico de Monterrey, Campus Monterey, Monterrey, Nuevo Leon, Mexico*

ABSTRACT: The present study traces the geochemical occurrence of arsenic and Heavy Metals along the active oxbow formations of river Ganga in MGP. Strong influence of silicate weathering and anthropogenic leaching of fertilizers is concurrently occurring with pyrite oxidation leading to high arsenic vulnerability. While most carbonate minerals are in a state of oversaturation but monsoon led bicarbonate enrichment play a major role via replacement of adsorbed arsenic under conditions suitable for microbial degradation. Further, the steep changes in river directional profile has caused high sediment deposition and compaction leading to increase in the soil bulk density, which interestingly traces the river flow path, coherent with the spatial distribution of soil organic carbon stock (socs). Further, localized concentration of arsenic matches the variation profile of clay content and iron concentration of the soil confirming the role of river led recharge in promoting oxidizing conditions within small stretch of study area. Low cation exchange capacity further supports our concern for groundwater vulnerability in terms of enhanced cation retention of the soil. Our study is a pioneering work in understanding arsenic spatial vulnerability under fluvial forcing through developing indicators which trace the arsenic mobilizing chemistry in a more coherent and direct ways.

1 INTRODUCTION

The younger deltaic alluviums of the Mid Gangetic Plains (MGP) have historically reported higher arsenic exposure, mainly through dissociation of silicate minerals containing arsenic. Some studies have earlier speculated the role of desorption process which is being mediated by microbial action under increased organic content of the soil as well as anthropogenically induced NO_3^- (Kinniburgh *et al.*, 2003; Kumar *et al.*, 2007; Ravenscroft *et al.*, 2001; Singh *et al.*, 2020). Few studies further suggested the role of large-scale irrigation pumping in flooding these plains with carbon supply leading to arsenic mobilization. However, Saunders *et al.* (2005), explained the situation from a geological evolution perspective, stressing the role of continental retreat, leading to a rise in sea level and subsequent deposition of alluvial, marsh and peats in the low-land Gangetic plains during the early and later Holocene period. Due to the low temperature during the Pleistocene period, most of the sediment derived in the lowland region were a product of physical weathering leading to unfavorable conditions for chemical weathering to occur (McArthur *et al.*, 2004). As a result, the arsenic derived in these floodplains are of secondary origin and have less association with primary arsenic bearing minerals, possibly hinting the combined role of fluvial processes and mineral weatherability.

WQI assimilates all the components of hydrochemistry causing its high correlation with arsenic, as it is representative of the underlying geological conditions. HPI, however, accounts more anthropogenic influences dominated by metal contamination. The objective of the present study is to associate the degree of agreement of both these indices in tracing the arsenic concentration in the MGP which is actively characterized by the landform deformation through fluvial processes. Secondary soil features such as organic and clay content play a significant role in mobilizing the arsenic from its mineral state under suitable soil pH and cation exchange capacity. It is therefore necessary to evaluate the role of soil characteristics in aiding the arsenic vulnerability. The aim of the study is threefold – (i) to trace the operative arsenic geochemistry in the region, (ii) to confirm the role of fluvial forcing in mediating the arsenic mobilization process and (iii) to investigate and correlate the sensitivity of mineral weathering with the arsenic enrichment process.

DOI: 10.1201/9781003317395-15

2 MATERIALS AND METHODS

2.1 *Study area*

Buxar district in Bihar covers an area of approximately 1700 km^2, covering almost 1.8% of the total area in Bihar. The geomorphology of the region is dominated by the alluvium depositions brought by the active floodplains during the course of its flow in the MGP. The terrain of the region is slightly sloping towards the northeast, tracing the flow direction of the river Ganga. While the vegetation cover is high for the region but the recent deforestation activities have brought a massive change in the soil and landform characteristics. As per the 2011 census the population of the district is more than 1.7 million with major concentration of settlement along the river Ganga (Figure 1).

2.2 *Sample collection and analysis*

A total of 31 samples were collected in 2015 during the post monsoon season to evaluate the concentration of arsenic, heavy metals and other major ions. Pre-washed polyethylene bottles were used for collecting the groundwater samples by first discarding the groundwater yield obtained during the initial 5–10 minutes. Further filtration was done using Milli-pore 0.45-μm of filter paper and some water sample was preserved at $pH < 2$ with 2M HNO_3 for heavy metal analysis. In-situ parameters were analyzed using Water Quality probe and Hanna Multiparameter (H19828). ICP-OES were used for estimating Na^+, K^+ Ca^{2+}, Mg^{2+}. Argentometric methods of titration were used for estimating Cl^- and UV Spectrophotometry was used for estimating anions such NO_3^-, PO_4^{3-} and SO_4^{2-}. Atomic Absorption Spectroscopy (AAS) method was used for estimating the arsenic concentration of the samples.

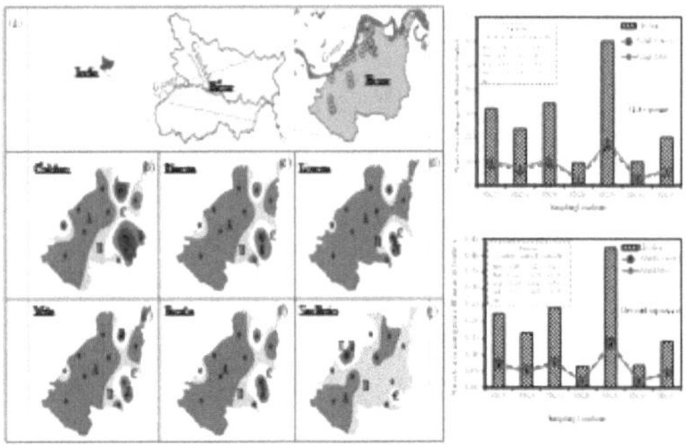

Figure 1. The study area around Buxar district, showing the sampling locations with yellow circles, situated adjacent to the River Ganga (a). The plus sign denotes the block headquarters (11) in the district. The population of children below 12 years of age, with the respective block counts given in brackets (b), A (5) > 30,000; B (1) – 25,000 – 30,000; C (0) – 20,000 – 25,000; D (1) – 15,000 – 20,000 and E (4) < 15,000; The illiterate and literate population of the district (c), (d), A > 65,000; B – 50,000 – 65,000; C – 35,000 – 50,000; D – 20,000 – 35,000 and E < 20,000. The male and female population of the district (e), (f), A > 65,000; B – 50,000 – 65,000; C – 35,000 – 50,000; B – 20,000 – 35,000 and A < 20,000. The sex ratio in the district (g), A > 940; B – 930 – 940; C – 920 – 930; B – 910 – 920 and A < 910. Non-carcinogenic Hazard Index via (i) Oral exposure and (ii) Dermal exposure.

2.3 *WoSIS based soil chemical profiling*

The data for soil and its secondary chemical characteristics including cation exchange capacity, pH, soil organic carbon content and soil fractions have been retrieved from International Soil Reference and Information Centre (ISRIC) database. The database includes around 1,96,000 in-situ collection points along with approximately six million soil profile records. Further remodelling at a spatial resolution of 250 m, enables us to understand the depth varying chemical evolution of the soil. For our study we have used the data for soil profile at the surface. These secondary features of the soil also help in deriving meaningful interpretations about the possible mineral weatherability and the natural potential of the soil to increase the fluvial vulnerability. The evaluation of vegetation type and extent cover was mapped using LANDSAT 8 satellite images. Further, NOAA Climate Data record of AVHRR has been considered to derive Leaf Area Index (LAI) and Photosynthetically Active Radiation (FPAR). Finally, the activeness of the microbial activity in Buxar district has been mapped using MODIS (MOD17A2H) derived Gross Primary Productivity (GPP).

3 RESULTS AND DISCUSSION

3.1 *Characterization of arsenic hydrogeochemistry*

The silicate weathering of the region is con-current with arsenic release into the groundwater. The two most abundant minerals of arsenic i.e. arsenolite (As_4O_6) and claudetite (As_2O_3) are extremely undersaturated with SI being −16.41 and −16.18 respectively (Figure 2). The groundwater conditions are both oxidizing and reducing with the change being brought in a very small span of area. The higher concentration of arsenic is however achieved under oxidizing conditions when pH lies between 8.1 to 8.6, hinting strongly to the possibility of pyrite oxidation. This is supported by the linear trend between Fe and As but for the mobilization to occur, strong microbial activity is needed under high organic content of the soil. High soil organic carbon stock (SOCS) ranging between 8–33 tons/ha and tracing trends of gross primary productivity and the spatial distribution

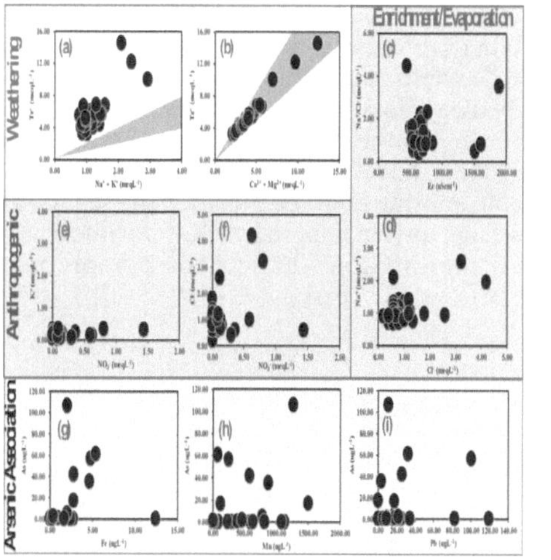

Figure 2. Hydrochemistry of the study region, delineating the geogenic and anthropogenic disturbances.

of arsenic confirms the role of pyrite oxidation and the subsequent release of arsenic under microbial degradation. The concentration of arsenic correlates with Fe (r = 0.32) and Mn (r = 0.34) signifying the dissociation of Fe from pyrite during microbial breakdown. The high vulnerability of arsenic could be due to the recharge observed during the preceding monsoon season as supported by the high correlation between HCO_3^- and NO_3^- (r = 0.72). According to Appelo *et al.* (2002), bicarbonates does have the ability to re-place arsenic from the adsorbed state to increase its availability in the groundwater leading to the active participation of rainfall recharge in increasing the arsenic vulnerability. However, the major source of bicarbonate is still biotite led silicate weathering as revealed by significantly high correlation of 0.86.

The fluvial processes have a dominant effect in bringing widescale landform deformation which is visible as increased bulk density of the soil, reaching to almost 1587 kg/m^3 near the meandering ox-bow formations. These regions are also characterized by pockets of high bedrock depth (156 m) possibly hinting at the enhanced role of depth-dependent organic recycling by the microorganisms in order to release arsenic from the mineral state. The trend of these parameters is consistent with the spatial distribution of arsenic which have highly vulnerable zones located near the mouth of the ox-bow formation. The high concentration of Fe in the pockets of arsenic vulnerable zones does suggests the role of a possible adsorption of arsenic on Fe-OOH surface however, the oxidative potential of the groundwater suggest the occurrence of As(V), which explains why low spatial hydrologic mobility of arsenic has occurred in the region. Along with positive ORP, the phosphate concentration is also high (r = 0.32) which mean that the arsenate uptake by aerobic micro-organisms will occur, leading to inhibition of oxidative phosphorylation and thereby disrupting the entire nutrient recycling in the near future. Weak correlation between phosphate, nitrate (r = −0.05) and bicarbonates (r = −0.11), clearly shows that phosphate is released in the sub-surface system with little contribution from surface recharge. Interestingly, lead has the highest correlation with phosphate (r = 0.36) among any other heavy metal. The trend becomes clearer when we see the symbiotic behavior of lead and iron (r = 0.45), which seems to suggest the natural occurrence of common minerals such as galena in the bedrock deposits. This also validates the correlation between arsenic and manganese (r = 0.34) as galena belongs to the sulfide group of minerals containing impurities such as iron sulfide and arsenide niccolite, which in our opinion explains their common origins and hence the correlations. However, witnessing the oxidation state of the groundwater, it is expected that galena will transform into anglesite (lead sulphate) or cerussite (lead carbonate), in the near future causing an enrichment of sulphate and bicarbonates through it, in addition to the effects induced by carbonate weathering. The cation exchange capacity of the soil at the groundwater sampling location lies in the range between 250–250 mmol.c/kg. This usually signifies presence of illite, as the dominant clay mineral, considering the fact that the concentration of clay lies between 18–40%.

Principal Component Analysis (PCA) provided a better conformity to our speculative understanding about the operative processes active at the sub-surface level. The common origin of manganese (PC2 = 0.675) and arsenic (PC2 = 0.732) was further validated by its relative lesser space distance occupied on the first two principal components (Figure 3). It could also be seen that the sample for which the arsenic concentration has reached above 100 μg/kg has exceptionally high concentration of manganese reaching to almost 1400 μg/kg. This explains the dominant signature of sulphide mineral dissociation. The significantly high correlation of pH with manganese (r = −0.52), explains the link of pH dependent manganese oxyhydroxides formation causing adsorption of arsenic which in turn depends on ORP for its enhanced enrichment. The PC1 has very accurately captured the variances among physical parameters such as TDS (PC1 = 0.95) and major ions such as Na + (PC1 = 0.85) and Mg^{2+} (PC1 = 0.88). However, the first principal component also explains a significant variance of NO_3^- (PC1 = 0.88) clearly signifying that the first principal component is explaining the surface recharge of anthropogenically contaminated river water into the groundwater system.

Close aggregation of iron and lead also signifies the role of galena as a dominant mineral with

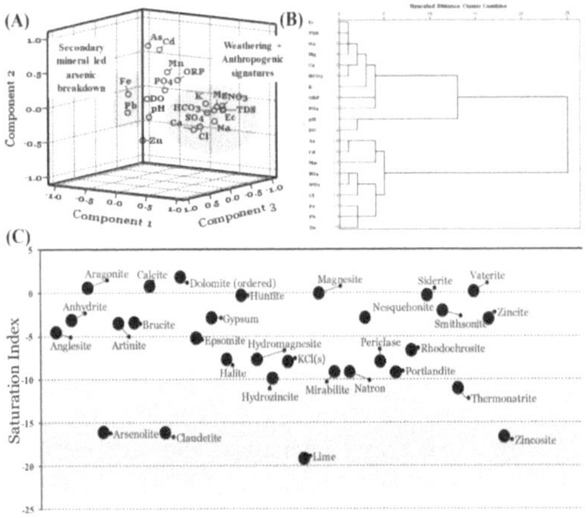

Figure 3. PCA and HCA clusters along with SI values of major minerals.

varying degrees of sulfide impurities. Further, K nearest neighbor clustering was done which revealed that the arsenic samples can be accurately grouped into three categories, which traced our three previously identified processes. Moreover, the heat map of the samples and parameters along with its Hierarchical Clustering Analysis (HCA) clearly establishes and links the TDS variations with arsenic variability of the samples, especially sample number 29 and 18, thereby conforming the role of river led TDS enrichment in having a direct causative relationship with arsenic mobilization. Interestingly, more than the pH of the groundwater, the pH of the soil shares a consistent behavior in mediating this enrichment process. Further, for both the samples, potassium and iron vary in a complementary fashion clearly revealing the role of silicate weathering to be active for the sample with highest arsenic vulnerability.

4 CONCLUSIONS AND RECOMMENDATIONS

The present study has encompassed all the major scientific tools including remote sensing, literature survey, modelled soil characteristics and in-situ groundwater sampling in order to interpret the vulnerable situation progressing in the MGP with regards to arsenic enrichment. We have presented a detailed discussion concerning the role of anthropogenically introduced fertilizers and heavy metals, on the corresponding sensitivity of the arsenic dissolution process in aquifer system. Further, we have demonstrated through our analysis that river morphology does have a significant role in aiding arsenic concentration through sediment deposition along the oxbow formations. Therefore, it has now become imperative to use more remote sensing and modelling driven approach for monitoring of situation in MGP as the situation changes nature with seasons and years.

ACKNOWLEDGEMENTS

This study is supported by DST-SERB Young Scientist Grant to MK.

REFERENCES

Appelo C.A.J. & Postma D. 2004. *Geochemistry, Groundwater and Pollution*. CRC Press.

Kinniburgh D.G., Smedley P.L., Davies J., Milne C.J., Gaus I., Trafford J.M., Burden S., Huq S.I., Ahmad N. and Ahmed K.M. 2003. *The Scale and Causes of the Groundwater Arsenic Problem in Bangladesh. In Arsenic in ground Water* (pp. 211–257). Springer, Boston, MA.

Kumar M., Kumari K., Ramanathan A.L. & Saxena R. 2007. A comparative evaluation of groundwater suitability for irrigation and drinking purposes in two intensively cultivated districts of Punjab, India. *Environ. Geol.* 53(3): 553–574.

McArthur J.M., Banerjee D.M., Hudson-Edwards K.A., Mishra R., Purohit R., Ravenscroft P., Cronin A., Howarth R.J., Chatterjee A., Talukder T. & Lowry D. 2004. Natural organic matter in sedimentary basins and its relation to arsenic in anoxic ground water: the example of West Bengal and its worldwide implications. *Appl. Geochem.* 19(8): 1255–1293.

Ravenscroft P., McArthur J.M. & Hoque B.A. 2001. Geochemical and palaeohydrological controls on pollution of groundwater by arsenic. In: W.R. Chappell, C.O. Abernathy & R. Calderon (eds.) *Arsenic Exposure and Health Effects IV*. Elsevier Science Ltd.:Oxford, pp. 55–77.

Saunders J.A., Lee M.K., Uddin A., Mohammad S., Wilkin R.T., Fayek M. & Korte N.E. 2005. Natural arsenic contamination of Holocene alluvial aquifers by linked tectonic, weathering, and microbial processes. *Geochemistry, Geophys. Geosystems* 6(4):1–7.

Singh A., Patel A.K., Deka J.P. & Kumar M. 2020. Natural recharge transcends anthropogenic forcing that influences arsenic vulnerability of the Quaternary alluviums of the Mid-Gangetic Plain. *NPJ Clean Water* 3(1): 1–12.

Arsenic in the Environment: Bridging Science to Practice for Sustainable Development –
van der Wal, Ahmad, Petrusevski, Weijma, Savic, van der Wens, Beerendonk, Bhattacharya, Bundschuh & Naidu (Eds)
 ISBN 978-1-032-32928-4

Source of arsenic based on geological and hydrogeochemical properties of geothermal systems: Case study of Anatolia (Turkey)

A. Baba[1] & T. Uzelli[2]
[1]*Department of International Water Resources, İzmir Institute of Technology, Urla-Izmir, Turkey*
[2]*Geothermal Energy Research and Application Center, İzmir Institute of Technology, Urla-İzmir, Turkey*

ABSTRACT: The arsenic (As) concentration in water resources is very crucial for human life. Arsenic values in water resources in Anatolia can reach important concentrations that need attention. The reason for this is that structural controls such as water-rock interactions and faults during the circulation of groundwater, generally of geogenic origin, affect the circulation of waters. The effect of geogenic As on waters can be seen mostly in monitoring studies in geothermal fluids in Anatolia. Significant As concentrations have been reached, especially in tectonically active regions such as the North Anatolian Fault and Aegean Extensional Fault Systems, and in volcanic fields. In this context, spatial correlation of high As values with structural and volcanic controls was made within the scope of this study.

1 INTRODUCTION

The presence of arsenic (As) in high concentrations in water resources has always been a risk factor. In addition to human activities, another factor that increases As concentrations in water resources is natural occurrence (Baba & Armannsson 2006; Baba 2010). Arsenic, a natural contaminant in aquifer systems used for drinking water supply, is involved in the hydrologic cycle through geologic units and structural controls. So, orogenic belts and active volcanism can create chemically unique environments. Structural controls such as faults, fractures, and fissures also directly affect element concentrations in aquifers and provide secondary permeability and porosity.

In tectonically active areas, surface and groundwater mix in aquifers as in the western Turkey with active extensional tectonics. Geothermal fluids reach the surface through these structural elements and geothermal production wells. Studies show that elements such as B, Li, F, and As are present in higher concentrations in the geothermal fluid and shallow-deep circulating waters interacting with the geothermal system.

Figure 1. Arsenic distribution map of Turkey (modified after Baba *et al.*, 2021).

While high concentrations in geothermal fluids interact with deeper rocks, high concentrations in shallow circulating waters are evidence that structural controls and geology influence the overall hydrogeologic system in the region. Geothermal systems develop in parallel with systems where hydrothermal and epithermal igneous deposits form. Alteration and mineral formation processes are the same or very similar. Mineral formations that develop from epithermal mineral deposits, ophiolitic melanges from old magmatic rocks, igneous rocks, and As concentrations in sedimentary basin sediments are significant. Arsenic is generally found in lithologic units in volcanic and tectonically inactive regions. While As occurs in units with deep geothermal systems in the structure of minerals such as pyrite and arsenopyrite, it is separated from the structure of As minerals by different natural and thermodynamic conditions.

As a result, physicochemical and biological degradation processes such as erosion and alteration, tectonic activities, and intense deformation in rock units carry As in the rock relatively strongly into the geothermal system (Baba & Sozbilir 2012; Birkle *et al.*, 2010; Bundschuh & Maity 2015; Lopez *et al.*, 2012; Mukherjee *et al.*, 2009). This study finds high concentrations of As in tectonically controlled basins, old massive masses, tectonically active contraction zones, old continental interaction zones, and geological units belonging to ancient ocean ruins that have closed and formed accumulation mélanges due to plate tectonics around Turkey.

DOI: 10.1201/9781003317395-16

2 MATERIALS AND METHODS

2.1 *Study area*

Turkey is located on Anatolian Plate between the Eurasian and Arabian-African plates in the Eastern Mediterranean seismic belt. The 'Neotectonic Period' in Anatolia is characterized by the compressional tectonic regime in Eastern Anatolia and the westward movement of the Anatolian Plate (Bozkurt 2001). Neotectonic deformation of the Anatolian Plate has formed different provinces. Geothermal systems, on the other hand, have developed under the control of tectonic and volcanic movements in different regions of Anatolia.

Along with the West Anatolian Extensional Fault Systems (WAES), the three most essential transform faults in Anatolia are (Sengor *et al.*, 1985): the North Anatolian Fault (NAF), the East Anatolian Fault (DAF) and the Northeast Anatolian Fault (NEAF). The NAF and EAF are the most prominent examples of the impact of the strike-slip fault systems (intra-continental transform fault) controlling the westward escapement of the Anatolian Plate.

2.2 *Spatial arsenic data collection and analysis*

All geothermal resources in the study area are dispersed under the control of active tectonics and volcanism. There are 415 geothermal fields defined throughout the country in 63 provinces in Turkey. According to Şener *et al.* (2023), more than 2000 geothermal springs (wells or hot springs) have been identified in Turkey. All these areas are affected by volcanism of different ages and fault zones. It is possible to make some definitions under the game type classification for areas without sharp differences between As distributions (Baba *et al.*, 2021).

Within the scope of this study, the general situation in the country was tried to be revealed by examining the As values determined in previous studies (Akkus *et al.*, 2005; Baba *et al.*, 2021; Baba & Armannsson 2006; Baba & Sozbilir 2012; Bundschuh *et al.*, 2013).

3 RESULTS AND DISCUSSION

3.1 *Geothermal fields of Western Anatolia*

The Western Anatolian Extensional Province (WAEP), where tectonically active and occasional Quaternary volcanism is observed, is known for the Neogene and Quaternary basins in approximately E-W and NE-SW directions. The essential morphological structures bounded by normal faults in the region are the Gediz and Büyük Menderes Grabens. Low and high-angle normal faults limiting these grabens and basins have caused the formation of critical geothermal systems. The reservoir of the system is the Menderes Massif units consisting predominantly of carbonates. Due to altered units of the Menderes Massif, volcano-sedimentary sediments with argillic, phyllic, and silica-hematitic hydrothermal alterations, high concentrations of geothermal fluids have been determined in some geothermal fields. Arsenic concentrations are reaching 4.1 mg/L in the geothermal fields in Denizli and 3.34 mg/L in Manisa. These values are due to geothermal fluids moving along high-angle normal faults and reaching deep.

3.2 *Geothermal fields of Eastern Anatolia*

The compression-dominated deformation that occurred along the collision zone known as the Bitlis-Zagros Suture Zone (BZSZ) of the Eurasian and Arabian Plates not only caused the Anatolian plate to rise but also caused the formation of a tectonic and volcanic region. Volcanic units (andesite, basalt, and rhyolite) have different spreading areas with different characters. Although there are few As studies in geothermal fluids in the region, high As values have been measured around tectonically controlled volcanic mountains with intense opening cracks. The highest As values were measured in the volcanic regions around Van, Sanliurfa, and Bingol at concentrations of 0.9 mg/L, 0.1 mg/L, and 0.06 mg/L, respectively.

3.3 *Geothermal fields of Central Anatolia*

The central part of Anatolia is located within a neotectonic regime area bounded by the NAF and EAF. While Ankara and its surroundings are an inactive volcanic region with the NAF, the southern parts are influenced by Quaternary-aged Cappadocia volcanism. While the As values in geothermal fluids can reach around 6.3 mg/L around the NAF (Çankırı region), it generally varies between 0.09 and 0.5 mg/L around Ankara and Bolu and between 0.02 and 0.18 mg/L around Niğde and Nevşehir (Cappadocia). In the vicinity of Kırşehir-Kayseri, As concentrations in granite, schist, and marble units within the Central Anatolian crystalline complex can rise to 0.3 mg/L.

4 CONCLUSIONS AND RECOMMENDATIONS

The structural control and physical properties of the rocks, as well as the accompanying volcanism, lead in part to differences in the physicochemical properties of the deep-circulating geothermal fluids. For this reason, taking into account all these factors, geothermal systems may contain fluids with different As concentrations. In accordance with the current situation in Turkey, As levels and chemical composition of geothermal resources, especially in the western regions, have been studied in detail in focal studies, while there are relatively few studies on water resources related to volcanism in other regions of Anatolia.

Considering the available databases, Turkey has geothermal fields that may contain high As concentrations, especially in the central and western parts, under the effect of the extensional tectonic regime and normal component strike-slip regime. Of course, this can be explained by the fact that the permeability values in the rocks and faults are higher than the other compressional tectonic regions of the Anatolian Plate. Apart from this, high As values were also reached in geothermal resources around the North Anatolian Fault, which is an intra-continental transform fault. The reason for this is that the planes of strike-slip faults are steeper, and the faults can reach deeper levels. Arsenic values are estimated to be high in other Anatolia areas, especially in regions with volcanic activity and alteration zones.

REFERENCES

Akkus I., Akilli H., Ceyhan S., Dilemre A. & Tekin Z. 2005. *Turkiye Jeotermal Kaynaklar Envanteri.* MTA Genel Mudurlugu Envanter Serisi-201, Ankara.

Baba A. & Ármannsson H. 2006. Environmental impact of the utilization of a geothermal area in Turkey. *Energ. Source Part B.* 1(3): 267–278.

Baba A. 2010. High arsenic levels in water resources resulting from alteration zones: a case study from Biga Peninsula, Turkey. In: J.-S. Jean, J. Bundschuh & P. Bhattacharya (eds.) *Arsenic in Geosphere and Human Diseases, (As2010).* CRC Press/Taylor and Francis (ISBN-13: 978-0-415-57898-1), pp. 18–20.

Baba A. & Sozbilir H. 2012. Source of arsenic based on geological and hydrogeochemical properties of geothermal systems in western turkey. *Chem. Geol.* 334: 364–377.

Baba A., Uzelli T. & Sozbilir H. 2021. Distribution of geothermal arsenic in relation to geothermal play types: a global review and case study from the anatolian plate (Turkey). *J. Hazard. Mater.* 414: 125510.

Birkle P., Bundschuh J. & Sracek O. 2010. Mech nisms of arsenic enrichment in geothermal and petroleum reservoirs fluids in Mexico. *Water Res.* 44 (19): 5605–5617.

Bozkurt E. 2001. Neotectonics of Turkey – a synthesis. *Geodin. Acta* 14: 3–30.

Bundschuh J., Maity J.P., Nanth B., Baba A., Gunduz O., Kulp T.R., Jean J.S., Kar S., Tseng Y., Bhattacharya P. & Chen C.Y. 2013. Naturally occurring arsenic in terrestrial geothermal systems of Western Anatolia, Turkey: potential role in contamination of freshwater resources. *J. Hazard. Mater.* 262: 951–959.

Bundschuh J. & Maity J.P. 2015. Geothermal arsenic: occurrence, mobility and environmental publications. *Renew. Sust. Energy. Rev.* 43: 1214–1222.

Lopez D.L., Bundschuh J., Birkle P., Armienta M.A., Cumbal L., Sracek O., Cornejo L. & Ormachea M. 2012. Arsenic in volcanic geothermal fluids of Latin America. *Sci. Total Environ.* 429: 57–75.

Mukherjee A., Fryar A.E. & O'Shea B.M. 2009. Major Occurrences of Elevated Arsenic in Groundwater and Other Natural Waters. In: K.R. Henke (ed.) *Arsenic—Environmental Chemistry, Health Threats and Waste Treatment.* John Wiley & Sons, Chichester UK, pp. 303–350.

Sengor A.M.C., Gorur N. & Saroglu F. 1985. Strike-slip faulting and related basin formations in zones of tectonic escape: Turkey as a case study. In: K.T. Biddle & N. Christie-Blick (eds.) *Strike-slip Faulting and Basin Formation* Society of Economic Paleontologists and Mineralogists, Special Publication, No. 37. Tulsa, Oklahoma, USA pp. 227–264.

Şener M. F., Uzelli T., Akkuş İ., Mertoğlu O. & Baba A. 2023. The potential, utilization, and development of geothermal energy in Turkey. *Bull. Miner. Res. Explor.* https://doi.org/10.19111/bulletinofmre.1229381

Arsenic in the Environment: Bridging Science to Practice for Sustainable Development – van der Wal, Ahmad, Petrusevski, Weijma, Savic, van der Wens, Beerendonk, Bhattacharya, Bundschuh & Naidu (Eds)
© 2023 the Author(s), ISBN 978-1-032-32928-4

Use of GIS and geospatial techniques to model spatial variability of geogenic arsenic in groundwater systems of Geita district in the Lake Victoria basin of western Tanzania

J. Ijumulana[1,2], R. Irunde[1,2], F.J. Ligate[1,2], V. Kimambo[1,2], P. Bhattacharya[2], J.P. Maity[3], A. Ahmad[4], F. Mtalo[1] & J. Mtamba[1]
[1]*DAFWAT Research Group, Department of Water Resources Engineering, College of Engineering and Technology, University of Dar es Salaam, Dar es Salaam, Tanzania*
[2]*KTH-International Groundwater Arsenic Research Group, Department of Sustainable Development, Environmental Science and Engineering, KTH Royal Institute of Technology, Stockholm, Sweden*
[3]*Department of Earth and Environmental Sciences, National Chung Cheng University, Min-Hsiung, Chiayi County, Taiwan*
[4]*SIBELCO Ankerpoort NV, Maastricht, The Netherlands*

ABSTRACT: For the past few decades, sporadic studies in African continent have reported elevated concentrations of arsenic (As) in several sources of drinking water of some countries posing a serious health risk mainly to rural communities. In Tanzania, As concentrations above WHO guideline (10 μg/L) for drinking water have been reported in parts of Lake Victoria basin where special interested has shifted to groundwater as a primary source of drinking water. In this study, GIS and geospatial methods were used to describe the spatial variability of As concentrations in the vicinity of gold mining areas. Around 92% of drinking water sources had As above 10 μg/L and the percentage varied between sources. Similarly, a significant positive global Moran's I indicated clustering of As concentrations in space with highest spatial variability in the rural settings posing a health risk to rural communities in Geita district.

1 INTRODUCTION

Elevated arsenic concentrations (As) in drinking water sources is an increasing water quality problem worldwide. In Tanzania, high levels of As have been reported around gold mining areas in Lake Victoria basin of north-western Tanzania (Ahoulé *et al.*, 2015; Kassenga & Mato 2008; Nyanza *et al.*, 2014). Mining activities, have been considered as primary triggers of As in most of drinking water sources (Almås & Manoko 2012; Nyanza *et al.*, 2014). This study quantifies the spatial variability of As in the amidst mining areas of Geita district by using modern geospatial methods and technologies.

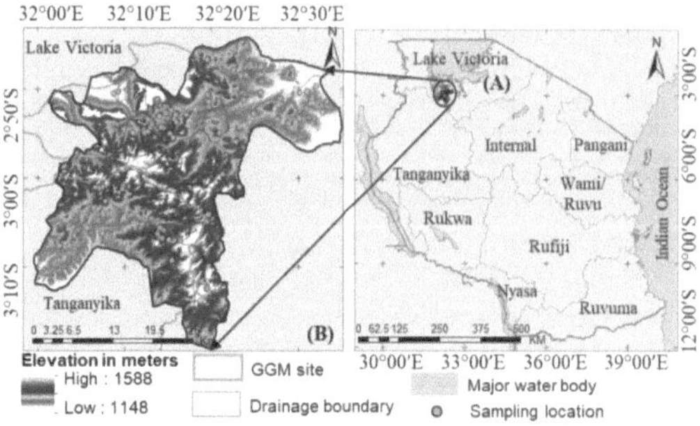

Figure 1. Study area in Lake Victoria basin (A) and water/rock/soil/sediment sampling locations in Geita district (B).

2 MATERIALS AND METHODS

2.1 *Study area*

This study was conducted in Geita district, south of Lake Victoria at the boundary between Lake Victoria and Tanganyika basins (Figure 1). The district is well endowed with precious metals and minerals that contribute significantly to the national economic development. Like in many places of Tanzania, particularly in central and northern development zones, groundwater contributes around 80% of water supply, especially in rural areas of Geita region. In these areas, substantial groundwater is hosted in Precambrian basement aquifers within the Tanzanian Craton. Access to groundwater is through shallow wells, boreholes and springs.

2.2 *Spatial data collection and analysis*

Both water and rock/soil/sediments were collected using standard methods during fieldwork campaigns during 2016 and 2018. A total of 130 ground water samples were collected from drinking water wells. The samples represented shallow wells (52 samples), boreholes (51 samples) and springs (21 samples). In

DOI: 10.1201/9781003317395-17

addition, water samples were collected from surface water including dams (3 samples) and rivers (3 samples). Rock/soil/sediment samples were collected strategically at or near the drinking water source. The position for each sample location was determined using geographical information systems (GIS) data collector (Global Mapper 20 series). Each water sample was analyzed in situ to determine total As concentration using As Hach Field Test Kit. Petrographic analysis was used to determine As concentrations in each rock/soil/sediment sample at the Geology laboratory of the University of Dar es Salaam. The Carl ZEISS made Primotech petrographic microscope was used in the analysis. For soils and sediments, the chemical analysis was done using an Olympus Delta handheld X-ray Fluorescence (XRF) Spectrometer. Spatial variability in As concentrations was studied using predicted values. The predicted values were generated using ordinary kriging method. The predictions were done at 0.25×0.25 km grid interval.

3 RESULTS AND DISCUSSION

3.1 *Arsenic concentrations in drinking water sources*

Overall, 92% of the water samples had As > 10 µg/L, an optimal recommended value for drinking water (WHO 2011). A significant positive global Moran's I statistic (I = 306, z-value = 5.96, p-value = 0.001) was obtained indicating clustering of As concentrations in space (Moran 1950). The concentration varied between sources with shallow wells varying between 0 and 280 µg/L while 96% of the samples had As concentration > 10 µg/L. For boreholes, As varied between 0 and 250 µg/L with ~ 76% of samples having As > 10 µg/L. For springs, ~ 93% of samples had As > 10 µg/L with range of between 0 and 290 µg/L.

3.2 *As concentrations in rocks, soils and sediments*

The ore petrographic analysis of the rock thin sections indicated dominance of arsenopyrites and other sulphide/ore minerals including pyrites and chalcopyrites. For soils and sediment samples, the XRF analysis results indicated high levels of As ranging between 47 and 270 mg/L at 6 locations while no As traces were detected at the other two locations.

3.3 *Spatial variability of As concentrations with respect to the GGM site*

Figure 2 indicates The spatial variation of As concentrations with distance from Geita Gold Mining (GGM) site. Although all the predicted As values were above 10 µg/L, the median values at an interval of 0.5 km slightly increased linearly with distance from the GGM site (Adjusted $R^2 = 0.101$, p-value = 0.002). Highest variability (many extreme values = outliers) in predicted As concentrations was observed far away from the GGM site, particularly in rural areas where artisanal and scale mining activities are in full swing.

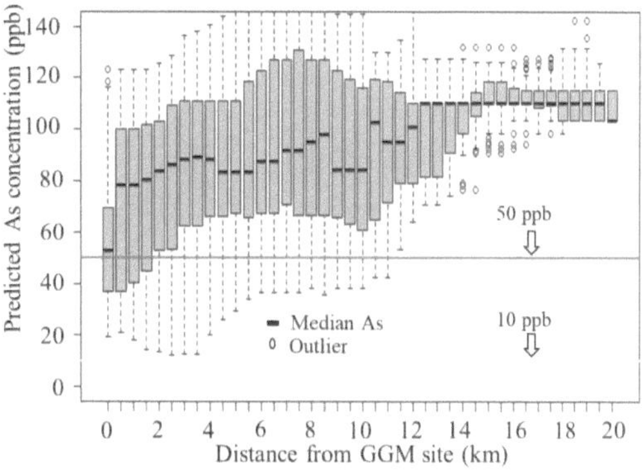

Figure 2. Spatial variability of As concentrations in Lake Victoria Gold Fields of Geita district.

4 CONCLUSIONS AND RECOMMENDATIONS

In this study the occurrence of As in drinking water sources was investigated. The elevated As concentrations are mobilized from oxidation of sulphide/ore minerals, particularly arsenopyrites, pyrites and chalcopyrites. The presence of elevated As concentrations in soils and sediment samples is an indication that As in drinking water sources is mobilized through water and rocks/soils/sediments interaction.

ACKNOWLEDGEMENTS

We acknowledge the Swedish International Development Cooperation Agency (Sida) for supporting the DAFWAT program (Contribution: 51170071).

REFERENCES

Ahoulé D.G., Lalanne F., Mendret J., Brosillon S., Maïga & A.H. 2015. Arsenic in African waters: a review. *Wat. Air Soil Pollut.* 226(9): 302.

Almås Å.R. & Manoko M.L.K. 2012. Trace element concentrations in soil, sediments, and waters in the vicinity of Geita Gold Mines and North Mara Gold Mines in Northwest Tanzania. *Soil Sed. Contamin.* 21(2): 135–159.

Kassenga G.R. & Mato R.R. 2008. Arsenic contamination levels in drinking water sources in mining areas in lake Victoria basin, Tanzania, and its removal using stabilized ferralsols. *Int. J. Biol. Chem. Sci.* 2(4): 389–400.

Mnali S. 2001. Assessment of heavy metal pollution in the lupa gold field, SW Tanzania. *Tanzania J. Sci.* 27(2): 15–22.

Moran P.A. 1950. Notes on continuous stochastic phenomena. *Biometrika* 37(1/2): 17–23.

Nyanza E.C., Dewey D., Thomas D. S., Davey M. & Ngallaba S.E. 2014. Spatial distribution of mercury and arsenic levels in water, soil and cassava plants in a community with long history of gold mining in Tanzania. *Bull. Environ. Contam. Toxicol.* 93(6): 716–721.

WHO 2011. Guidelines for Drinking-Water Quality, Fourth Edition, *WHO Chronicle* 38(4): 104–108.

Arsenic in the Environment: Bridging Science to Practice for Sustainable Development –
van der Wal, Ahmad, Petrusevski, Weijma, Savic, van der Wens, Beerendonk, Bhattacharya, Bundschuh & Naidu (Eds)
© 2023 the Author(s), ISBN 978-1-032-32928-4

Distribution of pollutants in water of the Bustillos Lagoon in Chihuahua, Mexico

J.M. Ochoa-Rivero[1], H.A. Fuentes-Hernandez[2], V.M. Reyes-Gomez[2] & B.A. Rocha-Gutierrez[3]
[1]*INIFAP. Instituto Nacional de Investigaciones Forestales, Agrícolas y Pecuarias, Campo Experimental La Campana, Chihuahua, México*
[2]*INECOL. Instituto de Ecología A.C., Red Ambiente y Sustentabilidad, Chihuahua, México*
[3]*Facultad de Ciencias Químicas, Chihuahua, UACH. Universidad Autónoma de Chihuahua, México*

ABSTRACT: The Bustillos Lagoon is a very important water body for the Chihuahua, Mexico, but water pollution by different sources have been previously reported in this area. The aim of this study was to determine the main physicochemical parameters and the occurrence of As in the water of the Bustillos Lagoon. Randomly, 22 points were selected by GIS, and samples were collected in the area to evaluate water quality, which is used for crops irrigation. The spatial distribution was performed by was Co-Kriging available in ArcMap© 10.4 software, he physicochemical parameters were quantifying in situ and the arsenic (As) analysis was performed by ICP-MS. In general, the physiochemical parameters value were within acceptable levels according the Mexican Regulations for aquatic life. The concentrations of As found in the study area exceed the national and international values established by drinking water consumption regulations in the 22 points of Lagoon WHO (0.010 mg/L and Mexican Regulations; 0.025 mg/L and freshwater range; 0.15–0.45 µg/L). It is concluded that water of the Bustillos ecosystem represent a hazard for aquatic life and a possible risk exposure if the water is used for crops irrigation and any domestic uses.

1 INTRODUCTION

Bustillos Lagoon is considered one of the most important natural resources in the state of Chihuahua.

Besides, it has been considered a priority wetland by international organizations, due to its migratory bird biodiversity. Unfortunately, anthropogenic activities have contributed to the deterioration of this aquatic ecosystem. For instance, water pollution by pesticides-insecticides, wastewater discharges, and inappropriate solid waste disposal, are the main source of pollution in this area. The main discharges to the Bustillos Lagoon are the runoff from an area of intensive agriculture, and industrial and domestic wastewater discharges. Water from the lagoon is used for crop irrigation and other agricultural application (Rubio-Arias *et al.*, 2005). Likewise, Ochoa-Rivero *et al.* (2017) reported high levels of organochlorine pesticides like DDT and its metabolites in water. However, the presence and distribution of arsenic in this area have not been evaluated. The objective was to determine physicochemical parameters an occurrence of As in the water of the Bustillos, Lagoon, Chihuahua, Mexico.

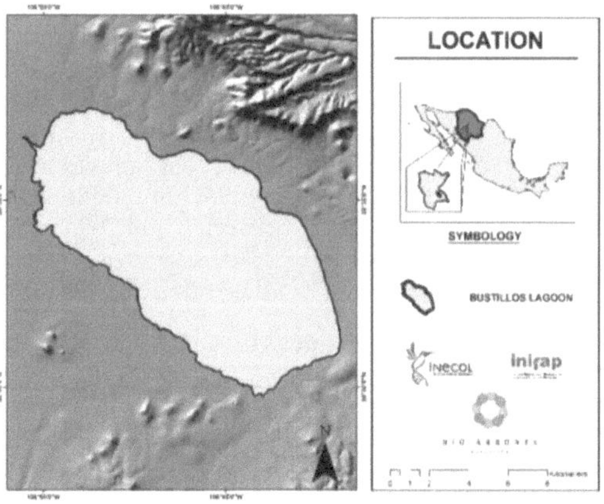

Figure 1. Map of the study area around the endorheic basin of Laguna de Bustillos in Chihuahua, Mexico.

2 MATERIALS AND METHODS

2.1 *Study area*

The study was carried out at the endorheic basin called "Laguna de Bustillos and/or Aquifer of Cuauhtémoc" in Chihuahua, Mexico (Figure 1). The basin belongs to Hydrological Region No. 34 called "Closed Basins of the North" The Bustillos Lagoon watershed is between 28° 13'19" and 28° 59'35"N and the length 106° 34'39" and 107° 10'33"O (Figure 1), and the total area is 3,288 km^2. This basin is about 90 km from the city of Chihuahua. The main activities in the basin are: agriculture, livestock (intensive and extensive), and the main crops in the area are corn, wheat, barley and apple, among others.

2.2 *Water sampling*

Sampling area to collect the water was obtained randomly. The area was divided into 4.34 km^2 quadrants using the geographic software Fishnet ArcMap© 10.4 software. Then, 22 quadrants were randomly designated as sampling sites. In each sampling site,

DOI: 10.1201/9781003317395-18

during the August 2018 water samples were collected at depth 0.50 m. A total of 22 water samples were obtained and analyzed following the Mexican Standandars for wastewater collection (NMX-AA-014-1980).

2.3 *Physical and chemical analysis*

The physicochemical parameters such as: pH, temperature (T), electrical conductivity (EC_w), redox potential (Eh), total dissolved solids (TDS), and temperature (°C) were measured *in situ* with Myron L® FS6FCE PoolPro handheld Pool Meter. The concentration of arsenic (As) in the water samples was analyzed by ICP-MS (Inductively coupled plasma mass spectrometry) Perkin Elmer Model ELAN 6100®.

2.4 *Creation of spatial database and data analysis*

ArcGIS software was used to create spatial database comprising the location and description of each water sample and As. The deterministic interpolation method was Co-Kriging available in ArcMap© 10.4 software.

3 RESULTS AND DISSCUSION

Table 1 shows the descriptive statistics of the physicochemical and As variables during August, 2018.

The physiochemical parameters in general are within acceptable for wastewater discharges in body waters (NOM-001-SEMARNAT-1996). Figure 2 shows spatial distribution of As in water based on WHO and Mexican Standards in drinking water (NOM-127-SSA1-2000). In Figure 2(a) the levels of As showed that 22 point exceeded the standards (0.010 mg/L – WHO; 0.025 mg/L Mexican) and ranged (0.15–0.45 μg/L) proposed by Bissen & Frimmel 2003) for freshwater (lakes and rivers). However, in Figure 2(b) the levels of As indicated that 21 point exceeded the Mexican standard. Only one site reported concentrations below the regulation.

The contribution of this problem can be attributed to the geological composition of the lagoon and to anthropic activities such as agricultural-pesticides in the area (Barrett *et al.*, 2019; Benavides *et al.*, 2008; Rubio *et al.*, 2006).

Table 1. Descriptive statistics of physiochemical parameters at 22 sampling sites in August 2018 at Bustillos Lagoon, Chihuahua, Mexico.

Variable	N	Mean	EE	SD
pH (Units)	22	8.6	0.060	0.31
CEw (mS)	22	1,271.40	18.900	88.80
ORP (mV)	22	74.82	8.720	40.91
TDS (mg/L)	22	254.08	3.770	17.68
T (°C)	22	19.89	0.750	3.54
As (mg/L)	22	0.060	0.002	0.01

*N = Sample size, EE = Experimental Error, SD = Standard Deviation

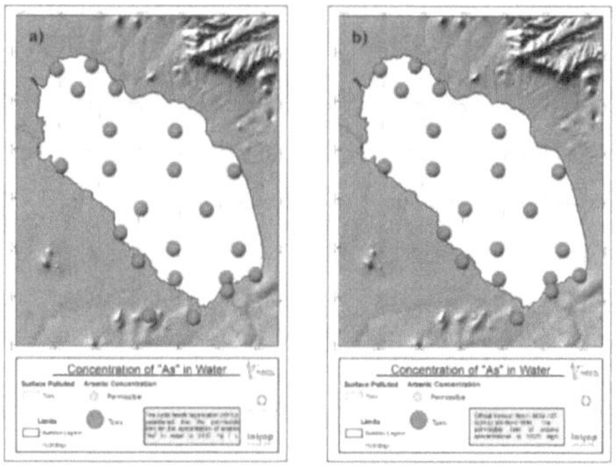

Figure 2. Spatial distribution of As in Bustillos Lagoon in Chihuahua, Mexico.

4 CONCLUSIONS AND RECOMMENDATIONS

Arsenic contamination in Bustillos Lagoon represents a risk exposure for aquatic life and humans due the irrigation use for crops. All the water samples exceeded the WHO guidelines. It is highly recommend to continue monitoring water and natural and anthropegenic sources of As and find a solution like bioremediation or other treatment to use the Bustillo's water safely.

ACKNOWLEDGEMENTS

This study was financially supported by The Fundación Gonzalo Rio Arronte I.A.P. "Manejo Integrado de la Laguna de Bustillos en Chihuhua (A333)".

REFERENCES

Barrett P.M., Hull E.A., Burkart K., Hargrave O., McLean J., Taylor V.F., Jackson B.P., Gawel J.E. & Neumann R. B. 2019. Contrasting arsenic cycling in strongly and weakly stratified contaminated lakes: evidence for temperature control on sediment–water arsenic fluxes. *Limnol. Oceanogr.* 64(3): 1333–1346.

Benavides A., Moreno M., Sosa M., Puga S., Alcalá J. & Quintana C. 2008. Evaluación de la calidad del agua en las principales lagunas del estado de chihuahua. *Revista Latinoamericana de Recursos Naturales* 4(2): 84–88.

Bissen M. & Frimmel F.H. 2003. Arsenic: a review. part i: occurrence, toxicity, speciation, mobility. *Acta Hydrochim. Hydrobiol.* 31: 9–18.

Ochoa-Rivero, J.M., Reyes-Fierro, A.V., Peralta-Pérez, M., Rosario, D., Zavala-Díaz De La Serna, F.J., Ballinas-Casarrubias, L., Salmeron, I., Rubio-Arias, H. & Rocha-Gutiérrez, B.A. 2017. Levels and distribution of pollutants in the waters of an aquatic ecosystem in northern mexico. *Int. J. Environ. Res. Public Health* 14(5): 456.

Rubio A. H., Saucedo R. A., Lara C. R., Wood K., & Jiménez J. 2005. Water quality in the laguna de Bustillos of Chihuahua, México. *WIT Transactions on Ecology And The Environment.* 80.

WHO. 2011. *Guidelines for Drinking Water Quality.* Volume 4. World Health Organisation, Geneva, pp 315–318.

Arsenic in the Environment: Bridging Science to Practice for Sustainable Development – van der Wal, Ahmad, Petrusevski, Weijma, Savic, van der Wens, Beerendonk, Bhattacharya, Bundschuh & Naidu (Eds)

Capillarity as a hydrologic control in the unsaturated zone drives arsenic transport from groundwater to soil solid phase in Matehuala (Mexico)

A. Gómez-Hernández[1], J.L. Hernández-Martínez[1], D. Meza-Figueroa[2], N. Martínez-Villegas[1] & B. SenGupta[3]
[1]*Instituto Potosino de Investigación Científica y Tecnológica, División de Geociencias Aplicadas, San Luis Potosí, México*
[2]*Departamento de Geología, Universidad de Sonora, Sonora, Mexico*
[3]*School of Energy, Geoscience, Infrastructure & Society, Heriot-Watt University. Water Academy, Scotland, UK*

ABSTRACT: Recent studies showed an arsenic concentration decrease from 91.5 to 11.3 mg/L in Matehuala (Mexico) groundwater, a semi-arid environment where calcite and gypsum predominate. Arsenic retention has been widely studied and explained in iron-rich systems, however, little is known about hydrological controls over arsenic attenuation in semi-arid soils rich in calcite and gypsum. In this study, mineralogy, water-rock interaction modeling soil fractionation have been used to explain arsenic transport from groundwater to soil solid phase and arsenic distribution in a low iron oxides content soil. In the alluvial-evaporitic geologic setting of the study area, arsenic wa distributed mainly in soluble fraction and adsorbed in iron oxy-hydroxides and bonded to no-crystalline pyrite Capillarity plays a key role in soluble arsenic retention from groundwater to soil solid phase, which under reducing conditions typical of aquifer systems, would spread due to As(III) high mobility.

1 INTRODUCTION

Natural arsenic (As) retention has been related to the surrounding mineralogy and mainly explained by adsorption phenomena on iron oxy-hydroxides (Zhao *et al.*, 2019), however, in scarce iron oxides environments, such geochemical control over As retention in soil could be insignificant; thus, As transport from soil system to hydrosphere, atmosphere and biosphere becomes an environmental concern issue. Traditionally As transport has been explained by mass transfer phenomena. Dissolution of As bearing minerals due to runoff infiltration reaching groundwater through soil and deeper subsoil layers, the As movement due to plant roots absorption, and As incorporation into the atmosphere by microbial As methylation processes that produce volatile methyl arsines are just some of natural phenomena that accounts for As transport (Khaska *et al.*, 2018). Nevertheless, the importance capillarity as driving force in the As retention in the soil solid phase due to capillary rise of contaminated groundwater from an aquifer has not been enough evaluated where studies are very scarce and only report the load of heavy metals and organic contaminants (Kurt *et al.*, 2016).

2 MATERIALS AND METHODS

2.1 *Study area*

Matehuala sub-basin is located to the NE of San Luis Potosí state in Mexico and flanked to the west by the Sierra de Catorce and the Sierra Azul to the east. A synclinal configuration is delimited by lower and upper Cretaceous limestones, where a large alluvial lacustrine package is transgressed by an evaporitic deposited to the NE. Agriculture and cattle maintenance depend exclusively from seasonal precipitations and groundwater exploitations, which in turn, are used for recreational activities (Martínez-Villegas *et al.*, 2018; Ruíz-Huerta *et al.*, 2017). Arsenic attenuation through natural mechanisms has been reported in the groundwater flow system where As decreased from 91.5 to 11.3 mg/L between two sampling groundwater locations 2 km distanced each other (Martínez-Villegas *et al.*, 2013).

2.2 *Soil sampling and laboratory analysis*

Soil type and soil profile horizons in the study area including the saturated zone were described, 22 soil samples were collected based on depth, the

DOI: 10.1201/9781003317395-19

mineralogy of each soil horizon was determined by powder X-ray diffraction (Rigaku Smartlab, copper anode and scanning range 0–90°). Total elemental concentrations in soil were determined by X-ray fluorescence using the USEPA 6200 method in a Thermoscientific Niton XLt analyzer. The organic and inorganic carbon content was determined in a Total Organic Carbon Shimadzu analyzer, the samples surface morphology was analyzed using a Scanning Electron Microscope (Qanta 250) and elemental analysis by dispersive energy of the most contaminated samples was obtained. Particle size distribution was determined by gravimetric method and soil moisture was measured with a Lutron PMS-714 device. A seven stages sequential chemical extraction procedure for the soil type was operatively designed from the Tessier method (Tessier *et al.*, 1979). Regarding groundwater, the solid-liquid equilibrium was modeled in PHREEQC with water quality (ICP-OES analysis).

3 RESULTS AND DISCUSSION

The results showed a profile of sandy soil and finer particles distributed in aggregates cemented by gypsum, 220 cm depth and four horizons (A, AB, B1 and B2), particle size distributed mainly in sands, followed by silts and finally, clays. Mineralogy results and As in soil solid phase analysis, showed that soil formation in Matehuala is mainly due to physical erosion processes where the hydraulic and eolian finer particles transport from the Sierra de Catorce are responsible for the minerals accumulation such as calcite, dolomite and orthoclase in the sedimentary sub-basin. At 220 cm, >70% of gypsum in soil suggests an erosion process of the evaporitic unit. Iron oxides were undetectable by X-Ray diffraction, suggesting that iron oxides content in soil corresponds mainly to non-crystalline ferrihydrite; higher aging iron minerals (e.g. hematite) are unlikely in carbonated soils.

On the surface, an impacted soil by anthropogenic activities is evidenced due to the presence of environmental concern elements like Pb and As, which are introduced into the soil by dissolution-infiltration processes secondary minerals as anglesite and pharmacolite. Arsenic and iron concentrations correlated in B2 horizon, suggesting that As is adsorbed by iron oxy-hydroxides. Furthermore, in B2 horizon the highest values of moisture were observed, showing that an appreciable As fraction in the soil solid phase would occur in soluble metal arseniates form, and dissolved As were transported in the contaminated groundwater rise by capillarity through soil porosity.

Soil formation processes and the As distribution in the soil showed congruence, thus: in capillary fringe, almost 40% of the total As was distributed mainly in the soluble fraction and the remaining 60% adsorbed in iron oxy-hydroxides and bonded to no-crystalline pyrite. In the intermediate zone, As was observed mainly bonded to non-crystalline pyrite, proving in this way, unweathered minerals transport from Sierra de Catorce and deposition and accumulation in the Matehuala Valley; and finally, in soil surface, traces of stabilization treatments sub-products of ancient metallurgical activities wastes.

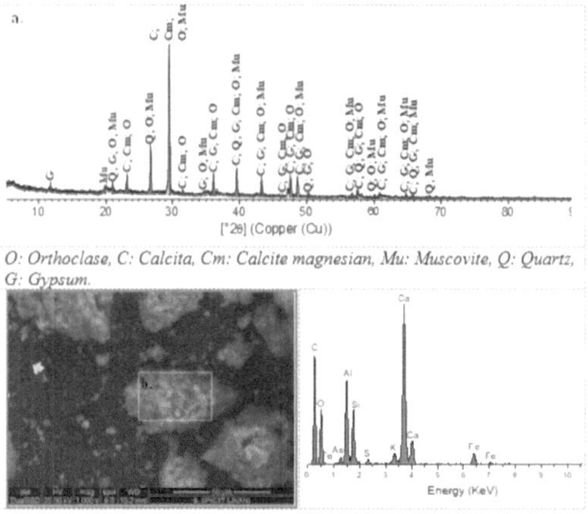

Figure 1. a) XRD of the most arsenic contaminated soil horizon. B) SEM and c) EDS of the most As contaminated soil horizon.

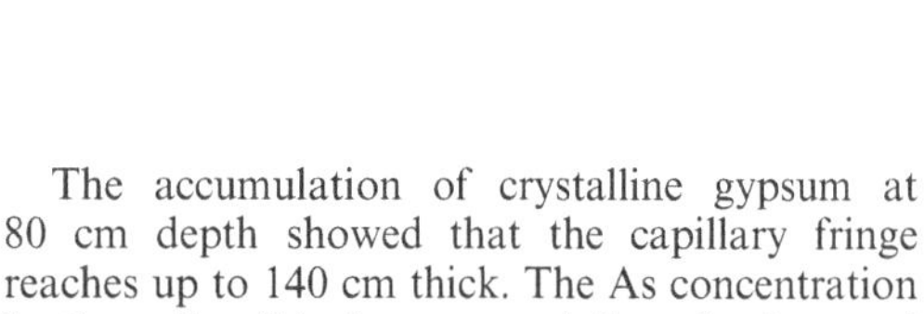

Figure 2. a) Arsenic and iron concentrations in soil. b) Arsenic distributed in soluble soil fraction.

The accumulation of crystalline gypsum at 80 cm depth showed that the capillary fringe reaches up to 140 cm thick. The As concentration in the soil solid phase ranged from background values (25 mg/kg) reported in the study area (AB and B1 horizons) to 359 mg/Kg on the B2 horizon overlaying contaminated aquifer. While gypsum dissolution, induces a high content of sulfates in the groundwater, which in turn, would control pharmacolite precipitation on B2 horizon, explaining in this way the observed As concentration in soil soluble fraction.

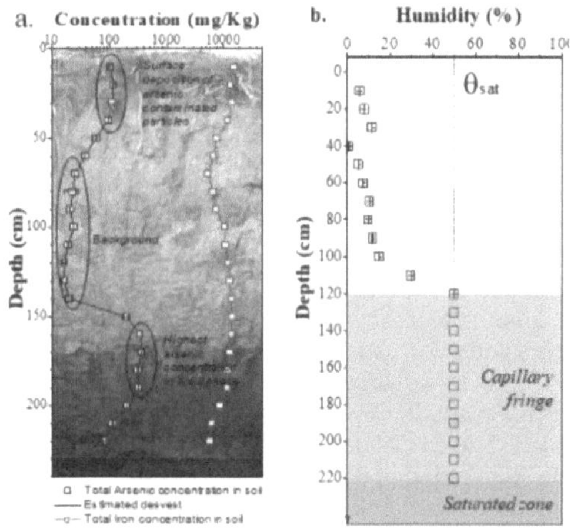

Figure 3. a) Arsenic and iron concentration along soil profile. b) Soil humidity variation to soil depth profile.

4 CONCLUSIONS AND RECOMMENDATIONS

The results of this study contribute to a better understanding of the importance of capillarity as a hydrological control that drives the As transport from liquid phase in a contaminated aquifer to the unsaturated zone of a soil rich in calcite, gypsum and low iron oxi-hydroxides content in a semi-arid environment impacted by mining-metallurgical activities. The importance of the capillary fringe as a pollutant sink requires its inclusion in the estimation of mass balances and in pollutants environmental fate modeling.

ACKNOWLEDGEMENTS

This study was funded by grant NA140182 Newton Advanced Fellowship from the Royal Society. NMV is thankful to CONACyT for Grant No. 7073. APG and JLHM are thankful to CONACYT for 588409 and 166112 graduate and postdoctoral fellowships, respectively.

REFERENCES

Huerta-Diaz M.A. & Morse J.W. 1992. Pyritization of trace metals in anoxic marine sediments. *Geochim. Cosmochim. Acta* 56: 2681–2702.

Khaska M., Le Gal La Salle C., Sassine L., Cary L., Bruguier O. & Verdoux P. 2018. Arsenic and metallic trace elements cycling in the surface water-groundwater-soil continuum down-gradient from a reclaimed mine area: isotopic imprints. *J. Hydrol.* 558: 341–355.

Kurt Z., Mack E.E. & Spain J.C., 2016. Natural attenuation of nonvolatile contaminants in the capillary fringe. *Environ. Sci. Technol.* 50, 10172–10178.

Martínez-Villegas N., Briones-Gallardo R., Ramos-Leal J.A., Avalos-Borja M., Castañón-Sandoval A.D., Razo-Flores E. & Villalobos M., 2013. Arsenic mobility controlled by solid calcium arsenates: a case study in Mexico showcasing a potentially widespread environmental problem. *Environ. Pollut.* 176, 114–122.

Martínez-Villegas N., Hernández A., Meza-Figueroa D. & Sen Gupta B., 2018. Distribution of arsenic and risk assessment of activities on soccer pitches irrigated with arsenic contaminated water. *Int. J. Environ. Res. Public Health* 15: 1060.

Ruíz-Huerta E.A., de la Garza Varela A., Gómez-Bernal J.M., Castillo F., Avalos-Borja M., SenGupta B. & Martínez-Villegas N. 2017. Arsenic contamination in irrigation water, agricultural soil and maize crop from an abandoned smelter site in Matehuala, Mexico. *J. Hazard. Mater.* 339: 330–339.

Schwertmann U. 1990. Some Properties of Soil and Synthetic Iron Oxides. In: M.F. De Boodt, M.H.B. Hayes, A. Herbillon, E.B.A. De Strooper & J.J. Tuck (eds.), *Soil Colloids and Their Associations in Aggregates, NATO ASI Series*. Springer US, Boston, MA, pp. 57–84.

Stumm W., Furrer G., Wieland E. & Zinder B., 1985. *The Effects of Complex-Forming Ligands on the Dissolution of Oxides and Aluminosilicates*, in: Drever, J.I. (Ed.), *The Chemistry of Weathering*. Springer Netherlands, Dordrecht, pp. 55–74.

Tessier A., Campbell P.G.C. & Bisson M., 1979. Sequential extraction procedure for the speciation of particulate trace metals. *Anal. Chem.* 51, 844–851.

Zhao C., Yang Jun, Zheng Y., Yang Junxing, Guo G., Wang J. & Chen T., 2019. Effects of environmental governance in mining areas: the trend of arsenic concentration in the environmental media of a typical mining area in 25 years. *Chemosphere* 235, 849–857.

Arsenic in the Environment: Bridging Science to Practice for Sustainable Development – van der Wal, Ahmad, Petrusevski, Weijma, Savic, van der Wens, Beerendonk, Bhattacharya, Bundschuh & Naidu (Eds)
© 2023 the Author(s), ISBN 978-1-032-32928-4

Arsenic mobility in hydrogeologic system of Bolivian Altiplano: Status and comparison of the Lower Katari and Southern Poopó Basins

I. Quino Lima[1,2], M. Ormachea Muñoz[2], O.E. Ramos Ramos[2], J. Quintanilla[2], J.P. Maity[3,4], A. Ahmad[1,5] & P. Bhattacharya[1,6]
[1]*KTH-International Groundwater Arsenic Research Group, Department of Sustainable Development, Environmental Science and Engineering, KTH Royal Institute of Technology, Stockholm, Sweden*
[2]*Laboratorio de Hidroquímica, Instituto de Investigaciones Químicas, Universidad Mayor de San Andrés, La Paz, Bolivia*
[3]*Department of Earth and Environmental Sciences, National Chung Cheng University, Min-Hsiung, Chiayi County, Taiwan*
[4]*School of Applied Science, KIIT University, Bhubaneswar, India*
[5]*SIBELCO Ankerpoort NV, Maastricht, The Netherlands*
[6]*KWR Watercycle Research Institute, Nieuwegein, The Netherlands*

ABSTRACT: Water resource in LKB (Lower Katari Basin) and SPB (Southern Poopó Basin) [TDPS (Lake Titicaca – Desaguadero River – Lake Poopó – Salt Flat of Coipasa) system] the southeastern part of Lake Titicaca and Lake Poopó in Bolivian Altiplano are contaminated with high concentrations of arsenic (As) (>10 µg/L) compared to the WHO and NB-512 guideline. Studied regions are characterized by a semiarid climate, slow hydrological flow, with geological formations of volcanic origin including brines and mineral deposits. The investigation focused on evaluating the present status and comparison of the geochemical processes of As in relation to the sources and mobilization in groundwater (GW) in LKB and SPB. Groundwater (GW), surface water (SW) and sediment samples were collected from both basins. The elevated concentration of As (LKB: 0.8–288 µg/L and SPB: 2.6–207 µg/L), boron (B) (LKB: 96–2473 µg/L and SPB: 507–4359 µg/L), manganese (Mn) (LKB: 0.6–7259 µg/L) and salinity (LKB: 125–11740 µS/cm) were observed in water compared to the WHO limit and is a serious concern about the GW quality for human consumption. The mineralization of GW is governing by the dissolution and exchange process of bases and anthropogenic generated solids and liquids effects the surface water quality in LKB as well as river bank communities. The spatial distribution of As was attributed to the geology of both the basins and the heterogeneously distributed evaporites in the sediments. The concentrations of As are found in alluvial sediments in northern region of LKB and "PACK belt" (an approximately 25 km long belt stretching along the southern shores of the Lake Poopó, between the villages of Pampa Aullagas and Condo K) in SPB, where iron (Fe) and aluminum (Al) oxides as well as hydroxides are the most predominant mineral phases as potential sorbents of As.

1 INTRODUCTION

The elevated concentrations of arsenic (As) are found in groundwater (GW) worldwide, especially in Argentina, Chile, Mexico, China, USA, Taiwan and Hungary; as well as in the state of West Bengal in India, Bangladesh and Vietnam (Lima *et al.*, 2021). The geogenic and anthropogenic As causes the toxicity to biotic life. The contamination is due to the release of As from sulphide-rich minerals formed due to extensive magmatic and hydrothermal activities, and the extensive volcanism along the Andean mountains (Lima *et al.*, 2021).

The present study investigates the geochemical processes of As-mobilization and other trace elements (TEs), geochemical characterization of GW, along with surface and subsurface water quality in the Lower Katari Basin (LKB) and Southern Poopó Basin (SPB), in the Bolivian Altiplano (Figure 1).

Figure 1. The TDPS system (Titicaca – Desaguadero – Poopó – Salt flat of Coipasa) in the Bolivian Altiplano. (a), Lower Katari Basin (b) and Southern Poopó Basin (c).

2 MATERIALS AND METHODS

2.1 *Sampling and analysis*

Groundwater, surface water and sediment (core sediment/river bed sediment) were sampled and analyzed during dry season from nineteen (LKB) and two (SPB) monitoring wells (piezometers) during August 2017 and November 2009 respectively. Major anions, cations, trace elements (Si, As, Al, B, Cd, Cu, Fe, Li, Mn, Pb, Sr, Cr, Ba and Zn) were analyzed was carried out in the ground- and surface water samples. The sequential extraction of 54 sediment core samples of the LKB using the BCR protocol (Quino Lima *et al.*, 2021).

3 RESULTS AND INTERPRETATIONS

The southeast of Lake Titicaca (LKB) and the south of Lake Poopó (SPB) are regions with a lack of good quality water. The elevated concentration of As (LKB: 0.8–288 µg/L and SPB: 2.6–207 µg/L), boron (B) (LKB: 96–2473 µg/L and SPB: 507–4359 µg/L), manganese (Mn) (LKB: 0.6–7259 µg/L) and salinity (LKB: 125–11740 µS/cm) is serious concern about the GW quality for human consumption. The GW mineralization in both the basins is controlled by dissolution of evaporitic/carbonate minerals, surface and by ion exchange processes.

Alluvial sediments are enriched with As in the northern region of LKB and PACK belt in SPB control the distribution of As in groundwater in the two basins (Figure 2a-d). The slow movement of GW facilitates the As release, probably due to the low surface slope and the large area of fine-grained sediments. Sequential extraction of sediments and geochemical modeling (SI) indicate that As is associated with Fe and Al oxides and hydroxides, which are probably the most important As adsorbents in sediments in the central and southern regions of both basins. Low As mobility in the central and southern regions of both basins is due to $CaCO_3$ as an important adsorbent mineral.

The spatial variability of As concentrations is controlled by the presence of As rich S minerals in SPB and by evaporites that are unevenly distributed in the sediments. The regions with volcanic formations in both basins, despite having lower mean concentrations of As in GW (LKB: 22 µg/L; SPB: 39 µg/L), remain detrimental to health. Another similarity between the two basins lies in the presence of As, B and Li both in SW and GW.

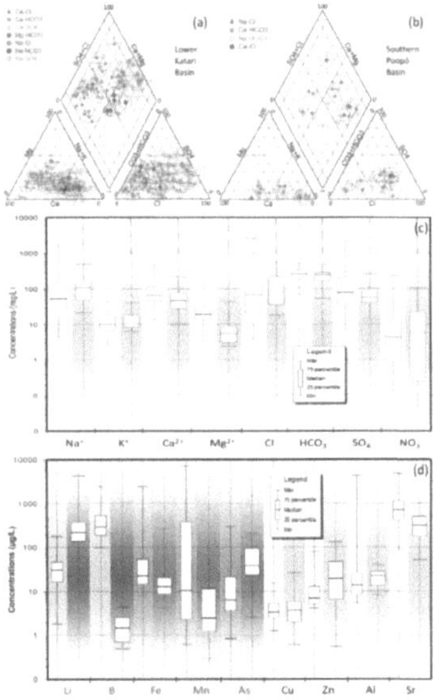

Figure 2. Piper plot of groundwater in a) LKB; and b) SPB (b), Box and Whisker plots for c) set of major ions in groundwater in LKB and SPB (right side) and d) trace elements in groundwater in LKB and SPB (right side). Redox sensitive (red color) and evaporative (blue color).

4 CONCLUSION

The water resources (GW and SW) are affected by high levels of As, B and salinity in both LKB and SPB, Mn and NO_3 in LKB, and Li in SPB at concentration levels higher than the WHO guidelines for drinking water. The spatial distribution of As is related to the geology and the heterogeneity in the occurrence of evaporites in the sediments. The concentrations of As in alluvial sediments in northern region of LKB and "PACK belt" in SPB, Fe and Al oxides and hydroxides are the predominant mineral phases as potential sorbents of As. Arsenic in the lowlands could be mobilized in GW due to increased salinity, increased residence time and other evaporation processes. Therefore, future research should focus on exploring the potential for deep aquifers to provide safe drinking water for both regions.

ACKNOWLEDGEMENTS

We gratefully acknowledge the financial support through Sida Contribution 75000553 for this study.

REFERENCE

Quino Lima I., Ormachea Muñoz M., Ramos O.E.R., Quintanilla Aguirre J., Maity J.P., Ahmad A. & Bhattacharya P. 2021. Hydrogeochemical contrasts in the shallow aquifer systems of the Lower Katari Basin and Southern Poopó Basin, Bolivian Altiplano. *J. South Amer. Ear. Sci.* 105: 102914.

Arsenic in the Environment: Bridging Science to Practice for Sustainable Development – van der Wal, Ahmad, Petrusevski, Weijma, Savic, van der Wens, Beerendonk, Bhattacharya, Bundschuh & Naidu (Eds)
© 2023 the Author(s), ISBN 978-1-032-32928-4

Factors controlling varying arsenic concentration in the Copiapó River, Atacama Region, Chile

K. Bieger[1] & M.A. Alam[2]
[1]*Departamento de Química y Biología, Facultad de Ciencias Naturales, Universidad de Atacama, Copiapó, Región de Atacama, Chile*
[2]*Departamento de Geología, Facultad de Ingeniería, Universidad de Atacama, Copiapó, Región de Atacama, Chile*

ABSTRACT: For the most part of the Copiapó River (between 20–100 km from the headwaters, corresponding to Lautaro Reservoir and Piedra Colgada sector respectively), Arsenic content is below the Chilean norm for potable water of 0.01 mg/L. However, in its catchment area located in the Andean Cordillera, some of the tributaries have relatively higher arsenic concentration, of both geogenic and anthropogenic origin, affirmed by the presence of arsenic bearing mineralization and mining activities in the area. A spike in the Arsenic concentration can be observed close to the city of Copiapó (Km. 100 from the headwaters) can be attributed to a local contamination source that not only contributes arsenic, but a wide spectrum of different metals. Downstream, Piedra Colgada onwards, arsenic concentration increases again. This increase however is due to contribution from very old (not dated, but chemically highly evolved) aquifer in that sector that has elevated arsenic concentration.

1 INTRODUCTION

Copiapó, capital of the Atacama Region in northern Chile, is located on the banks of the Copiapó River (Figure 1). Between 1997 and 2015, the river had dried due to the lack of precipitation and partly due to overexploitation. Heavy rains in the catchment area of the river on March 25 in 2015 (Barrett *et al.*, 2016; Bozkurt *et al.*, 2016; Jordan *et al.*, 2019; Rondanelli *et al.*, 2019; Wilcox *et al.*, 2016) not only caused the reappearance of the river, but also led to heavy flood that affected the entire region with contaminated water and mud from mineral-rich higher sectors, besides causing several casualties and substantial economic damage. Likely, potential arsenic (As) content of the dust caused by dried mud was one of the major concerns for the local population inhabiting several mining towns, including Copiapó. This study involved measurement of As, and other toxic element concentrations in the Copiapó River carried out in two different seasons.

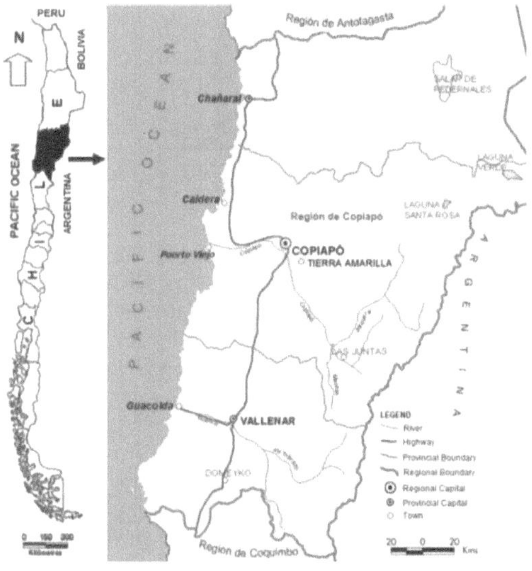

Figure 1. Study area, showing the Copiapó River, its tributaries, and major towns in the administrative region of Atacama, Chile.

2 MATERIALS AND METHODS

2.1 *Study area*

The Copiapó River (Figure 1) is located in the Atacama Desert's southern part – one of the driest places on the Earth. Its basin has a surface area of 18,400 km^2 (DGA 2004) and receives rare precipitation. It originates at the Jonquera and the Pulido rivers' confluence at La Junta, and is joined by the Manflas River at about 2.5 km downstream. It runs for about 160 km westwards before flowing into the Pacific Ocean close to the old port

DOI: 10.1201/9781003317395-21

town of Puerto Viejo. It initially flows northwestward through Quebrada Paipote (Paipote Ravine) for about 90 km; however, it flows roughly westward after passing through the regional capital Copiapó.

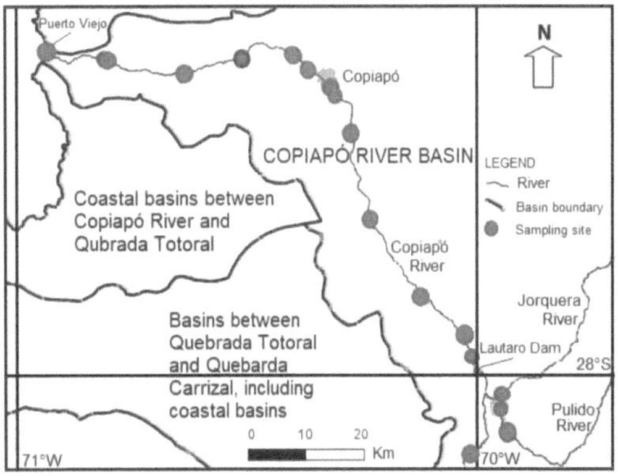

Figure 2. Sampling sites along the course of the Copiapó River. Also shown are the official hydrographic basin boundaries first established in 1978 (DGA 1978; 2013; 2014; SUBDERE 2013).

2.2 *Mud and dust sampling and analysis*

Samples of the mud spread throughout the city of Copiapó following the flood of the Copiapó River, and airborne dust of the dried mud was analyzed using X-ray fluorescence and atomic absorption techniques Universidad de Atacama. X-ray fluorescence analyses (XRF) were conducted in El Salvador in a private company, while Atomic Absorption Spectroscopy (AAS) analyses were carried out at the Pontificia Universidad Católica de Chile, Santiago.

2.3 *Water sampling and analysis*

Water samples were collected, and in situ measurements of physicochemical parameters were done during two sampling campaigns in October 2017 and June 2018. Water samples were taken from the origin of the river east of Laturao Dam at La Junta to Puerto Viejo. It falls into the Pacific every 10 or 20 km, depending upon the accessibility. It is important to mention that the first two data points belong to the Copiapó River's tributaries, and different tributaries (Manflas, Jorquera) were sampled in the two sampling campaigns.

Samples were analyzed using ICP-MS at Activation Geological Services, Coquimbo, for elemental chemistry, throughout the river course (Table 1).

Table 1. Variations in the concentration of arsenic along the course of the Copiapó River during the two sampling periods.

Sampled on 10/25/2017		Sampled on 06/01/2018	
Km*	As	Km*	As
0	2.3	0	13
10	15	12	4.1
20	11	20	9.1
30	7.9	38	6.5
60	7.1	60	6.1
80	6.7	80	6.3
95	7.3	95	7.4
100	7.6	100	8.8
115	8.9	115	7.1
130	12	130	8.6
150	12	150	9.6
160	5.5	160	11
170	3.6	170	8.3

Notes
*Distance from headwaters. As: Arsenic concentration in μg/L.

3 RESULTS AND DISCUSSION

3.1 *Arsenic concentration in mud and dust*

Arsenic concentration measured in the mud and dust samples using X-ray fluorescence was well below 100 mg/L. The results were corroborated further by atomic absorption analysis of the samples, which gave As concentration values of around 5 mg/L in all the cases. The difference in the values obtained using these two methods could be attributed to the low solubility of As-bearing minerals in the water. Although Chile does not have a norm establishing a permissible limit for toxic elements, including As, in sediments and soil, the obtained values for mud and soil are within the established legal limits in many countries (Teaf *et al.*, 2010 and references therein).

3.2 *Arsenic contamination in Copiapó River*

As concentration variation through the course of the river (Figure 3) shows that it is below the permissible limit for As in potable water of 10 μg/L in Chile (NCh 409 2006) in most of its course, specifically between 20 and 100 km, corresponding to Lautaro Dam and Piedra Colgada sectors respectively. However, in the headwaters section, its two tributaries show As concentration of up to 13 and 15 μg/L of both geogenic and anthropogenic origin, affirmed by the presence of arsenic bearing mineralization and mining activities in the area.

A spike observed close to the city of Copiapó (Km. 100) can be attributed to a local contamination source that not only contributes As, but a wide

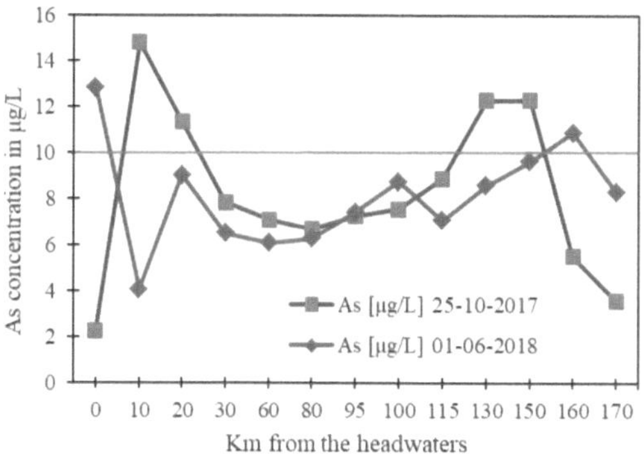

Figure 3. As concentration variation along the Copiapó River.

spectrum of different metals. Downstream, Piedra Colgada onwards, As concentration exceeds the permissible limit of 10 µg/L. This is due to the contribution from very old (not dated, but chemically highly evolved) aquifers in that sector that feed the river.

4 CONCLUSIONS AND RECOMMENDATIONS

Water quality monitored throughout the Copiapó River course shows that in most of the river sections, As concentration was below the Chilean guideline for drinking water of 10 µg/L (NCh 2005). However, in the upper section of the catchment within the Andean Cordillera, its tributaries have relatively higher As concentrations (>10 µg/L), of geogenic origin, affirmed by the presence of As-bearing mineralization in those areas. On the other hand, the aquifers feeding the river contribute As to it in the downstream area. It can be concluded that As contamination of the Copiapó River is mostly of geogenic origin.

Although As concentration in the mud and dust of the Copiapó city is not at a hazardous level, it needs to be monitored following flooding events in the future. Water quality of the Copiapó River should be monitored continuously at closer intervals throughout the river for toxic elements, including As to ascertain the point sources in order.

ACKNOWLEDGEMENTS

The authors thank the Universidad de Atacama for administrative and technical support to carry out this work. Klaus Bieger would like to thank the Faculty of Natural Sciences of the Universidad de Atacama for financial support for this project (Incuba 2018; 2019). A special thank is due to the undergraduate students of the Hydrochemistry course, who participated in some sampling and analysis sessions, Ricardo Santelices for help in XRF analysis, and Ady Giordano for the AAS analysis.

REFERENCES

Barrett B.S., Campos D.A., Veloso J.V. & Rondanelli R. 2016. Extreme temperature and precipitation events in March 2015 in central and Northern Chile. *J. Geophys. Res. Atmos.* 121: 4563–4580.

Bozkurt D., Rondanelli R., Garreaud R. & Arriagada A. 2016. Impact ofwarmer eastern tropical pacific SST on the March 2015 Atacama floods. *Monthly Weather Review* 144(11): 4441–4460.

DGA 1978. Clasificación de Cuencas Hidrográficas de Chile. *Dirección General de Aguas, Ministerio de Obras Públicas, Gobiernno de Chile.*

DGA 2013. Análisis Crítico de la Definición de Cuencas Del Banco Nacional de Aguas. *Dirección General de Aguas, Ministerio de Obras Públicas, Gobiernno de Chile.*

DGA 2014. Inventario de Cuencas, Subcuencas Y Subsubcuencas de Chile. *General de Aguas, Ministerio de Obras Públicas, Gobiernno de Chile.* DOI: 10.1002/2016JD024835

Jordan T.E., Herrera L.C., Godfrey L.V., Colucci S.J., Gamboa P.C., Urrutia M. J., González L. G. & Paul J.F., 2018. Isotopic characteristics and paleoclimate implications of the extreme precipitation event of March 2015 in Northern Chile. *Andean Geol.* 46(1): 1–31.

NCh 2005. Agua Potable – Parte 1. Requisitos. NCh 409/1 Of84. Instituto Nacional de Normalización, Santiago, Chile.

Rondanelli R., Hatchett B., Rutllant J., Bozkurt D. & Garreaud R. 2019. Strongest MJO on record triggers extreme Atacama rainfall and warmth in Antarctica. *Geophys. Res. Lett.* 46: 3482–3491.

Subdere 2013. *Guía Análisis Y Zonificación de Cuencas Hidrográficas Para el Ordenamiento Territorial.* Subsecretaría de Desarrollo Regional y Administrativo, Ministerio del Interior y Seguridad Pública, Gobierno de Chile. ISBN: 978-956-8468-42-2

Teaf C.M., Covert D.J., Teaf P.A., Page E. & Starks M.J. 2010. Arsenic Cleanup Criteria for Soils in the US and Abroad: Comparing Guidelines and Understanding Inconsistencies. *Proceedings of the Annual International Conference on Soils, Sediments, Water and Energy*, Vol. 15, Article 10.

Wilcox A.C., Escauriaza C., Agredano R., Mignot E., Zuazo V., Otárola S., Castro L., Gironás J., Cienfuegos R. & Mao L. 2016. An integrated analysis of the March 2015 Atacama floods. *Geophys. Res. Lett.* 43: 8035–8043.

Arsenic in the Environment: Bridging Science to Practice for Sustainable Development –
van der Wal, Ahmad, Petrusevski, Weijma, Savic, van der Wens, Beerendonk, Bhattacharya, Bundschuh & Naidu (Eds)
 ISBN 978-1-032-32928-4

Silcrete formations in black saline alkaline lakes of Pantanal da Nhecolândia. A natural mechanism for arsenic trapping

A. Hechavarría-Hernández[1], A.H. Fostier[1], K. Chacón-Madrid[1] & L. Barbiero[2]
[1]*Instituto de Química, Universidade Estadual de Campinas, Campinas, São Paulo, Brazil*
[2]*Institut de Recherche pour le Développement (IRD), GET, Toulouse, France*

ABSTRACT: Elevated Arsenic water concentrations have been reported in alkaline lakes of Pantanal da Nhecolândia, with extreme values reaching 3 mg L^{-1}. Silcrete occurrence, resulting from a currently forming process, is frequently observed in the beaches of black alkaline lakes. Two different silcrete types have been distinguished, namely light grey (GS) and black (BS) silcretes, with 3 different layers in each sample. The silcrete is formed and degraded on the beach of the lake according to the seasonal water regime. The silicification process alters various aspects of hydro-bio-geochemical functioning of the lakes and may also affect dissolved arsenic concentration. Arsenic contents increase from bottom to top layer, a trend which was less pronounced in BS samples due to high organic matter contents that can impact the arsenic trapping mechanism. About 71 to 76% of arsenic in silcrete samples were associated with silica, suggesting water arsenic immobilization during the silcrete formation. The nanoparticulate material, preserving arsenic within siliceous structures in surface waters, can come from the degradation of silcrete. Thus, the dissolved arsenic would be trapped during the formation of silcrete and released within small siliceous structures during its degradation.

1 INTRODUCTION

Arsenic (As) concentrations, up to 3 mg/L, have been reported in alkaline lakes of Nhecolâdia, a sub-region of the Brazilian Pantanal, the largest wetland in the word. These lakes usually show pH values ranging from 9 to 10.7 and have been grouped into 3 classes according to their biogeochemical functionings, namely "crystalline", "black" and "green" saline lakes. Silcrete formation evidences, can be observed in the beach of black lakes across the region. High pH values favor the solubility of silicon (Si) through dissociation of silicic acid and silcrete horizons are forming and frequently observed in the beach of black lakes across the region (Nash & McLaren 2007). Evaporation also results in high dissolved arsenic contents in the lakes and surrounding soils solution (Richter *et al.*, 2019). The silicification occurs on the sand beach of the lakes, because of silica oversaturation in subsurface water, coming from the lake. Despite these high levels, As is only slightly bioavailable in alkaline lakes (Fostier *et al.*, 2020). Here we study the influence of silcrete formation on the mobility and bioaccessibility of As in black alkaline lakes.

2 MATERIALS AND METHODS

Nhecolândia corresponds to the southern half of the megafan of the Taquari river. Samples were taken in and around a black saline lake (SR06) at São Roque farm, in the center of Nhecolândia (Figure 1). Two types of Silcrete samples were collected in the beach of the lake at the end of the dry season (October 2019), referred as light grey (GS) and black (BS) silcrete (BS). Because the water level was higher than normal, BS sample was partially covered by lake water, while GS was exposed. Water samples were also collected at the border and in the center of the lake.

Figure 1. Study area, black lake (SR06) in São Roque farm, Pantanal da Nhecolândia.

 DOI: 10.1201/9781003317395-22

2.1 *Laboratory analysis*

Water samples were sequentially filtered (0.22 µm) and ultrafiltered (3 kDa) to separate dissolved and colloidal fractions. As and Si concentrations in these samples were determined by inductively coupled plasma mass spectrometry (ICP-MS) and inductively coupled plasma optical emission spectroscopy (ICP-OES) respectively. The pH was measured in the field. The silcrete samples were ground and sieved (100 mesh) and digested using method (EPA 3052). As concentration in these samples were quantified through hydride generation atomic fluorescence spectroscopy (HG-AFS). A fractionation of As in silcrete samples was carry out using a sequential extraction method (Javed *et al.*, 2013). Total organic carbon (TOC) was determined using a TOC analyzer.

3 RESULTS AND DISCUSSION

3.1 *Silcrete analysis*

In silcrete samples, Si was the most abundant element. Three layers can be observed towards the depth, namely an indurated crust, a cemented intermediate horizon with incrustations of organic matter (OM), then a slightly cemented sandy material. The boundaries between these layers were diffuse in the BS samples due to the enrichment in OM by evapoconcentration. In the GS samples, the As concentration increases towards the surface, from GS3 to GS1 (Figure 2A), suggesting a relationship between the As content and the silicification process. This trend was not observed in the BS sample, probably due to the high OM contents (Figure 2B) which alters the As trapping mechanism. Whatever the sample, the sequential extraction shows that 71 to 76% of the As was associated with silica. This result confirms that the silicification process in black lakes plays a role in the trapping of dissolved As.

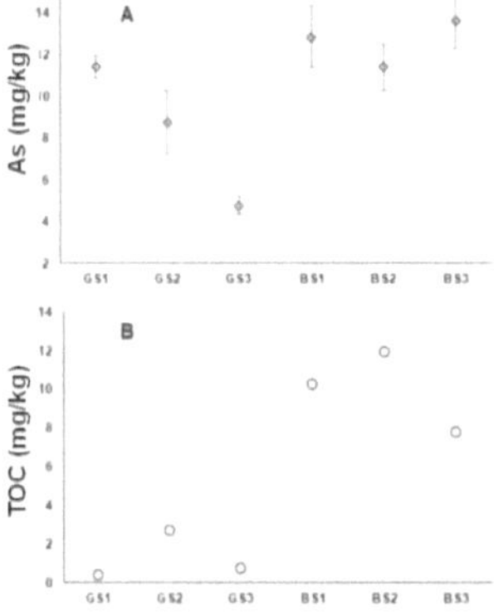

Figure 2. As (A) and TOC (B) concentrations in the three layers identified in silcrete samples, light grey GS and black BS (1 indurated crust, 2 intermediate horizon, 3 sandy material).

3.2 *Water analysis*

The lake water pH and Si content were 9.1 and the 117 ± 1 mg/L, respectively. These conditions favor the transport of silica and its precipitation within the beach by wicking effect (Barbiero *et al.*, 2016). At the same time, the total As content in lake water reached 123.1 µg/L. About 36% of As and 52% of Si are linked to suspended nanoparticles. Thus, arsenic could be trapped in the siliceous structure, just like in the silcrete within the beach, which would limit its mobility and bioaccessibility. This suspended matter could arise from the degradation and dismantling of the silcrete observed close to the lake shore.

4 CONCLUSIONS

The results highlight the link between local silica and arsenic cycles. They seem to indicate that the As cycle results, on the one hand, from a dissolved transfer to the beach in the dry season, and on the other hand, from a return to the lake within siliceous structures in the wet season. The balance between these two processes would control the mobility and bioaccessibility of As in this type of black alkaline lakes.

ACKNOWLEDGEMENTS

For the financial support to Fundação de amparo à pesquisa do estado de São Paulo (FAPESP 2016/14227-5) and Conselho Nacional de Desenvolvimento Científico e Tecnológico (CNPq 142323/ 2017-9). To the Spectrometry, Sample Preparation and Mechanization Group (GEPAM/ UNICAMP) for the analytical support.

REFERENCES

Barbiero L., Berger G., Filho A.T.R., Meunier J.-F., Martins-Silva E.R. & Furian S. 2016. Organic control of dioctahedral and trioctahedral clay formation in an alkaline soil system in the Pantanal wetland of Nhecolândia, Brazil. *PLoS ONE* 11(7): e0159972

Javed M.B., Kachanoski G. & Siddique T. 2013. A modified sequential extraction method for arsenic fractionation in sediment. *Anal. Chim. Acta* 787: 102–110.

Nash D.J. & Sue J.M. 2007. *Geochemical Sediments and Landscapes*. Wiley-Blackwell Publishing Ltd, UK. 488p.

Richter L., Hechavarría Hernández A., Pessôa G.S., Zezzi Arruda M.A., Rezende-Filho A.T. Bartimann de Almeida R., Menezes H.A., Valles V., Barbiero L. & Fostier A.H. 2019. Dissolved arsenic in the upper Paraguay river basin and Pantanal wetlands. *Sci. Total Environ.* 687: 917–28.

 ISBN 978-1-032-32928-4

Arsenic mobilization in aquifer sediments in the southwest USA

S. Bhattacharjee[1], Z. Anwar[1], L. Dunnican[1], F. Ahmed[1], S. Ghasemi[1], C. Rosales[1], T. Lehman[2], K. Millerick[1], K. Rainwater[1] & A. Deonarine[1]
[1]*Department of Civil, Environmental, and Construction Engineering, Texas Tech University, Texas, USA*
[2]*Department of Geoscience, Texas Tech University, Texas, USA*

ABSTRACT: The city of Midland, TX is prone to drought and has resorted to the use of groundwater to meet its water demand. However, due to naturally occurring arsenic in aquifer sediments, arsenic concentrations in the extracted groundwater are often above the US EPA maximum contaminant limit of 10 µg/L. The aim of this study is to identify biogeochemical factors which control the mobilization of arsenic from the sediment to the groundwater. Aquifer sediments were collected at a well site up to a depth of 643 ft. Geological and total elemental analyses using XRF and ICPMS indicated co-occurrence of As and Fe, along with S in certain sediment strata. Arsenic speciation and microbial community analyses will be performed to identify the driving factors of As mobilization.

1 INTRODUCTION

Arsenic (As) is a trace element carcinogen and a drinking water contaminant whose concentration in drinking water is regulated by the US EPA (10 µg/L) (US EPA 2001). Despite this, the U.S. Geological Survey and Centers for Disease Control and Prevention estimates that 21 million people are still exposed to elevated arsenic levels (>10 µg/L) in 25 states of United States (Figure 1) (Ayotte *et al.*, 2017), due to naturally occurring arsenic in the bedrock of groundwater aquifers. Arsenic is a problem in the state of Texas which is heavily reliant on its aquifers to meet the water demand for its 28 million inhabitants (Texas Demographic Center 2019). One such example is in the city of Midland where the depletion of surface reservoirs due to drought conditions led to the use of a groundwater source. Total arsenic concentrations have been measured continuously since the beginning of groundwater extraction and fluctuate from 2 to 36 µg/L (2013–2019 data).

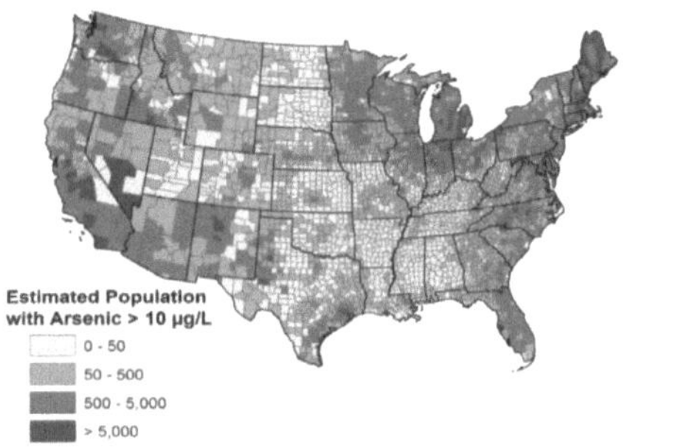

Figure 1. Map showing the estimated As concentrations in well water across the United States. Source: (Ayotte *et al.*, 2017).

This study aims to understand the biogeochemical controls on arsenic mobilization from aquifer sediments to the groundwater at the Midland well site. Specific objectives are to determine: (i) total concentration of arsenic in different sediment strata; (ii) arsenic species distribution (e.g., arsenothiols, arsenite, arsenate) in the aquifer sediments and groundwater; (iii) co-occurrence of arsenic with other elements; and (iv) how microbial activity might influence arsenic mobilization.

2 MATERIALS AND METHODS

An aquifer sediment core was collected from the Midland well field using a sonic drill that provided continuous samples to a depth of 643 ft. A 5 lb aliquot of the sediment material was collected in plastic bags approximately every 10 ft. Samples were transported to the lab and stored at 4°C until analysis. Based on geological examination, the various sediment strata and parent geological formations were identified (e.g., sandstone from the Gatuña Formation, TX). For elemental analysis, a portable X-ray fluorescence (pXRF) gun was used to measure total elemental concentrations for all 64 soil samples. Based on this data, 11 different sample depths were identified for further analysis. These samples were microwave digested with metal-free HNO_3 and H_2O_2 followed by ICP-MS to obtain total elemental concentration data. 16S rRNA gene amplicon surveys (Illumina) were performed on 4 sample depths to characterize bacteria and archaea within soils. Further analyses

 DOI: 10.1201/9781003317395-23

include arsenic speciation using HPLC-ICP-MS for the examination of arsenic species distribution in aquifer sediments and groundwater, as well as microbial community functional gene surveys to examine the possible influence of microorganisms on arsenic mobilization.

3 RESULTS AND DISCUSSION

The studied well field lies within the Monument Draw Trough (a Neogene alluvial aquifer) and penetrates a substantial thickness of Gatuña Formation which is composed of unconsolidated or semi-consolidated quartzose sand. Surface sediment includes the Wickett-Sharvana soil (fine to medium-grained quartzose sand) and wind-blown sand to a depth of 50 ft. From a depth of 50 to 75 ft, is the Mescalero Caliche (sandy pedogenic carbonate). The saturated zone of the aquifer is within the Gatuña Formation (from 75 ft to 566 ft) and underlying Dockum Group (from 566 to 643 ft). The Dockum Group consists of an upper mudstone and a lower sandstone interval.

pXRF analysis revealed the co-occurrence of arsenic (2 to 10 mg/L) and iron (4163 to 24902 mg/L) at multiple depths (Figure 2). The highest arsenic concentrations were observed in clay-rich intervals of the Gatuña Formation and Dockum Group. High sulfur concentrations (6723-125003 mg/L) were also observed at a depth of 581 to 620 ft, which also coincided with peaks in As and Fe concentrations. Our ICP-MS data confirmed the co-occurrence of As and Fe and detected elevated arsenic and iron concentration at similar depths to the pXRF data (Figure 3). Also of note was the co-occurrence of Mn and Fe, and the detection of Se, which has been shown to negate the effect of As on memory and depression (Hall 2012; Johnson *et al.*, 2013).

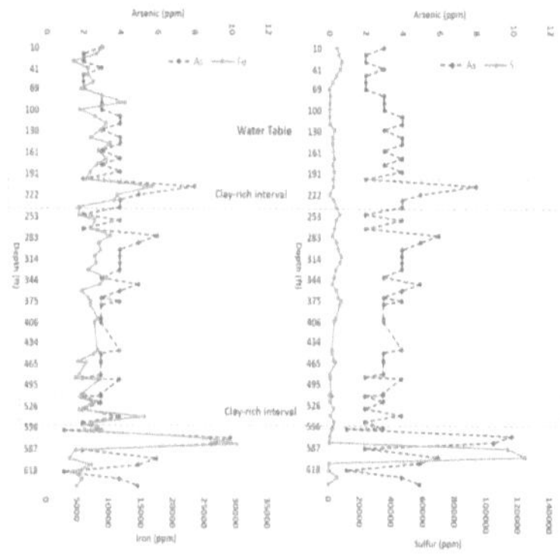

Figure 2. Concentration of As, Fe and S at different depth detected by pXRF.

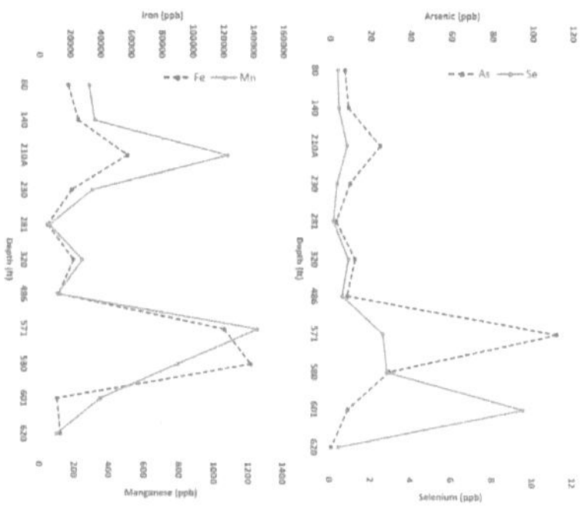

Figure 3. Concentration of As, Fe, Mn and Se at selected depth detected by ICP-MS.

4 CONCLUSIONS AND RECOMMENDATIONS

The studied well field is an important source of drinking water for the city of Midland. High arsenic concentration at different depths emphasises the importance of understanding the dynamics of arsenic mobilization in aquifer sediment. Therefore, it becomes indispensable to have in-depth arsenic analysis to understand the different driving factors behind the observed elevated As concentrations (>10 μg/L) in groundwater.

REFERENCES

Ayotte J.D., Medalie L., Qi S.L., Backer L.C. & Nolan T.B. 2017. Estimating the high arsenic domestic well population in the conterminous United States. *Environ. Sci. Technol.* 51: 12443–12454.

Hall J., Edwards M., Barber R., Johnson L., Gong G. & O'Bryant E. 2012. Higher groundwater selenium exposure is associated with better memory: a project frontier study. *Neuroscience & Medicine* 3: 18–25.

Johnson L.A., Phillips J.A., Mauer C., Edward M., Balldin V.H., Hall R.J., Barber R., Conger T.L., Ho E.J. & O'Bryant E. 2013. The impact of GPX1 on the association of groundwater selenium and depression: a project frontier study. *BMC Psychiatry* 13: 7.

Texas Demographic Center 2019. Estimated population of Texas, Its Counties, and Places. Available at https://demographics.texas.gov/Resources/publications/2019/20191205_PopEstimatesBrief.pdf (Retrieved on 16th February 2020)

US EPA 2001. National Primary Drinking Water Regulations. Available at https://www.epa.gov/ground-water-and-drinking-water/national-primary-drinking-water-regulations (Retrieved on 16th February 2020)

1.2 Biogeochemical processes controlling arsenic mobility and redox transformation

Arsenic in the Environment: Bridging Science to Practice for Sustainable Development –
van der Wal, Ahmad, Petrusevski, Weijma, Savic, van der Wens, Beerendonk, Bhattacharya, Bundschuh & Naidu (Eds)
© 2023 the Author(s), ISBN 978-1-032-32928-4

Optical and molecular signatures of dissolved organic matter in groundwater from the Hetao Basin, China

W. Qiao[1,2] & H.M. Guo[1,2]
[1]*State Key Laboratory of Biogeology and Environmental Geology, China University of Geosciences, Beijing, P.R. China*
[2]*School of Water Resources and Environment, China University of Geosciences, Beijing, P.R. China*

ABSTRACT: Recent studies have deduced that both recalcitrant and labile organic matters were implicated in As mobility in the Hetao Basin, China. We therefore monitored optical characteristics using fluorescence spectroscopy and molecular signatures of DOM by FT-ICR MS analysis in the As-prone aquifers. The results revealed that high-As groundwater was associated with more DOM with humic-like fluorescence, and with a large number of polycyclic aromatics and polyphenols compounds. The possible reason was that DOM with high bioreactivity was preferential consumed in high-As groundwater. Surface water contained much more abundant labile DOM formulas, which could act as electron donors for the reductive dissolution of Fe(III) oxides once introduced into groundwater. The humic-like substances showed remarkably association with polycyclic aromatics and polyphenols, while tryptophan-like substances were similar to phenolic and aliphatic compounds.

1 INTRODUCTION

Both recalcitrant and labile organic matters (OM) were reported to be implicated in As mobility in anoxic aquifers (Guo *et al.*, 2019; Mladenov *et al.*, 2010). The Hetao Basin, China, is a typical inland basin hosting high arsenic (As) groundwater. High As groundwater was coincidentally associated with high contents of humic substances (HS) and labile OM (Guo *et al.*, 2019), which mobilized As by forming ternary complexations of HS, As, and Fe colloids (Li *et al.*, 2019), and facilitating the reductive dissolution of As bearing Fe(III) oxides by labile organic matter (OM) degradation (Guo *et al.*, 2019). Linking the optical and molecular properties of OM with As levels may help in probing the mechanisms of As enrichment. The present study aims at (1) characterizing optical characteristics of dissolved organic matter (DOM) with different As concentrations, (2) investigating molecular signature of DOM and its relation to As mobility, and (3) understanding the linkage between optical characteristics and molecular signature of DOM, and thereafter evaluating their roles on As mobility with respect to chemical nature of DOM.

2 MATERIALS AND METHODS

2.1 *Study area*

The Hetao Basin is located in the middle Inner Mongolia, China, locally stretching north of the Yellow River and south of the Langshan Mountains. Quaternary lacustrine sediments are widely distributed across the basin, being mainly composed of Pleistocene to Holocene sand, sandy silt, and silt clay (Guo *et al.*, 2019). Groundwater flow is nearly stagnant, which favors As accumulating in groundwater. Climatic conditions in the basin is characterized as low precipitation and strong evaporation, which leads to the fact that local residents rely on groundwater resources for various use.

2.2 *Water sampling and analysis*

Eleven groundwater samples and two surface water samples were taken from the northern part of the Hetao Basin in 2017. Groundwater was sampled after continuous pumping for 20 min, and was measured for temperature, pH, EC, ORP, and alkalinity in the field site. Samples for DOC and FT-ICR MS analyses were acidified with HCl to pH < 2.0, while another aliquot was not acidified for optical spectroscopic analysis. Trace elements (e.g., As, Fe, Mn) were analyzed by an ICP-MS (7500CE, Agilent). DOC concentration was analyzed by a Shimadzu TOC-VCPH analyzer. Optical spectroscopic measurements were done using a UV-Visible spectrophotometer (UV8100, LabTech Ltd) and a Fluoromax-4 Spectrofluorometer (Jobin Yvon Horiba). Molecular compositions of DOM were measured by a Bruker Apex ultra FT-ICR MS equipped with a 9.4-T superconducting magnet and an Apollo II electrospray ion source.

2.3 *Data processing*

A set of optical parameters of DOM, including fluorescence index (FI), biological indexes (BIX), humification index (HIX), and $SUVA_{254}$, were obtained to monitor optical characteristics of DOM.

Based on our dataset, excitation emission matrices (EEMs) were used to establish a PARAFAC model. Four components were recognized, including two terrestrial humic-like components (C1, C2), microbial

DOI: 10.1201/9781003317395-24

humic-like components (C3), and tryptophan-like fluorescence (C4).

According to the results of FT-ICR MS analysis, DOM formulas in the van Krevelen diagram were divided into five groups based on the modified aromaticity index (AI_{mod}) and the H/C ratio, including polycyclic aromatics (M1; $AI_{mod} > 0.66$), polyphenols (M2; $0.5 < AI_{mod} < 0.66$), highly unsaturated and phenolic compounds (M3; $AI_{mod} \leq 0.5$ and H/C < 1.5), aliphatic compounds (M4; $1.5 \leq$ H/C < 2), and saturated compounds (M5; $AI_{mod} < 0.5$ and H/C > 2) (Seidel *et al.*, 2014).

3 RESULTS AND DISCUSSION

3.1 *Aqueous arsenic concentration and its relation to fluorescent characteristics of DOM*

Groundwater As concentration ranged from 1 to 850 μg/L, among which six samples had high As levels (>50 μg/L), and five samples had low As levels (<50 μg/L). Surface water samples had As concentration < 10 μg/L.

The PARAFAC results have recognized three recalcitrant DOM components (i.e., C1, C2, and C3) and one labile component (i.e., C4). In general, the fluorescence intensities of C1, C2, and C3 components were higher in high As groundwater, while C4 components exhibited relative higher intensities in low As groundwater and surface water (Figure 1). It showed that DOM in high As groundwater was more humified than that in low As groundwater. Surface water contained much more DOM with high bioreactivity.

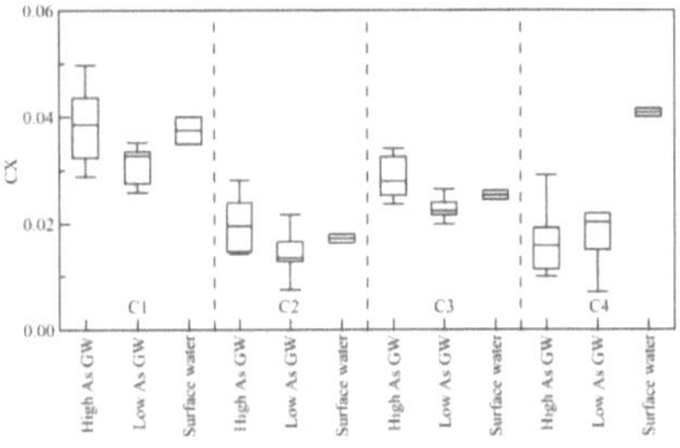

Figure 1. The fluorescence intensities of C1, C2, C3, and C4 components in high As groundwater, low As groundwater, and surface water.

3.2 *Molecular characteristics of DOM*

Generally, high As groundwater contained higher total numbers of formulas (average 4425) than low As groundwater (average 3832). Among the total number of DOM formulas, highly unsaturated and phenolic compounds were the most abundant compounds, which accounted for around 75%, 79%, and 72% in high As groundwater, low As groundwater, and surface water, respectively. Aliphatic compounds, which are considered as labile compounds, were enriched in surface water (17%) than those in low As groundwater (6.4%) and high As groundwater (6%). This indicated that surface water with abundant labile DOM may enhance As mobilization once introduced into groundwater (Harvey *et al.*, 2002).

3.3 *Relating optical properties to molecular characteristics of DOM*

Optical components of DOM recognized using PARAFAC analysis and its molecular compositions identified by FT-ICR MS were linked utilizing significant spearman correlations (Figure 2). The results indicated that the three humic-like components (C1, C2, and C3) had similar molecular-level compositions, and were coincidently associated with polycyclic aromatics and polyphenols (Figure 2). On the contrary, the C4 component was associated with phenolic and aliphatic compounds (Figure 2), which was regarded as labile compounds.

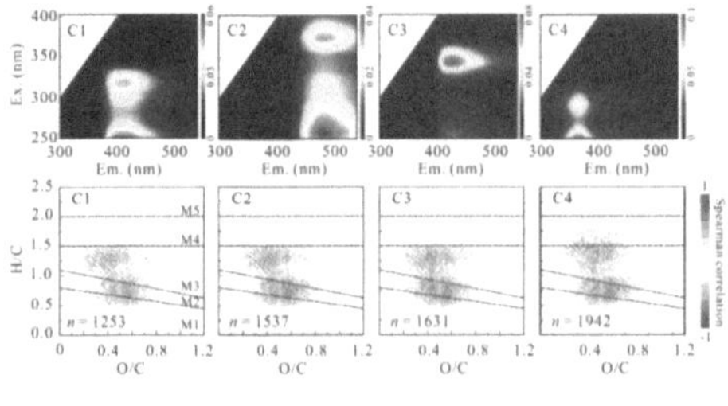

Figure 2. Significant spearman correlations ($p < 0.05$) between optical components (C1, C2, C3, and C4) and molecular compositions. Red points mean positive correlations, while blue points represent negative correlations.

ACKNOWLEDGEMENTS

The study was financially supported by the National Natural Science Foundation of China (grant No. 41825017 and 41672225), the Fundamental Research Funds for the Central Universities (grant No. 2652018197).

REFERENCES

Guo H.M., Li X.M., Xiu W., He W., Cao Y.S., Zhang D. & Wang A. 2019. Controls of organic matter bioreactivity on arsenic mobility in shallow aquifers of the Hetao basin, P.R. China. *J. Hydrol.* 571: 448–459.

Li X. M., Guo H.M., Zheng H., Xiu W., He W. & Ding Q., 2019. Roles of different molecular weights of dissolved organic matter in arsenic enrichment in groundwater: evidences from ultrafiltration and EEM-PARAFAC. *Appl. Geochem.* 104: 124–134.

Mladenov N., Zheng Y., Miller M.P., Nemergut D.R., Legg T., Simone B., Hageman C., Rahman M.M., Ahmed K. M. & McKnight D.M., 2010. Dissolved organic matter sources and consequences for iron and arsenic mobilization in Bangladesh aquifers. *Environ. Sci. Technol.* 44: 123–128.

Seidel M., Beck M., Riedel T., Waska H., Suryaputra I.G.N. A., Schnetger B., Niggemann J., Simon M. & Dittmar T. 2014. Biogeochemistry of dissolved organic matter in an anoxic intertidal Creek Bank. *Geochim. Cosmochim. Acta* 140: 418–434.

 ISBN 978-1-032-32928-4

Arsenic transformation and possible mobilization by indigenous microbes in hot spring environment

J.P. Maity[1], Y.H. Huang[1], G. Day[1], P. Banerjee[1], A.C. Samal[3], A. Ahmad[2], P. Bhattacharya[2] & C.Y. Chen[1]

[1]*Department of Earth and Environmental Sciences, National Chung Cheng University, Chiayi, Taiwan*
[2]*KTH-International Groundwater Arsenic Research Group, Department of Sustainable Development, Environmental Science and Engineering, KTH Royal Institute of Technology, Stockholm, Sweden*
[3]*Department of Environmental Science, University of Kalyani, Nadia, West Bengal, India*

ABSTRACT: Indigenous bacteria play an important role for arsenic (As) mobilization in aqueous environment. In present study, investigated reduction characteristics of arsenate by different indigenous bacterial isolates KTL (GU329910), CL (GU329907) and BL (GU329908), which was isolated from Kuan-Tzu-Ling (KTL), Chung-Lun (CL) and Bao-Lai (BL) hot springs (Taiwan) respectively. Morphological and 16S-rRNA analysis exhibits that the pure culture of isolate KTL (GU329910), CL (GU329907) and BL (GU329908) are similar (99% similarity) to *Bacillus pocheonensis*, *Desulfovibrio psychrotolerans* (sulfate reducing deltaproteobacterium), and *Clostridium sulfidigenes* (mesophilic, proteolytic, thiosulfate- and sulfur-reducing bacterium), respectively. *D. psychrotolerans* (GU329907) and *C. sulfidigenes* (GU329908) reduced SO_4^{2-} to S^{2-} and As(V) to As(III), efficiently. The growth rate and arsenic reduction was exhibited higher in presence of *D. psychrotolerans* (GU329907) compared to *C. sulfidigenes* (GU329908). Thus, the sulfate reducing bacteria contributes in arsenic mobilization process and forms more toxic As(III) species, which affects the biotic life.

1 INTRODUCTION

Arsenic have been found in geothermal systems of different geographical locations (e.g. USA, Turkey, Taiwan etc.), throughout the world. There are more than 100 different types of hot springs, cold spring, and mud spring, are located in Taiwan (Maity *et al.*, 2017).

In general, arsenic mobilization occurs through food chain or due to the oxidation and reduction process, through chemically and microbiologically in environment. Often, the reductive mobilization process plays an important role in arsenic transportation and mobilization at anaerobic environment.

Considering this background, the study was to investigate the arsenic and sulfate reduction characteristics of indigenous bacteria in hot springs in Taiwan, by which arsenic can mobilize to environment.

2 MATERIALS AND METHODS

2.1 *Study area and sampling*

Hot spring water were collected from Kuan-Tzu-Ling (KTL), Chung-Lun (CL) and Bao-Lai (BL) hot spring in Taiwan; and stored under anaerobic condition in laboratory at –20°C, after flushing with N_2. Environmental parameters were estimated in field.

2.2 *Isolation and identification of indigenous bacteria*

Indigenous bacteria were isolated and identified (16S-rRNA with $\geq$ 99% similarity) (CL, BL and KTL) using arsenic rich sulfate reducing media (serum bottle and dilution plate) (Hi-Media, India), in anaerobic environment. By the forward and reverse primer, such as 16S 27F (forward): 5' – AGAGTTTGAT CCTGGCTCAG-3' and 16S 1492R (reverse): 5' – GGTTACCTTGTTACGACTT-3' respectively.

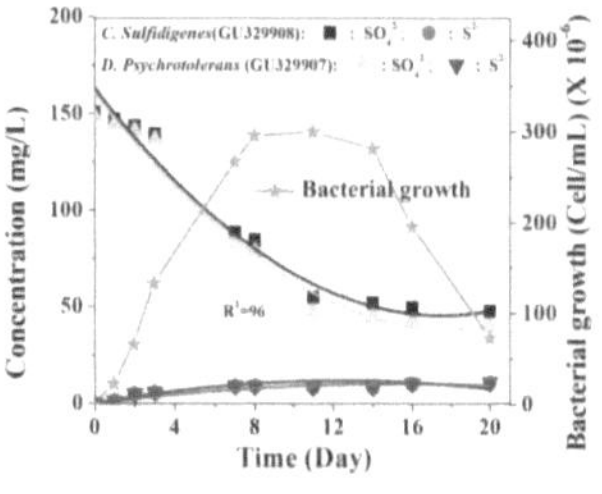

Figure 1. Sulfate reduction by indigenous bacteria.

2.3 *Analysis of arsenic and sulfate reduction*

Arsenic transformation [(0–2000 μg/L) As(V) to As(III)] and sulfate reduction (0–150 mg/L SO_4^{2-}) was performed using two efficient strain,

DOI: 10.1201/9781003317395-25

D. psychrotolerans (GU329907) and *C. sulfidigenes* (GU329908), at 28°C up to 20 days. SO_4^{2-} and S concentrations determined by IC and Microprocessor-Controlled Photometer-PC MultiDirect. Arsenic was measured with the ICP-MS (Hewlett-Packard 4500, Japan) and HPLC-AFS. Accuracy was checked by a certified material (TMDW; Lot # 623609, HPS, USA), within ± 5% of certified values (accuracy: ±5% and precision: ±2%).

3 RESULTS AND DISCUSSION

3.1 *Environmental parameter*

Temperature of the KTL, BL and CL hot springs was 82.2 ± 1.2°C, 62.0 ± 0.7°C and 50.0 ± 0.4°C, respectively. The pH of the KTL CL, and BL water noted 7.01 ± 0.02, 7.41 ± 0.02 and 8.00 ± 0.02 respectively. The Eh of KTL (−396 mV) and CL (−360 mV) water were low compared with BL (−185 mV), suggesting reducing nature of spring water.

3.2 *Identification of indigenous bacteria*

The nucleotides similarities (%) in between KTL (GU329910) and *B. pocheonensis* (FJ009384); *B. drentensis* (FJ009411) and *B. soli strain G8* (FJ009379) were found 99.0, 98.0, and 98.0. Therefore, bacterial isolates from KTL hot spring was assigned to be closely related to *B. pocheonensis*. The nucleotides similarities (%) in between BL(GU329908) and *C. sulfidigenes* (EF199998); *C. thiosulforeducens* (AY024332) and *C. subterminale* (AF241844) were 99.0, 99.0 and 98.0, respectively. Therefore, isolates from BL hot spring can be assigned to *C. sulfidigenes*. The nucleotides similarities (%) between CL (GU329907) and *D. psychrotolerans* (AM418397), *D. acrylicus* (NR025978) and *D. desulfuricans* (FJ655909) were noted as 99.0, 98.0, and 98.0, respectively. Therefore, isolates from BL hot spring were assigned to *D. psychrotolerans*.

3.3 *Sulfate reduction by indigenous bacteria in hot spring environment*

Hot spring water was rich in different types of reducing bacteria. The indigenous strain *D. psychrotolerans* (GU329907) and *C. sulfidigenes* (GU329908) were transformed the 150 mg/L of sulfate to sulfide with increasing time. The transformation rate for both the strain noticed higher order upto 11days; and after, the rate was observed slower (Figure 1). Therefore, the redox sensitive elements may release to hot springs water by the action of reducing bacteria (e.g. SO_4^{2-} transformed into S^{2-} by sulfate-reducing bacteria as *C. sulfidigenes* and *D. psychrotolerans*) and exposed to the environment.

3.4 *Arsenic reduction and possible mobilization by indigenous bacteria in hot spring environment*

The biotransformation of arsenic (As(V) to As (III)) by *D. psychrotolerans* (GU329907) and *C. sulfidigenes* (GU329908) are shown in Figure 2, with different initial concentrations of As(V) (0–2000 μg/L). The transformation results reflecting the relationship between incubation time with the concentration of As(V) and As(III) revealed that *D. psychrotolerans* (GU329907) and *C. sulfidigenes* (GU329908) transformed 2000 μg/L of As (V) to As(III) within 10 and 14 days, respectively. In contrast in between two strain, the *D. psychrotolerans* (GU329907) are more efficient than *C. sulfidigenes* (GU329908). So, the sulfate reducing bacteria transform the As(V) to As(III) in hot spring environment, which help to mobilized the arsenic in surface and subsurface environment.

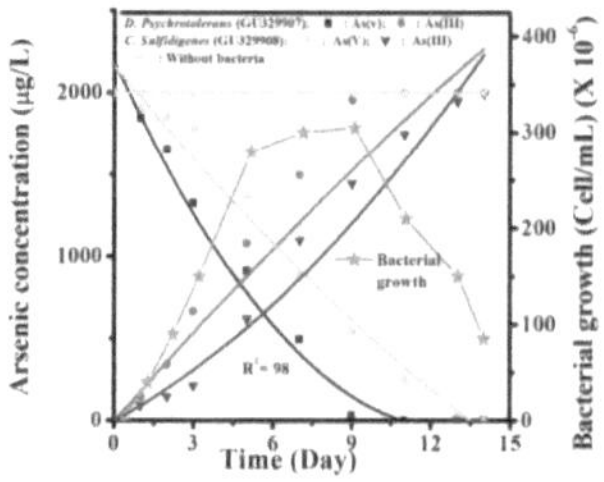

Figure 2. Arsenic reduction by indigenous bacteria.

4 CONCLUSIONS

Hot springs are rich with reducing bacteria. The indigenous sulfate reducing bacteria (*D. psychrotolerans* (GU329907) and *C. sulfidigenes* (GU329908)) transforms the As(V) to As(III) in hot spring environment, which help to mobilized the arsenic in surface and subsurface environment. However, different indigenous sulfate reducing bacteria exhibits the different rate of arsenic transformation from As(V) to As(III).

ACKNOWLEDGEMENTS

We acknowledge the Ministry of Science and Technology (Taiwan) for financial support (MOST 108-2811-M-194-510, MOST 107-2811-M-194-006 and MOST 108-2116-M-194-006).

REFERENCE

Maity J.P., Chen C.Y., Bundschuh J., Bhattacharya P., Mukherjee A. & Chang Y.F. 2017. Hydrogeochemical reconnaissance of arsenic cycling and possible environmental risk in hydrothermal systems of Taiwan. *Groundw. Sustain. Dev.* 5: 1–13.

 ISBN 978-1-032-32928-4

Arsenite and arsenate binding to ferrihydrite organo-mineral coprecipitate: Implications for arsenic mobility and fate in natural environments

H. Du
College of Resources and Environment, Hunan Agricultural University, Changsha, P.R. China

ABSTRACT: Arsenic mobility in soils, sediments and groundwater systems is strongly controlled by adsorption occurring at iron oxide/water interfaces, and the extent of this adsorption may be influenced by the presence of natural organic matter (NOM). We show that the coprecipitation of humic acid (HA) with ferrihydrite (Fh) can significantly reduce the retention of both As(III) and As(V) over a wide pH range (4–11), and with increased organic carbon loading, there is reduced arsenic adsorption. On pure Fh, As(III) is adsorbed to a greater extent than As(V) at pH > 6.5 (the crossover pH), whereas the crossover pH shifts to more acidic pH in the presence of HA, implying that the binding of As(III) is more favorable than As(V) in the presence of NOM. Both As(III) and As(V) are complexed with the ferric hydroxyl functional groups, and no ternary Fh-HA-As complexes are detected. We observed that ~40% of the adsorbed As(III) is oxidized to As(V) on pure Fh, compared to only ~29% of As(III) oxidation on the Fh-HA coprecipitate, indicating that NOM hinders As(III) oxidation on iron (hydr)oxide.

1 INTRODUCTION

In soils, sediments, and aquatic systems, natural organic matter commonly coexists with iron (hydr) oxides (Davis 1982). The acidic oxygen-containing functional groups, mainly carboxylic andphenolic, present in the NOM, can bind strongly with the hydroxyl groups of minerals, resulting in the formation of an organo-mineral composite (Kleber *et al.*, 2015). Natural organic matter thereby can influence the potential transport and fate of arsenic by altering the structure and properties of iron (hydr)oxides, or by direct competition with arsenic for the available adsorption sites. To date, however, very little information is available on the contrasting adsorption of As(III) and As(V) to iron (hydr)oxides in the presence of NOM, and there is also a need to clarify how NOM affects arsenic transformation and speciation at iron (hydr)oxide surfaces.

2 MATERIALS AND METHODS

2.1 *Sample preparation*

A commercial peat humic acid (HA) procured from the Sigma-Aldrich (H16752-2; Shanghai, China) was used as a representative of natural organic matter (NOM). Pure ferrihydrite (Fh) and the two Fh-HA coprecipitates (5 wt% and 15 wt% OC) were prepared by hydrolysis of a Fe^{3+} salt solution with and without HA, to pH ~7.5 with 1 M NaOH (Du *et al.*, 2018).

2.2 *Arsenic adsorption experimental designs*

Arsenic adsorption experiments were conducted in batch systems in 0.01M NaCl electrolyte at room temperature. For adsorption isotherms (pH 7 as a representative), the adsorption solution contained 0.1 g/L adsorbent and 0–0.8 mmol/L As(III) or As (V). Suspensions were immediately shaken for 24 h. All adsorption suspensions were separated by centrifugation into a residual thick paste for analyses by X-ray photoelectron spectroscopy (XPS) and a clear supernate for determination of total arsenic concentration using ICP-OES (Perkin Elmer Optima 8300).

2.3 *X-ray photoelectron spectroscopy experiments*

X-ray photoelectron spectroscopy (XPS) measurements were performed on a KRATOS Axis Ultra X-ray photoelectron spectrometer (Thermo Fisher Scientific, USA) with Al Ka radiation. The adsorption samples were kept in a sealed plastic bag under N_2 atmosphere before XPS measurements. For a high-resolution scan (O, C, Fe and As), the step size was 0.05 eV. All binding energies (BE) were referenced to the C 1s peak at ~284.8 eV. The XPS results were fit using a curve-fitting program XPSPEAK41, with the full width at half-maximum (FWHM) maintained constant for a particular spectrum.

DOI: 10.1201/9781003317395-26

3 RESULTS AND DISCUSSION

3.1 *Macroscopic phenomenon*

When Fh is coprecipitated with HA, the adsorption capacity of the resulting Fh-HA composite for both As(III) and As(V) decreases compared to that for pure Fh. The binding affinities further indicate that both As(III) and As(V) have a lower affinity with the Fh-HA coprecipitate than with pure Fh. Given the reduced adsorption capacity and binding affinity, it can be concluded that HA significantly inhibits arsenic adsorption to Fh surfaces.

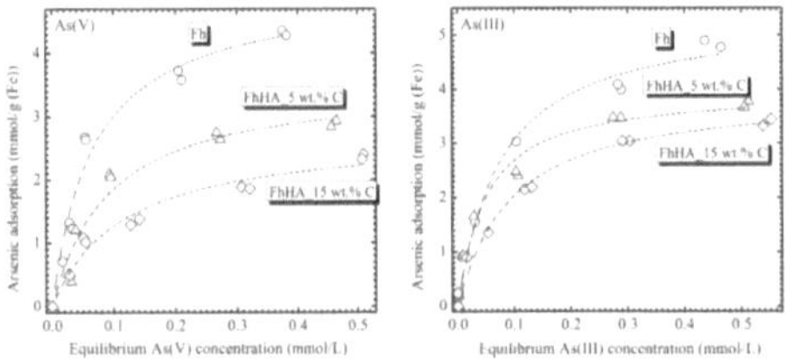

Figure 1. Adsorption isotherms of As(III) and As(V) on ferrihydrite and ferrihydrite-HA coprecipitates at pH 7. Experimental solutions contain 0.1 g/L adsorbent and 0–0.8 mmol/L adsorbate in 0.01M NaCl electrolyte. Lines are Langmuir fit.

3.2 *Molecular binding information*

Component at ~531.2 eV increases from ~29% to ~33% and ~48% after As(III) and As(V) adsorption, respectively (Table 2), which indicates that a proportion of the eOH groups interact with As anions, possibly through ligand exchange (eOeAs), and this new species contributes to an increase in the peak intensity. The surface ferric hydroxyl groups therefore play a key role in the formation of arsenic complexes on Fh surfaces.

Our XPS results show lower peak intensities (29%) for the pentavalent form in the As(III)-adsorbed Fh-HA coprecipitate than that (40%) in the As(III)-adsorbed Fh, suggesting that HA does not promote the oxidation but rather inhibits the oxidation of As(III) to As(V) on iron (hydr)oxide.

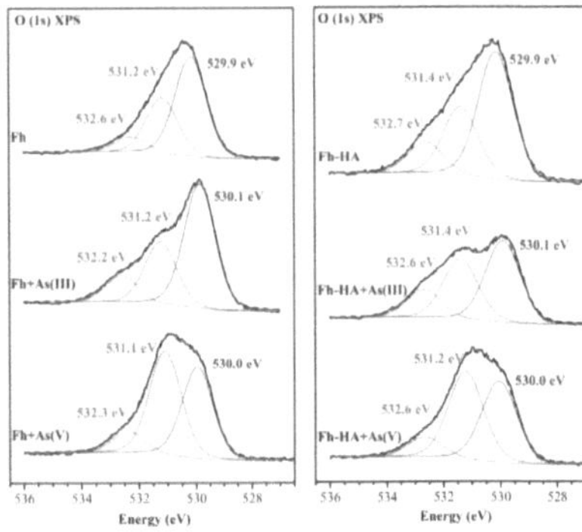

Figure 2. O (1s) XPS spectra of ferrihydrite and ferrihydrite-HA coprecipitate (15 wt% OC) before and after the binding of As(III) and As(V) at pH 7.

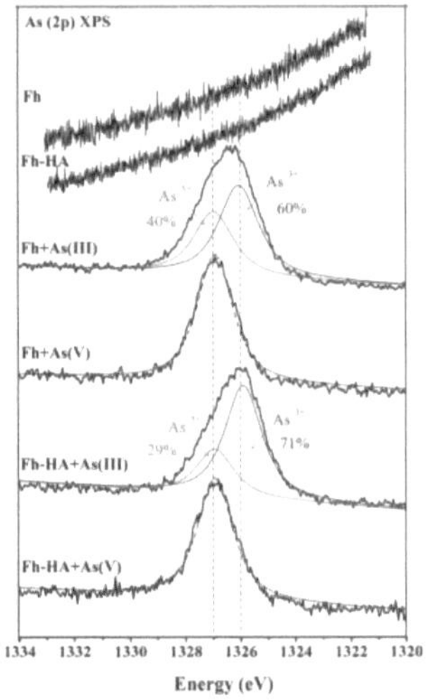

Figure 3. Arsenic (2p) XPS spectra of ferrihydrite and ferrihydrite-HA coprecipitate before and after the binding of As(III) and As(V) at pH 7.

4 CONCLUSIONS AND RECOMMENDATIONS

The presence of NOM decreases both arsenite and arsenate adsorption to ferrihydrite, and hinders As (III) oxidation on iron (hydr)oxide. The results of this study suggest that NOM interacts with arsenic in ways that promote arsenic mobility and especially promote the mobility of arsenate relative to arsenite, which is of great significance for evaluating the migration and bioavailability of arsenic in both natural and contaminated environments.

ACKNOWLEDGEMENTS

We acknowledge the National Natural Science Foundation of China (41907015).

REFERENCES

Davis J.A. 1982. Adsorption of natural dissolved organic matter at the oxide/water interface. *Geochim. Cosmochim. Acta* 46: 2381–2393.

Du H., Peacock C.L., Chen W. & Huang Q. 2018. Binding of Cd by Ferrihydrite organo-mineral composites: Implications for Cd mobility and fate in natural and contaminated environments. *Chemosphere* 207: 404–412.

Kleber M., Eusterhues K., Keiluweit M., Mikutta C., Mikutta R. & Nico P.S. 2015. Mineraleorganic associations: Formation, properties, and relevance in soil environments. *Adv. Agron.* 130: 1–140.

Arsenic in the Environment: Bridging Science to Practice for Sustainable Development –
van der Wal, Ahmad, Petrusevski, Weijma, Savic, van der Wens, Beerendonk, Bhattacharya, Bundschuh & Naidu (Eds)
© 2023 the Author(s), ISBN 978-1-032-32928-4

Responses of bacteria and genes to arsenite under nitrate-reducing conditions in a non-contaminated paddy soil

X. Li[1], S. Li[3], J. Qiao[2] & F. Li[2]
[1]*Guangdong Provincial Key Laboratory of Chemical Pollution and Environmental Safety, School of Environment, SCNU Environmental Research Institute, South China Normal University, Guangzhou, P.R. China*
[2]*Guangdong Key Laboratory of Sugarcane Improvement and Biorefinery, Guangdong Bioengineering Institute (Guangzhou Sugarcane Industry Research Institute), Guangzhou, P.R. China*
[3]*Key Laboratory of Integrated Agro-environmental Pollution Control and Management, Guangdong Institute of Eco-Environmental Science & Technology, Guangzhou, P.R. China*

ABSTRACT: Recent studies demonstrated that microbial arsenite (As(III)) oxidation associated with nitrate (NO_3^-) reduction might be an important process in diminishing arsenic bioavailability and toxicity to rice when paddy soils are contaminated by arsenic. However, the responses of bacterial communities and functional genes to As(III) under nitrate-reducing conditions are poorly understood in a non-contaminated soil. Our results revealed that microbial oxidation of As(III) to As(V) was substantially accelerated by nitrate addition in a paddy soil with low arsenic background content, while nitrate reduction was not affected by As(III) addition. Metagenomic binning revealed that *Pseudogulbenkiania* sp. was the main nitrate-reducing bacteria with *narG*, *nirS* and *norBC* genes, and *Azoarcus* sp. was the predominant As(III)-oxidizing bacteria with As(III) oxidase genes *aioA* and *aioB*. It was also confirmed that the abundance of *Azoarcus* spp. and *aioA* genes were enhanced in the presence of As(III) and nitrate. These findings suggest that non-contaminated paddy soils also harbor arsenic oxidizing bacteria and genes, which could be stimulated in the presence of As(III) and nitrate.

1 INTRODUCTION

As(III) oxidation is of significance to diminish the concentration of As(III) in flooded paddy soils (Khan *et al.*, 2010). NO_3^- application effectively decreased As accumulation, particularly As(III), in rice plants (Wang *et al.*, 2018). Denitrifying bacteria that have genes associated with As(III) oxidation are expected to be enriched during anaerobic As(III) oxidation and NO_3^- reduction in As-contaminated paddy soils. However, it remains poorly understood whether any microbial communities and genes related to As transformation can be stimulated by As(III) under denitrifying conditions in non-contaminated paddy soils. As(V)-resistant bacteria can be isolated even from soils with very low arsenic levels, suggesting that arsenic resistance is not confined to organisms inhabiting arsenic-laden environments (Jackson *et al.*, 2005). We hypothesized that non-contaminated paddy soils might also harbor arsenic oxidizing bacteria and genes. The present study aims to investigate the responses of bacteria and genes to As(III) under nitrate-reducing conditions in a paddy soil with low arsenic background content using metagenomic sequencing and real-time qPCR analysis.

2 MATERIALS AND METHODS

2.1 *Soil sampling and characterization*

The soil samples were collected from 3–5 sampling points of a flooded paddy field in South China Botanical Garden, Guangzhou, China (23°10‘38.26“N, 113°21’ 10.12” E). The soil contained 41.0 g/kg organic matter, 27.9 g/kg total iron, and 3.93 mg/kg total arsenic.

2.2 *Microcosm experiments*

Anaerobic microcosms were set up by adding 1 g of paddy soil (wet weight) into a serum vial containing 100 mL anaerobic, sterile bicarbonate-buffered medium (pH 7.0, 30 mM bicarbonate buffer, 5 mM sodium acetate, 1 mL/L vitamin solution and 1 mL/L trace element solution). Five microcosm treatments were prepared: (i) soil control; (ii) + As(III); (iii) + NO_3^-; (iv) + As(III) + NO_3^-; (v) sterilized soil + As (III) + NO_3^- (soil was sterilized by autoclaving for 20 min). The initial concentrations of $NaAsO_2$ and $NaNO_3$ were 100 μM and 10 mM, respectively. All of the vials were incubated at 30°C under an atmosphere of N_2/CO_2 (80:20, v/v) in the dark without shaking. The experimental treatments were conducted in triplicate.

2.3 *Determination of arsenic and nitrogen species*

For As species, the samples were measured using high performance liquid chromatography-atomic florescence spectroscopy (SA-20, Jitian Inc., Beijing, China). For NO_3^- and NO_2^- quantification, the samples were measured colorimetrically at 540 nm using a continuous flow analyzer (Skalar SAN++, The Netherlands).

DOI: 10.1201/9781003317395-27

2.4 *Metagenome sequencing*

Paired-end sequencing libraries with insert sizes of 350 bp were constructed using the total community genomic DNA samples extracted from the + NO_3^- and + As(III) + NO_3^- microcosms on day 8. Triplicate DNA extractions were performed and pooled for shotgun metagenome library construction. Binning of metagenomic contigs was carried out using CONCOCT software. The completeness/contamination calculation, essential single copy genes identification, and coverage distribution of the obtained genome bins were assessed using CheckM 0.9.7. The metagenomic reads have been deposited in the NCBI SRA under accession number SRP5177929.

2.5 *Quantification of Azoarcus spp. and aioA gene*

The quantitative PCR reactions were performed in triplicate on the CFX 384^{TM} Real-Time PCR Detection System (Bio-Rad Laboratories, USA) with SYBR Green I detection. The 10 μL qPCR reaction contained 5 μL of 2 × IQ^{TM} SYBR® Green Supermix (Bio-Rad), 0.1 μM (*aioA* and 16S rRNA gene) or 0.4 μM (*Azoarcus* spp.) of each primer, and 1 μL of DNA (1–10 ng).

3 RESULTS AND DISCUSSION

3.1 *Arsenic and nitrogen transformation*

More than 90% of the added As(III) was oxidized to dissolved As(V) in the presence of NO_3^- with an As (III) oxidation rate of 0.29 ± 0.09 per day (R = 0.81) in the + As(III) + NO_3^- microcosms. Less than 20% of the added As(III) was oxidized to As(V) at the end of the incubation period (day 8) in the + As(III) microcosms. The concentration of NO_3^- decreased over time to 5.1–5.5 mM in the + NO_3^- and + As(III) + NO_3^- microcosms. Concomitantly, the concentration of NO_2^- increased to 3.0–3.9 mM at the end of incubation in these two microcosms. No NO_3^- reduction was observed in the sterilized soil + As(III) + NO_3^- control during the entire incubation.

3.2 *Genome binning*

A total of 8 high quality draft genomes were successfully reconstructed. Phylogenetic analysis based on the deduced protein sequences showed that Bin18, Bin25, Bin46, Bin48, Bin54, Bin78, and Bin79 clearly clustered with the reference genomes of *Geobacter* sp. M18, Candiadatus *Lloydbacteria*, *Pseudogulbenkiania subflava*, *Paludibacter propionicigenes*, *Caulobacter segnis*, *Flavisolibacter ginsengisoli*, and *Azoarcus tolulyticus*, respectively.

For As(III) oxidation, the abundance of *aioA* and *aioB* genes were determined to be 0.03 and 0.003 FPKM in the + NO_3^- microcosms, while 0.09 and 0.008 FPKM in the + As(III) + NO_3^- microcosms, respectively. The abundance of the arsenic-related genes in each reconstructed bin indicated that the *aioB* gene was derived solely from the *Azoarcus* sp.-like genome (Bin 79) in both the + NO_3^- and + As (III) + NO_3^- microcosms.

For nitrogen transformation, the relative abundance of the denitrification associated genes (particularly *narG*, *nirS* and *norBC*) was higher, compared with the genes associated with DNRA and assimilatory nitrate reduction. The most abundant reads for *narG* (65–75%), *nirS* (71–89%) and *norBC* (76–81%) genes were derived from *Pseudogulbenkiania* sp. (Bin46) in both the + NO_3^- and + As(III) + NO_3^- microcosms.

3.3 *Quantification of Azoarcus spp. and aioA gene*

Copy numbers of the *Azoarcus* spp. increased over time to 6.7 ± 0.9 and 2.5 ± 0.2 (× 10^5 gene copies/g soil) on day 8 in the + As(III) + NO_3^- and + NO_3^- microcosms, respectively, which was an order of magnitude higher than those in the + As(III) and soil control microcosms (0.4–0.6 (× 10^5 gene copies/g soil)). An increase in copy numbers of the *aioA* gene was only observed in the + As(III) + NO_3^- microcosms, which reached the highest level of 5.4 ± 0.5 (× 10^5 gene copies/g soil) on day 8, compared to 0.30–0.33 (× 10^5 gene copies/g soil) in the + As(III), + NO_3^- and soil control microcosms.

4 CONCLUSIONS

Our results suggest that *Pseudogulbenkiania* spp. are mainly responsible for the sequential reduction of NO_3^- to NO_2^-, NO, and N_2O, while *Azoarcus* spp. principally serve as As(III) oxidizer. As previously reported, a high number of *ars* gene clusters was found in the genome of *Azoarcus* sp. CIB, and *Pseudogulbenkiania* spp. can also stimulate iron oxidation coupled to nitrate reduction. Together with our findings, both *Pseudogulbenkiania*- and *Azoarcus*-related spp. have high adaptability and a variety of physiological and ecological attributes for arsenic and nitrogen cycling, which could potentially be stimulated to tackle arsenic-contaminated sites.

ACKNOWLEDGEMENTS

This work was supported by National Natural Science Foundation of China (41877043 and 41907130) and Guangdong Natural Science Funds for Distinguished Young Scholars (2017A0303 06010).

REFERENCES

Jackson C.R., Dugas S.L. & Harrison K.G. 2005. Enumeration and characterization of arsenate-resistant bacteria in arsenic free soils. *Soil Biol. Biochem.* 37, 2319–2322.

Khan K.A., Stroud J.L., Zhu Y.G. *et al.* 2010. Arsenic bio availability to rice is elevated in bangladeshi paddy soils. *Environ. Sci. Technol.* 44(22): 8515–8521.

Wang X., Liu T., Li F. *et al.* 2018. Effects of simultaneous application of ferrous iron and nitrate on arsenic accumulation in rice grown in contaminated paddy soil. *ACS Earth Space Chem.* 2(2): 103–111.

Arsenic in the Environment: Bridging Science to Practice for Sustainable Development –
van der Wal, Ahmad, Petrusevski, Weijma, Savic, van der Wens, Beerendonk, Bhattacharya, Bundschuh & Naidu (Eds)
 ISBN 978-1-032-32928-4

Linking microbial community composition to arsenic mobilization in the western Hetao Basin: Potential importance of ammonium as an electron donor

W. Xiu[1,2], J. Lloyd[3], H.M. Guo[1,2], W. Dai[2], S. Nixon[3], N.M. Bassil[3], C. Ren[2], C. Zhang[2], T. Ke[2] & D.A. Polya[3]

[1]*State Key Laboratory of Biogeology and Environmental Geology, China University of Geosciences, Beijing, P.R. China*
[2]*School of Water Resources and Environment, China University of Geosciences (Beijing), Beijing, P.R. China*
[3]*Williamson Research Centre for Molecular Environmental Science, School of Earth and Environmental Sciences, The University of Manchester, Manchester, UK*

ABSTRACT: The western Hetao Basin was selected as the study area to address spatially-dependent correlations between microbial community composition and geochemical zonation, approximately along the groundwater flow path from a proximal fan (geochemical-group I: low As, oxidizing), through a transition area (geochemical-group II: moderate As, moderately-reducing) and then to a flat plain (geochemical-group III: high As, reducing). High-throughput Illumina 16S rRNA gene sequencing showed that the microbial community structure in the proximal fan included bacteria affiliated with organic carbon degradation and nitrate-reduction or even nitrate-dependent Fe(II)-oxidation, mainly resulting in As immobilization. In contrast, for the flat plain, high As groundwater contained Fe(III)- and As(V)-reducing bacteria, consistent with current models on As mobilization driven via reductive dissolution of Fe(III)/As(V) mineral assemblages. Interestingly, spearman correlations between hydrogeochemical data and microbial community compositions indicated that ammonium as a possible electron donor induced reduction of Fe oxide minerals, suggesting a wider range of metabolic pathways (including ammonium oxidation coupled with Fe(III) reduction) driving As mobilization in high As groundwater systems.

1 INTRODUCTION

Biogeochemical As cycling is commonly coupled to or affected by the biogeochemical cycling of iron (Fe), carbon (C), nitrogen (N) and sulfur (S) (Guo *et al.*, 2016; Islam *et al.*, 2004; Li *et al.*, 2017; Oremland & Stolz 2003; Zhu *et al.*, 2017). Hydrogeochemical conditions are vital in affecting microbial community structure (Al Lawati *et al.*, 2012; Dhar *et al.*, 2011; Hassan *et al.*, 2015; Li *et al.*, 2017; 2014; Mirza *et al.*, 2014). In western Hetao Basin, strong geochemical-microbiological interactions involving As, Fe, N, C and S in the aquifers were observed by previous hydrogeological, biogeochemical, microbiological and isotopic studies have demonstrated (Guo *et al.*, 2013, 2016; Jia *et al.*, 2014). However, the spatial distribution of microorganisms, especially with respect to geochemical zonation requires further examination. Therefore, the objectives of this study were to: i) characterize the microbial community structure along a representative groundwater flow path; ii) evaluate the relationship between hydrogeochemical conditions and the microbial community composition; and iii) identify the key microorganisms and processes affecting As behavior along the flow path.

2 MATERIALS AND METHODS

2.1 *Study area*

Our study area is located in the western Hetao Basin and extends from an alluvial fan (front of the Langshan Mountains) to a flat plain with an area of approximately 25 km^2. Several metre-thick clay layers occur at around 40 m BLS, dividing shallow and deep groundwater aquifers (Jia *et al.*, 2014). A representative groundwater flow path was selected from the alluvial fan to the flat plain (Figure 1). The hydro-geochemical zonation was established based on the hydrogeological setting, chemical and isotopic characteristics (Jia *et al.*, 2014).

DOI: 10.1201/9781003317395-28

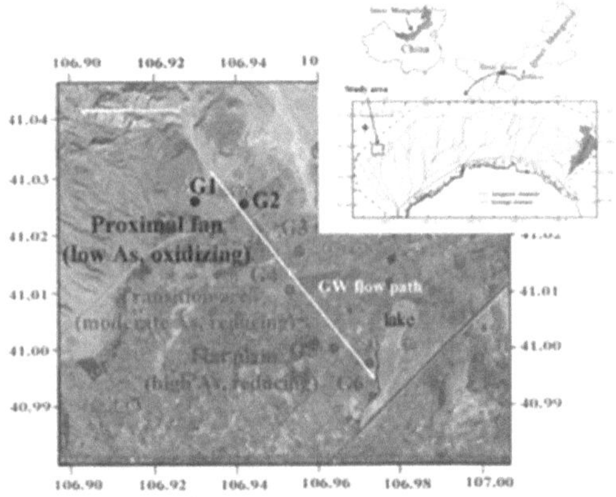

Figure 1. Study area and water sampling locations.

2.2 *Sample collection, measurements and data analysis*

Six groundwater samples (depths between 40 and 80 m BLS; G1 and G2 in the proximal fan, G3 and G4 in the transitional area, and G5 and G6 in the flat plain) were collected along the groundwater flow (Figure 1). Groundwater parameters such as, pH, temperature (T), electrical conductivity (EC), Fe(II), S(-II) and redox potential (Eh) were measured in the field. Major cations and trace elements were determined by ICP-AES and ICP-MS, respectively. Dissolved As species were analyzed by HPLC-HG-AFS. DOC was determined using a Shimadzu TOC-VCPH analyzer.

Chemical samples were filtered through 0.45-μm membrane filters while microbial samples were collected by filtration of 5–10 L of groundwater through 0.2-μm filters (Millipore). The biomass-containing filters were placed in a 50 mL sterile conical centrifuge tube and transported to the laboratory using dry ice, and subsequently frozen at −80°C until analysis within seven days.

DNA was extracted from each sample using the FastDNA® SPIN Kit for Soil (MP Biotechnology, Solon, OH, USA). The 16S rRNA gene was amplified from the genomic DNA using primers 338F and 806R targeting the V3-V4 region (Xu *et al.*, 2016). Raw FASTQ files were demultiplexed, quality-filtered using QIIME v.1.9.1 (Caporaso *et al.*, 2010). Data analysis were conducted using R and related packages.

3 RESULTS AND DISCUSSION

3.1 *Aquatic chemistry*

Along the groundwater flow path from the proximal fan to the flat plain, we observed a decrease of ORP and increases in total dissolved As (TAs), total dissolved Fe (TFe), dissolved Fe(II) and total dissolved Mn (TMn) (Figure 2). Spearman ranking correlations showed that Dissolved As species (TAs, As(III) and As(V)), Fe species (TFe and Fe (II)) and TMn were also correlated positively with each other ($p < 0.034$). Interestingly, ammonium levels showed strong positive relationships with TAs ($p = 0.019$) and As(III) ($p = 0.008$).

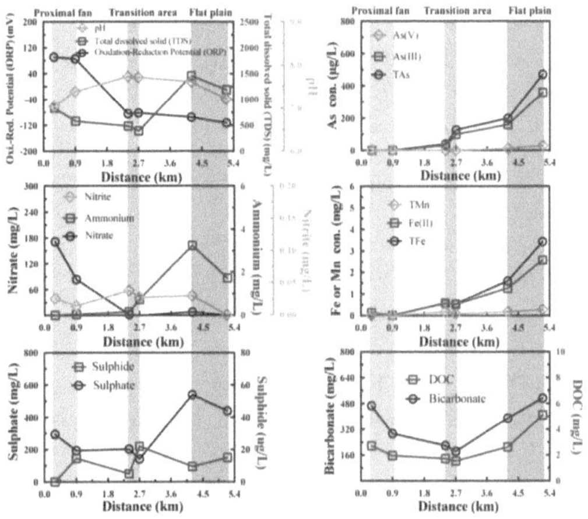

Figure 2. Changes of the groundwater parameters in samples along the groundwater flow.

3.1.1 *Microbial ecology*

A total of 291,007 16S rRNA gene sequences were obtained from six groundwater samples after removal of low-quality sequences and chimeras. Overall, the five most heavily represented classes (>1%) in the groundwater samples were Gammaproteobacteria, Alphaproteobacteria, Betaproteobacteria, Actinobacteria, Flavobacteria, and Deinococci. The genera observed suggested the presence of a diverse bacterial community able to catalyze a wide range of biogeochemical processes.

3.1.2 *Correlations between microbial community and groundwater chemistry*

PCA results suggested that, approximately along the groundwater flow paths, the changes of microbial community compositions were associated with three different groups of geochemical parameters (Figure 3). Correspondingly, top 50 genera were also clustered into three groups, as visualized in a heatmap based on Spearman correlations (Figure 4). Spearman ranking correlations showed that Microbial group I and Microbial group II were mainly associated with Geochemical-group I and Geochemical-group II, respectively. While Microbial group III was mainly associated with the Geochemical-group II and III data. Interestingly, five genera were simultaneously positively correlated with ammonium, Fe, and As, reported to participate in ammonium oxidation either partially

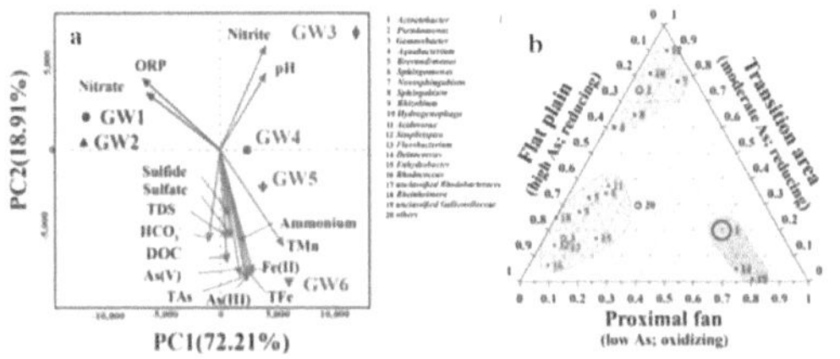

Figure 3. Principle component analysis (PCA).

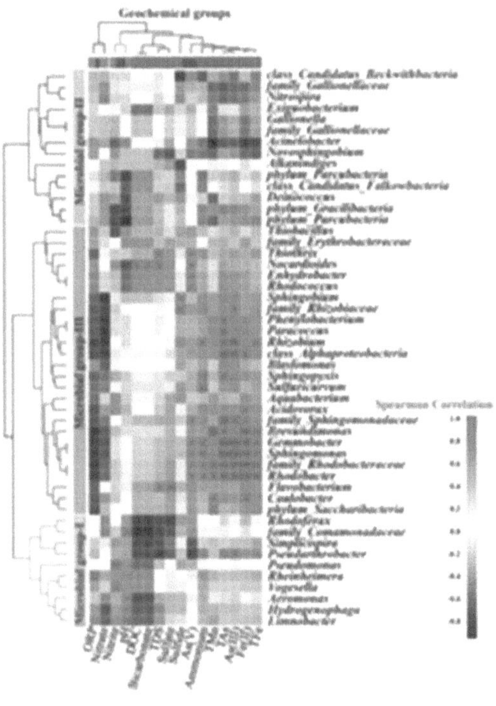

Figure 4. Correlations between microbial community and groundwater chemistry.

or completely (Erkal *et al.*, 2019; Fitzgerald *et al.*, 2015; Huang *et al.*, 2014; Kuykendall *et al.*, 2015; Liu *et al.*, 2017; Staley & Brenner 2015), indicating the potential importance of ammonium as an electron donor during As mobilization.

4 CONCLUSIONS AND RECOMMENDATIONS

In the present study, an obvious change of major genera in groundwater samples along the flow path was found corresponding to the geochemical evolution from the proximal fan (geochemical-group I: low As, oxidizing), through the transition area (geochemical-group II: moderate As, moderately-reducing), to the flat plain (geochemical-group III: high As, reducing). In the flat plain, high As groundwater contained Fe(III)- and As (V)-reducing bacteria, consistent with current models on As mobilization driven via reductive dissolution of Fe(III)/As(V) mineral assemblages. We provide the evidence for ammonium as electron donor, with potential ammonium-oxidizing, Fe(III))-reducing organisms detected, which may contribute significantly to the As enrichment in the reducing aquifers of the Hetao Basin. We suggest that ammonium oxidation coupled with Fe(III) reduction would widen the range of metabolic pathways driving As mobilization in high As groundwater systems.

ACKNOWLEDGEMENTS

The study is financially supported by the National Natural Science Foundation of China (Nos. 41702272, 41825017 and 41672225), the Fundamental Research Funds for the Central Universities (Nos. 2652017051, 2652017165), and the Natural Environment Research Council of United Kingdom (grant NE/P01304X/1).

REFERENCES

Al Lawati W.M., Rizoulis A., Eiche E., Boothman C., Polya D.A., Lloyd J.R., Berg M., Vasquez-Aguilar P. & van Dongen B.E. 2012. Characterisation of organic matter and microbial communities in contrasting arsenic-rich holocene and arsenic-poor pleistocene aquifers, Red River Delta, Vietnam. *Appl. Geochem.* 27(1): 315–325.

Caporaso J.G., Kuczynski J., Stombaugh J., Bittinger K., Bushman F.D., Costello E.K., Fierer N., Pena A. G., Goodrich J.K. & Gordon J.I. 2010. QIIME allows analysis of high-throughput community sequencing data. *Nature Methods* 7(5): 335.

Dhar R.K., Zheng Y., Saltikov C.W., Radloff K.A., Mailloux B.J., Ahmed K.M. & van Geen A. 2011. Microbes enhance mobility of arsenic in pleistocene aquifer sand from Bangladesh. *Environ. Sci. Technol.* 45(7): 2648–2654.

Erkal N.A., Eser M.G., Özgür E., Gündüz U., ErogluI. & Yücel M. 2019. Transcriptome analysis of rhodobacter capsulatus grown on different nitrogen sources. *Arch. Microbiol.* 1–11.

Fitzgerald C.M., Camejo P., Oshlag J.Z. & Noguera D. R. 2015. Ammonia-oxidizing microbial communities in reactors with efficient nitrification at low-dissolved oxygen. *Wat. Res.* 70: 38–51.

Guo H., Liu C., Lu H., Wanty R.B. *et al.* 2013. Pathways of coupled arsenic and iron cycling in high arsenic groundwater of the Hetao basin, Inner Mongolia, China: An iron isotope approach. *Geochim. Cosmochim. Acta* 112: 130–145.

Guo H., Zhou Y., Jia Y., Tang X., Li X., Shen M., Lu H., Han S. *et al.* 2016. Sulfur cycling-related biogeochemical processes of arsenic mobilization in the Western Hetao basin, China: evidence from multiple isotope approaches. *Environ. Sci. Technol.* 50 (23):12650–12659.

Hassan Z., Sultana M., van Breukelen B.M., Khan S.I. & Röling W.F.M. 2015. Diverse arsenic-and iron cycling microbial communities in arsenic contaminated aquifers used for drinking water in Bangladesh. *FEMS Microbiol. Ecol.* 91(4), fiv026.
Huang T., Gao B., Hu X.K., Lu X. *et al.* 2014. Ammonia-Oxidation as an engine to generate nitrous oxide in an intensively managed calcareous fluvo-aquic soil. *Sci. Rep.* 4: 3950.
Islam F.S., Gault A.G., Boothman C., Polya D.A., Charnock J.M., Chatterjee D. & Lloyd J.R. 2004. Role of metal-reducing bacteria in arsenic release from Bengal Delta sediments. *Nature* 430(6995): 68.
Jia Y., Guo H., Jiang Y., Wu Y. & Zhou Y. 2014. Hydrogeochemical zonation and its implication for arsenic mobilization in deep groundwaters near alluvial fans in the hetao basin, inner mongolia. *J. Hydrol.* 518: 410–420.
Kuykendall L.D., Young J.M., Martínez-Romero E., Kerr A. & Sawada H. 2015. Rhizobium. *Bergey's Manual of Systematics of Archaea and Bacteria.*
Li P., Jiang Z., Wang Y.H., Deng Y., Van Nostrand J. D., Yuan T., Liu H., Wei D.Z. & Zhou J.Z. 2017. Analysis of the functional gene structure and metabolic potential of microbial community in high arsenic groundwater. *Wat. Res.* 123: 268–276.
Li Y., Guo H. & Hao C. 2014. Arsenic release from shallow aquifers of the Hetao basin, Inner Mongolia: evidence from bacterial community in aquifer sediments and groundwater. *Ecotoxicology* 23(10): 1900–1914.
Liu H., Zhu L., Tian X. & Yin Y. 2017. Seasonal variation of bacterial community in biological aerated filter for ammonia removal in drinking water treatment. *Wat. Res.* 123: 668–677.
Mirza B.S., Muruganandam S., Meng X. *et al.* 2014. Arsenic (V) reduction in relation to Iron (III) transformation and molecular characterization of the structural and functional microbial community in sediments of a basin-fill aquifer in Northern Utah.. *Appl. Environ. Microbiol.* 80(10): 3198–3208.
Oremland R.S. & Stolz J.F. 2003. The ecology of arsenic. *Science* 300(5621): 939–944.
Staley J.T. & Brenner D.J. 2015. *Enhydrobacter. Bergey's Manual of Systematics of Archaea and Bacteria.*
Xu N., Tan G., Wang H. & Gai X. 2016. Effect of biochar additions to soil on nitrogen leaching, microbial biomass and bacterial community structure. *Eur. J. Soil Biol.* 74: 1–8.
Zhu Y., Xue X., Kappler A., Rosen B.P. & Meharg A.A. 2017. Linking genes to microbial biogeochemical cycling: lessons from arsenic. *Environ. Sci. Technol.* 51 (13), 7326–7339.

Arsenic in the Environment: Bridging Science to Practice for Sustainable Development –
van der Wal, Ahmad, Petrusevski, Weijma, Savic, van der Wens, Beerendonk, Bhattacharya, Bundschuh & Naidu (Eds)
© 2023 the Author(s), ISBN 978-1-032-32928-4

Investigation of biochemical properties of soil and groundwater in arsenic affected blocks of Murshidabad district and isolation of potential arsenic resistant bacteria

S. Ahmed, A. Basu, D. Mandal, I. Saha & M. Biswas
Sripat Singh College, Jiaganj, Murshidabad, India

ABSTRACT: Arsenic (As) is one of the major contaminants of soil and groundwater, responsible for a number of health hazards. Various blocks of Murshidabad district show As concentration above the maximum permissible limit in soil and water samples. In the present study, we have focused on some highly As contaminated regions of Murshidabad district. We have analyzed various biochemical parameters of soil and water samples of this district. The soils of these regions show high alkalinity and the groundwater samples also exhibits a basic pH. The total dissolved solids of these water samples varied from 200 mg to 300 mg. Also, the bacterial load in the water samples was extremely high. Since, groundwater is a source of drinking water in the blocks of Murshidabad district, consumption of such high amount of total dissolved solids would take a toll on the detoxifying and excretory system of the body i.e., on the hepatic and renal systems. The uncharacterized microorganisms in the soil could be faecal coliforms or other pathogenic bacteria, and their consumption would have serious health consequences. After serial dilutions (10^6 to 10^9 folds) of the groundwater and the soil samples, we could still isolate some bacteria thriving in these samples. Since, the soil and groundwater of these regions are highly As contaminated, bacterial colonies isolated after serial dilutions could be potentially As resistant bacteria.

1 INTRODUCTION

Arsenic (As) toxicity is one of the serious problems both from national and global perspective. Arsenic is a toxic element, which contaminates soil and groundwater. In India, As contamination of soil and groundwater is a severe problem in lower Gangetic plain and Ganga-Brahmaputra deltaic region (Mukherjee *et al.*, 2005). U.S. Environmental Protection Agency prescribed the maximum permissible limit of As in drinking water to be 10 µg/L (US EPA 2006). However, World Health Organization notified that in absence of any alternate source of drinking water, the maximum permissible limit of As would be 50 µg/L. In countries like Canada and Australia, this permissible limit is 5 µg/L and 7 µg/L, respectively. Arsenic is released in the soil by natural biogeochemical cycles, from there it leaches into the groundwater by natural processes. Arsenic enters the body through water and food. There are evidences that staple food crops like rice have accumulated As. This leads to biomagnification of As within the food chain.

West Bengal is one of the severely As affected states of India. Within West Bengal nine districts are affected by As toxicity. Murshidabad is one of the severely As affected districts of West Bengal. The river Ganga separates it from Bangladesh. Eastern bank of Bhagirathi river constitutes more As contaminated blocks of Murshidabad district (64.7% above 10 µg/L and 32.5% above 50 µg/L) compared to the western bank of the river (30.1% above 10 µg/L and 11.7% above 50 µg/L). In this study, our main aim is to analyze the conditions of soil and groundwater in highly As contaminated regions of Murshidabad district. For this purpose, we have qualitatively and quantitatively checked various parameters of soil & groundwater samples from these affected regions. Also we have attempted the isolation of potential As resistant bacteria from these regions.

2 MATERIALS AND METHODS

2.1 *Chemical analysis of soil and groundwater*

Soil samples were diluted to obtain a concentration of 1 mg soil /1 ml distilled water. Soil samples were mixed well with water by constant stirring so, that the soluble solutes i.e. ions and electrolytes come in the polar phase. The particulate matter in the soil (soil sediments) was allowed to settle down. The pH of the solution was measured at 25°C. Similarly, the pH of the water samples were measured at the same temperature.

2.2 *Estimation of bacterial load in the soil and groundwater samples and isolation of potential arsenic resistant bacteria*

Water samples, and soil samples dissolved in water was plated in a LB-agar plate to estimate the bacterial load in the soil and groundwater samples. Further, serial dilutions of water and soil samples were made and these diluted stocks were plated in LB- agar plates for isolation of single colonies of potential As resistant bacteria.

DOI: 10.1201/9781003317395-29

3 RESULTS AND DISCUSSION

3.1 *Chemical characterization of soil and groundwater*

Water samples were collected from Bhagobangola I, Hariharpara, Chunakhali and Asrampara, which are the regions amongst the severely As affected blocks of Murshidabad district. Soil samples were also collected from Bhagobangola I, Hariharpara and Chunakhali. pH of the soil and water samples were measured at 25°C. Both soil and water samples from the four As contaminated blocks were found to be alkaline in nature. Soil sample of Hariharpara showed maximum pH 8.37 amongst all the other soil samples. Among the water samples, water sample from Bhagobangola I showed the maximum pH 7.60. All the soil samples showed pH above 8. Total dissolved solids (TDS) in the water samples were found to be very high. TDS in water samples from Hariharpara was 200 mg/L. The other three sources showed TDS concentration of 300 mg/L.

Figure 1. Soil samples collected from a) Mahisasthali village.

3.2 *Isolation of potential arsenic resistant bacteria from the soil and groundwater samples*

Next, with an aim to isolate As resistant bacteria from these samples, soil and water samples from aforementioned As affected blocks were plated with and without dilutions on LB agar plate. Huge number of bacterial colonies was observed when undiluted soil and water samples were plated. As PHED had marked these blocks as severely As contaminated, we wanted to identify As resistant microorganisms from soil and water samples of these blocks. Therefore, we isolated bacteria from these soil and water samples by serially diluting the samples and spreading them on LB agar plate. Single bacterial colonies could be observed in the plates. And considering the huge As toxicity of the aforementioned regions, these colonies could be categorized as colonies of potential As resistant bacteria.

4 CONCLUSIONS AND RECOMMENDATIONS

We have concentrated our study on four blocks of Murshidabad district. These were Bhagobangola I, Hariharpara, Chunakhali and Asrampara. All of these blocks had been marked by PHED (Public Health and Engineering Department) as severely As affected blocks. We collected soil and water samples from these areas and measured the pH and total dissolved solids (TDS). Both soil and water samples of these areas were found to be very alkaline. Soil & water samples from the above mentioned As affected blocks were plated with and without dilutions on LB-agar plate. With serial dilutions single bacterial colonies could be obtained on the LB-agar plate. Considering the huge As toxicity of these regions, the soil and water microbiota obtained from these regions could be potentially As resistant. Huge number of bacterial colonies was observed when undiluted soil and water samples were plated. Therefore, people of these blocks were drinking water contaminated with millions of bacteria. As characterization of these bacteria was not done therefore, it could only be assumed that these water samples might be contaminated with toxic bacteria including fecal coliforms and other pathogenic bacteria.

ACKNOWLEDGEMENTS

The authors are thankful to Sripat Singh College for providing laboratory facilities & Department of Biotechnology, Government of West Bengal for the financial assistance.

REFERENCES

Hendryx M. 2009. Mortality from heart, respiratory and kidney disease in coal mining areas of Appalachia. *Int. Arch. Occup. Environ. Health* 82(2): 243–249.

Huq S.M.I., Joardar J.C., Parvin S., Correll R. & Naidu R. 2006. Arsenic contamination in food chain: transfer of arsenic into food materials through groundwater irrigation. *J. Health Popul. Nutr*, 24(3): 305–316.

Kapaj S., Peterson H., Liber K. & Bhattacharya P. 2006. Human health effects from chronic arsenic poisoning–a review. *J. Environ. Sci. Health A* 41(10): 2399–2428.

Mukherjee S.C., Saha K.C., Pati S., Dutta R.N., Rahman M. M., Sengupta M.K. *et al.* 2005. Murshidabad—one of the nine groundwater arsenic-affected districts of West Bengal, India. Part I: magnitude of contamination and population at risk. *Clin. Toxicol.* 43: 823–834.

Pais I. & Benton Jones J. 1997. *The Hand Book of Trace Elements.* Boca Raton Florida, 223p.

US EPA. 2006. *Drinking Water Requirements for States and Public Water System.* EPA, USA.

Arsenic in the Environment: Bridging Science to Practice for Sustainable Development –
van der Wal, Ahmad, Petrusevski, Weijma, Savic, van der Wens, Beerendonk, Bhattacharya, Bundschuh & Naidu (Eds)
© 2023 the Author(s), ISBN 978-1-032-32928-4

Understanding the interactions among dissolved organic matter, fecal contamination, and arsenic in the groundwaters of southeast Bangladesh

H.V. Kulkarni[1,3], S. Barua[1,2], M.G. Kibria[3], P. Bhattacharya[4] & S. Datta[1,3]
[1]*Department of Geology, Kansas State University, Manhattan, KS, USA*
[2]*School of Earth, Atmosphere and Environment, Monash University, Clayton, VIC, Australia*
[3]*Department of Geological Sciences, University of Texas at San Antonio, San Antonio, TX, USA*
Department of Earth and Environmental Sciences, University of Texas Arlington, Arlington, TX, USA
[4]*KTH-International Groundwater Arsenic Research Group, Department of Sustainable Development, Environmental Science and Engineering, KTH Royal Institute of Technology, Stockholm, Sweden*

ABSTRACT: Understanding the interactions among Dissolved Organic Carbon (DOC), arsenic and fecal contamination in the groundwater of southeast Bangladesh is crucial as groundwater is the most important drinking water sources in Bangladesh. Groundwater samples (13 – 236 m depths) from four piezometer nests and nearby surface water samples from Matlab (SE Bangladesh) were collected and analyzed for inorganic, organic and microbial constituents. The groundwaters contained total dissolved As (As_T) ranging from 0.2 to 407 µg/L, dissolved organic carbon (DOC) between 0.7 and 12 mg/L, molar C:N between 0.2 and 8, and total coliform (MPN) between 12 to 238 CFU. Absorbance and fluorescence spectroscopic analyses of dissolved organic matter (DOM) revealed that the groundwater contained aromatic humic-like fluorescence signature, characterized by humification index (HIX) of 10±7, and specific ultraviolet absorbance (SUVA) ranging from 0.3 to 4 L/mg.m. Surface water samples from the study area contained low As_T (2-11 µg/L) and very high total coliform (308 – 241,957 CFU). These results indicate that almost all the groundwater samples analyzed in this study contained unsafe levels of AsT (>50 µg/L) as well as presence of total coliform bacteria in detectable levels suggesting potential pathogens. Higher DOC, lower C:N and high humic-like fluorescence signature of groundwater DOM are consistent with arsenic mobilization by reductive dissolution of iron bearing aquifer sediments by microbes under reducing conditions and fueled by DOM. The presence of coliform bacteria in groundwater, even at >200 m depths, is alarming. The source of coliform bacteria in groundwater may be via infiltration of surface water that is contaminated by anthropogenic wastes, and further supported by higher DOC concentrations in the groundwater.

1 INTRODUCTION

Safe drinking water scarcity and water pollution are among the most significant environmental threats in the Bengal basin. Since the surface water sources were contaminated with pathogens causing a significant risk of water-borne diseases and infant mortality (Escamilla *et al.*, 2013), the drinking water supply in Bengal basin was shifted to shallow groundwater. Consequently, more than 10 million new hand-pumped wells were installed to provide drinking water for over 130 million people. Several decades ago, elevated dissolved arsenic (As) concentrations in groundwater were linked with chronic diseases such as arsenicosis and eventually turned into a massive humanitarian disaster. Arsenic concentrations in groundwater commonly exceed the regulatory limit of 50 µg/L beneath the Holocene floodplains within the alluvial and deltaic plains of the Bengal basin. The widely accepted mechanism of As release is the microbially-mediated reductive dissolution of As bearing iron (oxy)hydroxide (FeOOH) rich sediments fueled by dissolved organic carbon (DOC) (Nickson *et al.*, 1998). The presence of As in groundwater has been documented over decades. However, the probable nature and sources of DOC in groundwater and how these DOC influence As and coliform contamination have not been well understood in southeast Bangladesh groundwater. Thus, the present study investigates the potential role of DOC contributing to driving As release as well as the concentration of DOC available for coliform growth in groundwater.

2 METHODS

Nineteen groundwater samples (13 to 236 m depth) and nine surface water samples (eight ponds and one canal proximal to the piezometer nests) were collected between June and July 2014 from four piezometer nests in Matlab (southeast Bangladesh). Major cations and trace elements were measured in 0.45-µm filtered samples acidified with 0.2% v/v trace metal grade nitric acid using ion chromatography (IC) and high resolution inductively coupled plasma mass spectrometry (HR-ICP-MS), respectively. Major anions and stable isotopes of water were measured in 0.45-µm filtered and unacidified samples using IC and Picarro Cavity Ringdown Spectrometer, respectively. Bulk DOC and nitrogen were measured in 0.2-µm filtered samples acidified with 0.2% v/v hydrochloric acid by thermic oxidation using Shimadzu TOC/TN Analyzer. Absorbance and fluorescence spectroscopic properties of

DOI: 10.1201/9781003317395-30

DOM were measured by Horiba Aqualog Fluorometer in 0.2-μm filtered and unacidified samples. US EPA approved Colilert test was used to determine most probable number (MPN) of coliform bacteria in groundwater.

3 RESULTS AND DISCUSSION

Stable isotopic signature of groundwater samples showed $\delta^{18}O$ and $\delta^{2}H$ ranging from -2 to -4‰, and -10 to -20‰ respectively which plotted between local meteoric water line and pond evaporation line, indicating that the groundwater was recharged through moderately evaporated local precipitation. Analysis of chloride to bromide mass ratio (Cl:Br) in groundwater samples showed the signatures of evaporation (Cl:Br, 10-100), organic matter degradation (Cl:Br, <10), and sewage influence (Cl: Br, > 1,000) as described in Knappett *et al.* (2018).

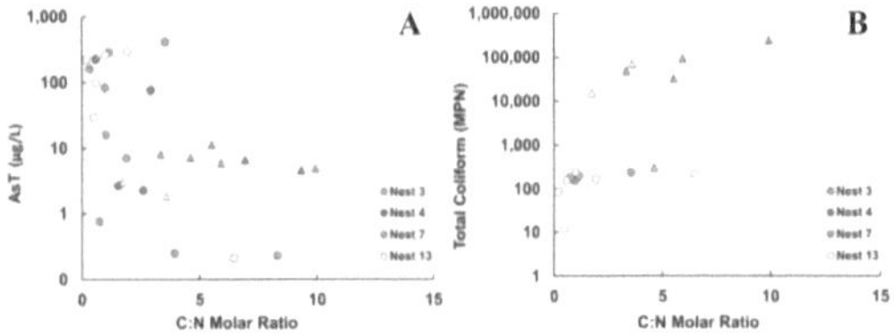

Figure 1. Bivariate plots show correlation of molar ratio of organic carbon to nitrogen (C:N Molar Ratio) with total dissolved arsenic (A), and with total coliform count (B). Circles and triangles represent groundwater and surface water samples from each piezometer nest area, respectively.

Groundwater samples contained total dissolved arsenic (As_T) of 114±130 μg/L and ranged between 0.2 and 407 μg/L, whereas in surface water samples As_T was only 6±3 μg/L. Total coliform count in groundwater samples ranged between 12 to 238 CFU whereas, it ranged between 308 and 241,957 CFU in surface water samples. Groundwater also contained DOC concentrations ranging between 3 and 9 mg/L, and C:N ratio that ranged from 2 to 10. Groundwater DOM was found to be highly humic-like, aromatic and terrestrially derived as indicated by humification index (HIX) values ranging between 4 and 11, and SUVA between 2 and 4 L/mg.m. Groundwater As_T concentrations appear to be negatively correlated with C:N ratio of groundwater DOM (Figure 1A) consistent with previous studies (Mladenov *et al.*, 2015). Aromatic and humic-like DOM has been linked with As mobilization as it serves as electron donor (McArthur *et al.*, 2001) for heterotrophic microbial metabolism (such as iron reduction), forms aqueous complexes with iron and arsenic and catalyzes arsenic mobilization via electron shuttling (Kulkarni *et al.*, 2018). Groundwater samples appeared to have higher total coliform count with increasing C:N ratio (Figure 1B) which indicates that carbonaceous organic matter may favor the growth of coliform bacteria in groundwater. Some prior studies have shown that the DOC from various sources (algal or terrestrially derived) can support the growth of coliform bacteria (Bouteleux *et al.*, 2005) and the DOC may favor the growth of coliform bacteria in groundwater.

4 CONCLUSIONS

Groundwater samples from Matlab (Bangladesh) exceeded national safe drinking water guideline of 50 μg/L at 13-100 m depths, whereas deeper groundwater >150 m contained less than 1 μg/L of As_T. However, all the groundwater samples including deep groundwater, contained detectable levels of total coliform bacteria which suggests that the groundwater that contain safe levels of As, contain unsafe levels of coliform bacteria. Our analyses indicate that while sedimentary trapped organic matter supports As mobilization, it may also support the growth of coliform bacteria in the aquifer. Groundwater is the most important drinking water source in Bangladesh, and the findings of this study provide valuable insights on presence of arsenic and coliform bacteria that pose risk to human health.

ACKNOWLEDGEMENTS

We sincerely thank SASMIT (Sustainable Arsenic Mitigation), KTH Royal Institute of Technology, and K-State for assistance in field sampling and laboratory analyses.

REFERENCES

Bouteleux C., Saby S., Tozza D., Cavard J., Lahoussine V., Hartemann P. & Mathieu L. 2005. Escherichia coli behavior in the presence of organic matter released by algae exposed to water treatment chemicals. *App. Environ. Microbiol.* 71: 734–740.

Escamilla V., Knappett P., Yunus M., Streatfield P.K. & Emch M. 2013. Influence of latrine proximity and type on tubewell water quality and diarrheal disease in Bangladesh. *Ann. Assoc. Amer. Geograph.* 103: 299–308.

Knappett P.S., Li Y., Hernandez H., Rodriguez R., Aviles M., Deng C., Piña V., Giardino J.R., Mahlknecht J. & Datta S. 2018. Changing recharge pathways within an intensively pumped aquifer with high fluoride concentrations in Central Mexico. *Sci. Total Environ.* 622: 1029–1045.

Kulkarni H., Mladenov N., Datta S. & Chatterjee D. 2018. Influence of monsoonal recharge on arsenic and dissolved organic matter in the holocene and pleistocene aquifers of the Bengal basin. *Sci. Total Environ.* 637–638: 588–599.

McArthur J.M., Ravenscroft P., Safiulla S. & Thirlwall M. F. 2001. Arsenic in groundwater: testing pollution mechanisms for sedimentary aquifers in Bangladesh. *Wat. Resour. Res.* 37: 109–117.

Mladenov N., Zheng Y., Simone B., Bilinski T.M., McKnight D.M., Nemergut D., Radloff K.A., Rahman M.M. & Ahmed K.M. 2015. Dissolved organic matter quality in a shallow aquifer of Bangladesh: implications for arsenic mobility. *Environ. Sci.Technol.* 49: 10815–10824.

Nickson R.T., McArthur J.M., Burgess W.G., Ravenscroft P., Ahmed K.M. & Rahman M. 1998. Arsenic poisoning of Bangladesh groundwater. *Nature* 395: 338.

Arsenic in the Environment: Bridging Science to Practice for Sustainable Development –
van der Wal, Ahmad, Petrusevski, Weijma, Savic, van der Wens, Beerendonk, Bhattacharya, Bundschuh & Naidu (Eds)
 ISBN 978-1-032-32928-4

Radiocarbon analysis of RNA, DIC, DOC and CH_4 to constrain the sustainability of pumping Pleistocene aquifers in Bangladesh

B.J. Mailloux[1], M.R. Mozumder[2], B.C. Bostick[2], T. Ellis[2], C. Harvey[3], G. Slater[4], E. Trembath-Reichert[5], I. Choudhury[6], K.M. Ahmed[6] & A. van Geen[2]
[1]*Environmental Sciences Department, Barnard College, Columbia University, New York, USA*
[2]*Lamont-Doherty Earth Observatory, Columbia University, Palisades, New York, USA*
[3]*Department of Civil and Environmental Engineering, Massachusetts Institute of Technology, Cambridge, Massachusetts, USA*
[4]*School of Geography and Earth Sciences, McMaster University, Hamilton Ontario, Canada*
[5]*Department of Geology, Arizona State University, Tempe, Arizona, USA*
[6]*Department of Geology, University of Dhaka, Dhaka, Bangladesh*

ABSTRACT: Pleistocene aquifers are key for lowering the chronic exposure of the rural population to arsenic. Too little is known, however, about the sources of reactive carbon that maintain reducing conditions in these low-arsenic aquifers. This matters as enhanced supply of reactive carbon due to perturbations in groundwater flow could potentially release arsenic to groundwater. To shed light on this process, our team measures the radiocarbon content of labile microbial matter (RNA) and compares it to the radiocarbon content of potential sources of reactive carbon. Results to date suggest that recent recharge supplies reactive carbon in the dissolved form rather than the sediment, possibly in part in the form of methane. This dynamic situation suggests arsenic concentration could vary of time and should be monitored in vulnerable areas.

1 INTRODUCTION

In Bangladesh, the groundwater pumped from deeper Pleistocene-aged aquifers is typically low in arsenic (As), in contrast to shallower sands deposited more recently during the Holocene. Wells that tap Pleistocene aquifers are widely used to mitigate the damaging health effects of chronic exposure to toxic levels of As from drinking shallow well water. However, deep wells tapping the Pleistocene aquifer contain elevated levels of As in some parts of the country (Khan *et al.*, 2019), including a limited number in Araihazar (Choudhury *et al.*, 2016), 25 km east of the capital Dhaka, where Columbia University and Dhaka University have been conducting hydrogeological and biogeochemical studies for almost two decades. Rather than regional patterns of elevated As at depth linked to broad-scale differences in hydrogeology across the country, this presentation focuses on the source of carbon fueling microbial respiration, and therefore the potential release of As to groundwater, in a Pleistocene aquifer that is generally low in As but already at risk because of increasing depressurization by massive pumping for the municipal water supply of Dhaka (Knappett *et al.*, 2016).

2 METHODS

Samples were collected in Araihazar for radiocarbon analysis of microbial RNA, building on the previous work of Mailloux *et al.* (2013; 2019) dissolved inorganic carbon (DIC), dissolved organic carbon (DOC), and methane (CH_4), along with basic groundwater chemistry including As, from Pleistocene aquifers in the 40-250 m depth range.

3 RESULTS AND DISCUSSION

For low-As aquifers >150 m deep, the DIC (n=13) ages shows an average $\Delta^{14}C$ of -698‰ which is equivalent to an average age of 10,700 years. This corresponds with the end of the Pleistocene, suggesting that the aquifer is sealed from recent recharge accelerated by Dhaka pumping. In contrast, the DOC (n=9) shows an average $\Delta^{14}C$ of -369‰ which is equivalent to an average age of 3,900 years. The young DOC age may be a more sensitive indicator that the aquifer is actually not so well sealed, and that heavy pumping around Dhaka is causing vertical drawdown of organic carbon from shallower depths.

Thus far, the carbon in microbial RNA from 3 deep low-As wells are well clustered at an average $\Delta^{14}C$ of -483‰, which is equivalent to 5,200 years. This new result suggests that even in deep, low-As Pleistocene wells, microbes are respiring the younger advected carbon rather than sedimentary carbon >12 kyr in this depth range. In addition, a radiocarbon content of RNA in a deep Pleistocene aquifer

DOI: 10.1201/9781003317395-31

that is intermediate between the radiocarbon content of DIC and DOC suggests microbial activity that combines heterotrophy and autotrophy.

We are also studying a set of wells >150 m deep that are elevated in As in two villages of Araihazar containing a cluster of deep high As wells and another set of shallower wells in the 40-80 m depth range in different set of villages where Pleistocene sands extend closer to the land surface. We do not have as of yet RNA radiocarbon data from the deep high-As well, but we see in those two villages a fairly consistent relationship of As and CH_4 concentrations increasing in tandem in the 0.1-100 µg/L and the 0.1-100 µmol/L ranges, respectively. The two villages are unusual also because of a particularly thick (>50 m) clay layer separating the Holocene and Pleistocene aquifer at this location and, as a result, an exceptionally large vertical difference of 6 m throughout the year conducive to downward flow. Both the DIC and CH_4 analyzed from one of these wells contain bomb-radiocarbon input dating from atmospheric of nuclear bombs in the 1950s and 1960s. We therefore provisionally attribute the findings of elevated As in deep Pleistocene aquifer in Araihazar to rapid downward flow of shallow DOC and/or CH_4 along the annulus of these deep wells, which in turns triggers the local release of As to deep groundwater.

The mechanism regulating groundwater As in the region of Araihazar where the Pleistocene aquifer is shallower and readily accessible to individual households appear to be different. The two RNA radiocarbon measurements of -450‰ and -110‰ available so far suggest an association between elevated As concentrations and relatively young reactive carbon, but other data show evidence also of the opposite. At one site, for instance, groundwater below a thick clay capping the Pleistocene aquifer is free of bomb-radiocarbon whereas more recently recharged water containing bomb radiocarbon is very low in As. The one relationship that is systematic within these relatively shallow aquifers, and consistent with observations in the deep aquifer, is the association of elevated As concentrations with elevated concentrations of CH_4. As in the case of the deep Pleistocene aquifer, the radiocarbon content of RNA in deep Pleistocene aquifer samples is intermediate between the radiocarbon content of DIC and DOC. The data suggest, in addition, for the one high-As sample to date for RNA, DIC, and DOC, and CH_4 were radiocarbon dated, a possible role for CH_4 as a driver of microbially-mediated Fe reduction and, therefore, of As release to groundwater.

4 CONCLUSION

If confirmed, our findings of relatively young DOC and RNA could have significant implications for the response of Pleistocene aquifers to pumping-induced perturbations, especially CH_4 plays a role as well as it is much less retarded if at all by adsorption to aquifers sands compared to DOC.

ACKNOWLEDGEMENTS

Columbia University and the University of Dhaka's research in Araihazar has been supported for 20 years by the US National Institute of Environmental Health Sciences Superfund Research Program and several grants from the US National Science Foundation.

REFERENCES

Choudhury I., Ahmed K.M., Hasan M., Mozumder M. R.H., Knappett P.S.K., Ellis T. & A. van Geen. 2016. Evidence for elevated levels of arsenic in public wells of Bangladesh due to improper installation. *Ground Water*. 54(6): 871–877.

Khan M.R., Michael H.A., Nath B., L Huhmann B.L., Harvey C.F., Mukherjee A., Chakraborty M., Choudhury I., Ullah M.S., Ahmed K.M. Goodbred Jr., S.L., Schlosser P., Bostick B.C., Mailloux B.J. & van Geen A. 2019. High-arsenic groundwater in the Southwestern Bengal basin caused by a lithologically controlled deep flow system. *Geophy. Res. Lett.* 46 (22): 12637–13609.

Knappett P.S.K., Mailloux B.J., Choudhury I., Khan M., Michael H., Barua S., Mondal D.R., Steckler M. Akter H., Ahmed K.M., Bostick B., Harvey C., Shamsudduha M., Mihajlov I., R Mozumder R. & van Geen A. 2016. Vulnerability of low-arsenic aquifers to municipal pumping in Bangladesh. *J. Hydrol.* 539: 674–686.

Mailloux B.J., Trembath-Reichert E., Cheung J., Watson M., Stute M., Freyer G.A., Ferguson A.S., Ahmed K.M., Alam M.J., Buchholz B.A., Thomas J., Layton A.C., Zheng Y., Bostick B.C. & van Geen A. 2013. Advection of surface-derived organic carbon fuels microbial reduction in Bangladesh Groundwater. *Proc. Natl. Acad. Sci. U.S.A.* 110: 5331–5335.

Mailloux B.J., Kim C., Kichuk T., Nguyen K., Precht C., Wang S., Jewell T.N.M., Karaoz U., Brodie E.L., Williams K.H., Beller H.R. & Buchholz B.A. 2019. Paired RNA radiocarbon and sequencing analyses indicate the importance of autotrophy in a shallow alluvial aquifer. *Sci. Rep.* 9: 10370.

*Arsenic in the Environment: Bridging Science to Practice for Sustainable Development –
van der Wal, Ahmad, Petrusevski, Weijma, Savic, van der Wens, Beerendonk, Bhattacharya, Bundschuh & Naidu (Eds)*
 ISBN 978-1-032-32928-4

Arsenic bioaccumulation in phytoplankton in Brazilian soda lakes

A.H. Fostier[1], A. Hechavarría-Hernández[1], J.S. Costa[2], M. Fiore[2], J.P. Leister[1],
A.T. Rezende-Filho[3] & L. Barbiero[4]
[1]*Institute of Chemistry, University of Campinas (UNICAMP), Campinas, São Paulo, Brazil*
[2]*Center for Nuclear Energy in Agriculture (CENA), University of Sao Paulo (USP), Piracicaba, SP, Brazil*
[3]*Federal University of Mato Grosso do Sul (UFMT), Campo Grande, Brazil*
[4]*Geoscience Environment Toulouse (GET), Institute of Research for Development (IRD), Toulouse, France*

ABSTRACT: The soda lakes waters of Nhecolândia, a subregion of the Pantanal wetlands in Brazil, are characterized by high dissolved arsenic (As) concentrations (up to ~3 mg/L) and extreme pH conditions (pH> 8.5), resulting from an evapo-concentration process. Albeit these conditions, abundant planktonic community develops in these lakes that tolerate or metabolize As. Nevertheless, As bioaccumulation had never been studied in these lakes. In this study carried out in six alkaline lakes over the two very contrasting and 2018 dry periods from 2017 and 2018, we show that bioaccumulation factors in phytoplankton are low.

1 INTRODUCTION

High concentrations of dissolved arsenic (As), ranging from 28.8 to 2916 μg/L in water samples collected during the 2017 and 2018 dry periods, have been reported in soda lakes waters of Nhecolândia, a subregion of the vast Pantanal wetlands in Brazil (Richter *et al.*, 2019). Albeit extreme pH conditions, abundant planktonic community develops in these lakes that tolerate or metabolize As, suggesting possible bio accumulation. The aim of this work was therefore to determine As concentrations in phytoplankton and to assess the environmental factors that can control As bioaccumulation in these organisms.

2 MATERIALS AND METHODS

2.1 *Study area*

Nhecolândia lies in the floodplain on the southern half of the Taquari fan (Figure 1). It comprises an area of approximately 24,000 km^2. In this biome, approximately 500 shallow and non-stratified soda lakes exist in proximity with around 15,000 freshwater lakes. The climate is classified as tropical humid with a short dry season (July to October). The mean annual precipitation is ~1100 mm, and the annual evapotranspiration is approximately 1400 mm, providing an hydrological deficit of about 300 mm. This process also originates the evapo-concentration of many elements, including As (Barbiero *et al.*, 2008). Electrical conductivity of saline-alkaline lakewaters generally ranges from 1500 to 15,000 μS/cm with pH > 8.5. The alkaline lakes have been classified in 3 different types depending on their biogeochemical functioning, namely green, black or crystalline water lakes. When intense cyanobacterial bloom can be observed in green lakes, low bacterial abundance is generally observed in the others.

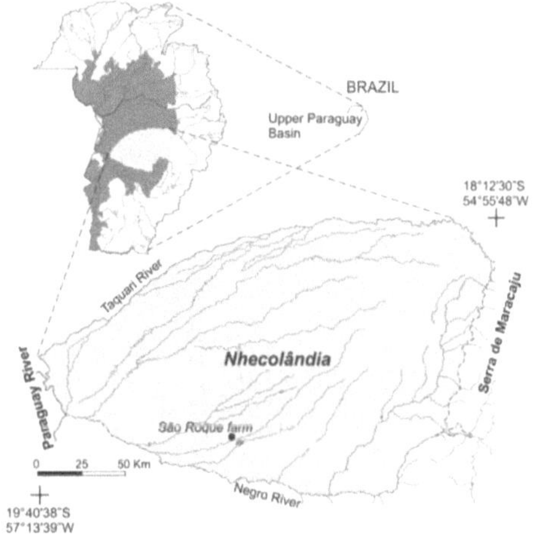

Figure 1. Location of the Pantanal wetland, Nhecolândia region and São Roque farm. (Adapted from Barbiero *et al.*, 2018).

2.2 *Water sampling and laboratory analysis*

Water and phytoplankton sampling was carried out at São Roque (SR) farm (Figure 1) in October 2017 and August 2018 in 6 lakes (green: SR01, SR04, SR05; SR08; black: SR06 and crystalline: SR07). Total dissolved As in water samples (filtered on 0.22 μm) was determined by hydride generation atomic fluorescence spectrometer (HG-AFS) (Millennium Excalibur 10.055, PSAnalytical). Accuracy (105%) and precision (4.5%) were assessed by analyzing the CRM NIST 1640a. Limits of detection (LD) and quantification (LQ) were generally lower than 0.05 and 0.15 μg/L, respectively. All the samples were analyzed in triplicate. Lyophilized phytoplankton samples were digested with HNO3/H2O2 in sealed flasks in microwave oven (Provecto Analítica DGT 100). Total As

DOI: 10.1201/9781003317395-32

was then determined by HG-AFS after being mixed with K2S2O8 and passing through a coupled UV-reactor. Accuracy (96%) and precision (7.3%) were assessed by analyzing the CRM BCR414. LD and LQ were 0.17 and 0.50 µg/L, respectively. The pH value and nutrients contents were also determined in water by classical method.

3 RESULTS AND DISCUSSION

3.1 *Arsenic concentrations in water and phytoplankton and bioaccumulation factor*

Arsenic concentrations in water and phytoplankton samples are presented in Table 1. Bioaccumulation factors (BF) were calculated as the ratio between As concentration in phytoplancton and As concentration in water (Figure 2).

Table 1. Mean arsenic concentrations in water (µg/L) and phytoplankton (µg/kg) and relative standard deviation (%).

	2017		2018	
	Water	Pkyto	Water	Pkyto
SR01	2916 (3.5)	–	64(2.7)	10078 (12.7)
SR04	929 (2.3)	31733(1.6)	83 (3.7)	6823 (8.8)
SR05	821 (2.7)	36522 (7.9)	26 (3.1)	3695 (19.3)
SR06	1323 (2.9)	39943 (8.8)	38 (4.1)	12256 (9.4)
SR07	53 (2.7)	22692 (4 8)	9.5 (4.6)	–
SR08	1721 (2.3)	36769 (2.9)	40 (9.0)	3673 (20.3)

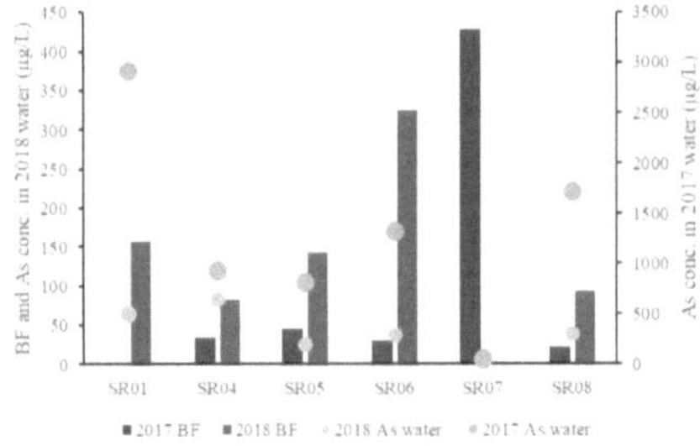

Figure 2. Arsenic concentration in water samples and bioaccumulation factor (BF) in phytoplankton samples collected in six alkaline lakes during the 2017 and 2018 dry seasons.

The levels of As measured in the waters (Table 1) were much higher in 2017, characterized by a very dry season, than in 2018, which on the contrary corresponded to a dry season with frequent rains. As a result, the waters were 6 to 45 times more concentrated in 2017 than in 2018. Similarly, As contents in phytoplankton were 3 to 10 times higher in 2017 than in 2018 for the same lakes. By contrast, calculated BF were higher in more diluted waters collected in 2018, as well as in the SR07 lake (Figure 2). The pH values were in the same order in 2017 and 2018, ranging from 8.6 to 10.2. In compensation, some parameters showed large differences, namely salinity 0.99 to 25.12 g/L in 2017 and 0.53 to 1.63 g/L in 2018; NO_3^-: 0.056 to 0.847 mg/L in 2017 and 0.014 to 0.249 mg/L in 2018; PO_4^{3-}: 0.42 to 86.92 mg/L in 2017 and 0.06 to 2.38 mg/L in 2018. SR07 always showed lower pH and dissolved load. In addition, BF was negatively correlated with water As contents (R^2 = 0.76). Recent studies revealed that As mainly occurs as As(V) in these alkaline waters (Richter *et al.*, 2019), which is the primarily As species accumulated by phytoplankton (Rahman *et al.*, 2012). However, when compared with BF values reported in As contaminated freshwater lakes (e.g. Caumette *et al.*, 2011), BF observed in alkaline lakes of Nhecolândia are low and increase with decreasing salinity. In both years, nutrients involved in microbial metabolism (C, N and P) where usually positively correlated with As concentration in phytoplankton (mainly NO_3^- and PO_4^{3-}), suggesting assimilation by the organisms. BF values were positively and negatively correlated with these nutrients in diluted waters and mineralized 2017 waters, respectively.

4 CONCLUSIONS AND RECOMMENDATIONS

The results clearly show the low bioavailability of dissolved As in these alkaline lakes, especially in the most alkaline ones. Such low availability could be due to different processes, such as complexation with organic matter, or sequestration by silica in black lakes preventing from As uptake by microorganisms. The silica As sequestration in silcrete formations in black lake is presented and discussed by Hechavarría-Hernández *et al.* (2023 this volume).

ACKNOWLEDGEMENTS

We acknowledge the São Paulo Research Foundation (FAPESP, N°2016/14227-5) and Ec2co-INSU (CNRS, France) for financial support, the Coordination for the Improvement of Higher Education Personnel (CAPES) and the National Council for Scientific and Technological Development (CNPq) for scholarships.

REFERENCES

Barbiero L., Rezende Filho A.T., Furquim S.A.C., Furian S., Sakamoto A.Y., Valles V., Graham R.C., Fort M., Ferreira R.P.D. & Neto J.P.Q. 2008. Soil morphological control on saline and freshwater lake hydrogeochemistry in the pantanal of Nhecolândia, Brazil. *Geoderma* 148: 91–106.

Caumette G., Koch I., Estrada E. & Reimer K.J. 2011 Arsenic speciation in plankton organisms from contaminated lakes: transformations at the base of the freshwater food chain. *Environ. Sci. Technol.* 45: 9917–9923.

Rahman M.A. & Hasegawa H. 2012. Arsenic in freshwater systems: influence of eutrophication on occurrence, distribution, speciation, and bioaccumulation. *Appl. Geochem.* 27: 304–314.

Richter L., Hernandez A.H., Pessôa G.S., Arruda M.A.Z., Rezende-Filho A.T., Almeida R.B., Menezes H.A., Valles V., Barbiero, L. & Fostier A.H. 2019. Dissolved arsenic in the upper Paraguay river basin and Pantanal wetlands. *Sci. Total Environ.* 687: 917–928.

Arsenic in the Environment: Bridging Science to Practice for Sustainable Development – van der Wal, Ahmad, Petrusevski, Weijma, Savic, van der Wens, Beerendonk, Bhattacharya, Bundschuh & Naidu (Eds)
© 2023 the Author(s), ISBN 978-1-032-32928-4

Environmental biochemistry of arsenic species in contaminated areas of Chile

I. Pizzaro & D. Román
Analytical Chemistry Laboratory, Universidad de Antofagasta, Antofagasta, Chilel

ABSTRACT: Antofagasta Region of Chile is worldwide recognized as severely arsenic (As) contaminated area. Due to its abundance, mainly as As (V) and As (III), certain microorganisms such as several types of bacteria have evolved. The aim of this work was to study the most relevant As resistant bacteria that exist in highly As contaminated sediments in El Tatio geyser field. Inductively coupled plasma mass spectrometry (ICP-MS) was employed to determine the concentration of As. Results showed a great accumulation of this element. The separation of bacterial cells into cytoplasmic and membrane fractions were carried out by differential centrifugation. The results demonstrated that, between 20-70% of arsenate was reduced to arsenite. Only in one case it was observed the presence of methylated species of As such as DMA and MMA.

1 INTRODUCTION

Arsenic (As) is an abundant element and worldwide distributed through the Earth crust. Volcanogenic activity, underground waters, thermal springs and mining or industrial activities release this toxic element and left a heritage of As in some well-known contaminated places such as New Zealand, Japan, Argentina, Bangladesh, etc. (Shakoor *et al.*, 2015). This is also the case of the geothermal springs Tatio Geysers field in Atacama Desert of the II Region of Chile that contain As concentrations in water of 22-60 mg/L. This high concentration and high temperature of water, about 60°C, together the high UV radiation, produce than only microbial communities able to resist these extreme conditions, could live in this area, probably developing different defence or energy uptake mechanisms in a similar way than reported in other world places. Since the first knowledge about As biotransformation through a series of reductions of pentavalent to trivalent arsenic species and subsequent oxidative methylation with the sulfur atom from S-adenosyl-methionine as redox partner.

2 MATERIALS AND METHODS

2.1 *Study area*

Samples were taken from Platform A1, belonging to the geothermal perforation project of El Tatio, 4 km down the Geyser El Tatio field. Four places very close (22°22′S; 67° 59′O) and above 3700 m of altitude were selected for sampling. Figure 1 shows the locations where sediments and water samples were taken. Three water samples were collected in each place in the board of affluent and the water surface and in a column of 10 cm. Three sediments were also taken in the fourth places. All samples were maintained in polyethylene containers at 4°C in an icebox before transportation to the laboratory and processing.

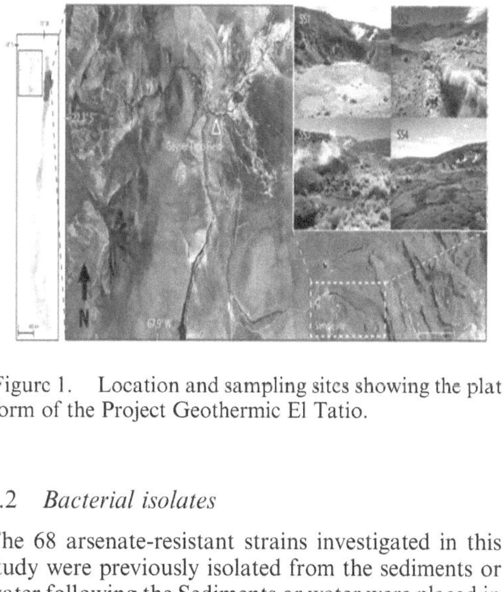

Figure 1. Location and sampling sites showing the plat form of the Project Geothermic El Tatio.

2.2 *Bacterial isolates*

The 68 arsenate-resistant strains investigated in this study were previously isolated from the sediments or water following the Sediments or water were placed in sterile flasks containing 12 mL of CDM medium (2% wt/vol) supplemented with 5 Mm of As(V) and incubated at 25±2°C for 48 h. Enrichment cultures were centrifuged at 5000 rpm for 15 min, washed twice with sterile deionized water and spread onto a new As (V)-supplemented CDM agar (2%) at 28±2°C for 10 days in the dark. The bacterial isolates were submitted to different experiments

2.3 *Tolerance level of As and determination of bacterial reduced activity of As(V)*

Bacterial isolates resistance to As(V) was evaluated by minimum inhibitory concentrations (MIC) tests under aerobic conditions. MICs of the isolated strains were assessed as previously described (Escalante *et al.*,

DOI: 10.1201/9781003317395-33

2009). Aliquots of bacterial cultures (OD600=0.3) were grown on CMD agar plates (2%) containing different concentrations of As(V) (2– 100 mM) and then incubated at 28 °C in oxic or anoxic conditions. The MIC is defined as the lowest concentrations of As(V) that completely inhibited bacterial growth on agar medium after 72 and 120 h of incubation. Control bacteria were obtained in the same CMD agar without As(V). The abilities of the obtained As-resistant bacteria to reduce As(V) were tested under aerobic conditions. The isolates were first tested using a qualitative $AgNO_3$ screening method (Simeonova *et al.*, 2004). The qualitative method to determine the transformation of original As(V) specie to As(III).

2.4 *Total arsenic and As speciation by ICP-MS*

Total As determination in the cytoplasm and plasmatic membrane after incubation to 5mM and 20 mM were performed by ICP-MS. Arsenic species were identified by comparison of the LC-ICP-MS chromatograms of the cytoplasm and citoplasmatic membrane with those provided by injection of standards solution of As(III), As(V), MMA, DMA, AsB and AsC. Furthermore, spiking the extracts with a specific As species were also performed to confirm the presence of this species.

3 RESULTS AND DISCUSSION

3.1 *Bacterial isolates and arsenic tolerance*

Only thirty-two isolates were resistant to As(V). Under aerobic growth conditions, all of the bacterial isolates exhibited resistance to As(V) with MICs exceeding 20 mM. Ten isolates (Hydrogenophaga sp. SS2, Brevundimonas sp. SS2, Pseudomonas sp. SS3, Acinetobacter sp. SS3, Rahnella sp. SS4, Pseudo-monas sp. SS4, Pseudomonas sp. SS4B, Pseudomo-nas sp. SS4C, Pseudomonas sp. SS4D, Enterobacter sp. SS4), in aerobic growth conditions, showed of 50 mM of As(V). Seven isolates (Aeromonas sp. SS1, Exiguobacterium sp. SS2, Streptomyces sp. SS2, Rhodobacter sp. SS2, Aeromonas sp. SS4, Myco-bacterium sp. SS4, Micromonospora sp. SS4 and Haemophilus sp. SS4) were even resistant to As(V) at 100 mM whereas the remaining isolates showed lower levels (<50 mM) of resistance to As(V).

3.2 *Total As concentration in cytoplasm and citoplasmatic membrane and bacterial reduction capacity of As(V)*

In the cytoplasm, both As(III) and As(V) species were be detected. From all the 28 strain tested, in the 15% of the bacterial from sediments and water, practically only As(III) is present. This is the case of Streptomyces sp. SS2, Exiguobacterium sp. SS4B, Pseudomonas sp. SS1 Band Enterobacter sp. SS4. 20% of the bacteria were able to reduce the 25-50% of As (V) to As (III), 15% of them were able to reduce the 75-100% the As(V) initial. The recovery of total As (Figure 2) is, in the most cases up to 80%, but not in all cases. In Pseudomonas sp. SS2 isolate, the recovery respect incubation is only 21%.

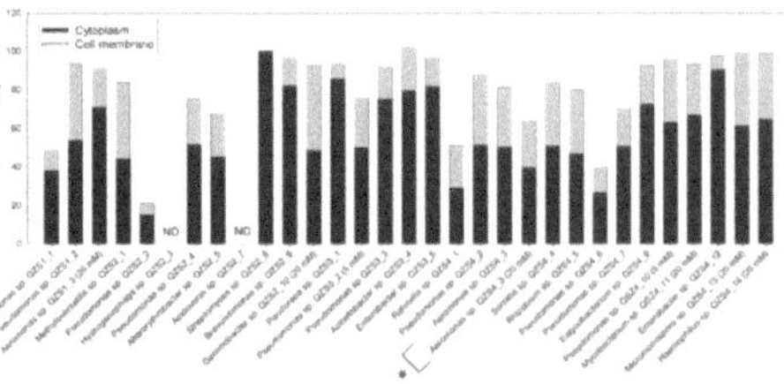

Figure 2. Distribution of total arsenic in cytoplasm and membrane.

4 CONCLUSIONS AND RECOMMENDATIONS

The microbial community under study shifted with very different phylogenetic composition. The naturally high As concentration, ranging up to 67 mg/L and 80 mg/L, of the place and the capacity to develop defence mechanisms determine the different communities and the different behaviour of the bacterial community towards As. One of the characteristics of all these bacteria is the lack of capacity to developed methylation mechanisms. Only As(III) and As(V) have been found as As species present in the cytoplasm and citoplasmatic membrane for all of them.

ACKNOWLEDGEMENTS

We acknowledge the Project VR007/11 Cooperación al Desarrollo de la UCM, Project MECESUP-UA VRI/2014.

REFERENCES

Escalante G., Campos V., Valenzuela E., Yañez J., Zaror C. & Mondaca M.A. 2009. Arsenic resistant bacteria isolated from arsenic contaminated river in the Atacama Desert (Chile). *Bull. Environ. Contam. Toxicol.* 83: 657–661.

Shakoor M.B., Niazi N.K., Bibi I., Rahman M.M., Naidu R., Dong Z., Shahid M. & Arshad M. 2015. Unraveling health risk and speciation of arsenic from groundwater in rural areas of Punjab, *Pakistan. Int. J. Environ. Res. Public Health* 12: 12371–12390.

Simeonova D., Lievremont D., Lagarde F., Muller D., Groudeva V. & Lett M. 2004. Microplate screening assay for the detection of arsenite-oxidizing and arsenate-reducing bacteria. *FEMS Microbiol. Lett.* 237: 249–253.

1.3 Arsenic and other trace elements in global groundwaters

Arsenic in the Environment: Bridging Science to Practice for Sustainable Development –
van der Wal, Ahmad, Petrusevski, Weijma, Savic, van der Wens, Beerendonk, Bhattacharya, Bundschuh & Naidu (Eds)
© 2023 the Author(s), ISBN 978-1-032-32928-4

Processes controlling arsenic distributions in groundwater from an inland basin, China

H.M. Guo[1,2], W. Xiu[1] & W. Qiao[2]
[1]*State Key Laboratory of Biogeology and Environmental Geology, China University of Geosciences, Beijing, P.R. China*
[2]*School of Water Resources and Environment, China University of Geosciences (Beijing), Beijing, P.R. China*

ABSTRACT: High arsenic (As) groundwater has been widely found in arid-semiarid basins, which poses negative effects on human health. Although As distribution in groundwater from flat plains of the basin is quite patchy, there are increasing trends along flow paths from alluvial fan to the plains near the mountains. Both hydrogeolocal and biogeochemical conditions control distributions of dissolved As in groundwater. In oxic conditions of alluvial fans, As is preferentially adsorbed on aquifer sediments with high contents of Fe(III) oxides. Along the flow paths, both dissolved organic matter (DOM) and sedimentary organic matter (SOM) increase, which consume dissolved oxygen (DO) and lead to reducing conditions. Labile SOM is believed to be the primary electron donor, which triggers reductive dissolution of Fe(III) oxides and/or sulfate reduction depending on availability of sulfate. In addition to reduction dissolution, sulfur disproportionation-induced Fe(III) reduction causes As mobilization.

1 INTRODUCTION

High concentrations of dissolved arsenic (As) in groundwaters have been reported in both inland basins and river deltas (Guo *et al.*, 2014). Those groundwaters are mostly hosted under reducing aquifers, excluding high As groundwater in South America (such as Argentina, Chile and Mexico) under oxic conditions (Smedley & Kinniburgh 2002). In both inland basins and river deltas, groundwater As is patchily distributed (Guo *et al.*, 2017; Wang *et al.*, 2019), which makes the location of safe drinking water wells and the prediction of groundwater As extremely difficult. Under reducing condition, reductive dissolution of Fe(III) oxides has been well consented to contribute to As enrichment in groundwater (Fendorf *et al.*, 2010), which would be triggered by surface-sourced DOM and/or SOM in term of carbon sources for microbes and complexation/competition agents (Guo *et al.*, 2019; Mladenov *et al.*, 2015). However, factors controlling patchy distribution of dissolved As in groundwater are poorly understood.

The objectives of this study are to 1) outline distribution of groundwater As in the Hetao basin, a typical inland basin hosting high As groundwater; 2) evaluate factors controlling As distributions in groundwater.

2 MATERIALS AND METHODS

2.1 *Study area*

The Hetao basin, being located in the northwest of Inner Mongolia China, is a typical inland basin hosting high As groundwater. It lies to the north of the Yellow river, to the south of the Langshan Mountains (Figure 1). Shallow groundwater at depths <40 below land surface (bls) is mainly hosted in Pleistocene and Holocene fine sand aquifers, which flows from northwest to southeast in the northern part and from southeast to northwest in the southern part. High As groundwater generally occurs in later Pleistocene lacustrine sediments with relatively high contents of organic matter (Guo *et al.*, 2011). In the northern part, groundwater is mainly used for irrigation, while in the southern part groundwater is mainly for drinking (Guo *et al.*, 2016).

2.2 *Sampling and analysis*

Groundwater and aquifer sediments were taken from the basin and a typical hydrogeological unit located near the mountain where groundwater flows from the alluvial fans to the flat plain. Analyses of groundwater and sediment samples can be found elsewhere (Guo *et al.*, 2016, 2019; Li *et al.*, 2014).

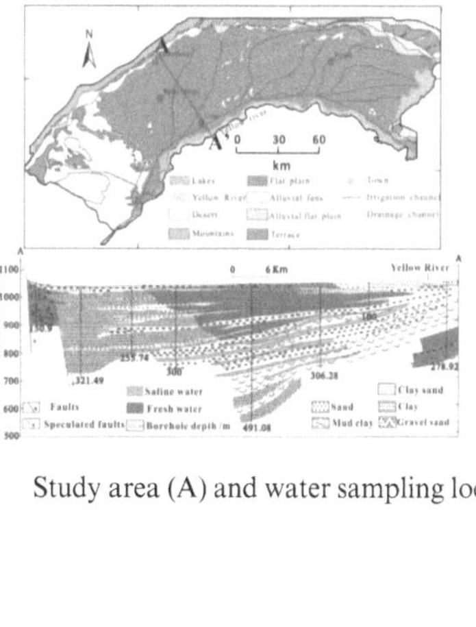

Figure 1. Study area (A) and water sampling locations (B).

3 RESULTS AND DISCUSSION

3.1 *Spatial distributions of dissolved As*

Groundwater As is patchily distributed at the basin scale (Guo *et al.*, 2011). High As groundwater occurs in the northern and eastern basin more frequently. Usually, As concentrations are quite different in wells with the same depth over distances from tens of meters in the hotspot of the flat plain. However, As concentrations generally show increasing trends along the flow paths near the mountains. This distribution pattern supports to locate safe wells with low As concentrations at the alluvial fans (Guo *et al.*, 2016).

3.2 *Processes controlling As concentrations*

In the alluvial fans, groundwater occurs under oxic conditions and contains low As concentrations (mostly <10 μg/L). Batch experiments shows that aquifer sediments adsorbed both As(V) and As(III) with capacities of 19.5 and 33.0 μg/g, respectively. Adsorption on Fe(III) oxides in aquifer sediments is the dominant process limiting dissolved As concentrations in this zone (Figure 2).

Along the flow path, the labile organic matter in groundwater system, especially in aquifer sediments, consumes oxidants (including oxygen and NO_3^-) and leads to reducing environments. Iron isotopic signature shows that microbially mediated reduction of Fe(III) oxides occurs (Guo *et al.*, 2013). The reduction of Fe(III) oxides releases the adsorbed As (Figure 2). In the transition area under moderately reducing conditions, the released As and Fe (II) would be partly readsorbed on the residual Fe(III) oxides. High As concentrations are also observed under reducing conditions with SO_4^{2-} reduction. Precipitation of Fe sulfides can not lead to low concentrations of dissolved As, although it decreases As levels.

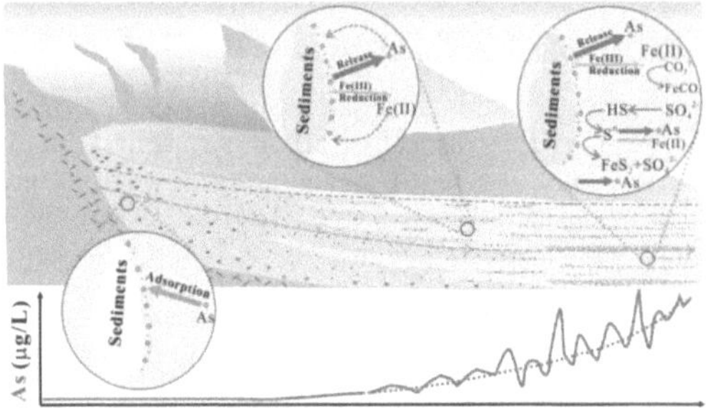

Figure 2. Processes controlling distribution of dissolved As.

Sulfur and oxygen isotopes of SO_4^{2-} indicate that microbially mediated reduction of SO_4^{2-} proceeds at low rates. In combination with the presence of elemental sulfur in aquifer sediments, the fractionation factors between $^{34}S_{SO4}$ and $^{18}O_{SO4}$ suggest that microbial disproportionation of elemental sulfur should take place (Guo *et al.*, 2016), which is coupled with reduction of Fe(III) oxides. The sulfur disproportionation-induced Fe(III) oxide reduction would also release adsorbed As (Figure 2).

4 CONCLUSIONS AND RECOMMENDATIONS

Arsenic concentration is patchily distributed in groundwater of the Hetao basin, and locally show increasing trends along flow paths near the mountains. Arsenic adsorption, Fe(III) oxide reduction, and sulfur dispropotionation are the major processes controlling As distribution, which are affected by the availability of DO, labile OM, Fe(III) oxides, and SO_4^{2-}. This study shows that groundwater in the alluvial fans has low As concentrations, and should be the suitable candidate for water supply.

ACKNOWLEDGEMENTS

The study was financially supported by National Natural science Foundation of China (Nos. 41825017 & 41672225).

REFERENCES

Fendorf S., Michael H.A. & van Geen A. 2010. Spatial and temporal variations of groundwater arsenic in south and Southeast Asia. *Science* 328: 1123–1127

Guo H.M., Li X.M., Xiu W., He W., Cao Y.S., Zhang D. & Wang A. 2019. Controls of organic matter bioreactivity on arsenic mobility in shallow aquifers of the Hetao basin, P.R. China. *J. Hydrol.* 571: 448–459.

Guo H.M., Zhou Y.Z., Jia Y.F., Tang X.H., Jiang Y.X., Lu H., Han S.B., Zheng H. *et al.* 2016. Sulfur cycling-related biogeochemical processes of arsenic mobilization in the Western Hetao basin, China: evidences from multiple isotope approaches. *Environ. Sci. Technol.* 50(23): 12650–12659

Guo H.M., Wen D.G., Liu Z.Y., Jia Y.F. & Guo Q. 2014. A review of high arsenic groundwater in Mainland and Taiwan, China: distribution, characteristics and Geochemical Processes. *Appl. Geochem.* 41: 196–217.

Guo H.M., Liu C., Lu H., Wanty R.B., Wang J. & Zhou Y. 2013. Pathways of coupled arsenic and iron cycling in high arsenic groundwater of the Hetao basin, Inner Mongolia, China: an iron isotope approach. *Geochim. Cosmochim. Acta* 112: 130–145.

Guo H.M., Zhang B., Li Y., Berner Z., Tang X.H. & Norra S. 2011. Hydrogeological and biogeochemical constrains of arsenic mobilization in shallow aquifers from the Hetao basin, Inner Mongolia. *Environ. Pollut.* 159: 876–883

Li Y., Guo H.M. & Hao C.B. 2014. Arsenic release from shallow aquifers of the hetao basin, inner mongolia: evidence from bacterial community in aquifer sediments and groundwater. *Ecotoxicology* 23: 1900–1914

Mladenov N., Zheng Y., Simone B., Bilinski T.M., McKnight D.M. *et al.* 2015. Dissolved organic matter quality in a shallow aquifer of Bangladesh: implications for arsenic mobility. *Environ. Sci. Technol.* 49: 10815–10824.

Smedley P.L & Kinniburgh D.G. 2002. A review of the source, behavior and distribution of arsenic in natural waters. *Appl. Geochem.* 17: 517–568

Wang Y. X., Pi K. F., Fendorf S., Deng Y.M. & Xie X. J. 2019. Sedimentogenesis and hydrobiogeochemistry of high arsenic late pleistocene-holocene aquifer systems. *Earth Sci. Rev.* 189: 79–98.

Arsenic in the Environment: Bridging Science to Practice for Sustainable Development –
van der Wal, Ahmad, Petrusevski, Weijma, Savic, van der Wens, Beerendonk, Bhattacharya, Bundschuh & Naidu (Eds)
 ISBN 978-1-032-32928-4

Competitive adsorption mechanism of As(III) and As(V) in shallow groundwater of Jianghan Plain

Y. Zhang, J.X. Wu, Q. Yu & H.P. Ye
College of Resources and Environmental, South-Central University for Nationalities, Wuhan, P.R. China

ABSTRACT: Inorganic arsenate and arsenite are the main forms of arsenic in groundwater. They have different mobility, toxicity and bioavailability in the natural environment. It is essential to clarify whether As(III) and As(V) will compete for active sites on certain minerals in sediments. A series of laboratory batch experiments in 10 mM PIPES and 25 mM NaCl mixed solution have been carried out to study the competitive adsorption between As(III) and As(V) on aquifer sediments. The results demonstrated that arsenic is more easily enriched in organic layer, iron-aluminum oxide and clay layer deposits with large specific surface area. The amount of As(III) and As(V) alone adsorbed on sediment was approximately the same. When the As(III)/As(V) molar ratio is very high, the adsorption of As(V) on the sediment can be inhibited by As(III) and there will be a more obvious inhibition with the increasing of As(III) concentration. However, the competitive adsorption between As(III) and As(V) on sediment can be neglected when the As(III)/As(V) molar ratio is low. As(V) seems to promote the adsorption of As(III) on sediment, especially when the concentration of As(V) is low.

1 INTRODUCTION

The high arsenic groundwater poses a great threat to human health and environmental problem in many countries around the world. Inorganic arsenic is more toxic than methylated arsenic, and As(III) is more toxic and labile than As(V) (Huang & Matzner 2007). Adsorption is one of the key processes controlling the mobility and availability of As(V) and As(III) in natural environments (Sø *et al.*, 2008). The mobility and the potential bioavailability of arsenic in natural environments are strongly affected by the interaction between arsenic species and mineral surfaces (Smedley & Kinniburgh 2002). It has been reported that arsenic in groundwater is mainly in the form of inorganic arsenate and arsenite. Arsenate and arsenite have different mobility, toxicity and bioavailability in the natural environment. Therefore, it is important to study the competitive adsorption behavior.

2 MATERIALS AND METHODS

2.1 *Site description*

Jianghan Plain is an alluvial plain formed by the Yangtze and Han rivers located in the Middle Reaches of the Yangtze River includes the central and southern regions of Hubei Province (Figure 1). It has a subtropical monsoonal climate and annual temperature ranges from 15°C to 17°C. The average annual precipitation is 1269 mm and the average annual evaporation is 1200 mm. The field monitoring site was constructed in Shahu Village. A total of 39 monitoring wells installed in 13 points has been distributed in this site (Figure 1), each with a nest of borehole at three different depths of 10 m, 25 m and 50 m.

2.2 *Sediment collection and analysis*

A Geoprobe DT21 rig was used to get continuous in-situ sediment samples from ZK01 (Figure 1) in May 2017 with depth below 30 m. The sediment (15 cm long) in sampling were cut according to its lithological changes and immediately wrapped with cling film and tin foil paper after the cores were brought out to the surface. Then the samples were put into 8 L anaerobic box with oxygen scavenging packs and anaerobic indicator (AnaeroPack, Mitsubishi, Japan) and immediately preserved at 4°C freezer.

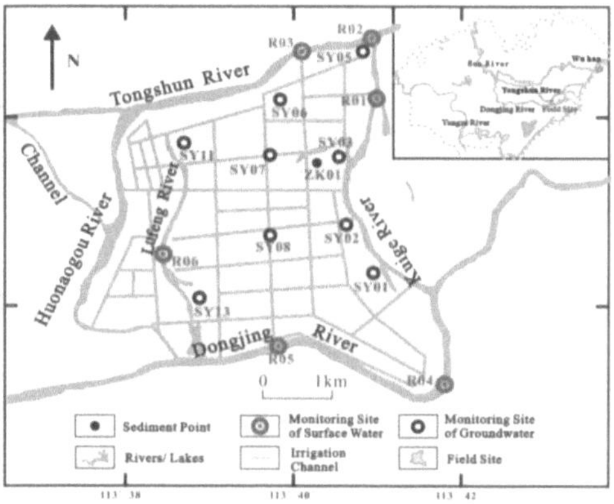

Figure 1. Study area and water sampling locations.

The sediment samples of 17.4 m was chosen as the representatives of the As-contaminated aquifer sediment for our anaerobic batch experiments, referred to hereafter as HJ17.4. Concentrations of As and Fe(T) were determined by atomic fluorescence spectrometry

DOI: 10.1201/9781003317395-35

(KDR-AFS1101Z, China) after digestion of sediments with HF-HNO_3 in a microwave digestion apparatus (Multiwave PRO, Anton paar). Total elemental compositions were determined by X-ray fluorescence spectrometry (XRF-Axios advanced). Total organic carbon in the sediment was determined by TOC analyzer (multi N/C 3100 TOC/TN) after removing inorganic carbon with 2 M HCl. The specific surface area of sediment was determined by five-point N_2-BET method (JW-BK132F). All measurements were repeated in triplicate.

2.3 *Experimental procedures*

The frozen sediments were thawed and homogenized in an N_2 filled glove box. The stock solutions of As(III) and As(V) were diluted with a mixture of 10 mM PIPES (1,4-piperazinediethanesulfonic acid) + 25 mM NaCl to prepare a solution of 1-60 μM As(V) and As(III). Different concentrations of As(V) and As(III) solution were mixed in 1:1 volume ratio to make mixed solution with As(III)/As(V) molar ratio ranged from 0.03 to 30. The initial concentrations of As(V) and As(III) and the ratio of As(V)/As(III) referred to the field groundwater data. The adsorption behaviors of As(III), As(V) and the mixed solution on sediment surface were conducted in 50 mL serum bottle in an anaerobic glove box in triplicate. The serum bottle which contained 1.0 g sediment sample and 25 mL correlation solution with a background electrolyte solution was placed in a shaking incubator with 25°C and shaken at 180 rpm for 48 h. After standing for 2 hours, the supernatant were collected to detect the concentrations of As(III) and As(V).

3 RESULTS AND DISCUSSION

3.1 *Sediment characterization*

The XRF results revealed that HJ17.4 sediment sample was mainly consisted of SiO_2, Al_2O_3 and Fe_2O_3. The lithology of HJ17.4 was clay. The As concentration of HJ17.4 sediment sample was 21.15 mg/kg and the TOC content and the specific surface area of HJ17.4 were 12.46 g/kg and 31.6 m^2/g, respectively, which all higher than those of HJ12, HJ20.1 and HJ26.1 sediment samples (not shown in this paper). It demonstrated that arsenic was more easily enriched in organic layer, iron-aluminum oxide and clay layer deposits.

3.2 *Competitive adsorption between As(III) and As(V)*

Figure 2 shows that the amount of As(III) and As(V) adsorbed by sediment was approximately the same. When the As(III)/As(V) molar ratio is in the range of 0.46-35.34, As(III) inhibits the adsorption of As(V) on sediment surface and the inhibition behavior becomes stronger with the increase of As(III) concentration (Figure 2a), which may be correlated to that As(III) can compete with As(V) for the active site on the iron-manganese (hydrogen) oxide surface. But, As(V) seems to promote the adsorption of As(III) on sediment surface and the lower concentration

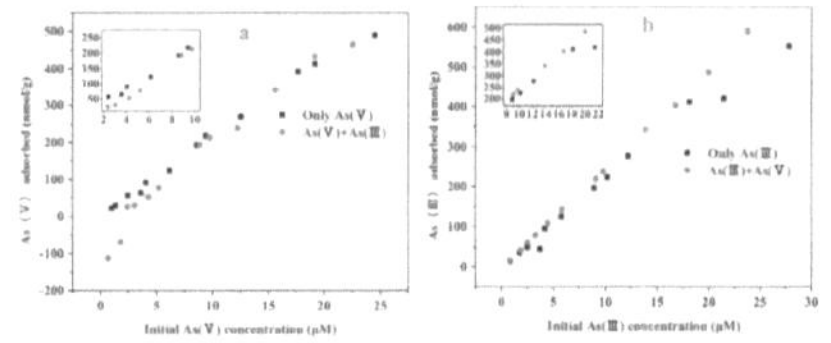

Figure 2. Adsorption of As(III) and As(V) on HJ17.4 sediment sample at Jianghan Plain in 10 mM PIPES buffer solutions with a 25 mM NaCl background. The pH was adjusted to 7.0.

of the As(V), the more obvious of promoting effect (Figure 2b). The competitive adsorption between As(III) and As(V) on sediment can be ignored when As(III)/As(V) is 0.04~0.46. When the initial molar ratio of As(III)/As(V) was 0~0.66, the amount of As(V) adsorbed on the sediment drops sharply, while the amount of As(III) adsorbed on the sediment rises sharply. However, when the initial molar ratio of As(III)/As(V) is 0.66~35.34, the amount of As(V) adsorbed by sediments decreases steadily, and the amount of As(III) adsorbed by sediments rises steadily (not shown).

4 CONCLUSIONS AND RECOMMENDATIONS

Arsenic is more easily enriched in organic layer, iron-aluminum oxide and clay layer deposits with large specific surface area. The amount of As(III) and As(V) alone adsorbed on the sediment surface was approximately the same. When the As(III)/As(V) molar ratio is very high, the adsorption of As(V) on the sediment can be inhibited by As(III) and there will be a more obvious inhibition with the increasing of As(III) concentration. However, the competitive adsorption between As(III) and As(V) on sediment can be neglected when the As(III)/As(V) molar ratio is low. As(V) seems to promote the adsorption of As(III) on sediment, especially when the concentration of As(V) is low.

ACKNOWLEDGEMENTS

We acknowledge the National Natural Science Foundation of China (Contribution: 41702245) and the Natural Science Foundation of Hubei Province (Contribution: 2017CFC862).

REFERENCES

Huang J.H. & Matzner E. 2007. Mobile arsenic species in unpolluted and polluted soils. *Sci. Total Environ.* 377: 308–318.

Smedley P.L. & Kinniburgh D.G. 2002. A review of the source, behaviour and distribution of arsenic in natural waters. *Appl. Geochem.* 17: 517–568.

Sø H.U., Postma D., Jakobsen R. & Larsen F. 2008. Sorption and desorption of arsenate and arsenite on calcite. *Geochim. Cosmochim. Acta* 72: 5871–5884.

Arsenic in the Environment: Bridging Science to Practice for Sustainable Development –
van der Wal, Ahmad, Petrusevski, Weijma, Savic, van der Wens, Beerendonk, Bhattacharya, Bundschuh & Naidu (Eds)
© 2023 the Author(s), ISBN 978-1-032-32928-4

Understanding the groundwater systems in the high and low arsenic zones across the Bhagirathi-Hooghly River, India

M. Chakraborty[1], A.K. Mishra[1] & A. Mukherjee[1,2,3]
[1]*Department of Geology and Geophysics, Indian Institute of Technology–Kharagpur, Kharagpur, India*
[2]*School of Environmental Science and Engineering, Indian Institute of Technology Kharagpur, Kharagpur, India*
[3]*Applied Policy Advisory to Hydrogeosciences Group, IIT Kharagpur, Kharagpur, West Bengal, India*

ABSTRACT: The groundwater arsenic (As) concentrations across the Bhagirathi-Hooghly (B-H) River show contrasting spatial variability. Here, we aim to understand this variability as a function of the differences in hydrostratigraphy and the groundwater evolution on the opposite banks of the B-H River. Our results indicate a strong control of aquifer architecture, hydrogeochemistry and redox drivers on the As distribution patterns on the opposite banks of B-H River.

1 INTRODUCTION

Arsenic (As) concentrations are found to vary widely (at local to regional-scales) across the extensively contaminated flood and delta plains of the Ganges River system. Such spatial variability in As concentrations is believed to be a function of various (hydro)geological, (hydro)geochemical and anthropogenic factors that dictates the mobilization (and sequestration) and distribution of As within the delta aquifers (Chakraborty *et al.*, 2015).

Across the Bhagirathi-Hooghly (B-H) River (a major distributary of the Ganges River), such sharp contrast in As concentrations are observed, wherein the eastern bank aquifer (EBA) are found to be extensively contaminated by As, while the western bank aquifer (WBA) are largely free of As (above permissible limit) (Mukherjee *et al.*, 2018; Sankar *et al.*, 2014). In this study, we investigate the differences in the groundwater system across the B-H River in terms of the hydrostratigraphy and the groundwater reaction pathways that results in such variability of As among the EBA and WBA.

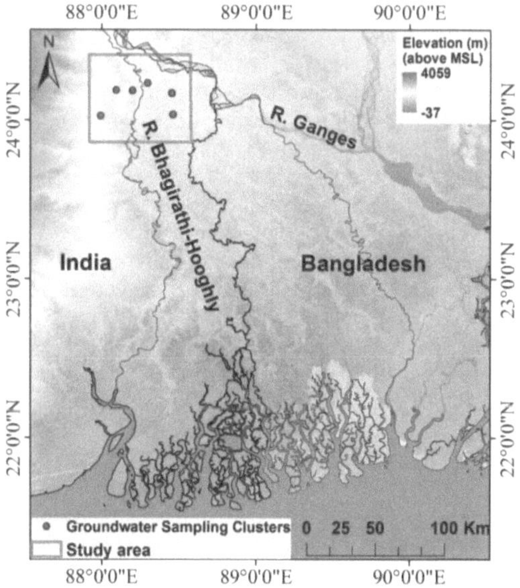

Figure 1. Study area map showing sampled water well clustrers.

1.1 *The study area*

The study area is located within the district of Murshidabad (West Bengal), and is broadly dissected by the B-H River, which flows due south. The area to the east of the river is marked by deltaic lowlands having a dense network of active river channels and loose alluvial deposits with abundant organic matter (OM). The area the west circumscribes the cratonic uplands (further west) and is characterized by a sharp increase in topography, fewer active channels, and older lateritic clays and alluvial sediments containing calcareous and ferruginous nodules.

1.2. *Methodology*

The work involves development of a hydrostratigraphic model for the study area in RockWorks 17 (RockWare, USA) using ~250 borewell lithologs collected from West Bengal Public Health Engineering Department (WBPHED), followed by development of minimal reaction-path (inverse) models between six well clusters (having a total of 62 groundwater sampled wells tapping upto a depth of 85 m below msl) using PHREEQC (version 3.6.2, Parkhurst and Appelo 2013).

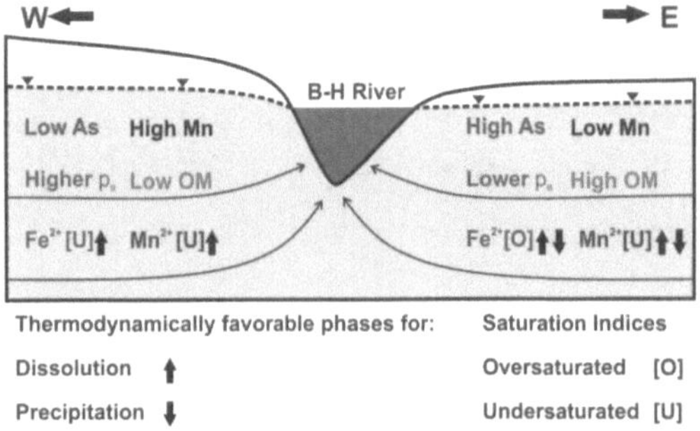

Figure 2. Conceptual diagram of the hydrogeochemical signatures along groundwater evolution pathways on the EBA and WBA across the B-H River.

2 RESULTS AND DISCUSSION

The hydrostratigraphic model shows that the WBA is mostly confined by a thick (upto ~30 m) surficial clay layer and is intercalated by discontinuous aquitard lenses, which possibly restricts any infiltration from the surficial/shallow depths into the aquifers. The EBA, however, is mostly unconfined with a very thin (<5 m) surficial clay layer and no intervening aquitard layers of sigificance. This possibly promotes deeper infiltration of high-As or OM-rich waters into the deeper aquifers. The infiltrational contribution from river water is found to be lower in the confined aquifers of WBA [median ~ 61.6%] than in th unconfined aquifers of the EBA [median ~ 33.3%].

Our geochemical modeling shows large-scale favoured dissolution of siderite in the EBA, wherein groundwater is already oversaturated with respect to siderite, which explains the high Fe content in these aquifers. In the EBA the groundwater redox potential is measured to be lower than the As (III)/As(V) redox barrier, and hence the dissolution of Fe is coupled with widespread liberation of sorbed As in these aquifers, which is the dominant mobilisation mechanism of As in these aquifers (Bhattacharya *et al.*, 1997). The WBA also dominantly shows favoured dissolution of siderite but the groundwater is still undersaturated with respect to the phase. Inspite of the suggested thermodynamically favoured ongoing dissolution of Fe in the region, the As co-liberation and retention in dissolved state is expected to be hindered because the groundwater redox potential in the WBA is higher than the As (III)/As(V) redox barrier.

3 CONCLUSION

This study was done to develop a comprehensive understanding on spatial variability of the groundwater As across the River B-H. The presence of an unconfined aquifer on the eastern side possibly permits a higher surficial infiltration unlike the confined system to the west. In the EBA, the groundwater redox potential has crossed the As(III)/As(V) barrier thus co-liberating As along with Fe. Whereas for WBA, mobilization of As is getting restricted in spite of Fe dissolution is due to higher groundwater redox potential that has not crossed the As(III)/As(V) barrier.

REFERENCES

Bhattacharya P., Chatterjee D. & Jacks G. 1997. Occurrence of arsenic contaminated groundwater in alluvial aquifers from delta plains, Eastern India: options for safe drinking water supply. *Int. J. Water Resour. Dev.* 13:79–92.

Chakraborty M., Mukherjee A. & Ahmed K.M. 2015. A review of groundwater arsenic in the Bengal Basin, *Bangladesh and India: from Source to Sink. Curr. Pollut. Rep.* 1: 220–247.

Mukherjee A., Fryar A.E., Eastridge E.M., Nally R.S., Chakraborty M. & Scanlon B.R. 2018. Controls on high and low groundwater arsenic on the opposite banks of the lower reaches of river Ganges, Bengal Basin, India. *Sci. Total Environ.* 645: 1371–1387.

Parkhurst D.L. & Appelo C.A.J. 2013. Description of input and examples for PHREEQC version 3—a computer program for speciation, batch-reaction, one-dimensional transport, and inverse geochemical calculations: U.S. geological survey techniques and methods, book 6, Chap. A43, 497 p. URL: http://pubs.usgs.gov/tm/06/a43/

Sankar M.S., Vega M.A., Defoe P.P., Kibria M.G., Ford S., Telfeyan K., Neal A., Mohajerin T.J., Hettiarachchi G.M., Barua S., Hobson C., Johannesson K. & Datta S. 2014. Elevated arsenic and manganese in groundwaters of Murshidabad, West Bengal, India. *Sci. Total Environ.* 488–489: 570–579.

Arsenic in the Environment: Bridging Science to Practice for Sustainable Development – van der Wal, Ahmad, Petrusevski, Weijma, Savic, van der Wens, Beerendonk, Bhattacharya, Bundschuh & Naidu (Eds)
© 2023 the Author(s), ISBN 978-1-032-32928-4

Arsenic contamination of groundwater from western part of the Bhagirathi River in the lower delta plain of West Bengal, India

B.A. Shah & S. Chatterjee
Department of Geological Sciences, Jadavpur University, Kolkata, India

ABSTRACT: The spatial distribution of groundwater arsenic is focused on the western part of the Bhagirathi River in Bengal Delta. About 21.26% of tubewells in Purbasthali Block contain arsenic above 10 μg/L (WHO guideline) and 41.73% of tubewells have iron above 1 mg/L (Bureau of Indian Standards guidelines is 0.3 mg/L). Maximum concentrations of arsenic and iron in tubewell samples are 374 μg/L and 17.6 mg/L, respectively. Mean value of arsenic and iron concentrations are 48 μg/L and 4.16 mg/L, respectively. It may be noted that 34.64% of tubewlls are located in the depth of 21 to 40 m. Most of the arsenic-contaminated tubewells are located within the depth of 21 to 40 m in the Holocene Newer Alluvium aquifers. Arsenic pollution in the Holocene aquifers is mostly confined in Katwa Formation. Sijua Formation is almost arsenic-safe in groundwater, which is in upper Pleistocene to lower Holocene in age. Most of the contaminated tubewells are located in Holocene Newer Alluvium and spatially associated with meander cut-offs, paleo-meander scars, abandoned channels and back swamp areas.

1. INTRODUCTION

Arsenic (As) contamination in soil and groundwater is one of the major environmental problems around the world (Ravenscroft *et al.*, 2009). It may be noted that nine districts of Bengal Delta, West Bengal (viz. Malda, Murshidabad, Nadia, North 24-parganas, Kolkata, South 24-parganas, Barddhaman, Hoogly and Howrah) have As contamination in groundwater (Acharyya & Shah 2010; Bhattacharya *et al.*, 1997; Chakraborti *et al.*, 2009; Mandal *et al.*, 1996). The upper permissible limit of As in drinking water is 10 μg/L as per WHO guidelines, which has been endorsed by Bureau of Indian Standards (BIS). In this study, groundwater As in tubewells is tested within the Holocene Newer Alluvium aquifers, as well as the Pleistocene Older Alluvium aquifers. The results on spatial variability of As occurrence with respect to local geological setting are presented from western part of the Bhagirathi River in the lower delta plain of West Bengal.

2 MATERIALS AND METHODS

2.1 *Study area*

The study area is located in the western bank of the Bhagirathi River in Barddhaman District belongs to 23°20′ N to 23°35′ N and 88°10′ E to 88°25′ E (Figure 1). Geographically, it is located from parts of the Damodar fan and floodplain of the Bhagirathi River. The western flank of Bhagirathi River has four major lithostratigraphic units, viz. (i) Lalgarh Formation represented by reddish brown lateratic soil (ii) Sijua Formation representing the older alluvial terrace (iii) Chinsurah Formation representing an assemblage of floodplain features and paleo-meander scars (iv) Katwa Formation representing present day channels and meander zone. The top most layers are represented by unconsolidated sand silt and clay of present day deposits (Mukhopadhyay & Ghosh 2010). The Quaternary Geology map (modified after Geological Survey of India 2014) also reveals four prime units of Quaternary formation (Figure 2).

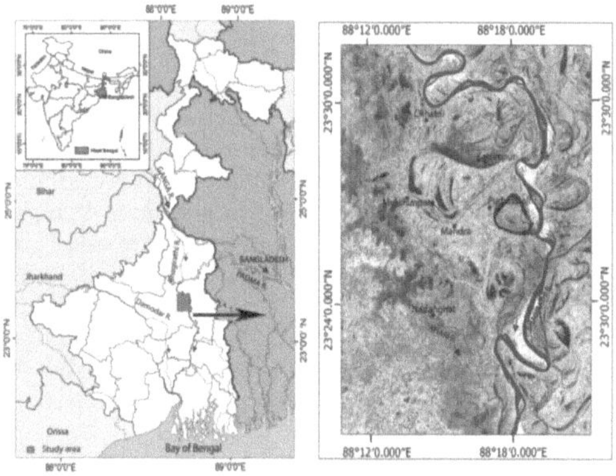

Figure 1. The study area is located in the lower delta plain of West Bengal.

2.2 *Water sampling and laboratory analysis*

The geographic locations of tubewell water samples were recorded by a hand-held Global Positioning System (Garmin eTrex Vista). Tubewell water samples are collected with 1 drop of dilute nitric acid (1:1) AR Grade. The information of tube well depth is acquired from local people. Total 127 tubewells were brought to the laboratory to test the level of arsenic (As) and iron (Fe) concentrations.

DOI: 10.1201/9781003317395-37

Arsenic analysis is done at the laboratory through flow injection hydride generation atomic absorption spectrometry (FI-HG-AAS) system of Perkin-Elmer Model 3100. Iron was analyzed by 1, 10 phenanthroline method by the use of UV spectro-photomet. ArcGIS software was used to create spatial distribution of arsenic in groundwater.

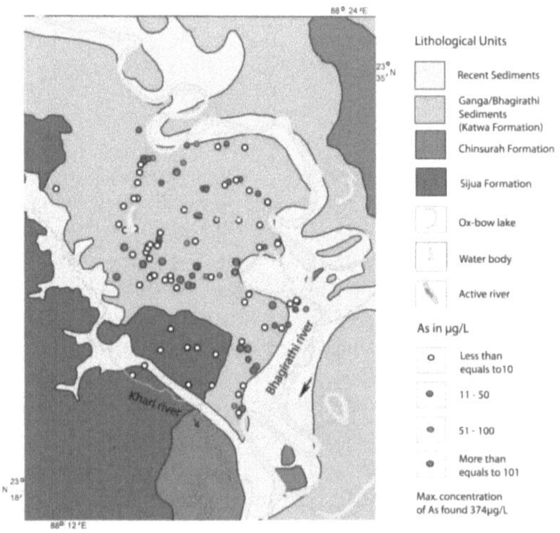

Figure 2. Spatial distribution of groundwater arsenic in the western bank of the Bhagirathi River in the lower delta plain of West Bengal.

3 RESULTS AND DISCUSSION

3.1 *Spatial distribution of arsenic in groundwater with local geological settings*

The individual tubewell based As concentrations are classified into four distinct groups. The tubewell water analysis report shows that about 50.40% of the samples have As concentration above WHO guidelines i.e. >10 μg/L. Mean value of arsenic is 48 μg/L. The most of the As-contaminated tubewells are located (Ganga-Bhagirathi River sediments) in the Katwa Formation (Figure 2). While almost all the tubewells located within Sijua Formation have low level of dissolved arsenic (As <10 μg/L). The Fe values are also classified into four distinct groups. It is observed that about 86.61% samples have Fe concentration above 1 mg/L. The mean value of Fe is 4.17 mg/L. It may be noted that the As-contaminated tubewells have high concentrations of Fe but high Fe concentrations spots are not always associated with elevated As value. However, the correlation coefficient (R) between As and Fe in tubewell water is 0.618, which is significantly positive correlation. It is observed that maximum number of tubewell of shallow depth $\leq$ 60 m has As concentration above WHO and BIS guidelines. The general trend of As concentration is decreasing with increasing depth; though some sporadic occurrences of As concentration at relatively deeper depths $\geq$ 61 m are also found. Interestingly, maximum percentages of tubewells have arsenic ranges 11–100 μg/L and are found in the depth of 21 to 40 m.

The As-contaminated aquifers were associated with grey to black coloured organic-rich argillaceous sediments. Arsenic releases in the aquifer of Bengal Delta by biomediated reductive dissolution of hydrated iron oxide with the presence of anaerobic heterotypic Fe^{3+} reducing bacteria (Islam *et al.*, 2004; Kinniburgh & Smedley 2001).

4 CONCLUSIONS AND RECOMMENDATIONS

Arsenic pollution in lower deltaic part in the western side of the Bhagirathi River are mostly confined in Katwa Formation, while Sijua Formation is almost arsenic safe in groundwater, as it belongs to upper Pleistocene to lower Holocene deposits. Arsenic-contaminated aquifers are mostly located within the middle to upper Holocene lowland organic-rich, fine sand to silt and clayey sediments in the Bengal Delta. Most of the As-contaminated tubewells are located within the depth of 21 to 40 m. The upland terraces mainly made up of the Pleistocene sediments topped by an oxidised impervious zone, are safe from groundwater arsenic problem.

ACKNOWLEDGEMENTS

The financial support of this study came from UGC Start-Up Grant and JU-RUSA 2.0 project, which is gratefully acknowledged.

REFERENCES

Acharyya S.K. & Shah B.A. 2010. Groundwater arsenic pollution affecting deltaic West Bengal, India. *Curr. Sci.* 99: 1787–1794.

Bhattacharya P., Chatterjee D. & Jacks G. 1997. Occurrence of arsenic-contaminated groundwater in alluvial aquifers from delta plains, Eastern India: options for safe drinking water supply. *Int. J. Water Res. Dev.* 13(1): 79–92.

Chakraborti D. *et al.* 2009. Status of groundwater arsenic contamination in the state of West Bengal, India: a 20-year study report. *Mol. Nutr. Food Res.* 53: 542–551.

Islam F.S., Gault A.G., Boothman C., Polya D.A., Charnock J.M., Chatterjee D. & Lloyd J.R. 2004. Role of metal-reducing bacteria in arsenic release from Bengal delta sediments. *Nature* 430: 68–71.

Kinniburgh D.G. & Smedley P.L. 2001. *Arsenic Contamination of Groundwater in Bangladesh.* British Geological Survey Report, WC/00/19, Dhaka.

Mandal B.K. *et al.* 1996. Arsenic in groundwater in seven districts of West Bengal, India – the biggest arsenic calamity in the world. *Curr. Sci.* 70: 976–986.

Mukhopadhyay D.K. & Ghosh G. 2010. Arsenic pollution of groundwater in parts of West Bengal – a case study. *Ind. Jour. Geosciences* 64: 41– 48.

Ravenscroft P., Brammer H. & Richards K. 2009. *Arsenic Pollution: A Global Synthesis.* Wiley-Blackwell, Chichester.

Arsenic in the Environment: Bridging Science to Practice for Sustainable Development –
van der Wal, Ahmad, Petrusevski, Weijma, Savic, van der Wens, Beerendonk, Bhattacharya, Bundschuh & Naidu (Eds)
© 2023 the Author(s), ISBN 978-1-032-32928-4

Arsenic contamination in groundwater within the central Gangetic plains of India: Sources and mobilization

A.L. Ramanathan[1], S.K. Yadav[1], M. Kumar[2] & P. Bhattacharya[3]
[1]*School of Environmental Sciences, Jawaharlal Nehru University, New Delhi, India*
[2]*Department of Environmental Science, Central University of Haryana, India*
[3]*KTH-International Groundwater Arsenic Research Group, Department of Sustainable Development, Environmental Science and Engineering, KTH Royal Institute of Technology, Stockholm, Sweden*

ABSTRACT: Groundwater pollution due to presence of naturally occurring arsenic (As) is reported in sedimentary aquifer worldwide and health related issues due to drinking of As contaminated water have been documented in many parts of the world. In recent year, high concentration of As in groundwater in several parts of India has become a major cause of concern. Late Quaternary stratigraphy and sedimentation in the Ganga Alluvial Plain have influenced As contamination of groundwater, although other environmental factor such as water interactions, ion exchange, and shift in the climatic conditions, redox processes and anthropogenic activities are also responsible for elevated concentration of As in groundwater of Gangetic plain. Arsenic was transported from weathered rock and adsorbed on dispersed iron-oxyhydroxide in the Ganga alluvial plain. Arsenic was released later to groundwater mainly by reductive dissolution of hydrated-iron-oxide and corresponding oxidation of sediment organic matter. Elevated concentration of As in groundwater has been younger alluvium (dark and grey sediment) of Holocene age deposits. Older Alluvium, comprising Pleistocene brownish yellow sediment, extending as deeper aquifers in Newer Alluvium areas, is low in groundwater As. Inorganic arsenite {As(III)} is more dominated on arsenate in the groundwater. The concentration of As(III) in agricultural soil samples varies from not detectable to 40 μg/kg and As (V) was observed as the major species while the total As concentration varied from 3528 to 14,690 μg/kg in our study area. Results of X-ray diffraction (XRD) and environmental scanning electron microscope (ESEM) revealed the presence of hematite and goethite throughout the vertical section below while magnetite was observed only in the upper oxidized layer of the core collected from the study area. Alteration of Fe-oxides and presence of fibrous goethite indicating presence of diagenetic sediment. Siderite plays a crucial role as sinks to the As in subsurface sediments. The study also concluded that decomposition of organic matter trigger mobilization of As into the groundwater.

1 INTRODUCTION

Groundwater pollution due to presence of naturally occurring arsenic (As) is reported in sedimentary aquifer worldwide and health related issues due to drinking of As contaminated water have been documented in many parts of the world. In recent year, high concentration of As in groundwater in several parts of India has become a major cause of concern (Hossain *et al.*, 2012; Rahman *et al.*, 2015). Arsenic pollution of groundwater and soil has become a major environmental issue worldwide in recent decades.There are numerous As compounds reported in the atmosphere and biological systems. Around the globe more than 100 countries has been recorded groundwater pollution due to As contamination (Barringer & Reilly 2013; Naujokas *et al.*, 2013). Natural As enters into the soil by weathering and transportation while anthropogenic As comes from coal burning, mining and applications of As based pesticides. Several studies from India and Bangladesh has been reported with elevated As in soil and water (Kumar *et al.*, 2016; Rahman *et al.*, 2015; Verma *et al.*, 2015). For more than two decades, As-rich groundwater has been used for irrigation in the middle Gangetic plain and putting agricultural soils in this plain at high risk of As pollution.

2 METHODS

Nineteen (19) groundwater samples were collected in Samastipur district by following the standard procedure of APHA (2008) during June 2015. For As(III) a disposable cartridge (Metal Soft Center, PA, USA) was used. Eleven (11) agricultural soil samples (0-15 cm depth) were collected from the study area. Beside water and soil samples two deep core (subsurface sediment) were also collected. One core was from older alluvium and another from younger alluvium. The hand-flapper drilling method was used to obtain core samples and drilling were done to depths of 46.2 and 33.5 m at Methrapur and Haral Chapar respectively. Digestion of soil and sediment samples were done by microwave assisted extraction technique along with 1M phosphoric acid (Rahman *et al.*, 2009). Analysis of water and digested soil samples were done on ICP-MS (Agilent Technologies, Japan 7500ce). Mineralogical investigation done by XRD 6000, (Shimadzu Corporation, Japan) analysis and environmental scanning electron microscopes (ESEM) (FEI Quanta 450 FEG) fitted with an EDAX Apollo X SDD EDX detector.

DOI: 10.1201/9781003317395-38

3 RESULTS AND DISCUSSION

In the study area, As concentrations ranged from 1.3 to 104.7 μg/L. The average As concentrations (20 g/L) are more than twice the WHO drinking water norm of 10 μg/L. From the Eh-pH diagram, it was observed that approx. 73% groundwater samples were As(III) dominated while 27% were As(V) dominated. Arsenic concentrations in groundwater usually vary from region to region based on regional geology (Rahman *et al.*, 2015). In central gangetic plain As(III) is dominated species.

The analytical result of As in SRM soil (n = 7) showed that experimental value (105 ± 10 μg/g) was almost similar to the certified value (105 ± 8 μg/g). The total As concentration in agricultural soils ranged from 3527 to 14,690 μg/kg (Table 1), which is generally 2 to 7 times higher than the world average value of 2000 μg/kg for igneous and sedimentary rocks.

Table 1. Concentration of inorganic As species with their respective extractable amount (%) in agricultural soil (n=11) (results expressed in μg/kg and are based on peak area calculation). As(ext): extractable As; As(tot): total As; nd: not detected.

S. No.	Lat.	Long.	As(III)	As(V)	As(ext)	As(tot)	% Ext.
1	25.57	85.67	nd	1050	1050	3528	29.8
2	25.57	85.64	nd	2413	2413	10,600	22.8
3	25.55	85.67	39	6835	6874	12,940	53.1
4	25.56	85.72	20	5913	5933	10,830	54.8
5	25.56	85.72	nd	3420	3420	9527	35.9
6	25.56	85.71	31	4257	4289	9525	45
7	25.54	85.64	40	5404	5444	9742	55.9
8	25.58	85.68	nd	4403	4403	12,410	35.5
9	25.55	85.66	0.1	4513	4513	11,630	38.8
10	25.61	85.67	12	3766	3779	13,490	28
11	25.55	85.67	nd	6152	6152	14,690	41.9

The percentage of extractable concentration of inorganic As with 1 M ortho-phosphoric acid ranged from 22.9 to 56. The concentration of As(III) and As(V) varied from not detectable to 40.1 and 1050 to 6835 μg/kg, respectively. Arsenate was detected in almost all agricultural soil samples while As(III) was detected only in 6 samples in a low spanning range, i.e. 0.1–40.1 μg/kg.

For both subsurface cores, the sediment was divided into five distinct lithofacies based on texture and lithology. A four-color version of the Munsell Color Chart has been used to identify the As concentration. A lithology for both the sediment core were plotted with the photographic view. At a depth of 21.3 mbgl, higher concentrations of As(tot), As(V), and As(III) were found in Methrapur, Dal Sing Sarai, with concentrations of 20,057, 6289, and 264 g/kg, respectively. At depth of 6.5 mbgl, dark grey miceaceous sand with organic matter was reported which was associated with higher As pollution.

It has been reported that the concentration of As varies according to the color and grain size of the sediment in the Gangetic plain (Hasan *et al.*, 2009). Brown to yellowish color with coarser sediment contains small amounts of As while having a grey to black color with fine grain size contains comparatively larger. A very good correlation (r^2=0.96) was observed between As in soil and sediment samples.

The statistical analysis shown As(V) positively associated with As(ext) and As(tot), indicate that the presence of As(V) and As(ext) was also higher in the As region. As(III) had a positive relationship with sand but a negative relationship with silt. In Harail Chapar, sand has a negative correlation with As(tot), while in Methrapur, there is no major correlation, suggesting that As is less associated with this fraction of sediment.

Mineralogical study suggested that hematite and goethite are the minerals available almost each depth of the sediment core. A very small XRD peak of siderite and magnetite were also observed in both the sediment cores. Another possible source of elevated As in the upper section may be the presence of chlorite, muscovite, and Fe-OH coated quartz in core sediments.

4 CONCLUSIONS

In our study area, As(III) is the most abundant element in groundwater, while As(V) is the most abundant element in agricultural field soil and subsurface sediment. However in subsurface sediment As(III) was dominant species. Inorganic As forms were linked to: first, the aquifer's redox conditions, which were represented in sediment color; and second, grain size, with dark sediment containing more As(III) and As(V). Siderite is a recognised As sink that can be found in the reduced portion of the sediments. The presence of fibrous goethite and the modification of Fe-oxides in this region indicate partial solubilization of oxyhydroxides. This studies did not provide any microbiological testing to see whether microbes are involved in As mobilisation in the middle Gangetic plain.

REFERENCES

Barringer J.L. & Reilly P.A. 2013. *Arsenic in Groundwater: A Summary of Sources and the Biogeochemical and Hydrogeologic Factors Affecting Arsenic Occurrence and Mobility*. INTECH Open Access Publisher.

Hasan M.A., von Brömssen M., Bhattacharya P., Ahmed K.M., Sikder A.M., Jacks G. *et al.* 2009. Geochemistry and mineralogy of shallow alluvial aquifers in Daudkandi Upazila in the meghna flood plain, Bangladesh. *Environ. Geol.* 57: 499–511.

Hossain E., Islam. K., Yeasmin F., Karim M.R., Rahman M., Agarwal S., *et al.* 2012. Elevated levels of plasma big endothelin-1 and its relation to hypertension and skin lesions in individuals exposed to arsenic. *Toxicol. Appl. Pharm.* 259(2): 187–194.

Kumar M., Ramanathan A.L., Rahman M.M. & Naidu R. 2016. Concentrations of inorganic arsenic in groundwater, agricultural soils and subsurface sediments from the middle gangetic plain of Bihar, India. *Sci. Total Environ.* 573: 1103–1114.

Naujokas M.F., Anderson B., Ahsan H., Aposhian H.V. *et al.* 2013. The broad scope of health effects from chronic arsenic exposure: update on a worldwide public health problem. *Environ. Health Perspect.* 121: 295.

Rahman M., Mamun A.A., Karim M.R., Islam K. *et al.* 2015. Associations of total arsenic in drinking water, hair and nails with serum vascular endothelial growth factor in arsenic-endemic individuals in Bangladesh. *Chemosphere* 120: 336–342.

Verma S., Mukherjee A., Choudhury R. & Mahanta C. 2015. Brahmaputra river basin groundwater: solute distribution, chemical evolution and arsenic occurrences in different geomorphic settings. *J. Hydrol. Regional Studies* 4: 131–153.

Arsenic in the Environment: Bridging Science to Practice for Sustainable Development –
van der Wal, Ahmad, Petrusevski, Weijma, Savic, van der Wens, Beerendonk, Bhattacharya, Bundschuh & Naidu (Eds)
© 2023 the Author(s), ISBN 978-1-032-32928-4

Distribution of arsenic and uranium in groundwater utilized as drinking water in Bihar, India

L.A. Richards[1], A. Kumar[2], P. Shankar[2], A. Gaurav[2], A.K. Ghosh[2] & D.A. Polya[1]
[1]*Department of Earth and Environmental Sciences and Williamson Centre for Molecular Environmental Science, The University of Manchester, Manchester, UK*
[2]*Mahavir Cancer Institute and Research Center, Phulwari Sharif, Patna, India*

ABSTRACT: Geogenic contaminants such as arsenic (As) and uranium (U) in groundwater can lead to dangerous health impacts. In this study, we have undertaken a groundwater survey across all districts of Bihar particularly in order to identify spatial distribution of trace contaminants (notably As and U) and the associated dominant geochemical conditions. Concentrations of As and U vary substantially and are strongly inversely correlated, consistent with the redox controls on As and U mobility. This work has important implications on understanding the spatial controls on As and U, remediation priorities and selection, and may be useful in informing further characterization and/or monitoring efforts.

1. INTRODUCTION

A major environmental and public health challenge is the presence of geogenic contaminants such as arsenic (As) and uranium (U) in groundwater. In Bihar, India, the presence of As (Chakraborti *et al.*, 2017; Chakraborti *et al.*, 2018; Ghosh *et al.*, 2012; Saha *et al.*, 2010), and to a lesser extent U (Kumar *et al.*, 2018, 2020) has been reported, although spatially representative, publically available and non-summarized data remains difficult to obtain, and thus wider understanding of the dominant geochemical processes and controls on spatial variability in the area remains limited. The aim of this study was to systematically obtain representative groundwater samples from across Bihar, in order to (i) understand the distribution of geogenic contaminants (particularly As and U) and (ii) characterize the geochemical processes expected to impact groundwater geochemistry in Bihar.

2 MATERIALS AND METHODS

2.1 *Field area*

The field area is in Bihar, northern India (Figure 1). Approximately ~ 150 field sites were targeted prior to sampling to encompass (i) representative locations in all 39 districts of Bihar; and (ii) logistic feasibility. Multiple depths were targeted at each location to capture multiple aquifers.

2.2 *Water sample collection*

Groundwater samples were collected from existing private and government wells (typically hand-pumps). Most sampling was completed in the pre-monsoon season between March and July 2019. In total ~ 270 samples were collected at an average sampling density of 2 depths (within ~ a maximum of 500 m from each other; often < 50 m apart) per ~ 680 km^2. All sampled sources were in frequent use and were flushed for ~ 1 – 2 minutes prior to sampling. Subsamples were collected for cation, anion and dissolved organic carbon analysis.

2.3 *Water sample analysis*

In-situ measurements of pH, oxidation-reduction potential (*Eh*), temperature and electrical conductivity (EC) were made. Cations and trace metal (loids) including As and U were measured using inductively coupled plasma atomic emission spectrometry (ICP-MS) and anions using ion chromatography (IC) at the University of Manchester. Alkalinity was estimated by charge balance. Quality assurance/quality control measures included certified reference materials, calibrations run every 10 samples and duplicate analysis of 15% of samples.

3 RESULTS AND DISCUSSION

3.1 *Dominant groundwater chemistry*

Groundwater was generally circum-neutral (median pH = 7.2), with a wide range of *Eh* (median = 200 mV) and EC (median = 0.65 mS/cm; median temperature = 27 °C). The majority of groundwater is the Na-HCO_3^- type (suggesting ion exchange), with some samples instead Ca-HCO_3^- or Na-Cl type.

3.2 *Arsenic and uranium in Bihar*

Concentrations of groundwater As range from < 1 to ~ 900 µg/L, with ~ 15% of samples from this study exceeding the WHO provisional guideline of

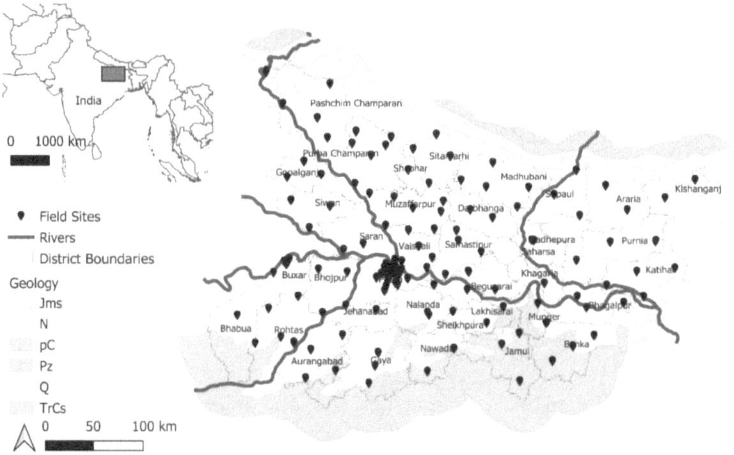

Figure 1. Field sites in Bihar, India. Maps were created using QGIS (Version 3.10 A Coruña; https://qgis.org/en/site/) with open access layers from Natural Earth and the US Department of the Interior.

10 μg/L (World Health Organization 2011), and ~4% exceeding 50 μg/L. Groundwater U concentrations range from < 1 to ~80 μg/L), with ~7% of samples exceeding the WHO provisional guideline of 30 μg/L. There is a strong and statistically significant inverse relationship between As and and U (t value = -2.7; $p < 0.05$; degrees of freedom = 267), consistent with redox controls on As and U mobility. Concentrations of As tend to be higher in areas near and to the north of the Ganga River, particularly between the Ghagara and Gandak tributaries; U concentrations are elevated mostly in a NW-SE band to the east of the Ghagara and running south of the Ganga. A strong relationship between As, Fe and depth is observed. In contrast U is correlated with *Eh*, HCO_3^- and NO_3^-. Detailed interpretation on the distribution and geochemical controls on groundwater chemistry in Bihar is the subject of ongoing work by co-authors.

4 CONCLUSIONS

Here we report the inorganic geochemical characterization (including As and U and associated parameters) spanning all districts in Bihar, India. Despite the relatively limited sampling density, this spatially comprehensive interpretation provides important information on the distribution and dominant geochemical controls on groundwater As and U in Bihar, with implications on remediation priorities and selection, and identifying areas where further sampling and/or monitoring may be beneficial.

ACKNOWLEDGEMENTS

This research was supported by the DST-Newton Bhabha NERC-EPSRC Indo-UK Water Quality Programme (NE/R003386/1 & DST/TM/INDO-UK/2K17/55(C) & 55(G); 2018 - 2021) and a Dame Kathleen Ollerenshaw Research Fellowship (LR). P. Lythgoe is thanked for analysis. Support of local authorities and landowners is greatly appreciated.

REFERENCES

Chakraborti D., Rahman M.M., Das B., Chatterjee A., Das D., Nayak B., Pal A., Chowdhury U.K., Sad Ahmed S., Biswas B.K., Sengupta M.K., Hossain M. A., Samanta G., Roy M.M., Dutta R.N., Saha K.C., Mukherjee S.C., Pati S.P., Kar P.B., Mukherjee A. & Kumar M. 2017. Groundwater arsenic contamination and its health effects in India. *Hydrogeol. J.* 25(4): 1165–1181.

Chakraborti D., Singh S.K., Rahman M.M., Dutta R. N., Mukherjee S.C., Pati S. & Kar P.B. 2018. Groundwater arsenic contamination in the Ganga river basin: a future health danger. *Int J Environ Res Public Health* 15(2): 180.

Ghosh A.K., Bose N., Kumar R., Bruining H., Lourma S., Donselaar M.E. & Bhatt A.G. 2012. Geological origin of arsenic groundwater contamination in Bihar, India. In J.C. Ng, B.N. Noller, R. Naidu, J. Bundschuh & P. Bhattacharya (eds.), *Understanding the Geological and Medical Interface of Arsenic (As2012)*. London: Taylor & Francis Group.

Kumar D., Singh A., Jha R.K., Sahoo S.K. & Jha V. 2018. Using spatial statistics to identify the uranium hotspot in groundwater in the mid-eastern gangetic plain, india. *Environ. Earth Sci.* 77: 702.

Kumar D., Singh A., Kumar P., Jha R.K., Sahoo S.K. & Jha V. 2020. Sobol sensitivity analysis for risk assessment of uranium in groundwater. *Environ. Geochem. Health* 42: 1789–1801.

Saha D., Sarangam S.S., Dwivedi S.N. & Bhartariya K. G. 2010. Evaluation of hydrogeochemical processes in arsenic-contaminated alluvial aquifers in parts of Mid-Ganga basin, Bihar, Eastern India. *Environ. Earth Sci.* 61: 799 – 811.

World Health Organization. 2011. Guidelines for Drinking-Water Quality: Fourth Edition. Geneva.

Arsenic in the Environment: Bridging Science to Practice for Sustainable Development –
van der Wal, Ahmad, Petrusevski, Weijma, Savic, van der Wens, Beerendonk, Bhattacharya, Bundschuh & Naidu (Eds)
 ISBN 978-1-032-32928-4

Pseudo contour maps from logistic regression modelling: Case study of groundwater arsenic distribution in Gujarat state, India

R. Wu[1], J. Podgorski[1,2], M. Berg[2] & D.A. Polya[1]
[1]*Department of Earth and Environmental Sciences, & Williamson Research Centre for Molecular Environmental Science, University of Manchester, Manchester, UK*
[2]*Eawag, Swiss Federal Institute of Aquatic Science and Technology, Dübendorf, Switzerland*

ABSTRACT: Logistic regression modelling is a widely used prediction method to generate probability maps of groundwater arsenic exceeding a selected threshold concentration. In this study, we converted logistic regression modelled probability maps to a pseudo contour map which comprehensively integrated important information of several probability maps of groundwater arsenic concentrations in Gujarat state, India. The pseudo contour map of groundwater arsenic concentrations clearly and intuitively indicates high groundwater arsenic concentrations (>10 μg/L) occurring in Kachchh District and Banas Kantha District and shows groundwater arsenic concentrations in northern and central Gujarat to be slightly higher than near the western and eastern borders of the state. The cutoffs, where sensitivity equals specificity, is one of the most appropriate criteria of diagnosis of continuous probability results, better ensuring the accuracy of the generation of a pseudo contour map. However, the probability cutoff of each contour is likely different, so that the generated map is pseudo-contour rather than a contour map.

1 INTRODUCTION

High arsenic groundwaters pose adverse health effects for hundreds of millions of people worldwide (Bhattacharya 2017). Rational prediction of the distribution of arsenic contamination in groundwaters is an important tool to managers to prevent people's exposure to arsenic. Logistic regression modelling has been widely used to explore the statistical relationship between dependent variables (groundwater arsenic) and other environmental independent variables in order to predict the distribution of arsenic in groundwaters. Probability maps are a common way to present the prediction results of logistic regression models. In this article, we develop a new way of integrating logistic regression model results to generate a pseudo contour map of groundwater arsenic concentrations, in this case, in Gujarat state, India.

2 MATERIALS AND METHODS

2.1 *Study area and groundwater arsenic*

The Gujarat State is situated between North latitudes 20°06' 00" to 24°42' 00" and East longitudes 68°10' 00" to 74°28' 00" (Figure 1), with an area of 196,024 km^2. (CGWB 2016). Groundwater arsenic data were obtained from CGWB (2016).

2.2 *Probability maps of different arsenic concentrations*

Five different thresholds for logistic regression models were set, including 10 μg/L, 5 μg/L, 4 μg/L, 3 μg/L, and 2 μg/L, since 10 μg/L is the WHO provisional guide value for arsenic in drinking water and most arsenic concentrations are below 10 μg/L in the dataset. Five probability maps, each with the different selected threshold was produced based on the best performing logistic regression models.

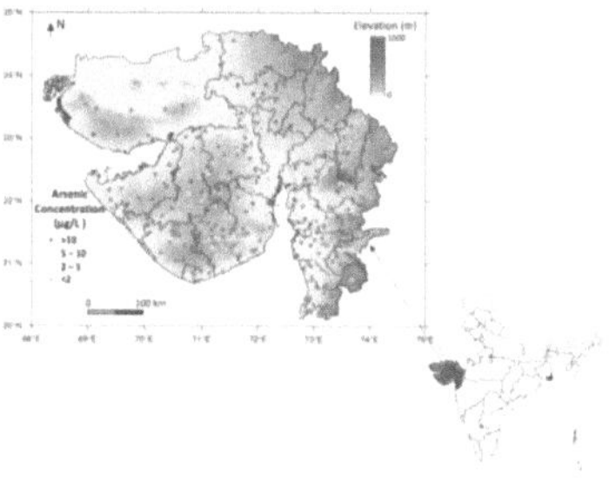

Figure 1. Study area and the distribution of groundwater arsenic concentrations. (data from CGWB 2016).

2.3 *Determining the cutoffs for converting probability maps to a pseudo contour map*

The sensitivity, accuracy, and specificity of the best performing logistic regression models were plotted against cutoff which is in range of 0 to 1. Sensitivity and specificity lines intersected at a certain cutoff which was then used to compare with the probabilities calculated by logistic regression in order to determine whether arsenic concentrations exceeded the given threshold (Podgorski *et al.*, 2017).

DOI: 10.1201/9781003317395-40

2.4 *Creation of pseudo contour map*

Converting probability maps of groundwater arsenic concentrations exceeding selected thresholds to a pseudo contour map is, to the best of our knowledge, a new method in this context. The areas whose probabilities exceeded the selected cutoff were considered as the zone of occurrence of groundwater arsenic concentrations exceeding the selected threshold concentration. Combining occurrence zones of different groundwater arsenic concentrations generated a pseudo-contour map of groundwater arsenic concentrations.

3 RESULTS AND DISCUSSION

3.1 *The cutoffs of converting probability maps to a pseudo contour map*

Probability maps of arsenic concentration exceeding 10 μg/L, 5 μg/L, 4 μg/L, 3μg/L and 2 μg/L are shown in Figure 2. The sensitivity, accuracy, and specificity of the best performing logistic regression models were plotted against cutoffs (Figure 3). The cutoffs (10 μg/L: 0.69, 5 μg/L: 0.66, 4 μg/L: 0.61, 3μg/L: 0.57 and 2 μg/L: 0.50) where sensitivity intersects with specificity were used to create a pseudo map of arsenic concentrations in groundwaters.

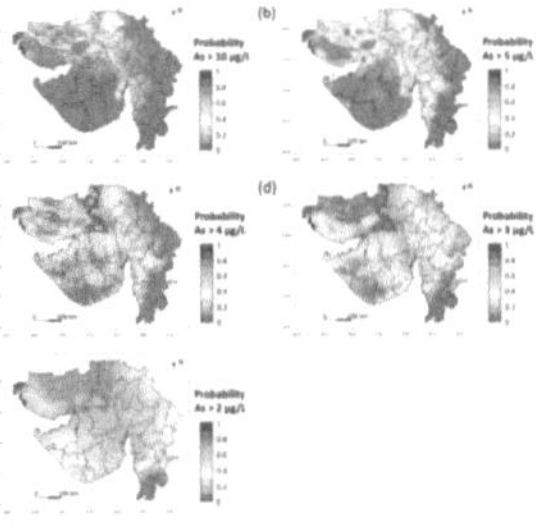

Figure 2. Probability map of arsenic concentration exceeding thresholds of (a) 10 μg/L, (b) 5 μg/L, (c) 4 μg/L, (d) 3 μg/L, and (e) 2 μg/L in Gujarat.

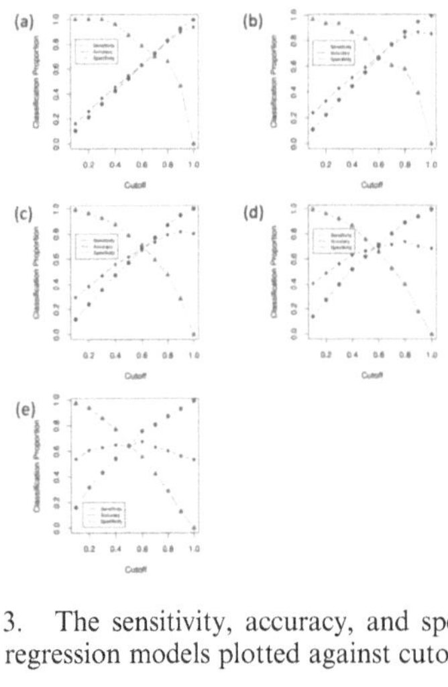

Figure 3. The sensitivity, accuracy, and specificity of logistic regression models plotted against cutoffs.

3.2 *Pseudo contour map of the distribution of groundwater arsenic in Gujarat state*

The pseudo contour map of arsenic concentrations in Gujarat groundwaters is shown in Figure 4. High groundwater arsenic concentrations (>10 μg/L) occur in Kachchh District and Banas Kantha District. Groundwater arsenic concentrations in northern and central Gujarat are slightly higher than in the western and eastern borders. The pseudo contour map shows the distribution of arsenic clearly and intuitively.

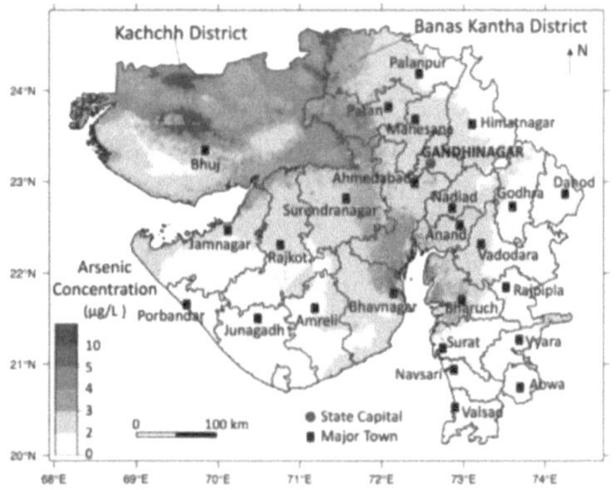

Figure 4. Pseudo contour map of arsenic concentrations in groundwaters in Gujarat. Raw data from CGWB (2016).

4 CONCLUSIONS

The pseudo contour map can integrate information from several probability maps using different thresholds in order to clearly display the distribution of different groundwater arsenic concentrations in one map. The cutoffs where sensitivity equals specificity is one of the most appropriate criteria of diagnosis of continuous probability results, better ensuring the accuracy of the generated pseudo contour map. However, the probability cutoff of each contour is likely different, so that generated map is not really a contour map but rather a pseudo-contour map. In this study, the pseudo contour map clearly and intuitively indicates the distribution of high groundwater arsenic concentrations (>10 μg/L) in a state, Gujarat, with largely low reported groundwater arsenic concentrations.

REFERENCES

Bhattacharya P., Polya D. & Jovanovic D. (eds.). 2017. *Best Practice Guide on the Control of Arsenic in Drinking Water*. IWA Publishing.

CGWB-Central Ground Water Board. 2016. Groundwater Year Book – 2015 – 2016 Gujarat state and UT of Daman & Diu.

Podgorski J.E., Eqani S.A.M.A.S., Khanam T., Ullah R., Shen H. & Berg M. 2017. Extensive arsenic contamination in high-ph unconfined aquifers in the Indus Valley. *Sci. Adv.* 3(8): e1700935.

Arsenic in the Environment: Bridging Science to Practice for Sustainable Development –
van der Wal, Ahmad, Petrusevski, Weijma, Savic, van der Wens, Beerendonk, Bhattacharya, Bundschuh & Naidu (Eds)
© 2023 the Author(s), ISBN 978-1-032-32928-4

Gold-mine related arsenic contamination of drinking water sources in Kolar Gold Fields, India

D. Ghosh[1], P.C. Arya[1] & P.K. Sarath[1,2]
[1]*Biogeochemistry Research Group, Centre for Earth Sciences, Indian Institute of Science, Bangalore, India*
[2]*LOCEAN, Sorbonne University, Paris, France*

ABSTRACT: The mining of Kolar Gold Fields (KGF), Karnataka, carried out for more than a century, has largely impacted the groundwater aquifers. In addition to the water logged sulfide-rich gold (Au) minerals, the leaching of mine tailing dumps had introduced hazardous elements and cyanide into the aquifer system. A detailed survey covering a radius of 10 km of the Kolar Gold Field, with samples from 53 borewells were collected during pre and post-monsoon seasons of 2018, and different parameters were assessed. Contradicting the previous reports, the mineral dissolution and contaminant transport was found to be the function of pH. Almost 80% of the water samples exceeded the regulatory limits of potable water criteria with excess arsenic (12-127 μg/L), fluoride (>1.5 mg/L), dissolved salts (>500 mg/L) and total hardness (>382–1027 mg/L). Additionally the nitrate and sulphates are also above the regulatory limits. A multiple regression model was developed to predict Water Quality Index (WQI) of groundwater from each borewell for the different seasons. From the hydrogeochemical facies, it can be concluded that the dissolution of sulfide bearing Au-rich minerals, releasing H3AsO+ at low pH is transported through, facilitating rock weathering, results in groundwater composition of Ca–HCO_3 type and Ca–Na–HCO_3 or Ca–Mg–Cl type.

1 INTRODUCTION

The contaminant transport from historically active abandoned gold mines, altering the fluxes of hazardous elements in the groundwater has been observed in many corners of the world (Chakraborti *et al.*, 2013; Mamatha & Rao 2010). The sulfide phases of inter-grown gold, leads the production of As-rich wastes (Langmuir *et al.*, 1999). Historically active mines have large metalloid site dumps.

Since there is no rehabilitation of such abandoned mining sites, or attenuation of As-rich wastes, many incidences of As related health conditions has been reported. Arsenic is a frequent component of gold mines as: arsenides, sulfides, and sulfosalts. In abandoned water logged mines, the reductive conditions result in weathering of such minerals, lowering of pH and mobilization of As and other elements.

2 MATERIALS AND METHODS

2.1 *Study area*

Kolar District consists predominantly of migmatitic gneiss and the younger granites. The Along the joints of the granitic gneiss fracture/fissure system facilitate groundwater circulation and also hold moderate quantities of water. The overall transmissivity of the formations vary from 2 to 1935 m^2/d. The general yield of the formations ranges from 0.8 to 30 L/s (Mamatha & Rao 2010).

The contact zones of the peninsular gneiss and granites have been reported to have high levels of fluoride (F). The source might be the weathering of fluoride bearing minerals, like fluorite, fluorapaptite and fluoride-rich micas. In Kolar, such contact zones between the litho-units of the taluks Bagepalli and Gudibanda with the occurrence of peninsular gneiss in Mulbagal taluk, as seen in the geological map of the district (Mamatha & Rao 2010).

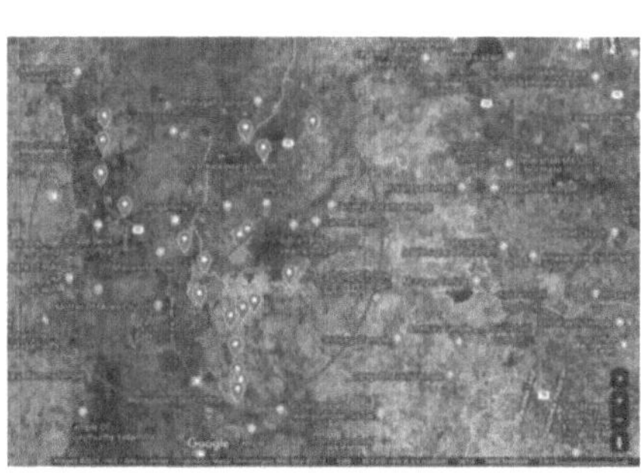

Figure 1. The study area showing a radius of 10 km around the Kolar Gold Field.

2.2 *Assessment of drinking water quality around KGF*

The potable water demand of the urban population in the district Kolar is met 80% by groundwater resources abstraction through installed boreholes and 20% through tapped Cauvey river water supply (seasonal). Almost 95% of the irrigation is done using groundwater sources. Most of these abstraction wells are linked to aquifers at the contact zones of granite and peninsular gneiss.

2.3 *Water sampling and analysis*

Water sampling was carried out during summer/dry season (Mar-Apr 2018) and post monsoon (Aug-Sept 2018). A total of 53 pre-installed bore-wells were targeted covering a radius of 10 kms from KGF (Figure 1). General physio-chemical parameters such as, pH, TDS, temperature (T), electrical conductivity (EC) and redox potential (Eh) were measured in situ. Major anions were analyzed by cholorometric analyses and the cations were quantified using inductively coupled plasma-optical emission spectrometry (ICP-OES). The Biochemical Oxygen Demand (BOD) was measured directly by BOD digital incubator and Chemical Oxygen Demand (COD) by digital reactor and photometer. Cyanide was quantified by colorimetric method and the coliform load was estimated by MPN test.

2.4 *Data analysis and water quality assesment*

ArcGIS software was used to create spatial distribution of mean As (µg/l). The Water Quality Index (WQI) was used to expresses overall water quality with a single digit, based on several water quality parameters.

$$WQI = \sum_{i=1}^{i} Q_i W_i / \sum_{i=1}^{i} W_i$$

where, W_i is unit weight of water quality and Q_i subject index or quality rating.

3 RESULTS AND DISCUSSION

3.1 *Water quality parameters*

Most of the water samples (94%) had a WQI > 75-100, indicating a very poor quality, and 6% were between 50-75 indicating poor quality. All the water samples though had a low BOD and MPN indicating no sewage/fecal contamination. The As heat map (Figure 2a) shows contaminant site and distribution. Increase in As load during dry season was observed, which reduces drastically during post monsoon due to dilution effect. The pH of water was found to be the key parameter controlling the distribution (Figure 2b). Significant doses of As-biomarkers are reported from Northern part

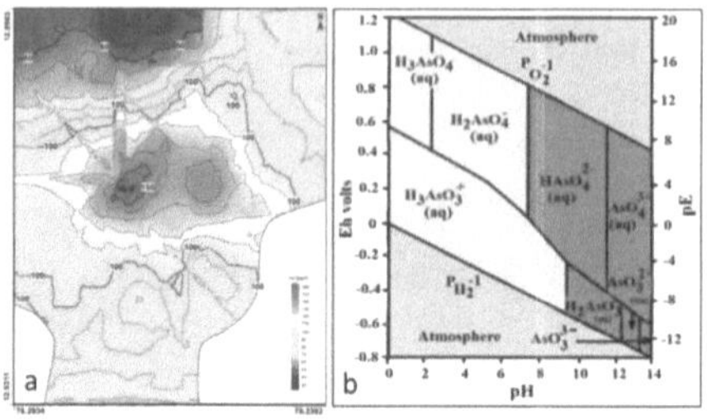

Figure 2. a) Distribution of As in and around KGF; b) The schematic representation of partioning of As as the function of pH and Eh volts, in and around ores.

of the state. Presently, many cases of arsenicosis and skin lesion were recognized by our research team. However, the documentation of these medical cases is lacking.

4 CONCLUSIONS

The long avoided As contamination of potable water sources from the abandoned gold mine in Kolar has now posed a threat to human health. The contamination of Fluoride and cycnide is affecting the population as well the agricultural yield. The Rehabilitation of the sites and supply of safe drinking water is an utmost need. This is a preliminary observation based on the small sample size, and the work is in progress to link the overall geology and mineral chemistry of the region with the release of As and F.

ACKNOWLEDGEMENTS

The work was supported by the DST Inspire Faculty grant [DST/INSPIRE/04/2015/002362], awarded to DG.

REFERENCES

Chakraborti D., Rahman M.M., Murrill M., Das R., Siddayya Patil S.G., Sarkar A., Dadapeer H.J., Yendigeri S., Ahmed R. & Das K.K. 2013. Environmental arsenic contamination and its health effects in a historic gold mining area of the Mangalur greenstone belt of Northeastern Karnataka, India. *J. Hazard. Mater.* 262: 1048–1055.

Langmuir D., Mahoney J., MacDonald A. & Rowson J. 1999. Predicting arsenic concentrations in the porewaters of buried uranium mill tailings. *Geochim. Cosmochim. Acta* 63(19–20): 3379–3394

Mamatha P. & Rao S.M. 2010. Geochemistry of fluoride rich groundwater in Kolar and Tumkur districts of Karnataka. *Environ. Earth. Sci.* 61:131–142. (and the references within).

Arsenic in the Environment: Bridging Science to Practice for Sustainable Development –
van der Wal, Ahmad, Petrusevski, Weijma, Savic, van der Wens, Beerendonk, Bhattacharya, Bundschuh & Naidu (Eds)
© 2023 the Author(s), ISBN 978-1-032-32928-4

Arsenic contamination in sedimentary groundwater basin of Mukunkan in northern Sri Lanka

M. Premathilaka[1] & R. Chandrajith[2]
[1]*National Water Supply and Drainage Board, Ratmalana, Sri Lanka*
[2]*Department of Geology, Faculty of Science, University of Peradeniya, Peradeniya, Sri Lanka*

ABSTRACT: Present study was carried out to assess the status of arsenic (As) in groundwater of Murunkan sedimentary basin. This basin supplies 10,000 m^3/day of groundwater as potable water to surrounding region. A total of 19 groundwater samples were collected from shallow wells and deep wells and analyzed for their contents of cations and anions, including trace elements. Highest As concentration recorded was 6.59 (μg/L) which is less than the recommended WHO limit. The mean of As concentration of shallow and deep wells were 4.26 (μg/L) and 2.04 (μg/L), respectively. It revealed that relatively high prevalence of As in the shallow groundwater compared to deep groundwater. Mean Fe contents in deep and pumping wells were also higher compared to shallow wells and non-pumping wells. The results also revealed that As and Fe contents behave in similar manner.

1 INTRODUCTION

Drinking water (10,000 m^3/d) to the Mannar town and surrounding region that located in the northern Sri Lanka is mainly supplied by deep bore holes in confined aquifer systems in the Murunkan sedimentary basin. The deep aquifer system mainly recharges from the monsoonal rain and also storage of *Yodawewa* Reservoir particularly during the dry season. During the drought period, over-extraction is very common and leads to deterioration of groundwater quality. Murunkan groundwater basin is predominantly located in Miocene limestone sequence and it is one of the high yielding groundwater resources in Sri Lanka (Thilakerathne *et al.*, 2015). Since the basin is located close to the sea, over extraction of groundwater and also predicted sea-level rise could make a serious impact on the groundwater quality.

An elevated contents of As in groundwater in some shallow wells has been reported in the costal coastal aquifers systems in Sri Lanka (Bandara *et al.*, 2018). For instance, 23% wells in Mannar Island showed arsenic contents higher than recommended WHO limit. Therefore, this study was carried out to monitor the As and other contaminants in Murunkan aquifer system since this aquifer provide drinking water for the nearby community.

2 MATERIALS AND METHODS

2.1 *Study area*

The Murunkan sedimentary aquifer is one of the deep confined aquifers which is located in the north-western coastal areas of Sri Lanka. The area covers about 400 km^2. The area receives an annual rainfall of 900 to 1500 mm. Malwathu Oya, irrigation channels, and *Yodawewa* reservoir are the main surface water bodies in the Murunkan sedimentary basin area (Figure 1).

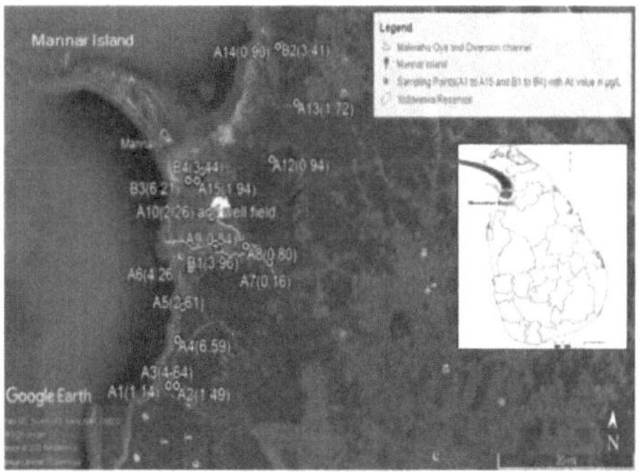

Figure 1. Study area and water sampling locations of study area.

2.2 *Hydrogeological and geological set up*

Geologically, the area is underlain by Miocene limestone and sandstone that overlaying Precambrian basement rock. Limestone layers are covered with superficial unconsolidated silty or sandy clay/clayey sand deposits of Recent age. Lower part of the overburden acts as unconfined aquifer while the deep layers formed confined aquifers. Thickness of unconsolidated formation varies from place to place with a maximum thickness of 10 m. All deep wells represents the confined aquifer while shallow wells represents both type of aquifers.

DOI: 10.1201/9781003317395-42

2.3 *Water sampling and analysis*

Water sampling was carried out during the rainy period and 19 water samples were collected from shallow (4) and deep wells (15) that drilled in the confined aquifer (Figure 1). The field measurement of pH, temperature (T), and electrical conductivity (EC) were carried out with field test kits. Anions (F^-,NO_3^-,SO_4^{2-} and PO_4^{3-}) in samples were measured by ion chromatography. Na^+ and Ca^{2+} and were measured using atomic absorption spectrophotometry and trace elements (Li, B, Al, Cr, Mn, Fe, Co, Ni, Zn, As, Se, Sr, Mo, Cd, Ba, Hg, Bi) were determined with a Inductively Coupled Plasma Mass Spectrometer (ICPMS).

3 RESULTS AND DISCUSSION

The analyzed parameters and their compositions are summarized in Table 1 and the As contents are summarized in Table 2.

Table 1. Summary statistics of chemical analysis of groundwater in Murunkan aquifer.

Parameter	Mean	Min	Max
Temp °C	29.6	27.1	34.3
pH	7.43	6.95	7.93
EC (μs/cm)	2710	866	7130
HD (mg/L)	541	236	1296
F^- (mg/L)	0.53	0.01	1.53
NO_3^- (mg/L)	1.7	0.5	7.7
PO_4^{3-} (mg/L)	1.2	0.37	6.1
SO_4^{2-} (mg/L)	52	6	170
Na^+ (mg/L)	48.3	46.8	48.3
Ca^{2+} (mg/L)	151	25.8	410
(in μg/L)			
Li	4.45	0.55	12.8
B	228	44.7	501
Al	18.0	3.1	122
Cr	1.04	0.11	2.91
Mn	152	1.48	1361
Fe	23.9	7.27	130
Co	0.37	0.03	1.32
Ni	1.97	0.46	6.33
Zn	629	15.5	58.3
Se	7.2	1.59	12.4
Sr	1371	595	2626
Mo	2.23	0.26	9.06
Cd	0.14	0.01	0.34
Ba	227	61.1	797

Electrical conductivity (EC) values in studied wells 866 to 7130 μS/cm with average of 2710 μS/cm. High EC points are located close to the coastline that decreased towards the landward direction. The fluoride values higher that 1 mg/L in some wells but average fluoride content (0.53 mg/L) is lower that that in the dry zone metamorphic aquifers (Rubasinghe *et al.*, 2015). The total Fe varied from 7.27 to 130 μg/L with a mean of 23.9 μg/L. Mean Fe contents in deep wells and pumping wells are higher compared to shallow and non-pumping wells. However, none of wells have exceeded the WHO recommended levels of iron for drinking water (0.3 mg/L).

Table 2. As and Fe contents (in μg/L) in shallow, deep, pumping, and non-pumping wells in the aquifer.

Parameter	Mean	Min	Max
As (all)	2.51	0.16	6.59
Fe (all)	23.9	7.27	130
As (shallow)	4.26	3.41	6.21
Fe (shallow)	13.4	7.27	18.4
As (deep)	2.04	0.16	6.59
Fe (deep)	26.7	9.13	130
As (pumping)	2.22	0.16	6.59
Fe (pumping)	26.2	7.27	130
As (non-pumping)	2.46	0.61	6.21
Fe (non pumping)	17.5	12.3	23.4

Measured heavy metals and metalloids in groundwater were lower than the WHO recommended levels for drinking water. Highest arsenic concentration recorded is 6.59 (μg/L), which is less than recommended WHO limit. The mean arsenic concentration of studied shallow and deep wells were 4.26 (μg/L) and 2.04 (μg/L), respectively. The results revealed that As and Fe concentrations behave in a similar manner in the Murunkan aquifer system.

4 CONCLUSIONS AND RECOMMENDATIONS

Studies revealed that highest arsenic concentration recorded is 6.59 (μg/L), which is less than recommended WHO limit and mean arsenic concentration in shallow well are higher than deep wells. Also study revealed that As and Fe concentrations behave in a similar manner.

ACKNOWLEDGEMENTS

We acknowledge the National Water Supply and Drainage Board and Department of Geology, University of Peradeniya for assisting the fieldwork and analysis.

REFERENCES

Bandara U.G.C., Diyabalanage S., Hanke C., van Geldern R., Barth J.A.C. & Chandrajith R. 2018. Arsenic-rich shallow groundwater in sandy aquifer systems buffered by rising carbonate waters: a geochemical case study from Mannar Island, Sri Lanka. *Sci. Total Environ.* 633: 1352–1359.

Rubasinghe R., Gunatilake S. & Chandrajith R. 2015. Geochemical characteristics of groundwater in different climatic zones of Sri Lanka. *Environ. Earth Sci.* 74, 3067–3076.

Thilakerathne A., Schuth C. & Chandrajith R 2014. The impact of hydrogeological setting on geochemical evolution of groundwater in karstified limestone aquifer basin in Northwest Sri Lanka. *Environ. Earth Sci.* 73(12); 8061–8073.

Arsenic in the Environment: Bridging Science to Practice for Sustainable Development –
van der Wal, Ahmad, Petrusevski, Weijma, Savic, van der Wens, Beerendonk, Bhattacharya, Bundschuh & Naidu (Eds)
© 2023 the Author(s), ISBN 978-1-032-32928-4

Arsenic occurrence in the groundwater of South Africa

T.A. Abiye[1] & P. Bhattacharya[2]
[1]*School of Geosciences, University of the Witwatersrand, Johannesburg, South Africa*
[2]*KTH-International Groundwater Arsenic Research Group, Department of Sustainable Development, Environmental Science and Engineering, KTH Royal Institute of Technology, Stockholm, Sweden*

ABSTRACT: Arsenic (As) in groundwater has variable concentration, which is controlled by the mineralization of crystalline rocks. Consequently, its concentration in the groundwater reaches up to 253 μg/L (Namaqualand), 6150 μg/L (west of Johannesburg), about 500 μg/L in the Karoo aquifers. Acid mine drainage is also found to be important source for As and other toxic metals in the environment.

1 INTRODUCTION

In regions that are characterized by climatic aridity with low rainfall occurrence, groundwater from weathered and fractured crystalline aquifers is a primary source of water supply for various activities. The development of groundwater resource from these aquifers has been regarded as critical driver for the agricultural sector in South Africa, because of its availability at shallow depth (≈40 m), besides highly productive aquifers to the depth of about 400 m located within the Karoo (Rosewarne *et al.*, 2013) and Transvaal meta-sedimentary rocks (Abiye *et al.*, 2011). The geochemical composition of an aquifer has direct control over the groundwater quality exacerbated by mining and other economic activities such as agriculture and industry, besides mineralization.

Being a well-known gold, uranium, PGE and coal producing country in the world, vast area of South Africa has a potential to release arsenic (As) into aquifers (Abiye & Bhattacharya 2019). Coal and gold are the primary economic minerals of the country. Coal is associated with carbonaceous and black shales of the Ecca Group (Karoo Supergroup. It is primarily mined through an open cast method where the overlying less mineralized rocks are dumped by exposing the sulphide rich coal layers to be oxidized, which could potentially mobilizes As in groundwater besides leaching from coal and mobilization through AMD (Figure 1). On average the As content in the coal seams falls between 0.9 and 8.2 mg/kg (Willis, 1983 *cited in* Wagner & Hlatshwayo 2005). On the other hand, gold occurs in quartz veins associated with mafic and ultramafic greenstone belts, conglomerates in the Witwatersrand basin that contains gold-uranium-pyrite ores. Arsenopyrite (FeAsS) is a widely available mineral in gold and coal mineralized zones of South Africa. Mining associated to gold, coal and uranium is the main activity that mobilizes As. The abandoned gold tailings dams in the Johannesburg region contain very high metal concentration that is released into surface and ground water resources, where acidic mine water often highly oxidising, continuously decants into streams (Abiye 2014; Abiye *et al.*, 2018; Winde 2006).

Figure 1. Acid mine drainage associated with gold mining in the Krugersdorp area (As is about 6150 μg/L).

2 METHOD

Extensive literature assessment was conducted in order to gather relevant information on the As and toxic metal concentration in the groundwater system of South Africa. In order to get a first-hand information on As, 24 groundwater samples from various aquifers and analyzed by IC-PMS. In this abstract we focus only on As. During sample collection, pH and ORP values were recorded in the field.

3 RESULTS AND DISCUSSION

The geochemical analysis of 2.5 m thick soil profile around Johannesburg revealed that the upper horizon contains high As concentration, which could be due to leaching from the gold tailings

DOI: 10.1201/9781003317395-43

dams and interaction with acid mine drainage (Abiye & Bhattacharya 2019). Up to the depth of 0.5 m, there is high As concentration (≈10 mg/L to 200. Below the depth of 0.5 m, there is an exponential decrease in the As concentration that falls between ≈5 mg/L to 50 mg/L, indicating the possibility of a decrease in the oxidation process.

In the gold producing region of South Africa, the average concentration of As in the pyrite within the Black Reef Formation of the Transvaal Supergroup is about 1394 mg/kg (Barton & Hallbauer, 1996 *cited in* Rosner 1999). Arsenic was also detected at high concentration in the gold tailings dams that ranged between 82.7 mg/km and 123.4 mg/kg and with the concertation of 0.12 mg/L in the groundwater impacted by mine water seepage (Rosner 1999). Even though pyrite (FeS_2) does not contain As, there is high possibility that niccolite (NiAs) and cobaltite (CoFeAsS) could be associated to mineralized zones, where gold mining is taking place.

In the Johannesburg region, As content in groundwater, ranges from 17 μg/L to 318 μg/L, related to sulphide oxidation due to the gold and uranium mining. In the northern section of Limpopo province, As concentration was reported by Nephalama & Muzerengi (2016) with the maximum and minimum concertation of 1553 mg/L and 0.83 mg/L, respectively, associated with the coal mining activity from the Karoo sediments. Owing to the release of As from rocks in groundwater, its concentration in the groundwater reaches up to 1000 μg/L (Sami & Druzynski 2003).

Based on the As and ORP relationship, slightly acidic to neutral groundwater occurs in oxidizing condition, while groundwater with an alkaline pH resides in reducing condition (Figure 2). In both cases, high As concentrations were recorded. In the condition where ORP is negative (reducing condition), As concentration ranges between <20 μg/L and 240 μg/L, while for positive ORP (oxidizing condition), the As concentration ranges between <20 μg/L and 253 μg/L showing the presence of different arsenate species. The overall pH values of the groundwaters are near neutral to slightly alkaline (6.0–8.84). Under these conditions As^{5+} is the most prevalent species (Muñoz *et al.*, 2016).

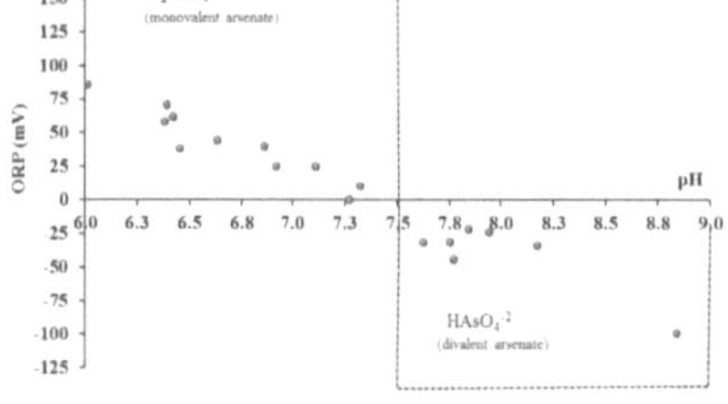

Figure 2. Arsenic distribution with measured ORP in groundwater.

4 CONCLUSIONS AND RECOMMENDATIONS

Both oxidizing and reducing conditions facilitate As mobilization, which could depend on the degree of mineralization of the host rock. Exceptionally high As with the ORP values of +10 mV and -34 mV has been attributed to pyrite and iron mineralization in granite. Mineralogical composition of the host rocks that have undergone metamorphism controls the concentration of As in groundwater. The high As concentration in different lithologies could be related to weathering of As containing minerals and subsequent leaching into groundwater. The most widespread source of As in South Africa is related to acid mine drainage associated with gold and coal ming. Due to the health related risk associated with As, it is essential to take precautionary measure on boreholes with high As content.

REFERENCES

Abiye T.A. Mengistu H. & Demlie M.B. 2011. Groundwater resource in the crystalline rocks of the Johannesburg area, South Africa. *J. Water Resour. Prot.* 3 (4): 199–212.

Abiye T.A., 2014. Mine water footprint in the Johannesburg area: a review based on existing and measured data. *S. Afr. J. Geol.* 117 (1): 87–96.

Abiye T.A., Mkansi S., Masindi K. & Leshomo J. 2018. Effectiveness of wetlands in retaining metals from mine water, South Africa. *J. Water Environ.* 32: 259–266.

Abiye T.A. & Bhattacharya P. 2019. Arsenic concentration in groundwater: archetypal study from South Africa. *Groundw. Sustain. Dev.* 9:100246.

Muñoz M.O., Aróstegui J.G., Bhattacharya P., Sracek O., Moreno M.E.G., Kohfahl C., Aguirre J.Q., Diaz J.H. & Bundschuh J. 2016. Geochemistry of naturally occurring arsenic in groundwater and surface-water in the southern part of the Poopó lake basin, Bolivian Altiplano. *Groundw. Sustain. Dev.* 2-3: 104–116.

Nephalama A. & Muzerengi C. 2016. Assessment of the influence of coal mining on groundwater quality: case of Masisi village in the Limpopo province of South Africa. In: C. Drebenstedt and M. Paul (eds.) Mining Meets Water – Conflicts and Solutions. Proceedings IMWA 2016, Freiberg/Germany. pp 430–438.

Rosner T. 1999. The environmental impact of seepage from gold mine tailing sams near Johannesburg, South Africa. Unpublished PhD Thesis. University of Pretoria, South Africa. 231p.

Rosewarne P., Woodford A., van Tonder G., Goes M., Esterhuyse C., Visser D., O'Brien R., Talma S. & Tredoux G. 2013. Karoo groundwater Atlas ume 2 Karoo groundwater expert group.

Sami K. & Druzynski A.L. 2003. Predicted spatial distribution of naturally occurring arsenic, Selenium and uranium in groundwater in South Africa – reconnaissance survey. WRC Report No. 1236/1/03.

Wagner N.J. & Hlatshwayo B. 2005. The occurrence of potentially hazardous trace elements in five highveld coals, South Africa. *Int. J. Coal Geol.* 63: 228– 246

Winde F. 2006. Challenges for sustainable water use in dolomitic mining regions of South Africa-a case study of uranium pollution: part 1 sources and pathways. *J. Phys. Geogr.* 27 (4): 333–347.

Arsenic in the Environment: Bridging Science to Practice for Sustainable Development – van der Wal, Ahmad, Petrusevski, Weijma, Savic, van der Wens, Beerendonk, Bhattacharya, Bundschuh & Naidu (Eds)
© 2023 the Author(s), ISBN 978-1-032-32928-4

Natural occurrence and the effects of the evaporation on the arsenic concentrations in a semi-arid area in the Mexican Altiplano

D. Cauich-Kau[1,2], A. Cardona-Benavides[2], J. Castro-Larragoitia[2] & T. Rüde[1]
[1]*Institute of Hydrogeology, RWTH Aachen University, Germany*
[2]*Earth Sciences, Autonomous University of San Luis Potosi, San Luis Potosí, México*

ABSTRACT: In the Mexican Altiplano in the last decades, several investigations were performed in order to have a better understanding of the sources, mobility and process that controls arsenic in the groundwaters. In the Cerritos Basin in the State of San Luis Potosí elevated concentrations of arsenic above the Mexican critical value (25 µg/L) were found being of concern to the public health. This study not only deals with the evaporation effect on arsenic concentrations but also for its natural occurrence in igneous rocks. This source is connected to the basin by a karst system pointing to the need of fully understanding all parts of groundwater systems.

1 INTRODUCTION

In the early 1960s, Mexican institutions reported health problems in arid to semi-arid areas inhabitants, because of groundwater consumption with high arsenic (As) concentrations (Armienta *et al.*, 2008; Morales *et al.*, 2015). In these regions, anthropogenic and geogenic sources are favoring As in groundwater. The former is due to the old mining waste and the latter is because the specific physicochemical conditions (arsenopyrite oxidation, climatic conditions, and low pH values). Recently, elevated As concentrations (max 226.4 µg/L) were reported in the central part of the San Luis Potosi State, particularly in the Cerritos Basin (Konefke 2018; Palm 2008;). There, the As concentrations were identified not only at mineralized areas but also in the shallow groundwaters in the basin fill sediments. The aim of this study is to get a deeper understanding on the link between the source rock weathering and the As concentrations in the groundwater of the Cerritos basin.

2 MATERIALS AND METHODS

2.1 *Study area*

Cerritos Basin is located at the central part of San Luis Potosí State in an area dominated by a karstified limestone sequence of Cretaceous age. Felsic rocks of Tertiary age, intruded the calcareous sequence, magmatic activity took place leaving a mineralized granite in the NW of the study area and some felsic volcanic rocks to the East (Figure 1). The average annual temperature and rain are 19°C and 662 mm respectively. The study area includes Cerritos, Guadalcázar, and Villa Juárez municipalities in which agriculture and livestock are the actual main economic activities. In the Guadalcázar area, several ancient mining activities are located associated with the intrusive body El Realejo. The study area belongs to the hydrological region of Pánuco and has only intermittent streams.

2.2 *Assessment of the sampling points*

The entire population of the study area depends on the groundwater resources. Shallow-dug (5-15 m

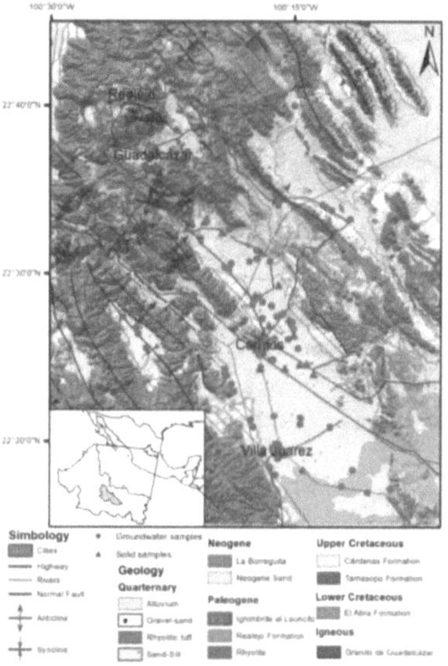

Figure 1. Study area, general geology, and sampling points.

deep) and deep (30-180 m) wells are the main source for groundwater in the valley. Geophysical investigations (TEM surveys) identified alluvial sediments with up to 300 m thickness. In Guadalcázar area the main groundwater source are springs but also shallow wells tapping the limestone units.

2.3 *Field campaigns and laboratorial analysis*

In total, 43 solid samples were taken: *i*) 9 rocks including the intrusive body El Realejo, and rhyolitic rocks, *ii*) 5 sediments downstream the intrusive body, *iii*) 23 soils in agricultural areas, and *iv*) 6 tailings at the ancient mining places. Water sampling was carried out in two seasons, the first in September 2018 and the second a year later. In all, 95 water samples were collected (including the samples collected by Palm (2008) and Konefke (2018)): 13 springs, 41 dug wells, 39 deep wells, 1 pond, and 1 drainage. The physicochemical parameters like pH, temperature (C), electrical conductivity (EC), dissolved oxygen (DO), and alkalinity were field measured. All the samples were prepared and analyzed for major and trace elements. The inductively coupled plasma-optical emission spectrometry and inductively coupled plasma mass spectrometry were used. The stable environmental isotope analyses (^{18}O and ^{2}H) were performed by absorption cell spectroscopy.

3 RESULTS AND DISCUSSION

3.1 *Solid samples*

The As concentrations found in the rocks were in the range of 1.8-50 (mg/kg), in which it is observed that some sites exceed the background values (1-30 mg/kg) (Kabata-Pendias 2011). In the felsic intrusive bodies and rhyolitic rocks, high As have been found in the range of 9-14 and 1.8-50 mg/kg respectively. These variations depend on the rock alteration, being the most altered with major concentrations; however, the tailings located at the NW Guadalcázar have the highest As (5,906 mg/kg) (Figure 2a). This elevated value could be the caused by the interaction with Fe hydroxides, they are likely to have an influence on the As dynamics, as these minerals represent important natural As sources and sinks. Furthermore, clay and silt are transported from NW by ephemeral streams and are deposited in a pond at the southern rim of the Guadalcázar. There, the surface water in contact with these soils, with high As (65 mg/kg) seeps into a karst system leaving the fine material as secondary sink and tends to increase the As. This water enriched with As (32 μg/L) discharge in a waterfilled doline which is the end of the drainage and is located in the limestones, there the As concentrations differs completally with the As found in the springs located in the limestones (see

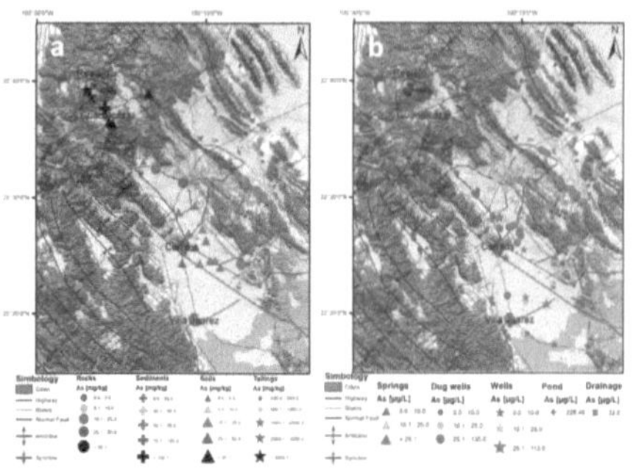

Figure 2. Arsenic concentrations in a) solid samples; b) groundwaters.

Figure 2b) as well. The water seeping this doline could be another As source in the shallow waters from the Cerritos valley. By the Cerritos area, all the soils samples are above the background values (10-26 mg/kg) (Kabata-Pendias 2011). These concentrations indicate an influence from the fine material coming from the Guadalcazar area, but also a possible anthropogenic contamination, because of the use of fertilizers. An evidence of it is observed in the soil (27 mg/kg) of the sample located at the south of Cerritos where the surrounding area is mainly formed by limestone units that does not contain As.

3.2 *Groundwater samples*

Most of the samples were collected from an aerobic water source (Table 1). In the shallow granular aquifer, only 74% of the sites were below the Mexican regulation (25 μg/L) (Figure 2b). Not only in the dug wells but also in the deep wells high As concentrations were found. In the first, the major density of "hotspots" are located in the valley northeast Cerritos in an agricultural area. There, the soils with As (average 15 mg/kg) and the climatic conditions (temperature and humidity) could enrich the shallow groundaters with As; however, three dug wells placed in the basin fill NE Cerritos showed a local efffect, this is, As concentrations in the range of 32.3-38 (μg/L) but with different physicochemical characteristics pH (7.07-7.78), T (16.7-22.3 °C), eC (603-712 μS/cm), and DO (6-6.5 mg/L). This indicates a different geochemical process in each dug well even they are located in a short distance. In the limestones feeding the springs, some dug wells, and some deep wells are located, low As were identified in the range of 0.15-9.52 (μg/L), these information corroborates that the limestones do not contribute with this element to the groundwater. In addition, the general flow direction (NW to SE) does not identify a trend, which indicates that there are other As mobility and enrichment controls. One of these controls could be the evaporation.

Table 1. Range of the field parameters and As concentrations.

Type	Springs	Dug wells	Wells	Pond	Drainage
As [µg/L]	0.01–9.52	1.03–127.85	1.22–113.20	226.40	32.97
T [°C]	19.1–29.9	14.9–23.8	17.5–28.7	34.50	23.20
pH	6.42–8.19	6.36–7.97	6.72–8.01	5.87	6.85
eC [µS/cm]	420–3690	180–3030	318–4770	265.00	1419.00
DO [mg/L]	3.04–10.84	0.40–8.39	0.27–10.80	8.63	0.84
HCO_3^-[mg/L]	192–469	107–584	165–1063	74.87	599.00
Water type	61% HCO_3^-–Ca 30% SO_4^{2-}–Ca	44% HCO_3^-–Ca 20% SO_4^{2-}–Ca	38% HCO_3^-–Ca 28% SO_4^{2-}–Ca	Mix–Ca	HCO_3^-–Mix

3.3 *Isotopic data*

The stable isotope data can be used for multiple kinds of interpretation, such as identification of possible recharge areas, loss through evaporation, and monitoring chemical processes through evaporation. The isotopic data was compared with the Local Meteoric Water Line (LMWL) (Figure 3) calculed by Calva (2011) for a better understanding of the climatic conditions in the study area. The slope of the LMWL is slightly smaller than the Global Meteoric Water Line (GMWL). This value (s=8.09) is attributed to the evaporation of water droplets during precipitation (secondary evaporation), which would affect the isotopic content of precipitation. According to Calva Hernández (2011) the major sources of water vapour in this region are the Gulf of Mexico and the Caribbean Sea. As previally mentioned, the springs, some dug wells, and some deep wells located in the limestonces are not influenced for the evaporation, however, the deep wells placed at the basin fill (Figure 2a) have some evaporation effects. This could happen because of a shallow depth and the design itself. For the dug wells, the results showed that 47% of them placed in the basin fill NE Cerritos are influenced by evaporation process (Figures 2b & 3) but also the

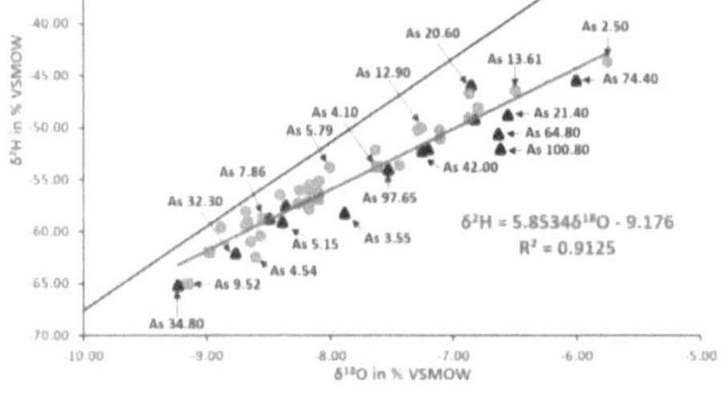

Figure 3. Isotopic data and As concentrations in µg/L.

As enrinchment could come from the fine material and the doline water as well indicanting that the material coming from the NW Cerritos have influence in groundwater sources.

4 CONCLUSIONS AND RECOMMENDATIONS

Geogenic As is a problematic issue in the groundwater supply for the population, since groundwater is the sole source for this vital resource in the study area. The evidence showed that not only the dug wells located in the basin fill are affected by elevated As concentrations above the Mexican regulations but also some deep wells located at the basin fill as well. One of the causes of these results could be the evaporation process but also the sediments derived from the mineralized zone in the intrusive body El Realejo that can be a secondary sink. For a future development in the area it is important to consider not to drill for wells in the basin fill, a better option is placing wells in the limestone units yet the groundwater there has concentrations below the critical value.

ACKNOWLEDGEMENTS

We acknowledge the Institute of Geology of the Autonomous University of San Luis Potosi for supporting the solid and groundwater analysis.

REFERENCES

Armienta M.A. & Segovia N. 2008. Arsenic and fluoride in the groundwater of Mexico. *Environ Geochem Health*. 30: 345–353.

Calva Hernández D. 2011. Caracterización Isotópica de la Precipitación en el Estado de San Luis Potosí. *Universidad Autónoma de San Luis Potosí, MSc Thesis.*

Kabata-Pendias A. 2011. *Trace Elements in Soils and Plants*. Taylor and Francis Group, LLC, USA.

Konefke L. 2018. *Causes for Elevated Concentrations of Arsenic and Uranium in the Northern Cerritos Basin*, San Luis Potosí, México. RWTH Aachen University, MSc Thesis.

Morales I., Villanueva-Estrada R.E., Rodríguez R. & Armienta M.A. 2015. *Geological, Hydrogeological, and Geothermal Factors Associated to the Origin Of Arsenic, Fluoride, and Grounwater Temperature in a Volcanic Environment "El Bajío Guanajuatense"*, Mexico. Environ. Earch Sci. 74: 5403–5415.

Palm J. 2008. *Hydrogeological Mapping of the Cerritos Aquifer System*. RWTH Aachen University, Diploma Mapping Thesis.

Arsenic in the Environment: Bridging Science to Practice for Sustainable Development –
van der Wal, Ahmad, Petrusevski, Weijma, Savic, van der Wens, Beerendonk, Bhattacharya, Bundschuh & Naidu (Eds)
 ISBN 978-1-032-32928-4

Presence of geogenic arsenic caused by thermal activity in the Celaya Valley Aquifer: Environmental implications

J.F.A. Landa-Arreguín[1], R.E. Villanueva-Estrada[2], J.E. Ortega-Gutiérrez[1],
J.I. Morales-Arredondo[3], B.S. Amézaga-Campos[1] & M.A. Armienta-Hernández[3]
[1]*Posgrado en Ciencias de la Tierra, Universidad Nacional Autónoma de México, Del. Coyoacán, Ciudad de México., México*
[2]*Instituto de Geofísica, Universidad Nacional Autónoma de México, Morelia, Mich., México*
[3]*Instituto de Geofísica, Universidad Nacional Autónoma de México, Ciudad de México, México*

ABSTRACT: In the municipalities of Juventino Rosas, Villagrán, Celaya and Apaseo el Grande located in the Celaya Valley Aquifer (CVA), high levels of arsenic (As) have been reported in different drinking water wells that exceed the maximum limits allowed by World Health Organization (WHO) (0.01 mg/L) and by the Official Mexican Standard of permissible limits for drinking water (0.025 mg/L), however the highest values in the region belong to two hot springs whose values are 0.3 mg/L. The origin of levels of arsenic is considered as geogenic and are related with thermal activity causing the dissolution of the volcanic glass of the ignimbrites of the stratigraphic sequence. Therefore, the regional geology may affect the arsenic concentrations of the CVA and the health of the inhabitants, through water ingestion, being currently a serious problem in the municipalities.

1 INTRODUCTION

Arsenic concentrations in natural surface drainage systems usually exceed 0.01 mg/L in areas of geothermal activity (Welch & Stollenwerk 2003). The arsenic origin in thermal systems is associated to rhyolitic rocks, particularly the chloride type fluid with neutral pH (Webster & Nordstrom 2003) and also found in volcanic glass (Bundschuh *et al.*, 2008). In various areas of Mexico and Latin America higher arsenic concentrations than the WHO guideline have been reported, so due to its toxicity, arsenic is considered a serious health threat to the population of those zones (Armienta & Segovia 2008; López *et al.*, 2012).

2 MATERIALS AND METHODS

2.1 *Study area*

The municipalities of Juventino Rosas, Villagrán, Celaya and Apaseo el Grande belonging to the Guanajuato state, are located in the north-central part of the Trans-Mexican Volcanic Belt (TMVB) physiographic province (Figure 1) and hydrologically correspond to the Celaya Valley Aquifer (CVA) located in the southeastern of the subprovince known as "Bajío Guanajuatense".The geology in the region is composed mainly for volcanic rocks of different compositions of Cenozoic age and below them there is a Cretaceous calcareous basement sequence, being the most ancient rocks of the region (Figure 1).

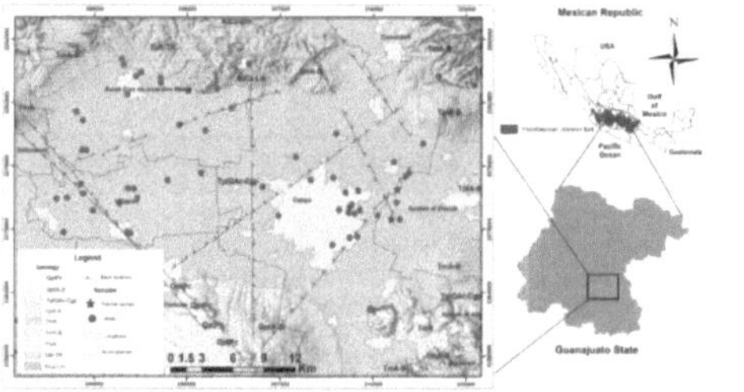

Figure 1. Map of the study area showing the geological characteristics and water sampling locations.

2.2 *Water sampling and laboratory analysis*

A total of 62 drinking water well samples were collected from CVA in two field campaigns. The first one was carried out during October 2016 in rainy season and were taken 17 groundwater samples from Juventino Rosas, the second one was in Villagrán, Celaya and Apaseo el Grande during February and March 2017 in dry season, of which 15 samples were collected from Villagrán water wells, 26 samples from Celaya, and finally, 4 samples from drinking water and 2 more from hot

 DOI: 10.1201/9781003317395-45

springs were collected from Apaseo el Grande municipality.

The physico-chemical parameters were measured in the field. Major ions collected were analyzed in the laboratory by ion chromatography, arsenic concentrations from Juventino Rosas and Villagrán were analyzed by flame atomic absorption spectrometer and from Celaya and Apaseo el Grande were measured by inductive coupled plasma mass spectrometry. The whole chemical analysis was carried out at the Geophysics Institute of the National Autonomous University of Mexico (UNAM).

3 RESULTS AND DISCUSSION

3.1 *Distribution of arsenic in the study area of CVA*

The four municipalities are affected by high concentrations of arsenic, therefore are susceptible to presenting health problems in the population.

However, the municipality of Celaya is most affected because the drinking water well with the highest concentrations of arsenic is found there and there are more wells that exceed considerably WHO and NOM-127 guideline (Figure 2).

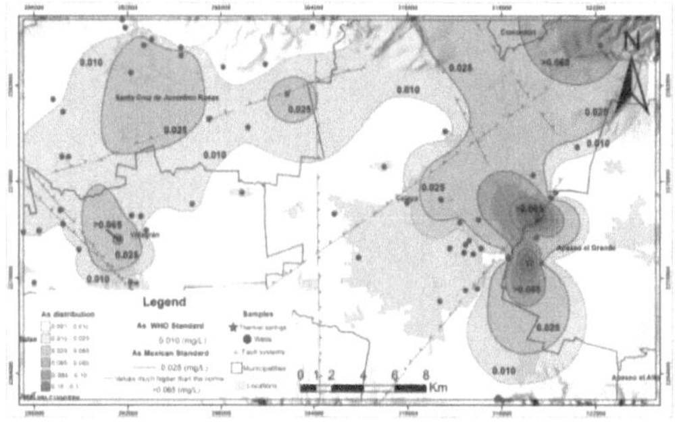

Figure 2. Distribution map of arsenic contaminated drinking water wells in CVA.

3.2 *High arsenic concentration in groundwater in CVA*

In the municipalities of Juventino Rosas, Villagrán, Celaya and Apaseo el Grande high levels of arsenic have were determined in 50% of the total wells of the study area (Figure 3) which exceed the maximum limits allowed by the WHO (0.01 mg/L) most of them have thermal conditions. According to the official Mexican drinking water standards (NOM-127 modified in 2000), 10 drinking water wells of the study area are contaminated with arsenic, 5 of them located in Celaya and the others in Juventino Rosas y Villagrán (Figure 3). However, the two thermal springs have the highest arsenic concentrations in the region with 0.3 mg/L As (Landa-Arreguín *et al.*, 2017).

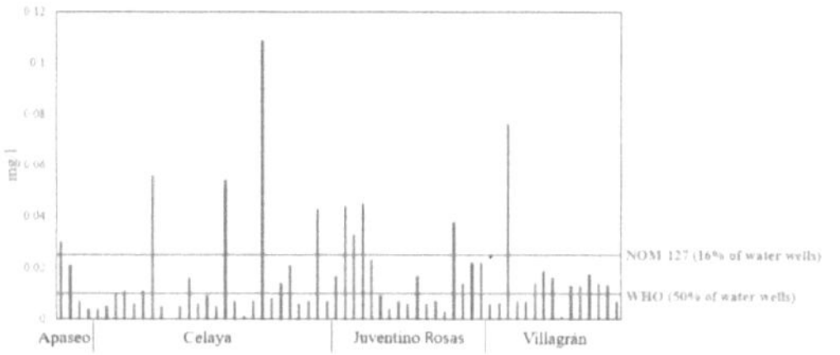

Figure 3. Arsenic concentrations in drinking water wells by municipalities and WHO and NOM-127 guidelines.

3.3 *Health and environmental implications of high arsenic concentrations in drinking water*

The demand and dependence of groundwater for development of economic activities in the region, has increased the risk for the inhabitants to be exposed to arsenic. Currently, it has become a serious health problem, as well as over-exploitation of the aquifers. A serious health and environmental problem are urban wells that supply water to the inhabitants whose concentrations also exceed the limits of the WHO and the Mexican standards. The origin of high As values are geogenic (Morales-Arredondo *et al.*, 2020). The main environmental risks associated with arsenic are the health affectations of the exposed population.

4 CONCLUSIONS AND RECOMMENDATIONS

The high population density and the development of economic activities in the region has increased the groundwater demand, causing an over-exploitation of water of the CVA. In the region, the municipalities have had to drill deep water wells for covering the demand, however some of them present thermalism. The presence of As has been detected in many of the wells in the area, most of the drinking water wells that exceed the WHO and NOM-127 permissible limits have thermal condition. The high levels of arsenic concentrations are due to geogenic processes related with the thermal activity in the area, causing devitrification of volcanic glass of the ignimbrite rocks then arsenic release. Arsenic represents a risk for ingestion of drinking water of the inhabitants of Juventino Rosas, Villa-grán, Celaya and Apaseo el Grande in the CVA being a serious health and environment problem in the region. It is recommended to continue monitoring water wells and to evaluate health and environmental risks in the municipalities.

ACKNOWLEDGEMENTS

Funding was provided by the Mexican Center for Innovation in Geothermal Energy (Centro Mexicano de Innovación en Energía Geotérmica [CeMIE-Geo]): CONACyT-SENER-Energetic Sustainability Sectoral Fund (Fondo Sectorial CONACyT-SENER-Sustentabilidad Energética), project # 207032-2013-04, "Map of geothermal provinces based on fluid geochemistry and aquifer distribution: A tool for the exploration and development of conventional geothermal resources" (*Mapa de provincias geotérmicas a partir de la geoquímica de fluidos y la distribución de acuíferos: herramienta para la exploración y desarrollo de los recursos geotérmicos convencionales*) and by DGAPA – PAPIIT: IA1010-19 project.

REFERENCES

Armienta M.A. & Segovia. N., 2008. Arsenic and fluoride in the groundwater of Mexico. *Environ Geochem Health* 30:345–353.

Bundschuh J., Pérez-Cabrera A. & Litter M.I., 2008. Distribución del Arsénico en la Regiones Ibérica e Iberoamérica.

Landa-Arreguín J.F.A., Villanueva-Estrada R.E., Rocha-Miller R.G., Rodríguez-Salazar M.T.J., Rodríguez-Díaz A.A. & Hernández-Mendiola E., 2017. Resultados Preliminares del Estudio Geoquímico de la Zona Geotérmica de Rancho Nuevo, Guanajuato. Memorias del XXIV Congreso Anual de la Asociación Geotérmica Mexicana, Morelia, Mich., 29–31 marzo.

López D.L., Bundschuh J., Birkle P., Armienta M.A., Cumbal L., Sracek O., Cornejo L. & Ormachea M., 2012. Arsenic in volcanic geothermal fluids of Latin America. *Sci Total Environ.* 429:57–75.

Morales J.I., Armienta M.A. Ortega-Gutiérrez J.E., Flores Ocampo I.Z. & Flores Vargas R., 2020. Evaluation of the carbono dioxide behavior in a termal aquifer located at Central Mexico and its relation to silicate weathering. *Int. J. Environ. Sci. Technol.* 17: 3411–3430.

Morales-Arredondo I., Villanueva-Estrada R.E., Rodríguez R. & Armienta M.A., 2015. Geological, hydrogeological and geothermal factors associated to the origin of arsenic, fluride, and groundwater temperature in a volcanic environment "el bajío guanajuatense" mexico. *Environ. Earth Sci.* 74: 5403–5415.

NOM-127-SSA1-1994-modificada 2000 Salud Ambiental, Agua Para Uso y Consumo Humano. Límites Permisibles de Calidad y Tratamiento a Que Debe Someterse el Agua Para Su Potabilización. Mexico.

Webster J.G. & Nordstrom D.K. 2003. Geothermal arsenic. In: A.H. Welch & K.G. Stollenwerk (eds.). *Arsenic in Ground Water: Geochemistry and Occurrence*. Springer Science and Business Media New York. pp. 101–125.

Welch A.H. & Stollenwerk K.G. (eds.). 2003. *Arsenic in Ground Water: Geochemistry and Occurrence.* Springer Science and Business Media.

WHO 2011. *Guidelines for Drinking-Water Quality*. vol. 4. World Health Organisation, Geneva, pp. 315–318.

Arsenic in the Environment: Bridging Science to Practice for Sustainable Development –
van der Wal, Ahmad, Petrusevski, Weijma, Savic, van der Wens, Beerendonk, Bhattacharya, Bundschuh & Naidu (Eds)
© 2023 the Author(s), ISBN 978-1-032-32928-4

Geochemical baseline of arsenic in surface water and sediments of Chile: Regional distribution and its relationship to geology and climate

J. Tapia
Departamento de Ciencias Geológicas, Facultad de Ingeniería y Ciencias Geológicas, Universidad Católica del Norte, Antofagasta, Chile

ABSTRACT: Despite the significant differences in regional geology and climate as a function of latitude in Chile, environmental regulations for arsenic (As) concentrations in water are based on a constant international value (WHO) and do not consider potentially high (natural) background values. Furthermore, there are no sediment and soil reference guidelines. This study presents the analysis of newly-collected surface water and sediment data along three E-W transects of distinct regions of Chile in order to evaluate the relationship between As content and the regional geological units, crustal thickness, and climate. Analyzed samples indicate that average As values vary substantially from the south (0.37 μg/L in water and 9 mg/kg in sediments) to the north (2873 μg/L in water and 153 mg/kg in sediments), and there exists a correlation between high arsenic values, low precipitation, increased crustal thickness, and specific volcanic rocks such as Neogene volcanics of the Central Volcanic Zone. Regionally-based limits for As concentrations will therefore be recommended and discussed.

1 INTRODUCTION

Historically, As has been a significant health concern for inhabitants of Chile: examples include elevated concentrations in the hair of the 7,000 year old Chinchorro mummies and more recent contamination of surface water in the Antofagasta Region from 1958 to 1970 (800 μg/L to 1300 μg/L of As), causing health issues for Antofagasta inhabitants. Furthermore, Chile represents the world's largest producer of Cu and high natural values of As in surface water and sediments have been mistaken for anthropogenic contamination (namely mining). As such, this study aims to: (i) present a new compilation of data on As in surface water and sediments from distinct regions of Chile (Figure 1); (ii) correlate concentration trends to the country's latitudinal differences in climate, geology, metallogeny, and crustal thickness; and (iii) discuss the regionally-based differences and recommendations for Chilean guidelines of As in water and sediments.

2 MATERIALS AND METHODS

2.1 *Study area*

Data was collected from three representative E-W transects that intersect outcrops spanning from the Andes to the northern, central, and southern Chilean coasts (Figure 1). Each transect is characterized by distinct natural factors such as geology and climate; for example, Neogene volcanics of the Central Volcanic Zone (northern transect) are thought to contain elevated As concentrations.

2.2 *Water and sediment sampling and analysis*

Sampling campaigns were performed in winter and summer in these three transects (Figure 1). Water samples were filtered (0.22 μm) and acidified in the field. Sediments were kept at 4°C, then were oven dried at 50°C and sieved into gravel (>2 mm), sand (2 mm - 63 μm), and silt-clay (< 63 μm) grain size in the laboratory.

The concentration of As in filtered water were determined by ICP-MS. In regards to sediments, 300 mg of the sediments were digested in 12 mL

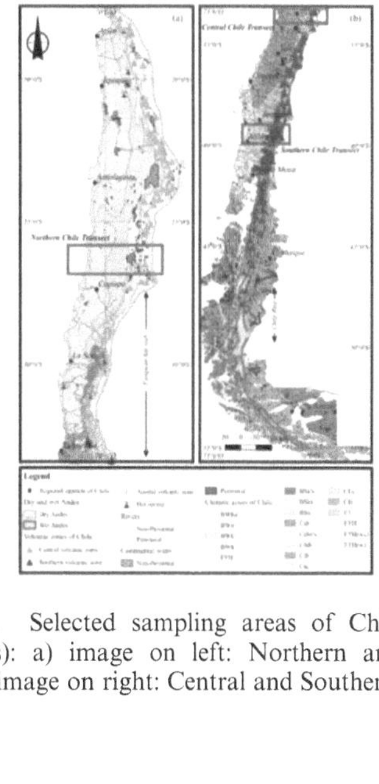

Figure 1. Selected sampling areas of Chile (in red rectangles): a) image on left: Northern and Central Chile; b) image on right: Central and Southern Chile.

HNO_3 + 4 mL HCl. The concentration of As in the digested samples were measured by ICP-MS. Solid state partitioning of As and the characterization of As-bearing minerals are currently in progress at the Universidad Católica del Norte.

2.3 *Water sampling and analysis*

Water sampling was carried out during summer/dry season (Mar-Apr 2018) and post monsoon (Aug–Sept 2018). A total of 53 pre-installed bore-wells were targeted covering a radius of 10 kms from KGF (Figure 1). General physio-chemical parameters such as, pH, TDS, temperature (T), electrical conductivity (EC) and redox potential (Eh) were measured in situ. Major anions were analyzed by cholorometric analyses and the cations were quantified using inductively coupled plasma-optical emission spectrometry (ICP-OES). The Biochemical Oxygen Demand (BOD) was measured directly by BOD digital incubator and Chemical Oxygen Demand (COD) by digital reactor and photometer. Cyanide was quantified by colorimetric method and the coliform load was estimated by MPN test.

3 RESULTS AND DISCUSSION

Analyzed samples show that in general, As values in sediments exceed that of the upper continental crust (5.2 mg/kg; Hu & Gao 2008). Furthermore, As concentrations generally increase from south to north (Table 1); As concentrations in water range over approximately four orders of magnitude.

Table 1. Average As values in water and sediments of the three studied transects (Figure 1).

Location	Average As in water [µg/L]	Average As in sediments [mg/kg]
El Salado River and salt flats (northern transect)	2873	153
Maipo River (central transect)	39	*
Los Ríos Region (southern transect)	0.37	9

*Currently being analyzed.

Furthermore, inspection of the spatial distribution shows that high As concentrations in Chile appear to be positively correlated to a higher crustal thickness, which has been also observed in Tibet due to factors such as the underlying magma chamber or high-temperature conditions generated by mantle upwelling or crustal thickening (Guo *et al.*, 2019). In addition, precipitation in northern Chile is lower than 20 mm·year-1; conversely, in southern Chile, lower As concentrations appear to be inversely proportional to precipitation due to dilution (Figure 2).

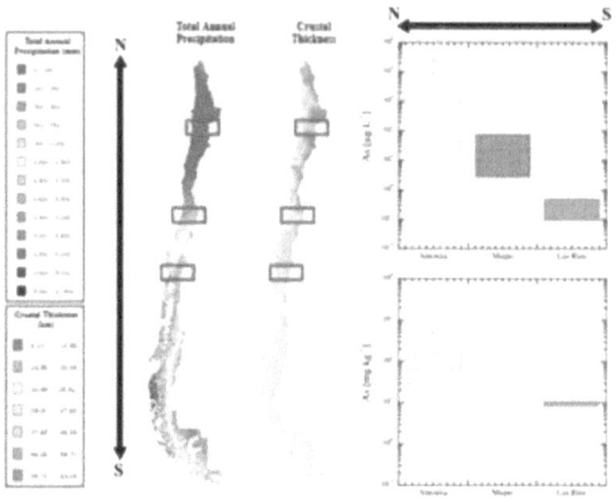

Figure 2. Chile transect showing total annual precipitation (mm), crustal thickness (km), and As concentrations in water and sediments.

4 CONCLUSIONS

Analyses conducted to this point indicate that As concentrations in water and sediments of Chile vary regionally and are correlated to various natural factors. Regional-based guidelines for As in water are recommended, instead of a constant international value (WHO), and standards are also needed for sediments. Results of this study could also have implications in other developing countries that currently lack sufficient environmental measures and are characterized by highly variable settings similar to Chile.

ACKNOWLEDGEMENTS

I acknowledge Brandon Schneider for his English improvement and Camila Lagomarsino, Cristóbal Quiroz and Brian Jones for the analysis of water and sediment samples of southern Chile.

REFERENCES

Guo Q., Planer-Friedrich B., Liu M., Yan K. & Wu G. 2019. Magmatic fluid input explaining the geochemical anomaly of very high arsenic in some southern tibetan geothermal waters. *Chem. Geol.* 513: 32–43.

Hijmans R.J., Cameron S.E., Parra J.L., Jones P.G. & Jarvis A. 2005. Very high resolution interpolated climate surfaces for global land areas. *Int. J. Climat.* 25: 1965–1978.

Hu Z. & Gao S. 2008. Upper crustal abundances of trace elements: a revision and update. *Chem. Geol.* 253: 205–221.

Laske G., Masters G., Ma Z. & Pasyanos M. 2013. Update on CRUST1.0 – A 1-degree global model of earth's crust, Geophys. Res. Abstracts, *15*, Abstract EGU 2013-2658.

Arsenic in the Environment: Bridging Science to Practice for Sustainable Development –
van der Wal, Ahmad, Petrusevski, Weijma, Savic, van der Wens, Beerendonk, Bhattacharya, Bundschuh & Naidu (Eds)
 ISBN 978-1-032-32928-4

Arsenic in different water sources from Arica and Parinacota, Chile

G. Pincetti-Zúñiga, L.A. Richards & D.A. Polya
Department of Earth and Environmental Sciences and Williamson Research Centre for Molecular Environmental Science, University of Manchester, Manchester, UK

ABSTRACT: In Arica and Parinacota, within the arid northern Chile, water quality of the scarce water sources is commonly affected by elevated arsenic (As) above 10 µg/L (WHO guideline). Other potentially harmful elements are also a concern. Unsafe waters, notably from cold springs, are still used for drinking by part of the rural population. An inorganic water quality survey was made to study the hydrochemistry of waters in different basins within the region, quantifying the levels of As, and studying its co-occurrence with other elements. From the 89 samples collected, 72% had elevated arsenic. From those 32 samples were indicated as drinking sources (max 223 µg/L). The highest As corresponds to thermal springs (21,300 µg/L), which contribute with As in the upper section of the basins. In addition, B, Se, Pb, Cd, and Cr were found above WHO guidelines, and were highly correlated with As.

1 INTRODUCTION

Naturally high concentrations of arsenic (As) in water sources in northern Chile has been of interest since the massive As poisoning in Antofagasta between 1958 and 1970 (Bundschuh *et al.*, 2012). Ever since, elevated As has been found in other areas, including the northernmost part of Chile, Arica and Parinacota region. In this region water sources are scarce, and their quality varies depending on the local geology of the river basins. Elevated concentrations of arsenic particularly affect extensive areas within the Lluta (Guerra *et al.*, 2016), and Camarones basins (Arriaza *et al.*, 2018). The water sources available in this region mostly occur as perennial and ephemeral streams, cold springs, and wells in lower areas of the valleys. Many of these sources are still used for drinking without previous treatment by rural communities, while surface waters are widely used for irrigation. The naturally high concentrations of arsenic and other metal/loids in waters from this region are widely attributed to the interaction with volcanic rocks, and thermal waters (Tapia *et al.*, 2019).

Through the sampling of surface, ground, and spring waters, and their (inorganic) hydrochemical characterization, we aim to perform a water quality assessment based on drinking guidelines, with focus in rural areas, and compare the hydrochemistry of waters from contrasting locations.

2 MATERIALS AND METHODS

2.1 *Study area*

The study area is enclosed within the Arica and Parinacota administrative region (18° 30' S 70° 20' W) (Figure 1). The local geography includes narrow valleys within the Atacama desert to the west, and to the east, the Andes Mountains and the Altiplano plateau. In the Atacama desert, rainfall is rare and typically <1 mm per annum. In the plateau the precipitations can reach >300 mm per annum, due to the summer rain associated with the *Altiplanic winter*. Thus, water sources within this region (Figure 1) strongly depend on the recharge areas upstream. Thermal springs are broadly found in high altitude areas, often near to active volcanoes.

2.2 *Sites selection and sampling*

Sampling sites selection was focused on (i) surface and (if present) groundwaters within Lluta and Camarones basin; (ii) accessible thermal springs, (iii) water sources of interest by local collaborators (e.g. tap water). The water sampling was based in standard protocols (Polya & Watts 2017) and was carried out in April 2018. A total of 89 water samples were collected from several locations (Figure 1). Water sources include streams and

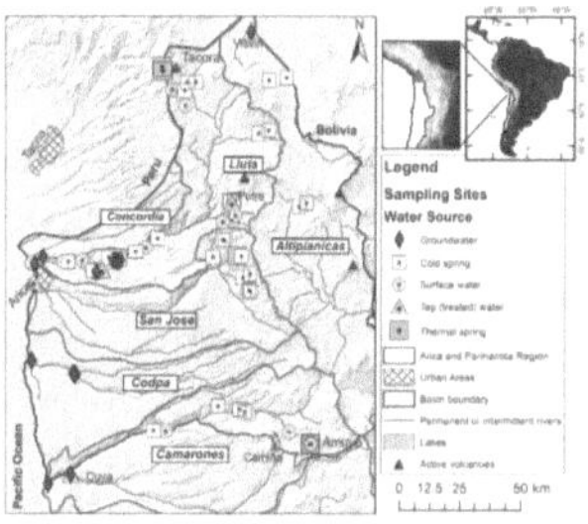

Figure 1. Area of study and sampling sites, including the main river basins and grouping of water sources.

DOI: 10.1201/9781003317395-47

irrigation canals (n=32), cold (n=24) and hot springs (n=5); dug wells (n=12), tube wells (n=3); and raw (n=8) and treated (n=5) tap water. Samples for chemical composition analysis were filtered (0.45 μm) upon collection. Samples for cations analysis were acidified to pH<2.

2.3 *Chemical data obtention and validation*

Physicochemical parameters were measured *in-situ* using a multi-meter including pH, oxidation-reduction potential (ORP), temperature, and electrical conductivity (EC). Alkalinity was measured with a field titration test kit. Chemical analyses were performed at the Manchester Geochemistry Analytical Unit. The analytical techniques used include ICP-AES, ICP-MS, and IC. Several samples were subject to dilutions before analysis. Certified reference materials were used and were broadly within 10%. Charge balance error was within 10% (all samples).

3 RESULTS AND DISCUSSION

3.1 *Inorganic hydrochemistry*

The studied waters had near-neutral pH, were oxidized, and slightly saline (median: 7.3, 197mV, and 2 mS/cm, respectively). An important variability in was observed between samples from different sources, notably the thermal springs. Waters near Tacora thermal springs had pH~1.5 and high electrical conductivity EC~20 mS/cm, whereas in Amuyo pH~7 and EC~14 mS/cm. Most of the samples with EC<0.5 mS/cm were from cold springs >3,000 m.a.s.l. within all basins.

The chemical composition of the studied waters was dominated by Na-Cl type by 51%, notably in Lluta and Camarones valleys; followed by 19% Ca (Na)-HCO_3 type, while 16% were Ca-SO_4 waters. Overall, the concentration range for Na was 5-1920 μg/L (median 234 μg/L), Ca 1-700 μg/L (median 110 μg/L), Mg 0.2-180 μg/L (median 22), and K 0.4-250 (median 16 μg/L). The anions ranged between 1-4300 μg/L for Cl (median 370 μg/L), SO_4 3-4460 μg/L (median 240 mg/L), HCO_3 0-430 ppm (median 60 μg/L), and NO_3 0-74 mg/L (median 3 mg/L). In general, waters from greater altitude areas without influence of thermal sources, had lower concentrations of all major elements.

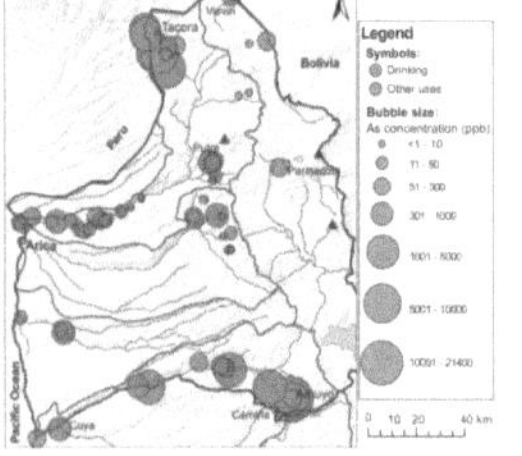

Figure 2. Arsenic concentrations and sources used for drinking.

3.2 *Arsenic occurrence and concentrations*

Overall, 72% of the samples (n=91) had As concentrations above the WHO guideline of 10 μg/L, with maximum values in Camarones (Amuyo springs: 21,300 μg/L) and Tacora area (5,300 μg/L) (Figure 2). Of the sources used for drinking (n=32), 44% had elevated As ranging between 0.4-223 μg/L. Of these samples 2 correspond to tap (treated) water (max 57 μg/L) and 5 to untreated tap water (max 97 μg/L). In rural areas it is typical to use cold springs as a drinking supply, using the natural gradient to drive water from higher ground areas.

3.3 *Other elements concentrations and associations*

Other potentially harmful elements above WHO guidelines were identified, notably in raw and treated sources used for drinking (n=32), including: B (n=11>2.4 μg/L), Cd (n=6>3 μg/L), Pb (n=3>50 μg/L), Se (n=1>40 μg/L), and Cr (n=1>50 μg/L). Strong and significant Pearson's correlations ($p < 0.05$, $r>0.5$) between As and Ca, Na, Cl, B, Li, Zn, Pb, and Cd were identified, suggesting a common source likely associated with the water rock interactions with volcanic rocks and thermal water mixing.

4 CONCLUSIONS

This survey identified areas where small rural communities are drinking water with elevated concentrations of As of up to 220 μg/L. The association of As with other potentially harmful elements suggests a common source likely from thermal waters and volcanic rocks.

ACKNOWLEDGEMENTS

GP was supported by CONICYT-PAI/INDUSTRIA-79090016, UoM SRF 2018, and GCRF (UK).

REFERENCES

Arriaza B., Amarasiriwardena D., Standen V., Yáñez J., Van Hoesen J. & Figueroa L. 2018. Living in poisoning environments: invisible risks and human adaptation. *Evol. Anthropol.* 27(5): 188–196.

Bundschuh J., Litter M.I., Parvez F., Román-Ross G., Nicolli H. B., Jean J.S., Liu C.W., López D., Armienta M.A., Guilherme L.R.G., Cuevas A.G., Cornejo L., Cumbal L. & Toujaguez R. 2012. One century of arsenic exposure in latin america: a review of history of history and occurrence from 14 countries. *Sci. Total Environ.* 429: 2–35.

Guerra P., Gonzalez C., Escauriaza C., Pizarro G. & Pasten P. 2016. Incomplete mixing in the fate and transport of arsenic at a river affected by acid drainage. *Water Air Soil Pollut.* 227: 73.

Polya D.A. & Watts M.J. 2017. Chapter 5. In: P. Bhattacharya, D. A. Polya & D. Jovanovic best practice guide on the control of arsenic in drinking water.

Tapia J., Murray J., Ormachea M., Tirado N. & Nordstrom D.K. 2019. Origin, distribution, and geochemistry of srsenic in the Altiplano-Puna plateau of Argentina, Bolivia, Chile, and Perú. *Sci. Total Environ.* 678: 309–325.

*Arsenic in the Environment: Bridging Science to Practice for Sustainable Development –
van der Wal, Ahmad, Petrusevski, Weijma, Savic, van der Wens, Beerendonk, Bhattacharya, Bundschuh & Naidu (Eds)*

Advances in improving the knowledge of geogenic arsenic distribution in the Salto aquifer, Uruguay

E. Alvareda[1], E. Abelenda[2], I. Machado[3], V. Bühl[3], N. Mañay[4], J. Ramos[1] & P. Gamazo[1]
[1]*Water Department, Centro Universitario Regional Litoral Norte, Universidad de la República, Rivera, Salto, Uruguay*
[2]*Dirección Nacional de Minería y Geología, Ministerio de Industria, Energía y Minería, Hervidero, Montevideo, Uruguay*
[3]*Analytical Chemistry, DEC, Faculty of Chemistry, Universidad de la República, Montevideo, Uruguay*
[4]*Toxicology, DEC, Faculty of Chemistry, Universidad de la República, Gral. Flores, Montevideo, Uruguay*

ABSTRACT: Total arsenic (tAs) levels above those recommended by WHO guidelines for drinking water (10 μg/L) were detected in Northern Uruguay's aquifers. However, the information about the natural baseline, distribution and hydrochemistry of geogenic metalloids from these hydrogeological formations, is scarce. Particularly, the Salto district on the northern littoral is located above the Salto-Arapey aquifer system and the Guarani aquifer system (GAS) the most important geothermal reservoir. Over the years, Salto aquifer is being used for irrigation, recreation and drinking water without any treatment, and it is considered as a safe source. Water quality is a concern and there are no epidemiological studies in Uruguay associated with environmental exposure on geogenic arsenic. For this purpose, we present 32 wells data from Northwest Salto with their hydrogeological origin and geo-referenced. Results have shown tAs range (1.0 - 58.9 μg/L), total phosphorus (tP) (50.0 – 178.5 μg/L) as well as other physicochemical. The highest tAs levels may belong to groundwater form Cretacic formations (Guichon and Arapey). Significant information is being emerged, promoting safety water management of Uruguay hydric resources, preserving human health.

1 INTRODUCTION

The Salto city, located above a rich area of aquifers (Salto, Arapey and Guarani), is stand out to be one of the main horticultural and thermal region of Uruguay. Surrounded by horticultural farms, Salto is the center of the citrus production. Thermal water from Guarani aquifer is relevant to tourism being significant in our foreign exchange earnings. Salto aquifer (SA) has been used for irrigation, recreation and drinking water without any previous treatment since it is considered to be a safe source. Previous researches show different total arsenic (tAs) levels from Arapey and Guarani aquifers (Abelenda 2016; Machado *et al.*, 2019) but there was a lack of data from SA. The present study sheds light to i) the tAs in the geological source and the spatial variability of tAs in the SA, and ii) the chemical groundwater characterization, to assess future human health risk.

2 MATERIALS AND METHODS

2.1 *Study area*

The study area covers approximately 500 km^2, being the Dayman River it southern limit, Salto Grande dam and Brazo del Itapebí Grande reservoir at the northern and the Uruguay River at western (Figure 1). The region has a humid subtropical climate, with rainfall throughout the year, and warm summers with temperatures above 40°C.

2.2 *Geology of the monitored aquifers*

The geology includes volcanic and sedimentary units, the 50% of monitoring wells belong to the Arapey aquifer (AA) from Arapey formation (ArF) from the Precambrian basement in the south and Paleozoic to Mesozoic sediments and basaltic spills at North of Uruguay (part of Paraná basin), including effusive rocks and basaltic spills with interstratified wind sandstone levels. Groundwaters from AA are depth located (~40 meters) in horizontal and vertical fissures of basalt with the presence of sandstones between castings, which porosity accumulates large amounts of water. The 19% of wells in SA (~20–30 meters) from Salto Formation constituted by alluvial deposits from Mio-Pliocene, with psamites and psefites, being a narrow strip parallel to Uruguay River. The SA is free to semi-confined shallow aquifer with a great extent. It has fine to very fine sandstones and river sedimentation gravels, is located above the clays of the Fray Bentos Formation (FBF). Its high hydraulic conductivities makes it suitable to be exploited for horticultural purposes. 12% of wells

DOI: 10.1201/9781003317395-48

correspond to the FBF from Oligocene represented by fine sandstones, limolites, loéssics, fangolites and conglomerate and reddish-pink-brown diamictites. 13% of wells belong to GAS from pores and fissures of sandstones formed during the Mesozoic and 6% of wells belong to the Guichon Formation, classified as reddish feldspathic wacke, with fine sandstones, and rounded quartz grains, with 90% of smectite (Abelenda 2016).

2.3 *Water sampling and laboratory analysis*

Groundwaters from 20 wells were sampled during April to October 2019. Physico-chemical parameters measured *in situ* were: temperature (T), pH, electrical conductivity (EC), redox potential (Eh) with HACH SensIon® portable multimeter and HACH® calibration solutions. Hardness (APHA 2340C) was analyzed in the Water Department Laboratory from Uruguay University. tAs and tP were determined in the Laboratory of the Energy and Mining National Bureau (DINAMIGE), Uruguay. The lasting data (12 wells) were collected from Abelenda (2016) and Machado *et al.* (2019) (Figure 1).

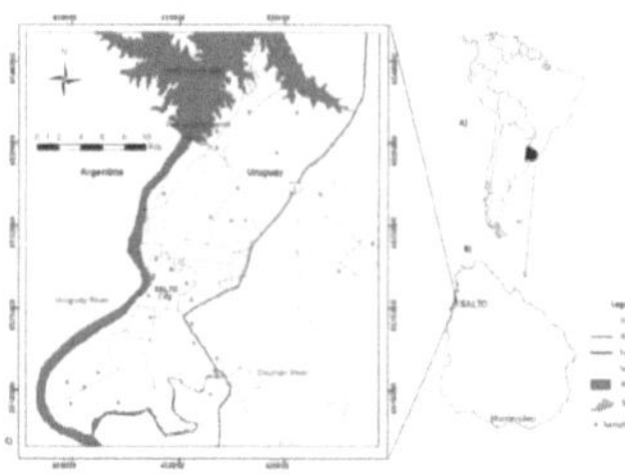

Figure 1. Study area A) Uruguay in South America B) Salto city and C) Sampling zones with green points. Images extracted from Abelenda (2016) and Google Images with modifications.

2.4 *Data analysis*

A spatial database including, geo-references, geology, physical and chemical data was generated. QGIS 2.18.23 software was used for mapping. Statistical analysis was performed in RStudio®.

3 RESULTS AND DISCUSSION

3.1 *Groundwater hydrochemistry*

Hydrochemical analyses of groundwater from Salto aquifer (SA) reveal: pH (7.4-7.8), Eh (106.9-122.1 mV), EC (274.15-579.10 μS/cm), Harness (62.53-336.35 mg/L $CaCO_3$), tAs (3.5 - 9.6 μg/L) and tP (50.0-178.5 μg/L). The highest EC (217.85 - 710.85 μS/cm), Hardness (154.2-474.3 mg/L $CaCO_3$) and As (14.7–49.3) were detected in Arapey aquifer (AA), however, tP range (50.0-78.0 μg/L) was lower than those from SA. Maximum As level (20.7–58.9 μg/L) were detected in SAG (Machado *et al.*, 2019).

3.2 *Geo-As-data analysis*

Two main hydrogeological units are distinguished in the study area. In a preliminary characterization, the SA composed by sandstone of fluvial sedimentary deposit originates mainly calcium-chloride type water. Meanwhile, groundwater from the fissured AA, with a secondary porosity and constituted by tholeiitic basalts, generates mainly bicarbonate-magnesium type. Besides, tAs from AA were higher than those from SA showing a maximum value of 49.3 μg/L in an area originated from fractured basalt. AA depth waters could be originated in geothermal systems adjacent to basaltic magmatism where is confined the Guarani aquifer.

4 CONCLUSIONS AND RECOMMENDATIONS

A preliminary study on the distribution of As and their potential geogenic sources from Salto region was carried out. tAs levels exceed those WHO recommended values from drinking water (10 μg/L) in the studied aquifers. The highest As levels were from fractured basalt of AA mainly. Even so, more studies are required to understand their mobilization mechanisms. These results could help consumers to be aware of this As exposure to preserve their health. Assessing tP levels, both SA and AA could be vulnerable to anthropogenic contamination considering the fact that this is an important region for horticulture and thermalism. Some specific risk management measures should be taken to minimize the exposure: good practice of groundwater management as efficient drilling systems to minimize environmental impact and health risk.

ACKNOWLEDGEMENTS

We acknowledge the National Agency for Research and Innovation (ANII) Uruguay for supporting the FMV_1_2017_536565 project.

REFERENCES

Abelenda E. 2016. Caracterización Hidrogeoquímica e Hidrodinámica y Estimación de la Recarga de Los Acuífero Salto y Arapey. *Degree Thesis, Licenciatura en Geología Facultad de Ciencias, Universidad de la Republica, Rivera Uruguay.*

Alvareda E. *et al.* 2019. Preliminary groundwater hydrochemistry studies grom Salto Northwest Uruguay as drinking water and their possible effects in human health. *Proceedings from MEDGEO 2018. The 8th International Conference on Medical Geology-Moscow*, Russia, p.115.

American Public Health Association (APHA). 2017. *Standard Methods for the Examination of Water and Wastewater.* Washington, DC: American Public Health Association.

Machado I., Bühl V. & Mañay N. 2019. Total arsenic and inorganic arsenic speciation in groundwater intended for human consumption in uruguay: correlation with fluoride, iron, manganese and sulfate. *Sci. Tot. Environ.* 681: 497–502.

QGIS Development Team. 2014. QGIS Geographic Information System. *Open Source Geospatial Foundation Project.*

Arsenic in the Environment: Bridging Science to Practice for Sustainable Development –
van der Wal, Ahmad, Petrusevski, Weijma, Savic, van der Wens, Beerendonk, Bhattacharya, Bundschuh & Naidu (Eds)

Arsenic levels in groundwater and its correlation with relevant inorganic parameters in Uruguay from the Medical Geology perspective

I. Machado[1], V. Bühl[1] & N. Mañay[2]
[1]*Analytical Chemistry, DEC, Faculty of Chemistry, Universidad de la República, Montevideo, Uruguay*
[2]*Toxicology, DEC, Faculty of Chemistry, Universidad de la República, Montevideo, Uruguay*

ABSTRACT: The aim of this work was to perform a groundwater monitoring in Uruguay, from the perspective of Medical Geology, focusing attention on arsenic levels and its correlations with other relevant inorganic parameters. For this purpose, a total of 46 groundwater samples from private wells, out of the scope of the state-run water utility company, were analyzed. The accuracy of the analytical methods was ensured by using certified reference materials. Arsenic concentration range was 1.72 – 120.5 μg/L, being half of the samples above the limit of 10 μg/L recommended by WHO for drinking water, with the corresponding risks for human health. Pearson correlations were performed, resulting in strong positive correlations for the pairs As/Cl^-, As/F^-, As/Na and As/V. These relationships between arsenic and other inorganic parameters in groundwater should be deeply studied, to prevent long-term health effects.

1 INTRODUCTION

Medical Geology is a growing field in Uruguay and groundwater quality has been the focus of research for multiple studies, being arsenic (As) exposure through drinking water, one of its main health risks concerns. Human exposure to As occurs through different pathways, but of all the potential sources, drinking water poses one of the greatest threats to human health as supported by the large number of documented cases around the world. Its mobilization in the environment occurs through complex natural biogeochemical reactions and human interactions, being the most recognized problems, the ones associated with the mobilization under natural conditions (Smedley & Kinniburgh 2013). The aim of this study was to perform a water monitoring of private wells destined to human consumption in Uruguay, that are out of the scope of the state-run drinking water supply company (OSE) and not systematically controlled, focusing attention on As levels and its correlation with relevant inorganic parameters such as chloride (Cl^-), fluoride (F^-), sodium (Na) and vanadium (V), from a Medical Geology perspective.

2 MATERIALS AND METHODS

2.1 *Samples*

A total of 46 groundwater samples were collected from private wells located at different localities of Uruguay, at depths between 30 and 50 m by means of pneumatic submersible pumps. Samples were collected in the autumn season between March and June 2018. The sampling points were randomly selected considering their spatial distribution in the area, in order to obtain an accurately representative hydrochemistry. All samples were stored in polypropylene bottles and kept at 4°C in the dark until analysis. Certified reference materials were used for accuracy and precision evaluation. Population density was also considered.

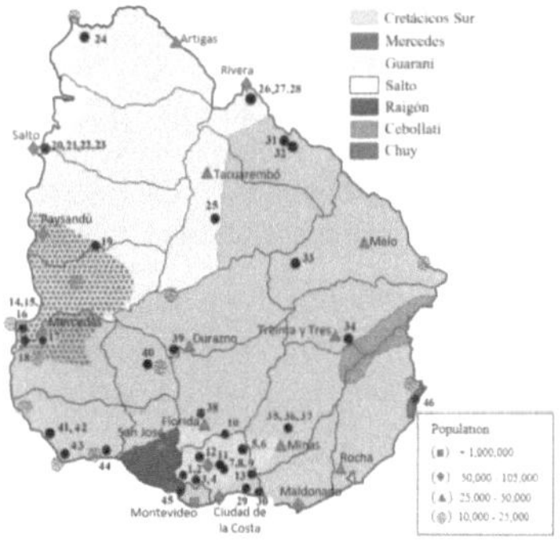

Figure 1. Main Uruguayan aquifers, most populated cities (INE 2020) and sampling points.

2.2 *Laboratory analysis*

Analysis of As and V were performed by electrothermal atomic absorption spectrometry (ETAAS), using a Thermo Scientific iCE 3500 spectrometer, equipped with a graphite furnace atomizer and employing Zeeman-based correction. Analysis of Cl^- and F^- were performed by ion chromatography (IC), using a Thermo Fisher-Dionex ICS5000+ HPIC system. Analysis of Na was performed by flame atomic absorption spectrometry (FAAS), using a Perkin Elmer AAnalyst 200 spectrometer.

2.3 *Statistical analysis*

Pearson correlation analysis were carried out using MS Excel® and differences at a 1% significance level ($p < 0.01$) were considered as statistically significant.

DOI: 10.1201/9781003317395-49

3 RESULTS AND DISCUSSION

3.1 *Arsenic levels*

Arsenic levels ranged from 1.7 to 120.4 μg/L, being half of the samples above the limit of 10 μg/L recommended by WHO for drinking water, with the corresponding risks for human health (WHO 2018). Furthermore, 10 samples were above the limit of 20 μg/L established by Uruguayan regulation.

The highest values corresponded to samples S7 and S8, 120.4 and 63.2 μg/L respectively, from Cretácicos Sur aquifer which is described as a porous aquifer composed of fine sandstones, with clay cement and low dissolved iron levels (Machado *et al.*, 2019). Samples S1 to S5 and S9, also abstracted from this area, presented As levels above those recommended by WHO between 10.3 and 17.9 μg/L. Samples S41 to S44 from the Thermal Corridor zone over Salto aquifer, ranged from 20.7 to 32.7 μg/L. These relatively high natural concentrations obtained may be only associated to the geological framework, since there is no industrial activity in this area. The results are in accordance with the pH values above 8.5, where As starts desorbing from the oxides' surfaces, increasing its concentration in solution. Sample S17 from Mercedes aquifer area presented an As concentration of 53.6 μg/L, being the third most concentrated sample. This area is composed by continental sediments constituted by conglomerates, sandstones and pelites (Machado *et al.*, 2019).

On the other hand, samples S24 to S28 from Guarani aquifer area, presented low As levels ranging from 1.80 to 2.39 μg/L. Similar values were obtained for sample S34 from Cebollati aquifer and sample S46 from Chuy aquifer, being the reported values 1.82 and 1.87 μg/L respectively.

3.2 *Correlations with inorganic parameters*

Fluoride levels ranged from 0.024 to 1.029 mg/L, being below the limit of 1.5 mg/L recommended by WHO. These results revealed a significant variability between samples that could be related to differences in the geochemical profiles of the sampling areas. Vanadium levels ranged from 3.0 to 167.2 μg/L, being the most concentrated samples S17 from Mercedes aquifer and S20 to S23 from Salto aquifer. These high concentrations were expected at the west side of the country, due to the presence of continental sediments containing volcanic ashes. Even though there is no recommended value by WHO in drinking water, these 5 samples exceed the maximum limit of 100 μg/L established by regional regulation. Regarding Cl^- and Na levels, national regulation has established limits of 250 and 200 mg/L respectively, mainly due to sensorial properties. Samples S7 and S9 from Cretácicos Sur aquifer exceeded the maximum limit for Cl^-, with values of 258.7 and 697.1 mg/L respectively, suggesting the possible presence of marine sediments. None of the analyzed samples exceeded the limit for Na.

Pearson analysis showed highly significant positive correlation coefficients: As/Cl^- (r = 0.3921), As/F^- (r = 0.5172), As/Na (r = 0.5549), As/V (r = 0.6201), Cl^-/Na (r = 0.4920), F^-/Na (r = 0.5801), F^-/V (r = 0.5687) and Na/V (r = 0.6300). The strong correlation As/F^- may be due to the dissolution of As-containing minerals facilitated by the alluvial quaternary basin, as well as ion-exchange with fluorite. This high correlation supports the hypothesis of a common source for these analytes, which could be related to the dissolution of volcanic glass, especially in the western part of the country. The strong correlation As/V is supported by the presence of continental sediments containing decimeter-thick volcanic ash layers intercalated, in some parts of the Raigón aquifer and other formations located at the west side of the country. The strong correlation V/F^- also demonstrates that the three elements are related and may have a probable common origin.

The distribution of these parameters in groundwater is heterogeneous, probably due to differences in the aquifer matrix that may change the geochemical conditions required for its mobilization.

4 CONCLUSIONS AND RECOMMENDATIONS

Results showed that several areas might be affected by dangerous As levels in groundwater. Canelones department, where the highest values were found, is the country's second most populated zone. Also, Salto and Mercedes, with high As levels, need to be addressed, considering their population of 104,000 and 42,000 inhabitants respectively. Chronic exposure to As levels above 10 μg/L in drinking water is a major cause of public health concern, supported by the substantial scientific evidence due to its association with several adverse effects on human health (WHO, 2018). However, effects of chronic exposure to As lower concentrations still remain unclear. For this reason, investing in research to better understand the fate and behavior of As in natural geochemical systems is of key importance, in order to keep As levels in drinking water as low as possible.

ACKNOWLEDGEMENTS

To ANII and PEDECIBA-Química for its support.

REFERENCES

Machado I., Bühl V. & Mañay N. 2019. Total arsenic and inorganic arsenic speciation in groundwater intended for human consumption in Uruguay: correlation with fluoride, iron, manganese and sulfate. *Sci. Total Environ.* 681: 497.

Smedley P. & Kinniburgh D. 2013. Arsenic in groundwater and the environment. In: O. Selinus (ed.) *Essentials of Medical Geology*. Springer, Dordrecht, pp. 279–310.

WHO. 2018. *Arsenic. Key Facts.* World Health Organization, Geneva, Switzerland.

Arsenic in the Environment: Bridging Science to Practice for Sustainable Development –
van der Wal, Ahmad, Petrusevski, Weijma, Savic, van der Wens, Beerendonk, Bhattacharya, Bundschuh & Naidu (Eds)
© 2023 the Author(s), ISBN 978-1-032-32928-4

Geologic/geomorphic controls on groundwater arsenic occurrence in a loess-type aquifer in southern Pampean plain, Argentina

L. Sierra[1], P. Weinzettel[1], S. Dietrich[1], S. Bea[1], E. Kruse[2], L.A. Richards[3], G. Pincetti-Zúñiga[3] & D.A. Polya[3]
[1]*IHLLA-Instituto de Hidrología de Llanuras "Dr. Eduardo J. Usunoff", Azul, Argentina*
[2]*CEIDE-Facultad de Ciencias Naturales y Museo, Universidad Nacional de La Plata, La Plata, Argentina*
[3]*DEES-Department of Earth and Environmental Sciences, University of Manchester, Manchester, UK*

ABSTRACT: The Chaco-Pampean plain is the most populated region in Argentina and is threated by highly variable As contamination in groundwater. We conducted an assessment of As distribution based on a detailed geologic/geomorphic and groundwater surveys in the Pampean aquifer in the southern Pampean Plain. Groundwater As increased from older hills, young loess hills to floodplains (F unit) as a result of variable reactive substrates and position on flow paths. Integrated assessment of local As contamination is essential to reduce exposure by selection of appropriate mitigation strategies.

1 INTRODUCTION

The Chaco-Pampean plain is probably the largest region in Argentina affected by arsenic (As) contaminated aquifers reducing social/economic development of 8 million inhabitants. This region is an extensive flatland with naturally high concentrations of toxic elements such as As, F, V and B in alkaline and oxidizing hydrogeochemical conditions. Several studies have identified volcanic glass shards (VGS) and minerals in the loessic aquifer as major primary As sources. Nevertheless, desorption process on Fe-oxyhydroxides are a common mobilization process. Local factors are important controlling As mobilization (Nicolli *et al.*, 2012).

2 MATERIALS AND METHODS

2.1 *Study area*

The study area is located in the southern rain-fed Pampean Plain in a temperate to subhumid climatic zone (P – PET = 83 mm) (Figure 1A). The study area (100 km^2) is located in the border of the Inter-ranges Pampas surrounding Tres Arroyos city (60 K inhabitants) in the Claromecó creek basin (Figures 1B & 1C). Locally, the 120 m thick Pampean aquifer is the most exploited aquifer (Weinzettel *et al.*, 2005). Recent survey reveals an enrichment of mobilizable-As in paleosoils and young deposits (Sosa *et al.*, 2019).

2.2 *Water sampling and laboratory analysis*

This study involved detailed surface mapping to characterize landscape patterns, as well as studies of the vadose zone and aquifer (new cored borehole). Groundwater samples were obtained from boreholes (n=38) by pumping after purging, with on-site measurement inside a flow-through cell according to Sierra (2019). Major ions were analyzed at the IHLLA laboratory and As by ICP-MS (ACME Lab).

3 RESULTS AND DISCUSSION

Upper section 70-m of the Pampean aquifer is composed by four lithological units (Sierra 2019). The deeper unit (LU1) (30 to 70 m depth) is composed of

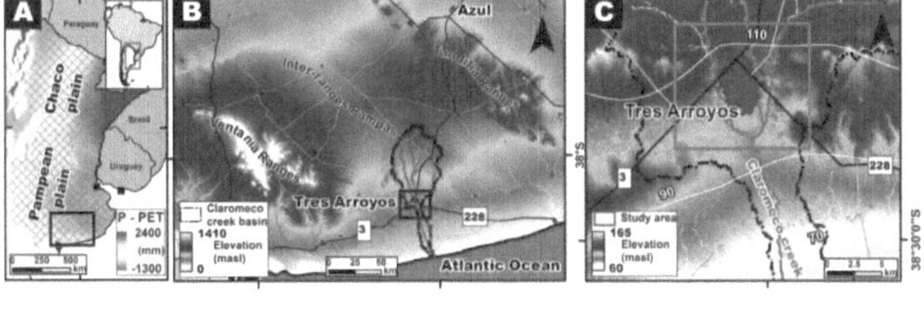

Figure 1. A) Precipitation & Potential Evapotranspiration balance (P – PET) in the Chaco-Pampean plain. B) Location of Claromecó creek basin. C) Physiography and regional phreatic surface of the study area based on Weinzettel *et al.* (2005).

DOI: 10.1201/9781003317395-50

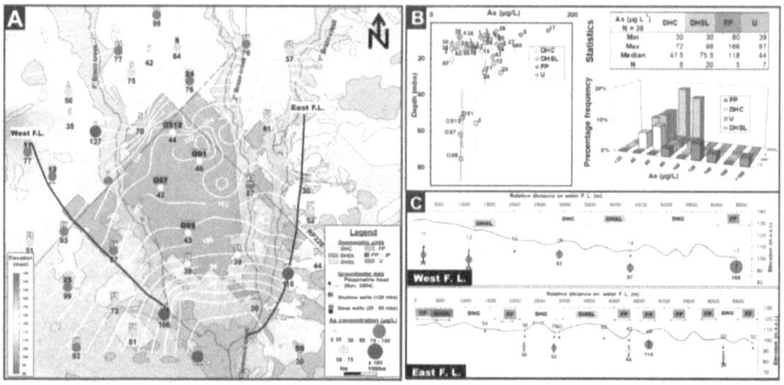

Figure 2. A) Groundwater As in surveyed wells overlain on geomorphic/piezometric map, B) As contents classified by geomorphic units: well screen interval depth distribution, descriptive statistics and frequency distribution, C) Flow paths as indicated in A.

upper Miocene pedosedimentary-calcareous sequences enriched with montmorillonitic clays and crystalline Fe-oxide. The shallower units (LU2, LU3, UL4) have variable characteristics. Main geomorphic units (four) include discontinuous hills floodplains with trellis-stream pattern (Figure 2A).

Hills are characterized by Mio-Pliocene calcretized matrix-supported conglomerates (LU2) and is recognized in three geomorphic units with increasing abundance of Holocene loess deposits (LU4): dissected hills capped by calcretes (DHC), dissected hills mantled by scarce loess (DHSL) and with deep loess (DHDL). Floodplain unit (FP) are filled with Late Pleistocene fluvial sediments (LU3) and reworked loess deposits. Minor units (three) are: urban areas (U) and perennial/intermittent ponds (PP, IP).

3.1 *Assessment of As concentrations patterns in geomorphic units*

Groundwater As concentrations in all samples exceeded the WHO provisional guide value (10 $\mu g\ L^{-1}$). Groundwater As in the DHC and DHSL units, shows lower values than in the FP unit (Figures 2A and 2 B). The former units are located in recharge areas and less reactive substrate (coarser and low-VGS sediments). Notwithstanding this, the DHSL unit shows an increase in groundwater As probably related to the presence of VGS enrichment in LU4. Thus, progressive increases in groundwater As ("West Flow Line", Figure 2C) may be related to hydrolysis of silicates and alkaline desorption of Fe-oxyhydroxides in paleosols (LU1) or calcite in LU2 (Sierra 2019). Meanwhile, "East Flow Line" shows high As lateral variability possibly related to local discharge flows from the hills to FP unit as a result of a concave downwards piezometric surface. The highest FP unit groundwater As are probably related to higher residence times in young and reactive beds such as high-VGS content LU4 and hydromorphic paleosoils in fluvial LU3 (Sosa *et al.*, 2019). Hydrolysis and desorption processes in young sediments favors As mobilization. Groundwater–surface water interaction and evapo-concentration may favor high-As natural waters in discharge areas.

4 CONCLUSIONS

Detailed geologic and geomorphic mapping of shallow aquifer is consistent with groundwater As mobilization in alkaline and oxidizing conditions. This may be useful in informing reduction of human exposures to high As by well switching or drilling.

ACKNOWLEDGEMENTS

This research was supported by PICT1805/14 (ANPCyT) and PID0075/11 (COHIFE) projects and HELAR1902 grant (British Council). We are grateful to landowners and IHLLA´s technicians. LR acknowledges a Leverhulme Trust Fellowship and DP an EPSRC IAA award (UoM).

REFERENCES

Nicolli H., Bundschuh J., Blanco M., Tujchneider O. & Dapeña C. 2012. Arsenic & associated trace elements in groundwater from the chacopampean plain, Argentina: 100 years of research. *Sci. Total Environ.* 429: 36–56.

Sierra L. 2019. *Nuevas Metodologías en la Exploración del Agua Subterránea en Base a la Investigación del Origen y la Movilidad del Arsénico en el Acuífero Pampeano.* FCEFyN, Universidad Nacional de Córdoba. Doctoral Thesis.

Sosa N.N., Kulkarni H.V., Datta S., Beilinson E., Porfido C., Spagnuolo M. & Surber J. 2019. Occurrence and distribution of high arsenic in sediments and groundwater of the Claromecó fluvial basin, Southern Pampean plain (Argentina). *Sci. Total Environ.* 695: 133673.

Weinzettel P., Varni M. & Usunoff E. 2005. Caracterización hidrogeológica del Área Urbana y periurbana de la ciudad de Tres Arroyos, Buenos Aires. In: Proceedings of IV Hydrogeology Argentine Congress. Rio Cuarto. 171–180.

Arsenic in the Environment: Bridging Science to Practice for Sustainable Development –
van der Wal, Ahmad, Petrusevski, Weijma, Savic, van der Wens, Beerendonk, Bhattacharya, Bundschuh & Naidu (Eds)
© 2023 the Author(s), ISBN 978-1-032-32928-4

Geochemistry of arsenic in surface and groundwaters of Los Pozuelos basin, Puna region, NW Argentina

J. Murray[1], D.K. Nordstrom[2], B. Dold[3] & A. Kirschbaum[1]
[1]*Instituto de Bio y Geo Ciencias del NOA (IBIGEO), CONICET-Universidad Nacional de Salta, Rosario de Lerma, Salta, Argentina*
[2]*United States Geological Survey, Boulder, CO, USA*
[3]*Division of Geosciences and Environmental Engineering, Luleå University of Technology, Luleå, Sweden*

ABSTRACT: Los Pozuelos is a closed basin in the Puna region of NW Argentina. In this semi-arid region closed basins are the most important feature for the hydrologic systems. The basin center is occupied by a fluctuating playa lake called Los Pozuelos lagoon, which is a UNESCO Biosphere Reserve. Local original communities use groundwater for drinking and cooking. Lowest concentrations of As and dissolved solids are in the rivers (1.46–27 μg/L) and the highest concentrations are in the lagoon (43.7–200.3 μg/L). In groundwater, arsenic also reach high concentrations (3.82–113 μg/L), especially during the dry season where Na/K mass ratios (0.2 and 0.3), and formation of Na-rich efflorescent salts suggest that high evaporation rates increase As concentration. As(V) is the dominant species in all the water types, except for the lagoon, where As(III) occasionally dominates because of organic matter buildup. There are four potential sources of As: i) oxidation of As sulfides in Pan de Azúcar mine wastes, and discharge into the basin; ii) weathering and erosion of mineralized shales; iii) weathering of volcanic eruptive non-mineralized rocks, and iv) weathering of gold mines. Because it is a closed basin, the arsenic released from the natural and anthropogenic sources is transported in solution and in fluvial sediments and finally accumulates in the center of the basin where the concentration in water increases by evaporation with enhancement by organic matter reducing conditions. Remediation methods should be applied in wells with high As concentrations.

1 INTRODUCTION

High levels of arsenic (As) have been detected in all the water types in the Altiplano-Puna (Tapia *et al.*, 2019). However, only a few publications report As speciation as well as a complete physical chemical characterization of the waters that help to understand the mobility of As in the closed basins (Tapia *et al.*, 2019). Extensive studies related to the geogenic and anthropogenic origin of As in the region are also scarce. Los Pozuelos, located in NW Argentina (Figure 1), is one of the most populated closed basins in the Argentinian Puna with a total population of about 3500 inhabitants. The local original communities use groundwater for drinking and cooking. Previous studies indicate the presence of As in surface and groundwater of the basin associated to different anthropogenic and geogenic sources (Murray *et al.*, 2016). The aim of this work is to understand the origin of As, its mobility, and cycling in surface and groundwater of Los Pozuelos basin. To achieve our objective, we analyzed and interpreted long term data (2011 to 2017) that includes As concentration, arsenic redox speciation, and the geochemistry (major ions and physical-chemical parameters) of surface and groundwaters of the basin. The effect of the geomorphological closed hydrologic system and semi-arid conditions on As mobility, concentration, and cycling are discussed. This work contributes to the larger framework of As sources, mobility, and cycling in the closed basins of the Altiplano-Puna.

2 MATERIALS AND METHODS

2.1 *Water sampling and analysis*

A total of 120 water samples were obtained from rivers (n=32), lagoon (n=6) and groundwater (n=28), which were sampled during 6 field campaigns between 2011 and 2017. Samples were filtered and preserved in the field. Cations were determined by inductively coupled plasma - optical emission spectrometry (ICP-OES) and by inductively coupled plasma-mass spectrometry (ICP-MS). Anion concentrations were determined by ion chromatography (IC). Arsenic redox species were determined at the U. S. Geological Survey (Boulder, Colorado) laboratories. Total dissolved arsenic (As(T)) and dissolved arsenite (As(III)) concentrations were determined with a Perkin Elmer (FIAS 100) hydride generation atomic absorption spectrometry system (HG-AAS).

3 RESULTS AND DISCUSSION

3.1 *Total As and As speciation in surface and groundwater of Los Pozuelos basin*

Rivers are mainly Na-HCO_3 water types and the total As concentrations vary in a range from 1.46 to 27 μg/L. The concentration of As in rivers increased from

DOI: 10.1201/9781003317395-51

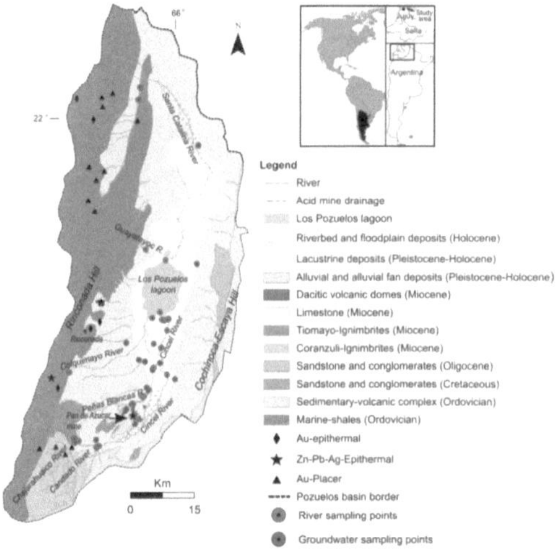

Figure 1. Study area and water sampling locations. (adapted from Murray *et al.*, 2019).

the headwaters to the center of the basin. The highest value was observed in Peñas Blancas River after acid mine drainage discharge (27 μg/L) and in Cincel River in the discharge to the lagoon (17.8–20.3 μg/L).

Arsenate (As(V) ≫ As(III)) is the dominant species in all river water samples. Oxidation promotes higher As (V) concentration in rivers and also sorption in fluvial sediments. The composition of groundwater is mainly $Ca\text{-}HCO_3$ and $Na\text{-}HCO_3$. Total dissolved As in groundwater ranges between 3.82 and 113 μg/L where As (V) is the dominant species. The shallow unconfined aquifer conditions promote oxidation and higher As (V) concentrations. 70% of the wells of the basin are not suitable for drinking purposes since their As concentrations are above the World Health Organization limit value (10 μg/L). During the dry season, As concentration increases in groundwater due to strong evaporation. This is supported by the composition of groundwater in some wells in the center of the basin that change to $K\text{-}HCO_3$ in dry season with Na/K ratio of 0.2 and 0.3 and thenardite ($Na_2(SO_4)$) + halite (NaCl) + eugsterite ($Na_4Ca(SO_4)_3 \cdot 2(H_2O)$) precipitates. However, $\delta^{18}O_{H2O}$, δD_{H2O} analyses are needed to better understand As mobility and its behavior under semi-arid conditions (Murray *et al.*, 2019).

In the lagoon, the concentrations of total dissolved As vary between 43.7 and 200.3 μg/L with As(V) the dominant species. The higher As concentration was observed during the dry season of May-2015 where the concentration increases up to 149.4 and 200.3 μg/L with As(III) the dominant species. High evaporation and the closed system effect most probably favor high As concentrations in the lagoon. However, the abrupt increase of total dissolved As and the relative increase of As(III) during the 2015 period suggests a process that could be related to the biological activity in the lagoon. The arrival of large numbers of flamingos increase bird guano and organic matter producing reductive conditions that could favor reduction of As(V) to As(III).

4 CONCLUSIONS AND RECOMMENDATIONS

In Los Pozuelos basin the As concentrations increases from the upstream areas to the center of the basin in both surface and groundwater with the highest values observed in Los Pozuelos Lagoon. Arsenate is the dominant species in all surface and groundwater, except for the lagoon where variations between As(V) and As(III) species were observed and associated with organic matter and reductive conditions in Los Pozuelos Lagoon. There are four potential sources for As in surface and groundwater i) acid mine drainage discharging into the basin; ii) weathering and erosion of mineralized shales; iii) weathering of volcanic eruptive non-mineralized rocks, and iv) weathering of gold mines. Because it is a closed basin, the arsenic released from the natural and anthropogenic sources is transported in solution, and in fluvial sediments, and finally accumulated in the center of the basin where the concentrations increase because streams and sediment coalesce, and because of high evaporation. Remediation methods should be applied in wells with high As concentrations.

ACKNOWLEDGEMENTS

We thank local inhabitants for allowing sampling and the National Park Administration for the assistance with the field work. The Environmental and Mining Secretaries of Jujuy Province authorized permits to develop this study. This research was financed by the Argentinian projects CIUNSa 2262, 1674, PICT 2015- 1069, PIP 201101-189 and PIP 201101-133 and the USGS National Research Program.

REFERENCES

Murray J., Nordstrom D.K., Dold B. & Kirschbaum A. 2016. Distinguishing potential sources for as in groundwater in Pozuelos basin, Puna region Argentina. In: P. Bhattacharya, M. Vahter, J. Jarsjö, J. Kumpiene, A. Ahmad, C. Sparrenbom, G. Jacks, M.E. Donselaar, J. Bundschuh & R. Naidu (eds.) *Arsenic Research and Global Sustainability (As2016)*, Taylor & Francis Group, pp. 94–96.

Murray J., Nordstrom D.K., Dold B., Romero Orué M. & Kirschbaum A. 2019. Origin and geochemistry of arsenic in surface and groundwater of Los Pozuelos basin, Puna region Argentina. *Sci. Total Environ.* 697: 134085

Tapia J., Murray J., Ormachea M., Tirado N. & Nordstrom D.K. 2019. Origin, distribution, and geochemistry of arsenic in the Altiplano-Puna plateau of Argentina, Bolivia, Chile, and Perú. *Sci. Total Environ.* 678: 309–325.

1.4 Arsenic mobility and fate in soils, sediments and mining wastes

Arsenic in the Environment: Bridging Science to Practice for Sustainable Development –
van der Wal, Ahmad, Petrusevski, Weijma, Savic, van der Wens, Beerendonk, Bhattacharya, Bundschuh & Naidu (Eds)
 ISBN 978-1-032-32928-4

Tracing the dynamic changes of arsenic species across the interfaces in water-soil-plant system

Z.-F. Yuan[1,2], W. Gustave[1,2] & Z. Chen[2]
[1]*Department of Environmental Science, University of Liverpool, Brownlow Hill, Liverpool, UK*
[2]*Department of Health and Environmental Sciences, Xi'an Jiaotong-Liverpool University, Jiangsu, P.R. China*

ABSTRACT: In flooded soils, soil-water interface (SWI) and plant rhizosphere are the key zone controlling biogeochemical dynamics. Arsenic (As) species and concentrations vary greatly at micro- to cm-scales along the redox gradient. However, the As pattern in the narrow interfaces is rare described, which is mainly due to the lack of proper techniques. Here, with the newly developed Integrated Porewater Injection (IPI) sampler, which is designed for soil porewater sampling with minimum disturbance to saturated soil environment, we successfully revealed the distribution of arsenate (As(V)) and arsenite (As (III)) across SWI and rice rhizosphere in every 2 mm. The results showed the oxygen in surface water and oxygen loss from rice roots strongly impact the dynamic change of As(III) and As(V). With growing rice, iron and As were linearly decreased across 0–10 mm distance from root surface, coinciding with the formation of redox zone by radial O_2 loss from root. The oxidizing zone can retain most of mobile As (40 - 80%) from bulk soils. We present a model of As transport from soil to rice, linking the temporal and spatial dynamic regulation of As in paddy soils.

1 INTRODUCTION

Rice (Oryza sativa L.) is the staple food in many regions of the world, but its quality is threatened by arsenic (As) in paddy soil (Zhu *et al.*, 2008). Arsenic releasing from Release of As in paddy soil is stimulated by anaerobic condition induced by periodic flooding during rice production. Along flooding, the release of As changes as a result of the interplay between the mobilization and immobilization of As driven by varying biogeochemical processes. In addition, immobilization of As could be promoted in the rhizosphere due to the radial O_2 loss (ROL) from rice root, which would significantly affect As release across the rhizosphere (spatial). However, the spatiotemporal regulation of As is not well resolved in paddy soil.

Recently, a technique, called Integrated Porewater Introduction (IPI) sampler, is developed in our group to *in situ* measure fine-scale (mm) total aqueous Fe, As and their species repeatedly aross soil-water interface(Yuan *et al.*, 2019). With IPI sampler, we tried to study spatiotemporal dynamics of As species across SWI and rice rhizosphere.

2 MATERIALS AND METHODS

2.1 *Soil and rice properties*

The soil sample with high As concentration (146 mg/kg) was collected from paddy fields in Shaoguan (SG, 25°6′N, 113°38′E), China. Indica rice (O. sativa L.) hybrid, Yliangyou-1 (YLY-1), was used.

2.2 *Porewater sampler preparation and application*

The IPI sampler used in this study shares the same design as reported in previous report (Yuan *et al.*, 2019). The IPI sampler employs the hollow fiber membrane tube (1.7 mm in diameter) to passively sample porewater in its surroundings (Figure 1).

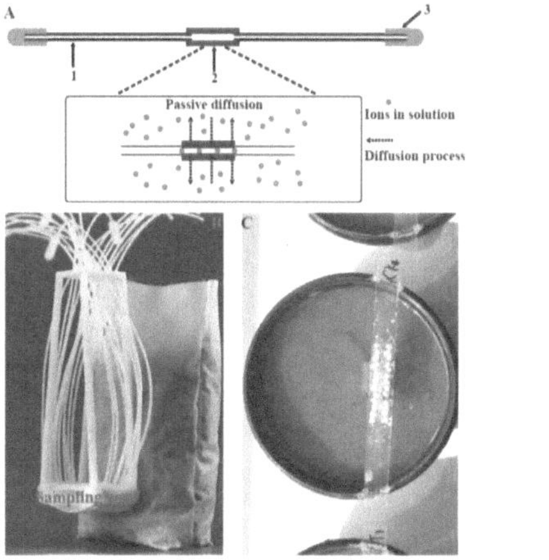

Figure 1. Schematic diagram of IPI sampler. (A) The design and work mechanism of IPI sampler; (B) photo of Rhizon profiler; (C) photo of Rhizon profiler deployed in soils.

DOI: 10.1201/9781003317395-52

When the equilibrium state was reached, the solution inside the tube was sampled for downstream instrumental analysis.

3 RESULTS AND DISCUSSION

3.1 *As species across SWI*

Two inorganic arsenic species, including As(III) and As(V), were detected in soil porewater. (Figure 2) The concentrations of As(III) and As(V) remained almost undetectable in the oxic zones but increased rapidly to up to 2.2 and 0.75 μM/L in reducing soils. Based on the changing gradient, As(III) proportion profile was divided into three stages: 1) a rapid decrease from 100 to 80.6% in the first 10–20 mm top soil; 2) a slow decrease from 80.6 to 75.5% in 20–30 mm soil; 3) a stable value (~75%) in deep soils (30–50 mm). Surprisingly, the initial release of As along SWI is almost all belonging to As(III).

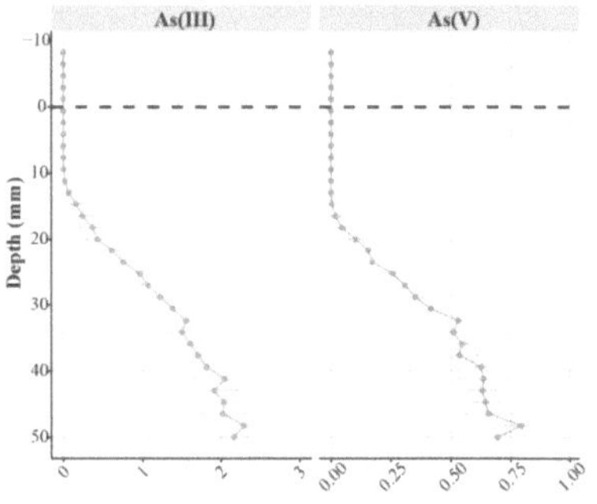

Figure 2. Profile of As species (μM/L) and arsenite [As (III)] proportion across SWI in Shaoguan (SG).

3.2 *Heterogeneous distribution of As species across the rhizosphere*

Based on spatiotemporal changes of As across the rhizosphere, we demonstrated the key role of the redox zone to immobilize mobile As pool around root. In order to mechanically interpret the potential biogeochemical processes regulating As redox in that zone, As species variation across the rhizosphere was measured (Figure 3).

Both As(III) and As(V) were significantly immobilized in the redox zone (Figure 3). For the control group without plant, sum of As(III) and As(V) remained around 90.9 μg/L across the rhizosphere, with an As(III) proportion of 60%. By contrast, sum of As(III) and As(V) decreased linearly from 101 μg/L at 10-20 mm distance from rice root to as low as 23.5 μg/L at 0–2 mm around root, accompanied by an As(III) proportion increasing from 66.0 to 84.3%.

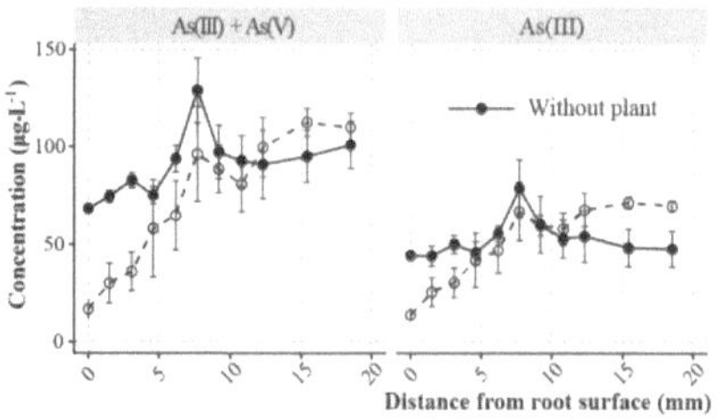

Figure 3. Spatial changes of As species across the rhizosphere (0-20 mm) at DAT 40. Arsenic species included arsenite [As(III)] and arsenate [As(V)].

An opposite trend was observed for As(III) proportion. Figure 3 clearly showed that the redox zone retained a relatively high As(III), while As(V) was almost depleted within 0-5 mm zone.

4 CONCLUSIONS AND RECOMMENDATIONS

To the best of our knowledge, it is the first time that we present a fine mapping of As(III) and As (V) in SWI and rice rhizosphere. Our results showed IPI sampler is a powerful to study the As species in soil system.

ACKNOWLEDGEMENTS

This work was supported by the National Science Foundation of China (41977320, 41571305) and XJTLU RDF-15-01-39.

REFERENCES

Yuan Z.-F., Gustave W., Bridge J., Liang Y., Sekar R., Boyle J., Jin C.-Y., Pu T.-Y., Ren Y.-X. & Chen Z. 2019. Tracing the dynamic changes of element profiles by novel soil porewater samplers with ultralow disturbance to soil–water interface. *Environ. Sci. Technol.* 53(9): 5124–5132.

Zhu Y.G., Williams P.N. & Meharg A.A. 2008. Exposure to inorganic arsenic from rice: a global health issue? *Environ. Pollut.* 154(2): 169–171.

Arsenic in the Environment: Bridging Science to Practice for Sustainable Development – van der Wal, Ahmad, Petrusevski, Weijma, Savic, van der Wens, Beerendonk, Bhattacharya, Bundschuh & Naidu (Eds)
© 2023 the Author(s), ISBN 978-1-032-32928-4

Arsenic(III) adsorption and oxidation by a Pleistocene brown sediment from the Hetao Basin, China

Z.P. Gao[1,2] & H.M. Guo[1,2]
[1]*State Key Laboratory of Biogeology and Environmental Geology, China University of Geosciences, Beijing, P.R. China*
[2]*School of Water Resources and Environment, China University of Geosciences (Beijing), Beijing, P.R. China*

ABSTRACT: The ability of As(III) adsorption and oxidation by the Pleistocene brown sediment from inland basins still remains unclear. To address this issue, we investigated As(III) adsorption and oxidation by a Pleistocene brown sediment from the Hetao Basin in the presence of competing anions (e.g., OH^-, PO_4, and HCO_3). The results show that the brown sediment contained around 4 mmol/kg extractable Mn(IV) oxides, and was capable of oxidizing dissolved As(III). However, percentage of dissolved As(V) reached a maximum of around 20% with increasing initial As(III) concentration up to 25 μmol/L, indicating its limited ability of oxidizing As(III). Therefore, adsorbed As was believed to be the sum of adsorbed As(III) and adsorbed As(V). Arsenic adsorption exhibited a nonlinear isotherm. Maximum As adsorption occurred at around pH 7.2. Oxidation of As(III) was likely enhanced by the increasing pH. Phosphate competed stronger with As(V) than with As(III), while effect of HCO_3 competition was greater on As(III) adsorption.

1 INTRODUCTION

Low arsenic (As) groundwater in aquifers with Pleistocene brown sediment is potentially threatened by high-As groundwater intrusion in the Hetao Basin, induced by increasingly groundwater extraction (Zhang *et al.*, 2018). Arsenic in high-As groundwater under reducing conditions are generally dominated by As(III) species. The vulnerability of the low-As zone to As(III) contamination was thus influenced by the retardation of As transport onto the oxidized brown sediments (van Geen *et al.*, 2013). Dissolved As(III) may be oxidized by Mn(IV) oxides once introduced into the low-As zone. Arsenic(III) were completely oxidized to As(V) by the Pleistocene brown sediments from Bangladesh (Stollenwerk *et al.*, 2007), whereas As(III) was partially oxidized by the brown sediment from the Red River floodplain only when organic pH-buffers were added (Rathi *et al.*, 2017). The capability of As(III) adsorption and oxidation by the brown sediments in the Hetao Basin still remains unclear. Therefore, the present study aims to (1) characterize geochemical compositions of the oxidized brown sediment, (2) investigate As(III) adsorption and oxidation by the brown sediment in the presence of competing anions (e.g., OH^-, HCO_3, and PO_4), and (3) evaluate the vulnerability of the low-As zone to As contamination.

2 MATERIALS AND METHODS

2.1 *Study area*

The Hetao Basin, China, is located north of the Yellow River and south of the Langshan Mountains. Aquifers in the alluvial fan near the mountains hosting low-As groundwater are characterized by the oxidized brown sediments, while high-As groundwater are found in the flat plain with gray sediments. Groundwater initially flowed from the mountain to the flat plain, while the flow direction was lately reversed because of heavy groundwater extraction, leading to the intrusion of high-As groundwater into the low-As zone (Zhang *et al.*, 2018).

2.2 *Sediment Characterization*

A brown fine sediment named Fan#1-18, located at a depth of around 18 m below land surface near the mountain, was selected to perform As(III) adsorption experiments. Contents of Fe(III)/Mn(IV) oxides in the sediment were quantified with different extraction procedures. Contents of Fe(III) oxides were determined from the difference in Fe contents between 0.1 mol/L ascorbic acid + 0.2 mol/L ammonium oxalate extract and HCl extract at pH 3. The difference in Mn contents between 0.1 mol/L $NH_2OH{\cdot}HCl$ + 0.01 mol/L HNO_3 and HNO_3 extract at pH 2 was considered as Mn(IV) oxide contents.

2.3 *Arsenic(III) adsorption experiments*

Batch experiments of As(III) adsorption onto Fan#1-18 were conducted in anoxic conditions, including As(III) adsorption kinetics, As(III) adsorption isotherm, pH effects, and PO_4 and HCO_3 competitions. The experiments were all conducted with 1 g wet sediment (0.84 g dray sediment) and 20 mL 20 mmol/L NaCl solutions. Batch experiments of kinetic As(III) adsorption, pH effects, and PO_4 and HCO_3 competitions were all conducted with 10 μmol/L As(III), while experiments for As (III) adsorption isotherm were carried out with As(III)

DOI: 10.1201/9781003317395-53

concentrations from 0.5 to 25 µmol/L. Blank experiments without As addition were also performed. Solution pH was adjusted using 1 mol/L (0.1 mol/L) HCl or NaOH in pH effect experiments, while pH in other batches was undisturbed and measured at the end of the experiments. After shaken for around 5 days, samples were taken and detected by ICP-AES (iCAP6300, Thermo) for dissolved As, Fe, Mn, and P, by HPLC-HG-AFS (AFS-9130, Titan) for As species, and by Alkalinity Test (MColortest TM, HC605827) for dissolved HCO_3.

3 RESULTS AND DISCUSSION

3.1 *Sediment characteristics*

No easily-dissolved Fe(II) minerals were extracted in HCl solution at pH 3. Total extracted Fe(III) oxides were around 123 mmol/kg, being comparable to those in the brown sediments from Bangladesh with around 151 mmol/kg (Stollenwerk *et al.*, 2007).

Content of sedimentary Mn(II) extracted by HNO_3 at pH 2 was around 0.14 mmol/kg, while total extracted Mn(IV) oxides were around 4 mmol/kg, very close to those in the sediment from Bangladesh with around 4.6 mmol/kg (Stollenwerk *et al.*, 2007).

3.2 *Arsenic(III) adsorption and oxidation*

Arsenic(III) was partially oxidized to As(V) when interacting with the brown sediment Fan#1-18 (Figure 1), being possibly related to the high content of Mn(IV) oxides. Therefore, the amount of adsorbed As in Figure 1 was the sum of adsorbed As(III) and As(V). In the kinetic As(III) adsorption experiments (Figure 1a), adsorbed As reached an plateau after three days, while As(III) oxidation and As(V) production were equilibrated after five-day reaction. It indicates that As(III) oxidation rates were slower than the adsorption rates. In the concentration-dependent As (III) adsorption experiments (Figure 1b), As adsorption exhibited nonlinear adsorption isotherm. With increasing addition As(III) concentrations, proportion of As (V) concentration in dissolved As gradually decreased to around 20%, indicating limited ability of Fan#1-18 on oxidizing As(III). Maximum As adsorption occurred at around pH 7.2 (Figure 1c). With increasing pH from 6.7 to 8.6, proportion of As(V) gradually increased from around 10% to 30%, being mainly due to increases in As (III) oxidation and/or As(V) desorption. Phosphate acted as a stronger competitor for As adsorption onto Fan#1-18 (Figure 1d). With increasing PO_4 concentration from around 0.5 to 25 µmol/L, both dissolved As (III) and As(V) concentrations increased, but percentage of As(V) increased from around 22% to 28%. It indicates relative stronger competition of PO_4 with As (V) than with As(III). Bicarbonate showed visible competitions with As adsorption onto Fan#1-18 (Figure 1e). Although relative abundance of As(V) kept relatively stable at around 0.9 µmol/L with increasing HCO_3 concentrations from 2 to 12 mmol/L, As(III) proportion increased from around 75% to 85%, possibly indicating that HCO_3 competed stronger with As(III) than with As(V).

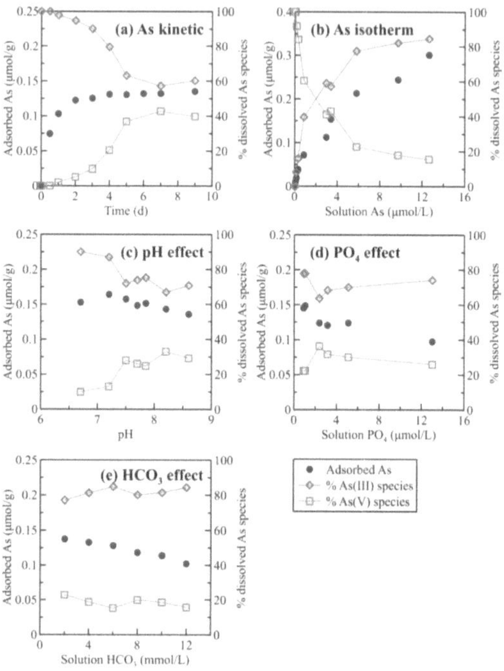

Figure 1. As(III) adsorption onto the Pleistocene brown sediment Fan#1-18. (a) Kinetic As(III) adsorption; (b) As (III) adsorption isotherm; (c) pH effects on As(III) adsorption; (d) PO_4 effects on As(III) adsorption; (e) HCO_3 effects on As(III) adsorption. Initial As(III) = 10 µmol/L in (a), (c), (d), and (e).

ACKNOWLEDGEMENTS

The study was financially supported by the National Natural Science Foundation of China (grant No. 41825017), the Fundamental Research Funds for the Central Universities (grant No. 2652018189).

REFERENCES

Rathi B., Neidhardt H., Berg M., Siade A. & Prommer H., 2017. Processes governing arsenic retardation on pleistocene sediments: adsorption experiments and model-based analysis. *Water Resour. Res.* 53: 4344–4360.

Stollenwerk K.G., Breit G.N., Welch A.H., Yount J.C., Whitney J.W., Forster A.L., Uddin M.N., Majumder R. K. & Ahmed N., 2007. Arsenic attenuation by oxidized sediments in Bangladesh. *Sci. Total Environ.* 379: 133–150.

van Geen A., Bostick B.C., Pham T.K.T., Vi M.L., Nguyen N.M., Phu D.M., Pham H.V., Radloff K., Aziz Z., Mey J.L., Stahl M.O., Harvey C.F., Oates P., Weinman B., Stengel C., Frei F., Kipfer R. & Berg M., 2013. Retardation of arsenic transport through a pleistocene aquifer. *Nature* 501: 204–207.

Zhang Z., Guo H.M., Zhao W.G., Liu S., Cao Y.S. & Jia Y. F., 2018. Influences of groundwater extraction on flow dynamics and arsenic levels in the Western Hetao basin, Inner Mongolia, China. *Hydrogeol. J.* 26(5): 1499–1512.

Arsenic in the Environment: Bridging Science to Practice for Sustainable Development – van der Wal, Ahmad, Petrusevski, Weijma, Savic, van der Wens, Beerendonk, Bhattacharya, Bundschuh & Naidu (Eds)
© 2023 the Author(s), ISBN 978-1-032-32928-4

Effects of phosphate on distribution of arsenic at the paddy soil-water interface: Experiments and modeling

L. Weng[1,2] & Y. Deng[1]
[1]*Agro-Environmental Protection Institute, Ministry of Agriculture and Rural Affairs, Tianjin, P.R. China*
[2]*Department of Soil Quality, Wageningen University, Wageningen, The Netherlands*

ABSTRACT: Contamination of arsenic (As) in paddy soils has drawn much attention in the past decades. Phosphate based fertilizers are commonly used in paddy field. Elevating speciation and distribution of As is essential to assess the risk of As in paddy soils. In this study, we investigated effects of phosphate on the solubility of As (arsenite and arsenate) in two paddy soils over a wide pH range, by combing the Ligand and Charge Distribution (LCD) model to reveal the influence mechanisms. Contrasting effects of phosphate on As speciation were observed in the suboxic and anoxic soil. Under suboxic conditions, phosphate inhibited Fe and As reduction probably due to stabilization of Fe-(hydr)oxides, but increased soluble arsenate concentration as a result of competitive adsorption between arsenate and phosphate. In an anoxic soil, phosphate stimulated Fe and As reduction and caused increases of arsenite in soil solution under both acidic and neutral/alkaline pH. The understanding obtained in this work can be used in future to explain and predict the environmental process of anionic pollutants such as arsenate, arsenite and phosphate in paddy fields as well as in natural soil/sediment-water systems.

1 INTRODUCTION

In paddy soils, which is often under flooded condition, both the solubility and bioavailability of As are high (Takahashi *et al.*, 2004). Phosphate based fertilizers are commonly used in paddy field. The geochemical behaviors of phosphate and arsenate are similar and both can be adsorbed by Fe-(hydr)oxides, which are important adsorbents in soils (Yamaguchi *et al.*, 2011). In addition to competition for adsorption, phosphate may influence As speciation via influencing the reduction-oxidation reactions. The behavior of As in soils is very sensitive to the redox potentials. For paddy soils under reduced conditions, both As reduction and Fe-(hydr)oxides dissolution would affect the mobility and toxicity of As, but the relative contribution of these two mechanisms is difficult to determine. However, a good understanding and reliable prediction of As behavior in paddy soils is crucial in risk assessment and remediation in polluted areas so as to reduce effects of As on ecosystem and human health.

2 MATERIALS AND METHODS

Two paddy soil samples were collected from the top soils (0–20 cm) of paddy fields from Liaoning (N1), and Hunan (N2) province of China. The soil samples were dried at room temperature and sieved over 2 mm.

Batch adsorption experiments were carried out in 50 mL polyethylene centrifuge tubes, to which 2.0 g soil samples and appropriate amounts of stock solutions were added. For each soil, the final soil-solution ratio (SSR) was 1:10, the final concentration of $CaCl_2$ was 10 mM, and the final concentration of arsenate was 0.25 mM. For phosphate treatment, the final concentration of PO_4^{3-} was 0.25 mM. The prepared suspensions were adjusted with acid and base to a pH range between 3 and 9, and were shaken horizontally (50 rpm) at room temperature for 7 days. After 7 days shaking, the end pH and Eh of the suspension were measured, then centrifuged and filtered through 0.22 μm membranes. The filtrates were analyzed on HPLC-HG-AFS for As speciation and concentration, on TOC analyzer for concentrations of DOC, and on ICP-OES for Fe, Al and Ca concentrations. The concentrations of phosphate were measured with the molybdate blue method.

The LCD modeling approach followed is similar to that of Cui & Weng (2013). The LCD model used the Charge Distribution and MUlti-SIte Complexation model (Hiemstra & Van Riemsdijk 1999) to describe the reactions of adsorbates with soil mineral surfaces, and used the Non-Ideal Competitive Adsorption model (Kinniburgh *et al.*, 1999) to calculate the speciation and charge of NOM adsorbed. The amount of NOM adsorbed was optimized for each soil and was kept constant for the same soil. Goethite was used as the representative material for reactive surfaces in soils. The LCD model calculations were performed using the computer program ORCHESTRA (Meeussen 2003).

3 RESULTS AND DISCUSSION

In soil N1 without addition of phosphate, total soluble As concentrations range from 0.2 μM to 4 μM in 10 mM $CaCl_2$ (Figure 1a). Arsenite dominates at acidic pH, whereas at high pH arsenate is the major species in soil solution. Both the dissolution of Fe-(hydr)oxides and the reduction of arsenate consume H^+, therefore, higher amount of soluble arsenite was observed under acidic conditions. Upon addition of 0.25 mM phosphate, the concentration of soluble arsenate increased especially for intermediate and alkaline pH. For example, the concentration of soluble arsenate increased from 0.4 μM to 1.8 μM at pH 6.0 with phosphate added in soil N1. It can be attributed to the competition between arsenate and phosphate in their adsorption to soil minerals. The LCD model calculations showed that arsenite in the adsorbed phase decreased as well, indicating that arsenate reduction was inhibited with phosphate addition in soil N1. This conclusion is in line with the decrease of soluble Fe concentration in the soil N1 when phosphate was added. With the modelling, speciation of As in the solid phase could be calculated. Based on the calculation, 0.8-2% of arsenate in soil N1 was reduced to arsenite without phosphate addition, whereas only <0.3% of arsenate was reduced when phosphate was added (Figure 1b).

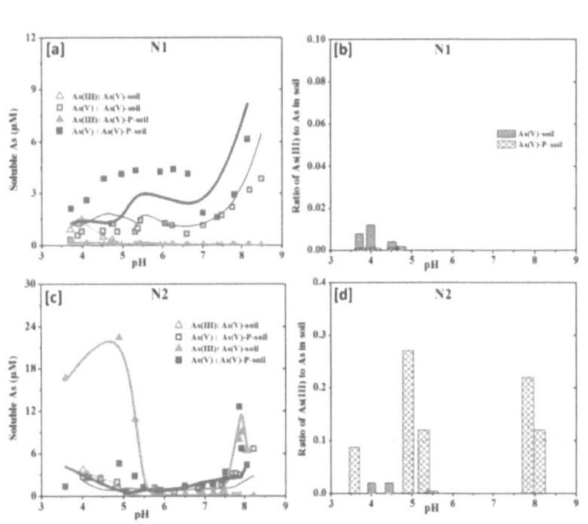

Figure 1. Concentration and speciation of As in soil solution (a, c) and the LCD model calculated ratio of total amount of reduced As(V) (solid and solution phase) to total amount of As(V) added (0.25 mM) in soils (b, d). Symbols are experimental results, and lines are LCD model calculations.

In soil N2 without phosphate addition, total concentrations of soluble As range from 0.4-7 μM in 10 mM $CaCl_2$ (Figure 1c). Arsenite dominates at low pH, whereas arsenate is more important at high pH in soil solution, which is similar to that in soil N1. However, the relative importance of arsenite is larger compared to soil N1, suggesting a stronger reduction of arsenate into arsenite in soil N2, in accordance with the higher soluble Fe concentration in soil N2 than in soil N1. With addition of 0.25 mM phosphate, arsenite became an important As species in soil N2 solution not only under acidic pH, but also under alkaline pH. For example, at pH 8.0 the concentrations of soluble arsenite were >70 times higher than that without phosphate addition. According to the LCD model calculations, with phosphate addition, the fraction of arsenate reduced to arsenite in soil N2 increased from <0.01% to 20% at pH around 8.0, and it increased from 1% to 27% at pH about 5.0 (Figure 1d). It can be seen that addition of phosphate led to a strong increase of arsenate reduction, causing a significant increase of soluble arsenite at both acidic and especially alkaline pH.

4 CONCLUSIONS AND IMPLICATIONS

The current study revealed that phosphate fertilization may lead to very different impact on As solubility in different paddy soils. Under (sub)oxic conditions, addition of phosphate results in stabilization of Fe-(hydr) oxides and decreases of As reduction, leading to a decrease of arsenite but an increase of arsenate in solution as a result of competitive adsorption with phosphate. On the contrary, under anoxic conditions phosphate fertilization in phosphorus deficient soils stimulates microbial reduction of Fe-(hydr)oxides and arsenate. Consequently, soluble arsenite concentration increases strongly not only at acidic pH, but also at alkaline pH, making arsenite a dominant soluble As species over a wide pH range. For paddy soils under the period of flooding, the pH of the soil tends to become neutral. In this case, phosphorus fertilization may strongly increase As concentration in soil solution.

ACKNOWLEDGEMENTS

This work was supported by the National Natural Science Foundation of China (No. 41771277).

REFERENCES

Cui Y.S. & Weng L.P. 2013. Arsenate and phosphate adsorption in relation to oxides composition in soils: LCD modeling. *Environ. Sci. Technol.* 47(13): 7269–7276.

Hiemstra T. & Van Riemsdijk W.H. 1999. Surface structural ion adsorption modeling of competitive binding of oxyanions by metal (hydr)oxides. *J Colloid Interface Sci.* 210 (1): 182–193.

Kinniburgh D.G., van Riemsdijk W.H., Koopal L.K., Borkovec, M., Benedetti M.F. & Avena M.J. 1999. Ion binding to natural organic matter: competition, heterogeneity, stoichiometry and thermodynamic consistency. *Colloids Surf. A: Physicochem. Eng. Asp.* 151(1–2): 147–166.

Meeussen J.C.L. 2003. ORCHESTRA, a new object-oriented framework for implementing chemical equilibrium models. *Environ. Sci. Technol.* 37: 1175–1182.

Takahashi Y., Minamikawa R., Hattori K. H., Kurishima K., Kihou N. & Yuita K. 2004. Arsenic behavior in paddy fields during the cycle of flooded and non-flooded periods. *Environ. Sci. Technol.* 38(4): 1038–1044.

Yamaguchi N., Nakamura T., Dong D., Takahashi Y., Amachi S. & Makino T. 2011. Arsenic release from flooded paddy soils is influenced by speciation, Eh, pH, and iron dissolution. *Chemosphere* 83(7): 925–932.

Arsenic in the Environment: Bridging Science to Practice for Sustainable Development – van der Wal, Ahmad, Petrusevski, Weijma, Savic, van der Wens, Beerendonk, Bhattacharya, Bundschuh & Naidu (Eds)
ISBN 978-1-032-32928-4

Arsenic in Chianan Plain sediments from SW Taiwan: Causes of concentration and potential of release to groundwater

H.-J. Yang & C.-Y. Lou
Department of Earth Sciences, National Cheng-Kung University, Tainan, Taiwan, Republic of China

ABSTRACT: A total of 510 samples collected from three sediment cores at east, central, and west parts of the Chianan Plain in SW Taiwan were analyzed for bulk arsenic concentration and 75 of them were subjected to sequential extraction to determine arsenic distribution among phases. The results showed that arsenic was concentrated during deposition in shallow marine environment and has a tendency of decreasing abundance from east to west, reflecting the control of transportation distance. Since a large portion of arsenic was hosted by secondary phases, including clay minerals, Mn-(hydro)oxide and iron-oxyhydroxide, it appears that arsenic was concentrated during sediment transportation and deposition when the primary arsenic-containing phases, mainly sulfides, were decomposed and released arsenic to the secondary phases. The secondary phases are more fragile to changes in Eh and Ph. They tend to breakdown during interaction with groundwater, which were consequently enriched in arsenic.

1 INTRODUCTION

The Chianan Plain in SW Taiwan is one of the worldwide known areas of high arsenic in groundwater. While arsenic cycling in the Chianan Plain have been addressed to some extents, the sources of arsenic in groundwater remain unclear. Hydrothermal veins, mud volcanoes, offshore basalts, authigenic sulfides, and coal/peat beds all have been considered (Lewis *et al.*, 2007). Being closely associated with groundwater, the sediments, however, have not been considered in a great extent as a possible source of groundwater arsenic. Here, we present arsenic concentration data from bulk analysis and sequential extraction from two sediment cores recovered from the coastal and foothill areas of the Chianan Plain, the Jin-Hu and Tsung-Shi cores, respectively. The data are combined with those from the Hsin-Shen core (Yang *et al.*, 2016) at the centre of the plain to reveal the spatial and temporal distribution of arsenic and then discuss the causes of arsenic concentration and the potential of arsenic release from sediments to groundwater.

2 MATERIALS (SAMPLES)

Each of the Tsung-Shi, Hsin-Shen, and Jin-Hu cores recovered ~250 meters' of sediments. These sediments were deposited as clay, silt, and sand layers with thickness vary from several tens of cm to 35 m. At least two samples were collected from each layer with more from thicker ones. Each sample weighed at least 60 g. More than 150 samples were collected from each of these three cores for chemical analyses. It is considered that the compositions of these ~500 samples are representative of the spatial and temporal variations of the Chianan Plain sediments.

3 ANALYTICAL METHODS

Arsenic concentrations were determined using an ICP-MS. For bulk arsenic, sediment samples were pulverized with a motor and tar. Sample powders were dissolved in Teflon beakers with a mixture of HF and HNO_3 followed by conversions to chloride form and then to nitrate form. Nitric solutions of a 5000x dilution from samples were subjected to ICP runs. About 25 samples from each core were selected for sequential extraction analysis. Arsenic occurred as adsorbate and those incorporated in amorphous and crystalline FeOOH, silicates, sulfides, and oxides were extracted following the procedure of Keon *et al.* (2001). The extractants were converted to nitric solutions for ICP runs.

4 RESULTS AND DISCUSSION

4.1 *Extent and causes of arsenic concentration*

The arsenic concentration and grain size of the sediments are apparently related. The sand-sized sediments that indicate deposition in coast environments contain more quartz and are characterized by invariably low arsenic concentrations, generally < 6 ppm (Figure 1). The silt-sized sediments were composed of higher proportions of clay and contain arsenic ranging from ~ 12 ppm up to over 50 ppm (Figure 1). Indicating deposition in greater depths, the thicker > 30 meters' clay-sized sediments generally have arsenic concentrations of 10–30 ppm (Figure 1). The unusually high arsenic concentrations of > 30 ppm all occur in thin < 10 meters' clay-silt layers that intercalated with slightly coarser layers of similar thickness (Figure 1). Reflecting shallow marine environments, such intercalations were mainly deposited during 7–9 and >15 kya (Figure 1).

DOI: 10.1201/9781003317395-55

The highest arsenic concentration decreases from ~250 ppm in eastern Tsung-Shi core to ~120 ppm in the central Hsin-Shen core and then to the ~90 ppm to the western Jin-Hu core. The number of high arsenic samples (>20 ppm) also decreases from 18 from the Tsung-Shi core to 12 from the Hsin-Shen core to 7 from the Jin-Hu core. Apparently, arsenic concentration in Chianan Plain sediments decreases from east to west. Such a systematic variation is considered to reflect the control of transport distance of the sediments. Since these sediments were derived from the Central Mountain Range at the east of the Chianan Plain, the eastern sediments are closer to the sources that supplied arsenic-containing minerals, mainly sulfides. Being transported to west, sulfides were decomposed and/or deposited, reducing the arsenic concentration in the sediments.

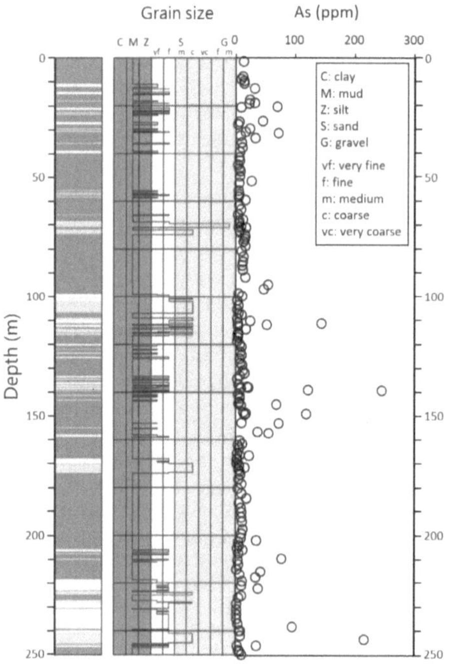

Figure 1. The relationship between depth, arsenic concentration and grain size of the Tsung-Shi core sediments.

4.2 *Arsenic release potential*

The high and variable arsenic concentrations reflect the control of different arsenic hosts. Sequential extraction revealed that 20–40% of arsenic in these sediments occurred as adsorbate on clay minerals as well as Mn and Fe oxides or hydroxides, whereas iron-oxyhydroxides incorporated other 30–50% of bulk arsenic (Figure 2). The rest of < 20% was distributed among silicates, sulfides, oxides (Figure 2). Since more than 80% of bulk arsenic was associated with the secondary phases, it appears that arsenic concentration occurred during transportation and deposition through decomposition of primary arsenic hosts, mainly sulfides. The arsenic hosted by the secondary phases is relatively

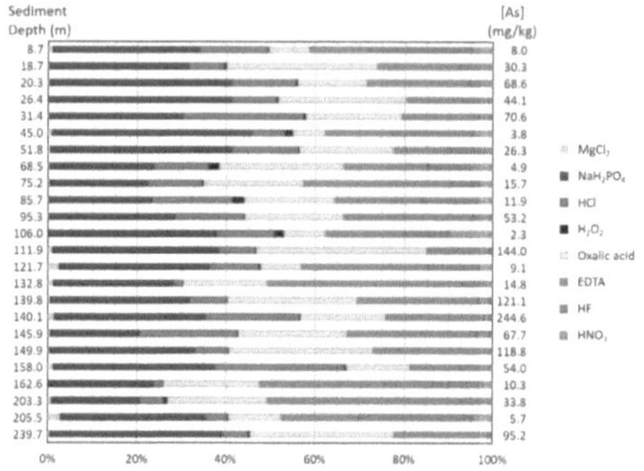

Figure 2. Arsenic content of the Tsung-Shi core sediments as percentages of total arsenic content in different extractants.

mobile. They could be released into groundwater upon changes in pH and/or Eh conditions. This inference needs to be confirmed by water-sediment interaction experiments and compositions of groundwater.

5 SUMMARY

Arsenic in the Chianan Plain was mainly concentrated in several ~20 meter thick depth ranges, where fine-grained layers with high arsenic were intercalated with slightly coarse ones. The westward decrease of arsenic content is attributed to the dominant control of transportation distance with the deposition sites near the sources containing more arsenic. During transportation and deposition, arsenic was adsorbed or incorporated into secondary phases, which are vulnerable to changes in pH-Eh condition and then release arsenic into groundwater.

ACKNOWLEDGEMENTS

We acknowledge the Central Geological Survey of Taiwan for providing sediment samples and Ministry of Science and Technology of Taiwan for funding supports.

REFERENCES

Keon N.E., Swartz C.H., Brabander D.J., Harvey C. & Hemond H.F. 2001. Validation of an arsenic sequential extraction method for evaluating mobility in sediment. *Environ. Sci. Technol.* 35: 2778–2784.

Lewis C., Ray D. & Chiu K.K. 2007. Primary geological srouces of arsenic in the Chianan plain (blackfoot disease area) and the Lanyang plain of Taiwan. *Int. Geol. Rev.* 49: 947–961.

Yang H.J., Lee C.Y., Jiang Y.J., Jean J.S., Shau Y.H., Takazawa E. & Jiang W.T. 2016. Distribution and hosts of arsenic in a sediment core from the Chianan plain in SW Taiwan: implications on arsenic primary source and release mechanisms. *Sci. Total Environ.* 569–570: 212–222.

Arsenic in the Environment: Bridging Science to Practice for Sustainable Development –
van der Wal, Ahmad, Petrusevski, Weijma, Savic, van der Wens, Beerendonk, Bhattacharya, Bundschuh & Naidu (Eds)
© 2023 the Author(s), ISBN 978-1-032-32928-4

The oxidation of As(III) by α-FeOOH with different structural defects and its implication for paddy soil remediation

L. Fang[1,2], Z. Hong[1,2] & F. Li[1,2]
[1]*Guangdong Key Laboratory of Integrated Agro-environmental Pollution Control and Management, Guangdong Institute of Eco-Environmental Science & Technology, Guangzhou, P.R. China*
[2]*National-regional Joint Engineering Research Center for Soil Pollution Control and Remediation in South China, Guangzhou, P.R. China*

ABSTRACT: Exploring the role of structural defects of iron oxides on the generation of ROS near the rhizosphere zone is essential for understanding As geochemical process and its regulation in paddy soils. Herein, the adsorption and oxidation of As(III) on the surface of goethite with different defects (i.e. oxygen vacancy) in the presence of Fe(II) under oxic conditions have been examined. Results indicate that Fe(II) rapidly triggers the As(III) oxidation, and the oxidation is enhanced with increasing of OVs. Both pH and Fe(II) content play an important role in As(III) oxidation, while the overall adsorption of As is barely affected. Importantly, superoxide radical and H_2O_2 are found responsible for As(III) oxidation. The finding of this work provides an insight into understanding the As(III) reaction process on iron oxides in paddy soils.

1 INTRODUCTION

The contamination with arsenic (As) in paddy soils is still a global challenge, threating food safety and human health. Studies show that arsenite [i.e. As(III)] has a relatively higher bioavailability and toxicity compared to arsenate [As(V)], which is preferentially formed under anoxic conditions determining the uptake rate of As by rice plants (Gorny *et al.*, 2015). It suggests that the process of radial oxygen loss (ROL) in the vicinity of the rhizosphere zone of rice plants plays an important role in the formation of iron plaques and subsequently the redox status of As (Mei *et al.*, 2009). Recently, a few studies reported that the microbially generated Fe(II) can effectively adsorb on the surface of iron oxides and react with oxygen molecules to form reactive oxygen species (ROS) that is likely to modulate the As species in soils (Elsner *et al.*, 2004). However, there is a lack of direct evidence in association with As speciation. More importantly, there is also little known how the structure of iron plaques (e.g. goethite) affects the reaction process of As redox process and its stability on the surface of iron oxides. Exploring the role of structural defects of iron oxides on the generation of ROS near the rhizosphere zone is essential for understanding As geochemical process and its regulation in paddy soils.

Herein, the present study mimicked reaction process of ROS generation near rhizosphere zone mediated by ROL. In particular, the potential effects of structural defects on the surface of the iron oxide (i.e. goethite) on the catalytic effects of adsorbed Fe(II) on ROS generation, and subsequently the oxidation and adsorption process of As(III) species were systematically investigated. Finally, the underlying reaction mechanisms were determined.

2 MATERIALS AND METHODS

2.1 *Synthesis and characterization of goethite with different defects*

Goethite was prepared using a hydrothermal method. In brief, a proper amount of Fe(NO3)3 was added into a KOH solution, and aging at 70°C. Goethite incorporated different levels of defects were manipulated by doping 1.0–3.0% Cu (w/w) as suggested by Jin *et al.* (2017). The as-prepared goethite samples were characterized by solid-state techniques.

2.2 *Synthesis and characterization of goethite with different defects*

A mixture of Fe(II) and goethite with a proper ratio was prepared under anoxic conditions prior to adding aliquots of As(III) stock solutions and exposing to air under atmospheric conditions. Effects of Fe (II) concentrations, goethite dose and pH were accordingly investigated. the ROS identification was conducted by spiking a suitable scavenger into the above solutions (i.e. isopropanol for ·OH, 1,4-Benzoquinone for O_2*-) respectively. The adsorption of total As and oxidation of As(III) were determined accordingly.

2.3 *Analytical methods*

The As(III) and As(V) concentrations were analyzed on a LC-HG-AFS (AFS-820; Jitian Instruments), while the Fe(II) concentration was measured using a Phenanthroline-based colorimetric method. Electron spin resonance was applied to confirm the possible radicals in reaction systems.

DOI: 10.1201/9781003317395-56

3 RESULTS AND DISCUSSION

3.1 *As(III) oxidation by goethite with defects*

Figure 1a shows that the incorporation of Cu does not significantly vary the crystalline structure of goethite, while the ESR spectra suggests the increasing spinning of paramagnetism of goethite due to the increased oxygen vacancies (OV; Figure 1b). The goethite with higher Cu contents will generally leads to a higher removal of As(III) (Figure 1c). More importantly, the presence of Fe(II) significantly leads to a higher As(III) oxidation, and the oxidation extent is highly related to the OV contents (Figure 1d), indicating the important active role of OV on As(III) oxidation.

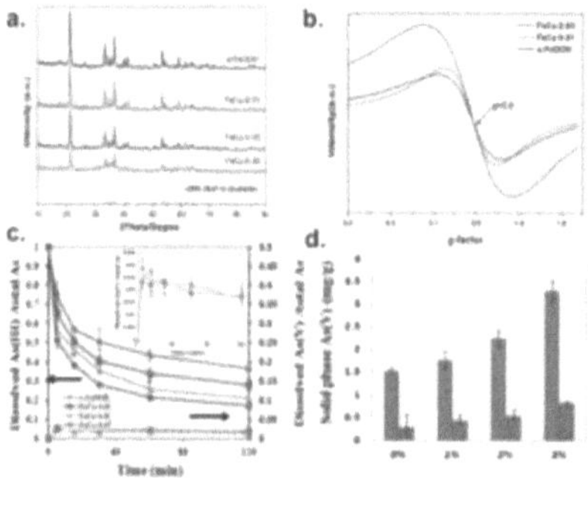

Figure 1. XRD (a) and ESR (b) results of the goethite with different defects; and removal kinetics of As(III) in solutions (c) and formation of As(V) with and without Fe (II) at pH 7.0.

3.2 *Effects of Fe(II) and pH on As(III) oxidation*

Both Fe(II) contents and pH play important roles in As (III) oxidation on the surface of goethite. Figure 2a suggests that the formation of As(V) is enhanced by increasing Fe(II) contents in solutions, while the overall As adsorption is little affected. Despite the rate of adsorbed As(III)/As(total) declines at lower pH conditions, the As(V) contents are higher (Figure 2b). This is likely due to the low pH favors the generation of ROS.

3.3 *Effects of Fe(II) and pH on As(III) oxidation*

To identify the possible ROS species and their respective contribution to the overall As(III) oxidation, we

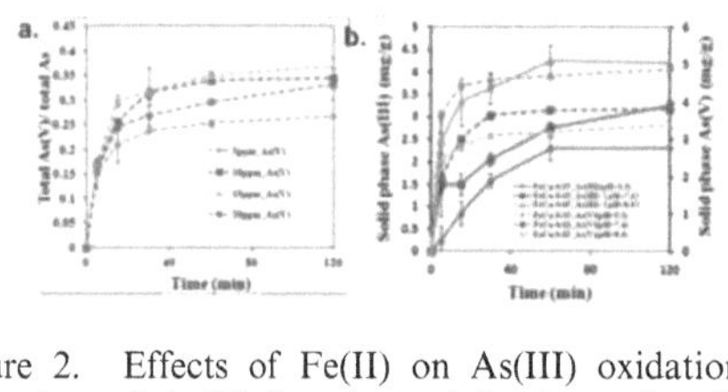

Figure 2. Effects of Fe(II) on As(III) oxidation. a) formation of As(V) is enhanced by increasing Fe(II) contents in solutions, while the overall As adsorption is little affected. b) the rate of adsorbed As(III)/As(total) declines at lower pH conditions.

further applied quenching experiments using four different reagents. As illustrated in Figure 3a, results indicate that superoxide (O_2*-) and H_2O_2 are the main ROS species for As(III) oxidation, with a contribution of 65∼80%, and 40∼50%, respectively. Herein, we draws a picture that the electron transferred from the adsorbed Fe(II) species will favor the generation of ROS, and the following the oxidation of As(III). The presence of OVs will accelerate this process as shown in Figure 3b.

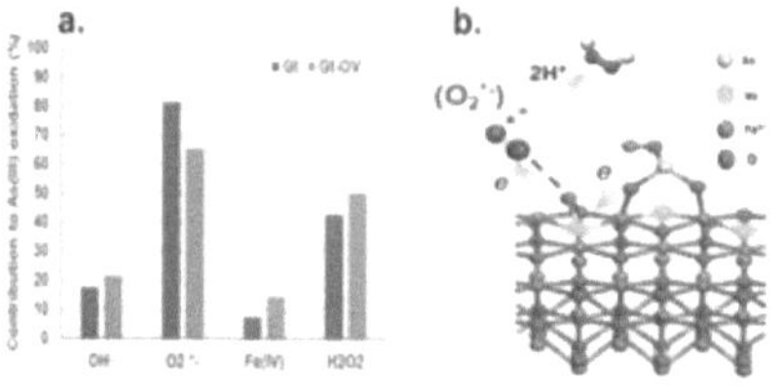

Figure 3. Contribution of ROSs to As(III) oxidation (a), and schematic diagram of As(III) oxidation on goethite (b).

4 CONCLUSIONS

The incorporation of Cu leads to isomorphism of Fe with Cu atoms, creating significant amounts of OVs. The presence of Fe(II) can result in the oxidation of As(III) on the surface of goethite under oxic conditions, and this oxidation process can be enhanced by the contents of OVs and Fe(II) contents, even though the overall adsorption of As is insignificantly affected. Strong evidence indicates that superoxide radicals and H_2O_2 are responsible for As(III) oxidation. Such process provides a new insight into in As(III) fate near the rhizosphere zone and possibly developing Fe mineral based materials by tuning structure of iron oxides for As(III) oxidation.

ACKNOWLEDGEMENTS

The current work was financially supported by the National Natural Science Foundation of China (Grant No. 21876161).

REFERENCES

Elsner M., Schwarzenbach R.P. & Haderlein S.B. (2004). Reactivity of Fe(II)-bearing minerals toward reductive transformation of organic contaminants. *Environ. Sci. Technol.* 38(3): 799–807.

Gorny J., Billon G., Lesven L., Dumoulin D., Madé B. & Noiriel C. (2015). Arsenic behavior in river sediments under redox gradient: a review. *Sci. Total Environ.* 505: 423–434.

Jin H., Tian X., Nie Y., Zhou Z., Yang C., Li Y., & Lu L. (2017). Oxygen vacancy promoted heterogeneous fenton-like degradation of ofloxacin at pH 3.2–9.0 by Cu substituted magnetic Fe_3O_4@ FeOOH nanocomposite. *Environ. Sci. Technol.* 51(21): 12699–12706.

Mei X., Ye Z., & Wong M.H. (2009). The relationship of root porosity and radial oxygen loss on arsenic tolerance and uptake in rice grains and straw. *Environ. Pollut.* 157 (8–9): 2550–25574.

Arsenic in the Environment: Bridging Science to Practice for Sustainable Development – van der Wal, Ahmad, Petrusevski, Weijma, Savic, van der Wens, Beerendonk, Bhattacharya, Bundschuh & Naidu (Eds)
© 2023 the Author(s), ISBN 978-1-032-32928-4

Occurrence of arsenic and its potential for mobility in soils in Muthurajawela marsh, Sri Lanka: Relationship with acid sulfate soil materials

C.L. Vithana[1,2], P.A.K. Ulapane[1], R. Chandrajith[1], L.A. Sullivan[3], J. Bundschuh[4], N. Toppler[5], N.J. Ward[2] & A. Senaratne[1]
[1]*Department of Geology, University of Peradeniya, Palle, Sri Lanka*
[2]*Southern Cross Geoscience, Southern Cross University, East Lismore, NSW, Australia*
[3]*University of Canberra, Bruce, Australia*
[4]*University of Southern Queensland, Toowoomba, QLD, Australia*
[5]*Environmental Analysis Laboratory, Southern Cross University, East Lismore, NSW, Australia*

ABSTRACT: Muthurajawela marsh in Sri Lanka is an ecologically and economically significant, unique ecosystem with high biodiversity. Acid sulfate soil (ASS) soil materials in Muthurajawela soil are a good source and sink of trace elements. Previous studies have shown an occurrence of higher amounts of trace elements such as Cu, Zn and Cr in Muthurajawela marsh. However, the occurrence of arsenic (As) and its potential for mobility in soils are unknown. In this study we estimated the arsenic content partitioned into different soil fractions down the soil profiles in three different locations of the marsh. The sulfidic soil material content was also determined as chromium reducible sulfur (CRS) in the same soil profiles. Soils collected from northern and middle segments of the marsh contained substantial amount of As exceeding the natural background level (40 mg As kg^{-1}) set for soils. More than 75% of the total As in these two soil profiles was partitioned into ASS materials (e.g. pyritic or iron oxyhydroxy sulfate minerals) indicating a close relationship between As and ASS materials.

1 INTRODUCTION

Muthurajawela has a very sensitive ecosystem and plays dual roles as a wildlife habitat and floodwater retention area (Master Plan of Muthurajawela & Negombo Lagoon 1991). The marsh is the only peat bog in Sri Lanka and comprises of substantial amount of sulfidic soils materials underneath (Dissnayake 1984). Sulfidic soil materials (e.g. pyrite, iron monosulfides) and their oxidized products such as iron oxy/hydroxy sulfate minerals (e.g. schwertmannite and jarosite) collectively form acid sulfate soil (ASS) materials (Sullivan *et al.*, 2018). Arsenic is a critically important trace metalloid that can pose a threat to humans and ecosystems. In ASS environments, the mobility of As is strongly controlled by the ASS soil materials. Arsenic in Muthurajawela is highly likely to be associated with ASS materials. Arsenic associated with those minerals can be mobilized when they are subject to various processes including transformation, dissolution and precipitation which is controlled by surrounding environmental conditions (i.e. pH, Eh, organic matter etc.). For a long time, the surrounding environmental conditions have been greatly altered due to natural (prolonged flooding, drought) and anthropogenic activities (land clearing, fragmentation, authorized/unauthorized residential/industrial projects) in Muthurajawela. These activities can affect the stability of ASS materials and the mobility of As. However, the availability of As and its mobilization potential in soils in Muthurajawela marsh are unknown. The main objective of this study was to understand the occurrence of arsenic and its bioavailability in Muthurajawela marsh.

2 MATERIALS AND METHODS

2.1 *Study area*

Muthurajawela marsh is located on the western coastal belt of Sri Lanka about 10 km north of Colombo City. It covers approximately 3100 ha and connects to Negombo lagoon from its northern end (Figure 1). The western border of the marsh is the Hamilton Canal and southern and eastern sides are bordered by heavily urbanized and industrialized highlands. The marsh receives freshwater mainly from rainfall (via south east and inter-monsoons) and streams which enter to the northern segment of the marsh. The average rainfall of the marsh is 2000-2500 mm and daily maximum temperature is 31.5°C (Dissanayake 1984). The soil profile in Muthurajawela is mainly comprised of peat, clay and some sand.

2.2 *Soil sampling, field tests and soil drying procedure*

Soil profiles were collected from northern (Site 1), middle (Site 2) and southern (Site 3) segments of the marsh using a Russian-D-corer up to 1.5 m depth. Each soil profile was sliced into different depths based on the colour and placed in labelled sealable plastic bags with

DOI: 10.1201/9781003317395-57

air extruded. In the field, the pH of soil in water (pH_F) and 30% H_2O_2 (pH_{FOX}) were measured for all depths in each soil profile. Collected soils were immediately stored at 4°C until brought to the laboratory. Soil samples collected for chromium reducible sulfur (CRS) method were placed in an oven and dried at 85°C to a constant weight. The oven dried soils samples were ground to powder using a ball mill and stored in sealed plastic bags at room temperature. The moisture content at 105°C was determined to report the extracted arsenic concentrations in different soil fractions.

Figure 1. Study area and sampling locations.

2.3 *Determination of Chromium Reducible Sulfur (CRS) content and arsenic in different soil fractions*

CRS content in each depth of soil profiles were determined according to Burton *et al.* (2008). Arsenic partitioned in different soil fractions were determined by following the sequential extraction procedure described in Claff *et al.* (2010).

3 RESULTS AND DISCUSSION

3.1 *CRS content in soil profiles*

All three soil profiles contained substantial amount CRS within 1.5 m depth exceeding the threshold limit (> 0.06%) set for medium textured clay soils indicating the occurrence of hypersulfidic soil materials. The CRS data were consistent with the field data (pH_F and pH_{FOX}).

3.2 *Arsenic fractions in soil profiles*

Arsenic concentrations in Site 1 and Site 2 soil profiles were substantially higher and ranged from 5 – 470 mg/kg indicating enrichment of As in soils at the two sites. In Site 1, As in surface horizons (upto 50 cm) was largely present in the residual fraction and below that more than 95% was in pyritic fraction. A contrasting As partitioning behavior was shown in Site 2 where > 75% of total As was associated in the crystalline oxide fraction (Figure 2). The

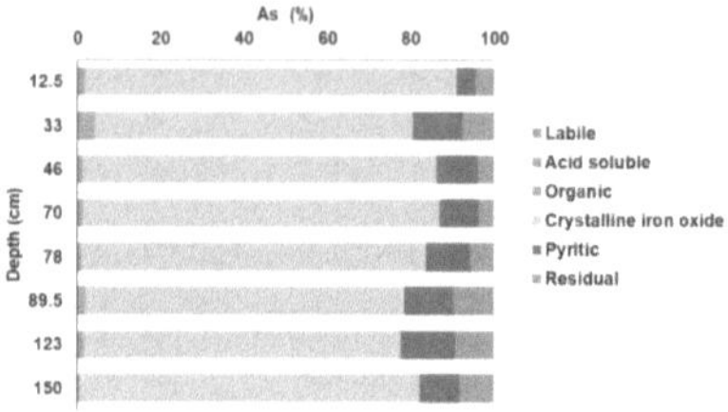

Figure 2. Arsenic partitioning pattern in Site 2 (middle) in Muthurajawela marsh.

experiment on assessing the As partitioning behavior in Site 3 (southern) is ongoing.

4 CONCLUSIONS AND RECOMMENDATIONS

The data indicate that soils in northern and middle segments of Muthurajawela marsh are highly likely to be contaminated by arsenic. Almost all As in soils in Muthurajawela marsh can be associated with ASS materials. Consequently, the mobility of As is strongly dependent on the stability of those ASS minerals. Given the higher amount of natural and anthropogenic activities in the marsh, As in those ASS minerals may release to the surrounding water bodies especially to the groundwater which can be transported to other areas via aquifers. Therefore, future studies should focus on investigating As concentrations in the groundwater in the marsh and potential impact of As in the groundwater of Muthurajawela.

ACKNOWLEDGEMENTS

We acknowledge the funding given by National Science Foundation (NSF) Sri Lanka (Grant: NSF/PDRS/2018/01).

REFERENCES

Burton E., Sullivan L., Bush R., Johnston S. & Keene A. 2008. A simple and inexpensive chromium-reducible sulfur method for acid-sulfate soils. *Appl. Geochem.* 23(9): 2759–2766.

Claff S., Sullivan L., Burton E. & Bush R. 2010. A sequential extraction procedure for acid sulfate soils: partitioning of iron. *Geoderma* 155(3-4): 224–230.

Dissanayake C. 1984. Geochemistry of the Muthurajawela peat deposit of Sri Lanka. *Fuel* 63(11): 1494–1503.

Master Plan of Muthurajawela and Negombo Lagoon. 1991. *Greater Colombo Economic Commission: Euroconsult*, The Netherlands.

Sullivan L., Ward N., Toppler N. & Lancaster G. 2018. *National Acid Sulfate Soils Guidance: National Acid Sulfate Soils Sampling and Identification Methods Manual.* Water Quality Australia.

Arsenic in the Environment: Bridging Science to Practice for Sustainable Development – van der Wal, Ahmad, Petrusevski, Weijma, Savic, van der Wens, Beerendonk, Bhattacharya, Bundschuh & Naidu (Eds)

Impact of tropical monsoon climate on arsenic enrichment in Holocene sediments of Southeastern Sri Lanka

K.M. Premaratne[1], R. Chandrajith[1], N. Ratnayake[2,3], K. Gayantha[4] & L. Si-Liang[5]
[1]*Department of Geology, Faculty of Science, University of Peradeniya, Palle, Sri Lanka*
[2]*Ocean University of Sri Lanka, Colombo, Sri Lanka*
[3]*Department of Earth Resources Engineering, Faculty of Engineering, University of Moratuwa, Moratuwa, Sri Lanka*
[4]*MaxPlank Institute of Biogeochemistry, Jena, Germany*
[5]*Institute of Surface Earth System Science, Tianjin University, Tianjin, P.R. China*

ABSTRACT: A 5.1 m sediment core extracted from Pottuvil lagoon in southeast coastal lowlands of Sri Lanka was used to trace the temporal variability of As enrichment in coastal sediments in response to dynamic monsoon climate throughout the Holocene. Geochemical study along with AMS 14C dating and Bacon Age Depth modeling suggested that As enrichment in lagoon sediments is favored by strong monsoon activity phases and As depletion is occurred during weak monsoon periods. Most interestingly, As accumulation declined during 5.2 and 4.2 Ka Bond events in response with the abrupt weakening of monsoon climate. This study reveals that As enrichment in Holocene sediments of Southeastern Sri Lanka is significantly affected by changing tropical monsoon climate over past 6000 yrs BP in millennial and centennial time scale. Chronology of major climate shifts detected in Pottuvil proxy record is comparable with regional records from Indian monsoon regime.

1 INTRODUCTION

Arsenic (As) is a commonly found element in the Earth's crust with an average concentration of ~5 mg kg^{-1} (Flora 2015). In sediments, As is relatively enriched in metal oxide rich layers (Chaillou *et al.*, 2003). However, in quaternary time scale, detailed studies on paleoclimate and arsenic accumulation in sediments are rare in Sri Lanka. Located in the core of Indian monsoon domain, Sri Lanka is having a classic 'tropical monsoonal' climate which provides the ideal stage for-paleo studies. Therefore, this study aimed to comprehensive exploration of Holocene climate-arsenic link with respect to tropical monsoon climate of Sri Lanka using lagoon sediments from southeastern coastal lowlands of Sri Lanka.

2 MATERIALS AND METHODS

2.1 *Study area*

Pottuvil lagoon is a brackish water coastal lagoon located in Ampara District of Eastern Province of Sri Lanka, extending from 6°57′00″N and 81°46′15″E to 6°50′00″N and 81°51′30″E which is spread over an area of 24.49 km^2 in southeastern coastal plain. Climatically, the lagoon belongs to Dry Zone of Sri Lanka. The lagoon receives freshwater mainly from direct precipitation of northeast monsoon and from several minor streams including Dooviligal Oya and Karanda Oya. The mean annual rainfall of the lagoon area ranges from 700-1000 mm (IUCN SL & CEA 2006; Silva *et al.*, 2013).

2.2 *Sediment core collection and laboratory analysis*

Coring was carried-out in Pottuvil lagoon using a rotary drilling equipment fixed in a floating barge which can acquire sediments in polyvinylchloride (PVC) tubes of 1.0 m length and 2.5″ diameter. A 5.1 m sediment core was extracted from the lagoon and the core was logged for its major and trace element contents, Total Organic Matter content (OM)% and Magnetic Susceptibility (MS) in selected subsections at a 5 cm interval. Mollusks shells and bulk sediment samples from selected depths of the sediment core were sent to Radiocarbon facility at Max-Planck Institute of Biogeochemistry, Gena, Germany and Radiocarbon Laboratory of Institute of Surface Earth System Science (ISESS), Tianjin University, China for AMS 14C analysis. All the radiocarbon dates were calibrated using Northern Hemisphere terrestrial calibration curve (IntCal13) (Reimer *et al.*, 2013) and age-depth model for the sediment core was developed using Bacon 2.2 software (Blaauw & Christen 2011).

2.3 *Data processing*

Origin2018 and SPSS16 were used in data analysis and graphical illustration and CorelDRAW software was used for image processing. Based on ICP-MS analysis, As, Fe, Al, K, Rb, Mg and Ti concentrations are presented as oxide concentrations. Other proxies including Rb/Sr ratio, Total REE, La/Yb, La/Lu ratio, Light Rare Earth (LREE) Enrichment %, Middle Rare Earth (MREE) Enrichment %, Heavy Rare Earth (HREE) Enrichment % and OM (weight percentage).

DOI: 10.1201/9781003317395-58

3 RESULTS AND DISCUSSION

3.1 *Temporal variability of arsenic accumulation in sediments*

The sediment core presents temporal variability of geochemical proxies of Holocene epoch from 5561 BP to 2313 BP. Pottuvil record shows drastic changes in Arsenic enrichment in the sediments throughout the Holocene. As concentration in the sediments fluctuated from 0.1389 mg/kg to 5.3313 mg/kg during this period. Our results show a strong positive correlation between As concentration and Fe_2O_3, Al_2O_3 & OM% for the whole length of the core which implies the role of iron & clay minerals, Organic matter as As carriers to the lagoon. Prolonged period of As enrichment is detected from 5141 BP to 4229 BP with short other peaks at 5409-5305, 4050-3885, 3228-3158 and 2988-2581 BP intervals. As is significantly depleted during 5305-5141, 4229-4050, 3885-3228, 3158-2988 and 2581-2313 BP.

3.2 *Impact of monsoon behavior on arsenic accumulation*

In our results, active wet monsoon phases were characterized by elevated levels of K_2O, Rb_2O, MgO, TiO_2, ΣREE, MS and low values of Rb/Sr, La/Lu, La/Yb, LREE enrichment (%) and weak monsoon periods behave the other way. Peaks of As enrichment in Pottuvil sediments coincides with strong monsoon periods as detected by above monsoon proxies. Similarly, As depletion is recorded in dry, weak monsoon activity phases. Further, all monsoon proxies show abrupt response linked to monsoon weakening during 5.2 Ka and 4.2 Ka Bons events detected in the Pottuvil core. Most interestingly, As accumulation declined during 5.2 and 4.2 Ka Bond events detected in the core (Figure 1). This observation is consistent with other proxies as well. However, the climate signal of 2.8 Ka Bond event is decoupled with which of 4.2 Ka and 5.2 Ka events and As accumulation increased accordingly.

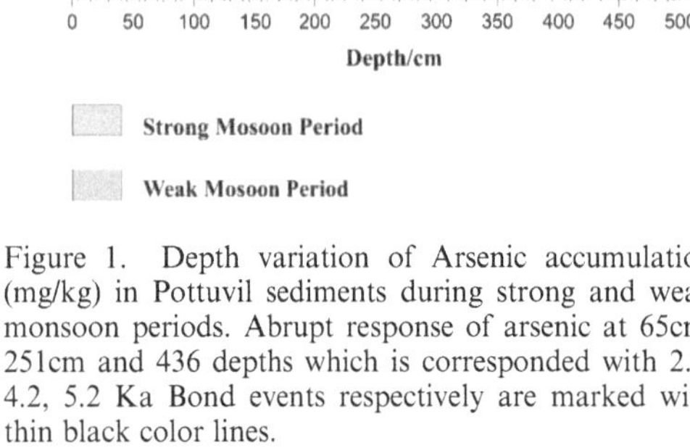

Figure 1. Depth variation of Arsenic accumulation (mg/kg) in Pottuvil sediments during strong and weak monsoon periods. Abrupt response of arsenic at 65cm, 251cm and 436 depths which is corresponded with 2.8, 4.2, 5.2 Ka Bond events respectively are marked with thin black color lines.

4 CONCLUSIONS AND RECOMMENDATIONS

Based on geochemical and physical proxies along with Bacon Age-depth modelling, this study reveals that arsenic enrichment in Holocene sediments of Southeastern Sri Lanka is significantly affected by changing tropical monsoon climate over past 6000 yrs BP in millennial and centennial time scale. Arsenic enrichment is favored by active wet monsoon periods and arsenic depletion in sediments is concurred at weak monsoon phases. Chronology of major climate shifts detected in Pottuvil proxy record is comparable with regional records from Indian monsoon regime. In a next step, arsenic in sediments need to be studied in present day monsoon scenario.

ACKNOWLEDGEMENTS

Financial assistance from National Science Foundation of Sri Lanka (Grant No. NSF/SCH/2018/06) and from National Natural Science Foundation of China (Grant No. 41861144026) is highly acknowledged.

REFERENCES

Blaauw M. & Christen J.A. 2011. Flexible paleoclimate age-depth models using an autoregressive gamma process. *Bayesian Analysis* 6(3): 457–474.

Chaillou G., Schäfer J., Anschutz P., Lavaux G. & Blanc G. 2003. The behaviour of arsenic in muddy sediments of the bay of Biscay (France). *Geochim. Cosmochim. Acta* 67 (16): 2993–3003.

Flora S.J.S. 2015. Handbook of arsenic toxicology, handbook of arsenic toxicology cited in costa l., mirlean n., quintana g., adebayo s. and johannesson k. 2019. distribution and geochemistry of arsenic in sediments of the world's largest choked estuary: the Patos Lagoon, Brazil. *Estuaries and Coasts* 42(7): 1896–1911.

IUCN Sri Lanka and the Central Environmental Authority 2006. *National Wetland Directory of Sri Lanka*, Colombo, Sri Lanka. ISBN: 955-8177-54-7.

Reimer P.J., Bard E., Bayliss A., Beck J.W., Blackwell P.G., Ramsey C.B., Buck C.E., Cheng H., Edwards R.L., Friedrich M. & Grootes P.M. 2013. IntCal13 and marine13 radiocarbon age calibration curves 0–50,000 years cal bp. *Radiocarbon* 55(4):1869–1887.

Silva E.I.L., Katupotha J., Amarasinghe O., Manthrithilake H. & Ariyaratna R. 2013. *Lagoons of Sri Lanka: from the Origins to the Present*. Colombo, Sri Lanka: International Water Management Institute (IWMI).

Sources and mobilization of geogenic arsenic in groundwater in a sedimentary terrain of Sri Lanka

R. Chandrajith[1], S. Diyabalanage[2] & C.B. Dissanayake[1]
[1]*Department of Geology, Faculty of Science, University of Peradeniya, Peradeniya, Palle, Sri Lanka*
[2]*Instrument Centre, Office of the Dean, Faculty of Applied Sciences, University of Sri Jayewardenepura, Nugegoda, Sri Lanka*

ABSTRACT: Although arsenic (As) in groundwater is not recorded in many parts of Sri Lanka, excessive levels were noted in the sedimentary terrains that are predominant in the north and north-western part of the island. The groundwater from these sedimentary terrains showed levels up to 44 µg/L arsenic. The maximum As level observed in the metamorphic terrain was 4.6 µg/L with an average of 0.3 µg/L. Bore-hole sediments collected from the high As region depicted very high affinities of both As and Fe, particularly in the finer fraction of the aquifer materials. The toxicity effects of high arsenic in such terrains remain a neglected health concern that needs greater attention.

1 INTRODUCTION

Arsenic (As) contamination of groundwater is one of the most discussed problems in "Medical Geology" since it leads to adverse health effects. The presence of As in groundwater notably in South and Southeast Asia has become a major health issue. The prevalence of skin disorders, cancer and cardiovascular diseases among others, is caused mainly by the consumption of As-laden drinking water and also food grown in As rich soils (Ahmed *et al.*, 2004; Bhattacharya *et al.*, 1997). Sri Lanka is an Island that has some unique geological, and geochemical characteristics. Geologically, over 90% of the island is made up of high grade metamorphic while the rest is mainly sedimentary sequences (Figure 1).

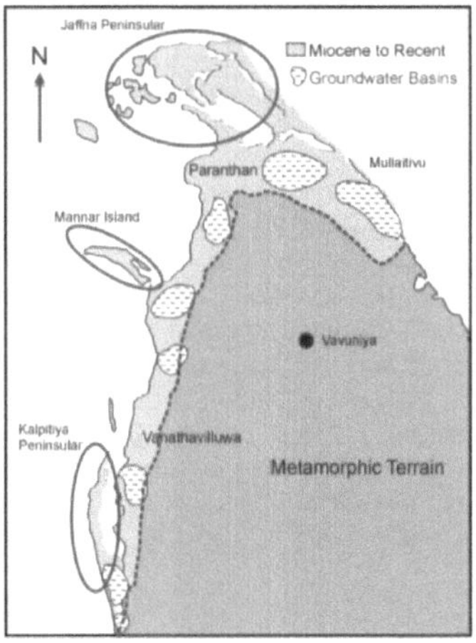

Figure 1. Sedimentary terrain of Sri Lanka and sampling areas.

Although As is not a common constituent in groundwater in metamorphic aquifers, elevated levels were reported in sedimentary aquifer systems (Bandara *et al.*, 2018; Chandrajith *et al.*, 2016; Weragoda & Kawakami 2017). For instance, over 30% of wells in the Mannar island exceeded the WHO recommended limits of As (Bandara *et al.*, 2018). It has been noted that As enriched coastal sand aquifers are confined to spits and bars, raised coastal beaches and in sand dunes. In this study, groundwaters from coastal sedimentary trains were investigated for As contamination and compared with the As contents in adjacent metamorphic terrains.

2 MATERIALS AND METHODS

During the present study, 177 groundwater samples were collected from sedimentary aquifers extending from the northeast (Mullaitivu) to western (Puttalam) coast including Kalpitiya and Jaffna Peninsulas, and Mannar Island (Figure 1). All groundwater wells in the region are shallow and constructed on sand dunes. For comparison purposes, 216 samples were collected from the metamorphic terrain. Trace element contents of samples were determined with a ICP-MS. In Mannar Island where the highest groundwater As was reported, an experimental piezometer was constructed and sediment and water samples were collected at different intervals. Both sediment and water were analysed for As and other trace elements.

3 RESULTS AND DISCUSSION

3.1 *Arsenic in sedimentary terrains*

Most groundwater samples from the sedimentary aquifers belonged to the Ca–HCO_3 type while some

DOI: 10.1201/9781003317395-59

of them were a mixture in the Ca–Mg–Cl type. The groundwater from sedimentary terrains showed up to 44 μg/L As out of which 10% contained higher than 10 μg/L. 26% samples had As above 5 μg/L (Table 1). All wells with high As concentrations (>25 μg/L) showed near-neutral pH conditions. Fe and Ba concentrations in groundwater indicated their coexistence with As. Sediment samples collected from high As wells showed the presence of Fe-oxide coated sands in the aquifer material that may also retain inorganic As. Sediment analysis depicted very high affinities of both As and Fe, particularly in the finer grain size fraction.

Table 1. Arsenic levels in sedimentary regions of Sri Lanka (in μg/L).

Region (# of samples)	Mean	Min.	Max.
Jaffna (35)	1.94	0.07	15.1
Mullaitivu (27)	0.42	0.03	2.73
Paranthan (34)	2.44	0.06	21.6
Mannar Island (43)	11.7	0.60	43.8
Wanathavilluwa (29)	1.51	0.10	6.50
Kalpitiya (43)	5.16	0.25	46.9

3.2 *Arsenic in metamorphic terrain*

The geochemical composition of groundwater from metamorphic aquifers mainly depends on the mineralogical composition of underlain rocks and climatic factors. Weathering of silicate minerals and ion exchange processes are mainly responsible for the hydrogeochemistry of groundwater in hard rock terrains while most water in the dry zone regions are Ca-Mg rich water. This may be due to dissolution of mineral phases such as pyroxene and amphiboles that dominate in aquifer rocks. However, in the wet zone, groundwater is predominantly bicarbonate type with Na-K rich phases (Rubasinghe *et al.*, 2015). In contrast to the sedimentary terrain, maximum As levels observed in the metamorphic terrain was 4.6 μg/L with an average of 0.3 μg/L (Table 2).

Table 2. Arsenic levels in metamorphic terrain of Sri Lanka (in μg/L).

Region (# of samples)	Mean	Min.	Max.
Yan Oya River Basin (77)	0.21	0.01	1.14
Malala Oya River Basin (25)	0.25	0.07	0.60
Wilgamuwa (46)	0.28	0.01	1.80
Monaragala (60)	0.24	0.02	1.00

4 CONCLUSIONS

The occurrence of As rich groundwater is a notable feature in the sedimentary aquifers in the northern part of Sri Lanka, particularly in the Mannar Island. In this region, aquifer materials are enriched with Fe-coated sand and As can then be released into the groundwater through reductive dissolution processes. Although iron oxide coated sand can act as a geochemical barrier that can retain or accumulate metals and metalloids, cyclic fluctuation of the water table and resultant changes in the redox conditions could increase the release of As from aquifer.

ACKNOWLEDGEMENTS

National Research Council (NRC), Sri Lanka Grant no. TO 14-05 is acknowledged.

REFERENCES

Ahmed K.M., Bhattacharya P., Hasan M.A., Akhter S. H., Alam S.M.M., Bhuyian M.A.H., Imam M.B., Khan A.A. & Sracek O. 2004. Arsenic enrichment in groundwater of the alluvial aquifers in Bangladesh: an overview. *Appl. Geochem.*19: 181–200.

Bandara U.G.C., Diyabalanage S., Hanke C., van Geldern R., Barth J.A.C. & Chandrajith R. 2018. Arsenic-rich shallow groundwater in sandy aquifer systems buffered by rising carbonate waters: a geochemical case study from Mannar island, Sri Lanka. *Sci. Total Environ.* 633: 1352–1359.

Bhattacharya P., Chatterjee D. & Jacks G. 1997. Occurrence of arsenic-contaminated groundwater in alluvial aquifers from delta plains, Eastern India: options for safe drinking water supply. *Int. J. Wat. Res. Dev.* 13: 79–92.

Chandrajith R., Diyabalanage S., Premathilake K.M., Hanke C., van Geldern R. & Barth J.A.C. 2016. Controls of evaporative irrigation return flows in comparison to seawater intrusion in coastal karstic aquifers in Northern Sri Lanka: evidence from solutes and stable isotopes. *Sci. Total Environ.* 548: 421–428.

Rubasinghe R., Gunatilake S. & Chandrajith R. 2015. Geochemical characteristics of groundwater in different climatic zones of Sri Lanka. *Environ. Earth Sci.* 74: 3067–3076.

Weragoda S.K. & Kawakami T. 2017. *Evaluation of Groundwater Quality in 14 Districts in Sri Lanka: A Collaboration Research Between Sri Lanka and Japan.* In: F. Kurisu, A.L. Ramanathan, A.A. Ahmad Kazmi & M. Kumar (eds.), *Trends in Asian Water Environmental Science and Technology*. Springer International Publishing, New Delhi, India, pp. 151–155.

Mineralogical and elemental composition in sediments of upper Siang River basin, Northeastern India

A. Dixit[1], S. Sathe[1,2], C. Mahanta[1], S. Kumar[1] & S. Chaudhuri[1]
[1]*Department of Civil Engineering, Indian Institute of Technology, Guwahati, Guwahati, India*
[2]*Department of Civil Engineering, Rajarambapu Institute of Technology, Sakharale, Maharashtra, India*

ABSTRACT: Arsenic (As) contamination in the Brahmaputra basin has been extensively studied, with reports of contamination in Yarlung Tsangpo, the Brahmaputra floodplain, and the large delta in Bangladesh. However, the presence of arsenic in the eastern Himalayan region is still not fully understood. In this study, we analyzed the mineralogical and elemental composition of sediment samples and found the presence of As-bearing minerals, particularly realgar and orpiment. The mineralogical composition further corroborated the elemental composition, with the presence of As, Fe, and S in the sediment samples. These findings highlight the pres-ence of arsenic in the remote upper Siang basin in the eastern Himalaya, thereby providing important insights into the trail of As contamination from Tibet to Bangladesh.

1 INTRODUCTION

The Yarlung Tsangpo basin is a crucial catch-ment area in the Brahmaputra basin, both in terms of sediment quality and quantity. Previous studies have shown that 50% of sediment in Brahmaputra floodplains originates from the Yarlung Tsangpo basin (viz. primarily from Siang and Yamne River) (Singh & France-Lanord 2002). Additionally Li *et al.*, 2011 reported the presence of arsenic (As) in river sediment at Yarlung Tsangpo basin. The Siang and Yarlung rivers are still relatively pristine, such contamination is likely due to geogenic sources. Our study focuses on the mineralogy and elemental composition, as well as the presence of As in river bank sediment collected from remote locations in the Yarlung Tsangpo basin. Our aim is to con-tribute to a better understanding of the potential sources and distribution of arsenic minerals in the Brahma-putra basin.

2 MATERIALS AND METHODS

2.1 *Study area*

The study area for this research is the Siang and Yamne River basin, which originates as the Yarlung Tsangpo before traveling through deep gorges near Namcha Barwa gorge (Figure 1). Upon entering India, it is known as the Siang River and becomes a major tributary of the Brahmaputra in the floodplains.

We collected sediment samples from three locations: Yammeng River (L1), a tributary of Yamne, Siang River in Yingkiong (L2), and Siang River at Pasighat (L3) (Figure 1). The sediment samples were collected from the river bank during September 2019.

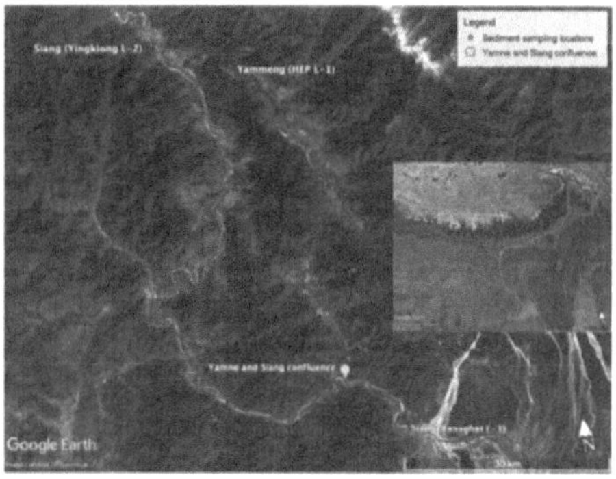

Figure 1. Study area and sediment sampling location in the Siang basin of Brahmaputra.

2.2 *Mineralogical and elemental analysis*

X-ray fluorescence (XRF) analysis was conducted for determination of elemental concentrations in air-dried sediment samples (approximately 0.5 g). The analysis was carried out over PANalytical Epsilon 3 instrument, operated at 20 kV voltage and 0.1 mA current. Mineralogical composition was determined by X-ray diffractometer (XRD) on air-dried, gently pulverized sediment samples. Data was collected for 2θ range of 3° to 70°, using Copper (CuKα radiation = 1.5418 Å) as a target material for single-crystal diffraction. Qualitative

DOI: 10.1201/9781003317395-60

identification of phases was performed on the Profex and X'Pert HighScore software.

3 RESULTS AND DISCUSSION

3.1 *Mineralogy*

XRD analysis shows the presence of major phases typically found in fluvial detrital sediments such as quartz, feldspar, and sheet silicates (biotite, muscovite, clay minerals). However minor phases of amphibole and As-bearing mineral such as realgar and orpiment are also observed. As-bearing phases were more distinct in Yingkiong sample (L2) compared to the Yammeng and Pasighat samples.

3.2 *Elemental composition*

The elemental analysis by XRF of the sediment revealed the presence of As in the L2 sample (Table 1). The lower concentration of As in sediment may be due to the inference of other trace metals in sediment. The presence of As in a pristine location (i.e. L1 and L2) suggests a geogenic source. The below detection limit is abbreviated as 'bdl' in the table below.

Table 1. Elemental concentration of the sediments based on the XRF analyses.

Elements \ Sample ID	L1	L2	L3
Fe %	11.5	18.4	17.9
Mn %	0.2	0.3	0.3
S (ppm)	377.1	2000	3000
P %	0.8	0.7	0.6
As (ppm)	bdl	23	bdl

4 CONCLUSIONS AND RECOMMENDATIONS

Although there are report of As occurrence of sediment in Yarlung Tsangpo (Li *et al.*, 2011). Unlike what we believe that may be some As is generated in the floodplain, this paper clearly opens up the possibility that As contamination at Tibetan plateau is finding its tail all the way down to enter India mainland through the entry point of Tibet to Yingkiong.

Additionally, it is established that total 1Bt of sediment is entering to Indian ocean through Ganges-Brahmaputra basin (Milliman & Farnsworth 2011). However, the As trail from Tibet to Indian Ocean has never been established before. Here we are trying to weave them together. With the study by Li *et al.* (2011) in Tibet, this study in Himalayan hillslope at entry point from Tibet to India, study by Mahanta *et al.* (2012); Mahanta *et al.* (2016); Sathe *et al.* (2016); Sathe *et al.* (2020) and Sathe *et al.* (2021) for Brahmaputra floodplain and Chakraborty *et al.* (2015) for Bangladesh delta, now we can say that the loop of As trail from Tibetan plateau to coastal Indian Ocean is established for the first time.

ACKNOWLEDGEMENTS

We acknowledge the Department of Civil Engineering and Central Instruments Facility, IIT Guwahati for providing financial and instrumental facilities for this study.

REFERENCES

Chakraborty M., Mukherjee A. & Ahmed K.M. 2015. A review of groundwater arsenic in the Bengal basin, Bangladesh and India: from source to sink. *Curr. Pollut. Rep.* 1(4): 220–247.

Li C., Kang S. & Zhang Q. 2011. Heavy metals in sediments of the Yarlung Tsangbo and its connection with the arsenic problem in the Ganges – Brahmaputra basin. *Environ. Geochem. Health* 33: 23–32.

Milliman J.D. & Farnsworth K.L. 2011. *River Discharge to the Coastal Ocean: A Global Synthesis*. Cambridge University Press, Cambridge, UK. 392 p.

Mahanta C., Sathe S.S. & Bhattacharya P. 2012. Grain size, mineralogy and sediment composition as factors controlling release and mobilization of arsenic in parts of the Brahmaputra floodplains, Northeastern India. *Geol. Soc. Amer., Abstracts with Programs* 44(7): p. 155.

Mahanta C., Sathe, S.S. & Mahagaonkar A. 2016. Morphological and mineralogical evidences of arsenic release and mobilization in some large floodplain aquifers. In: P. Bhattacharya, M. Vahter, J. Jarsjö, J. Kumpiene, A. Ahmad, C. Sparrenbom, G. Jacks, M.E. Donselaar, J. Bundschuh & R. Naidu (eds.) *Arsenic Research and Global Sustainability (As 2016)*. CRC Press/Taylor and Francis (ISBN 978-1-138-02941-5), pp. 66–67.

Sathe S.S., Goswami L., Mahanta C. & Devi L.M. 2020. Integrated factors controlling arsenic mobilization in an alluvial floodplain. *Environ. Technol. Innov.* 17: 100525.

Singh S.K. & France-Lanord C. 2002. Tracing the distribution of erosion in the Brahmaputra watershed from isotopic compositions of stream sediments. *Earth Planet. Sci. Lett.* 202(3–4), 645–662.

Arsenic in the Environment: Bridging Science to Practice for Sustainable Development –
van der Wal, Ahmad, Petrusevski, Weijma, Savic, van der Wens, Beerendonk, Bhattacharya, Bundschuh & Naidu (Eds)
© 2023 the Author(s), ISBN 978-1-032-32928-4

Arsenic occurrence in soil and sediments in Mewat, Haryana, India

G. Krishan[1] & P. Bhattacharya[2]
[1]*Groundwater Hydrology Division, National Institute of Hydrology, Roorkee, Uttarakhand, India*
[2]*KTH-International Groundwater Arsenic Research Group, Department of Sustainable Development, Environmental Science and Engineering, KTH Royal Institute of Technology, Stockholm, Sweden*

ABSTRACT: Arsenic (As) a natural component of earth's crust, is contaminating soil sediments, transported into surface water and groundwater and may poison crops and vegetables. Communities affected by As poisoning are generally not aware of its contaminations in the soil sediments. In the present study, 3 typical soil profiles were chosen to collect 15 soil sediment samples from the depth range of 0–150 cm. Samples were analysed for As, manganese and iron. Average concentrations of As, iron and manganese were found 27, 512 k and 568 k mg/kg. However, WHO (1993) has not specified any toxicity limits of As, iron and manganese concentrations in agricultural soils but Grsichek *et al.*, 1993 have specified limits as <1 k, <1000 k and < 10 K for As, iron and manganese, respectively in sediments excavated during drilling of RBF wells. As per these limits concentrations of iron and manganese were found exceeding the toxicity levels.

1 INTRODUCTION

Arsenic (As) contaminations cause a major threat to the health of inhabitants living in the fertile lands of Indo-Gangetic basin (IGB) and the risks are further increased by the conditions under which As is released from fluvial sediments from Himalayas to other natural resources (Bhattacharya *et al.*, 1997, 2001; Mukherjee *et al.*, 2006). Arsenic has been extensively encountered in groundwater in all parts of IGB (Ahamed *et al.*, 2006; Bhattacharjee *et al.*, 2005, 2011; Chakraborti *et al.*, 2003, 2008; Datta & Kaul 1976; Kumar *et al.*, 2017) but not much work has been done for determination of As in soil sediments. The higher concentrations of As in deeper layers of soils can contaminate the groundwater. Mewat, a semi-arid region in Haryana has been chosen for this study due to scarce perennial surface water resources and rely on groundwater for domestic and agriculture requirements (Krishan *et al.*, 2020, 2021a,b). Pollution of agricultural soil with heavy metals like As, iron and manganese may lead to various health impacts in the study area. Keeping this in view, 3 sites were selected for soil sediment sampling from a depth range of 0–150 cm to assess the concentrations of As, iron and manganese.

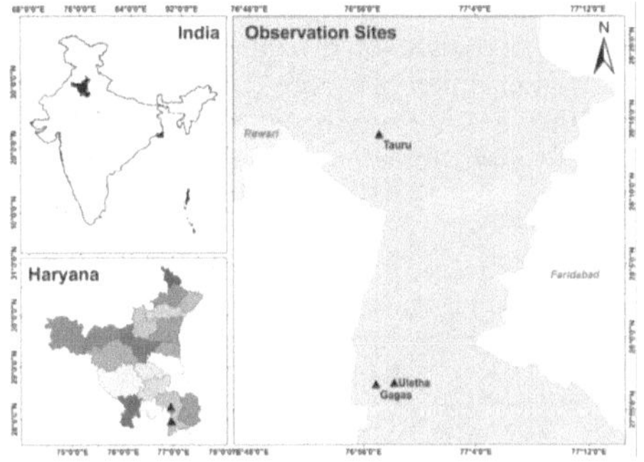

Figure 1. Study area showing soil sediment sampling locations.

2 MATERIALS AND METHODS

2.1 *Study area*

Mewat district a semi-arid region in Haryana (Figure 1) that covers about 1500 sq. km. geographical area. Farming is the main occupation of the community. Average annual rainfall is 500 mm.

2.2 *Soil sediment sampling and analysis*

The samples were collected from Ulheta, Tauru and Karhera sites from 5 different depths as 0–30 cm, 30–60 cm, 60–90 cm, 90–120 cm, 120–150 cm in Mewat district of Haryana. The soil samples were air dried, grinded and the texture was loamy sand to sandy loam at Ulheta; sandy loam to silty loam at Karhera/Ghagas and Sandy loam at Tauru. Grain size fraction of < 1 mm of all sediment samples was extracted for analysis of As, Fe, Mn using 0.005 M Diethylenetriamine pentaacetate

(DTPA) as per the methods specified by Lindsay and Norwell (1978). The extracts were analysed for As, Cd, Cr, Cu, Fe, Mn, Ni, Pb and Zn using a inductively coupled plasma emission spectrometer (ICPE 9000 series, Shimadzu) at the Central Analytical Facility, CSSRI (ICAR), Karnal.

3 RESULTS AND DISCUSSION

The average concentrations of As in soil sediments in Ulheta, Karhera and Tauru were found 31.6, 28.2 and 21.6 mg/kg, respectively while the average concentration of iron and manganese in soil sediments were found 442 k and 580 k, respectively in Ulheta, 566 k and 359 k, respectively in Karhera and 527 k and 765 k, respectively in Tauru (Table 1). Depth wise distribution of heavy metals at all the 3 sites is shown in Figure 2. At Karhera higher concentrations of As were found in depth range of 90–120 cm but almost constant concentrations were observed at Ulheta and Tauru. Variations in concentrations are found to be influenced by topographic differences along with the local vegetation and climate (Srivastava *et al.*, 2015). In addition to this, the sources and distribution of As in IGB are largely controlled by the distribution of organic matter, oxic-anoxic conditions, and indigenous microbial flora (Dutta *et al.*, 2011). WHO (1993) has not specified any toxicity limits of As, iron and manganese concentrations in agricultural soils but Grsichek *et al.*, 1993 have specified limits as <1 k, <1000 k and < 10 K for As, iron and manganese, respectively in sediments excavated during drilling of RBF wells. As per these limits concentrations of iron and manganese were found exceeding the toxicity levels.

Table 1. Summary of concentrations (mg/kg) of arsenic, iron and manganese in Mewat, Haryana.

	Ulheta			Karhera			Tauru		
	As	Fe*	Mn*	As	Fe*	Mn*	As	Fe*	Mn*
Min	22.5	334	468	13.3	529	216	17.8	404	667
Max	41.9	530	716	52.6	614	442	24.4	645	888
Avg	31.6	442	580	28.2	566	359	21.6	527	765

As = Arsenic; Mn = Manganese; Fe = iron, * 1000

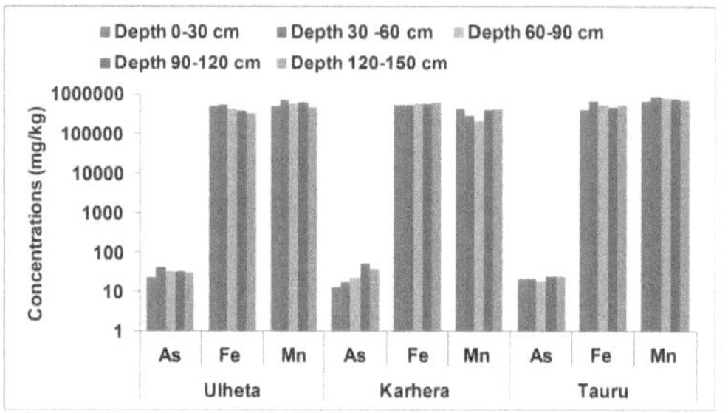

Figure 2. Depth wise distribution of arsenic, iron and manganese in Mewat, Haryana.

4 CONCLUSIONS AND RECOMMENDATIONS

Depth wise (0–150 cm) analysis of soil sediments at 3 sites in Mewat, Haryana show that concentrations of As in soil sediments were less than their toxicity levels (WHO 1993) while concentrations of iron and manganese are found exceeding the toxicity levels (Grischek *et al.*, 1993). This is concluded from the substantial low As in upto 60 cm depth of the soil sediments where the soil is used for growing crops and vegetables there might be chances of accumulation of these elements in the crops. There is an urgent need to carry out a detail study in such areas with the soil sediments impacted by these heavy metals.

ACKNOWLEDGEMENTS

GK acknowledge the funding received World Bank Funded National Hydrology project.

REFERENCES

Ahamed S., Sengupta M.K., Mukherjee A. *et al.* 2006. Arsenic groundwater contamination and its health effects in the state of Uttar Pradesh (UP) in upper and middle Ganga plain, India: a severe danger. *Sci. Total Environ.* 370: 310–322

Bhattacharjee S., Chakravarty S., Maity S. *et al.* 2005 Metal contents in the groundwater of Sahebgunj district, Jharkhand, India, with special reference to arsenic. *Chemosphere* 58: 1203–1217

Bhattacharya P., Chatterjee D. & Jacks G. 1997. Occurrence of arsenic contaminated groundwater in alluvial aquifers from delta plains, Eastern India: options for safe drinking water supply. *Int. J. Wat. Res. Manag.* 13(1):79–92.

Bhattacharya P., Mukherjee A. & Mukherjee A.B. 2011. Arsenic in groundwater of india. In: J.O. Nriagu (ed.) *Encyclopedia of Environmental Health*, vol. 1, pp. 150–164 Burlington: Elsevier.

Chakraborti D., Mukherjee S.C., Pati S., *et al.* 2003. Arsenic groundwater contamination in middle Ganga plain, Bihar, India: a future danger? *Environ. Health Perspect.* 111: 1194

Chakraborti D., Singh E.J. *et al.* 2008. Groundwater arsenic contamination in Manipur, one of the seven north-eastern hill states of India: a future danger. *Environ. Geol.* 56: 381–390

Datta D, & Kaul M. 1976 Arsenic content of drinking water in villages in Northern India. a concept of arsenicosis. *J. Assoc. Physicians India* 24(9): 599–604

Datta R., Quispe M.A. & Sarkar D. 2011. Greenhouse study on the phytoremediation potential of vetiver grass, *Chrysopogon Zizanioides* L., in Arsenic-contaminated Soils. *Bull. Environ. Contam. Toxicol.* 86: 12–128.

Gaur S., Joshi M., Jos H. & Saxena S. 2013. Analytical study of water safety parameters in ground water samples of Uttarakhand in India. *Ind. Public Health Res. Dev.* 4: 185–189

Grischek T., Dehnert J., Nestler W., Treutler H.C. & Freyer K. 1993. Description of System Conditions During Bank Filtration Based on Investigation of Bore Cores. *Proc. 2. Dresdner Grundwasser for schungstage*, 25.03.1993, Dresden, pp. 207–220 (in German).

Hazarika S. & Bhuiyan B. 2013. Fluoride, arsenic and iron content of groundwater around six selected tea gardens of Lakhimpur district, Assam, India. *Arch. Appl. Sci. Res.* 5: 57–61.

Krishan G., Ghosh N.C., Kumar C.P. *et al.* 2020. Understanding stable isotope systematics of salinity affected groundwater in Mewat, Haryana, India. *J. Earth Syst. Sci.* 129: 109.

Krishan G., Prasad G., Anjali Kumar C.P., *et al.* 2020. Identifying the seasonal variability in source of groundwater salinization using deuterium excess- a case study from Mewat, Haryana, India. *J. Hydrol. Reg. Stud.* 31: 100724.

Krishan G., Sejwal P., Bhagwat A. *et al.* 2021. Role of ion chemistry and hydro-geochemical processes in aquifer salinization—a case study from a semi-arid region of Haryana, India. *Water* 13(5): 617.

Lindsay W.L. & Norwell W.A. 1978. Development of DTPA of soil test for Zn, Fe, Mn and Cu. *J. Amer. Soil Sci.* 42: 421–428.

Mukherjee A.B., Bhattacharya P., Jacks G. *et al.* 2006. *Groundwater Arsenic Contamination in India: Extent and Severity*. In: R. Naidu, E. Smith, G. Owens, P. Bhattacharya & P. Nadebaum (eds.) *Managing Arsenic in the Environment: From Soil to Human Health*. CSIRO Publishing, Melbourne, Australia, pp. 533–594. (ISBN: 0643068686)

Shrivastava A., Ghosh D., Dash A. *et al.* 2015. Arsenic contamination in soil and sediment in india: sources, effects, and remediation. *Curr. Pollut. Rep.* 1: 35–46.

WHO 1993 *Standard Maxima for Metals in Agricultural Soils*. WHO Press, Geneva, Switzerland.

Arsenic in the Environment: Bridging Science to Practice for Sustainable Development – van der Wal, Ahmad, Petrusevski, Weijma, Savic, van der Wens, Beerendonk, Bhattacharya, Bundschuh & Naidu (Eds)
© 2023 the Author(s), ISBN 978-1-032-32928-4

Environmental implications of arsenic presence in volcanic and sedimentary rocks along the southeastern of El Bajío Guanajuatese, Guanajuato, Mexico

J.I. Morales-Arredondo[1], M.A. Armienta-Hernández[1], A.E. Lugo-Dorantes[2], F. Juárez-Aparicio[3], F. Romero[1], L.G. Martínez-Jardines[1] & I.Z. Flores-Ocampo[3]
[1]*Departamento de Recursos Naturales, Instituto de Geofísica, Universidad Nacional Autónoma de México, Col. Copilco Universidad, Delegación Coyoacán, México*
[2]*Earth Sciences Student, México, Universidad Nacional Autónoma de México, Col. Copilco Universidad, Delegación Coyoacán, Mexico*
[3]*Earth Sciences Postgraduate Program, Universidad Nacional Autónoma de México, Col. Copilco Universidad, Delegación Coyoacán, México*

ABSTRACT: Recent studies on arsenic (As) occurrence particularly in The Celaya Valley Aquifer (CVA) waters show that several sources of drinking water have elevated concentrations above national and international guidelines. In Villagrán and Juventino Rosas Municipalities, elevated concentrations of As above WHO guideline (10 µg/L) In both sites depending on groundwater as drinking water sources. Anthropogenic sources have been discarded to cause As presence, that may be due to water-rock interaction processes. However, the natural arsenic content in the geological environment is unknown. In this study, an assessment of arsenic was realized for volcanic and sedimentary rocks, collected from 14 sampling places along the southeastern of El Bajío Guanajuatese, Guanajuato, Mexico. These geological materials are representative of the hydrogeological environment. The high arsenic content occurs in volcanic rocks, this information will be used to identify if the proposed hydrogeochemical processes are related to the release and remotion of arsenic in the aquifer.

1 INTRODUCTION

High levels of arsenic (As) have been reported in groundwater in several Mexico (Armienta & Segovia 2008). The presence of this element in aqueous environment is associated with geogenic processes mainly (ASTDR 2008; Morales-Arredondo *et al.*, 2016), as geothermal environment. Elevated concentrations of As has been reported in Villagrán and Juventino Rosas, Municipalities, Guanajuato, México, (Morales-Arredondo *et al.*, 2018). However, the natural arsenic content in the geological environment from study area is unknown. In this study, an assessment of arsenic was realized for volcanic and sedimentary rocks, collected from 14 sampling places along the southeastern of El Bajío Guanajuatese, Guanajuato, Mexico during 2015 and 2017. The results have helped to define the rock type that contributes to the high arsenic content in the groundwater, the information will be used to identify if the proposed hydrogeochemical processes are related to the release of arsenic in the aquifer. Beside, sedimentary rocks (limestones) with low As contents have been used to carry out experiments to remove arsenic from contaminated water.

2 MATERIALS AND METHODS

2.1 *Study area*

The municipalities are located in the state of Guanajuato and belong to the hydrogeological province of Celaya Valley Aquifer (Figure 1), which is found between the physiographic regions of the Trans-Mexican Volcanic

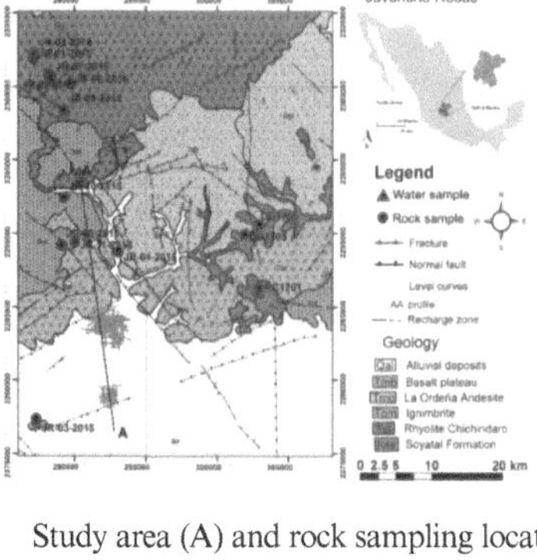

Figure 1. Study area (A) and rock sampling locations (B).

DOI: 10.1201/9781003317395-62

Belt and the Central Plateau. Lithologically, the area is formed by a sedimentary basin that overlays volcanic emissions of basic composition from the Pliocene-Pleistocene Era as well as acidic volcanic emissions from the early Oligocene to the late Mesozoic Eras. The basement is composed of limestones with low-grade metamorphism (Morales-Arredondo *et al.*, 2016). In the north, the Sierra de Guanajuato, which is mainly composed of rhyolite-ignimbrite rocks, is found. To the south, monogenetic volcanoes belonging to the Michoacán-Guanajuato field are present, and this region also forms part of the Trans-Mexican Volcanic Belt.

The groundwater of JR is influenced by a geological environment containing volcanic rocks of both acidic and basic composition and by a sedimentary basin filled with sediments of lacustrine-alluvial origin (Morales-Arredondo *et al.*, 2018). More than 90% of rural and urban population in CVA depends on groundwater resources abstracted through deep wells. More than 80% of groundwater wells target aquifers composed of volcanic rocks and sediments of lacustrine-alluvial origin.

2.2 *Water sampling and laboratory analysis*

Rock sampling was carried out at 2015 and 2017. A total of 29 water samples were collected, of which 14 samples were taken from volcanic rock sources and 3 samples from sedimentary rock sources were collected (Figure 1). The mineralogical analysis was determined with XRD and SEM-EBS at Laboratorio Nacional de Geoquímica y Mineralogía in the Instituto de Geología, UNAM and at Laboratorio Universitario de Petrología in the Instituto de Geofísica, UNAM respectively; Total arsenic contents were quantified with Hydride Generation Atomic Absorption Spectroscopy after acid digestion of the rock samples at Laboratorio de Química Analítica in the Instituto de Geofísica, UNAM.

2.3 *Creation of spatial database and data analysis*

ArcGIS software was used to create spatial database comprising the location each rock sample, and As concentration. The data analysis part involved calculating and mapping of summary statistics, i.e. minimum, maximum, average and standard deviation.

3 RESULTS AND DISCUSSION

3.1 *Mineralogical settings of sampled rocks*

The mineralogical analysis of representative volcanic rocks of the study area indicated that the rhyolites and ignimbrites were composed of a vitreous matrix containing mainly quartz phenocrysts, lithic fragments, feldspars, and iron oxides in addition to some plagioclases and accessory minerals of less than 5 μm in size. The basalt contained a crystalline matrix of plagioclases, pyroxenes, feldspars, quartz, opaque minerals, iron oxides, and iddingiste. The mineralogical analysis of representative sedimentary rocks indicate that limestones were composed of $CaCO_3$, SiO_2, traces of phyllosilicates, probably of the smectite or chlorite type and muscovite. The proportion of $CaCO_3$ defines a rock as a limestone or metamorphic limestone, which in turn presents minerals in smaller proportions such as quartz and clays.

3.2 *The concentration of arsenic in volcanic and sedimentary rocks from study area*

The arsenic contents in volcanic rocks varied from 0.64 mg/kg to 25.66 mg/kg with an average value of 5.32 ± 7.23 mg/kg. Meanwhile, the arsenic contents in sedimentary rocks varied from 1.29 mg/kg to 5.58 mg/kg with an average of 2.82 ± 2.4 mg/kg. These geological materials are representative of the hydrogeological environment.

4 CONCLUSIONS AND RECOMMENDATIONS

Arsenic contamination in groundwater of study area is a really problem in drinking water sources. The most targeted aquifers composed of rhyolites and ignimbrites seem to have high levels of arsenic exceeding WHO guideline. The communities are exposure to arsenic toxicity along of the CVA. The behaviour of excess arsenic is related to interaction between geothermal groundwater and volcanic rocks. These results have helped to define the rock type that contributes to the high arsenic content in the groundwater, the information will be used to identify if the proposed hydrogeochemical processes are related to the release of arsenic in the aquifer. Besides, sedimentary rocks (limestones) with low As contents have been used to carry out experiments to remove arsenic from contaminated water.

ACKNOWLEDGEMENTS

The research was financed partially by PAPIIT-DGAPA (Contribution IA-101019). We extend our gratitude to O. Cruz, A. Aguayo, N. Ceniceros, and A. Vasquez Salgado for performing the chemical determinations, to A. Rodríguez-Díaz for carrying out the mineralogical determinations through SEM-EDS, to R. Flores-Vargas and J.F. Landa-Arreguin for support during field work and rock sampling in the study area.

REFERENCES

Armienta M.A. & Segovia N. 2008. Arsenic and fluoride in the groundwater of Mexico. *Environ. Geochem. Hlth.* 30: 345–352.

ASTDR 2008. *Public Health Assesment Guidance Manual.* Agency for Toxic Substances and Disease Registry.

Morales-Arredondo I., Rodríguez R., Armienta M.A. & Villanueva Estrada R.E., 2016. The origin of groundwater arsenic and fluorine in a volcanic sedimentary basin in Central Mexico: A hydrochemistry hypothesis. *Hydrogeol. J.* 1: 1–10.

Morales-Arredondo J.I., Armienta Hernández M.A., Hernández-Mendiola E., Estrada-Hernández R.E. & Morton Bermea O. 2018. Hydrochemical behavior of uranium and thorium in rock and groundwater samples from southeastern of El Bajío Guanajuatense, Guanajuato, Mexico. *Env. Earth Sci.* 77:567

Arsenic in the Environment: Bridging Science to Practice for Sustainable Development –
van der Wal, Ahmad, Petrusevski, Weijma, Savic, van der Wens, Beerendonk, Bhattacharya, Bundschuh & Naidu (Eds)
 ISBN 978-1-032-32928-4

Arsenic contamination on sediments in the North and Tamiahua Beaches, Gulf of Mexico: Environmental implications

I.Z. Flores-Ocampo[1] & J.S. Armstrog-Altrin[2]
[1]*Postgraduate in Earth Sciences, National Autonomous University of Mexico, Mexico City, Mexico*
[2]*Institute of Marine Sciences and Limnology, National Autonomous University of Mexico, Mexico City, Mexico*

ABSTRACT: Recent studies, particularly in sediments of the Gulf of Mexico, show that there are high concentrations of arsenic (As). In Norte Beach sediments, concentration of As varies from 3.55 to 25.7 mg/kg, with an average value of 11.641 mg/kg. Similarly, the As content in Tamiahua Beach varies between 0.8 and 2.10 mg/kg (avg. 1.36 mg/kg). Different indices were used to infer the enrichment, toxicity and risk to the aquatic environment. The enrichment factor (EF) and geo-accumulation index (I_{geo}) obtained for As in the North Beach were 6.7–53.1 and 2.16–5.14, respectively. However, in Tamiahua beach sediments As content EF and I_{geo} values vary between 1.8 to 4.57 and 0.26–1.61, respectively. The Canadian Environmental Quality Guide (CEQG) divides concentrations into 2 levels: threshold effect (TEL) and probable effect (PEL), these levels define three probable biological effects. The As concentrations of the two beaches were between the TEL (threshold effect levels) and PEL (probable effect level). High concentration of As is due to the anthropogenic contamination derived from different industries located along the Gulf of Mexico coastal areas, which increase the possibility for adverse effect towards living organisms.

1 INTRODUCTION

The source, distribution and mobilization of arsenic (As) in the beach environment differ according to the location and may be related to natural or anthropogenic processes. Sediments are the environmental destination of many chemical substances and elements (including As) in aquatic ecosystems that act as sinks and, subsequently, as a source of the aquatic environment. Many aquatic organisms may be exposed to chemical substances through their immediate interactions with sediments.

For this study the beach areas are located the Gulf of Mexico, one located in the northern part, in the state of Veracruz, known as Tamiahua Beach and another to the south, in the state of Campeche, in Ciudad de Carmen, known as Playa Norte. To perform the analyses, 20 sediment samples were taken from each beach to determine the origin and degree of As contamination.

To infer the degree of As contamination in beach sediments, the enrichment factor (EF; Birth 2003), geoaccumulation index (I_{geo}; Müller 1969) and Canadian Environmental Quality Guide (CEQG) were calculated. The EF and I_{geo} allow to differentiate the origin of sediment contamination, whether it is due to anthropogenic or natural source. The CEQG define the probability of biological effects on the ecosystem.

The present study aims to investigate: a) the degree of As contamination, and b) source of the As in the sediments.

2 METHODS

The sediments sampled were treated as follows: 0.5 g of each sample was digested with aqua regia for 2 hours at 95°C. The samples were then cooled and diluted with deionized water. The samples were analyzed using a Varian ICP-MS. Quality Control for the digestion was 15% for each batch, 2 method reagent blanks, 6 in-house controls, 8 sample duplicates and 5 certified reference materials.

3 RESULTS AND DISCUSSION

The results of the study (Table 1) revealed that the As content in the North Beach is higher than in the Tamiahua Beach sediments, which varies from 3.55 to 25.7 mg/kg, with an average value of 11.641 mg/kg. Similarly, the As content in Tamiahua Beach varies between 0.8 and 2.10 mg/kg (avg. 1.36 mg/kg).

The EF value obtained for As in the North Beach ranges from 6.7 to 53.1, which is greater than the Sutherland scale (2000). According to the Sutherland scale, an EF value > 5 indicates an anthropogenic source. Hence, the As content in the North Beach sediments (>5) indicates an

DOI: 10.1201/9781003317395-63

Table 1. Arsenic concentrations (in mg/kg) in sediments of Norte beach (PN) and Playa Tamiahua (T).

Sample	As	Sample	As
PN 1	10.9	T 2	2.1
PN 3	6.0	T 3	1.8
PN 5	7.4	T 4	1.1
PN 7	25.7	T 8	1.3
PN 9	8.5	T 10	1.0
PN 11	17.0	T 11	0.8
PN 13	17.0	T 13	1.4
PN 15	11.6	T 14	1.0
PN 17	8.8	T 16	1.7
PN 19	3.5	T 19	1.4

Table 2. Values of the EF and Igeo index.

Sample	EF	Igeo
PN1	23.76	3.99
PN3	12.53	3.06
PN5	17.16	3.52
PN7	53.05	5.14
PN9	18.64	3.64
PN11	32.22	4.42
PN13	34.69	4.53
PN15	17.67	3.56
PN17	17.62	3.55
PN19	6.7	2.16
T2	4.57	1.61
T3	4.16	1.47
T4	2.49	0.73
T8	2.7	0.85
T10	2.25	0.58
T11	1.8	0.26
T13	3.08	1.04
T14	2.2	0.55
T16	3.26	1.12
T19	2.85	0.93

anthropogenic origin. On the other hand, the As content in the Tamiahua beach sediments are < 5, which ranges from 1.8 to 4.57. This indicates the absence of As contamination in the Tamiahua beach and the As content in sediments is due to natural origin.

The I_{geo} results were compared with the classes proposed by Müller (1981). North Beach revealed high level of anthropogenic contamination and the Igeo values (~2.16–5.14) are within the categories proposed by Müller (I_{geo} class = 5 and 6). Class 6 reflects the enrichment factor 100 times above the background values. Tamiahua Beach has Igeo values between 0.26 and 1.61, which indicates that the sediments of this beach are poorly or not contaminated.

In addition, the As content was compared with the Canadian standard CEQG (Canadian sediment quality guidelines for the protection of aquatic life) to assess whether the As contents represent a risk to living organisms that live in the beach area or not. The maximum values for TEL are 7.24 mg/kg and for PEL it is 41.6 mg/kg. The As concentrations of the two beaches are between the TEL (threshold effect levels) and PEL (probable effect level). This indicates the possibility for adverse effect towards living organisms. Additionally, there are several toxicological studies carried out in marine fauna, which verify contamination by As (Aguilar-Ucán *et al.*, 2014; Botello 2005).

Finally, it is identified that there is a high correlation between As and vanadium, this behaviour indicates a common origin of both elements in the North Beach, which is probably related to the influence of oil industry.

4 CONCLUSIONS

The results of this study indicate that the enrichment of As content in the Tamiahua Beach sediments is due to a natural origin, whereas As content in the North Beach is due to an anthropogenic origin and indicates the influence of oil industry with high probability that As concentrations are toxic to the ecosystem.

REFERENCES

Aguilar-Ucán C.A., Montalvo-Romero C., Cerón-Bretón J.G. & Anguebes-Fransesch F. 2014. Levels of heavy metals in marine species: Oyster (Crassostrea virginica), Jaiba (Callinectes sapidus) and Camarón (Litopenaeus setiferus), from Ciudad del Carmen, Campeche, Mexico. *Rev. Lat. Rec. Na.* 10(1): 9–17.

Birth G. 2003. *A Scheme for Assessing Human Impacts on Coastal Aquatic Environments Using Sediments*. In: C.D. Woodcoffe & R.A. Furness (eds.) *Coastal GIS 2003*. Wollongong University Papers in Centre for Maritime Policy, Australia.

Botello A.V., Soto L.A., Ponce-Vélez G. & Susana Villanueva F. 2015. Baseline for PAHs and metals in NW Gulf of Mexico related to the deepwater horizon oil spill. *Estuar. Coast. Shelf Sci.* 156: 124–133.

Canadian Council of Ministers of the Environment 1995. Protocol for the Derivation of Canadian Sediment Quality Guidelines for the Protection of Aquatic Life. CCME EPC-98E.

Flores-Ocampo I.Z. 2020. Sediment Geochemistry and U-Pb Geochronology of Detrital Zircons from Playa Norte and Playa Tamiahua of the Gulf of Mexico: Implication of Provenance. MSc Thesis, UNAM, Mexico City.

Müller G. 1981. Die schwermetallbelastung der sedimente des neckars und seiner nebenflusse: Eine bestandsaufnahme. *Chem. Ztg.* 105: 157–164.

Sutherland R.A., Tolosa C.A., Tack F.M.G. & Verloo M.G. 2000. Characterization of selected element concentrations and enrichment ratios in background and anthropogenically impacted roadside areas. *Arch. Environ. Contam. Toxicol.* 38: 428–438.

Zhang Z., Wang J.J., Ali A. & DeLaune R.D. 2016. Heavy metals and metalloid contamination in Louisiana Lake Pontchartrain Estuary along I-10 bridge. *Transp. Res. D: Transp. Environ.* 44: 66–77.

Arsenic in the Environment: Bridging Science to Practice for Sustainable Development –
van der Wal, Ahmad, Petrusevski, Weijma, Savic, van der Wens, Beerendonk, Bhattacharya, Bundschuh & Naidu (Eds)
© 2023 the Author(s), ISBN 978-1-032-32928-4

Arsenic speciation in dust dispersion from mining to Mount Isa city

J. Zheng[1], B.N. Noller[1], T. Huynh[1], R. Taga[1,2], V. Diacomanolis[1,2], J.C. Ng[2], J. Aitken[3] &
H.H. Harris[4]
[1]*Centre For Mined Land Rehabilitation, The University of Queensland, St Lucia, QLD, Australia*
[2]*Queensland Alliance for Environmental Health Sciences, The University of Queensland, Woolloongabba, QLD, Australia*
[3]*School of Chemistry, The University of Sydney, Sydney, NSW, Australia*
[4]*Department of Chemistry, University of Adelaide, SA, Australia*

ABSTRACT: The study has sought to understand why higher % bioaccessibilities (BAc's) were observed for the risk assessment of As exposure to the population of Mount Isa for surface tailings and city fallout dust. The current study examined the speciation of arsenic (As) using arsenic K-edge to record XANES spectra and linear combination fitting of sample spectra with model compounds. The study identified that As speciation is not showing a clear difference with As compound composition, although small differences are apparent of probably from other minor constituents and the presence of very fine particles that have enhanced solubility of the compounds present.

1 INTRODUCTION

The Lead Pathway Air Study (2007–2012) (Noller *et al.*, 2017) examined arsenic (As) and undertook assessment of human exposure at Mount Isa from mining and processing activities (Zheng *et al.*, 2018). If not properly managed, dispersion of metals and metalloids from mine activities may cause adverse effects on the environment and human health (enHealth 2012). The study identified that As exposure to the population of Mount Isa is low but bioaccessibilities (BAc's) for surface tailings (31 ± 28% n = 7) and fall out (city; 47 ± 23% n = 15) may be a higher risk compared with garden soil (15 ± 8% n = 92; < 250 μm fraction) (Zheng *et al.*, 2018). The current study examines the speciation of As to further understand the higher %BAc's.

2 MATERIALS AND METHODS

Samples were collected from mining, mineral processing activities and the city residential area at Mount Isa, northern Australia. Samples initially sieved to <2 mm were ground to <20 μm for X-ray analysis. Total As concentrations were determined by ICP-MS on aqua regia digests of the samples. Arsenic K-edge XANES spectra were recorded from model compounds and samples at Beamline 20B Australian National Beamline Facility (ANBF) Photon Factory, KEK, Tsukuba, Japan (see models in Noller *et al.*, 2016). The spectra were recorded at room temperature in fluorescence mode using a thirty-six-element germanium-array detector. BL–20B was equipped with a channel-cut Si (111) monochromator which was detuned 50% to reject harmonics. Sodium arsenate was used as an internal standard for energy calibration, with the first peak of the first derivative assumed to be 11,873.6 eV. Solid models diluted in boron nitride (ground to <20 μm) were pressed into aluminium spacers then covered between two 63.5-μm Kapton tape windows (window size, 2*10 mm). Data analysis, including calibration, averaging and background subtraction of all spectra and principal component analysis (PCA), target and linear regression analyses of XANES spectra were performed using the EXAFSPAK software package (George & Pickering 2000). The accuracy of fitting was determined to be ~10% based on analyses of control model compound mixtures.

3 RESULTS AND DISCUSSION

Table 1 gives compound fitting of XANES spectra according to 3 categories of samples, namely copper processing and background mineralisation, tailings and city fallout, and city residential area samples. Thus, arsenic is generally associated with copper ore and its processing at Mount Isa (Noller *et al.*, 2017). Copper ore concentrate shows the presence of arsenopyrite (FeAsS) and As sulfides (III and V). Ferric arsenate is present in tailings and fall out, but Ca arsenate is only present in fallout. Calcium arsenite is only present in copper

 DOI: 10.1201/9781003317395-64

slag, which has a glassy matrix and is not easily mobilised as dust. Background mineralisation is also associated with Ca and Fe arsenates having low %BAc's Arsenic in roof gutter dust and garden soil are both similar in composition but carpet dust is more like surface tailings material. The high volume air (HVA) particulates have metal arsenates exclusively with no As sulfides present. Thus As sulfides are only present in carpet dust and is a residue from earlier times when workers wore their work clothes off-site to home.

The observation of higher %BAc's in tailings surface dust and city fallout (Zheng *et al.*, 2018) is not showing a clear difference with As compound composition, although small differences are apparent.

Thus, it is more likely that the observed increases in %BAc's for As in surface tailings and city fallout are from minor constituents and the presence of very fine particles that give enhanced solubility of the compounds present.

Table 1. Compound Fitting from XANES spectra.

Copper processing and background mineralisation				
	Sample			
Compound fraction	Copper ore concentrate	Copper smelter reclamation	Copper slag	Natural mineralisation
Copper processing and background mineralisation				
AsIII sulfide	0.28	–	–	–
Caarsenite	–	–	0.90	–
Ca arsenate	–	<0.1	–	0.50
Fe arsenate	–	0.72	0.11	0.56
FeAsS	0.69	–	–	–
Other	–	0.15 ($NaAs^{III}$)	–	–
Total	0.97	0.95	1.01	1.05
Residual	0.003	0.001	0.003	0.007
Tailings and city fallout				
	Sample			
Compound fraction	Surface tailings (TD5)	Surface Tailings (TD8)	Fallout City Station Street	Fallout City K-Oval
AsIII sulfide	0.39	0.45	–	–
AsV sulfide	0.13	–	–	–
Ca arsenate	–	–	0.76	0.61
Fe arsenate	0.23	0.31	0.16	0.32
FeAsS	0.25	0.20	–	–
Other	–	–	0.15 (α-As)	0.12 (α-As)
Total	1.01	0.98	1.06	1.05
Residual	0.0007	0.0005	0.006	0.004
City residential area				
Compound fraction	Roof gutter dust	Garden soil	Carpet dust	HVA Particulates (PM^{10})
AsIII sulfide	–	–	0.29	–
AsV sulfide	0.14	0.09	–	–
Ca arsenate	<0.1	0.10	–	0.59
Fe arsenate	0.84	0.5	0.45	0.5
Other	–	0.24 ($NaAs^{V}$)	0.26 ($NaAs^{V}$)	<0.1 ($NaAs^{V}$)
Total	1.02	0.96	1.01	1.09
Residual	0.0006	0.004	0.0006	0.007

4 CONCLUSIONS AND RECOMMENDATIONS

The observation of higher %BAc in tailings surface dust and city fallout (Zheng *et al.*, 2018) is not showing a clear difference with As compound composition, although small differences are apparent of probably from other minor constituents and the presence of very fine particles that have enhanced solubility of the compounds present.

ACKNOWLEDGEMENTS

This work was performed at the BL20B Beamline Facility Photon Factory (PF), KEK Tsukuba-Japan with support from the Australian Synchrotron Research Program, funded by the Commonwealth of Australia under the major National Research Facilities Program. Queensland Alliance for Environmental Health Sciences is a partnership between Queensland Health and the University of Queensland. Glencore-Mount Isa Mines funded the study.

REFERENCES

enHealth 2012. *Environmental Health Risk Assessment. Guidelines for Assessing Human Health Risks from Environmental Hazards.* The Environmental Health Committee.

George G.N. & Pickering I.J. 2000. *EXAFSPAK: A Suite of Computer Programs for Analysis of X-ray Absorption Spectra.* Stanford Synchrotron Radiation Laboratory: Stanford, CA.

Noller B.N., Ng J.C., Aitken J.B. & Harris H.H. 2016. Arsenic fate following mining of sulfide ore at mine sites and significance of the reduced state. In: P. Bhattacharya, M. Vahter, J. Jarsjö, J. Kumpiene, A. Ahmad, C. Sparrenbom, G. Jacks, M.E. Donselaar, J. Bundschuh & R. Naidu (eds.). *Arsenic Research and Global Sustainability (As2016).* CRC Press/ Balkema, Taylor and Frances Group, London, UK. pp 189–190.

Noller B., Zheng J., Huynh T., Ng J., Diacomanolis V., Taga R. & Harris H. 2017. *Lead Pathways Study – Air. Health Risk Assessment of Contaminants to Mount Isa City.* 7 February 2017 Mount Isa Mines Limited, Mount Isa. pp 1–414.

Zheng J., Noller B.N., Huynh T., Taga R., Ng J.C., Diacomanolis V. & Harris H.H. 2018. Health risk assessment of arsenic dispersion from mining in mount isa. In: Y.G. Zhu, H.M. Guo, P. Bhattacharya, A. Ahmad, J. Bundschuh & R. Naidu (eds.) *Environmental Arsenic in a Changing World: Proceedings of the 7th International Congress and Exhibition on Arsenic in the Environment (As2018)*, CRC Press, Taylor and Frances Group, London, UK. pp 372–373.

Arsenic in the Environment: Bridging Science to Practice for Sustainable Development –
van der Wal, Ahmad, Petrusevski, Weijma, Savic, van der Wens, Beerendonk, Bhattacharya, Bundschuh & Naidu (Eds)
© 2023 the Author(s), ISBN 978-1-032-32928-4

Leaching behavior of arsenic from Upper Cretaceous karst infillings materials in a fractured limestone aquifer (HESP, France)

R. Mhanna[1], A. Naveau[1], F. Battaglia[2], H. Thouin[2], M. Bueno[3], C. Fontaine[1], G. Porel[1], J. Bassil[4] & L. Caner[1]
[1]*CNRS/IC2MP, Université de Poitiers, Poitiers, France*
[2]*BRGM, Water, Environment and Ecotechnology Division, Orléans, France*
[3]*CNRS/IPREM, Université Pau & Pays de l'Adour, Pau, France*
[4]*L2GE, Lebanese University, Faculty of Sciences 2, Fanar, Matn, Lebanon*

ABSTRACT: Arsenic (As) mobility and transport into groundwater are highly dependent on the nature of the sedimentary matrix and the groundwater parameters. In this study, a combination of chemical fractionations, mineralogical characterizations, elemental mappings, and leaching experiments investigated the distribution of As in karst infillings materials from the Hydrogeological Experimental Site of Poitiers (HESP). The influence of pH and Eh conditions, contact time and the presence of bacteria on As mobility were then evaluated. The strong association between iron and As in the matrix is consistent with the low mobility of As in circumneutral pH conditions imposed by the nature of the limestone aquifer. In addition, the alternations of redox conditions tend to increase As release whereas the contribution of the bacteria present in the groundwater seems to regulate the mobilization.

1 INTRODUCTION

Geogenic arsenic (As) contamination in groundwater is a worldwide environmental issue. In this context, it is important to understand the biogeochemical processes involved in the mobilization of As from sedimentary materials into groundwater (Gillispie *et al.*, 2016).

In the present study, the behavior of karst argillaceous infillings materials, with relatively high contents of both organic matter and pyritic minerals, was investigated to understand the mechanisms responsible for retaining As in the Dogger aquifer. This aquifer exhibits high amounts of soluble selenium (Bassil *et al.*, 2016) but no detectable amount of As, in contrast to the associated sedimentary materials, which contain about fourfold higher As than selenium.

For this purpose, the distribution of As was first determined by a combination of elemental mappings and chemical fractionations in the clayey infillings. Then, batch leaching tests were performed to define the impact of pH, redox conditions, and the presence of bacteria on As mobilization.

2 METHODS

2.1 *Sampling*

The HESP is a research platform that aims to develop new approaches in the characterization and modeling of groundwater flow and solute transfer. It consists of an area of 10 ha with 37 wells cored through two aquifers: the Middle Jurassic (Dogger) and the infra-Lias aquifers separated by the marls of the supra-Lias. Large paleokarst cavities and smaller modern karst cavities are developed in the Dogger limestone, both partially infilled with dark argillaceous sediments.

Karst infillings samples used in the present study were collected in 2012 through non-destructive drilling at a depth between 66 m and 71 m.

2.2 *Laboratory analysis*

Chemical quantifications were performed by ICP-MS in an accredited laboratory (CRPG, Nancy, France). Powder samples (infra 50 µm fraction) were analyzed by X-ray diffraction using Phillips analytical X'Pert Pro apparatus in the same experimental conditions as described by Bassil *et al.* (2016). Standard thin sections were observed using a scanning electron microscope Jeol JSM-7900F equipped with a Bruker EDX spectrometer for chemical microanalyses (IC2MP, Poitiers).

A previously developed six-steps protocol of sequential extractions (Mhanna *et al.*, 2019), was used to define the distribution of As, Se, and Fe.

Batch leaching experiments were performed at 20°C in borosilicate flasks with suspensions of 30 g/L. The anaerobic condition was imposed by the addition of a mixture of gases. In situ water

DOI: 10.1201/9781003317395-65

was collected from a HESP well containing no detectable soluble As.

The quantifications of As and Se were performed by ICP-MS (Agilent 7500ce, IPREM, France), and Fe by ICP-OES (Agilent Tech 5110, IC2MP, France).

3 RESULTS AND DISCUSSION

Given the chemical and mineralogical composition of the sample (Table 1), investigations were primarily focused on the distribution of As, selenium, and iron.

Table 1. Main composition of the studied material.

Sample	SiO_2 %	Al_2O_3 %	Fe_2O_3 %	FeO %	Corg %	S tot %	As ppm	Se ppm
C5-Bj	40.6	30.4	1.8	1.1	5.68	0.57	22.4	6.3
Sample	Depth m	Calcite %	Pyrite %	Quartz %	Feld-K %	Anatase %	Clays %	SSA m^2/g
C5-Bj	67.9	4	1.4	9.8	8.1	5.3	71.4	52.2

Petrographic observations identified two pyrite morphologies: euhedral and framboidal, but only the former was rich in As. Around the euhedral crystals, diffuse halos of iron oxide are also enriched in As, but to a lesser extent. Hence, it could be suggested that a part of As coprecipitated or sorbed onto secondary iron oxides, derived from weathering of primary pyrite.

Chemical fractionations supported this assumption since the fractions associated with iron sulfide minerals (35%) and amorphous iron oxides (13%) represented an important part of As (Table 2). The exchangeable fraction was also significant (around 13%) whereas, the soluble As was the least abundant fraction with only 1% of total As.

Table 2. As, Se, and Fe distributions (chemical fractionation).

	As	Se	Fe
Soluble	1.5	4.7	0.2
Exchangeable	13.3	0.7	0.3
Carbonated	8.5	<LD	0.4
Amorphous Fe/Al	13.0	2.2	7.4
Alkaline Soluble	4.8	17.8	1.3
Pyritic	35.4	38.9	82.8
Refractory	23.4	36.3	7.5

Batch leaching studies as a function of pH (Figure 1) showed that As is highly dependent on pH. The negligible leaching in circumneutral pH is consistent with the chemical fractionations and

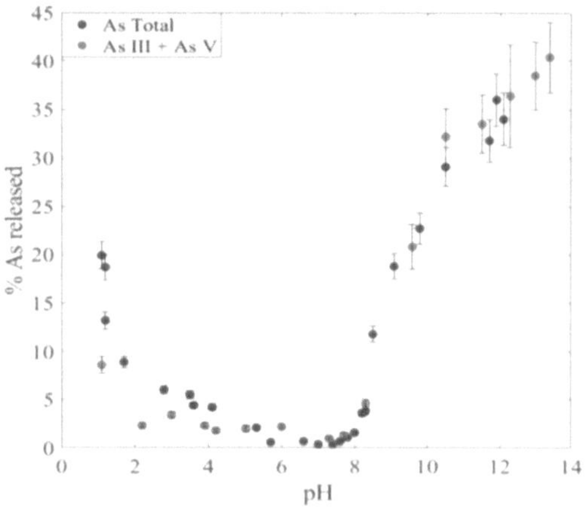

Figure 1. As leaching behavior as a function of pH.

can explain the absence of detectable As in the groundwater of the HESP. The increase of As mobility as pH decreased is attributed to the dissolution of calcite (around pH 4.5), of amorphous iron oxides (around pH 3) and finally of pyritic minerals at very low pH (McKibben *et al.*, 1986).

Leaching experiments as a function of time (Figure 2) showed that the alternation between anaerobic (in grey) and aerobic (in white) conditions favored the mobilization of As. Furthermore, the contribution of the bacteria present in the water tended to limit the release of As during anaerobic conditions. The chemical composition of the water also played a major role in the release of As: the carbonated facies of the HESP water strongly limited the mobilization of As, probably by buffering the pH conditions.

Under alkaline conditions, As mobilization drastically increased with the pH increase as

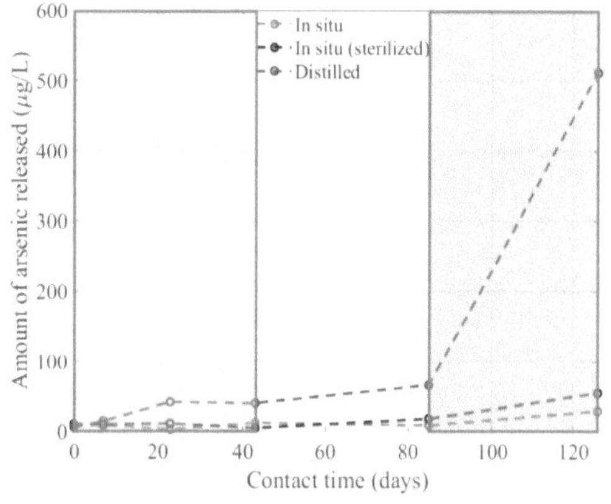

Figure 2. As leaching behavior as a function of time. The grey zone represents anaerobic conditions and the white zone represents aerobic conditions.

previously observed for selenium, iron, and TOC (Bassil *et al.*, 2018), suggesting the presence of ternary complexes between As oxyanions, iron, and humic-like organic matter (Martin *et al.*, 2017).

4 CONCLUSIONS

Elemental mappings combined with chemical fractionations and leaching experiments showed a strong association between As and Fe distributions: After partial alteration of the pyritic materials, the associated As was strongly retained by amorphous iron oxides and continental organic matter complexes. Moreover, the released As was very low in circumneutral pH conditions (as observed in the carbonated aquifer) but highly increased under pH variation. The mobilization of As also increased with the alternations of redox conditions, as observed during the fluctuations of the piezometric levels. Finally, the bacteria seem to play an important role in regulating As mobility under reducing conditions.

ACKNOWLEDGMENTS

We acknowledge the financial support from the European Union, the 'Région Nlle Aquitaine' and the 'Agence de l'Eau Loire-Bretagne'. We also acknowledge B. Nauleau and D. Paquet for their technical support in the HESP.

REFERENCES

Bassil J., Naveau A., Beuno M., Di Tullo P., Grasset L., Kazpard V. & Razack M. 2016. Determination of the distribution and speciation of selenium in an argillaceous sample using chemical extractions and post-extractions analyses: application to the hydrogeological experimental site of poitiers. *Environ. Sci. Pollut. Res.* 23: 9598–9613.

Bassil J., Naveau A., Beuno M., Razack M. & Kazpard V. 2018. Leaching behavior of selenium from the karst infillings of the hydrogeological experimental site of poitiers. *Chem. Geol.* 483: 141–150.

Gillispie E.C., Andujar E. & Polizzotto M.L. 2016. Chemical controls on abiotic and biotic release of geogenic arsenic from pleistocene aquifer sediments to groundwater. *Environ. Sci.: Processes Impacts* 18: 1090–1103.

Lowers H.A., Breit G.N., Foster A.L., Whitney J. & Yount J. 2007. Arsenic incorporation into authigenic pyrite, Bengal basin sediment, Bangladesh. *Geochim. Cosmochim. Acta.* 71: 2699–2717.

Martin D.P., Seiter J.M., Lafferty B.J. & Bednar A.J. 2017. Exploring the ability of cations to facilitate binding between inorganic oxyanions and humic acid. *Chemosphere* 166: 192–196.

McKibben M.A. & Barnes H.L. 1986. Oxidation of pyrite in low temperature acidic solutions: rate laws and surface textures. *Geochim. Cosmochim. Acta.* 50: 1509–1520.

Mhanna R., Naveau A., Beuno M., Caner L. & Bassil J. 2019. Distribution of arsenic in an argillaceous sample using sequential chemical extractions: application to the hydrogeological experimental site of poitiers (HESP). *E3S Web Conf.*, 98.

1.5 Advances and challenges in arsenic analysis in solid and aqueous media

Arsenic in the Environment: Bridging Science to Practice for Sustainable Development –
van der Wal, Ahmad, Petrusevski, Weijma, Savic, van der Wens, Beerendonk, Bhattacharya, Bundschuh & Naidu (Eds)
© 2023 the Author(s), ISBN 978-1-032-32928-4

Ultra-High Performance Liquid Chromatography (UHPLC) hyphenated to Inductively Coupled Plasma Mass Spectrometry (ICP-MS) for fast analysis of arsenic species in environmental and biological media

I. Herath[1], J. Bundschuh[1,2] & P. Bhattacharya[3]
[1]*Faculty of Health, Engineering and Sciences, University of Southern Queensland, West Street, Toowoomba, Queensland, Australia*
[2]*UNESCO Chair on Groundwater As within the 2030 Agenda for Sustainable Development, University of Southern Queensland, West Street, Toowoomba, Queensland, Australia*
[3]*KTH-International Groundwater Arsenic Research Group, Department of Sustainable Development, Environmental Science and Engineering, KTH Royal Institute of Technology, Stockholm, Sweden*

ABSTRACT: Longer retention time resulting in a slow separation is the major drawback of existing methodologies based on inductively coupled plasma mass spectrometry (ICP-MS) coupled to high performance liquid chromatography (HPLC-ICP-MS) and ion chromatography (IC-ICP-MS) for speciation of arsenic (As). This study developed a fast analytical protocol based on Ultra-High Performance Liquid Chromatography (UHPLC) hyphenated to ICP-MS detection for simultaneous separation and quantification of arsenite (As(III)), arsenate (As(V)), dimethylarsonate (DMA(V)) and monomethylarsenate (MMA(V)). A mixed mobile phase containing 8.5 mM of $NH_4H_2PO_4$ and NH_4NO_3 (1:1) at pH 6.0 achieved the separation of As species within 5 min which is almost a double saving in analysis time per sample compared to the existing methods (9-15 min). Unlike HPLC, UHPLC did not generate a higher column back pressure with increasing flow rate up to 2.5 mL/min resulting in a faster separation with excellent resolution of peaks. Limits of detection for As species were in the range of 0.3-0.5 μg/L. Faster separation achieved from the proposed method is indeed analytically cost effective in terms of ICP-MS running cost and energy consumption.

1 INTRODUCTION

The determination of total arsenic (As) concentrations is not always adequate under various environmental scenarios as the toxicological nature of As depends on its chemical speciation. Inductively Coupled Plasma Mass Spectrometry (ICP-MS) hyphenated to High Performance Liquid Chromatography (HPLC) and Ion Chromatography (IC) are widely used for simultaneous speciation of arsenic (As) at present (Chen *et al.*, 2006; Rahman *et al.*, 2009). However, longer retention time resulting in a slow separation is the major drawback of these existing analytical methods. In addition, fast separations achieved from HPLC based methods have always resulted in poor resolution and baseline separation between peaks (Li *et al.*, 2019). Therefore, a faster separation with excellent resolution of As species using the existing analytical techniques has still remained as a challenging task. Baseline separation and resolution between peaks need to be achieved perfectly in order to establish an accurate and precise analytical protocol for real sample applications. Hence, this study aimed to improve the existing HPLC related approaches in order to develop a fast analytical protocol for simultaneous separation and quantification of As species, including arsenite (As(III)), arsenate (As(V)), dimethylarsonate (DMA(V)) and monomethylarsenate (MMA(V)) using an UHPLC hyphenated to ICP-MS detection as a modern advancement for the analysis of environmental, and biological samples.

2 MATERIALS AND METHODS

2.1 *Optimization of mobile phase*

Two types of eluents based on different compositions of ammonium salts containing $NH_4H_2PO_4$ and NH_4NO_3 were examined for the speciation of As. The first mobile phase (M1) was prepared with $NH_4H_2PO_4$ and the second mobile phase (M2) was prepared with a mixture of $NH_4H_2PO_4$ and NH_4NO_3 (1:1). The concentration and pH of both eluents were optimized in the range of

DOI: 10.1201/9781003317395-66

8.5–30.0 mM and pH 5-6.6, respectively through the isocratic elution to select the best mobile phase to achieve an excellent separation of the targeted As species.

2.2 *Instrumentation of UHPLC coupled with ICP-MS*

For the first time, a modern UHPLC system (Flexar, Perkin-Elmer, USA) was coupled with ICP-MS detection for the quantification of the targeted As species with an anion exchange column; PRP-X100, 250 × 4.1 mm, 10 μm (Hamilton, USA). The targeted As species present in mixed standard solutions were freshly analysed and quantified by UHPLC-ICP-MS following the six point calibration (5-25 μg/L) to cover the entire concentration range of interest. The elemental As was detected by ICP-MS at m/z 75. To validate the accuracy and the reliability of the concentrations of As species measured from the proposed method, a certified rice reference material (ERM® – BC211) was tested under the same conditions.

3 RESULTS AND DISCUSSION

3.1 *Effects of eluent pH on resolution of peaks*

Figures 1 and 2 depict the variation of retention time in the separation of the targeted As species with changing the pH of M1 (30 mM) and M2 (8.5 mM).

The best peak resolution for all As species was achieved by M2 at pH 6.0.

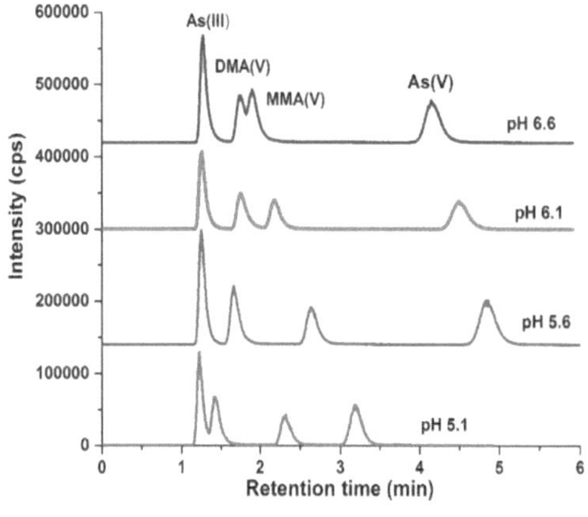

Figure 1. Effects of eluent pH on the resolution of As species through isocratic elution of 30 mM, $NH_4H_2PO_4$ (M1).

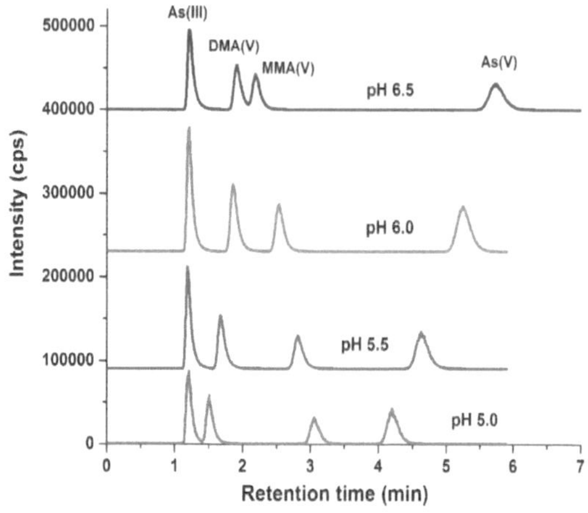

Figure 2. Effects of eluent pH on the resolution of As species through isocratic elution of 8.5 mM, $NH_4H_2PO_4$ and NH_4NO_3 (M2).

3.2 *Effects of eluent concentration on resolution of peaks*

The 8.5 mM of M2 at pH 6.0 was selected as an optimum mobile phase to separate the targeted As species with regard to high peak intensity, excellent peak resolution and fast analysis (Figure 3).

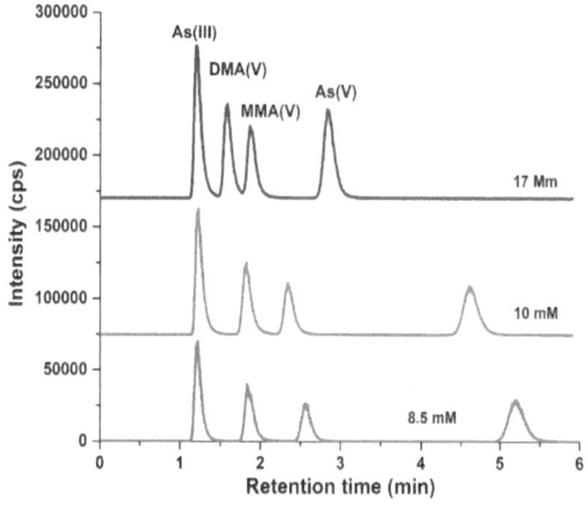

Figure 3. Effects of eluent concentration on the resolution of As species with M2 at pH 6.0.

3.3 *Overall performance of the method*

Figure 4 illustrates the full separation of the targeted As species with the optimized UHPLC conditions and the mobile phase containing 8.5 mM, $NH_4H_2PO_4$ and NH_4NO_3 at pH 6.0, flow rate – 2.5 mL/min and injection volume – 50 μL.

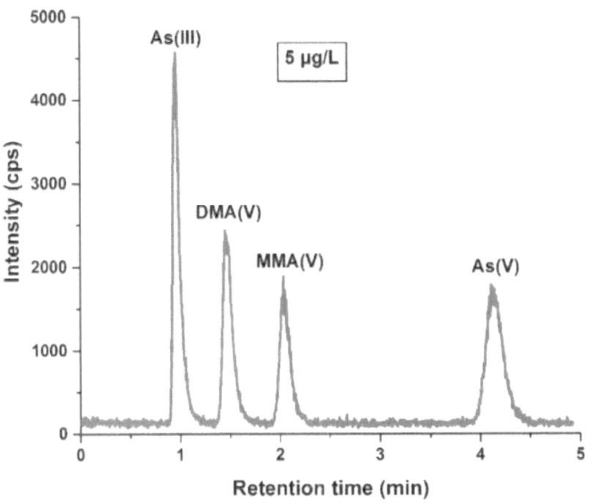

Figure 4. Chromatogram of a mixture of As species at concentrations of 5 µg/L with the optimized UHPLC (flow rate – 2.5 mL/min) and mobile phase conditions.

4 CONCLUSIONS

Results demonstrated that the mixed mobile phase containing 8.5 mM of $NH_4H_2PO_4$ and NH_4NO_3 at pH 6.0 (M2) is the best eluent for the separation of the targeted As species. Optimized chromatographic conditions in UHPLC achieved the separation of targeted As species with an excellent resolution within 5 min which is a double saving in analysis time per sample compared to the existing HPLC-ICP-MS approaches (9-15 min).

REFERENCES

Chen Z., Akter K.F., Rahman M.M. & Naidu R. 2008. The separation of arsenic species in soils and plant tissues by anion-exchange chromatography with inductively coupled mass spectrometry using various mobile phases. *Microchem. J.* 89(1): 20–28.

Li P., Pan Y., Fang Y., Du M., Pei F., Shen F., Xu B. & Hu Q. 2019. Concentrations and health risks of inorganic arsenic and methylmercury in shellfish from typical coastal cities in China: a simultaneous analytical method study. *Food Chem.* 278: 587–592.

Rahman M.M., Chen Z. & Naidu R. 2009. Extraction of arsenic species in soils using microwave-assisted extraction detected by ion chromatography coupled to inductively coupled plasma mass spectrometry. *Environ. Geochem. Health* 31(1): 93–102.

Arsenic in the Environment: Bridging Science to Practice for Sustainable Development –
van der Wal, Ahmad, Petrusevski, Weijma, Savic, van der Wens, Beerendonk, Bhattacharya, Bundschuh & Naidu (Eds)
© 2023 the Author(s), ISBN 978-1-032-32928-4

As(V) determination using bienzimatic biosensor AuNPs-screen-printed electrode

C. Núñez[1], V. Arancibia[1], V. Serafín[2], L. Agui[2] & P. Yáñez-Sedeño[2]
[1]*Chemistry Faculty, Pontificia Universidad Católica de Chile, Santiago, Chile*
[2]*Departamento de Química Analítica, Facultad de CC. Químicas, Universidad Complutense de Madrid, Madrid, Spain*

ABSTRACT: In this work a biosensor developed using AuNPs-SPCE by amperometry for determination of As(V) based on the enzymatic activity of two enzymes, AcP and Tyr. The functioning principle is the decreased sustrate amperometric signal by the inhibits enzymatic activity of AcP after the addition of As(V). On this way is possible to determined arsenic concentration. The sensitivity of the analytical method whit incorporation of AuNPs was 0.01 μA μmol/L (RSD = 5%) with a detection limit of 1.23 μmol/L. The linear response was in the range 1.0 to 20.0 μmol/L using buffer acetate 0.10 mol/L pH 6.0.

1 INTRODUCTION

Arsenic contamination has been identified as a public health problem worldwide because it has serious toxic effects even at low exposure levels. Natural processes and anthropogenic activities are triggers of arsenic emission and that may be mobilized in the surface. Intense efforts are devoted to developed methods for the detection of arsenic species and mitigation of effect depends on chemical forms and oxidation states. Electroanalytical techniques have advantage above the traditional techniques, as spectrometry and chromatography, due to its sensitivity, low cost and direct applicability to determine As(III) and As(V) without previous separation. Usually As(III) is determinate by electrochemical techniques due to As(V) is electrochemically inert, therefore its determination is performanced by reduction to As(III) form.

Cosnier *et al.* (2006) developed an amperometric bio-sensor based on simultaneous entrapment of acid phosphatase (AcP) and polyphenol oxidase (PPO) into anionic clays. The biosensor performed was based on the inhibition of acid phosphatase (AcP) caused for As(V) addition.

For this work, was used the principle described for Cosnier *et al.* Included different modified electrodes with aim of designed a electrode surface and analytic method able to determined As(V) in complex matrices.

2 MATERIALS AND METHODS

2.1 *Apparatus and equipment*

Amperometric measurements were made with a BAS Epsilon with a modified screen-printed electrode. The reference and auxiliary electrodes were an integrated system included in electrode Solutions were stirred during accumulation steps. pH was measured with an Orion model 430 pH meter.

2.2 *Amperometric method*

Measurements were performed in 10 mL of acetate buffer 0.1 M pH 6.0. The biosensor was performed incubating each enzyme during 15 minutes to room temperature.

The calibration curves were obtained and linear regression and detection limits were calculated. The detection limit (DL) was calculated from $y_{DL} = a + 3\sigma_{x/y}$ and $y_{DL} = a + bx_{DL}$, where a is the intercept, $\sigma_{x/y}$ is the random error in x and y, and b is the slope.

3 RESULTS AND DISCUSSION

3.1 *Operational parameters optimization*

Gold nanoparticles (AuNPs) due to their high electrical conductivity and surface area have been employed as modifier screen-printed electrode also improved the electronic transfer in enzymatic activity. On the other hand, electronic conductive polymers like nPEDOT (nanoparticles Poly(3,4-ethylenedioxythiophene)) have proven useful in the development of electrochemical biosensors because they provide a suitable means for immobilization of biomolecules while improving the speed of electronic transfer.

In order to attain high a sensitivity and precision of the method, was studied different modified electrode platform. The Figure 1 resume the electrodes modified. The electrode surface was characterized

DOI: 10.1201/9781003317395-67

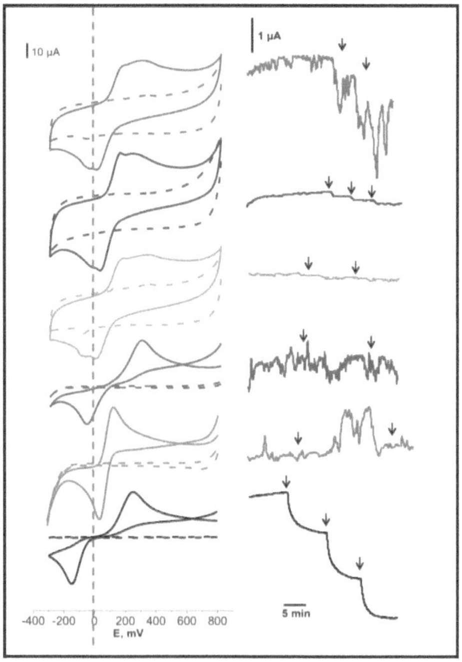

Figure 1. Cyclic voltammetry and amperometric measurements carried out for the study of different electrochemical platform. Conditions CV: HQ = 50,00 mmol L^{-1} en CH_3COOH = 0,10 mol L^{-1}; pH = 6,00; v = 0,05 V. Amperometrics conditions: 1-NP = 10,00 µmol L^{-1}; CH_3COOH = 0,10 mol L^{-1}; pH = 6,00; E = −0,20V. © (■)nPEDOT/AuNPs-SPCE; (■)GA/nPEDOT/SPCE; (■) nPEDOT/SPCE; (■)AuNPs (commercial)/SPCE; (■) AuNPs(synthesized)/SPCE (■) AuNPs-SPCE.

by means of cyclic voltammetry and amperometric detection showed that a significant difference in the platform used, indicated that the AuNPs-SPCE was the more biocomplatible surface.

The major parameters such as pH buffer and potential reduction were optimized. The results show that the As(V) signal decreased when the pH buffer is less ad greater than 6.00. The potential reduction was studied in the range −0.25 to 0.05 V were optimal potential was −0.20 V.

3.2 *Study of the inhibition of the enzymatic reaction by modifications of As(V) and validation method.*

With the optimized experimental variables, the detection of As(V) was carried out based on the decrease of the current intensity in the steady state corresponding to the reduction of quinone to catechol.

The results obtained are shown in Figure 2. It can be seen the decreased substrate signal after the addition of As(V). This allows a better detection of

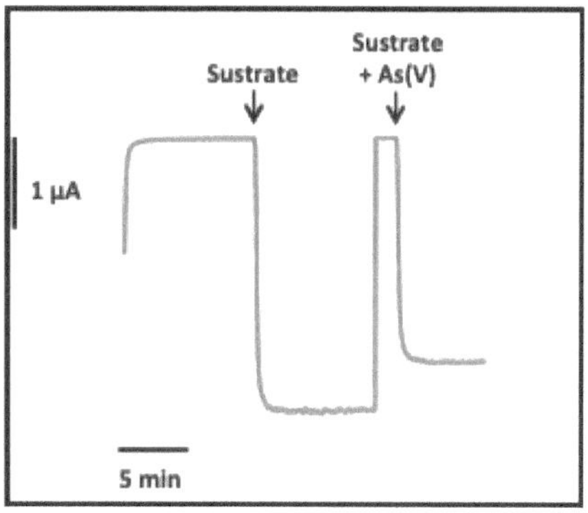

Figure 2. Inhibition of signal of sustrate after the addition of As(V).

the current variation related to the concentration of As (V).

Under the optimum analytical conditions calibration was obtained. The decreased in the signal current is proportional to As(V) concentration from 1.0 to 20.0 µmol/L, the DL was found to be 1.23 µmol/L. The correlation coefficient and slope were 0.9993 and 0.011 µA µmol/L. On the other hand, stability studies was carried out showed that is possible to realized ten consecutive measure of As(V) without decrease steady-state current and inhibition was completely reversible.

4 CONCLUSIONS AND RECOMMENDATIONS

Using the association of AcP and PPO for the specific amperometric determination of As(V) descri-bed for Cosnier *et al.* (2006), it was possible to modify a gold nanoparticles screen-printed electrode and develop a sensitive and stable method capable to determined arsenic. More studies are required to determine the application of the method in complex matrices. The positive results open the possibility of developments in this field.

ACKNOWLEDGEMENTS

Financial support by FONDECYT under Regular Project 1171189, CONICYT through a doctoral Scholarship N°21130492 and *Tesis en el sector productivo* N°7815120004.

REFERENCE

Cosnier S., Mousty C., Cui X., Yang X. & Dong S. 2006. Specific determination of As(V) by an acid phosphatase-polyphenol oxidase biosensor. *Anal. Chem.* 78(14): 4985–4989.

Arsenic in the Environment: Bridging Science to Practice for Sustainable Development –
van der Wal, Ahmad, Petrusevski, Weijma, Savic, van der Wens, Beerendonk, Bhattacharya, Bundschuh & Naidu (Eds)
 ISBN 978-1-032-32928-4

Arsenic determination in urine samples by anodic stripping voltammetry using different electrodes

J. José Triviño[1], C. Nuñez[2] & V. Arancibia[2]
[1]*Facultad de Ciencias Biológicas, Pontificia Universidad Católica de Chile, Santiago, Chile*
[2]*Facultad de Química y de Farmacia, Pontificia Universidad Católica de Chile, Santiago, Chile*

ABSTRACT: Four methods for the determination of As(III) in urine were developed, validated and applied to real samples using modified screen printed electrodes: Graphene and gold nanoparticles (GH-SPCE; AuNPs-SPCE); and gold modified Boron Doped Diamond (BDDE) and Edge Plane Pyrolytic Graphite (PGE). To obtain sensitive and selective methods, the effects of various parameters such as pH, accumulation potential, and time (Eacc, tacc) were optimized in the presence and absence of urine. The determination of As, in all four cases, was performed by ASV and the following LDs were obtained: 0.28 (GH-SPCE); 42.0 (AuNPs-SPCE / urine) µg/L. Whereas with BDDE and PGE LDs were 2.3 and 13.2 µg/L. Methodologies with GH-SPCE and AuNPs-SPCE/urine electrodes were applied to determine As(III) and As(T) in human urine samples from workers at the arsenic abatement plant. While BDDE and PGE were used to determine the concentration of As(III) in doped urine and water samples, because its reproducibility was low.

1 INTRODUCTION

Arsenic is one of the most toxic elements found in nature, and it constitutes one of the main concerns in relation to human health. This one can be found in drinking water, in the air as volatile arsines, and in soil. Arsenic enrichment in water in the Antofagasta Region of Chile is well known. Here the streams are characterized by a high As content (100–1000 mg/L), mainly associated with the quaternary volcanic and present-day geothermal activity in the Andes mountains. In 1962 arsenic enrichment in drinking water triggered in the Second Region of Chile an intake of this metalloid at concentrations greater than 500 µg/L. It has been estimated that 7% of all deaths occurring in Antofagasta between 1989 and 1993 were due to past exposure to arsenic in drinking water. The quality of water from the Loa River is poor due to high salinity and high dissolved arsenic and boron, because its main tributary (El Salado) has its origin in the geothermal field of El Tatio. Different analytical methods are used to determine trace levels of arsenic in a variety of samples. Electroanalytical techniques are adequate because of their sensitivity and low cost of instrumentation and supplies and they can be used to determine As(III) and As(V) without prior separation. Another advantage is the portability of the instrumentation and its ability to perform real time analysis.

In this work electroanalytical methodology was optimized to determine arsenic and then apply it in the determination of this analyte in environmental and/or biological samples. For this, several electrodes were used: hanging drop mercury electrode (HMDE), carbon paste electrodes with and without modifying (CPE, MCPE) and screen-printed electrodes of carbon, graphene and with carbon and gold nanotubes. These studies were performed in the presence and absence of ligands.

2 MATERIALS AND METHODS

2.1 *Reagents and materials*

All solutions were prepared in deionized water (Merck). Standard solutions of arsenic were prepared by diluting a stock solution of 1000 mg L^{-1} As_2O_3 (Fluka). Acetic acid, phosphoric acid, boric acid, nitric acid, hydrochloric acid, and sodium hydroxide were suprapur grade from Merck.

2.2 *Instrumentation and electrodes*

Screen–printed carbon electrodes modified with graphene and gold nanotubes were purchased from DropSens. Stripping voltammograms were obtained using a CHI 852E. Electrochemical Analyzer (CHI Instruments, Austin, TX), with CHI–852D software, using a two–electrode arrangement consisting of SPCE as working electrode, with a combined ORP (HANNA) reference/auxiliary electrode. Au-BDD and Au-PGE electrodes were immersed into $HAuCl_4$ solution and deposition was conducted at the optimized potential for a fixed time. The modified electrode was prepared *in situ* and *ex situ.*

2.3 *General voltammetric procedure*

In the electrochemical cell (10.0 mL) containing 0.01 mol/L. Britton Robinson buffer, the working electrodes together with the combined reference/

 DOI: 10.1201/9781003317395-68

auxiliary electrode were introduced. Arsenic was accumulated to adequate Eacc during for tacc with constant stirring. Then, after a 10 s equilibration time, stripping voltammograms (ASV) were recorded. After choosing the optimal parameters, calibration curves were made to obtain linear ranges, detection limits, and sensitivity. The standard addition method was used for the validation and analysis of samples. The DL was calculated using the approximation of Miller and Miller for calibration curves with yDL = a + 3sx/y and yDL = a + bxDL, where a is the intercept, sx/y is standard deviation of the calibration curve and b is the slope.

3 RESULTS AND DISCUSSION

Preliminary experiments were carried out to identify the general features that characterize the behavior of the As in different electrodes. At the beginning of the sweep to negative potentials the deposit occurs and in the reverse reaction the respective oxidation. To find the optimum experimental conditions, the effect of chemical and instrumental factors on the intensity and shape of the stripping peak current of As was studied; the results are reported in the following section. On the other hand, for gold modified Boron Doped Diamond (BDDE) and Edge Plane Pyrolytic Graphite (PGE) the study in function of gold in solution was carried out previously.

3.1 *Effect of pH*

To perform the study based on pH, 10 mL of Britton Robinson buffer 0.1 mol/L (or H_3PO_4, HCl, HNO_3 solution) was added to the electrochemical cell. the anodic voltamperogram was plotted. Subsequently, aliquots of NaOH solution 30% were added, measuring the pH after each addition and plotting the respective voltamperograms. With all the electrodes, the optimum pH was between 1.0 – 3.0.

3.2 *Effect of Eacc and tacc*

The effect of Eacc was studied in all potential range, positive and negative with the scan to positive and negative potential too using a solution of As(III) 5.0 to 10.0 μg/L. The signal of As in all the cases was about −0.2 to + 0.2 V. On the other hand, the effect of deposition time on the peak current of As(III) and As(V) was studied. The peak current of As with all the electrodes increased with increasing deposition time from 00 to 120 s and then decreased sharply for higher times. For the stability and reproducibility of electrodes short times were used.

3.3 *Calibration curves and analysis of samples*

In the best conditions for the all electrodes calibration curves were obtained. Figure 1 shows the plots obtained with the different electrodes. LDs obtained were: 0.28 (GH-SPCE); 42.0 (AuNPs-SPCE/urine) μg/L. Whereas with BDDE and PGE LDs were 2.3 and 13.2 μg/L. The method was

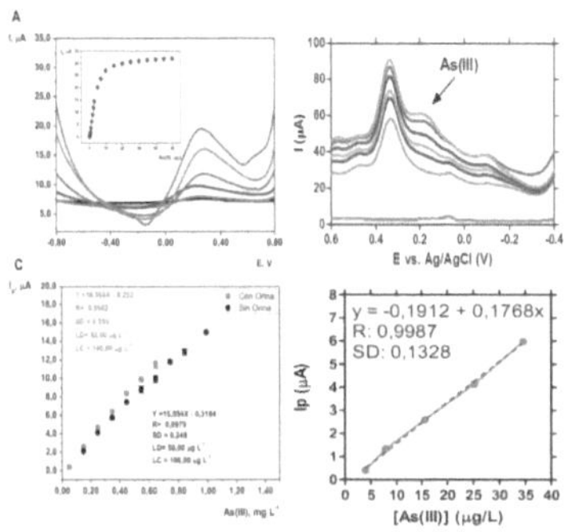

Figure 1. (A) Voltammograms obtained with GH-SPCE. (B) Voltammograms of urine sample obtained with Au-BDDE. (C) Calibration curve obtained with AuNPs-SPCE. (D) Calibration curve of a urine sample obtained with Au-BDDE.

validated with spiked urine, spiked tap water and reference material TMDA-62.2. Finally were applied to determination of As(III) and As_{total} in human urine samples from workers at the mineral industry. Others samples were Loa river and Tap water. The obtained results were compared with measurement made by an external analytical laboratory determined arsenic by ICP–MS.

4 CONCLUSIONS

According to the results obtained, GH-SPCE electrode was the most suitable for the determination of As in the urine of people contaminated with different levels without the need for prior treatment. The AuNPs-SPCE electrode has the advantage of its direct use as a sensor, however the detection limit is 42.0 therefore it is only suitable for urine samples with high levels (found in our analysis). The Au-BDDE electrode is also suitable for determining As in urine samples, however it is necessary to previously perform a digestion with acid and peroxide. The values of As obtained with urine samples cannot be published.

ACKNOWLEDGEMENTS

Financial support by FONDECYT Regular Project 1171189 is gratefully acknowledged.

REFERENCES

Núñez C., Arancibia V. & Triviño J.J. 2018. A new strategy for the modification of a carbon paste electrode with carrageenan hydrogel for a sensitive and selective determination of arsenic in natural waters. *Talanta* 187: 259–264.

Nuñez C., Arancibia V. & Gómez M. 2016. Determination of arsenic in the presence of copper by adsorptive stripping voltammetry using pyrrolidine dithiocarbamate or diethyl dithiophosphate as chelating-adsorbent agents. *Effect of CPB on the Sensitivity of the Method Microchem. Jour.* 126: 70–75.

Arsenic in the Environment: Bridging Science to Practice for Sustainable Development –
van der Wal, Ahmad, Petrusevski, Weijma, Savic, van der Wens, Beerendonk, Bhattacharya, Bundschuh & Naidu (Eds)
© 2023 the Author(s), ISBN 978-1-032-32928-4

Novel biosensor field kit for trace arsenic analysis in field samples

E. Hicks, M. McDonald, C. Dalton, D. Silver & R.M. Mayall
FREDsense Technologies Corp., Calgary, Alberta, Canada

ABSTRACT: Rapid on-site measurements of arsenic (As) are essential for the timely remediation of As-contaminated groundwater for both municipal and emergency response applications. Current field tests either suffer from complicated end-user instructions (e.g., arsine gas generation tests) or a lack of accuracy and specificity (e.g., colourimetric strip tests). The FRED-As system combines genetically modified bacteria that relieve the burden of the end-user with an electrochemical measurement system that provides enhanced accuracy and signal intensity compared to traditional bacterial detection approaches. This system was tested against commercial field kits and gold-standard ICP-MS performed at an off-site third-party analytical laboratory. The FRED-As tests demonstrated excellent specificity and sensitivity down to <2 ppb As under lab conditions, as well as outperforming the commercial field tests. The FRED-As system was then validated in field trials with municipal and emergency response partners, providing on-site data within ±2 ppb of the off-site analytical laboratory.

1 INTRODUCTION

Despite the high toxicity of arsenic (As) at low levels, measuring it accurately in the field remains a challenge. The WHO has set a limit for arsenic at 10 parts per billion (ppb) for As in drinking water. This low limit makes field analysis and timely data challenging, as the resultant tests need to be extremely sensitive. Currently, laboratory analysis is the primary method of choice for measurement of As, where inductively coupled plasma mass spectrometry (ICP-MS) is typically used. While some field kits are commercially available, they generally suffer from a trade-off of either a lack of usability or a lack of accurate measurements.

Herein, we demonstrate a novel method for the field analysis of arsenic. Coupling a genetically modified bacterial strain with screen-printed electrodes, we have developed the FRED-As system, a rugged bio-sensor capable of field operation. The sample can either detect As(III) directly or total dissolved As via a simple reduction reaction (Zhang *et al.*, 2004). Water samples are injected along with a small volume of the bacterial strain into a plastic cartridge. This is then incubated for one hour in a battery-operated device before analysis is performed. The FRED-As system was experimentally validated under laboratory conditions before being transitioned to field trials with partner organizations.

2 MATERIALS AND METHODS

2.1 *Determination of sensitivity under laboratory conditions*

Samples were prepared using As(III) ICP-MS analytical standard (Sigma-Aldrich) diluted in ultrapure water. 100 μL each of known concentrations of As(III) were mixed with 100 μL of an electrochemical substrate and 50 μL of the biosensor strain. Samples were prepared in triplicate and incubated at 37°C for one hour. Cyclic voltammetry (CV) was used to measure the output of the bacterial strain. The resultant electrochemical signal was taken as an indirect measurement and correlated to the amount of arsenic present in the sample.

2.2 *Determination of sensitivity under laboratory conditions*

Specificity was determined by comparing signals obtained from As(III) to signals obtained from other metals and inorganic compounds. 100 ppb standards were freshly prepared for each sample in ultrapure water. Similar to the sensitivity tests, 100 μL of each sample was mixed with 50 μL of the biosensor strain and 100 μL of the electrochemical reaction mixture in triplicate. Samples of 0, 5 and 10 ppb As(III) were also tested to serve as comparisons. All samples were incubated at 37°C for 1 hour prior to measurement via CV.

2.3 *Field pilot on drinking water samples*

Samples were collected from a municipal water system, both before and after arsenic remediation, as well as directly from various wells at a contaminated site in Bemidji Minnesota from a historical oil spill. A previously published As(V) reduction procedure (Zhang *et al.*, 2004) was used to measure total As in the samples (As(V) + As(III)) as well as As(III) individually. 800 μL of each sample was injected into the middle chamber of a FRED-As cartridge. 800 μL of ultrapure water was injected into the first control chamber (i.e., 0 ppb As) while a known 10 ppb sample prepared from an analytical standard was injected into the second control chamber to calibrate each sample. 200 μL of the biosensor strain was also injected into each cartridge chamber. Cartridges were placed into a FRED unit that performs a 1-hour incubation prior to measurement via CV. Samples were run in duplicate.

DOI: 10.1201/9781003317395-69

3 RESULTS AND DISCUSSION

3.1 *Initial validation of the FRED-As system*

Initial tests have shown that the FRED-Arsenic test is both sensitive and specific. Sensitivity tests showed clear detection down to 2 ppb As(III), substantially below the WHO limit of 10 ppb (Figure 1A). The test showed a linearity to 50 ppb (Figure 1B), after which a plateau was observed. The error associated with the As(III) measurements was low ($< \pm 2$ ppb) for As(III) concentrations below 20 ppb, but increased substantially near the plateau of 50 ppb. This could be due to the inherent toxicity of As(III) to the bacterial cells, or an artefact of measuring signals approaching the saturation point of the system of the bacterial strain. The resultant electrochemical signal was taken as an indirect measurement and correlated to the amount of arsenic present in the sample.

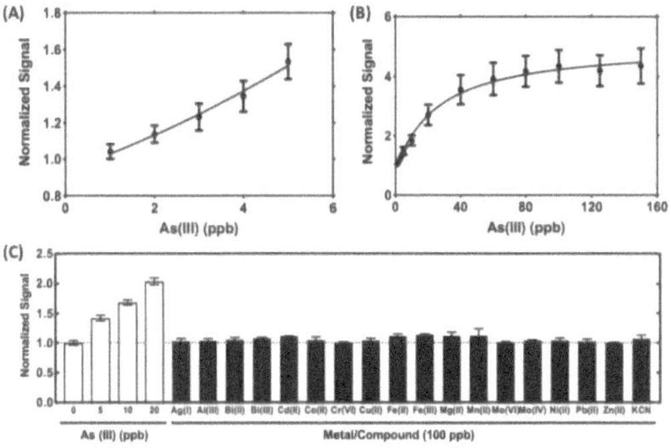

Figure 1. (A) limit of detection for the FRED-As system. (B) Dynamic range of the FRED-As system. (C) Specificity of the FRED-As system towards 100 ppb interferents.

The specificity of the system was tested by exposing the biosensor strain to 100 ppb samples of various metals and inorganic compounds. The reported signal from the FRED-As biosensor strain for these possible interferents was quite low, typically within 10% of the normalized signal for ultrapure water (Figure 1C). This is in comparison to the ca. 40% increase seen for 5 ppb As(III). When compared to commercially available field tests, results were quite similar at low (<5 ppb) levels of As(III). However, differences were observed at intermediate levels of As(III). When samples of 20 ppb As(III) in ultrapure water were tested, the FRED-As system reported a value of ca. 21 ppb, while a commercially available arsine gas test reported ca. 7 ppb and a colourimetric strip test reported a concentration between 10–30 ppb of As.

3.2 *Field trails of FRED-As system and comparison to off-site analytical laboratory results*

Field trials were then conducted with partner organizations who coordinated independent analytical laboratory analysis of the samples. Figure 2 shows the comparison for the FRED-As detection of As(III) and the total As content (after a As(V)→ (III) reaction) compared to the third-party results. While differences were observed at extreme concentrations of As, the data was comparable for low and intermediate levels (i.e., < 40 ppb). Similar results were obtained in municipal drinking water treatment plant samples and environmental samples, providing confidence in the lack of matrix effects on the validity of these results.

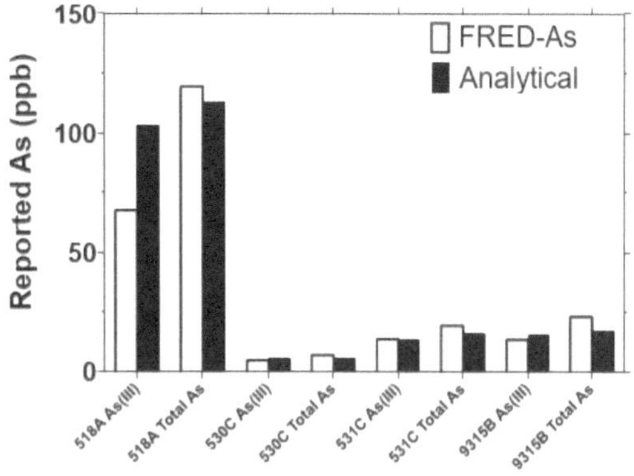

Figure 2. Comparison of the FRED-As system versus third-party ICP-MS analysis of arsenic levels in field samples.

4 CONCLUSION AND DISCUSSION

FRED-As represents a novel method for measuring trace arsenic in the field. Its sensitivity, specificity and ease of use together make it a powerful tool for water operators, emergency responders, communities and even consumers. Future work will focus on further refining the FRED-As system and consolidating further field trials on diverse water matrices.

ACKNOWLEDGEMENTS

We would like to thank our municipal pilot partners for site access and access to real world water samples. We would also like to acknowledge Dr. Cozzarelli for coordinating third-party laboratory analysis.

REFERENCE

Zhang Q., Minami H., Inoue S. and Atsuya I. 2004. Differential determination of trace amounts of arsenic(III) and arsenic(V) in seawater by solid sampling atomic absorption spectrometry after preconcentration by coprecipitation with a nickel–pyrrolidine dithiocarbamate complex. *Anal. Chim. Acta* 508(1): 99–105.

Section 2: Arsenic in agriculture and food production

2.1 Processes and pathways of arsenic in agricultural ecosystems

Arsenic in the Environment: Bridging Science to Practice for Sustainable Development –
van der Wal, Ahmad, Petrusevski, Weijma, Savic, van der Wens, Beerendonk, Bhattacharya, Bundschuh & Naidu (Eds)
© 2023 the Author(s), ISBN 978-1-032-32928-4

Using machine learning to map arsenic in rice in Bangladesh

J. Podgorski[1,2], S. Islam[1,3] & D.A. Polya[1]
[1]*Department of Earth & Environmental Sciences, The University of Manchester, Manchester, UK*
[2]*Eawag, Swiss Federal Institute of Aquatic Science and Technology, Department Water Resources and Drinking Water, Dübendorf, Switzerland*
[3]*Department of Soil Science, Bangladesh Agricultural University, Mymensingh, Bangladesh*

ABSTRACT: Groundwater arsenic contamination is a public health disaster in many areas of Bangladesh. Elevated concentrations of arsenic (largely as inorganic arsenic) have also been reported in rice with concentrations variably above guideline values. Consumption of rice is therefore one of the major exposure routes of arsenic to humans. We have used random forest modelling to identify areas of high arsenic-in-rice hazard. Datasets of spatially continuous environmental parameters, including climate, soil and modelled groundwater arsenic along with six crop-specific *in-situ* variables, were used to map the probability of total arsenic in rice exceeding 120 μg/kg. Results shows a 4 ± 3% greater probability of high arsenic rice with shallow groundwater irrigation. Greater probabilities of high rice arsenic are also evident in High Yielding Varieties in areas with high groundwater arsenic. The models may be used to inform land and resource management strategies to reduce human exposure to arsenic.

1 INTRODUCTION

Geogenic arsenic contamination of groundwater in Bangladesh is a public health disaster due to the drinking of this arsenic contaminated water. Consumption of inorganic arsenic containing rice is another channel of exposure. Rice is grown as a lowland crop in flooded paddy soils under reducing conditions, which greatly enhances the bioavailability of arsenic to rice. Grain arsenic concentrations are also impacted by the combined influences of soil characteristics, environmental conditions, and management strategies. Arsenic contents in rice vary widely, with most reported concentrations in the range 20 to 900 μg/kg (Islam *et al.*, 2017). In Bangladesh, *boro* (winter) rice may be of greater concern than *aman* (monsoon) rice given that approximately 70% of the area under cultivation with *boro* rice is irrigated with groundwater (BBS 2018) with expected arsenic concentrations similar to those reported by BGS/DPHE (2001). Duxbury *et al.* (2003) observed that *boro* rice had a significantly greater arsenic concentration than rain-fed *aman* rice and that the arsenic concentrations corresponded 'reasonably well' with the map of arsenic in drinking water wells. This is consistent with the strong positive correlation observed in greenhouse studies between arsenic in rice and arsenic in soil water (Abedin *et al.*, 2002) and the accumulation of arsenic in paddy field soils arising from irrigation with high arsenic water. Notwithstanding that, other multiple interacting environmental and chemical factors affect arsenic availability in soil and its uptake by plants (Duxbury *et al.*, 2003). The present study uses a machine learning, specifically random forest, approach to provide a hazard assessment of arsenic contamination of rice grown in Bangladesh.

2 MATERIALS AND METHODS

2.1 *Data collection and preparation*

Model groundwater arsenic data were obtained from Podgorski & Berg (2020). Rice arsenic data were obtained from the Islam *et al.* (2017) study, in which 965 rice samples were collected from a range of geographic locations through household surveys from 73 upazila's in 20 districts selected to represent severely, moderately and less affected arsenic in drinking water areas and to covering most of the regions of Bangladesh except hilly areas. The collected rice samples were grown in both the dry (*boro* Rice) and wet seasons (*aman* rice) and include most local, high yielding variety (HYV) and hybrid varieties. The irrigation database compiles crop-specific data on the area irrigated by different sources of water (e.g. groundwater, surface water) in each upazila.

2.2 *Modeling*

We utilized existing country-wide datasets of soil and climate parameters as well as groundwater arsenic hazard and six crop-specific parameters to model the probability of the total concentration of arsenic in rice exceeding120 μg/kg. The model was trained on 965 samples from 73 upazilas of Bangladesh. Arsenic concentrations from the same location were averaged, resulting in a 75–90% reduction in the number of unique data rows available for modelling. The random forest's out-of-bag (OOB) error statistic was used to assess a model's performance.

3 RESULTS AND DISCUSSION

Figure 1 displays the arsenic in rice hazard map (120 μg/kg level, OOB error 27.8%) of Bangladesh based

DOI: 10.1201/9781003317395-70

on just geospatial parameters (i.e. no crop-specific variables). The distribution of high arsenic hazard, particularly in the southwest, matches well with the distribution of high groundwater arsenic. Exceptions in the south-east region have a low grain arsenic concentration but the highest groundwater arsenic: this may be due to the surface water source of irrigation water used in this south-east region.

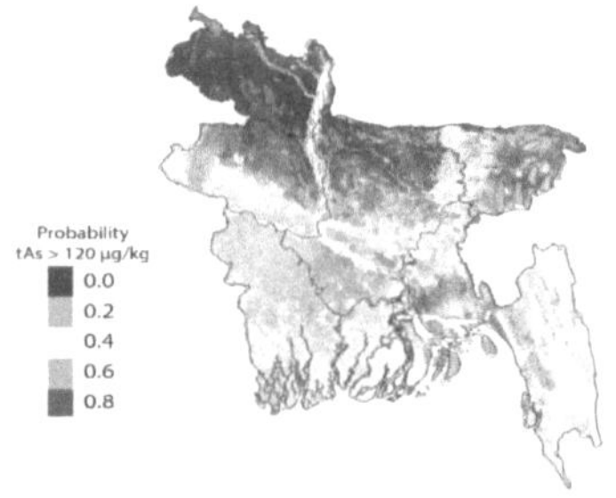

Figure 1. Model probability of high (> 120 µg/kg) total arsenic in Bangladesh grown rice based only on geospatial variables.

Irrigation by groundwater (with typically more arsenic than surface or rainwater) is associated with higher arsenic-in-rice hazard (Figure 2).

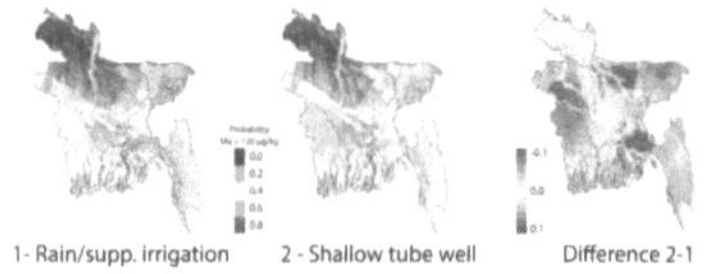

Figure 2. Model probability of high (> 120 µg/kg) total arsenic in Bangladesh grown rice based on geospatial variables and irrigation source. Model 4 ± 3% greater probability of high arsenic rice is found in areas with shallow groundwater irrigation.

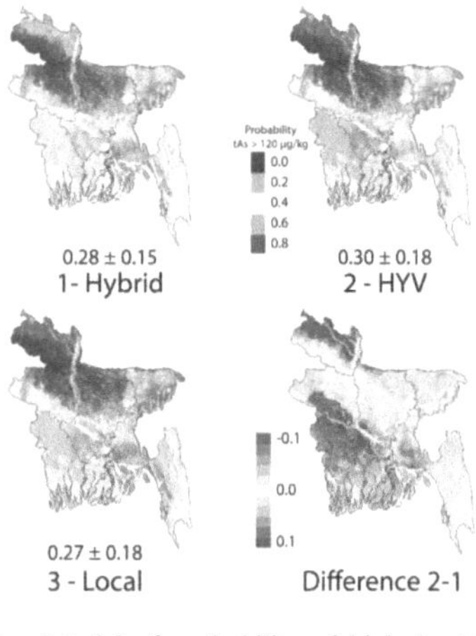

Figure 3. Model of probability of high (> 120 µg/kg) total arsenic in Bangladesh grown rice based on geospatial, irrigation source and rice type variables. HYV is a high yielding variety.

Poor model fits (data not shown) were obtained when only crop-specific variables were used in the modelling. The lowest OOB error (18.0%) was observed for a model involving all geospatial variables, irrigation source and grain size. Greater probability of high arsenic in type HYV was observed in areas with elevated groundwater arsenic hazard (Figure 3).

Study limitations include (i) the requirement for more extensive and more widely distributed arsenic-in rice data and arsenic speciation data, (ii) potential materially significant degree of incidental or otherwise association between the spatial distributions of rice types, irrigation methods and soil parameters; and (iii) difficulty in modelling the large number of rice varieties with different soil-plant-grain arsenic transfer characteristics.

4 CONCLUSIONS

Geospatial, irrigation source and rice type variables are collectively useful in predicting total arsenic concentrations in Bangladesh grown rice. The modelling suggests the importance of management of paddy soils and rice varieties to limit (inorganic) arsenic accumulation in rice.

ACKNOWLEDGEMENTS

We are grateful to the University of Manchester for a Global Challenge Research Fund Fellowship to SI providing a valuable opportunity to network and building long-lasting relationship with researchers there; and an EPSRC IAA award to DP funding a fellowship to JP. We thank the Computer & GIS Unit of the Bangladesh Agricultural Research Council (BARC) for GIS files for soil mapping.

REFERENCES

Abedin M.J., Cotter-Howells J. & Meharg A.A. 2002. Arsenic uptake and accumulation in rice (Oryza sativa L.) irrigated with contaminated water. *Plant and Soil* 240: 311–319.

BBS 2018. *The Yearbook of Agricultural Statistics of Bangladesh*. Bangladesh Bureau of Statistics, Dhaka, Bangladesh.

BGS/DPHE 2001. *Arsenic Contamination of Groundwater in Bangladesh, Final Report*. In: D.G. Kinniburgh & P.L. Smedley (eds.), British Geological Survey Report WC/00/19.

Duxbury J.M., Mayer A.B., Lauren J.G. & Hassan N. 2003. Food chain aspects of arsenic contamination in Bangladesh: effects on quality and productivity of rice. *J. Environ. Sci. Health A Tox. Hazard. Subst. Environ. Eng.* 38(1): 61–69.

Islam S., Rahman M.M., Islam M.R. & Naidu R. 2017. Geographical variation and age-related dietary exposure to arsenic in rice from Bangladesh. *Sci. Total Environ.* 601–602: 122–131.

Podgorski J.E. & Berg M. 2020, Global threat of arsenic in groundwater. *Science* 368(6493): 845–850

 ISBN 978-1-032-32928-4

Speciation of arsenic in groundwater, sediment and plant leaves in North-western Bangladesh: Implications for arsenic mobilization

A.H.M. Selim Reza[1], C.-C. Brombach[2], H. Fröllje[2] & T. Pichler[2]
[1]*Department of Geology and Mining, University of Rajshahi, Rajshahi, Bangladesh*
[2]*Department of Geosciences, Geochemistry and Hydrogeology, University of Bremen, Bremen, Germany*

ABSTRACT: Arsenic (As) contamination in groundwater of Chapai-Nawabganj district, Bangladesh is of increasing environmental concern to the local community. In this study, the current As contamination status in groundwater, sediment, and plant leaves was investigated. High-performance liquid chromatography (HPLC) coupled with inductively coupled plasma mass spectrometry (ICP–MS) and hydride generation atomic fluorescence spectrometry (HG-AFS) was applied to the speciation of As in groundwater, sediment and plant leaves samples collected from As contaminated area, Chapai-Nawabganj, Bangladesh. Elevated levels of As in the groundwater of up to 1492 μg/L and high concentrations of arsenite As (III) (1009 μg/L) were detected in groundwater. Speciation of As in groundwater samples (n = 20) showed the presence of inorganic As only; arsenite (As(III)) contained 55%–100% of total As and arsenate (As(V)) ranged from 0% to 45%. The relationship between δD and $\delta^{18}O$ of the groundwater samples indicates that the groundwater originate from meteoric water. A sediment profile of As from the study area showed higher As concentrations at depth of 65 ft compared with data from deeper and shallower depth of the same core at Chougachi. An extraction procedure revealed that As could be found in all four phases (weak acid soluble, reducible, oxizable and aqua regia soluble As). The bioavailable As in the sediment was in the range of 25 to 7448 μg/kg. There was a significant correlation between As in groundwater and As in plants. The buildup of As in plants was directly related to the As concentration of tube well waters. Accumulation of As was observed in plant leaves collected from plants in this area. Arsenic concentrations in plant leaves samples ranged from 96 to 589 μg/L. Only a small part of the As (25%) in the plant leaves was extractable with ethanol-water (1:3), and most of this As (95%) was inorganic As (arsenite 5–80 μg/L and arsenate 10 –69 μg/L). Organic As compound like dimethylarsenic acid was detected at trace levels in plant leaves.

1 INTRODUCTION

Arsenic (As) contamination in groundwater aquifers is a global health concern because As is a toxic element and human carcinogen (Ahmed *et al.*, 2004). The toxicity of As largely depends on its chemical species, with its inorganic forms (arsenite and arsenate) being more toxic than the organo-As compounds (e.g., monomethyl arsenate (MMA)) (Rahman *et al.*, 2009).

By investigating the total As content and quantifying the As species in drinking water, sediments and plant leaves, we can further assess the human health risk from intake of water and investigate the impact of contaminated water on plants'As budget. This is critically-important for resolving the emerging issue of As contamination in Chapai Nawabganj. The primary and specific objectives of the present study were to: (i) determine the total As content and species of As in groundwater, sediment and plant leaves; (ii) evaluate the pathways of As in a highly contaminated terrestrial environment.

2 MATERIALS AND METHODS

2.1 *Study area*

The study area is situated in the stable part of Bengal basin which represents a depression resulted from the uplift of the Barind Tract (Figure 1) in the eastern margin.

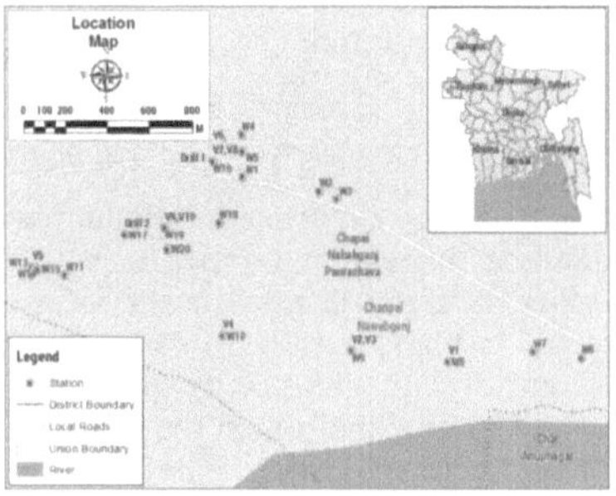

Figure 1. Study area (A) and water sampling locations (B).

2.2 *Analysis of major cations and trace elements*

Major cations and As were analyzed by inductively coupled plasma optical emission spectrometry (ICP-OES) using the PerkinElmer Optima 7300 DV. Calibration was done using single-element calibration standards. Inductively coupled plasma mass spectrometry (ICP-MS) was used for the determination of trace elements in all water samples.

DOI: 10.1201/9781003317395-71

2.3 *Analysis of As species in groundwater*

For speciation analyses, the water samples were filtered in the laboratory with 0.45 μm filter discs. Samples high in As were diluted with MilliQ water to fit into the calibration range of 0 to 50 μg/L for the different As species. The speciation analysis was done via hydride generation atomic fluorescence spectrometry (HG-AFS; PS Analytical, UK) coupled to a HPLC pump (Knauer K-100, Germany) (HPLC-HG-AFS).

2.4 *Extraction of As species from plant leaves*

The plant leaf digests were analyzed by ICP-OES for their major cation concentrations. The international reference material Dorm-2 (National Research Council Canada; dogfish muscle) and sample duplicates were used to control reproducibility of our measurements, which is generally good (deviation <10%) except for Na (~20%). Two procedural blanks were measured accordingly to the samples.

3 RESULTS AND DISCUSSION

3.1 *Arsenic speciation in groundwater samples*

In order to understand the current status of As contamination in the groundwater of Chapai-Nawabganj, water samples collected from an As rich aquifer were analyzed for total As and As species. The results demonstrate that the Chapai Nawabganj region is still highly contaminated with As with total As concentrations up to 1492 μg/L. We observe a large range of As concentrations, with 4 samples being below or close to the WHO limit (samples W7, W11, W14, W15). However, average (median) As concentrations of all samples are 400 μg/L (242 μg/L), and thus significantly higher than the WHO and Bangladesh limit for drinking water.

3.2 *Arsenic and arsenic species in sediments*

Arsenic occurs in the soil through a natural geogenic source and the interaction of the As between the water – sediment phase depends in the local hydrology, geology and land use. The two sediment cores from different drilling sites should give more information on the As binding form in the sediment. Since the depth profile of As in the sediment reflects the variation in As input into groundwater, the sediment core was sectioned every 10 ft (with 5 ft increments at the depth of 60 to 80 ft). At drilling station 1, the concentration of As in the sediment decreases from 7.8 mg/kg at 10 ft to 2.5 mg/kg at 40 ft. The concentration of As increases then again to a maximum concentration of 28.7 mg/kg As at a depth of 65 ft and decreases subsequently with depth to 1 mg/kg As at 120 ft. Drilling station 2 shows the maximum concentration of As at 10 ft depth and decreases with depth to 1.1 mg/kg at 110 ft.

3.3 *Arsenic and arsenic species in plant leaves*

Ten plant leaves samples (bean, mango, arum greens, guava, drumstick, star fruits, mango leaves, pumkin, bean and mango sampled at different places) were investigated. These plant leaves are the most common species in the Chapai-Nawabganj and grown soil watered from groundwater with elevated As concentrations ranging from 96–549 μg/L (Table not shown). Total As concentrations and extraction efficiencies using ethanol– water are summarized (table not shown).

As concentration in the plant leaves was 96–549 μg/L whereas drumstick contained 549 μg/L As in the leaves. Drumstick consisted of 80 μg/L of AsIII and 40 μg/L of AsV in the study area. These results indicate that the plant leaves could uptake different As species. The results showed that the ethanol method was effective for plant leaves.

High concentrations of As were detected in the drumstick plant leaves, indicating uptake of As from their surrounding water in Chapai-Nawabganj.

4 CONCLUSIONS AND RECOMMENDATIONS

The results showed that 85% groundwater samples collected from various sites of Chapai-Nawabganj are unfit for drinking purpose. The As concentration decreases with depth in groundwaters, and is characterized by As(III) as the major As species, possibly due to microbial activity. Arsenic species in groundwaters are about 0–45% As(V), and 55–100% As(III). The methylated species DMA and MMA were not detected in the groundwater.

The bioavailable portion of As in sediments ranges from 25 to 7448 μg/kg. Arsenic is bioavailable by two major pathways throughout study area: 1) easily-exchangeable As from sediments to groundwater, and 2) in groundaters, which may allow for biological uptake by plants and transfer up the food web.

Positive correlation between As in groundwater and plant leaves reveled that plant can uptake As directly from irrigated tube well water.

ACKNOWLEDGEMENTS

We acknowledge German Academic Exchange for supporting this work.

REFERENCES

Ahmed K.M., Bhattacharya P., Hasan M.A., Akhter S.H., Alam S. M.M., Bhuyian M.A.H., Imam M.B., Khan A. A. & Sracek O. 2004. Arsenic enrichment in groundwater of the alluvial aquifers in Bangladesh: an overview. *Appl. Geochem.* 19: 181–200.

Rahman M.M., Ng J.C. & Naidu R. 2009. Chronic exposure of arsenic via drinking water and its adverse health impacts on humans. *Environ. Geochem. Health* 31: 189–200.

Arsenic in the Environment: Bridging Science to Practice for Sustainable Development – van der Wal, Ahmad, Petrusevski, Weijma, Savic, van der Wens, Beerendonk, Bhattacharya, Bundschuh & Naidu (Eds)
© 2023 the Author(s), ISBN 978-1-032-32928-4

Rice seeds (IR64) priming with potassium humate enhances germination and growth under arsenic stressed condition

D. Mridha, A. De, A. Das & T. Roychowdhury
School of Environmental Studies, Jadavpur University, Kolkata, India

ABSTRACT: Rice nursery bed preparation was heavily affected by the arsenic (As) present in the soil and water in West Bengal as majority of the rice field is arsenic affected in exposed areas. Arsenic toxicity reduced the germination percentage and reduced plant vigour, beneficial and nutritional constituents. Humic acid (HA) is already rendered the quality of soil condition and plant growth modulator, which encourage to choose HA as material for priming agent. The priming experiment with HA showed 15% higher germination percentage and reduced As concentration in root (25.36–27.97%) and shoot (79.62–84.81%) compare to unprimed seeds under As stress condition. Arsenic toxicity reduced the overall seedling vigour compare to control and As also greatly affects the root development due to its primarily accumulation largely in root. HA priming also showed good result in root development under different As stress.

1 INTRODUCTION

Rice is a major staple crop for the people living in South-East Asia and a large amount of rice grain is still cultivated in Arsenic (As) contaminated fields. Arsenic enters in the human body through mainly water and food chain. Arsenic, which is a level 1 carcinogen molecule, produces several adverse effects on plant, animal and human. Till date, researchers are working extensively on As mitigation from rice and other crops. Arsenic stress reduces the germination and growth of germinated seedlings. The seed priming is a novel approach to enhance the germination and growth of plant seedling under different level of abiotic stress. Humic acid (HA) helps to increase seed germination but its role in As stress condition is not evaluated. HA application increases the germination percentage (Tejada & Gonzalez 2003) and promotes the plant health in oxidative stress condition. This study aims to unveil: 1) the potential of HA as a priming material for rice, 2) to study the effect of HA in seed germination under different inorganic As (iAs) stressed condition.

2 MATERIALS AND METHODS

2.1 *Experimental design and seed priming*

IR 64 rice variety was selected for this experiment. The seeds were first sterilized with 70% alcohol for two times and then washed thoroughly with double distilled water. The priming was done with 10 ppm solution of potassium humate for 24 hours in dark. After priming was done, the seeds were taken out from the priming solution and dried properly before sowing. Arsenic stress was applied with two different doses of sodium arsenate (Na_2AsO_4) solution (6 and 8 ppm). The experimental design was consisted of control; 6 ppm Na_2AsO_4 + unprimed seeds; 8 ppm Na_2AsO_4 + unprimed seeds; 6 ppm Na_2AsO_4 + primed seeds and 8 ppm Na_2AsO_4 + primed seeds.

2.2 *Final germination percentage*

All the petri plates with soaked tissue paper in double distilled water followed by two arsenic dose variant were kept in dark inside the incubator at $24 \pm 0.5°C$ with 70% humidity for germination. The final germination was calculated after 3 days from sowing. The seed was counted as germinated when the radical was $\geq$ 0.05 cm.

2.3 *Determination of physiological growth parameters and vigour*

The 10 days seedlings were collected from each treatment sets and cleaned with 70% ethanol followed by thorough washing with double distilled water and surface dried properly before taking the weight of seedling shoot, root length and seedling fresh and finally seedling vigour was calculated (Abdul-Baki & Anderson 1973).

2.4 *Determination of total arsenic*

0.1 g oven dried root and shoot tissues were digested for overnight with HNO_3 and H_2O_2 in 2:1 volumetric ratio. Total arsenic was quantified according to the prescribed protocol by Roychowdhury *et al.*, 2002 using flow injection hydride generation atomic absorption spectrometry (FI-HG-AAS) method.

3 RESULTS AND DISCUSSION

3.1 *Effect of As and Humic acid on seed germination and plant vigour*

Germination of the rice seeds was primarily inhibited by As. The priming of seeds with HA shows a promising

DOI: 10.1201/9781003317395-72

result in terms germination under As stressed condition. 100% germination (Figure 1) was found in control after 3 days of sowing, compared to 50% and 60% of germination in 8 ppm and 6 ppm As stressed condition, respectively. Though the germination was 65% and 75%, respectively for HA primed seeds. HA primed seeds showed a yield of 15% in germination than unprimed seeds under different As stress. The germination percentage of HA primed seeds was increased due to its growth modulation and metal chelation property (Tsang *et al.*, 2014). The seedling vigour for primed seeds was increased up to 1.27–1.32 folds than unprimed seeds under As stress condition.

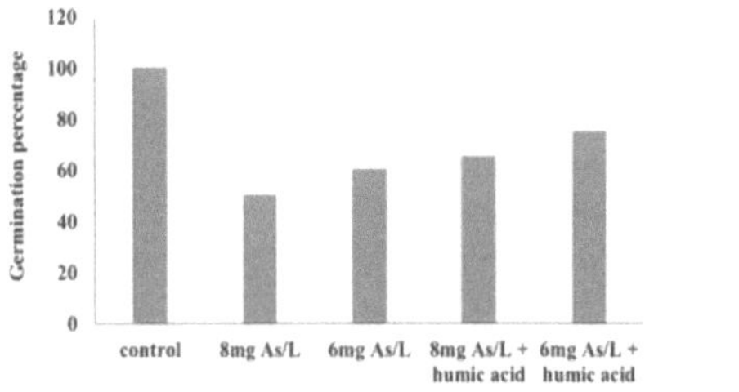

Figure 1. Germination percentage for primed and unprimed seeds under different As stress after 10DAS.

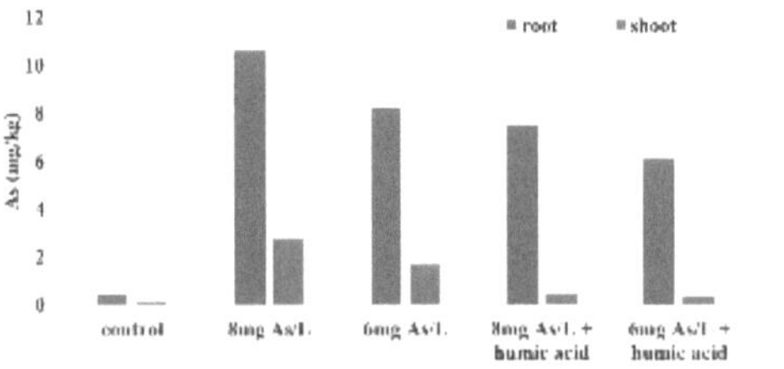

Figure 2. Arsenic content in root and shoot part of 10 DAS seedlings under different treatments.

3.2 *Arsenic concentration in root and shoot part of the seeding*

Control set without any arsenic stress showed less As content in root (0.45 mg/kg) and shoot (0.14 mg/kg) than the rest of the treatments. The unprimed seeds germinated under various doses of As, accumulated greater amount of As than the primed seeds (Figure 2). Unprimed seeds germinated under 8 mg/L As stress had 10.56 mg/kg and 2.7 mg/kg of As in root and shoot, respectively, these numbers were somewhat higher than the root (8.2 mg/kg) and shoot (1.62 mg/kg) for unprimed seeds germinated under 6 mg/L As stress. Apparently, the HA primed seeds under 8 mg/L As stress found to accumulate 7.5 mg/kg and 0.41 mg/kg of As in root and shoot, respectively portion. The similar trends of As accumulation found in HA primed seeds under 6 mg/L As stress whereas reduced amount of As in root (6.12 mg/kg) and shoot (0.33 mg/kg) have been observed for 10DAS seedling possessed. Root had the higher accumulation than the shoot in each treatment, this might be due to direct exposure of root to the medi arsenic solution. The primed seeds also had greater vigour compare to unprimed seeds under As stress (Figure 3).

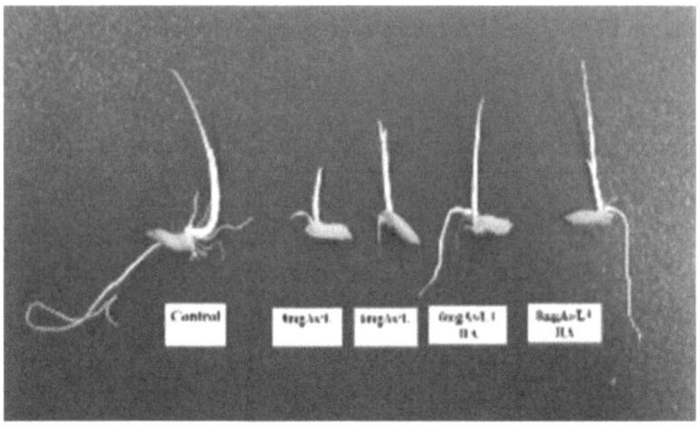

Figure 3. Germinated seeds in different treatments after 5DAS.

4 CONCLUSIONS

This experiment presented the role of HA as promoter of seed germination and plant vigour under As stress condition. The results showed that HA primed seeds were much more resistant to arsenic stress. Though the role of HA under As stress condition till graining or maturity stage was not performed in this experiment. The results also suggested that the chelation activity was highest in root portion which restricted the entry of As in shoot portion. Thus, HA priming should be encouraging to promote as priming agent for rice nursery preparation and cultivation under As contaminated areas in West Bengal and India.

ACKNOWLEDGEMENTS

We acknowledge 'School of Environmental Studies', JU for providing necessary facility for conducting this experiment and UGC for providing fellowship to DM (First Author).

REFERENCES

Abdul-Baki A.A. & Anderson J.D. 1973. Vigour determination in soybean by multiple criteria. *Crop Sci.* 10: 31–34.

Roychowdhury T., Uchino T., Tokunaga H. & Ando M. 2002. Survey of arsenic in food composites from an arsenic-affected area of West Bengal, India. *Food Chem. Toxicol.* 40(11): 1611–1621.

Tejada M. & Gonzalez J.L. 2003. Effects of Foliar Application of a by product of the two-step olive oil mill process on maize yield. *Agronomie* 23: 617–623.

Tsang D.C. & Hartley N.R., 2014. Metal distribution and spectroscopic analysis after soil washing with chelating agents and humic substances. *Environ. Sci. Pollut. Res.* 21 (5): 3987–3995.

Arsenic in the Environment: Bridging Science to Practice for Sustainable Development –
van der Wal, Ahmad, Petrusevski, Weijma, Savic, van der Wens, Beerendonk, Bhattacharya, Bundschuh & Naidu (Eds)
© 2023 the Author(s), ISBN 978-1-032-32928-4

The association of the biogeochemical processes between arsenic and organic substances in paddy soils

G.L. Duan[1,2], Y.P. Yang[1,2], X.Y. Yi[1,2] & Y.G. Zhu[1,2,3]
[1]*State Key Lab of Urban and Regional Ecology, Research Center for Eco-Environmental Sciences, Chinese Academy of Sciences, P.R. China*
[2]*University of Chinese Academy of Sciences, P.R. China*
[3]*Key Laboratory of Urban Environment and Health, Institute of Urban Environment, Chinese Academy of Sciences, P.R. China*

ABSTRACT: Soil organic substances play an important role in global carbon cycle. It is widely recognized that application of organic substances to soil is benefit for carbon sequestration and enhancing soil fertility. However, when the soil is contaminated with arsenic (As), the effects of organic substance application on As mobility in soil and accumulation in rice plants should not be ignored. To understand how organic substances affect As transformation and translocation in the soil-microbe-rice system, both pot and microcosms experiments were carried out, including different dosages of rice straw, straw biochar and humic substances. Results showed that the application of organic substances significantly increased As mobility, enhanced As methylation and volatilization, elevated the accumulation of organic As in rice grains. Our study demonstrated that organic substances could regulate As mobility and transformation in paddy soil through regulating soil microbial community, and the biogeochemical processes of As in paddy soils is associated with iron and organic substances through microorganisms.

1 INTRODUCTION

Arsenic (As), a human carcinogen, tops the U.S. Environmental Protection Agency's Superfund List of Hazardous Substances. Food crops grown in As-containing soil contain high level of As. In particular, rice is a natural As accumulator. More than half of the world's population depends on rice as the major source of nutrition, and rice consumption has become the main route of dietary As intake for populations depend on rice as the staple food (Li *et al.*, 2011). Organic substances are recommended to be added to the soil for carbon sequestration and soil fertility. Recently, the application of organic substances, such as rice straw, has been demonstrated to remarkably increase As mobility in soil, enhanced As methylation and increased As accumulation in rice grains (Jia *et al.*, 2012; Yang *et al.*, 2018; Yi *et al.*, 2019). Therefore, it is of great importance to fully understand how organic substances affect As mobility and transformation in paddy soil, what are the mechanisms governing the interactions between organic substances and As, as well as what microorganisms are involved in these processes.

2 RESULTS AND DISCUSSION

2.1 *Straw incorporation increased As mobility and enhanced As methylation in paddy soil, also elevated As accumulation in rice plants*

Straw incorporation is widely performed in As contaminated paddy fields. To understand how straw and straw biochar incorporation affect As transformation and translocation in the soil-microbe-rice system, a pot experiment was carried out with different dosages of rice straw and straw biochar application (Yang *et al.*, 2018). Results showed that both straw biochar and straw application significantly increased As mobility. Straw biochar mobilized As mainly through increasing soil pH and DOM content. Straw incorporation mainly through enhancing As release from iron (Fe) minerals and arsenate (As(V)) reduction to arsenite (As(III)). Straw biochar didn't significantly affect As methylation, while straw incorporation significantly enhanced As methylation, elevated dimethylarsenate (DMA) concentration in soil porewater and increased As volatilization. Straw biochar didn't significantly change total As accumulation in rice grains, but decreased As(III) accumulation by Si inhibition. Straw incorporation significantly increased DMA, but decreased As(III) concentration in rice grains. After biochar application, dissolved As was significant positive correlated with the abundance of *Bacillus*, indicating that *Bacillus* might be involved in As release, and As(III) concentration in polished grains was negatively correlated with Si concentration. The significant positive correlation between dissolved As with Fe and the abundance of iron-reducing bacterium suggested the coupling of As and Fe reduction mediated by iron-reducing bacterium. The significant positive correlation between DMA in rice grains and the abundance of methanogenic bacteria indicated that methanogenic bacteria could be involved in As methylation after straw application. These results demonstrated how rice straw incorporation affects As fate in soil-microbe-rice system, and would provide some guidance to straw incorporation at As contaminated paddy soil. In addition, some functional microorganisms were revealed in the soil environment that can dominate As mobility and transformation after straw incorporation.

2.2 *The biogeochemical processes of arsenic in paddy soil is coupled with iron and humic substances*

Rice straw can be degraded to humic substances (HS) by microorganisms in the soil. Humic acid (HA) and fulvic acid (FA) are dominating humic substances (HS) in soil. Therefore, the effects of HA and FA addition (0.2%–1.5%) on As mobility and microbial community composition in paddy soil were investigated (Yi *et al.*, 2019). FA significantly increased the concentrations of As (12-fold), iron (Fe; 20-fold), manganese (Mn; 3-fold) and acetic acid (3-fold) in soil porewater, and also caused significant enrichment of *Desulfitobacterium* (41-fold). Furthermore, the FA addition significantly increased the relative abundance of *Bathyarchaeota* (4-fold), a microorganism that is suggested to be important for FA degradation. In contrast, HA slightly increased As (1.2-fold) in porewater, had little effect on Fe, Mn and acetic acid, and 1.5% HA addition significantly decreased As in porewater at day 14 (45%). Both HA and FA addition promoted As methylation. HA increased dimethylarsenate concentration and FA increased monomethylarsenate concentration in porewater. These results highlight the contrasting effects of different (HA vs. FA) organic substances on As fate in paddy soil and advance our understanding of the associations among As, Fe and organic substances through microorganisms in paddy soil.

2.3 *The effects of straw and biochar amendments on arsenic biotransformation microbes*

It has been demonstrated that straw biochar and straw application to paddy soil dramatically altered arsenic (As) biogeochemical cycling in soil-rice system, but it remains unknown how As biotransformation microbes (ABMs) contribute to these processes. Therefore, terminal restriction fragment length polymorphism (T-RFLP) analysis and clone library were performed to characterize ABMs. Through linear discriminant analysis (LDA) effect size (LEfSe) and correlation analysis, results revealed that *arrA*-harbouring iron-reducing bacteria (e.g., *Geobacter* and *Shewanella*) and *arsC*-harbouring *Gammaproteobacteria* (e.g., fermentative hydrogen-producing and lignin-degrading microorganisms) potentially mediated arsenate (As(V)) reduction under biochar and straw amendments, respectively. Methanogens and sulfate-reducing bacteria (SRB) carrying *arsM* gene might regulate methylated As concentration in soil-rice system. Network analysis demonstrated that the association among ABMs in rhizosphere was significantly stronger than that in bulk soil. Arsenite (As(III)) methylators carrying *arsM* gene exhibited much stronger co-occurrence pattern with *arsC*-harbouring As(V) reducers than with *arrA*-harbouring As(V) reducers. This study would broaden our insights for the dramatic variation of As biogeochemical cycling in soil-rice system after straw biochar and straw amendments through the activities of ABMs, which could contribute to the safe rice production and high rice yield in As-contaminated fields.

3 CONCLUSIONS AND RECOMMENDATIONS

Arsenic mobility and speciation in paddy soil are key factors regulating As accumulation and toxicity in rice grains, which are important issues for food safety. Our results showed that the application of organic substances significantly increased As mobility and enhanced As methylation and volatilization in paddy soils, also elevated the accumulation of organic As in rice grains. Our studies demonstrated that organic substances could regulate As mobility and transformation in paddy soil not only by physi-chemical processes, such as adsorption, complexation, etc., but also through regulating soil microbial community, and the biogeochemical processes of As in paddy soils is associated with iron and organic substances through microorganisms.

ACKNOWLEDGEMENTS

These studies were supported by the National Natural Science Foundation of China (grant no. 41977323 and 21677157), and National Key Research and Development Program of China (grant no. 2018YFD0800202 and 2016YFD0 800402).

REFERENCES

Li G., Sun G.X., Williams P.N., Nunes L. & Zhu Y.G., 2011. Inorganic arsenic in Chinese food and its cancer risk. *Environ. Int.* 37: 1219–1225.

Jia Y., Huang H., Zhong M., Wang F.H., Zhang L.M. & Zhu Y.G., 2013. Microbial arsenic methylation in soil and rice rhizosphere. *Environ. Sci. Technol.* 47: 3141–3148.

Yang Y.P., Zhang H.M., Yuan H.Y., DuanG.L., Jin D.C., Zhao F.J. & Zhu Y.G. 2018. Microbe mediated arsenic release from iron minerals and arsenic methylation in rhizosphere controls arsenic fate in soil-rice system after straw incorporation. *Environ. Pollut.* 236: 598–608.

Yi X.Y., Yang Y.P., Yuan H.Y., Chen Z., Duan G.L. & Zhu Y.G. 2019. Coupling metabolisms of arsenic and iron with humic substances through microorganisms in paddy soil. *J. Hazard. Mater.* 373: 591–599.

Arsenic in the Environment: Bridging Science to Practice for Sustainable Development –
van der Wal, Ahmad, Petrusevski, Weijma, Savic, van der Wens, Beerendonk, Bhattacharya, Bundschuh & Naidu (Eds)
 ISBN 978-1-032-32928-4

Effect of inorganic selenium in soil on the translocation of arsenic from soil to rice plant

G.D. Yang, G.R. Pokhrel, K.T. Wang & H.M. Zhuang
Fujian Provincial Key Laboratory of Agroecological Processing and Safety Monitoring, College of Life Sciences, Fujian Agriculture and Forestry University, Fuzhou, P.R. China

ABSTRACT: Selenium can regulate arsenic toxicity by strengthening antioxidant potential, but the antagonism between selenite or selenate nutrient and the translocation of arsenic species from paddy soil to different rice organs are poorly understood. In this study, a pot experiment was designed to investigate the effect of selenite or selenate on arsenite or arsenate toxicity to two indica rice cultivars (namely Ming Hui 63 and Lu You Ming Zhan), and the uptake and translocation of arsenic species from paddy soil to different rice organs. The results showed that selenite or selenate could significantly decrease the arsenate concentration in pore water of soils, and thus inhibited arsenate uptake by rice roots. However, the existence of selenite or selenate did not decrease arsenate concentration in rhizosphere pore water of two indica rice cultivars. There existed good antagonistic effect between selenite or selenate and the uptake of arsenite and arsenate in rice plant in the case of low arsenic paddy soil. However, this antagonism depended on rice cultivars, arsenic species and arsenic level in soil. Both selenite and selenate are all effective to decrease the translocation of inorganic arsenic from the roots to their aboveground rice organs in arsenite/arsenate-spiked paddy soil, but selenate had stronger inhibiting effect on their transfer factors than selenite.

1 INTRODUCTION

Arsenic (As) is a non-essential and toxic element, and it can hinder the growth of rice plant and inhibit rice's photosynthesis and yield (Yang *et al.*, 2015). Inorganic species of As and selenium (Se) elements, namely arsenite [As(III)], arsenite [As(V)], selenite [Se(IV)] and selenite [Se(VI)], are the major forms present in flooded soil (Elrashidi *et al.*, 1987), their close position in the periodic table makes them show similar characteristics in their uptake and transportation from soil to plants, but the competitive uptake and transportation from paddy soil to rice grain between inorganic As and inorganic Se was also poorly understood. In this study, two indica rice cultivars, namely Ming Hui 63 (MH63) and Lu You Ming Zhan (LYMZ), were used as the materials and a pot experiment was designed to investigate the effect of inorganic Se [Se(IV) and Se(VI)] regulation on inorganic As [As(III) or As(V)] toxicity to different rice cultivars, and the uptake and transportation of As species from paddy soil to different rice grain. On the other hand, the effect of inorganic Se regulation on As species in flooded soil without rice plants was also investigated to further obtain the effect of rice cultivars on the As species in flooded soil.

2 MATERIALS AND METHODS

2.1 *Plant materials and growth conditions*

Pot cultivation was carried out in the experimental field of Fujian Agriculture and Forestry University, Fuzhou, Fujian province of China, during rice growth period of June to October 2017. The day and night temperature was at the range of 25–37°C, the humidity was at the range of 68–90%, and day length was about 13 h. Two indica rice cultivars (MH63 and LYMZ) were used in this study, from the Rice Research Institute, Academy of Agriculture Sciences, Fujian of China.

2.2 *Arsenic and selenium treatments*

The rice seedling of cultivar MH63 and LYMZ (at the 4-leaf stage) was transferred into sandy loam soil (4 seedlings each pot) as described in Materials and methods 2.1 and grown in pot soil (12 kg soil per pot) within nine different treatments (4 replications each), respectively. Treatment 1 (CT00), no As or Se was spiked into the soil; Treatment 2 (CT04), 4.00 mg Se (IV)was spiked into per kg soil; Treatment 3 (CT06), 4.00 mg Se(VI) was spiked into per kg soil; Treatment 4 (AT30), 50.00 mg As(III) was spiked into per kg soil; Treatment 5 (AT34), 50.00 mg As(III) and 4.00 mg Se (IV) were spiked into per kg soil; Treatment 6 (AT36), 50.00 mg As(III) and 4.00 mg Se(VI) were spiked into per kg soil; Treatment 7 (AT50), 50.00 mg As(V) was spiked into per kg soil; Treatment 8 (AT54), 50.00 mg As(V) and 4.00 mg Se(IV) were spiked into per kg soil; Treatment 9 (AT56), 50.00 mg As(V) and 4.00 mg Se (VI) were spiked into per kg soil.

2.3 *Collection of rhizosphere pore water and pore water of soils*

About 40.00 mL of rhizosphere pore water was collected from the rooting zone using 10.00 cm capillary

DOI: 10.1201/9781003317395-74

sampler at tillering stage, booting stage, grouting stage and maturing stage of cultivars MH63 and LYMZ, respectively. Similarly, about 40.00 mL of pore water of soils was also collected from the central zone of plastic buckets using 10.00 cm capillary sampler at the corresponding stages without rice plants. Then the collected water sample was filtered through 0.22 μm polypropylene membranes, kept in −20°C until further determination of As species.

2.4 *Extraction and determination of arsenic species*

Arsenic speciation in roots, stems, sheaths, leaves, kernel and bran was extracted by a microwave assisted extraction (MAE) method and As species in rhizosphere pore water, porewater of soils and the extract of roots, stems, sheaths, leaves, kernel and bran, was determined using ion chromatography inductively coupled plasma mass spectrometry (IC-ICP-MS), based on our previous description (Li *et al.*, 2018).

2.5 *Statistical analysis*

Each sampling was conducted in 4 biological replications and their average values were used for statistical analysis. Software DPS7.5 was used to conduct the significant test by least significant difference (LSD) among different treatments.

3 RESULTS AND DISCUSSION

3.1 *Effect of selenium regulation on the total concentration of each arsenic species*

There existed good antagonistic effect between the addition of Se(IV)/Se(VI) and the uptake of As(III)/As(V) by rice organs, but this antagonism depended on rice cultivars and As species in soil (Figure 1).

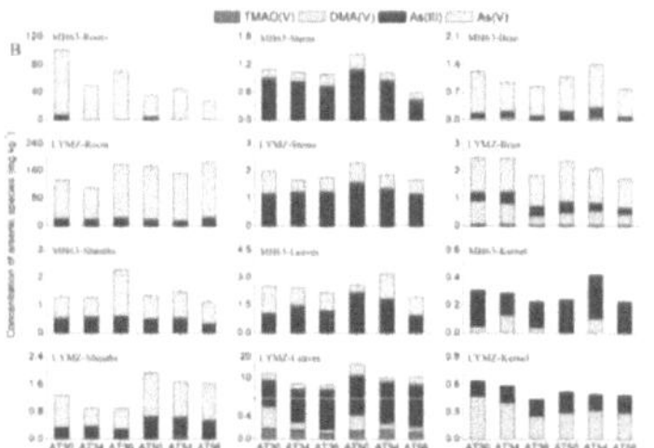

Figure 1. Effect of Se on As species in different organs of rice plants at maturing stage. Results are presented as mean ± standard deviation; n = 4; AT30, AT34, AT36, AT50, AT54, AT56, different As and Se treatments, their detail description could be seen in Section 2.2.

3.2 *Effect of selenium regulation on inorganic arsenic translocation*

The addition of Se(IV) significantly ($p < 0.01$) decreased the iAs transfer factor of leaf-to root in cultivar MH63 in As(III)-spiked paddy soil, decreased the iAs transfer factor of roots to the above-ground organ in cultivar LYMZ except that of bran-to-root in As(III)-spiked paddy soil, and decreased the iAs transfer factor of stem-to-root, sheath-to-root and leaf-to-root in cultivar MH63 in As(V)-spiked paddy soil, its inhibiting effect on the iAs transfer factor of roots to the aboveground rice organs was lower than Se(VI). Our results were opposite to the short-term hydroponic experiments of Hu *et al.* (2014) and Camara *et al.* (2018), which reported that Se (IV) was more effective to mitigate As translocation from rice roots to the above-ground rice organs than Se(VI).

4 CONCLUSIONS AND RECOMMENDATIONS

The existence of Se(IV) or Se(VI) could significantly decrease the As(V) concentrations in pore water of soils, and thus inhibited the As(V) uptake by rice roots. However, the existence of Se(IV) or Se(VI) didn't decrease the As(V) concentration in rhizosphere pore water of cultivar MH63 or LYMZ. The experimental results also indicated that there are good antagonistic effect existing between the addition of Se(IV)/Se(VI) and the uptake of As(III)/As(V) by rice plants in low As paddy soil, but the antagonism between Se(IV) or Se(VI) nutrient and the uptake of As(III) and As(V) depended on rice cultivars, As species and As concentration level in soil.

ACKNOWLEDGEMENTS

We acknowledge the Natural Science Foundation of China (grant numbers 21677033 and 21677034), Fujian Provincial Project of Science and Technology (grant numbers 2017Y0002), Project of China Tobacco Yunnan Industrial Co. Ltd. (grant number 2016YL03) and Project of Yunnan Company of China Tobacco Corporation (grant number 2017YN06 and 2016YN28) for this research.

REFERENCES

Camara A.Y., Wan Y., Yu Y., Wang Q. & Li H. 2018. Effect of selenium on uptake and translocation of arsenic in rice seedlings (*Oryza sativa* L.). *Ecotoxicol. Environ. Saf.* 148(2): 869–875.

Elrashidi M.A., Adriano D.C., Workman S.M. & Lindsay W.L. 1987. Chemical equilibria of selenium in soils: a theoretical development. *Soil Sci.* 144(1): 141–152.

Hu Y., Duan G.L., Huang Y.Z., Liu Y.X. & Sun G.X. 2014. Interactive effects of different inorganic As and Se species on their uptake and translocation by rice (*Oryza sativa* L.) Seedlings. *Environ. Sci. Pollut. Res.* 21(5): 3955–3962.

Qiu Z.Q., Lv Z.M., Wang K.T., Lan Y., Yang X.J., Rensing C., Fu F.F. & Yang G.D. 2018. Species distribution characteristics of arsenic in shellfish seafood collected from fujian province of China. *J. Food Compos. Anal.* 72 (1): 132–140.

Yang G.D., Xie W.Y., Zhu X., Huang Y., Yang X.J., Qiu Z. Q., Lv Z.M., Wang W.N. & Lin W.X., 2015. Effect of Arsenite-oxidizing bacterium B. laterosporus on arsenite toxicity and arsenic translocation in rice seedlings. *Ecotoxicol. Environ. Saf.* 120(1): 7–20.

*Arsenic in the Environment: Bridging Science to Practice for Sustainable Development –
van der Wal, Ahmad, Petrusevski, Weijma, Savic, van der Wens, Beerendonk, Bhattacharya, Bundschuh & Naidu (Eds)*
 ISBN 978-1-032-32928-4

Simultaneous application of iron-modified biochar and foliar silica sol in an arsenic-contaminated paddy field for safe production of rice

D. Pan[1,2], C. Liu[2], H. Yu[2], F. Li[2] & X. Li[1]
[1]*SCNU Environmental Research Institute, Guangdong Provincial Key Laboratory of Chemical Pollution and Environmental Safety, School of Environment, South China Normal University, Guangzhou, P.R. China*
[2]*Guangdong Key Laboratory of Integrated Agro-environmental Pollution Control and Management, Guangdong Institute of Eco-Environmental Science & Technology, Guangzhou, P.R. China*

ABSTRACT: Rice plants have a high ability to acquire arsenic (As), from flooded paddy soils and accumulate As in rice grains, posing harmful consequences on human health. In the present study, iron-modified biochar (Fe-BC) was used with foliar silica sol (Si) in an As-contaminated paddy soil in order to reduce As accumulation in rice grains for safe production. The results showed that the amendments of Fe-BC and Si significantly increased grain yields ($P < 0.05$), and decreased As concentration in brown rice ($P < 0.05$) after application for 2 years. The amendment of Fe-BC + Si resulted in the largest reductions in As accumulation in brown rice with an average reduction of 39-48% for four growing seasons, which led to the As concentration in brown rice below 0.20 mg/kg after 2-year practice. The pH value in the soil solution increased from 4.6 to 5.1, which was significantly negative correlated with soil available As. In conclusion, co-application of iron-modified biochar and foliar silica sol could be used as an effective strategy to reduce the As contents in rice grains.

1 INTRODUCTION

Rice plants have a high ability to acquire arsenic (As) from paddy soils which eventually lead to an increase in phytotoxicity and accumulation in rice grains, posing harmful consequences on populations consuming rice as a principal staple food. Iron oxides/hydroxides can facilitate immobilization of As in flooded paddy soil (Fan *et al.*, 2014). Biochar, known as an excellent soil amendment for its high surface area, pore size distribution and ion-exchange capacity, and zero valent iron (ZVI) and biochar mixture as a soil amendment can reduce As content in rice grains in pot experiment (Qiao *et al.*, 2017), and the reduction effect should be tested in field experiment. Therefore, iron-modified biochar (Fe-BC) was investigated in a two-year field experiment to identify As mobility in soil and As accumulation in rice grains. Joint remediation technologies can overcome the shortcomings of using a single remediation method, improve the remediation efficiency of pollution and reduce the cost of remediation. For example, ZVI-biochar mixture can reduce As content in rice grains by increasing As immobilization and thus decreasing availability in soil; foliage fertilizers by reducing As uptake and/or translocation to shoots and grains. Previous pot experiments found that foliar application of silica sol (Si) could reduce As accmulation in rice grains.Therefore, we also investigated the effectiveness of foliar silica sol on As reduction in rice grains in this field experiment. We hope that the joint application of Fe-BC and Si can decrease As content in rice grains by reducing As bioavailability in soil and As uptake and translocation in rice plants, simultaneously.

2 MATERIALS AND METHODS

2.1 *Preparation of Fe-BC and silica sol*

Both Fe-BC and silica sol were developed by Guangdong Institute of Eco-Environmental Technology (Guangzhou, China), which were produced by Tianjin Biomass Environmental Protection Technology Co., Ltd. (Tianjin, China) and by Iron Man Environmental Technology Co., Ltd. (Foshan, China), respectively. The preparation technology of Fe-BC has been described in the patent of Li *et al.* (2017). Before plowing cropland, the Fe-BC was mixed thoroughly with the top soil (0-20 cm depth) manually. The particle size distribution (PSD) of the silica sol was in a range of 4-15 nm.

2.2 *Field experiments*

The field experiments were carried out in a rice paddy at Shuitou Town, Fogang County, Qingyuan City, Guangdong Province, China (23°86′N 113°66′E). The variety was hybrid rice Tianyou 998. The total As content of this paddy soil was 22.6 mg/kg, which is lower than the limit from the Environmental Quality Standards (grade II) of As in paddy soil in China, but the As concentration in rice grains harvested in the soil exceeded the maximum allowable limit of China (0.2 mg/kg). The treatments included (1) Fe-BC application into the soil at a dose of 1.5 t/ha, (2) foliar application of silica sol (Si) (5 mM) on the rice leaf surface at a dose of 750 L/ha, (3) simultaneous applications of Fe-BC plus Si (Fe-BC + Si), and (4) the control (CK) without any amendment. Periodic irrigation and agricultural treatments such as insecticide

DOI: 10.1201/9781003317395-75

and weed killing were applied during every growing season when necessary. The experiments were conducted during March 2013 to November 2014, including four growing seasons.

2.3 *Plant materials and sampling*

The rice grains were harvested after maturity (approximately 83 days in early season and 66 days in late season after transplantation), rinsed with deionized water more than three times, and dried at 65°C in an oven for 48 h. The husk of the grains was removed. The brown rice was ground into flour with a squeezer for further chemical analysis. Meanwhile, the topsoils (0-20 cm) were sampled to the laboratory for subsequent analysis.

2.4 *Arsenic concentration determination*

Approximately 1 g of brown rice was digested in 10 mL of HNO_3 and $HClO_4$ (10:1, v/v) at 120-130°C until a clear solution was obtained. The digested solution was diluted with 1% HNO_3 to 50 mL and filtered with 0.45 μm filter. Concentration of As was determined by hydrogen generation-atomic fluorescence spectrometry (AFS-933, Beijing Titan Instruments Co., Ltd., China).

3 RESULTS AND DISCUSSION

3.1 *The rice grain yields*

The grain yields in Fe-BC, Si, and Fe-BC + Si treatments were obviously higher ($P < 0.05$) than the yield of the control through the 2-year field trial and especially in 2014. Moreover, the Fe-BC + Si treatment had the highest grain yields in all growing seasons except for the late season in 2013. These results indicate that iron-modified biochar and silica sol applications are conducive to increasing grain yields, and the combined applications are most effective in production enhancement.

3.2 *As concentration in brown rice*

Compared with the control, the Fe-BC and Si treatments significantly reduced the total As concentration in brown rice by 16-33% and 26-37%, respectively. The largest reductions in As concentration in brown rice were found in the Fe-BC + Si treatment with the reduction of 39%, 42%, 42%, and 48% for the four growing seasons, respectively. In particular, the mean total As concentration in brown rice dropped from 0.29-0.30 mg/kg in the CK to below 0.20 mg/kg in the treatment of Fe-BC + Si in 2013 and late season in 2014, less than the allowable limit of As in brown rice specified by China (0.2 mg/kg) in the form of inorganic As. By contrast, in the early season of 2014, the total grain As ranged from 0.29 to 0.33 mg/kg in the Fe-BC + Si treatment, which was probably related to the relatively high As accumulation in this growing season (grain As in the CK 0.45 mg/kg). Foliar application of Si reduced As accumulation in rice grains, possibly by reducing the As uptake and transport in rice plants (Duan *et al.*, 2014).

3.3 *Available As and pH in soil*

Fe-BC and Fe-BC + Si treatments significantly increased the soil pH in the four growing seasons except for the early season in 2013. In addition, significantly negative correlation between pH and $NH_4H_2PO_4$-extractable As was found. Accordingly, with the increase in pH after iron-modified biochar application, the concentrations of available As in soils decreased. On the other hand, significantly positive correlation between soil available As and As in brown rice was found. Hence, Fe-BC and Fe-BC + Si treatment significantly decreased available As in soil and finally further decreased grain As.

Therefore, the As reduction in rice grains in Fe-BC amendment may well result from the decrease in As available in soil (Qiao *et al.*, 2017).

4 CONCLUSIONS

Our results show that many measures have to be carried out to simultaneously decrease the concentration of available As in paddy soil, and reduce the As accumulation in rice grains. Our results from field experiments confirmed that combined applications of iron-modified biochar and foliar silica sol in the As-contaminated paddy soil was an excellent strategy for alleviating the accumulation of As in rice plants.

ACKNOWLEDGEMENTS

This work was supported by the National Natural Science Foundation of China (41877043).

REFERENCES

Duan G.L., Hu Y., Liu W.J., Kneer R., Zhao F.J. & Zhu Y. G. 2011. Evidence for a role of phytochelatins in regulating arsenic accumulation in rice grain. *Environ. Exp. Bot.* 71(3): 416–421.

Fan J.X., Wang Y.J. & Liu C. 2014. Effect of iron cxide reductive dissolution on the transformation and immobilization of arsenic in soils: new insights from X-ray photoelectron and X-ray absorption spectroscopy. *J. Hazard. Mater.* 279: 212–219.

Li F., Cui J., Liu C., & Liu C. 2017. *Iron-based Biochar Material, Preparation Therefore and Use Thereof in Soil Pollution Control.* United States US15518180.

Qiao J.T., Liu T.X., Wang X.Q., Li F.B., Lv Y.H., Cui J. H. & Liu C.P. 2018. Simultaneous alleviation of cadmium and arsenic accumulation in rice by applying zero-valent iron and biochar to contaminated paddy soils. *Chemosphere* 195: 260–271.

2.2 Dynamics of arsenic in rhizosphere

Arsenic in the Environment: Bridging Science to Practice for Sustainable Development –
van der Wal, Ahmad, Petrusevski, Weijma, Savic, van der Wens, Beerendonk, Bhattacharya, Bundschuh & Naidu (Eds)
 ISBN 978-1-032-32928-4

Effects of selenite on arsenic uptake and translocation in rice seedlings (*Oryza sativa* L.)

K. Wang[1], Y.N. Wan[1,2] & H.F. Li[1]
[1]*Beijing Key Laboratory of Farmland Soil Pollution Prevention and Remediation, College of Resources and Environmental Sciences, China Agricultural University, Beijing, P.R. China*
[2]*Institute of Agricultural Resources and Regional Planning, Chinese Academy of Agricultural Sciences, Beijing, P.R. China*

ABSTRACT: Arsenic (As) as a kind of carcinogen, raises a worldwide concern due to its threats to plants and human health via the food chains. In this research, the effects of selenite [Se(IV)] on arsenic uptake and translocation in rice seedlings (*Oryza sativa* L.) were investigated through hydroponic experiments in greenhouse. The results showed that compared to the As(III) or As(V) treatment alone, application of Se(IV) significantly increased the As content in rice roots by 94.9% or 54.2%, while dramatically decreased its content in shoots by 48.9% or 52.4%, respectively. Accordingly, As accumulation in above-ground parts of rice dropped significantly when Se(IV) presented in the culture solution. Our results demonstrated that Se(IV) could mitigate As translocation from roots to shoots in rice plants.

1 INTRODUCTION

Arsenic (As) is a ubiquitous contaminant which can enter the environment from anthropogenic activities. Accumulation of As in soil can not only inhibit the plants growth, but also pose a threat to human health via the food chains due to its high toxicity (Garg & Singla 2011). On the other hand, selenium (Se) is beneficial to plants if applied at proper dosage (El-Ramady *et al.*, 2016) and could cause antagonism with some heavy metals (such as As) via some relevant process (Camara *et al.*, 2018). Therefore, it is of great importance to understand the mechanism of Se on mitigation of As toxicity within plants.

In the present study, the effects of selenite [Se(IV)] on arsenic uptake and translocation in rice seedlings (*Oryza sativa* L.) were investigated through hydroponic experiments in greenhouse. Our findings would provide some theoretical basis for ensuring the safety of grain production and making possible for utilizing Se(IV) to regulate As in rice plants.

2 MATERIALS AND METHODS

2.1 *Experimental design and implementation*

The 30-day-old rice seedlings (*Oryza sativa* L., Zhunliangyou 608) were transferred to 2.5 L 1/2-strength Kimura solution (pH 5.5) containing 2 mM MES with four treatments: T1, 10 μM As(III); T2, 10 μM As(III) + 30 μM Se(IV); T3, 10 μM As(V); T4, 10 μM As(V) + 30 μM Se(IV). The treated rice seedlings were harvested after 72 h. Each treatment was replicated three times (two plants per pot). In this trial, As (III), As(V), and Se(IV) were supplied as $NaAsO_2$, $HAsNa_2O_4 \cdot 7H_2O$, and Na_2SeO_3, respectively.

2.2 *Determination of As content*

The contents of As in rice tissues were determined by a hydride generation atomic fluorescence spectrometer (HG-AFS) after digested (8 mL concentrated HNO_3, GR) using a microwave oven digestion system (CEM, MARS5).

3 RESULTS AND DISCUSSION

The As content in rice shoots was significantly decreased by 48.9% in As(III) + Se(IV) treatment (T2) or 52.4% in As(V) + Se(IV) treatment (T4); however, its content in roots was dramatically enhanced by 94.9% in T2 or 54.2% in T4, compared with its corresponding control treatment (Figure 1). Our results was in accord with the previous study on rice in which the As(III) uptake by rice roots was significantly

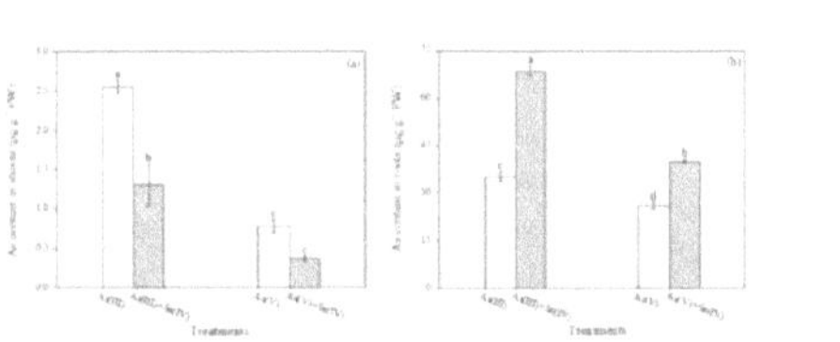

Figure 1. As content in (a) shoots and (b) roots of rice seedlings under the different treatments. Different letters indicate a significant difference between the treatments at $P < 0.05$.

DOI: 10.1201/9781003317395-76

elevated at the presence of Se(IV) (Camara *et al.*, 2018). However, other researchers reported Se(IV) invariably reduced the As(V) uptake by Chinese brake fern (Feng *et al.*, 2009). The variation in these observations could be ascribed probably to the different plant species, culture conditions, and even elements doses.

Regardless of As(III) or As(V) treatments, the translocation of As from roots to shoots was reduced obviously in rice seedlings with the lower transfer factor when Se(IV) presented in culture solution (Table 1). Thus, the proportion of As in above-ground parts dropped compared to its corresponding control treatment (Figure 2). A hydroponic experiment was conducted by Hu *et al.* (2014), who demonstrated that the As translocation from rice roots to above-ground parts decreased after application of Se(IV). Similarly, Wan *et al.* (2018) demonstrated the addition of Se (IV) inhibited the proportion of As in rice shoots and depressed the As translocation within rice seedlings. Evidence has shown that Se could enhance the formation of apoplastic barriers in endodermis, which was conducive to restrict the transportation of heavy metals within plants.

Table 1. Accumulation and translocation of As in rice plants under the different treatments.

	Total As (μg)		
Treatment	Shoots	Roots	Transfer factor
T1	32.7 ± 2.5 a	166.0 ± 16.0 b	0.072 ± 0.001 a
T2	15.9 ± 3.1 b	289.7 ± 9.5 a	0.019 ± 0.004 c
T3	9.9 ± 1.2 bc	117.4 ± 9.4 c	0.030 ± 0.003 b
T4	4.4 ± 0.2 c	173.9 ± 8.9 b	0.009 ± 0.001 d

Data are means ± SE (n = 3). Different letters in same column indicate a significant difference between the treatments at $P < 0.05$ according to Duncan's test.

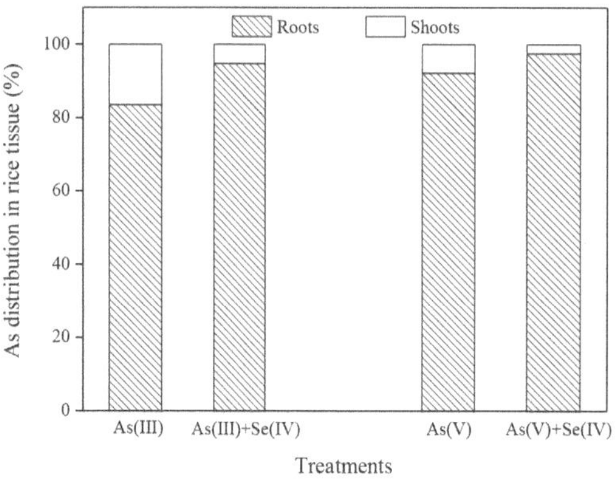

Figure 2. As distribution in rice seedlings under the different treatments.

Moreover, the total As in rice tissue in As(III) treatments was higher than that of in As(V) treatments (Table 1). It has been reported that As(III) enters into rice mainly through silicon (Si) transporter OsNIP2;1 (Ma *et al.*, 2008) with the fact that rice can absorb Si preferential due to its physiological characteristics. Therefore, the As(III) influx into rice roots was much more abundant relative to that of As(V).

4 CONCLUSIONS AND RECOMMENDATIONS

The study demonstrated that the application of Se (IV) significantly increased the As content in roots while decreased it in shoots, regardless of As(III) or As(V) treatments. Therefore, the As accumulation in above-ground parts of rice was dropped with the lower transfer factor. Further, the transformation of As speciation in rice plants after the inorganic Se applications should be investigated in future research.

ACKNOWLEDGEMENTS

This work was supported by the National Natural Science Foundation of China (No. 41907146).

REFERENCES

Camara A.Y., Wan Y., Yu Y., Wang Q. & Li H. 2018. Effect of selenium on uptake and translocation of arsenic in rice seedlings (*Oryza sativa* L.). *Ecotoxicol. Environ. Saf.* 148: 869–875.

El-Ramady H., Abdalla N., Taha H.S., Alshaal T., El-Henawy A., Faizy S.E.D.A., Shams M.S., Youssef S. M., Shalaby T., Bayoumi Y., Elhawat N., Shehata S., Sztrik A., Prokisch J., Fári M., Domokos-Szabolcsy É., Pilon-Smits E.A., Selmar D., Haneklaus S., & Schnug E. 2016. Selenium and nano-selenium in plant nutrition. *Environ. Chem. Lett.* 14(1): 123–147.

Feng R., Wei C., Tu S. & Sun X. 2009. Interactive effects of selenium and arsenic on their uptake by *Pteris vittata* L. under hydroponic conditions. *Environ. Exp. Bot.* 65(2–3): 363–368.

Garg N. & Singla P. 2011. Arsenic toxicity in crop plants: physiological effects and tolerance mechanisms. *Environ. Chem. Lett.* 9: 303–321.

Hu Y., Duan G., Huang Y., Liu Y. & Sun G. 2014. Interactive effects of different inorganic As and Se species on their uptake and translocation by rice (*Oryza sativa* L.) seedlings. *Environ. Sci. Pollut. Res. Int.* 21(5): 3955–3962.

Ma J.F., Yamaji N., Mitani N., Xu X.Y., Su Y.H., McGrath S.P. & Zhao F.J. 2008. Transporters of arsenite in rice and their role in arsenic accumulation in rice grain. *Proc. Natl. Acad. Sci. U. S. A.* 105(29): 9931–9935.

Wan Y., Camara A.Y., Huang Q., Yu Y., Wang Q. & Li H. 2018. Arsenic uptake and accumulation in rice (*Oryza sativa* L.) with selenite fertilization and water management. *Ecotoxicol. Environ. Saf.* 156: 67–74.

Arsenic in the Environment: Bridging Science to Practice for Sustainable Development –
van der Wal, Ahmad, Petrusevski, Weijma, Savic, van der Wens, Beerendonk, Bhattacharya, Bundschuh & Naidu (Eds)
© 2023 the Author(s), ISBN 978-1-032-32928-4

Biogeochemistry of arsenic in a highly polluted agricultural soil: Influence of water saturation and organic substrate supply

H. Thouin, N. Devau, M. Charron, F. Battaglia-Brunet, C. Joulian & J. Hellal
BRGM, Water, Environment, Process Development and Analysis Division, Orléans, France

ABSTRACT: Some agricultural soils was historically polluted by the destruction of chemical shells. The present work used laboratory experiments aiming to determine the influence of water saturation and organic substrate supply on the speciation and mobility and speciation of arsenic (As). A column filled with soil was fed with synthetic water, transiently supplied with organic substrates, during 320 days. Important quantities of As were leached. The initial As(V) present in soil was reduced to As(III) during the periods of organic substrates supply. Most of the As initially present in the solid phases was reduced at the end of experiment. Active As-transforming microbes were quantified in the solids and the evolution of their relative concentrations seemed to be linked to As speciation. The proportion of *aio*A genes encoding an arsenite oxidase, expressed as the ratio of *aio*A/16S genes abundances, tended to increase with the proportion of As(V) in the solids. The microbial activities played a major role in the evolution of As speciation, probably influencing As mobility.

1 INTRODUCTION

Arsenic (As) is a toxic element whose transfer from soil to agricultural products may induce health is-sues (Upadhyay *et al.*, 2019). The bio-availability and mobility of As depends on its speciation, itself influenced by the redox conditions that could substantially vary in soils according to water saturation. These conditions limit O_2 diffusion in pore space implying that reductive redox couples become predominant. If these anaerobic conditions are maintained during several days, reduction of oxidized As could occur. These reactions are microbially driven and imply organic substrates as electron donors. The reduced As species are more mobile, promoting their transfer either to plant or groundwater. The influence of anaerobic periods on the biogeochemistry of As needs to be evaluated extensively in agricultural areas affected by As pollution or high As geochemical background. Here, the effect of anaerobic periods on As biogeochemistry was examined applying complementary geochemical and microbiological monitoring methods.

2 MATERIALS AND METHODS

2.1 *Experimental methods*

The soil was sampled in an agricultural site polluted by the destruction of As-containing chemical shells near Verdun, France. It contained 983 ppm total As, mainly in the As(V) form. A glass column (30 cm high, 3.5 cm inner diameter, water jacket maintained at 20°C) was filled with polluted soil sieved at 2 mm and up-flow fed with oxygenated synthetic water (Werber 1991) during 320 days. Anaerobic conditions were induced in the column by supplementing the inlet water with organic substrates (acetate 10 mM, lactate 10 mM, formate 10 mM and glucose 2 mM).

2.2 *Analysis*

As speciation in water was determined by AAS after separation of As(III) and As(V) on resins. As peciation in solids was determined by HPLC-ICP-MS after extraction with H_3PO_4. MPN of As(III)-oxidizing and As(V)-reducing microbes (Thouin *et al.*, 2019) and quantification of 16S rRNA, *aio*A and *ars*B genes (Fernandez-Rojo *et al.*, 2017; Poirel *et al.*, 2012) were performed on the soil samples at the end of the experiment.

3 RESULTS AND DISCUSSION

3.1 *Evolution of As speciation and concentration in the column outlet*

Concentration of As increased from 5 mg/L to 30 mg/L at the beginning of the experiment (Figure 1) and then decreased between day 40 and day 170. After 160 days, As concentration fluctuated slightly around 5 mg/L. At the end of the experiment, 120 mg of As had been leached from the column. Until day 40, As (V) was the dominant As species.

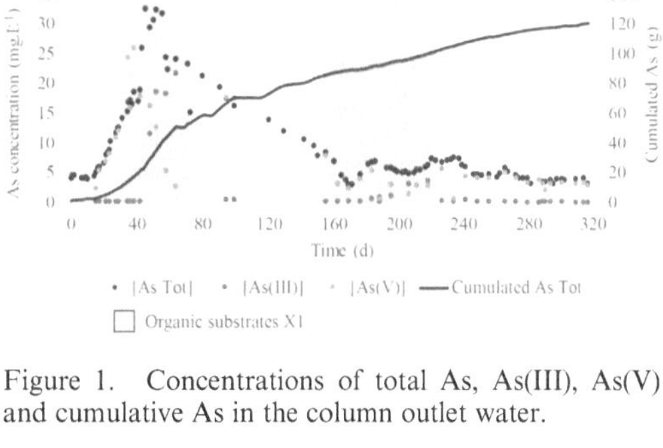

Figure 1. Concentrations of total As, As(III), As(V) and cumulative As in the column outlet water.

With the organic substrate addition, As(V) decreased and As(III) became the main species. During the period with no organic substrate, As(III) in the outlet decreased below 0.5 mg/L.

3.2 *Final analyses of the solids*

As speciation was determined in soil samples of the bottom (outlet) and the top (inlet) of the column at the end of the experiment (Figure 2). As(III) was the main As species in the two samples, and largely preponderant in the bottom zone, whereas As(V) was the main species in the initial soil. As(V) increased from the bottom (60 mg/kg) to the top (174 mg/kg). As(III) also increased but to a lesser extent (from 165 mg/kg to 204 mg/kg).

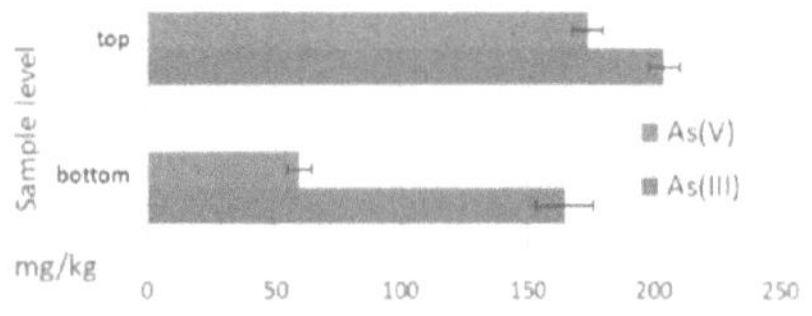

Figure 2. Distribution of As species in the soil at the end of experiment.

The range of numbers of As(III)-oxidizing microorganisms was around 10^5/g of dry soil except at the top of the column (Figure 3). The number of As(V)-reducing microorganisms was the highest in the bottom of the column with about 10^6 microorganism per gram of dry soil and then decreased toward the top of the column. The predominance of As(V)-reducing microbes in the bottom of the column might have contributed to the strong reduction of As(V) in the solid phases, resulting in 75% of As(III) in the solids sampled in this zone.

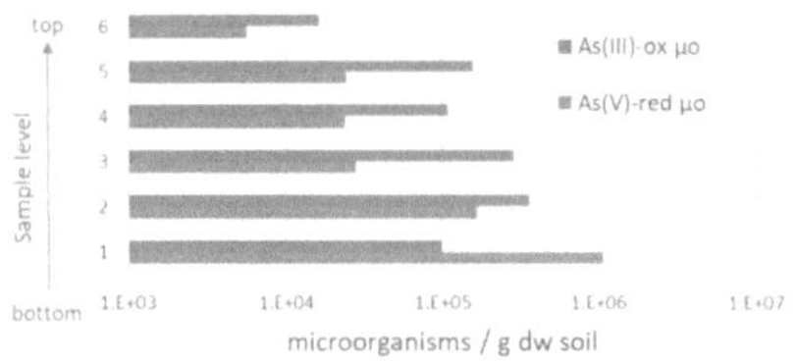

Figure 3. MPN of As(III) oxidizing and As(V) reducing microorganisms in the solids from the column at the end of experiment.

The abundance of *aio*A and *ars*B genes, respectively involved in As(III) oxidation and As(V) reduction, in the soil at the end of the experiment, is given in Figure 4, as the ratio between the functional genes and the 16S rRNA gene. Bacteria owning functional genes related to As transformation were detected in all samples. The proportion of *aio*A genes involved in As(III) oxidation varied irregularly inside the column, however it tended to increase from the bottom (with less As(V)) to the top (presenting a higher proportion of As(V)). The proportion of *ars*B genes also increased from the bottom to the top, but they were 10 times less abundant than *aio*A genes.

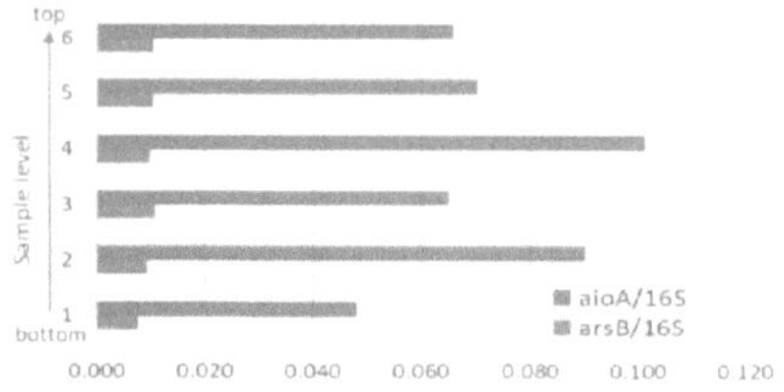

Figure 4. Abundance of genes involved in bacterial As transformation in the soil samples from the column at the end of experiment.

4 CONCLUSIONS

Column experiments performed in in alternated aerobic and anaerobic conditions showed strong mobility of As and important change of As speciation, both in water and solid phases. An effect of redox conditions on the distribution of As-transforming microbes was also noticed as an original result. These microorganisms could contribute to modifying As speciation so to the distribution of this toxic element between solid and aqueous phases in agricultural soils submitted to transient water saturation events if concentration of electron donor compounds is sufficient.

ACKNOWLEDGEMENTS

AgriAs is co-funded by EU and the Academy of Finland, L'Agence nationale de la recherche, Bundesministerium für Ernährung und Landwirtschaft and Forskningsrådet FORMAS under the ERA-NET Cofund WaterWorks2015 Call. This ERA-NET is an integral part of the 2016 Joint Activities developed by the Water Challenges for a Changing World Joint Programme Initiative (Water JPI).

REFERENCES

Fernandez-Rojo L. *et al.* 2017. Biological attenuation of arsenic and iron in a continuous flow bioreactor treating acid mine drainage (AMD). *Water Res.* 123: 594–606.

Poirel J., Joulian C., Leyval C. & Billard P. 2013. Arsenite-induced changes in abundance and expression of arsenite transporter and arsenite oxidase genes of a soil microbial community. *Res. Microbiol.* 164: 457–65.

Thouin H. *et al.* 2019. Microcosm-scale Biogeochemical Stabilization of Pb, As, Ba and Zn in mine tailings amended with manure and ochre. *Appl. Geochem.* 111: 104438

Upadhyay M.K., Shukla A., Yadav P. & Srivastava S. 2019. A review of arsenic in crops, vegetables, animals and food products. *Food Chem.* 276: 608–618.

Werber C. 1991. *Methods for Measuring the Acute Toxicity of Effluents and Receiving Waters to Freshwater and Marine Organisms.* 5th Ed., U.S. Environnemental Protection Agency, Washington, D.C.

Arsenic in the Environment: Bridging Science to Practice for Sustainable Development –
van der Wal, Ahmad, Petrusevski, Weijma, Savic, van der Wens, Beerendonk, Bhattacharya, Bundschuh & Naidu (Eds)
© 2023 the Author(s), ISBN 978-1-032-32928-4

Monothioarsenate uptake and reduction in *Arabidopsis thaliana*

A. Monroy-Licht[1,2], C. Rafferty[3], C.F. Kerl[4], B. Planer-Friedrich[4] & S. Clemens[3]
[1]*Department of Chemistry and Biology, Universidad del Norte, Barranquilla, Colombia*
[2]*Environmental and Computational Chemistry Group, School of Pharmaceutical Sciences, Universidad de Cartagena, Cartagena, Colombia*
[3]*Department of Plant Physiology, University of Bayreuth, Bayreuth, Germany*
[4]*Environmental Geochemistry, Bayreuth Center for Ecology and Environmental Research (BayCEER), University of Bayreuth, Bayreuth, Germany*

ABSTRACT: Monothioarsenate (MTA) is a newly discovered arsenic species, formed under sulfur-reducing conditions in paddy soil pore waters. It is structurally analogues of arsenate-As(V). Given the food safety risks associated with As accumulation in rice it is important to understand how thiolated As species are taken up and metabolized by plants. In this work, MTA reduction was evaluated using the model system *Arabidopsis thaliana*. The analysis was carried out in the accessions Col-0 (Wild type = WT) as well as in two *hac1* null alleles Krefeld (Kr-0) and a AtHAC1 mutant. Total arsenic accumulation and speciation assays in shoots for those plants showed that *hac1* null alleles plants accumulated more arsenic than wild type (Col-0) under As(V) conditions. HAC1 represents the known arsenate reductase. However, HAC1 mutant plants did not show this phenotype under MTA exposition. In spite of this, Shoot-As-speciation showed mainly As(III), followed by As(V) and few amounts of MTA, which could indicate that most of the MTA that is translocated to the shoots is first reduced to As(III), although a minor part remains as MTA. Arsenic speciation in roots showed that the dominant As species was arsenite for both plants in MTA/As(V) treatments. Nevertheless, a reduction in the percentage of the total accumulated arsenic present as arsenite was observed, hac1 null alleles has lower As(III) amount than Col-0, under As(V) conditions. Similarly, in plants exposed to MTA, the dominant As species in the roots was arsenite for all genotypes, followed by As(V) and a lower MTA proportion. The reduction of MTA to arsenite in roots gave evidence that MTA is actively taken up by plants and reduced inside the plant to As(III). A. thaliana exposed to MTA showed higher rates of As translocation from roots to shoots than plants under As(V) treatment. A better understanding of thioarsenate translocation will be especially important in order to prevent/reduce the T-As accumulation in edible plant parts.

1 INTRODUCTION

Arsenic is a toxic metalloid that is found ubiquitously in the environment (Farooq *et al.*, 2016), which is considered non-essential for plants and other organisms (Khalid *et al.*, 2017). This element causes skin diseases (Chowdhury *et al.*, 2000) and is classified as a class-1 carcinogen by the International Agency of Research on Cancer (IARC) (Niazi *et al.*, 2018). Contamination of As is a challenging issue, that it is not limited to water sources, indeed, soil pollution with this element has become a serious environmental hazard, particularly in agricultural areas (Mishra *et al.*, 2014). Arsenic accumulates in edible plant parts after being taken up from the soil. As(V) and As(III) are present in the soil solution, but those are not the only species to consider as As bioavailable fraction; several methylated species with variable proportions are also present (Zhao *et al.*, 2010) as well as Thioarsenic species, predominant in sulfidic environments (Planer-Friedrich *et al.*, 2007). Thioarsenate information about toxicity and mobility in plants is increasing (Kerl *et al.*, 2018; Planer-Friedrich *et al.*, 2017). However, important questions remain to be addressed. Because of this, the aim of this study was to evaluate the reduction of Monothioarsenate and As(V) in the model system *Arabidopsis thaliana*.

2 MATERIALS AND METHODS

2.1 *Plant material*

The experiments were set using *Arabidopsis thaliana* Col-0 (Wild type), Krefeld (0), which has loss-of-function alleles of HAC1 (Chao *et al.*, 2014) and a *hac1* mutant; HAC1 is responsible for arsenate reductase activity (Chao *et al.*, 2014). The T-DNA insertion mutant (SALK_058016C) was obtained from the Nottingham Arabidopsis Stock Centre (NASC). Homozygous line was identified.

2.2 *MTA and As(V) accumulation treatments and analysis of total As in roots and shoots in Arabidopsis*

Plants (Col-0, Kr-0 and *hac1* mutant) were grown hydroponically for 6 weeks according. After this time, addition of As started at concentrations of 10 μM As (V) and 10 μM MTA. The treatments were set for 48 hours. After that, the determination of total As was carried out for roots and shoots. For quantification, pooled root samples and pooled leaf samples, including shoots were separated and labeled according which each treatment. Shoots were placed in paper bags. Roots were carefully washed according to Planer-Friedrich *et al.*, 2017. Then roots and shoots were dried at 60°C for 3 days).

Approximately 100 mg of sample dried were used to determine the total As concentration. The sample was digested in 4 mL of concentrated HNO3 and 2 mL of 30% H2O2 in a microwave digestion system (Microwave System START 1500 MLS GmbH with a HPR-1000/10 S high pressure segment rotor). Digested samples were analyzed for total As by inductively coupled plasma optical emission spectrometry ICP-OES (icap 6000 series, Thermo Scientific). The internal standard was the multielement atomic spectroscopy standard solution V by Fluka Analytical. To the Standard 1:10000 was added the arsenic atomic absorption standard solution by Sigma-Aldrich. The microwave blank (HNO3/H2O2) was subtracted from all concentrations, and the results are reported as average concentrations with the standard deviation of three biological replicates. Three independent repetitions were conducted for each experiment.

2.3 *MTA and As(V) accumulation treatments and analysis of Arsenic speciation in roots and shoots in Arabidopsis*

Plants (Col-0, Kr-0 and *hac1* mutant) were grown hydroponically for 6 weeks. After this time, addition of As species started at concentrations of As(V) 10 μM and MTA 10 μM; the treatments were set for 48 hours. After that, pooled root and pooled leaf samples, including shoots were separated and labeled in accordance which each treatment. Roots were carefully washed according to Kerl *et al.*, 2018. After that, roots and shoots were flash-frozen, and ground in liquid N2. The extraction process was carried out following the Kerl *et al.* (2018) protocol. After that, samples were analyzed for As-speciation by inductively coupled plasma mass spectrometry (ICP-MS) Retention times of arsenite, arsenate, and MTA were determined using individual standards (Kerl *et al.*, 2018; Suess *et al.*, 2009).

3 RESULTS AND DISCUSSION

3.1 *Analysis of Total As in roots and shoots for Col-0, Kr-0 and hac1 mutant under MTA and As(V) treatments in Arabidopsis*

Hydroponic experiments showed that total As concentrations in the roots lower upon exposure to MTA compared to exposure to As(V). In the same way, Planer-Friedrich *et al.* (2017) found differences in total arsenic accumulation in roots around 4–5 times lower upon exposure to MTA versus As(V). The reason for a potential difference in uptake between MTA and As(V) is unclear yet. Statistical analysis of Total arsenic concentrations in roots, showed not significant differences between the genotypes under MTA treatment, but under As(V), the three lines tested are significantly different between each other, the *hac1* null alleles plants accumulated more arsenic than the WT (Figure 1A). On the other hand, in plants exposed to MTA conditions the T-As concentration in shoots was not significant between the genotypes (Figure 1B). Nevertheless, the translocation factor (TF) was higher in comparison to As(V) values, and significantly different between both. The highs TF value was found for *hac1* mutant (0.684) followed by Col-0 (0.17) and Kr-0 (0.09). Regarding to the shoots total arsenic concentration under As(V) condition. The *hac1* null alleles plants accumulated more arsenic than the WT, showing clear differences. *HAC1* is responsible for arsenate reductase activity in both the outer layer of the root (epidermis) and the inner layer adjacent to the xylem (pericycle). In its absence, the roots return less arsenic to the soil and the shoots accumulate up to 300 times more arsenic (Chao *et al.*, 2014). Nevertheless, disrupted *HAC1* plants did not show this phenotype under MTA exposition.

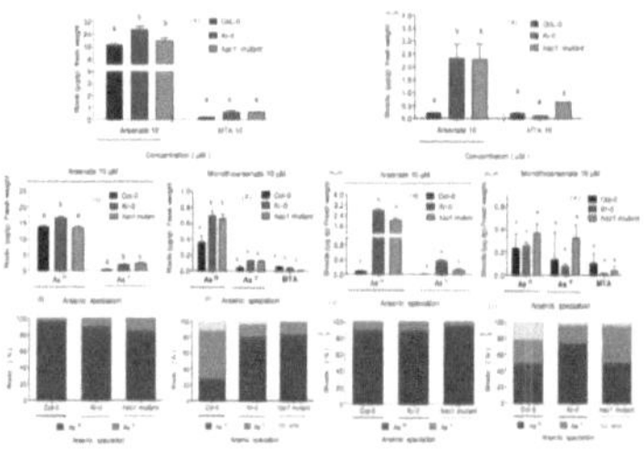

Figure 1. Total arsenic accumulation in roots (A) and shoots (B) of A. thaliana wild-type Col-0, hac1 null alleles and Kr-0, grown in hydroponic culture in one-tenth-strength Hoagland medium for 6 weeks. After this time, plants were transferred into 50 mL tubes for 48 h, under 10 μM arsenate or 10 μM monothioarsenate. Shoots were flash-frozen, and ground in liquid N2. For total arsenic analysis, digested samples were analyzed by ICP-OES. Arsenic speciation in roots (C & D) and in shoots (G and H) was carried out using PBS-extraction. Samples were analyzed immediately by IC-ICP-MS. In addition, the ratio of arsenic species was calculated for each treatment roots (E & F) and r shoots (I and J). Letters above bars indicate statistically different groups within treatments. Significance is based on pairwise analysis using post hoc analysis (Tukey's multiple comparisons test) following two-way ANOVA (significance defined as $p < 0.05$). Data represent average ± S.D (n = 3) of two independent assays.

3.2 *Arsenic speciation in roots and shoots in Arabidopsis under MTA and As(V) treatments*

Arsenic speciation in shoots showed that the dominant As species was arsenite in both treatments (94-84% in As(V)-exposed and 73-49% in MTA-exposed plants, respectively) (Figures 1J and 1H); with minor contributions of arsenate (10–5% and 44–21%, respectively) and under MTA treatment (21-4%). The absolute arsenite concentrations in shoots of the MTA-Col-0-exposed plants were higher than the same genotype under As(V) conditions, but for both *hac1* nullalleles, the concentrations in MTA-exposed plants were lower than in the As(V)-treatment (Figure 1H).

In plants exposed to As(V), the results between WT and both *hac1* loss-of-function alleles (Kr-0 and *hac1* mutant), who have increased the accumulation of arsenite and arsenate compared to Col-0 showed significant differences (Figure 1G). The total accumulated arsenic is present mainly as arsenite, and this phenomenon is explained due the capacity to convert arsenate to arsenite is reduced, then plants increase significant arsenic amount, primarily as arsenite, in the shoots (Chao *et al.*, 2014). This in line with the enormous differences found in the arsenic shoot concentrations between WT and *hac1* null alleles plants. The result of this study is in good agreement with data reported by Chao *et al.* (2014). On the other hand, the arsenic speciation in shoots under MTA conditions, showed not significance differences between Col-0 and two *hac1* null alleles plants for As(III), As(V) and MTA concentrations respectively. However, concentrations were decreasing in the tree genotypes followed the order of As(III) >As (V) > MTA. This could indicate that most of the MTA that is translocate to the shoots is reduced to As(III), although a minor part remains as MTA. Regarding to the analysis in roots, the dominant arsenic species was arsenite with minor contributions of arsenate in both treatments. On the other hand, the speciation data from plants grown under As(V), showed significant differences between As(III) and As(V) in both *hac1* null alleles plants, who had increased arsenic accumulation in comparison with WT (Figure 1C). This observation was also pointed out by Chao *et al.* (2014); who investigated the distribution of arsenic in roots in the *hac1* mutant.

In plants exposed to MTA, the dominant As species in the roots was arsenite for all genotypes. The results were: Col-0 79.9%, 8.9% and 11.1%; *hac1* mutant 83.2%, 15.5% and 1.4% and Kr-0 80.6%, 15.2% and 4.2% for As(III), As(V) and MTA series respectively (Figures 1E and 1F). The results showed significant differences between WT and both *hac1* loss-of-function alleles (Kr-0 and *hac1* mutant), who have increased the accumulation of arsenite. For arsenate and MTA species there are not significant differences. This results are in line with data reported on rice arsenic speciation for plants exposed to 10 μM MTA and 10 μM As(V), where mainly arsenite was found like a dominant species, followed by arsenate (Kerl *et al.*, 2018). On the other hand, the root to shoot translocation factor values were higher for plants exposed to MTA (TF values between 3.34-0.45) than them under As(V) conditions, (TF values between 0.19-0.02) which means that MTA is rapidly translocate to the shoots, where most is reduced to As(III). This observation is comparable to that determined in a previous study in A. thaliana ecotype Col-0 (Planer-Friedrich *et al.*, 2017) as well as the same tendency was identified in rice under 10 μM MTA conditions. In addition, it was found a close relation-ship with the higher As concentrations presented in the xylem (Kerl *et al.*, 2018).

4 CONCLUSIONS AND RECOMMENDATIONS

Nearly one hundred arsenic species have been detected in the environment. MTA is a newly discovered specie. It is formed under sulfur-reducing conditions in paddy soil pore waters. MTA as well as As(V) is taking up by plants. The arsenic accumulation and speciation analysis showed that hac1 null alleles plants accumulated more arsenic than the WT under As(V) conditions. The shoot-As- speciation was found mainly As(III), followed by As(V) under As(V) treatment. The same pattern was found for MTA conditions, but few amount of MTA was also observed; which could indicate that most of the MTA that is translocate to the shoots is reduced to As(III), although a minor part remains as MTA. Regarding to the root-As-speciation in plants exposed to MTA as well as under As (V), the dominant As species was arsenite for all genotypes followed for As(V) and MTA (under MTA treatment). The reduction of MTA to As(III) in roots gave evidence that MTA is actively taken up by plants and reduced inside the plant. Plants exposed to MTA, compare to As (V), showed higher rates of As translocation from roots to shoots. This observation requires further detailed investigations on passage to the grain; to determine whether MTA contributes to As accumulation in grains. The mechanisms for the high MTA translocation are not clear yet. However, this study contributed to support previous data. On the other hand, HAC1 experiments confirmed that this arsenate-reductase does not play a role as MTA-reductase. Nevertheless, the reduction to As(III) was evident in many experiments. The enzymes responsible for this process remain still unknown.

REFERENCES

Chao D.Y., Chen Y. *et al.* 2014. Genome-wide association mapping identifies a new arsenate reductase enzyme critical for limiting arsenic accumulation in plants. *PLoS Biol.* 12(12): e1002009.

Chowdhury U.K., Biswas B.K. *et al.* 2000. Groundwater arsenic contamination in Bangladesh and West Bengal, India. *Environ. Health Perspect.*108(5):393–397.

Farooq M.A., Islam F., Alia B. *et al.* 2016. Arsenic toxicity in plants: cellular and molecular mechanisms of its transport and metabolism. *Environ. Exp. Bot.* 132: 42–52.

Kerl C, Rafferty C *et al.* 2018. Monothioarsenate uptake, transformation and translocation in rice plants. *Environ. Sci. Technol.* 52 (16): 9154–9161.

Khalid S., Shahid M., Niazi N.K. *et al.* 2017. Arsenic behaviour in soil-plant system: biogeochemical reactions and chemical speciation influences (Chapter 4). In: N.A. Anjum, S.S. Gill & N. Tuteja (eds.) *Enhancing Cleanup of Environmental Pollutants. Volume 2.* Springer: Berlin, Germany.pp. 97–140.

Mishra B.K., Dubey C.S. *et al.* 2014. Concentration of arsenic by selected vegetables cultivated in the yamuna flood plains (YFP) of Delhi, India. *Environ. Earth Sci.* 72(9), 3281–3291.

Niazi N.K., Bibi I., Shahid M. *et al.* 2018. Arsenic removal by perilla leaf biochar in aqueous solutions and groundwater: an integrated spectroscopic and microscopic examination. *Environ. Pollut.* 232: 31–41.

Planer-Friedrich B., Kuhnlenz T., Halder D. *et al.* 2017. Thioarsenate toxicity and tolerance in the model system arabidopsis thaliana. *Environ. Sci. Technol.* 51(12): 7187–7196.

Planer-Friedrich B., London J. *et al.* 2007.Thioarsenates in geothermal waters of Yellowstone National Park: determination, preservation, and geochemical importance. *Environ Sci Technol.* 41(15): 5245–5251.

Suess E., Scheinost A.C., Bostick B.C. *et al.* 2009. Discrimination of thioarsenites and thioarsenates by X-ray absorption spectroscopy. *Anal. Chem.* 81(20): 8318–8326.

Zhao F.J., McGrath S.P. & Meharg A.A. 2010. Arsenic as a food chain contaminant: mechanisms of plant uptake and metabolism and mitigation strategies. *Annu. Rev. Plant Biol.* 61: 535–559.

Arsenic in the Environment: Bridging Science to Practice for Sustainable Development –
van der Wal, Ahmad, Petrusevski, Weijma, Savic, van der Wens, Beerendonk, Bhattacharya, Bundschuh & Naidu (Eds)
© 2023 the Author(s), ISBN 978-1-032-32928-4

Groundwater arsenic contamination with special reference to its accumulation in rice grain and additional entry during domestic scale post harvesting in Bengal delta

T. Roychowdhury, A. Das, N. Roy Chowdhury, M. Joardar, D. Mridha, A. De, U. Lama, P. Priyadarshni, K. Majhi, S. Chakraborti, D. Kumar & A. Devansh
School of Environmental Studies, Jadavpur University, Kolkata, India

ABSTRACT: Arsenic contamination of groundwater is a crucial water quality issue leading to various health exposures and sufferings in rural people of the world, particularly in Ganga-Meghna-Brahmaputra plain in Indian sub-continent. Arsenic is a notorious metalloid, which moves freely under the soil system, and also evenly percolates when comes into the nature while withdrawing groundwater. Cultivation of staple crops like paddy on these contaminated soils is one of the major routes for dietary exposure of arsenic. Arsenic uptake property of paddy plants was found to be dependent on the variety of cultivar rather than arsenic levels in groundwater or soil. Higher accumulation of arsenic has been observed in parboiled rice grain due to parboiling with arsenic-contaminated water.

1 INTRODUCTION

In rural areas of West Bengal, the arsenic (As) contaminated groundwater is not only the elevated source of As exposure through ingestion of drinking water and cooking purposes, but also is directly responsible for food chain contamination, since contaminated ground water is mainly used for irrigation (Roychowdhury 2008). Irrigation practices of agricultural fields with As-contaminated groundwater has led to As build-up in soil, with subsequent elevation of As by the process of translocation in food crops grown on these agricultural soils (Chowdhury *et al.*, 2018a). The present study is highlighted on the As accumulation in different parts of paddy plants including rice grain throughout its pre- (boro) and post-monsoon cultivation period. Further attention has been given on additional entry of As during domestic scale post harvesting technology used by the farmers for conversion of raw (sunned) rice to parboiled rice grain.

2 MATERIALS AND METHODS

2.1 *Study area, sample collection and preparation*

Samples were collected from As-affected blocks like Deganga, Gaighata located in North 24-Parganas district of West Bengal. Water samples were collected from shallow, large-diameter tubewells used for domestic purposes and agricultural irrigation in the paddy fields and about 0.1% v/v nitric acid was added as preservative. Paddy plant samples were collected throughout the cultivation period (vegetative, reproductive and ripening phases) during pre- and post-monsoon cultivation. Four kinds of whole grains (WG) i.e. raw WG, half boiled WG (wet), full boiled WG (wet) and parboiled WG (dry) along with As contaminated water samples were collected at 3 stages (raw water, half boiled water and full boiled water) during domestic scale post harvesting of paddy grain.

2.2 *Chemicals and reagents*

All the reagents were of analytical grade. Distilled deionized water was used throughout. Solid samples were digested in a Teflon bomb at about 100°C for 6–8 h by using a mixture of concentrated HNO_3 (96%) and H_2O_2 (30% v/v) in 2:1 ratio which were procured from Merck (Mumbai, India) (Roychowdhury 2008). About 10% of KI solution (10% in aqueous solution) and 8% of concentrated HCl (Merck, Mumbai, India) were added during sample preparation (5–10 ml) and allowed to settle for 45 mins. prior to analysis of As. A solution of 1.25% $NaBH_4$ (Merck, Mumbai) in 0.5% NaOH and 5.0 M HCl were used for Hydride Generation-Atomic Absorption Spectrometry (FI-HG-AAS) method. Standard stock solution of arsenate (1000 mg/L as As, traceable to SRM from NIST H_3AsO_4 in HNO_3, 0.5 M) from Merck, Darmstact, Germany was used for As estimation.

2.3 *Analysis*

An Atomic Absorption Spectrometer (Varian model AA 140, USA) was used for total As estimation by hydride generation method. Details of instrumentation, optimization conditions and methodology of HG-AAS system were as reported earlier (Chowdhury *et al.*, 2018a; Roychowdhury 2008).

DOI: 10.1201/9781003317395-79

3 RESULTS AND DISCUSSION

3.1 *Arsenic translocation in paddy plant parts and variation throughout the cultivation period*

Arsenic uptake property of paddy plants collected from 20 fields from the studied areas was found to be dependent on the variety of paddy plant rather than arsenic levels in groundwater (0.05–0.35 mg/L) or soil (20–50 mg/kg). Arsenic is translocated from root to aerial parts in descending order (Figure 1). Leaf, stem, root, root soil and surface soil showed a similar trend in their change of arsenic concentration throughout the cultivation period. Arsenic concentration was highest in vegetative phase; sharply declined in reproductive phase; followed by moderate increase in ripening phase (Chowdhury *et al.*, 2018a). The young root tissues in vegetative (primary) phase could uptake arsenic at a much faster rate than the older tissues in later phases. With the growth of the plant, formation of iron plaques on the surface of the root which sequester arsenic and prevented its uptake by plants in the reproductive phase. Finally, co-precipitation of arsenic with iron released from crystallized iron plaques results in loosening of the iron plaques from root surface which in turn contributes to increased concentration of arsenic in soil and plant parts.

3.2 *Variation in As concentration in paddy whole grains during various stages of post harvesting*

Considerable high concentrations of As (mean: 200 µg/kg) were found in about 90% of parboiled rice grain samples compared to raw or non-parboiled rice grain samples (54 µg/kg in 80% samples). This observation instigated to study the additional entry of As at various stages of parboiling for producing parboiled rice grain from raw (sunned) rice grain. A maximum increase of 180% of As content in parboiled rice grain was observed compared to raw rice grain (Chowdhury *et al.*, 2018b). Arsenic content in individual whole grain samples ar various phases of parboiling is shown in Figure 2. Significant increase in parboiled whole grain As concentration was dependent upon the large difference between As concentrations of the water and the raw whole grain used for parboiling.

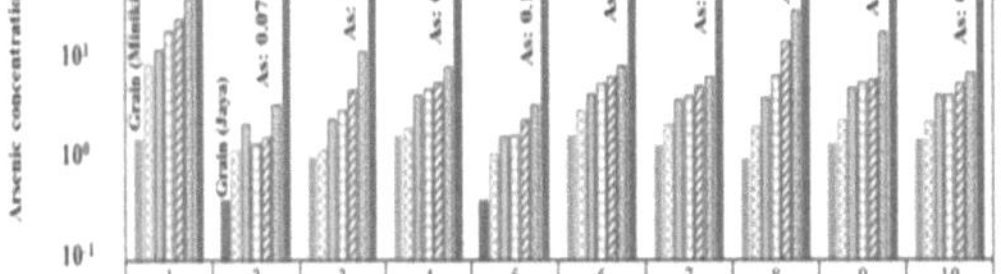

	1	2	3	4	5	6	7	8	9	10
Grain	1.46	0.37	0.97	1.56	0.36	1.56	1.27	0.93	1.28	1.43
Pedicle	7.59	1.12	1.13	1.85	1.03	2.72	1.95	1.88	2.18	2.10
Leaf	10.5	1.99	2.19	3.75	1.50	3.84	3.38	3.50	4.37	3.74
Stem (8-12)	16.1	1.29	2.70	4.32	1.53	4.83	3.70	5.83	4.98	3.72
Stem (4-8)	20.6	1.49	4.20	4.98	2.19	5.76	4.58	12.53	5.32	4.82
Stem (0-4)	37.8	3.04	10.1	7.13	2.97	7.20	5.57	25.0	15.4	6.22
Root	380	38.6	193	193	122	351	406	195	349	103

* Arsenic concentration of ground water in mg/l

Figure 1. As concentrations (mg/kg) in various parts of paddy plant at final phase before harvesting.

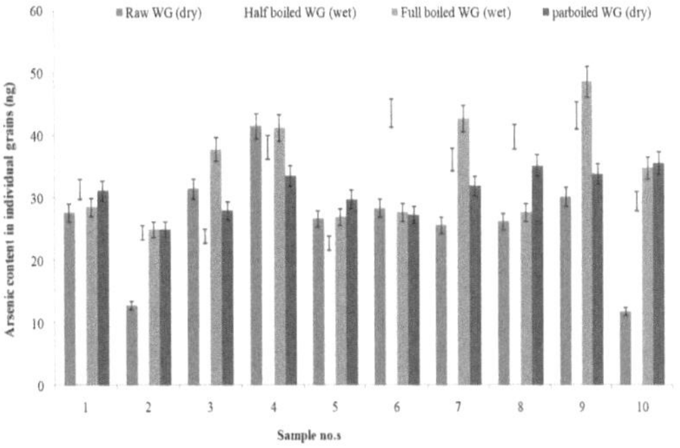

Figure 2. Histogram showing As content of individual whole grain samples at various stages of parboiling.

4 CONCLUSIONS AND RECOMMENDATIONS

A considerable amount of decrease in As concentration has been observed from root to grain of paddy plant. The rice grain As had little to no correlation with surface soil, root soil and irrigation water As concentration in all the three phases of growth. Arsenic uptake probably is related to the rice variety, i.e. the particular cultivar rather than As levels in groundwater or soil. Investigation on the role of parboiling procedure of rice being cultivated and processed in As prone areas suggested that parboiling of rice with As-contaminated groundwater contributes majorly in the increment of As concentration in parboiled rice and its by-products. Different rice cultivar seems to have different rates of arsenic uptake.

ACKNOWLEDGEMENTS

Financial support from 'Department of Science & Technology', Govt. of West Bengal for providing research project grant is highly acknowledged

REFERENCES

Chowdhury N.R., Das R., Joardar M., Ghosh S., Bhowmick S. & Roychowdhury T. 2018a. Arsenic accumulation in paddy plants at different phases of premonsoon cultivation. *Chemosphere* 210: 987–997.

Chowdhury N.R., Ghosh S., Joardar M., Kar D. & Roychowdhury T. 2018b. Impact of arsenic contaminated groundwater used during domestic scale post harvesting of paddy crop in West Bengal: arsenic partitioning in raw and parboiled whole grain. *Chemosphere* 211: 173–184.

Roychowdhury T. 2008. Impact of sedimentary arsenic through irrigated groundwater on soil, plant, crops and human continuum from Bengal delta: special reference to raw and cooked rice. *Food Chem. Toxicol.* 46(8): 2856–2864.

2.3 Microbial ecology and molecular mechanisms of arsenic biotransformation in soils and plant uptake

*Arsenic in the Environment: Bridging Science to Practice for Sustainable Development –
van der Wal, Ahmad, Petrusevski, Weijma, Savic, van der Wens, Beerendonk, Bhattacharya, Bundschuh & Naidu (Eds)*
© 2023 the Author(s), ISBN 978-1-032-32928-4

Silica nanoparticles inhibit arsenic uptake into rice suspension cells via improving pectin synthesis and the mechanical force of the cell wall

J. Cui & F. Li
Guangdong Key Laboratory of Integrated Agro-environmental Pollution Control and Management, Guangdong Institute of Eco-environmental Science & Technology, Guangzhou, P.R. China

ABSTRACT: Our previous studies indicated that the foliar application of silica nanoparticles (SiO_2 NPs) could obviously reduce arsenic (As) accumulation in rice. However, the mechanism underlying this effect at the single-cell level has not been reported. In this study, we investigated for the first time the effects of SiO_2 NPs on inhibiting As uptake into rice using individual rice cells. The results indicated that the addition of SiO_2 NPs could enhance the proportion of live cells by weakening oxidative stress upon As exposure. Compared to the treatment of cells with As only, treatment with SiO_2 NPs could maintain the integrity of the cell, increase the thickness of the cell wall (77.4%) and the ratio of As in the pectin (19.6%). In addition, the pectin content, cation exchange capacity (CEC) and pectin methylesterase (PME) activity were also increased in the SiO_2 NPs-pretreated cells, leading to a decreased degree of pectin methylesterification and an improved mechanical force of the cell walls. Furthermore, in the SiO_2 NPs-pretreated rice cells, the expression of the OsLis1 and OsLis2 genes was lower, whereas the expression of the OsNIP1;1 and OsNIP3;3 genes was higher than that of the As-only group.

1 INTRODUCTION

Arsenic (As) is a toxic element that is a non-threshold carcinogen and is distributed widely in the environment. It has been reported that the pollution of paddy soil with As by industrial activities, irrigation with As-polluted groundwater, and the use of arsenic-based fertilizer and pesticide resulted in a further increase in As concentrations in rice. Rice is a staple food crop worldwide, especially in Southeast Asian countries. When rice is cultivated in As-contaminated paddy soil, As can be easily taken up by rice, leading to the inhibition of plant growth, nutrient imbalances and the reduction of yield (Meharg *et al.*, 2009). The accumulation of As in human bodies through the food chain may pose detrimental effects to human health, including denaturation of proteins and enzymes, various types of cancer and impairment of the immune response. Our previous studies have also indicated that most SiO_2 NPs accumulate in the rice suspension cell walls. In this abstract, our aim was to investigate how SiO_2 NPs inhibit the cellular uptake and transport of As using individual rice cells.

2 MATERIALS AND METHODS

2.1 *Cell viability assay*

The rice cells were harvested by centrifugation and thoroughly washed with buffer to remove the cell medium. Subsequently, the samples were pre-fixed in 2.5% glutaraldehyde for 5 h, thoroughly rinsed with 0.1 M phosphate buffer solution and dehydrated in an ascending ethanol series (30, 50, 70, 90, 95 and 100% v/v). Then, the dehydrated cells were embedded in spur resin and cut into ultrathin sections using a microtome (Leica, Germany). Finally, the obtained samples were observed using Transmission Electron Microscope (TEM) equipped with an Oxford INCA Energy TEM 200 EDX system. The cell viability, reactive oxygen species (ROS) and mitochondrial transmembrane potential (MMP) assays were detected by flow Cytometry.

2.2 *Analysis of As and Si contents*

The concentration of As was measured using the method in our previous study.

2.3 *Atomic Force Microscope (AFM) imaging and measurement of mechanical properties*

The Lyophilized cell wall was treated with 0.02% Poly-L-Lysine for 10 min and then suspended in ultrapure water for 1 h. Subsequently, the obtained sample was washed with ultra-high purity water, dripped onto completely clean coverslips and then allowed to air dry overnight. The coverslip was glued to a glass slide and placed on a piezoelectric scanning tube of a MultiMode VIII AFM (Bruker, Santa Barbara, CA, USA). AFM imaging was performed with a Bruker ScanAsyst-Air probe (tip radius = 10 nm, spring constant = 5 N/m) and a silicon nitride cantilever. Before each measurement, the spring constant of the probe was calibrated. All images were obtained at a low scan rate (1 Hz). Measurement of the mechanical properties of the cell wall was performed after each imaging. The Poisson's ratio of the sample was assumed to be 0.3. The force-distance curve was continuously recorded by tapping the

DOI: 10.1201/9781003317395-80

sample surface in a scan range of 2 × 2 μm in a grid of 16 × 16 points. Young's modulus was calculated using Nanoscope analysis software based on the Hertz-Sneddon model.

3 RESULTS AND DISCUSSION

3.1 *Effects of SiO2 NPs pretreatment on alleviating As toxicity and oxidative stress*

To quantify cell viability, the rice cell survival percentage was detected using flow Cytometry. The results showed that the survival percentage of the As-only treated cells decreased obviously with increasing As concentration. However, compared with treatment with As alone, the survival percentage of SiO2 NPs-pretreated cells at concentrations of 10, 20 and 40 μM As increased by 92.5%, 85.4% and 83.6%, respectively. To correlate the effect of SiO2 NPs on As toxicity, we further measured the level of ROS. The results indicated that the fluorescence intensity of SiO_2 NPs-pretreated cells at concentrations of 0, 10, 20 and 40 μM As decreased by 4.5-, 6.2- and 4.8-fold, respectively, compared with treatment with As alone. Furthermore, we found that the level of MMP in the SiO_2 NPs-pretreated cells was decreased drastically in comparison to that in the As-only treated cells. These results suggested that the addition of SiO_2 NPs could enhance As detoxification and significantly decrease oxidative stress upon As exposure.

3.2 *Effect of SiO_2 NPs pretreatment on the properties of cell wall*

The pectin content may influence the structural and mechanical properties of the cell wall. In this study, AFM was applied to directly measure the surface morphology and Young's modulus of the rice cell wall. The three-dimensional topography clearly indicated that the SiO_2 NPs-pretreated and untreated cell walls had large spatial mechanical differences. The surfaces of the cell walls treated with As alone were rough (Figure 1D). However, the surface of the cell wall became flat with the addition of SiO_2 NPs (Figures 1E, F and G). Meanwhile, the mechanical properties of cell walls were measured by AFM force curves following imaging, and Young's modulus was calculated according to the Nanoscope Analysis software (Figures 1C', D', E', F' and G'). The AFM-based elastic modulus mapping indicated that the mechanical strength (Young's modulus) of cell walls was 1.99 ± 0.43 GPa naturally, whereas the Young's modulus of cell walls with sole As treatment was significantly decreased to 0.26 ± 0.02 GPa. Compared with the cells cultured without the addition of SiO_2 NPs, the average Young's modulus values of SiO_2 NPs-pretreated cell walls under As stress at 0.1, 1 and 10 mM increased by 4.5-, 9.9- and 23.8-fold, respectively (Figure 1B). This result suggested that the presence of SiO_2 NPs significantly improved the mechanical properties of the cell wall compared with treatment solely with As.

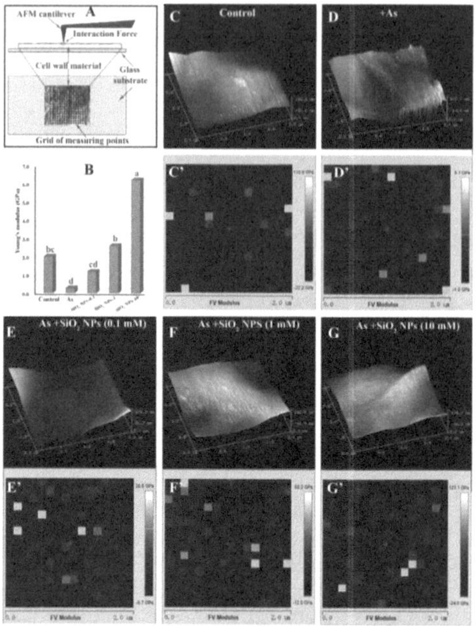

Figure 1. Schematic illustration to evaluate the mechanical properties of the cell wall using AFM (A); Young's modulus of the cell wall (B); three-dimensional topography of the cell wall pretreated with or without various concentrations of SiO_2 NPs (C, D, E, F and G); elasticity topography of the cell wall pretreated with or without various concentrations of SiO_2 NPs (C', D', E', F' and G').

4 CONCLUSIONS AND RECOMMENDATIONS

The data demonstrated that treatment with SiO_2 NPs decreases the cellular ROS and MMP levels and increases the pectin content under As stress, leading to an improved mechanical force of the cell wall. Additionally, the SiO_2 NPs treatment could regulate the expression of As-related genes upon exposure to As. Thus, As was mainly deposited into the cell walls, and pectin was the major contributor to the increased As retention in the cell wall.

ACKNOWLEDGEMENTS

We acknowledge National Natural Science Foundation of China (No. 41877137).

REFERENCE

Meharg A.A., Williams P.N., Adomako E., Lawgali Y. Y., Deacon C., Villada A. & Yanai J. 2009. Geographical variation in total and inorganic arsenic content of polished (white) rice. *Environ. Sci. Technol.* 43(5): 1612–1617.

Arsenic in the Environment: Bridging Science to Practice for Sustainable Development –
van der Wal, Ahmad, Petrusevski, Weijma, Savic, van der Wens, Beerendonk, Bhattacharya, Bundschuh & Naidu (Eds)
 ISBN 978-1-032-32928-4

Soil arsenic dynamics in differentially microbial propagated paddy fields of Bengal delta basin

A. Majumdar[1], M.K. Upadhyay[2], A.K. Srivastava[3], S. Srivastava[2], M.K. Jaiswal[1] & S. Bose[1]
[1]*Department of Earth Sciences, Indian Institute of Science Education and Research (IISER) Kolkata, Mohanpur, West Bengal, India*
[2]*Plant Stress Biology Laboratory, Institute of Environment & Sustainable Development, Banaras Hindu University, Varanasi, India*
[3]*Nuclear Agriculture and Biotechnology Division, Bhabha Atomic Research Centre, Mumbai, India*

ABSTRACT: Soil arsenic (As) determination is crucial for the assessment of arsenic flux from agricultural soil as focused in this study. Total content of arsenic flux has been measured by using two parametric equations that has been developed considering seasonal variation as well as irrigation practices namely conventional flooding (CF) and periodical flooding (PF). Depending on soil physic-chemical parameters, soil nutritional profiling has been done to know the difference between these two differentially irrigated fields. Soil microbial diversity has been identified using metagenomics and arsenic resistant (As^R) gene specific microbial community was checked via high-throughput sequencing. Altered microbial community with differential gene expression were found that suggested a change in arsenic content under these field conditions and how varied As^R microbes can propagate bioavailable soil As. CF field soil contains an average As concentration of 61.33 ± 0.96 mg/kg whereas reduction in As content found to be 2.5 fold in PF fields resulting in 26.14 ± 0.77 mg/kg. The study, here, shows positive correlation between microbial redox change of As and water management practices making lesser bioavailable soil As.

1 INTRODUCTION

Arsenic contamination from soil to different crops specifically in rice has been proved as a serious health issue to its consumers worldwide (Bhattacharya *et al.*, 2007; Upadhyay *et al.*, 2019). Determination of As dynamics in rhizospheric soil reveals the degree of contamination and elemental behavior with soil physic-chemical parameters (Kabata-Pendias 2000). With gradual deposition of As on top soil, extracted from contaminated groundwater (Meharg & Rahman 2003), resulted in arsenic resistance in microbes that involves different strategies to mitigate the toxic effect of As on cellular activities. Bioavailability (BA) of soil As can potentially affect rice plant ultra-structure (Majumdar *et al.*, 2019) and implementation of altered irrigation practices can change soil nutrient profiling which in turn can alter microbial diversity that might show active arsenic resistance gene expression. Present study has made a focus on proving a correlation at field based experiments.

2 MATERIALS AND METHODS

2.1 *Experimental set up and samplings*

Fields for rice cultivation was designed based on a conventional flooding (CF) and periodical flodding (PF) patterns. Samplings were done at three time point of rice cultivation and soil samples were taken from 0–20cm depth using PVC pipes, dried, homogenized, and pellets were made for total elemental analysis in X-ray fluorescence (XRF) spectroscopy and BA elemental analysis in inductively coupled plasma optical emission spectrscopy (ICP-OES). Collection of fresh soil for microbial isolatoin ad metagenomic analysis were done at the begining and end of cultivation under such irrigaton practices to check diversity alteration.

2.2 *Sample processing and instrumentation*

Soil samples were completely dried, finely powdered and screened through 2mm sieve. Soils were checked for pH, ORP, salinity, organic matter for further analysis with pourbiax diagram. Flux equations developed based on several parameters. 1:4 ratio of boric acid powder and soil taken to make pellet for XRF analysis. Sequential extraction process was followed for soil digestion using a series of oxidants and neutralizers for BA assessment in ICP-OES as per Majumdar *et al.* (2018). Plant parts (root, shoot, whole grains) were collected in respective growth phases and digested as mentioned by Majumdar *et al.* (2018). Plant internal ultra-structures were observed using field emission scanning electron microscope (FE-SEM) by making a very thin section of plant's shoot and node with further coating with ionized spray of gold particles to make the sample more prominent for visualization.

2.3 *Soil for metagenomics and bacterial characterization*

Rhizospheric rice plant root soil samples were collected fresh from both the fields at three growth phases for bacterial isolation for As tolerance assessment and for metagenomics, beginning and end soil samples were used to identify the difference between As responsive gene expressions. For bacterial As tolerance, a series of growth

DOI: 10.1201/9781003317395-81

media were used supplementing up to 2000 mg/kg of arsenite and 3000 mg/kg of arsenate with subsequent making of pure cultures of selected As^R colonies. Further those were tested for As oxidation-reduction change, biochemical reactions of IMViC test, and tests for plant growth promoting substances. Selected bacterial colonies were identified using 16SrRNA sequencing that proved to be useful for plant growth promotion in hydroponic cultivation of rice seedlings.

3 RESULTS AND DISCUSSION

3.1 *Arsenic content in soil-plant system at two different fields*

Total As concentrations were found to be 61.33 ± 0.96 mg/kg in average from CF soils and PF soils contained 26.14 ± 0.77 mg/kg. BA soil As **(a)** and plant As contents **(b)** were measured as in Figure 1.

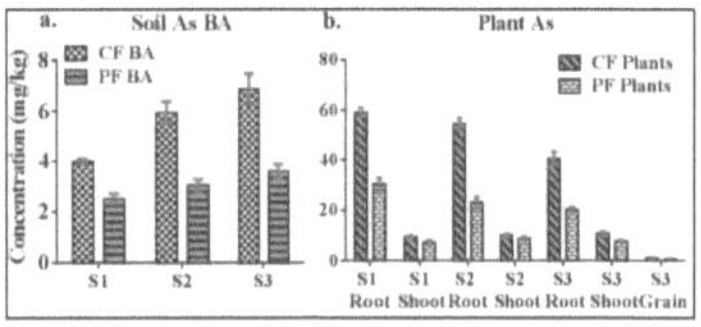

Figure 1. Arsenic BA in two differentially irrigated field soil and respective plant parts.

Developed equations are mentioned below:

$$AdC_S^{a2} = Ah^aC_W^2 + AdC_S^{a1} + W^aC_r^a(\text{Aman})$$
$$C_w^2(DP_s) = \left[AdC_s^{b1} + Ah^bC_w^1\right] - \left[AdC_s^{b2} + A\left(h^b - h^a\right)C_w^1\right](\text{Boro})$$

These two equations determines the net content of As, getting percolated to the soil layers or flushed out from the field.

3.2 *Ultra-structure enhancement under altered irrigation practice and microbial amelioration*

Rice plant vascular system has been found much more rigid retaining structural integrity under PF condition with As^R bacterial presence as in Figure 2.

Figure 2. Xylem, phloem and bundle sheath observation in FE-SEM and plant growth comparison in hydroponics.

Application of two specific bacterial strains in hydroponics medium for growth observation as in Figure 2 showed clear enhancement of plant's growth in terms of biomass by minimizing As stress to plant cell compared to stressed plants. Stress responsive marker and anti-oxidant enzyme assessment of these plants supported the fact of bacterial usefulness in As remediation and plant growth promotion.

3.3 *Metagenomics of As responsive microbes and specific identification*

Cultured bacteria were selected having 1100 mg/kg arsenite and 2000 mg/kg arsenate resistance. Microbial community was found to be changed under differential irrigation mode showing differential AsR gene expression accordingly. arsC, aioA, aioB, arrA, arrB, aioX and arsM genes were targeted for diversity assessment that showed distinct differences in two altered irrigated fields.

4 CONCLUSIONS

Arsenic stress on rice plant's internal ultra-structure was assessed under two irrigation practices that also changed microbial community showing differential As^R gene expression from these fields. Specific identified bacterial cultures was sucessfully proved to be plant growth enhancer by minimzing As stress.

ACKNOWLEDGEMENTS

Authors are thankful to MoES, Govt. of India (MoES/P.O. (Geosci)/56/2015), IISER Kolkata, BHU and BARC facilities for providing research grant and instrumental support.

REFERENCES

Bhattacharya P., Welch A.H., Stollenwerk K.G., McLaughlin M.J., Bundschuh J. & Panaullah G. 2007. Arsenic in the environment: biology and chemistry. *Sci. Total Environ.* 379(2–3): 109–120.

Kabata-Pendias A., 2000. *Trace Elements in Soils and Plants*. CRC Press.

Majumdar A., Barla A., Upadhyay M.K., Ghosh D., Chaudhuri P., Srivastava S. & Bose S. 2018. Vermiremediation of metal (loid)s via eichornia crassipes phytomass extraction: a sustainable technique for plant amelioration. *J. Environ. Manage.* 220: 118–125.

Majumdar A., Upadhyay M.K., Kumar J.S., Barla A., Srivastava S., Jaiswal M.K. & Bose S. 2019. Ultrastructure alteration via enhanced silicon uptake in arsenic stressed rice cultivars under intermittent irrigation practices in Bengal delta basin. *Ecotoxicol. Environ. Saf.* 180: 770–779.

Meharg A.A. & Rahman M.M. 2003. Arsenic contamination of bangladesh paddy field soils: implications for rice contribution to arsenic consumption. *Environ. Sci. Technol.* 37(2): 229–234.

Upadhyay M.K., Majumdar A., Barla A., Bose S. & Srivastava S. 2019. An assessment of arsenic hazard in groundwater–soil–rice system in two villages of Nadia district, West Bengal, India. *Environ. Geochem. Health* 41: 2381–2395.

Arsenic in the Environment: Bridging Science to Practice for Sustainable Development – van der Wal, Ahmad, Petrusevski, Weijma, Savic, van der Wens, Beerendonk, Bhattacharya, Bundschuh & Naidu (Eds)
 ISBN 978-1-032-32928-4

Thiourea (TU) mediated impact on selected transporters and arsenic content in rice (*Oryza sativa* L.) grown in West Bengal, India

M.K. Upadhyay[1], S. Srivastava[1], A. Majumdar[2], A. Barla[2], S. Bose[2], A.K. Srivastava[3] & P. Suprasanna[3]

[1]*Institute of Environment and Sustainable Development, Banaras Hindu University, Varanasi, Uttar Pradesh, India*

[2]*Department of Earth Sciences, Indian Institute of Science Education and Research Kolkata (IISER-K), Mohanpur, West Bengal, India*

[3]*Nuclear Agriculture and Biotechnology Division, Bhabha Atomic Research Center, Mumbai, Maharashtra, India*

ABSTRACT: Arsenic (As) is a ubiquitously present carcinogen and its contamination through food is major threat to the large population of the world. The problem is severe in South East Asia including West Bengal India. Thiourea (TU) is a redox active thiol (-SH) based molecule and found positive against variety of stresses, including As stress in lab condition. It can also improve the redox environment of the plant system which is the key component of As toxicity in rice in lab based study. In this work, we, evaluated the impact of TU on selected transporter expression ($\log_2$ fold) in developing grains of gosai cultivar of rice. The gosai rice cultivar was selected on the basis of largely consumption pattern in the area, grown in three fields (Control, Experimental field 1 and Experimental field 2) in Nadia district, West Bengal, India. The result indicated impact of TU supplementation on As accumulation pattern in gosai rice grain. Further As concentration was also found to be reduced at grain level by 17.24%. Thus TU can act as sustainable solution to tackle this problem by giving quick response in real field scale. This result could also be useful to understand the mechanism of As reduction in rice in sustainable manner.

1 INTRODUCTION

Arsenic (As) is considered as potential carcinogenic metalloid which is classified in class 1 carcinogen by IARC. 150 million people around the world are affected by As hazard. The condition is severe in South East Asian countries including Bengal delta basin region. There are many districts identified so far in West Bengal, where groundwater As concentration is >10 ppm and >50 ppm including Nadia which is permissible limit of As in in drinking water given by WHO and Indian standard. This contaminated groundwater only is used for irrigation in rice fields too as a part of intensive agriculture in Gangetic basin since long time. This the main reason of As accumulation in agricultural crops in this belt. Arsenic accumulates in significant concentrations due to its promising reducing environment (anoxic; flooded) cultivation, various suitable transporters. Further it will diminish the rice yield too (Huhmann *et al.*, 2019). Rice is one of the staple crop across the world, therefore, key source of exposure pathway to majority of population of the world. Further, import and export of rice can may also enhanced the chance of exposure even in no contaminated region of the globe. Redox imbalance, generation of reactive oxygen species (ROS) are important symptoms of As toxicity (Srivastava *et al.*, 2014). Sulfur is also important for plant growth which is also reduced by As toxicity to plant. TU is also sulfur containing ROS scavenger and redox regulator which has already applied in variety of stresses in plants (Srivastava *et al.*, 2014). We applied TU and assessed its impacts on As accumulation in rice grains, transporter expression analysis and yield pattern in three selected As contaminated field.

2 METHODS/EXPERIMENTAL DESIGNING

2.1 *Site description*

The study was accompanied in one of the remote village named Sarapur (23°01′07.8″ N-88°39′43.1″E-23°01′14.3″N 88°38′24.7″E), in Nadia district of West Bengal, India. This area is identified as prolonged use of As contaminated groundwater for irrigation purpose in various crops, vegetables and oil seed crops including rice. Agriculture is the chief profession of the native people. So, this region represented as As contaminated site for agronomy based study

2.2 *Growth of plants, collection and processing the samples*

We choose the rice cultivar Gosai which are the local, predominant and high consumption rate in our designated area. The cultivation of rice was performed in winter (Boro, Jan-April) season at three distinct places

DOI: 10.1201/9781003317395-82

viz- Control, Experimental field 1 (38.74 ± 0.129 mg/kg) and Experimental field 2 (84.77 ± 0.512 mg/kg) presence/absence of thiourea (TU) application (500 ppm for seed soaking followed by two foliar spray application; one at the flowering stage and second at the grain filling stage). The area of per selected field was 100 ft^2. The developing grain samples were dip into liquid nitrogen (Liq N_2) immediately after cutting and chill pack in dry ice. Further we have taken out the samples in lab by preserve in ice filled box and stored at −80 degree. The samples were analyzed for log2 fold expression pattern analysis after crushing in Liq N_2. Similarly, the dehusked ripened grain (brown rice) samples cut washed thrice, air dried and preserved for evaluating the As concentration in it. Further, they were separated and after digestion it has been used for ICP-MS analysis. The soil samples from these three field also evaluated for XRF analysis to know the As concentration. Lastly manually harvested paddy was dehusked and yield data was taken.

3 RESULTS AND DISCUSSIONS

3.1 *Expression profiling of three As transporters in developing rice grains*

The *LSi1* gene was found to be downregulated in both the experimental field but the in experimental field 2 it was more in comparison to other field. However in presence of TU it was downregulated in experimental field 2 that shows TU can only act after some threshold level of stress. *Lsi2* was found to upregulated in presence/absence of TU in experimental field 1 whereas in experimental field 2 it was found to be downregulated in presence/absence of TU. *LSi6* was found to be downregulated in all the condition and in both the experimental field.

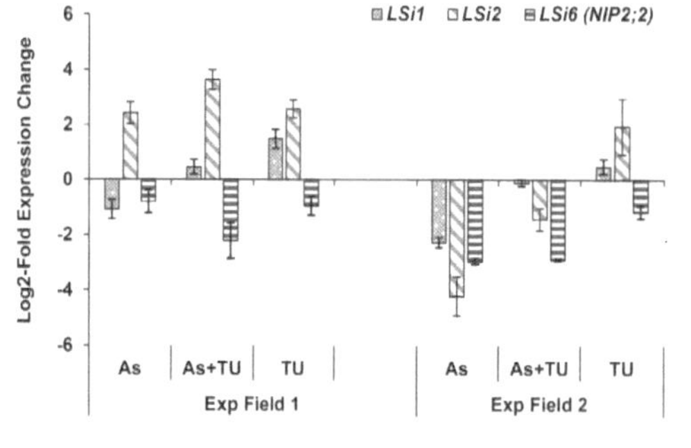

Figure 1. The log 2-fold change in three transporters in comparison with control field.

3.2 *Arsenic (As) concentration in rice grains*

In Boro (winter), the As concentration (mg/kg) in rice grains was found 0.62 ± 0.06, 0.55 ± 0.05 in control field, 0.68 ± 0.04, 0.58 ± 0.04 in experimental field 1 and 0.72 ± 0.04, 0.61 ± 0.04 in experimental field 2 respectively (Figure 2). All the values are in mg/kg. These values are in mg/kg ± SD. The accumulation of As was significantly reduced at grains level in TU treated samples in experimental field 1 only. Therefore, with this result we can conclude that for better result TU may need some threshold level of stress.

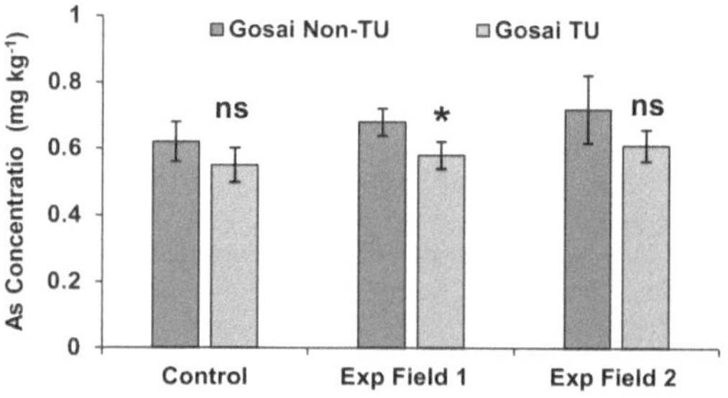

Figure 2. Arsenic (As) in mg/kg in gosai rice rice cultivar with/without thiourea (TU). All values are means of triplicates ± SD. The value found to be significant in Exp field 1 by analysing t-test.

4 CONCLUSIONS

This study specifies that TU can significantly reduce the As load in rice grain by modulating different transporters. Further, TU supplementation also improved growth of plants along with yield (data not shown here). Hence, TU treatment by seed soaking/ priming and spraying at the time of flowering/grain filling can be an achievable plan to mitigate As loading into the rice grains especially in contaminated areas because data showed our that TU actually needs minimal threshold level of As stress to give the full fledge result. However more study also would be needed to standardize the study in holistic manner.

ACKNOWLEDGEMENTS

The authors are thankful to IESD-BHU, Varanasi and BARC Mumbai, India for the lab facility. The first and presenting author is thankful to CSIR-New Delhi, India for Senior Research Fellowship (SRF) to pursue Ph.D in India.

REFERENCES

Huhmann B., Harvey C.F., Uddin A., Choudhury I., Ahmed K.M., Duxbury J.M. *et al.* 2019. Inversion of high-arsenic soil for improved rice yield in Bangladesh. *Environ. Sci. Technol.* 53: 3410–3418.

Srivastava S. & D'souza S.F. 2009. Increasing sulfur supply enhances tolerance to arsenic and its accumulation in *Hydrilla verticillata* (L.f.) royle. *Environ. Sci. Technol.* 43: 6308–6313.

Arsenic in the Environment: Bridging Science to Practice for Sustainable Development – van der Wal, Ahmad, Petrusevski, Weijma, Savic, van der Wens, Beerendonk, Bhattacharya, Bundschuh & Naidu (Eds)
© 2023 the Author(s), ISBN 978-1-032-32928-4

Unveiling arsenic tolerance potential of different rice (*Oryza sativa* L.) genotypes in paddy soil conditions of Pakistan

M.M. Hussain[1], I. Bibi[1], N.K. Niazi[1] & M. Shahid[2]
[1]*University of Agriculture Faisalabad, Faisalabad, Pakistan*
[2]*COMSATS University Islamabad, Vehari Campus, Vehari, Pakistan*

ABSTRACT: Rice is prone to high arsenic (As) accumulation under reduced paddy soil conditions which depends on genotypic variation, although it is poorly understood for Pakistan's scenario. This study aimed to distinguish As-safe (tolerant) genotypes from those which are As-unsafe (accumulating) in a pot experiment utilizing twenty different genotypes. Arsenic in soil (25 mg/kg) showed a significant effect on plant morphological parameters (number of tillers, plant height, root and shoot dry weights, number of spikelets, grain dry weight). Under As stress, Shaheen Basmati resulted in maximum number of tillers (8/pot) while Bas-370 has the minimum number of tillers (3/pot); Bas-2000 produced maximum shoot dry weight (2.5 g/pot) and Kainat genotype showed minimum shoot dry weight (0.12 g/pot). Total As concentration in root and shoot of fine varieties IR-6 and KSK-282 were (44 and 22 mg/kg DW) higher than that of coarse genotypes (3 and 2 mg/kg DW). Grain As concentration was found to be the lowest in IR-36, Noor Basmati, Shaheen Basmati and PB-95 (0.08, 0.10, 0.18, and 0.16 mg/kg DW). This study highlights that IR-36, Noor Basmati, Shaheen Basmati and PB-95 genotypes has grain As concentration below the FAO/WHO safe limit of As in rice grain (0.2 mg/kg DW) and should be considered for safe rice consumption and minimize As exposure to humans via rice consumption.

1 INTRODUCTION

Arsenic (As) contamination of paddy soils, both geogenically and anthropogenically, has led to serious food safety and health issues worldwide, particularly due to significantly higher As uptake in rice grain than that of other crops like wheat, maize, etc. Arsenic has not been recognized as an essential nutrient, but it has been reported to be accumulated in significant amounts in paddy rice (root, shoot, grain) under reduced soil conditions due to the reduction of arsenate (As(V)) form of As to arsenite (As(III) species, which is 100 times more mobile and bioavailable than As(V) under reduced paddy soil environments (Hussain *et al.*, 2019; Liao *et al.*, 2016).

Rice cultivars with high porosity in roots tend to result in more radial oxygen loss (ROL) from their roots, leading to higher As tolerance and lower As levels in their grain and straw. It has been reported that the porosity of root differs markedly among rice varieties and can be further enhanced by the formation of aerenchyma, flooding or deoxygenated nutrient solution. Recently, 13 cultivars grown at two field sites each in Bangladesh, India and China were evaluated for grain As. There was significant site, genotype, site by genotype interaction for total grain As. Correlations were observed only between sites in Bangladesh and India and not between countries. For seven cultivars, the As was speciated which revealed significant effects of site, genotype and site by genotype interaction for percentage of As (Tuli *et al.*, 2010). However, no research has been done so far on As tolerance among the rice genotypes in Pakistan. The objective of this study was to screen different rice genotypes to evaluate the As accumulation and/or tolerance under paddy soil conditions in a pot experiment.

2 MATERIALS AND METHODS

2.1 *Pot experiment*

Soil was collected from the surface layer (0–20 cm) of a paddy field having organic matter content of 0.66%, pH 7.71, available Na, P and K concentrations of 19, 7.7, 206 mg/kg, respectively. The soil was air dried, sieved through a 2 mm size sieve and mixed well prior to filling the pots (2 kg per pot). Soil was contaminated with sodium arsenate at 25 mg/kg level and left for 5 months incubation period in the laboratory to allow redistribution of As in different pools, possibly mimicking natural As contaminated soil. Total 20 rice genotypes (coarse and fine) were selected and grown under ambient conditions. All the 20 rice

DOI: 10.1201/9781003317395-83

varieties were screened by growing in As-contaminated soil.

2.2 *Arsenic analysis in rice shoot, root and grain*

For As analysis, root, shoot and grain samples were harvested after 100 days of growth, oven-dried and ground to fine powder prior to single acid digestion. Shoot, root and grain samples were digested using nitric acid and hydrogen peroxide at 180°C. Arsenic content in plant tissue was determined using a hydride generation-atomic absorption spectrometer (HG- AAS; Agilent AA240 with VGA 77, USA).

Analysis of variance (ANOVA) was used to determine significant differences between treatments following Tukey's test at $p < 0.05$.

3 RESULTS AND DISCUSSION

Results showed that for all 20 genotypes number of tillers, plant height, fresh and dry weight of grains were in range of 3–8 tillers/pot, 27–46 cm, 9–23 g/pot and 3–14 g/pot, respectively. While in As-contaminated soil, number of panicles ranged from 3–12/pot for all the screened 20 rice genotypes. The mean grain dry weight of different rice genotypes was in the range of 13–34 g/pot for rice plants in uncontaminated soil. Results showed that the trend in morphological and agronomic data was partially in agreement with Syu *et al.* (2015).

3.1 *Arsenic concentration in rice root, shoot, husk and grain*

Arsenic concentration in different parts of rice plants in both type of genotypes showed a decreasing trend in the order of root > shoot > grain. The mean shoot and root As concentration for all the twenty genotypes ranged from 1.53–8.70 and 101–386 mg/kg DW, respectively; and for rice grain mean As concentration spanned 0.12–0.76 mg/kg DW for the genotypes investigated in this study under As tress (Table 1). The maximum mean grain As content was observed in Basmati–385 and minimum in Noor Basmati ($P < 0.05$). Significantly, mean grain As concentration for IR-36, Noor Basmati, Shaheen Basmati and PB-95 was below safe limit of As in rice grain (0.2 mg/kg DW) set by WHO/FAO (Irem *et al.*, 2019).

Table 1. Total arsenic (As) concentration (mg/kg DW) in root, shoot and grain of 20 different rice genotypes in As-contaminated soil.

	Root As in As-contaminated soil		Shoot As in As-contaminated soil		Grain As in As-contaminated soil	
Varieties	Mean	S.D ±	Mean	S.D ±	Mean	S.D ±
PK-386	121 e	24	3.63 e	0.12	0.48 c	0.14
PS-2	220 b	42	4.78 d	0.08	0.36 d	0.19
NIAB IRRI-9	202 b	44	2.56 f	0.14	0.59 c	0.21
IR-36	386 a	21	3.69 e	1.56	0.19 f	0.14
IR-6	284 b	18	4.18 d	1.46	0.37 d	0.21
KSK-434	133 e	39	2.8 f	0.21	0.22 ef	0.05
KSK-282	365 a	33	1.53 g	0.36	0.32 d	0.13
KSK-133	324 a	47	1.87 g	0.21	0.45 c	0.13
Basmati 2000	134 e	18	3.76 e	0.46	0.24 e	0.11
Noor Basmati	190 c	15	3.64 e	1.2	0.12 f	0.03
Shaheen Basmati	178 c	15	4.86 d	1.1	0.16 f	0.09
Basmati-515	165 d	28	6.35 b	1.73	0.29 e	0.1
NIAB-2013	154 d	42	5.78 c	1.09	0.44 c	0.09
Basmati-198	101 f	11	6.82 b	2.28	0.56 b	0.11
Super Basmati	105 f	4.6	2.99 f	2.11	0.23 e	0.1
Basmati-385	170 c	6.4	8.70 a	0.16	0.76 a	0.14
Basmati-370	190 c	6.6	2.76 f	0.79	0.36 d	0.15
Kainat	226 b	24	6.85 b	1.62	0.59 b	0.14
PB-95	201 b	26	3.89 e	1.68	0.15 f	0.02
IR-1561	187 c	29	5.36 c	2.69	0.13 f	0.03

4 CONCLUSIONS

Arsenic concentration in root, shoot and grain was significatlly lower in coarse rice genotypes (Bas-198, Su-Bas, Bas-385 and NI-9) compared to the fine (Kainat, NI-9, IR-6, KSK-434 and KSK-282) genotypes. Arsenic contamination reduced the growth of different genotypes, although some rice genotypes were tolerant to As uptake from soil via root and its subsequent transfer to grains. Arsenic concentration in different rice parts was in the order of root > shoot > grain. Some rice genotypes accumulated more As in grains than the permissible limit of FAO. Basmati-385 is considered to be an As-unsafe genotype, while IR-36, Noor Basmati, Shaheen Basmati and PB-95 are identified as As-safe genotypes in terms of high As tolerance and reduced accumulation in grain. The IR-36, Noor Basmati, Shaheen Basmati and PB-95 contained grain As concentration below the safe limit of As in rice grain (0.2 mg/kg DW) set by WHO/FAO, hence these genotypes could be recommended for cultivation in As-conatminted areas of Pakistan. The data imply that As-safe rice genotypes should be promoted for cultivation in Pakistan, in order to minimize As induced health risk which could incur, nationally and globally, via consumption of As-rich rice grain. There is a need to investigate the transport pathways, uptake mechanisms and the detoxification mechanisms which could exist in different rice genotypes and establish new knowledge for As mitigation programs.

ACKNOWLEDGEMENTS

The authors are thankful to Higher Education Commission (6425/Punjab/NRPU/ R&D/HEC/ 2016 and 6396/Punjab/NRPU/R&D/HEC/2016), Pakistan for financial support. Thanks to M. Waseem Akhtar for contributing in pot experiment.

REFERENCES

Hussain M.M., Bib.i I., Shahid M., Shaheen S.M., Shakoor M.B., Bashir S. *et al.* 2019. Biogeochemical cycling, speciation and transformation pathways of arsenic in aquatic environments with the emphasis on algae. (Chapter Two). In: A.C. Duarte & V. Reis (eds.). *Comprehensive Analytical Chemistry*. 85, pp. 15–51.

Irem S., Islam E., Maathuis FJ.M., NiaziN.K., Li T. 2019. Assessment of potential dietary toxicity and arsenic accumulation in two contrasting rice genotypes: effect of soil amendments. *Chemosphere* 225: 104–114.

Liao G, Wu Q., Feng R., Guo J., Wang R., Xu Y. *et al.* 2016. Efficiency evaluation for remediating paddy soil contaminated with cadmium and arsenic using water management, variety screening and foliage dressing technologies. *J. Environ. Manag.* 170: 116–122.

Syu C.-H., Huang C-C, Jiang P-Y, Lee C-H, Lee D-Y. 2015.Arsenic accumulation and speciation in rice grains influenced by arsenic phytotoxicity and rice genotypes grown in arsenic-elevated paddy soils. *J. Hazard. Mater.* 286: 179–186.

Tuli R., Chakrabarti D., Trivedi P.K., Tripathi R.D. 2010. Recent advances in arsenic accumulation and metabolism in rice. *Molecular Breeding* 26: 307–323.

Arsenic in the Environment: Bridging Science to Practice for Sustainable Development –
van der Wal, Ahmad, Petrusevski, Weijma, Savic, van der Wens, Beerendonk, Bhattacharya, Bundschuh & Naidu (Eds)
 ISBN 978-1-032-32928-4

Arsenic content in a micro-basin soil and its effects on soil biodiversity

F.H. dos Santos[1], E.B. da Silva[2] & L.R.F. Alleoni[1]
[1]*ESALQ/USP Soil Science Department, College of Agriculture, University of São Paulo, São Paulo, Brasil*
[2]*UF/ IFAS Soil and Water Sciences Department, University of Florida, Gainesville, USA*

ABSTRACT: Soil contamination by potential toxics elements (PTE's) can change the ecosystem structure, its function and cause risk to human health. Soil is one of the main pathways of PTE's exposure to humans, so that is important to evaluate its content and distribution in the micro basins in order to assess the human health risk. In this study, we evaluated the Arsenic (As) content in 15 soils from Guamium micro-basin, located in Piracicaba, Sao Paulo, Brazil. The effect of As rates on collembola reproduction was also evaluated in both natural samples from Guamium soils and artificial soils (ArtS). As and Cu had the highest contents in natural soils, and their concentration were close or higher than prevention value established by national and international standards. As availability was high in the ArtS and it decreased the collembola reproduction. The Lowest Observed Effect Concentration (LOEC) on ArtS were 0.25 mg/kg, while No Observed Effect Concentration (NOEC) were less than 0.25 mg/kg. Conversely, the LOEC were 8.41 mg/kg in the natural soil, but it was not possible to define the NOEC.

1 INTRODUCTION

Soil is one natural resource responsible for the sustainability of food supply and environmental quality (Brevik & Burgees 2014). Changes in land use mainly due to deforestation affect their ability to purify and immobilize contaminants, such as arsenic (As). As is one of the most studied elements worldwide, recognized by its toxicity towards human and other forms of life. The changes in land use and grow up of content of arsenic in Guamium hydrographic basin soils, in Piracicaba, Brazil, has offered risk to the local population. The present study aims to investigate: i) the concentration of arsenic in soil samples from the Guamium River basin; and ii) evaluate the ecotoxicity of As in collembolas. In this abstracts results on actual situation, ecotoxic effects and health risk are presented.

2 MATERIALS AND METHODS

2.1 *Study area*

The Guamium hydrographic basin is located in Piracicaba, São Paulo (SP), Brazil, 47° 34′ 20.85″O 22° 32′ 18.64″S (Figure 1). It has an area of 7,051 ha, of which 491 ha are forest and only 243 ha of permanent preservation area. The remaining area is divided in both urban area (481 ha) and rural use (5,477 ha, representing 78% of the total), cultivated mainly with sugar cane. The last ones mentioned, are highly degraded due to contaminants present in inputs used in agriculture or those released by industries present in the region.

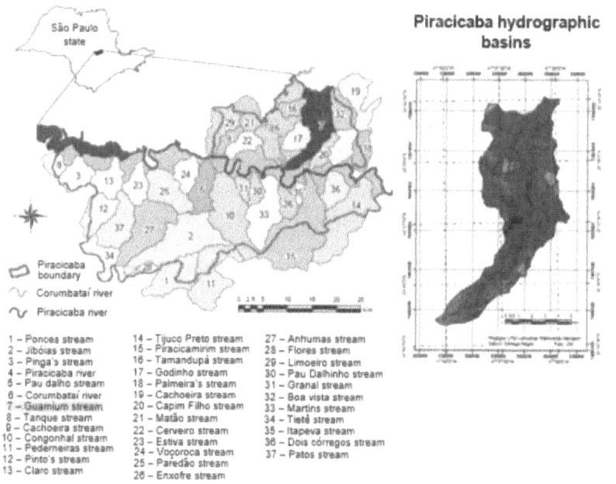

Figure 1. Study area.

2.2 *Soil sampling and laboratory analysis*

A total of 15 soil samples were collected at depths of 0 to 1.2m in each 10 cm. The parameters such as, pH, organic carbon and major inorganic ions were analyzed. The determination of PTE's was accomplished by 3051A method USEPA. As was determined by Atomic Fluorescence Spectrometry (AFS). The recovery of the extraction was verified through the inclusion of certified soil samples (Montana Soil, NIST, SRM 2711).

DOI: 10.1201/9781003317395-84

2.3 *Acute toxicity tests (ACT)*

Bioassays are used to assess acute toxicity and to determine the appropriate concentrations for testing sub-lethal effects in species, being described in normative protocols. Artificial soils (ArtS) composed of fine sand (70%), kaolinite clay (20%) and coconut fiber powder (10%), as well as samples of natural soil (Typic Hapludox Rhodic) were used to run the tests with five rates of arsenic (2.5; 5; 15; 45 and 135 mg kg^{-1}) that were added in the form of $Na_2HAsO_4.7H_2O$. The specie *Candida folsomia* was select since it is commonly used to assess the risk of pollutants because of sensitivity to the presence of EPT's (Phillips *et al.*, 2002).

3 RESULTS AND DISCUSSION

3.1 *Soil quality parameters*

The levels of PTE's of the soils of the Guamium basin were lower than the Quality Reference Value (QRV), with the exception of As and Cu. The average levels of As in the soils of the Guamium Basin in the superficial layer (0 – 0.2 m) ranged from 3.7 to 17.6 mg/kg, with an average of 7.7 mg/kg. All soils showed higher values than VRQ. The Guamium basin is located under three major geological formations: Serra Geral Formation (igneous rocks), Corumbataí Formation (sedimentary rocks) and Irati Formation (sedimentary rocks) (Figure 2). As there was no correlation with depth indicating a low influence of anthropic actions, this result may be an indication of the contribution of the geological material.

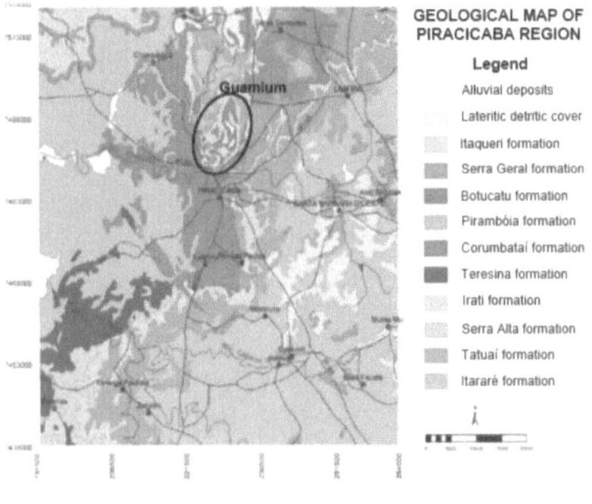

Figure 2. Geological map of Piracicaba, SP.

3.2 *Ecotoxicological test on collembola*

The reproduction of the collembola was not affected by the As doses in the natural soil. On the other hand, there was a significant reduction in the reproduction of collembola in the ArtS. This difference between soils can be explained by the greater availability of As in the ArtS compared to natural soil. The lower availability of As in natural soil was related to the higher levels of clay – iron and aluminum oxides and hydroxides present in this soil – in comparison to artificial soil. The greater availability of As in the artificial resulted in a low value for LOEC, whereas in natural soil it was not possible to observe this concentration. Moreover, the NOEC of the natural soil was the highest concentration extracted by the method USEPA 3051a.

4 CONCLUSIONS AND RECOMMENDATIONS

The high levels of arsenic present a serious threat to agriculture in Guamium basin soils, as well as to local residents. The most significant source of arsenic is the geology of the site. Due to the high degree of weathering and high anion retention capacity, As does not significantly affect the reproduction and development of collembola. In order to define the LOEC value in the ArtS, a new experiment would be necessary with As concentrations less than 0.25 mg/kg.

ACKNOWLEDGEMENTS

The authors are grateful for the financial support provided by the National Council for Scientific and Technology Development (CNPq) n° 132590/2019-0 and by São Paulo Research Support Foundation (FAPESP) n° 11/04453-4.

REFERENCES

Brevik E.C. & Burgess L.C. 2014. The influence of soils on human health. *Nat. Educ. Knowl.* 5(12): 1.

Phillips C.T., Kuperman R.G. & Checkai R.T. 2002. Toxicity of chemical-warfare agent HD to *Folsomia Candida* in different soil types. *Eur. J. Soil Biol.* 38: 281–285.

2.4 Arsenic in aquatic environment, speciation and toxicity

Arsenic in the Environment: Bridging Science to Practice for Sustainable Development –
van der Wal, Ahmad, Petrusevski, Weijma, Savic, van der Wens, Beerendonk, Bhattacharya, Bundschuh & Naidu (Eds)
 ISBN 978-1-032-32928-4

Arsenic transfer and metabolism in marine herbivorous fish, *Kyphosus* spp.: A research proposal

T. Siregar[1,3], T. Sanchez-Palacios[1], A. Gunn[1,2], T. Kelly[1,2] & S. Foster[1]
[1]*Omic Research Group, Laboratory of Analytical Chemistry, Faculty of Science and Technology, University of Canberra, Canberra, Australia*
[2]*National Centre for Forensic Studies, Faculty of Science and Technology, University of Canberra, Australia*
[3]*Research Center for Marine and Fisheries Post-Harvest, Ministry of Marine and Fisheries, Central Jakarta, Indonesia*

ABSTRACT: Arsenic in marine environment is ubiquitous, yet it can be toxic to living organisms. Arsenic enters the marine food chain through algae. Algae convert the accumulated inorganic containing As molecules into methylated forms and more complex molecules such as carbohydrates (i.e. arsenoribosides), fatty acids and other organic species. Arsenoribosides can make up 70% of the total accumulated As in algae. When transferred to herbivorous fish such as members of the family Kyphosidae, arsenoribosides can be converted into less toxic forms such as thio-, glycerol-, phosphate- and methylated-arsenoribosides or substantially converted into arsenobetaine. Much remains to be understood however regarding the transfer and metabolism of arsenoribosides in marine organisms. In this study, we proposed to conduct As-speciation analyses in the intestine system of *Kyphosus* spp., exposed to As-containing seaweed. Extraction and separation analytical techniques for As-speciation such as HPLC-HG-AFS, HPLC-ICPMS and LC-MS-MS will be employed for As containing metabolites determination. It is hypothesised that changes in proportion of As metabolites will be observed along the intestine of *Kyphosus* spp.

1 INTRODUCTION

Arsenic in the environment is ubiquitous (> 1μg/L) and it can be found in high concentrations, associated mainly to metal ores such as copper, lead and gold (Oremland & Stolz 2003). Arsenic concentration in seawater can range from 1 to 8 μg/L, and is mostly found as inorganic As (iAs) in two different oxidative stages, reduced (AsIII) and oxidized (AsV) (Mandal & Suzuki 2002). Accumulation of iAs in seawater depends on seasonal conditions, including sea depth and biological activities. In addition to that, antrophogenic activities such as mining, smeltering, combustion and pesticides in agriculture have contributed to the increasing levels of iAs in marine environments.

Inorganic As (arsenate/AsV and arsenite/ AsIII) in the water column and sediments can be taken up by seaweed. Inside, iAs is transformed into less toxic and more complex forms such as carbohydrate and ribose forms. The metabolism of iAs in marine primary producers occurs in a series of reduction, oxidation and addition reaction steps (Edmonds & Francesconi 2003). Interestingly, arsenoribosides (arsenosugars) are the most abundant form of As in seaweed, whereas in seaweed eating fish, arsenobetaine (AsB) is the most predominant form (Figure 1).

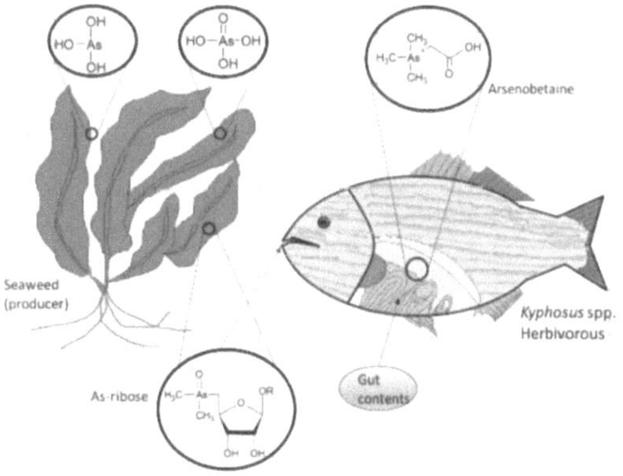

Figure 1. Arsenic uptake and transfer at the primary levels of the food chain in marine environment.

Three pathways for the transformation of arsenoribosides and methylated forms of As into AsB have been proposed (Caumette *et al.*, 2012) (Figure 2). The general consensus is that AsB biosynthesis occurs in situ, most likely inside the fish guts. However, more research is needed to determine whether there is one or several metabolic pathways involved in the biosynthesis of AsB in fish.

In the last decade, use of separation analytical methods such as high-performance liquid chromatography in combinaton with inductively couple plasma mass spectrometry (HPLC-ICP-MS) and liquid chromatography tandem mass spectrometry (LC-MS-MS) have been employed to determine the biochemical transformations of As in seaweed and fish. Most recently, the use of cost effective separation methods such as high-performance liquid chromatography hydrate generator atomic fluorescence spectrometry (HPLC-HG-AFS) has been adopted with promising outcomes. In this study our aim is to use widely adopted As-especiation techniques to (i) determine As species in the intestinal tract of two marine

DOI: 10.1201/9781003317395-85

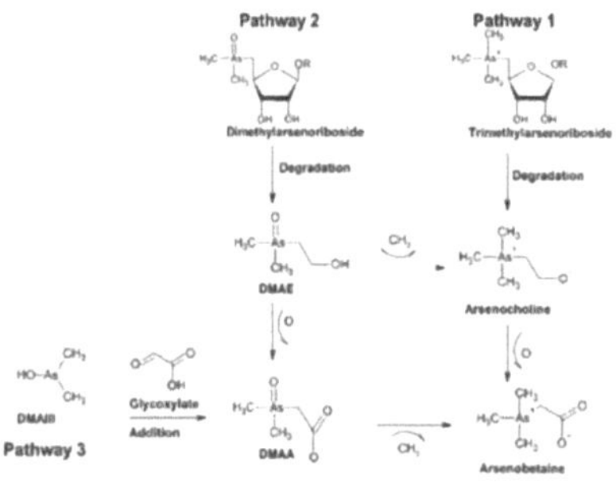

Figure 2. Possible pathways of Arsenobetaine formation, adapted from Caumette *et al.* (2012).

herbivorous fish from Indonesia, K. vaigiensis and K. cinerascens and (ii) investigate in particular the biosyntesis of AsB within these marine fish species.

2 MATERIALS AND METHODS

2.1 *Sample collection and storage*

Sampling of herbivorous fish *Kyphosus* spp. was undertaken in July 2019 in the Belitung Island catchment, Indonesia. Collected samples were kept on ice and transported to the Laboratory, Ministry of Marine and Fisheries, Jakarta. Fish were washed with DI water, total length and mass measured and dissected. Fish sample were divided into three groups (i.e. small, medium and big) based on their mass and length. The fat surrounding the instestine was removed to minimise cross contamination of As containing fat. Cleaned intestines were further divided into five gut segments as described by Mountfort *et al.* (2002). All intestine sections were placed in acid washed plastic vials and subsequent freeze-dried to preserve the identity of As forms present in the sample (LaboGene, ScanVac). Freeze-dried samples were sent to the analytical laboratory at the University of Canberra for further As-speciation analyses.

2.2 *Extraction and separation methods*

Speciation of arsenic species will be done using hypenated instrument HPLC-ICPMS (Perkin Elmer, USA) and HPLC-HG-AFS (Simadzu and PS Analytical). Standards (Sigma Aldrich and in-house Synthesis) and reference material were prepared prior to analysis. Methylated arsenic forms can be found as monomethyl, dimethyl or trimethyl arsenic as part of metabolic processes. The purpose of analysis using HPLC-GF-AFS is to see the process of methylation in both algae as a primary producer and fish as a primary consumer. This analysis will focus on reduction or oxidation processes and whether they are influenced by enzymatic processes. Other As metabolites such as arsenoribosides, arsenobetaine, As-containing lipid will be analysed using HPLC-ICPMS or LC-MS-MS when analytical standards not available.

Samples will be fractionated in three different extraction layers i.e. DCM/MeOH for lipid soluble, MeOH/H2O for water soluble and H2O for residue. Lipid soluble species including As containing fatty acids and As containing hydrocarbons will separated using C18, reverse phase column. Methanol/water soluble species will be separated using both anion and cation exchange columns. Cation species i.e. arsenobetaine, arsenocholine, ribosides and other complex methylated products will be separated using Supelco SCX column (buffer 20 mM pyridine, pH 2.6).

Inorganic and simple methylated As species will be separated using anion exchange column PRPX100 (buffer 20 mM $(NH_4)_3PO_4$, pH 5.6) as described in previous studies conducted in our research group (Foster *et al.*, 2007) with several adaptation (Figure 3).

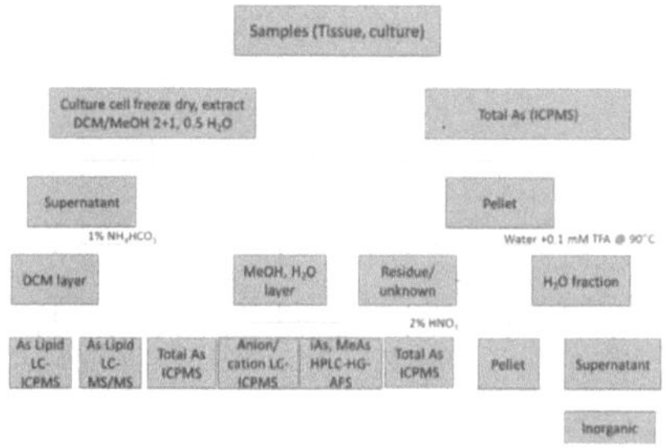

Figure 3. Arsenic extraction and analysis method.

3 EXPECTED OUTCOMES RESULTS AND PREDICTIONS

It is hypothesised that changes in proportion of As metabolites will be observed along the intestine of Kyphosus spp., such as As-ribosides, As methylated, Arsenobetaine and other arsenic-containing metabolite products.

ACKNOWLEDGEMENTS

We thank the Australian Government for providing a PhD scholarship for T. Siregar. Special acknowledge ments are made to Ministry of Marine and Fisheries for providing samples and lab facilities in Jakarta, In donesia.

REFERENCES

Caumette G., Koch I. & Reimer K.J. 2012. Arsenobetaine formation in Plankton: a review of studies at the base of the aquatic food chain. *J. Environ. Monit.* 14(11): 2841–2853.

Edmonds J.S. & Francesconi K.A. 2003. Organoarsenic compounds in the marine environment. In: P.J. Craig (ed.) *Organometallic Compounds in the Environment.* John Wiley and Sons, New York. p. 196–222.

Foster, S., Maher, W., Krikowa, F. & Apte, S. 2007. A Microwave-assisted sequential extraction of water and dilute acid soluble arsenic species from marine plant and animal tissues. *Talanta* 71(2):537–549.

Mandal B. & Suzuki K. 2002. Arsenic round the world: a review. *Talanta* 58(1): 201–235.

Mountfort D., Campbell J. & Clements K. 2002. Hindgut fermentation in three species of marine herbivorous fish. *Appl. Environ. Microbiol.* 68(3): 1374–1380.

Oremland R. & Stolz J. 2003. The ecology of arsenic. *Science* 300(5621): 939–944.

Arsenic in the Environment: Bridging Science to Practice for Sustainable Development – van der Wal, Ahmad, Petrusevski, Weijma, Savic, van der Wens, Beerendonk, Bhattacharya, Bundschuh & Naidu (Eds)
© 2023 the Author(s), ISBN 978-1-032-32928-4

Total, inorganic and bio-accessible arsenic in children diets of Bangladesh: Exposure assessment

M.M. Rahman[1,2], M. Alauddin[3], A.B. Siddique[1,2], M.R. Islam[1,2], S.K. Shaha[4] & R. Naidu[1,2]
[1]*Global Centre for Environmental Remediation (GCER), Faculty of Science, The University of Newcastle, Callaghan, NSW, Australia*
[2]*Cooperative Research Centre for Contamination Assessment and Remediation of the Environment (CRC CARE), The University of Newcastle, Callaghan, NSW, Australia*
[3]*Department of Chemistry, Wagner College, Staten Island, NY, USA*
[4]*Bangladesh Institute of Research and Rehabilitation in Diabetes, Endocrine and Metabolic Disorders (BIRDEM), Shahbag, Dhaka, Bangladesh*

ABSTRACT: This study investigated the potential exposure of arsenic (As) in diets for children in two villages of As-affected areas of Bangladesh. Children meals from lunch were collected and analyzed for total, inorganic As as well as bioaccessible As (BA). The results indicated that the concentrations of As in children meals were elevated which pose significant risk to the study group.

1 INTRODUCTION

Drinking water and food are two major exposure routes of arsenic (As) to human. Health effects due to the exposure of As-contaminated water and food has been reported from many regions including Taiwan, Chile, Argentina, Bangladesh, India where millions are exposed to As and suffering from various As-related diseases (Rahman *et al.*, 2009a). In a study, it was reported that consumption of cooked rice with >200 μg/kg may cause DNA damage in a section of population from As-affected area of West Bengal, India (Banerjee *et al.*, 2013). In our previous study, we reported that the female groups are more susceptible than male groups while the 2–5 yrs and 6–10 yrs age groups are supposed to be more exposed to As from rice consumption (Islam *et al.*, 2017). The results are frightening considering the future generation is at higher risk of As toxicity from the consumption of rice, which is the major staple food of the population in Bengal delta regions of Bangladesh and India.

Thus, it is crucial to estimate the actual exposure of As from diets for children in Bangladesh and the consequence risk by consuming food. In this study, we report exposure of As from diet (especially lunch) from two As-contaminated villages of Bangladesh. The objectives of this study were to investigate the concentrations of total and inorganic As in children diet collected during lunch time. The bioaccessible As (BA) in diets using simulated in vitro digestion method including the gastric (G) and gastrointestinal (GI) phases were also assessed.

2 METHODS

For this study, 14 diet samples from lunch menu were collected from 14 households in two As-contaminated villages (Shahpur and Sursui) of Bangladesh during 2018. The households were selected as random with at least 2 children. Altogether 31 children including both boys and girls were selected for this study and their portion size was weighted from lunch to determine the daily dietary intake rates of As. The samples were stored in zip-log bags, which were kept in the fridge at 4°C until processing. Food samples were dried in an oven at 65°C and then at homogenized by grinding. The samples were then transported by courier to the University of Newcastle. Digestion of the meal samples for total As analysis was carried out employing the method used by (Rahman *et al.*, 2009b). Inductively coupled plasma mass spectrometer (ICP-MS) (Agilent Technologies, Tokyo, Japan), was used to determine the amount of total As. The extraction of food samples for As speciation was conducted using (Signes-Pastor *et al.*, 2016). A liquid chromatography system (model 1260, Agilent Technologies, Tokyo, Japan) equipped with a guard and Hamilton PRP-X100 separation column, coupled with ICP-MS was used for As speciation analysis. The bioaccessible As (BA) was determined as per the method from the previously described studies (Kafaoglu

DOI: 10.1201/9781003317395-86

et al., 2016; Llorente-Mirandes *et al.*, 2016; Zhuang *et al.*, 2016).

3 RESULTS AND DISCUSSION

This study included 31 children (16 boys and 15 girls). Their age, body weight as well as daily amount of food consumption (rice and curry) was collected. The age and weight of children was 8.3 yrs (range 2–15 yrs) and 26 kg (range 10–56 kg). The average daily food consumption was 304 g (range: 85–563 g) (Table 1).

The mean, median and range of As in food (dry wt.) were 1072 µg/kg, 839 µg/kg and (586–1975) µg/kg, respectively, dry weight. Considering on an average 80% moisture contents in cooked food in this study, the mean, median and range of As in food (wet wt.) were 214 µg/kg, 168 µg/kg and (117–395) µg/kg, respectively. The total As concentration in children diets from Sonora, Mexico was ranged from 50 to 1150 µg/kg, dry wt. (García-Rico *et al.*, 2012), which is lower than this study.

Table 1. General information of study participants.

Gender	n	Age (yrs) Mean (range)	Weight (kg) Mean (range)	Daily consumption rate (g) Mean (range)
Boys	16	8.5 (3–14)	25 (13–43)	303 (148–563)
Girls	15	8.1 (2–15)	26 (10–56)	306 (985–489)
Total	31	8.3 (2–15)	26 (10–56)	304 (85–563)

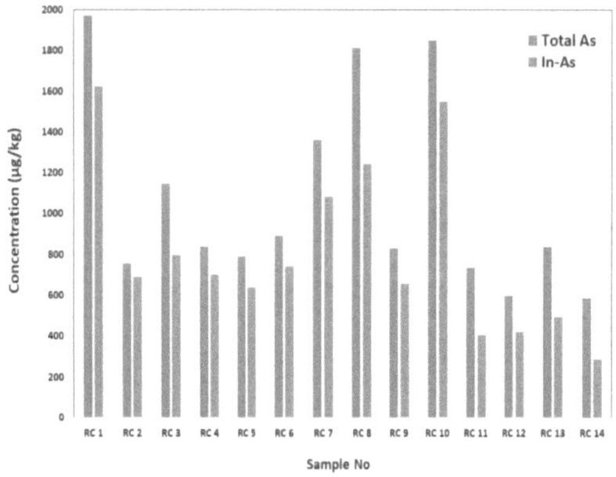

Figure 1. Total and inorganic As in children diets from Bangladesh.

From the analytical results, it appears that inorganic As was the predominant species present in food samples in this study. DMA(V) was also present in all food samples except one. This study could not detect any MMA(V) in food samples. The percentage of inorganic As in food was varied from 49% to 91%. The present of high percentage of inorganic As in diets pose significant risk to the study children.

The range of As extracted from the G and GI phases were 68–105% (mean 91 ± 13) and 72–117% (mean 98 ± 11), respectively. No significant differences was observed in the bio-accessible As between G and GI phases although bio-accessibility of As was slightly increased in the GI phase.

4 CONCLUSIONS

The study reveals that diet is crucial for the exposure of As in the children of As-affected areas, which poses a significant risk to the local inhabitants. Further work is necessary to reduce the As levels from diet as well as to minimize the human exposure risk and potential cancer risk from consuming As-contaminated food.

REFERENCES

Banerjee M., Banerjee N., Bhattacharjee P., Mondal D., Lythgoe P.R., Martínez M. *et al.* 2013. High arsenic in rice is associated with elevated genotoxic effects in humans. *Sci. Rep.* 3: 2195.

García-Rico L, Tejeda-Valenzuela L., Vélez D. & Montoro R. 2012. Content of selenium, total and inorganic arsenic and bioaccessibility of arsenic in children diets of Mexico. *J. Sci. Food Agric.* 92: 1725–1731.

Islam S., Rahman M.M., Islam M. & Naidu R. 2017. Geographical variation and age-related dietary exposure to arsenic in rice from Bangladesh. *Sci. Total Environ.* 601: 122–131.

Kafaoglu B, Fisher A., Hill S. & Kara D. 2016. Determination and evaluation of element bioaccessibility in some nuts and seeds by in-vitro gastro-intestinal method. *J. Food Compos. Anal.* 45: 58–65.

Llorente-Mirandes T, Llorens-Muñoz M., Funes-Collado V., Sahuquillo À. & López-Sánchez J.F. 2016. Assessment of arsenic bioaccessibility in raw and cooked edible mushrooms by a PBET method. *Food Chem.* 194: 849–856.

Rahman M.M., Ng, J.C. & Naidu R. 2009a. Chronic exposure of arsenic via drinking water and its adverse health impacts on humans. *Environ. Geochem. Health* 31: 189–200.

Rahman M.M., Owens G. & Naidu R. 2009b. Arsenic levels in rice grain and assessment of daily dietary intake of arsenic from rice in arsenic-contaminated regions of Bangladesh—implications to groundwater irrigation. *Environ. Geochem. Health* 31: 179–187.

Signes-Pastor A.J., Carey M. & Meharg A.A. 2016. Inorganic arsenic in rice-based products for infants and young children. *Food Chem.* 191: 128–134.

Zhuang P., Zhang C., Li Y., Zou B., Mo H., Wu K. *et al.* 2016. Assessment of influences of cooking on cadmium and arsenic bioaccessibility in rice, using an in vitro physiologically-based extraction test. *Food Chem.* 213: 206–214.

Arsenic in the Environment: Bridging Science to Practice for Sustainable Development – van der Wal, Ahmad, Petrusevski, Weijma, Savic, van der Wens, Beerendonk, Bhattacharya, Bundschuh & Naidu (Eds)

Uptake and dietary toxicity of arsenic in rice genotypes: Effect of organic and inorganic amendments

M.M. Hussain[1], N.K. Niazi[1,2], I. Bibi[1], M. Shahid[3], S. Bashir[4], M.F. Nawaz[5] & Z. Aslam[6]
[1]*Institute of Soil and Environmental Sciences, University of Agriculture Faisalabad, Faisalabad, Pakistan*
[2]*School of Civil Engineering and Surveying, University of Southern Queensland, Toowoomba, Queensland, Australia*
[3]*Department of Environmental Sciences, COMSATS University Islamabad, Vehari Campus, Vehari, Pakistan*
[4]*Sub-campus Depalpur, University of Agriculture Faisalabad, Okara, Pakistan*
[5]*Department of Forestry & Range Management, University of Agriculture Faisalabad, Faisalabad, Pakistan*
[6]*Department of Agronomy, University of Agriculture Faisalabad, Faisalabad, Pakistan*

ABSTRACT: This study examined the impact of various types of amendments (farmyard manure (FYM), cow dung (CD), biogas slurry (BGS), biomaterial (Biom), gypsum (GYP) and lignite (LIG)) on arsenic (As) accumulation by the two contrasting rice genotypes in a pilot-scale experiment. Arsenic-contaminated irrigation water (i.e., 45 mg/L) was applied at three distinct growth stages of rice plants at transplanting, tillering and milking stages for both the genotypes, i.e., Kainat and Basmati-385 (fine and coarse, respectively). Results showed that a significant increase in shoot, root and grain dry weight, plant height, number of tillers per plant where organic amendments were added with maximum grain weight yielded for FYM, CD and Biom treatments. Arsenic concentration in shoot and grain significantly reduced compared to the control for the respective genotypes. The shoot As concentration ranged from 3.1–28 mg/kg DW for Kainat as for Basmati-383 it was 1.7–16 mg/kg DW for all amendments and minimum was found where CD and Biom was applied. The average grain As concentration for Kainat genotype was higher in plants with GYP and LIG treatments compared to CD and Biom treatments (0.29 and 0.04 mg/kg DW). In the case of Basmati-385, the mean average grain As concentration was 0.24 and 0.09 mg/kg As DW for CD and Biom treatments, respectively. mean grain As concentration for both the rice genotypes in CD and Biom amendments was below safe limit of As in rice grain (0.2 mg/kg DW) set by WHO/FAO. This study highlights that organic amendments, CD and Biom, could bind As from soil and decrease As contents in rice shoot and grain, and CD and Biom could possibly enhance the growth and yield of two contrasting rice genotypes.

1 INTRODUCTION

Geogenic As contamination of groundwater is a global public health, agricultural and environmental issue. In contrast to other crops, As accumulation in rice is high under the reduced paddy soil conditions due to transformation of As(V) to As(III), which is more mobile and toxic than As(V) (Hussain *et al.*, 2019; Shakoor *et al.*, 2018). Rice is cultivated in paddy soils receiving As-contaminated irrigated water could be inevitably accumulated in its grain and shoot, thus increasing As exposure risk to the humans.

There exists abundant evidences that As negatively interferes with several biochemical and physiological processes inside plant causing reduced plant growth and yield (Abbas *et al.*, 2018). It is imperative to evaluate efficiency of different organic and inorganic amendments for reducing As uptake by rice grain. In this study, we examined impact of various organic and inorganic amendments on As uptake in rice grain.

2 MATERIALS AND METHODS

2.1 *Pot experiment*

Soil was collected from the surface layer (0–20 cm) of a paddy field having organic matter content 0.66%, pH 7.71, available Na, P and K concentrations of 19, 7.7, 206 mg/kg, respectively. The soil was air dried, sieved through a 2 mm size sieve and mixed well prior to filling the pots (2 kg per pot). Two rice (*Oryza sativa* L.) genotypes, Kainat and Basmati-385, (fine and coarse, respectively) were transplanted in each pot (2 seedlings per pot). Arsenic-contaminated irrigation water (concentration?) was applied at three intervals, i.e., at transplanting, tillering and milking stages. Different amendments were applied in soil, which included farmyard manure (FYM), cow dung (CD), biogas slurry (BGS), biomaterial (Biom), gypsum (GYP) and lignite (LIG) @ 2% wt/wt.

Various morphological parameters of rice plants were determined after forty days of transplantation. Chlorophyll a and b contents, hydrogen peroxide, and lipid peroxidation were also measured in thirty days old rice plants of both genotypes.

2.2 *Arsenic analysis in shoot, root and grain*

For As analysis, root, shoot and grain samples were harvested after four month of growth, oven-dried and ground to fine powder prior to acid digestion. Shoot, root and grain samples were digested using nitric acid and hydrogen peroxide at 180°C. Arsenic content in

DOI: 10.1201/9781003317395-87

plant tissue was determined using a hydride generation-atomic absorption spectrometer (HG-AAS; Agilent AA240 with VGA 77).

Analysis of variance (ANOVA) was used to determine significant differences between treatments following Tukey's test at $p < 0.05$.

3 RESULTS AND DISCUSSION

Results showed that increase in number of tillers and tiller length ranged from 28–51% and 28–50%, respectively, with the maximum values obtained for CD and FYM treatments over their respective control for Basmati-385 genotype. In the case of Kainat genotype, number of tillers and tiller length spanned 41–50% and 18–33%, respectively, with the maximum values attained for CD treatment compared to control. The application of CD and Biom reduced production of lipid peroxidation upto 96% in Kainat genotype with increase in chlorophyll contents upto 98% compared to control while in Basmati-385 CD and FYM reduced the lipid peroxidation production upto 87% with significant increase in chlorophyll contents upto 96% as compare to their respective control. The hydrogen peroxide production was reduced by 29% in Kainat by the application of CD and Biom and its reduction was upto 35% in Basmati-385 with the application of CD and Biom as compare to their respective control (Shahid *et al.*, 2017).

Table 1. Shoot and grain As concentrations in Kainat and Basmati-385 rice genotype.

		Shoot As concentration (mg/kg)		Grain As concentration (mg/kg)	
Sr. No	Treatment Genotypes	Kainat	Basniati-385	Kainat	Basinati-385
1	Control	3S.3 a	45.2 a	1.2 ab	1.2 ab
2	Gypsum	30.5 a	17.5 b	0.5 be	0.9 b
3	Farmyard manure	2S.S ab	18.0 b	0.7 b	0.7 c
4	Cow dung	10.0 d	15.5 c	0.2 c	0.1 d
5	Lignite	44.3 a	18.7 b	1.1 ab	1.4 a
6	Biogas slurry	12.6 d	13.0 c	1.6 a	1.0 ab
7	Biomaterials	16.9 c	5.1 d	0.04 d	0.09 d

3.1 *Arsenic concentration in rice root, shoot and graim*

Arsenic concentration in different parts§ of rice plants in both type of genotypes showed a decreasing trend in the order of root > shoot > husk > grain. The shoot As concentration ranged from 3.1–28 mg/kg DW for Kainat for all the applied amendments. In contrast, low shoot As content was observed for Basmati-383 which spanned 1.7–16 mg/kg DW. The average grain As concentration for Kainat genotype was higher in plants with GYP and LIG treatments compared to CD and Biom treatments (0.29 and 0.04 mg/kg DW). In the case of Basmati-385, the mean average grain As concentration was 0.74, 0.24 and 0.09 mg/kg As DW for FYM, CD and Biom treat-ments, respectively. Significantly, mean grain As concentration for both the rice genotypes in CD and Biom amendments was below safe limit of As in rice grain (0.2 mg/kg DW) set by WHO/FAO (Irem *et al.*, 2019). Chlorophyll contents of both rice genotypes were significantly ($P < 0.05$) higher in plants where FYM and CD were added compared to other amendments and their respective controls. With the application of CD and Biom the lipid peroxidation was significantly ($P < 0.05$) reduced both the genotypes.

4 CONCLUSIONS AND RECOMMENDATIONS

In conclusion, As concentrations were high in the fruit bodies of *Tricholoma matsutake*, *Boletus aereus* and *Collybia albuminosa*. Consumption of these species may pose health risks to human health.

ACKNOWLEDGEMENTS

The authors are thankful to Higher Education Commission (6425/Punjab/NRPU/ R&D/HEC/2016 and 6396/Punjab/NRPU/R&D/HEC/2016), Pakistan for financial support.

REFERENCES

Abbas G., Murtaza B., Bibi I., Shahid M., Niazi N.K., Khan M.I. *et al.* 2018. Arsenic uptake, toxicity, detoxification, and speciation in plants: physiological, biochemical, and molecular aspects. *Int. J. Environ. Health Res.* 15: 59.

Hussain MM, Bibi I., Shahid M., Shaheen S.M., Shakoor M.B., Bashir S. *et al.* 2019. Biogeochemical cycling, speciation and transformation pathways of arsenic in aquatic environments with the emphasis on algae (Chapter Two). In: A.C. Duarte & V. Reis (eds.) *Comprehensive Analytical Chemistry* 85, Elsevier, pp. 15–51.

Irem S., Islam E., Maathuis F.J.M., Niazi N.K. & Li T. 2019. Assessment of potential dietary toxicity and arsenic accumulation in two contrasting rice genotypes: Effect of soil amendments. *Chemosphere* 225: 104–114.

Shahid M., Rafiq M., Niazi N.K., Dumat C., Shamshad S., Khalid S. *et al.* 2017. Arsenic accumulation and physiological attributes of spinach in the presence of amendments: an implication to reduce health risk. *Environ. Sci. Pollut. Res.* 24: 16097–16106.

Shakoor M.B., Bibi I., Niazi N.K., Shahid M., Nawaz M.F., Farooqi A. *et al.* 2018. The evaluation of arsenic contamination potential, speciation and hydrogeochemical behaviour in aquifers of Punjab, Pakistan. *Chemosphere* 199: 737–746.

*Arsenic in the Environment: Bridging Science to Practice for Sustainable Development –
van der Wal, Ahmad, Petrusevski, Weijma, Savic, van der Wens, Beerendonk, Bhattacharya, Bundschuh & Naidu (Eds)*

Arsenic in wild growing edible mushrooms and the associated human health risk assessment

X. Liu[1,2], Y. Gao[1,2] & L.Q. Ma[1,2]
[1]*Institute of Environment Remediation and Health, Southwest Forestry University, Kunming, Yunnan, P.R. China*
[2]*Institute of Ecology and Environment, Southwest Forestry University, Kunming, Yunnan, P.R. China*

ABSTRACT: Wild growing mushrooms have been consumed as a delicacy in Europe and Asia due to their savorous taste, aroma and species heterogeneity as well as therapeutic effect in preventing diseases such as hypertension, hypercholesterolemia and cancer. Statutory limits for arsenic (As) in wild edible mushrooms has been established at 3.0 mg/kg DM. The health risks of metals depend more on bioavailability instead of total content, however, limited studies elucidated the health risk for humans. In addition, limited information about As speciation and distribution in mushroom was available. Moreover, more attention was put on cultivated mushrooms, less focus on the wild growing ones. Commonly, element contents in mushrooms are species- and/or growing stages-dependent, thus waiting for investigated.

1 INTRODUCTION

Food safety attracted increasing attention due to the growing environmental pollution/soil contamination. Soil heavy metals and metalloids including mercury (Hg), cadmium (Cd), lead (Pb) and arsenic (As) pose potential risks to human health via food chains. Mushroom is famous for its particular flavour and rich in valuable nutritious such as proteins, minerals and vitamins (Blagodatski *et al.*, 2018). However, mushrooms can accumulate soil toxic heavy metals to high concentrations, thus their potential in threatening food safety and human health deserves great attention (Bernaś *et al.*, 2006).

Arsenic is widely dispersed in the environment naturally or anthropogenically by agricultural and industrial activities. Once absorbed, it is being persistent in human body with long excretion half lives for decades and thus classified as potentially toxic elements. Food consumption was the main pathway for human exposure to toxic elements compared to inhalation and dermal contact. Thus the accumulation of toxic elements in the environment was of increasing concern due to food safety issues and potential health risks (Falandysz *et al.*, 2013). Wild growing mushrooms have been consumed as a delicacy in Europe and Asia due to their savorous taste, aroma and species heterogeneity as well as therapeutic effect in preventing diseases such as hypertension, cancer, and hypercholesterolemia (Falandysz & Rizal 2016).

Therefore, the aims of the study were to: 1) analysis As contents in wild edible mushrooms growing in naturally high background substrates; 2) determine As content in multiple species and changes during growing stages; 3) examine As speciation and distribution; and 4) propose numeric contributive information on the proportion of As speciation from total content on bioavailability and human health risk.

2 MATERIALS AND METHODS

2.1 *Study area*

The wild edible mushroom and the uppermost layer of 10 cm depth soil were sampled in Yunnan, China.

2.2 *Arsenic concentration in soils and mushrooms*

Plants or soil debris were removed from fresh fruit bodies using a plastic knife. Mushrooms (separately caps and stipes)and soil samples were examined in pools. The mushrooms and soil samples were freeze-dried to constant weight. The mushroom samples were pulverised to fine powder using a grinder, and the soil samples were crushed to pass a 0.149 mm sieve for total arsenic analysis (Giannaccini *et al.*, 2016).

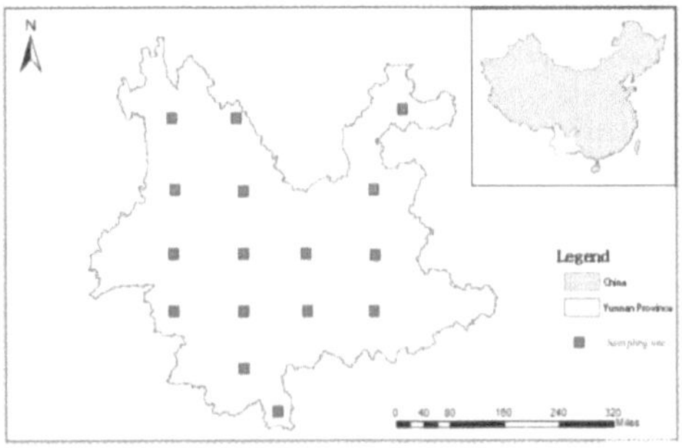

Figure 1. Study area and mushroom sampling locations.

DOI: 10.1201/9781003317395-88

2.3 *Arsenic soil-mushroom bioconcentration factor (BCF) analysis*

Samples were analyzed for trace elements. The wild edible mushrooms of different growth stages will analyzed for trace elements total content, speciation, and distribution. The bioconcentration factor (BCF) will be calculated to express the ability of mushrooms to accumulate elements and elements soil-to-mushroom bioavailability.

2.4 *Human health risk assessment by in vivo analysis*

In vitro and in vivo experiments was involved to examine mushroom elements bioavailability to human and to calculate the numeric contributive of the proportion of element speciation from total content in bioavailability. Basing on this, credible human risk assessment was proposed.

3 RESULTS AND DISCUSSION

3.1 *Arsenic concentration and distribution in six wild edible mushrooms*

Except two wild edible mushrooms *cyanobacterium* and *Cephalosporium*, whose As concentrations were lower than the standard value, the content of As in other wild mushrooms exceeded the national standard in varying degrees. The content of As in *Tricholoma matsutake* is much higher than that of other wild mushrooms.

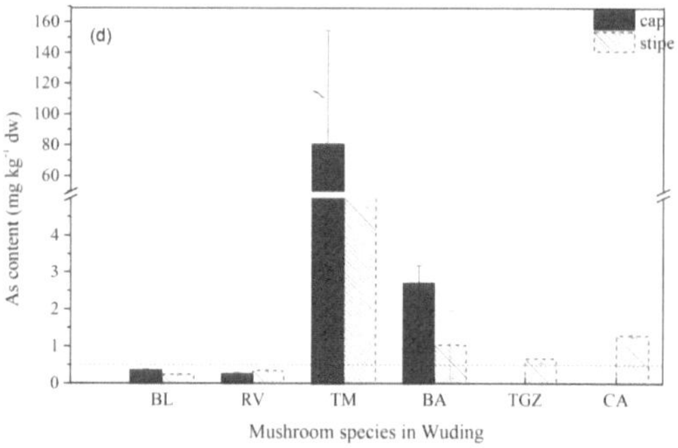

Figure 2. Arsenic concentration and distribution (cap and stipe) in six wild edible mushrooms.

3.2 *Human health risk assessment*

In order to determine the content of heavy metals in wild edible fungi, the mathematical model is used to evaluate whether the heavy metals in edible fungi will cause potential health hazards to human body.

The results showed that: 1) most edible fungi will not cause harm to human health; 2) compared with children, heavy metals in edible wild mushroom are more likely to cause potential harm to adults.

4 CONCLUSIONS AND RECOMMENDATIONS

In conclusion, As concentrations were high in the fruit bodies of *Tricholoma matsutake*, *Boletus aereus* and *Collybia albuminosa*. Consumption of these species may pose health risks to human health.

ACKNOWLEDGEMENTS

This work was supported in part by the National Natural Science Foundation of China, the Scientific Research Foundation of Southwest Forestry University, the Scientific Research Foundation of Ecology and Soil & Water Conservation Department, and Yunnan Education Department Science Research Foundation.

REFERENCES

Bernaś E., Jaworska G. & Lisiewska Z. 2006. Edible mushrooms as a source of valuable nutritive constituents. *Acta Sci. Pol. Technol. Aliment.* 5(1): 5–20.

Blagodatski A., Yatsunskaya M., Mikhailova V., Tiasto V., Kagansky A. & Katanaev V.L. 2018. Medicinal mushrooms as an attractive new source of natural compounds for future cancer therapy. *Oncotarget* 9 (49): 29259–29274.

Falandysz J. & Borovicka J. 2013. Macro and trace mineral constituents and radionuclides in mushrooms: health benefits and risks. *Appl. Microbiol. Biotechnol.* 97(2): 477–501.

Falandysz J. & Rizal L.M. 2016. Arsenic and its compounds in mushrooms: a review. *J Environ. Sci. Health C Environ. Carcinog. Ecotoxicol. Rev.* 34(4): 217–232.

Giannaccini G., Betti L., Palego L., Mascia G., Schmid L., Lanza M. & Lucacchini A. 2012. The trace element content of top-soil and wild edible mushroom samples collected in Tuscany, Italy. *Environ. Monit. Assess.* 184(12): 7579–7595.

Arsenic in the Environment: Bridging Science to Practice for Sustainable Development – van der Wal, Ahmad, Petrusevski, Weijma, Savic, van der Wens, Beerendonk, Bhattacharya, Bundschuh & Naidu (Eds)

Role of arsenic contaminated water during cooking of rice grain: An inverse relation

A. Das, M. Joardar, N. Roy Chowdhury, D. Mridha & T. Roychowdhury
School of Environmental Studies, Jadavpur University, Kolkata, India

ABSTRACT: Arsenic (As) contamination in both groundwater and rice grain is a wide ranged health hazard in south Asian countries especially in West Bengal, India. Rice being the staple food is a proficient method of As intake into human body. The study involves cooking of sunned and parboiled variety of rice cultivar Minikit and Ranjit with five different As contaminated waters (<3, 32, 45, 67 and 104 µg/l). Result shows a unidirectional flow of As from rice to water. The raw rice As concentration is high in parboiled one than in the sunned grains irrespective of the cultivars. The reduction rate of As in cooked rice from raw rice grain is inversely proportional to increasing As concentration in cooking water and the gruel As content is directly proportional with cooking water As concentration.

1 INTRODUCTION

Arsenic (As) contamination in groundwater is a yearly known hazard in the countries of south and south-east Asia. Apart from drinking water, being a group 1 carcinogen, its accumulation in rice grain through cultivation followed by contaminated cooked rice is a serious health risk for the population especially in the Bengal delta where people are immensely dependent on rice as their regular staple food (Roychowdhury 2008). This problem is observed not only in the zones with As affected soil and groundwater, even in the apparently non-contaminated sites like Kolkata, where population faces threat only through the consumption of rice as it is transported after being cultivated in the As endemic areas (Biswas *et al.*, 2019). Arsenic accumulation in rice grain depends on the type of cultivar as well as its variety, sunned or parboiled. Parboiled rice grain is procured after a two way boiling process of immediately cultivated paddy grain which sometimes gives an additional As burden as this post-harvesting procedure includes As contaminated water once again (Chowdhury *et al.*, 2018).

This study presently comprises cooking of both varieties of two different cultivars of rice grain (Minikit and Ranjit) with five different concentrations of As contaminated water (<3, 32, 45, 67 and 104 µg/L) in domestic level. The aim of the study is to investigate the role of As contaminated water in cooked rice with respect to the variety of rice grain. The current study is also an assessment towards the effect of consumption of As contaminated cooked rice while compared between their varieties, sunned and parboiled.

2 MATERIALS AND METHODS

2.1 *Study area*

The present study area is Gaighata, a block under North 24 Paraganas district in West Bengal, India. The district falls under the Gangetic delta plain reported with high As contaminated soil sediment (Chakraborti *et al.*, 2001; Roychowdhury 2010).

2.2 *Sample collection, preparation & preservation*

Raw rice grain (sunned and parboiled) and groundwater samples were collected from families residing in the above study area. Cooked rice samples were collected after preparation in a single house with five different concentrations of As contaminated waters maintaining the general ratio of rice: water (1:3). After cooking, the excess water was collected in each case which is known as gruel or total discarded water (TDW). Raw groundwater samples were preserved with addition of 0.1% (v/v) concentrated HNO_3 and stored at 4°C. Approximately 0.2 g of raw rice grain was taken for digestion whereas a substantial part of cooked rice was directly processed for digestion. The samples were digested with concentrated HNO_3 and H_2O_2 (2:1) in a hotplate at about 90°C for 1 h. The sample volumes were then made up to 5 ml with double distilled water and filtered through a suction filter (diameter 0.45 µm). The samples thus prepared for analysis.

2.3 *Analysis of samples*

For total As estimation, a Varian model AA140 (USA) was used which involves FI-HG-AAS method (Roychowdhury 2008). Quality control tests were performed by replicate analysis, calculating recovery of spiked digested samples as well as SRM sample analysis.

3 RESULTS AND DISCUSSION

3.1 *Arsenic concentration in raw rice grain*

In both the cultivars, the parboiled rice grain showed a prominently higher As concentration than the

DOI: 10.1201/9781003317395-89

sunned one. The Minikit cultivar holds As concentration 152 µg/kg in sunned variety whereas in parboiled it was 384 µg/kg. Similarly, the Ranjit cultivar showed As concentration as 79 µg/kg and 127 µg/kg in sunned and parboiled respectively. This values clearly state that the raw rice grain As concentration is cultivar dependent.

3.2 *Arsenic concentration in cooked rice and TDW*

Wet cooked As concentration depends on both the cooking water and raw rice As concentration. While cooking with non contaminated water (<3 µg/L), the As concentration in cooked rice had a sharp decrease from raw rice grain. The results were very surprising as in each case the wet cooked rice showed a reduction in As concentration from the raw rice grain. The interesting observation is that with increase in cooking water As, there is a decreasing trend in reduction of cooked rice As as well as an increasing trend in the gruel As content (Figure 1). Therefore, it is well understood that the As flow from water to rice is solely dependent on water As concentration in which the cooking has been done.

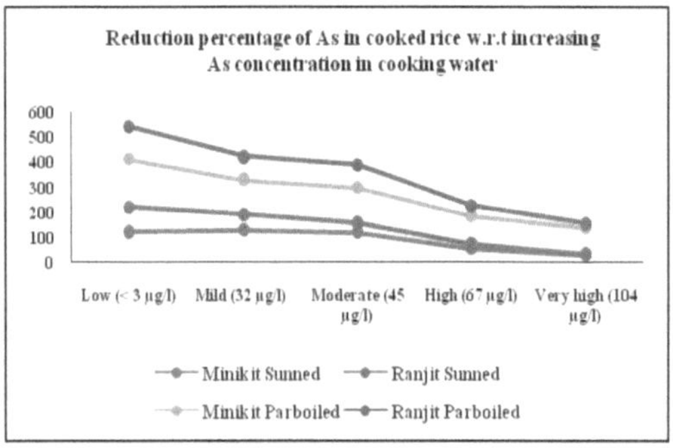

Figure 1. Relation between cooking water arsenic concentration (µg/L) and cooked rice arsenic concentration (µg/kg).

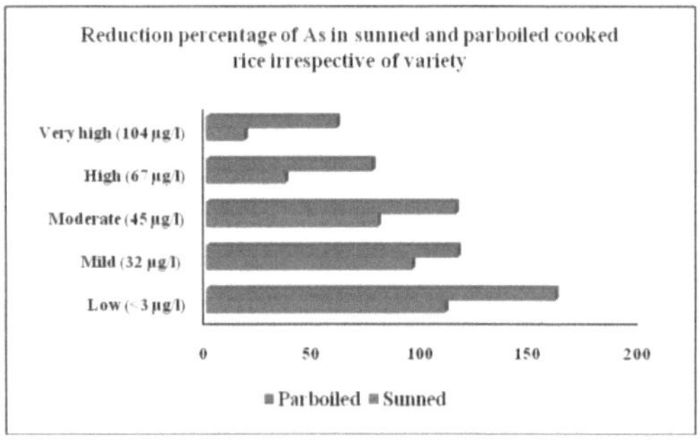

Figure 2. Arsenic reduction percentage in cooked rice from raw rice grain compared between sunned and parboiled rice variety.

3.3 *Reduction of arsenic in cooked rice with respect to rice variety*

Rice variety is an important phenomenon for this study. While cooking with five different concentrations of As contaminated water the reduction percentage of As concentration in cooked rice from raw rice grain was significantly high in parboiled variety than the sunned one (Figure 2). This is because the parboiled rice grain is already passed through a boiling process which results the additional As being loosely bound. This mobile As is then easily released during cooking procedure.

4 CONCLUSIONS AND RECOMMENDATIONS

The study includes five specific concentrations of As contaminated cooking water (range: <3–104 µg/L) for preparation of rice. Irrespective of cultivar and variety, there was observed a unidirectional flow of As from rice to water. It is always advised to cook rice with low or non contaminated water as the raw rice grain holds indigenous As concentration in the rural area. The flow of As from rice to water increases with increasing As concentration of cooking water. Therefore, intake of gruel with rice is never recommended for the rural health.

ACKNOWLEDGEMENTS

We acknowledge the Department of Science & Technology, West Bengal for financial support regarding our work and the villagers residing in the mentioned study area for sample collection.

REFERENCES

Biswas A., Swain S., Chowdhury N. R., Joardar M., Das A., Mukherjee M. & Roychowdhury T. 2019. Arsenic contamination in Kolkata metropolitan city: perspective of transportation of agricultural products from arsenic-endemic areas. *Environ. Sci. Pollut. Res.* 26: 22929–22944.

Chakraborti D., Basu G.K., Biswas B.K., Chowdhury U.K., Rahman M.M., Paul K. & Ray S.L. 2001. Characterization of arsenic bearing sediments in Gangetic Delta of West Bengal-India. In: W.R. Chappell, C.O. Abernathy & R.A. Calderon (eds.) *Arsenic Exposure and Health Effects.* Elsevier Science, UK, 27–52.

Chowdhury N.R., Ghosh S., Joardar M., Kar D., Roychowdhury T. 2018. Impact of arsenic contaminated groundwater used during domestic scale post harvesting of paddy crop in West Bengal: Arsenic partitioning in raw and parboiled whole grain. *Chemosphere* 211: 644.

Roychowdhury T. 2008. Impact of sedimentary arsenic through irrigated groundwater on soil, plant, crops and human continuum from Bengal delta: special reference to raw and cooked rice. *Food Chem. Toxicol.* 46(8): 2856–2864.

Roychowdhury T. 2010. Groundwater arsenic contamination in one of the 107 arsenic affected blocks in West Bengal, India: status, distribution, health effects and factorsresponsible for arsenic poisoning. *Int. J. Environ. Health Res.* 213: 414–427.

Arsenic in the Environment: Bridging Science to Practice for Sustainable Development – van der Wal, Ahmad, Petrusevski, Weijma, Savic, van der Wens, Beerendonk, Bhattacharya, Bundschuh & Naidu (Eds)
© 2023 the Author(s), ISBN 978-1-032-32928-4

Arsenic in cattle: Evaluation of possible exposure biomarkers

C.V. Alvarez-Gonçalvez[1], F.E. Arellano[1,2], A. Fernández-Cirelli[1,2] & A.L. Pérez-Carrera[1,2]
[1]*Universidad de Buenos Aires, Facultad de Ciencias Veterinarias, Centro de Estudios Transdisciplinarios del Agua (CETA), Buenos Aires, Argentina*
[2]*Universidad de Buenos Aires-CONICET, Instituto de Investigaciones en Producción Animal (INPA), Buenos Aires, Argentina*

ABSTRACT: Arsenic pollution is naturally present in groundwater and soils of the main agriculture production areas in Argentina. Arsenic can be taken up and bioaccumulated by forages, and may be biotransferred to animals, food and humans. Blood and hair are generally used in humans as biomarkers but there is a lack of information in livestock. This study shows that As levels in cattle hair from As affected area were significantly higher than in the control area. In analyzed blood samples no significant differences were found between control and As affected areas. Blood appear not to be an effective biomarker in cattle such as it has been determinate for human but hair may indicate chronic exposure in cattle.

1 INTRODUCTION

Arsenic (As) pollution affects several countries around the world. One of the most affected countries is Argentina, where As is naturally present in groundwater and soils of the main agriculture production areas (Bundschuh *et al.*, 2012). Arsenic can be taken up and bioaccumulated in plants and livestock, which means a potential risk to human health through agri-food.

Presence of toxic substances in blood is a known exposure biomarker (Lowry *et al.*, 1989). On the other hand, in the case of As, its levels in cattle hair might be used as another biomarker exposure. Even though, it is commonly used for humans, there is a severe lack of knowledge for other animal species. The aim of this study is to assess total arsenic levels in soils and drinking water from cattle farms; and to analyze the relationship with blood and hair As content in cattle.

2 MATERIAL AND METHODS

2.1 *Study area*

This study was carried out in the Pampean Plain (Argentina). The study area is one of the most important livestock production regions. The samples zones were located in Buenos Aires province (Control area) and in the southeast of Cordoba province (As affected area).

2.2 *Water sampling and laboratory analysis*

Soil and drinking water samples were taken up from livestock production systems and dairy farms. Soil samples correspond to surface soil (0 to 20cm deep). Samples were collected along transect in the field at regular intervals. Arsenic determination was according to ISO 11466 (1995). Soils were digested using a mixture of nitric and hydrochloric acids (1:3) at 120°C during 2h with reflux. Water samples correspond to groundwater used for cattle. They were collected in polypropylene tubes, acidified at 20% with HNO_3 and conserved refrigerated at 4°C. Blood and hair samples were collected from adult and female Holando cows. The samples were digested with HNO_3 at hot plate. Arsenic in all samples were determined by inductively coupled plasma optical emission spectrometry (ICP-OES Optima 2000, Perkin Elmer).

3 RESULTS AND DISCUSSION

Usually, arsenic in soils ranged between 1 to 95 mg/kg (Kabata Pendias & Pendias 2001), and in groundwater (main water source for animal drinking in the Pampean Plain) arsenic is documented in the literature up to values of 5000 μg/L (Ravenscroft *et al.*, 2009). Maximum total As content for animal drinking water and soil destined to agricultural activities in Argentina are 0.5 mg/L and 20 mg/kg respectively (Law 24.051), but the international guidelines recommend a maximum level of As in animal drinking water 200μg/L (FAO 1985) and in agricultural soil 17 mg/kg (CCME 2001). The As levels for all samples analyzed in this study are shown in Table 1.

It is known that drinking water is the main As source for cattle (Perez Carrera *et al.*, 2016), the levels obtained through forage and fodder or

DOI: 10.1201/9781003317395-90

Table 1. Arsenic levels (mín-máx.) for water, cattle's hair and blood in affected and control areas.

Area	Water µg/L	Hair As ng/g	Blood As µg/L
Affected area	51-268	< LD-1520	<LD-75
Control area	<LD-50	< LD-520	<LD

accidental soil take up is relatively low. In the analyzed soil samples, As levels were below the detection limit (LD = 3.5 mg/kg).

Regarding to As level in hair, in all samples from control area, were below 1000 ng/g. In affected areas As hair levels were between <LD-1520. In cattle there are no enough data to determine a confident baseline value. These results are according to previous studies (Pérez Carrera & Fernandez Cirelli 2012).

Arsenic levels in cattle blood were between <LD-75 µg/L. These values are below than those previously reported by Rana *et al.* (2008, 2010) in West Bengal, India, that inform mean levels in blood of 284 µg/L. Relationship between total As level in drinking water and As levels in blood and hair should be of relevance. However, no significant correlation ($p > 0.05$) was observed. Results are shown in Figure 1.

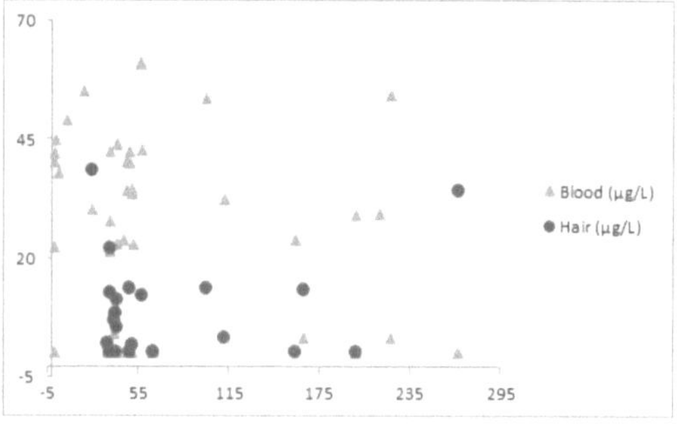

Figure 1. Relationship between total arsenic (µg/L) in blood and hair with water As concentration. X axis show As levels in water while Y axis shows As levels in blood and hair.

4 CONCLUSIONS

Total As concentration in 50% of water samples exceeded the recommended value for cattle (200µg/L) in the affected area. Blood appear not to be an effective biomarker for chronic exposure in cattle. Regards hair values in affected area were significantly higher than values determined in the control area and may be used as exposure biomarkers.

ACKNOWLEDGEMENTS

The authors are grateful to the University of Buenos Aires and CONICET for financial support.

REFERENCES

Bundschuh J., Litter M.I., Parvez F., Román-Ross G., Nicolli H.B., Jean J.S. *et al.* 2012. One century of arsenic exposure in Latin America: A review of history and occurrence from 14 countries. *Sci. Total Environ.* 429: 2–35.

CCME. 2001. Canadian soil quality guidelines for the protection of the environmental and human health: Arsenic (Inorganic) (1997). *Updated In: Canadian Environmental Quality Guidelines, 1999, Canadian Council of Ministers of the Environment, Winnipeg.*

FAO. 1985.*Water Quality for Agriculture.* Rome: Food and Agriculture Organization of the United Nations, 1985.

ISO 11466. 1995. Soil Quality – Extraction of Trace Element Soluble In Aqua Regia.

Kabata Pendias A. & Pendias H. 2010. *Trace Elements in Soils and Plants. Third edition.* CRC Press, Boca Raton, Ann Arbor, London.

Law. 24051. Argentina

Lowry L.K. 1995. Role of biomarkers of exposure in the assessment of health risks. *Toxicol. Lett.* 77(1–3): 31–38.

Perez-Carrera A.L. & Fernandez-Cirelli A. 2012. Arsenic exposure in cows from hight contaminated area in the Chaco Pampean Plain, Argentina. In: J.C. Ng, B.N. Noller, R. Naidu, J. Bundschuh & P. Bhattacharya (eds.) *Understanding the Geological and Medical Interface of Arsenic (As2012).* CRC Press/Taylor and Francis, pp. 201–202.

Pérez-Carrera A., Alvarez-Gonçalvez C.V. & Fernández-Cirelli A. 2016. Transference factors as a tool for the estimation of arsenic milk concentration. *Environ. Sci. Pollut. Res.* 23(16): 16329–16335.

Rana T., Bera A.K., Das S., Bhattacharya D., BandyopadhyayS., Pan D. & Das S.K. 2010. Effect of chronic intake of arsenic-contaminated water on blood oxidative stress indices in Cattle in an arsenic-affected zone. *Ecotoxicol. Environ. Saf.* 73(6): 1327–1332.

Rana T., Sarkar S., Mandal T. K., Bhattacharyya K. & Roy A. 2008. Arsenic residue in blood, urine and faeces samples from Cattle in the Nadia district of West Bengal in India. *Internet J. Vet. Med,* 4(1).

Ravenscroft P., Brammer H., Richards K. 2009. *Arsenic Pollution: A Global Synthesis.* John Wiley & Sons, West Sussex, UK.

Arsenic in the Environment: Bridging Science to Practice for Sustainable Development –
van der Wal, Ahmad, Petrusevski, Weijma, Savic, van der Wens, Beerendonk, Bhattacharya, Bundschuh & Naidu (Eds)
© 2023 the Author(s), ISBN 978-1-032-32928-4

Arsenic risk assessment through dairy products ingestion

F.E. Arellano[1,2], A. Fernández-Cirelli[1,2], S. Braeuer[3], W. Goessler[3] & A.L. Pérez-Carrera[1,2]
[1]*Universidad de Buenos Aires, Facultad de Ciencias Veterinarias, Centro de Estudios Transdisciplinarios del Agua (CETA), Buenos Aires, Argentina*
[2]*Universidad de Buenos Aires-CONICET, Instituto de Investigaciones en Producción Animal (INPA), Buenos Aires, Argentina*
[3]*Institute of Chemistry, University of Graz, Graz, Austria*

ABSTRACT: The aim of the present study was to investigate: i) the presence of As in raw milk and dairy products from Argentina; ii) the As daily intake for different consumers separated in age range and iii) the risk assessment trough milk consumption due to long term ingestion of inorganic As. The mean As concentration in raw milk was below than 20 ng/g (LMR, Decree No. 14, 2013), but in skimmed powdered milk and infant formula the mean exceed this LMR. Taking into account the ADDI obtained, and according to the reference value by oral exposure of 0.3 μg/kg*day (IRI, USEPA-2016), the infant group is the most exposed to As intake, followed by the children. Despite this, the risk of toxicity and target risk (TR) did not exceed the threshold values.

1 INTRODUCTION

Milk and dairy products are essential in human diet, being an important source of nutrients and minerals. However, in some cases, certain contaminants may accumulate in concentrations that could affect human health. In Argentina Arsenic (As) distribution is related to the geological origin, present mainly in groundwater (Pérez-Carrera A 2002). The As presence in drinking water and animal fodder could affect animal health and milk production (Pérez-Carrera *et al.*, 2016a,b). Also, As could be accumulated in dairy products and the ingestion of contaminated food may involve different risks to human health (Bargellini *et al.*, 2016; Carignan *et al.*, 2015). The present study aims to investigate: i) the presence of As in raw milk and dairy products; ii) the As daily intake for different consumers separated in age range and iii) the risk assessment trough milk consumption due to long term ingestion of As.

2 MATERIALS AND METHODS

2.1 *Study area*

Bovine, caprine and ovine raw milk samples were collected in 37 dairy farms from the southeastern part of the Province of Córdoba and the northern region of the province of Buenos Aires, Figure 1. Commercial milk (n: 36) and cheeses (n: 14) were obtained from different markets in the study areas. All samples were lyophilized, and were digested (three replicates per sample) with 5 ml of nitric acid 65% (Carl-Roth, Karlsruhe, Germany purified via sub-boiling in-house, Germany) in a microwave-heated digestor (MLS GmbH, Leutkirch, Germany). Then they were diluted with ultrapure water (18.2 MΩ*cm) to 10% v/v acidity.

Figure 1. Study area and raw milk sampling locations a. Cordoba province and b. Buenos Aires province.

2.2 *Determination of element concentrations*

Arsenic (As) was determined in all samples with Inductively Coupled Plasma Mass Spectrometry (ICPMS, Agilent ICP-MS 7500ce, Waldbronn, Germany). As (m/z 75) was measured without collision or reaction gas.

2.3 *Statistical analysis*

Data were analyzed using Infostat software, Version 2016I. Statistical differences ($p \leq 0.05$) were tested using one way ANOVA (Kruskal-Wallis. Risk assessment was estimated with the @Risk program versión 5.5 (Palisade Co.). The

DOI: 10.1201/9781003317395-91

target risk (TR) values > 1 involve a health risk prone to carcinogenic problems.

3 RESULTS AND DISCUSSION

3.1 *As presence in raw milk and dairy products*

Arsenic concentration in raw milk, commercial fluid milk and cheese is showed in Table 1. According with the obtained results, 9.52% of the bovine milk, 8% of the goat milk and 7.55% of the sheep milk samples exceeded the maximum tolerance limit (LMR) of As (20 ng/g) mentioned by the Mercosur Technical Regulation (Decree No. 14 2013). The rest of the raw milk samples showed As levels below the LOD (10 ng/g).

Table 1. As concentration determined in raw milk from cow, goat and sheep, commercial milk (powdered and fluid) and cheese. Reference material NIST- 8435.

	As (ng/g)	
	Mean±EE	Range
Goat raw milk (N:25)	< LOD	< LOD-70
Sheep raw milk (N:40)	< LOD	< LOD-80
Cow raw milk (N:42)	< LOD	< LOD-110
Commercial milk (N:36)	26.7 ± 4.78^{A}	< LOD-167
Cheese (N:14)	20.6 ± 12.3^{B}	< LOD-142
Reference Material	70.0 ± 65.0.	

Table 2 shows As concentration present in infant formula, whole and skimmed powdered milk. Mean As concentration present in skimmed powdered milk and infant formula, exceeded the LMR of 20 ng/g previously mentioned. Although, As concentrations found in this study were similar to those reported in other milk international production (Dobrzański *et al.*, 2005; Miedico *et al.*, 2016).

Table 2. Mean±EE and range As concentration determined in infantil formula, whole and skimmed powdered milk.

	Infant formula	Skimmed milk	Whole milk
As ng/g	20,50 ± 6,31	26,00 ± 2,72	16,30 ± 3,13
Range	< LOD-36	< LOD-38	< LOD-44
Reference	<LOD-77		

3.2 *Daily intake and risk assessment trough dairy products ingestion*

Table 3 shows the daily intake rate (ADDI) through As ingestion in infants (0 to 24 months), children (2 to 9 years), pregnant (15 to 40 years) and adults (9 to 70 years).

Table 3. As daily intake rate, median and range, for Infants, children, pregnant and adults through commercial milk (fluid and powdered) and cheese.

	Commercial Milk		Cheese	
	Median	Range	Median	Range
Infants	0.30	0-2.29	0.14	0-1.95
Children	0.08	0-0.59	0.04	0-0.50
Pregnant	0.03	0-0.27	0.02	0-0.23
Adults	0.02	0-0.14	0.01	0-0.12

Taking into account the ADDI obtained and according to the reference value of 0.3 μg/kg*day established by IRI (USEPA 2016), the infant group is the most exposed to As intake, followed by the children. Also, it was observed that the infant group is more exposed to this element through the daily intake of milk than cheese. No values of the TR estimated (Figure 2) were exceed the threshold of 1. Although, in the case of whole powdered milk (TR=0.95), skimmed powdered milk (TR= 0.92) and commercial fluid milk (TR= 0.97), the TRs were close to 1.

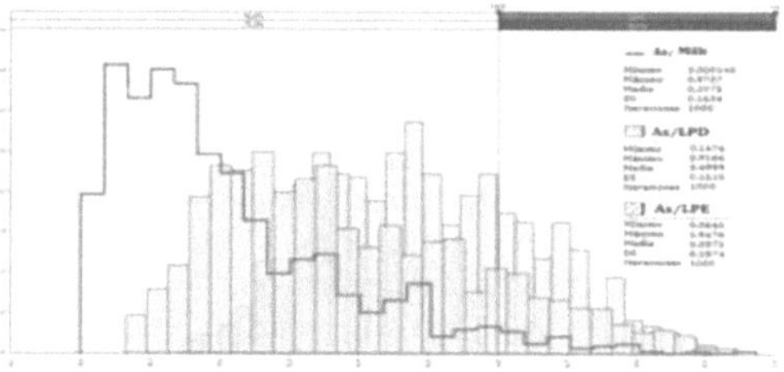

Figure 2. TR values for commercial fluid milk, whole powdered milk (LPE) and skimmed powdered milk (LPD).

4 CONCLUSIONS AND RECOMMENDATIONS

Arsenic concentrations present in the different types of dairy products and raw milk were higher than the LMRs suggested by the Mercosur Technical Regulation (decree no. 14 2013). In addition, the ADDI for infants and children were higher than those suggested by the IRI (USEPA 2016). Despite this, the risk of toxicity and TR did not exceed the threshold values. Also, it is necessary to emphasize that the risk analysis carried out takes into account the values determined in this work and ensures that within the obtained As concentrations, the dairy products consumption is relatively safe.

ACKNOWLEDGEMENTS

We acknowledge to the University of Buenos Aires and to CONICET (National Research Council) for financial support.

REFERENCES

Bargellini A., Venturelli F., Casali E., Ferrari A., Marchesi I. & Botella P. (2016). Trace elements in starter infant formula: dietary intake and safety assessment. ecotoxicology and toxicology: problems and decisions, *Environ Sci Pollut R* DOI: 10.1007/s11356-016-8290-9

Carignan C.C., Cottingham K.L., Jackson B.P., Farzan S.F., Gandolfi A.J., Punshon T., Folt C.L. & Karagas M.R. (2015). Estimated exposure to arsenic in breastfed and formula-fed infants in a United States cohort. *Environ Health Persp* 123:500–506. http://dx.doi.org/10.1289/ehp.1408789

Dobrzański Z., Kolacz R., Górecka H., Chojnacka K. & Bartkowiak A. 2005. The content of microelements and trace elements in raw milk from cows in the Silesian region. *Pol J Environ Stud* 14: 685–689

Miedico O., Tarallo M. & Pompa C y A.E. Chiaravalle. (2016). Trace elements in sheep and goat milk samples from apulia and basilicata regions (Italy): Valuation by multivariate data analysis. *Small Ruminant Research* 135: 60–65

Ministerio de Salud Pública, 2013. Decreto 14/013, *Modificación y Adaptación al Reglamento Técnico Mercosur Sobre Límites Máximos de Contaminantes Inorgánicos en Alimentos, Resolucion* 12/011.

Pérez Carrera, Alejo, 2002. *"Problemática del Arsénico en la Llanura Sudeste de la Provincia de Córdoba".* VIII Taller de Evaluación y Manejo de Riesgos por Exposición a Arsénico en Agua de Consumo. Maciel, Santa Fe, Argentina.

Pérez-Carrera A., Alvarez-Gonçalvez C.V. & Fernández-Cirelli A. (2016a). Transference factors as a tool for the estimation of arsenic milk concentration. *EnvironSci and Pollut* R17. DOI: 10.1007/s1135601667310.

Pérez Carrera A., Arellano F., & Fernández Cirelli A. (2016b). Concentration of trace elements in raw milk from cows in the southeast of Córdoba province, Argentina. *Dairy Sci Technol.* DOI 10.1007/s1359401602905

USEPA-IRIS. 2016. *IRIS Summary of Inorganaic Arsenic and others Trace Elements.* Retrieved November 2016, from Integrate Risk Information System: https://cfpub.epa.gov/ncea/iris/iris_documents/documents/subst/0278_summary.pdf

Section 3: Health impacts of arsenic

3.1 Exposure and epidemiology of arsenic and impacts on human health

Arsenic in the Environment: Bridging Science to Practice for Sustainable Development –
van der Wal, Ahmad, Petrusevski, Weijma, Savic, van der Wens, Beerendonk, Bhattacharya, Bundschuh & Naidu (Eds)
© 2023 the Author(s), ISBN 978-1-032-32928-4

In vitro assessment of arsenic release and transformation from As(V)-sorbed goethite and jarosite: The influence of human gut microbiota

N.Y. Yin & Y.S. Cui
College of Resources and Environment, University of Chinese Academy of Sciences, Beijing, P.R. China

ABSTRACT: The importance of arsenic metabolism by gut microbiota has been evidenced in risk characterization from As exposures. In this study, we evaluated the metabolic potency of human gut microbiota toward As(V)-bearing goethite and jarosite, presenting different behaviors of As release, and the solid-liquid transformation and partitioning. The release of As occurred mainly in the small intestinal phase for jarosite and in the colon phase for goethite, respectively. We found higher degree of As(V) and Fe(III) reduction by human gut microbiota in the colon digests of goethite than jarosite. Speciation analysis using high-performance liquid chromatography coupled with inductively coupled plasma mass spectrometry and X-ray absorption near-edge spectroscopy, revealed that 43.2% and 8.5% of total As was present as As(III) in the liquid and solid phase respectively after goethite incubation, while almost all generated As(III) was in the colon digests of jarosite. Therefore, As bioaccessibility in human gastrointestinal tract was predominantly contributed to Fe(III) dissolution in jarosite, and to microbial reduction of Fe(III) and As(V) in goethite. It expanded our knowledge on the role of Fe minerals in human health risk assessment associated with soil As exposures.

1 INTRODUCTION

Arsenic (As) as a class 1 carcinogen is ubiquitous in the natural environment, which presents a significant threat to human health (ATSDR 2015). For human exposure to contaminated soil and dust, inadvertent oral ingestion becomes one main exposure pathway, especially for children through outdoor hand-to-mouth behavior. In recent twenty years, *in vitro* methodologies have been developed to simulate the gastrointestinal tract, and As bioaccessibility is applied to evaluate risks to human associated with contaminated soils by many countries and regions (Bradham *et al.*, 2018).

The importance of As metabolism by human gut microbiota has been evidenced in contaminated soil and food (Calatayud *et al.*, 2018; Yin *et al.*, 2015). Speciation analysis must be considered an essential part of risk assessment associated with soil As exposures. There is a general lack of knowledge on the intervention of gut microbiota on As bioaccessibility and As(V)-bearing Fe (hydr)oxides. The objective of this study was to evaluate the metabolic potency of human gut microbiota toward As(V)-bearing Fe minerals, goethite and jarosite. Therefore, we determined As release and transformation during the incubation with human gut microbiota, and further partitioning between the liquid and solid phases.

2 MATERIALS AND METHODS

2.1 *Preparation of As(V)-bearing goethite and jarosite*

The synthesis of goethite and Jarosite was carried out according to methodologies suggested by Schwertmann & Cornell (2000), and Baron & Palmer (1996), respectively. Two levels of initial As(V) solutions (6.67 mM and 0.67 mM) were mixed with goethite and jarosite (30 g/L) at pH 4.6 and 9.2.

2.2 *Dynamic SHIME*

The colon microbial community was *in vitro* cultured in a dynamic human gastrointestinal simulator (SHIME), which consisted of five compartments to simulate stomach, small intestine, ascending colon, transverse colon and descending colon, respectively. Briefly, fresh fecal microorganisms obtained from a 28-year-old Chinese male volunteer (no antibiotic treatment for at least six months before the study), were inoculated into the three different colon compartments. The nutrition for the colon microbiota was provided by three times per day. The SHIME reactor was continuously stirred and regularly flushed with nitrogen gas to ensure anaerobic environment. The temperature (37 °C) and pH were operated automatically. The stable microbial community was expected after 4 weeks of adaptation.

2.3 *Metabolic potency of colon microbiota toward As(V)-bearing goethite and jarosite*

The physiologically based extraction test (PBET) combined with the SHIME was used. In brief, As-bearing goethite and jarosite was added to centrifuge tubes with gastric digest at the solid/solution ratio of 1:100. After the gastric phase, the pH was adjusted from 1.5 to 7.0 with $NaHCO_3$, and the bile salt and pancreatin were added. Following transition from the small intestinal to the colon phase, the digests were transferred without loss into anaerobic serum bottles with active descending colon solution at a solid/solution ratio of 1:200. The entire *in vitro* digestion process was maintained at 37°C and on a shaker at 150 rpm. Samples taken from at the end of each phase were centrifuged, and supernatants were filtered and stored at -80 °C until analysis.

2.4 *Sample analysis*

For liquid samples, As concentrations were determined using ICP-MS, and speciation analysis was conducted using HPLC-ICP-MS. For solid samples, X-ray absorption spectra of As were collected on beamline 1W1B (multi-pole wiggler) at the Beijing Synchrotron Radiation Facility (BSRF) in China.

3 RESULTS AND DISCUSSION

3.1 *As release and bioaccessibility*

In the small intestinal phase, the bioaccessible As was below detection limit of GL1 and GL3; and for goethite (GH2 and GH4) with high concentrations of As (V), the bioaccessibility was just 6.0% and 7.3%. But higher As bioaccessibility in jarosite was found from 32.3% to 45.6%.

Following the transition from small intestinal to colon phase, we observed the intense dissolution from As(V)-bearing goethite with the bioaccessibility of 52.6-74.6%. For jarosite, the bioaccessible As of JH2 and JH4 increased up to 1.4-fold in the colon phase ($P < 0.05$), but As bioaccessibility in JL1 and JL3 was congruent with that of small intestinal phase.

3.2 *As speciation and partitioning*

The predominant form in the colon digests was inorganic As (97.5-100%). For goethite, As(III) was predominantly present in colon digests, being 59.5-76.7% (2.0-13.5 mg/L). The colon digests of jarosite also displayed a reduction percentage of 45.4-65.6% (0.9-5.6 mg/L). However, we found an extremely low degree of methylation (0-2.5%) for the colon digests.

Arsenic K-edge XANES spectra indicated that As (III) had a great increase for As(V)-bearing goethite, and the content ranged from 31.2 mg/kg (5.6%) to 379.4 mg/kg (21.2%) after 48 h of incubation. In contrast, As(III) was just found in JH2, displaying 135.2 mg/kg (12.0%). Finally, no significant change was observed in other jarosite samples.

In comprehensive consideration of speciation analysis, As partitioning between the solid-liquid phase was shown in Figure 1. For As(V)-bearing goethite, 48.6% of

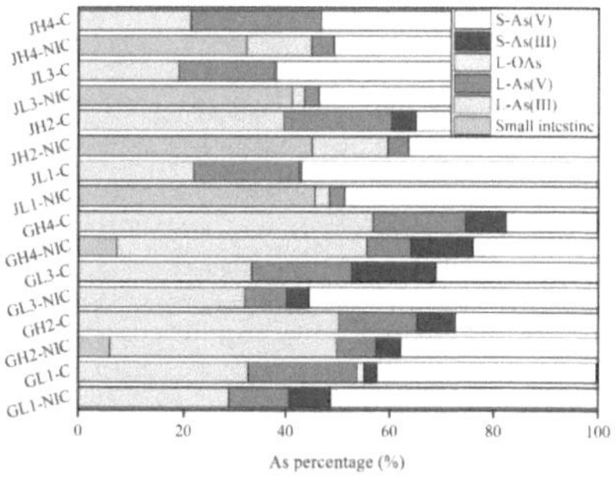

Figure 1. Percentages (%) of As(III) and As(V) between the solid and liquid phases. L = liquid phase; S = solid phase; O-As = methylated As species.

As(V) was reduced to As(III) (35.2-64.4%), and 16.0% of generated As(III) was detected on residual solids. For jarosite, 17.3% of As(V) was reduced and present in the colon digests.

4 CONCLUSIONS AND RECOMMENDATIONS

Human gut microbiota can increase As dissolution from Fe minerals, especially for goethite, while As release behavior presented mainly in small intestinal phase for jarosite. We found a high degree of As(V) reduction. About half of As(V) associated to goethite was reduced to As(III) with high toxicity, with almost 80% of generated As(III) present in the colon digest. The presentation of As transformation and fate in human gastrointestinal tract, is involved in some concurrent processes including As reduction and methylation, Fe(III) reductive dissolution, and As sequestration on biogenic Fe minerals. Information on As release and transformation from As(V)-bearing Fe minerals will result in more accurate risk assessment associated with soil As exposures.

ACKNOWLEDGEMENTS

We acknowledge the National Natural Science Foundation of China (grant no. 41877501) and the project of National Postdoctoral Program for Innovative Talents funded by China Postdoctoral Science Foundation (grant no. BX20180299).

REFERENCES

Agency for Toxic Substances and Disease Registry (ATSDR). 2015. *Toxicological Profile for Arsenic. U.S.* Department of Health and Human Services, Washington, D.C.

Baron D. & Palmer C.D. 1996. Solubility of jarosite at 4-35°. *Geochim. Cosmochim. Acta* 60: 185–195.

Bradham K.D., Diamond G.L., Burgess M., Juhasz A., Klotzbach J.M., Maddaloni M., Nelson C., Scheckel K., Serda S.M. & Thomas D.J. 2018. In vivo and in vitro methods for evaluating soil arsenic bioavailability: relevant to human health risk assessment. *J. Toxicol. Environ. Health B Crit. Rev.* 21(2): 83–114.

Calatayud M., Xiong C., Du Laing G., Raber G., Kevin Francesconi K. & van de Wiele T. 2018. Salivary and gut microbiomes play a significant role in in vitro oral bioaccessibility, biotransformation, and intestinal absorption of arsenic from food. *Environ. Sci. Technol.* 52:14422–14435.

Schwertmann U. & Cornell R.M. 2000. *Iron Oxides In The Laboratory: Preparation and Characterization* (2nd ed.). Wiley-VCH: Weinheim, Germany.

Yin N.Y., Zhang Z.N., Cai X.L., Du H., Sun G.X. & Cui Y.S. 2015. In vitro method to assess soil arsenic metabolism by human gut microbiota: arsenic speciation and distribution. *Environ. Sci. Technol.* 49(17): 10675–10681.

Arsenic in the Environment: Bridging Science to Practice for Sustainable Development –
van der Wal, Ahmad, Petrusevski, Weijma, Savic, van der Wens, Beerendonk, Bhattacharya, Bundschuh & Naidu (Eds)
© 2023 the Author(s), ISBN 978-1-032-32928-4

Chronic arsenicosis among different castes in a village of Patna district, Bihar, India

A.H. Jeelani
Centre of Social Medicine and Community Health, School of Social Sciences, Jawaharlal Nehru University, New Delhi, India

ABSTRACT: Arsenic (As) contamination of groundwater and its health ramifications in different states of India is an emerging public health concern. Among the affected region, states along lower and middle Gangetic plain are highly affected with elevated concentration of As in the groundwater. For this interdisciplinary study, by using Concurrent Quantitative Dominant Mix Method, the researcher attempts to understand health implications among different social groups in As exposed Haldi Chhapra village of Patna district, Bihar, India. The study found 3.65% individuals have manifestations of As poisoning. Among the affected individuals 5.44% belong to Schedule Castes (SC), 3.89% to Extremely Backward Class (EBC), 2.57% to Backward Class (BC) and 3% to Other Class (OC).

1 INTRODUCTION

The arsenic (As) contamination of groundwater has found in almost all the continents. However, South Asian countries, particularly the Ganga, Meghana, and Brahmaputra (GMB) plains due to geophysical structure and flood plain, are profoundly affected regions. In India, the lower and middle Gangetic plain has reported the elevated concentration of As in West Bengal and Bihar in 1984 and 2003 respectively (Chakraborti *et al.*, 2003; Garai *et al.*, 1984). Since the first report of arsenicosis in Bihar, around 10 million people in 17 districts have exposed to As beyond the permissible level (Singh 2015). The consumption of As contaminated water manifest into arsenicosis from skin discoloration that may be viewed as a cosmetic issue, painful skin lesion that can be disabling to the involvement of various internal organs. Rarer, but most serious, is the fact that cancer has been linked causally to chronic arsenicosis. The present study aims to investigate the health implications of chronic inorganic As consumption in a village of Patna district, Bihar, India among the different social stratifications. In this abstracts, results on the emerging public health concern of chronic arsenicosis have presented through an interdisciplinary approach.

2 MATERIALS AND METHODS

The interdisciplinary study has used Concurrent Quantitative Dominant Mix Method (QUANT +Qual) based on the time period and paradigm emphasis decision.

2.1 *Study area*

Haldi Chhapra, a village in Patna district of Bihar, India has been selected purposively based on the selection criteria: i) the available biomedical and environmental studies, ii) the concentration of the arsenic in the groundwater of the village (Figure 1). A study found 231 μg/L average concentration of arsenic in the groundwater; however the highest level of concentration was found 498 μg/L in the village. Along with the arsenic concentration in the groundwater 100% hand pumps were found affected and 577μg/L, 1224 μg/L, 1351 μg/L and 1468 μg/L per capita consumption of As through water was found among children, youth, adult and elderly respectively in the village. Besides this, 2.3% of individuals have been found affected by skin pigmentation (Singh & Ghosh 2012).

2.2 *Sample and sample selection technique*

The population of the village was recorded 8438 resides in 1427 households, according to ward wise voter summary report 2016. For this study, 130 households have been considered for the survey. The sample selection technique for household survey was stratified systematic random sampling where the whole village was stratified into four categories-Schedule Castes (SC), Extremely Backward Class (EBC), Backward Class (BC) and Other Class (OC). The castes were included in these four categories

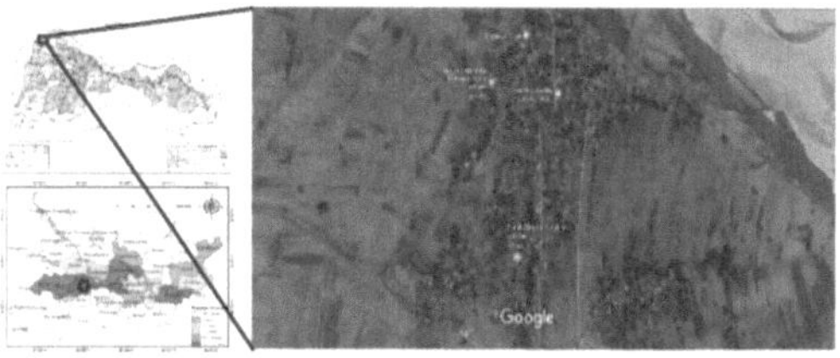

Figure 1. Location of study area in the district and state map.

based on the Bihar list of castes. For qualitative data, the researcher, has purposively selected two experts from Mahaveer Cancer Institute, Patna and five key informants along with affected individuals from the village.

2.3 *Research tools*

The initial days of the research were spent to know the social groups, water and health resources through *social* mapping of the village with community participation and transect walk. However, the researcher had employed household survey with a structured questionnaire to seek mainly three categories of questions-general information about the household, water-related information and arsenic-based queries. An in-depth interview with a semi-structured interview schedule has also been administered with the experts and key informants.

2.4 *Identification of arsenicosis and data analysis*

The cases of arsenicosis were identified either through the prescription of a physician if the patients have or confirmation from experts at Mahaveer Cancer Institute, Patna, if the affected individual has yet unidentified.

The quantitative data were entered into an Excel sheet and analyzed however, the interviews were transcribed or translated and use to develop a descriptive understanding of the arsenicosis.

3 RESULTS AND DISCUSSION

3.1 *The village and social groups*

Haldi Chhapra is a village of Patna district in Bihar, India, situated on the confluence of Sone and Ganga believed to be male and female rivers, respectively. The village is comprised of 15 different castes falling in the four categories of SC, EBC, BC and OC. Among these castes, SC and EBC have the highest percentage of landlessness (32.93% & 36.59%) and lowest ownership of water resources (60%) and income quartile. Rajput and Yadav are dominant castes in the village. Most of the social resources like the public distribution system, water filter and schools are in close proximity to the dominant castes.

3.2 *Arsenicosis: Manifestation and distribution*

The study found 3.65% of individuals have common and rare manifestations of arsenicosis. Among the affected individuals, 77.14% have skin pigmentation, and 14.28% have keratosis however, the rest have either combinations of symptoms or rare manifestation. The finding specific to children (between the age group of 0 to 18 years) is showing 3.06% have manifestations.

The arsenicosis among the four categories as it came through the result is: 5.46% among the SC, 3.89% in EBC, 2.57% belongs to BC, and 3% among the OC. Three clusters of the cases have been found as mentioned in the map along with few sporadic cases when putting all the cases manually on the social map. The red circle is the area where SC resides, a yellow circle where EBC lives however, blue circle belongs to the BC and OC (Figure 2).

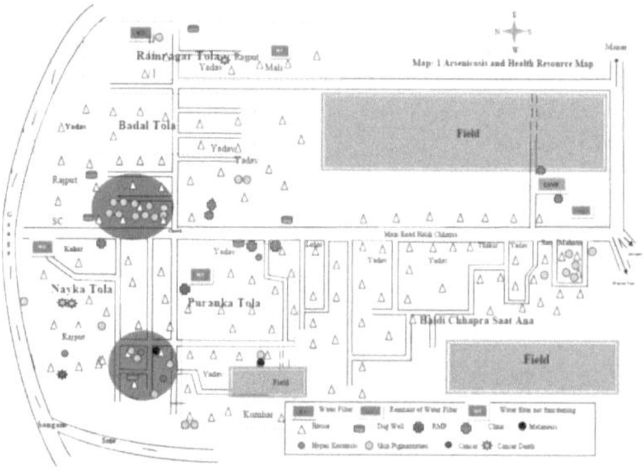

Figure 2. Arsenicosis map showing three clusters (red, blue, and yellow circle) of cases along with sporadic instances.

4 CONCLUSIONS AND RECOMMENDATIONS

The result of the research is showing 3.65% of arsenicosis however earlier study is showing 2.3% which means within six years of span, the cases have increased 1.6% in the village and particularly, children are at high risk. A follow-up study or surveillance system should be set up in the village to track the extent and nature of the problem. Another critical aspect of the finding is the caste system emerged as a social determinant of arsenicosis. The policy makers should consider caste in mitigation programs for ensuring equal access.

ACKNOWLEDGEMENTS

The researcher acknowledges Mahaveer Cancer Institute, Patna, villagers, and M.Phil guide Prof. Ritu Priya Mehrotra.

REFERENCES

Chakraborti D., Mukherjee S.C., Pati S., Sengupta M.K., Rahman M.M., Chowdhury U.K., Lodh D., Chanda C. R., Chakraborti A.K. & Basu G.K. 2003. Arsenic groundwater contamination in middle Ganga plain, Bihar, India: A future danger. *Environ. Health Perspect.* 111(9): 1194–1201.

Garai R., Chakraborty A.K, Dey S.B. & Saha K.C. 1984. Chronic arsenic poisoning from tube well water. *J. Indian Medical Assoc.* 82(1): 34–35.

Singh S.K. & Ghosh A.K. 2012. Health risk assessment due to groundwater arsenic contamination: children at high risk. *Hum. Ecol. Risk Assess. Int. J.* 18(4): 751–766.

Singh S.K. 2015. Groundwater Arsenic contamination in the middle-gangetic plain, Bihar (India): The danger arrived. *Int. J. Environ. Sci.* 4(2): 70–76.

Arsenic in the Environment: Bridging Science to Practice for Sustainable Development –
van der Wal, Ahmad, Petrusevski, Weijma, Savic, van der Wens, Beerendonk, Bhattacharya, Bundschuh & Naidu (Eds)
© 2023 the Author(s), ISBN 978-1-032-32928-4

Assessment of arsenic toxicity in the human blood samples from Simri and Tilak Rai ka Hatta village through haematological, hormonal and free radicals study

M.S. Rahman[1], S.K. Singh[1], A. Kumar[2] & A.K. Ghosh[2]
[1]*A.N. College (Magadh University, Bodh-Gaya) Bihar, India*
[2]*Mahavir Cancer Institute and Research Centre, Bihar, India*

ABSTRACT: Recent studies on arsenic (As) occurrence particularly in African waters show that several sources of drinking water have elevated concentrations above national and international guidelines. In Tanzania, elevated concentrations of As above WHO guideline (10 μg/L) in Lake Victoria Gold fields is emerging as a threat to public health depending on groundwater and surface water as drinking water sources. In this study, spatial statistics and GIS tools have been used to delineate the relationship between As occurrence and local geological settings. Among the 12 mapped local geological units, the most targeted aquifers for potable water are characterized by granitoids, migmatite, mafic and ultramafic meta-sediments (~50% of water points). The probability of having As levels above WHO guideline was 0.71 and 0.33 for surface water and groundwater systems respectively.

1 INTRODUCTION

The toxic metalloid arsenic is well known for its toxic effects and groundwater arsenic is emerging as a major problem in floodplains of the river Ganga. The large scale shifting of the water resource allocation from surface water to groundwater in India exposed millions of people to the arsenic. Presently, arsenic contamination in the groundwater of Bihar is reported from 18 districts, threatening more than 10 million people in the state (Hassan 2018; Kumar *et al.*, 2016; Rahman *et al.*, 2019a). Arsenic introduces oxidative damage which further initiates cytotoxicity. Oxidative stress arises when reactive oxygen species (ROS) are generated, which can react with cellular constituents such as thiols and lipids that results in the alteration of the antioxidant defense systems and cell membrane disruption. Inorganic arsenic interacts with protein sulfhydryls (ProSH) (cell membrane, cytoplasmic proteins, etc.) as well as with nonprotein sulfhydryls (glutathione, cysteine, lipoic acid, acetylCoASH, etc.) in biological systems. Arsenic toxicity affects human blood cells and causes megaloblastic dyserythropoietic anemia, pancytopenia and neutropenia (Rahman *et al.*, 2019b). The present study is aimed to investigate about the toxic effect of arsenic on human health by exploring few free radical parameters, hematological and hormonal assessment.

2 MATERIALS AND METHODS

2.1 *Study area*

The study was conducted in India in Simri and Tilak Rai ka Hatta village of Buxar district of Bihar (25°38′17.6″N/84°06′49.4″E and 25°41′36″N/84°07′51″E respectively). The population of the Simri village is approximately 17,670 in 2011 and 2621 households. The population of the Tilak Rai Ka Hatta village is approximately 5,348 and 340 households in 2011. The Simri village is situated approximately 1.65 km away from the banks of the river Ganga. Tilak Rai Ka Hatta village is a flood plain region of the river Ganga. Phulwarisharif, Patna (25°34′33″N 85°4′6″E) was selected as the control area.

2.2 *Free radicals assessment*

Thiobarbituric acid reactive substances (TBARS), as a marker for Lipid Peroxidation (LPO), were determined by the double heating method. This method is a spectrophotometric measurement of the color produced during the reaction to thiobarbituric acid (TBA) with malondialdehyde (MDA). Glutathione Peroxidase (GPx) activity was measured by the kinetic assay using the "continuous spectrophotometric rate determination reaction" and expressed as unit per liter in plasma (U/L). One unit of GPx represents 1μmol oxidized NADPH per min.

2.3 *Hematological parameters assessment*

The red blood cells (RBC) and white blood cells (WBC) counts were determined by the improved Neubauer hemocytometer method. The hemoglobin (Hb) concentration was determined by using the Sahli's hemoglobinometer. Hematocrit (HCT) was determined directly by microhematocrit tubes. Mean corpuscular volume (MCV), mean corpuscular hemoglobin (MCH) and mean corpuscular hemoglobin concentration (MCHC) was computed according

DOI: 10.1201/9781003317395-94

to Dacie and Lewis (1991). All tests were performed in duplicates.

2.4 *Hormonal assessment*

Using the ELISA method, Thyrotropin (TSH), Testosterone and Estradiol (E2) kit of AccuBind, Monobind, Lake Forest, CA, USA was utilized for the experiment.

2.5 *Statistical analysis*

Data were analyzed using statistical software package (GraphPad Prism 5) and values were expressed as Mean ± SEM.

3 RESULTS AND DISCUSSION

The complete results of all hematological parameters, free radicals and hormonal assessment in comparison with blood arsenic level is presented in Table 1.

Table 1. Showing the complete results of all hematological parameters, free radicals and hormonal assessment in comparison with blood arsenic level.

Parameters	Control area*	Test subjects from both villages**
Blood Arsenic Level (µg/L)	2.26 ± 0.6(n=20)	81.45 ± 5.93(n=375)
MDA Level (nmol/ml)	2.24 ± 0.3(n=20)	9.41 ± 0.62(n=375)
GPx (U/L)	227.50 ± 5.44(n=20)	304.47 ± 9.20(n=375)
Hb (g/dL)	14.11 ± 0.34(n=20)	9.86 ± 0.62(n=375)
RBC (x 10^6 µL)	4.29 ± 0.08(n=20)	2.95 ± 0.11(n=375)
HCT (%)	40.33 ± 0.53(n=20)	27.79 ± 1.77(n=375)
MCV (fL)	101.63 ± 3.47(n=20)	95.64 ± 5.44(n=375)
MCH (pg)	37.64 ± 1.25(n=20)	33.98 ± 1.91(n=375)
MCHC (g/dL)	37.65 ± 0.64(n=20)	35.58 ± 0.08(n=375)
WBC (x 10^3 µL)	7.38 ± 0.28(n=20)	6.67 ± 0.80(n=375)
TSH Level (µlU/ml)	2.92 ± 0.42(n=5)	7.23 ± 2.77(n=101)
Testosterone Level (ng/ml)	6.87 ± 0.27(n=5)	3.72 ± 0.62(n=87)
Estrogen Level (pg/ml)	26.66 ± 4.41(n=5)	158.11 ± 29.11(n=28)

*Mean Arsenic level in groundwater of control area = 4.30 ± 0.45 µg/L;
**Mean Arsenic level in groundwater of both villages = 252.02 ± 12.48 µg/L; P<0.01, Values are mean ± S.E.

4 CONCLUSIONS AND RECOMMENDATIONS

Arsenic contamination along the river Ganga is a major problem in drinking water extracted mainly through the handpumps. The increased MDA and GPx level shows the oxidative stress among the arsenic affected population. The decreased hemoglobin, RBC and hematocrit levels indicate the disturbances in the hematological parameters of the subjects. The increased TSH and Estrogen level, and the decreased Testosterone level shows the hormonal alterations among the test subjects belongs from both villages. The abnormal estrogen and testosterone levels provide the basis of infertility. The study reveals that arsenic poisoning is leading to severe hematological, hormonal and free radicals abnormalities in the population of arsenic hit areas of Buxar district of Bihar. Thus, it concludes that arsenic exposure to the population residing in the arsenic hit area leads to hematological changes and low immunity in them and indicating that they are at very high risk and alternate source of arsenic free drinking water is recommended. The proper mitigation and regular medical aids should also recommended to the affected population.

ACKNOWLEDGEMENTS

We acknowledge Anugrah Narayan College, Patna and Mahavir Cancer Institute and Research Centre, Bihar, India for the entire infrastructural facilities.

REFERENCES

Dacie J.V. & Lewis S.M. 1991. *Practical Haematology*. 7th Edition. *ELBS with Churchill Livingston, England*. p. 37–85.

Hassan M.M. 2018. *Arsenic in Groundwater: Poisoning and Risk Assessment, 1st Edition. Boca Raton: Taylor & Francis*, CRC Press. ISBN 9781439839270.

Kumar A., Rahman M., Iqubal M., Ali M., Niraj P.K., Anand G., Prabhat K., Abhinav & Ghosh A.K. 2016. Ground water arsenic contamination: a local survey in India. *Int. J. Prev. Med.* 7:100.

Rahman M.S., Kumar A., Kumar R., Ali M., Ghosh A. K. & Singh S.K. 2019a. Comparative quantification study of arsenic in the groundwater and biological samples of Simri village of Buxar district, Bihar, India. *Indian J. Occup. Environ. Med.* 23:126–32.

Rahman M.S., Kumar A., Kumar R., Ali M., Ghosh A. K. & Singh S.K. 2019b. Hematological and free radicals changes among people of arsenic endemic region of Buxar district of Bihar, India. *Int. J. Pub. Health Safe.* 4:178.

Arsenic in the Environment: Bridging Science to Practice for Sustainable Development –
van der Wal, Ahmad, Petrusevski, Weijma, Savic, van der Wens, Beerendonk, Bhattacharya, Bundschuh & Naidu (Eds)
© 2023 the Author(s), ISBN 978-1-032-32928-4

Arsenic exposure and health risk assessment – A study from upper Brahmaputra floodplain, Assam, India

R. Goswami[1] & M. Kumar[2,3]
[1]*Department of Environmental Science, The Assam Royal Global University, Guwahati, India*
[2]*Discipline of Earth Sciences, Indian Institute of Technology, Palaj, Gandhinagar, India*
[3]*Sustainability Cluster, School of Engineering, University of Petroleum & Energy Studies, Dehradun, Uttarakhand, India*

ABSTRACT: Widespread utilization of groundwater, containing high arsenic (As) levels, the Brahmaputra fllodplain, India raises serious health concerns and demands risk evaluation. Considering the present scenario, this study was planned with a holistic correlation groundwater As with the body tissues and consequent health risk assessment. Demographic information and other factors used in risk calculation were addressed through questionnaire based survey. Results exhibited High groundwater As along with significant as body burden. The maximum value for the Hazard Index (HI) indicated severe non-carcinogenic health impacts with high cancer risk. The future risk evaluation projection showed adverse health outcomes due to the overexposure.

1 INTRODUCTION

Illustrated by our earlier work, the inhabitants of northeastern part of India are particularly vulnerable to groundwater arsenic (As) contamination (Goswami *et al.*, 2014; Kumar *et al.*, 2016). Moreover, with the exception recent reports from our laboratory, there is a paucity of work on hazard indices related to consumption of groundwater with As levels in in this part above the WHO drinking water guideline (WHO 2011). Thus, there persists a clear lacuna in the evaluation of drinking water risk exposure, among private well users in As endemic areas of Brahmaputra floodplain. Further, our preliminary findings from this region indicate that, there has been little effort in identifying the relationship between magnitudes of As contamination and people's perception/understanding of the dangers of the issue. The central focus is a holistic correlation of groundwater As status with the consequent health risk assessment of the population of the upper Brahmaputra floodplain with concurrent questionnaire survey and biological samples analysis. The risk assessment study was designed to evaluate the maximum Hazard Index (HI) for both adults and children.

2 MATERIALS AND METHODS

2.1 *Study area*

The main area of focus was the Lakhimpur district in the Assam, India lying on the north bank of the Brahmaputra River (Figure 1). The Brahmaputra River controls the main drainage system in the district and the Subansiri–Ranganadi–Dikrong river system debouches in the Brahmaputra forming an intricate drainage network in the district (Central Ground Water Board 2013). Physiographically the study area can be divided broadly into three parts, i.e., the hilly tract comprising of the Siwalik sediments of lesser Himalayas, the foothill region and the extensive flood plain created by the river Brahmaputra and its tributaries in southern part (Central Ground Water Board 2013).

2.2 *Sampling and analysis of groundwater and biological samples*

Sample collection, preservation, digestion and analysis have been conducted using the methods illustrated by APHA (2005).

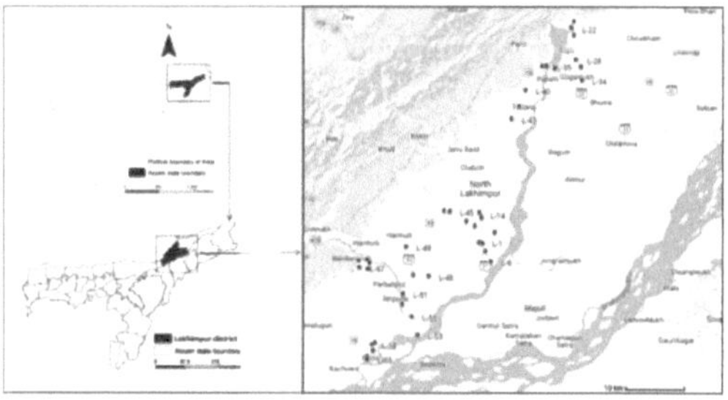

Figure 1. Map depicting study area location and sampling sites on Assam province of Indian Peninsula along with major rivers.

2.3 *Questionnaire survey*

A structured questionnaire was designed to get comprehensive idea about the magnitude of the problem, people's perception aiming at evaluating the societal outcome of the As problem and correlate its significance with people's understanding of the problem.

2.4 *Health risk assessment*

Risk characterization based on U.S.EPA framework has been carried out to find out the carcinogenic risk

DOI: 10.1201/9781003317395-95

and non carcinogenic effect among children and adult (EHRA 2004).

3 RESULTS AND DISCUSSION

3.1 *Chemistry of groundwater and sediments*

Major ion compositions of the analyzed groundwater samples showed remarkable spatial variations (Table 1) with dominance in the order of $Ca^{2+} > Na^{+} > Mg^{2+} > K^{+}$. Similar variation was recorded in distribution of major anions, with relative dominance trend of $HCO_3^{-} > SO_4^{2-} > Cl^{-}$. High groundwater arsenic concentrations was found (average=137) with 80% having As levels > WHO limit (0.01 mg/L).

Table 1. As conc. in hair, nail and urine of exposed adult and children of Majuli (the units are µg/L for urine and water and µg/kg for hair and nail).

Parameter!	Children Eanaa	Avs± SD	Adult Range	Aye± SD
Ase (Yte)	(2-12)	H.d±2_7	2 5-50	40.5±7.S
Gr-sim d water	{20-3 sa;.	152±B7_5		
Hair	<251- 30 Si)	9P0±BOO	{220- 5461)	173S±]27B
Kail	{125- 5Z51)		{426- 11725)	2702±2737
Urine	{54.4- GP7 Si	j33±10S	{43.1- 304.3}	142±B0_B

3.2 *Hydogeochemical facies*

The major ion compositions were plotted on Piper diagram (Figure 2) to comprehend the probable hydrogeochemical regime depicting the dominating water type of the study area. Figure 2 reveals that all the analyzed samples belong to Ca-HCO_3 type water implying that alkalinity favours the As dissolution. Although the sample size for the water type analysis was lower (n=20), the dominance of Ca-HCO_3 indicated homogeneity in the water type and good recharge, thus showing high groundwater sediment interaction in the studied location.

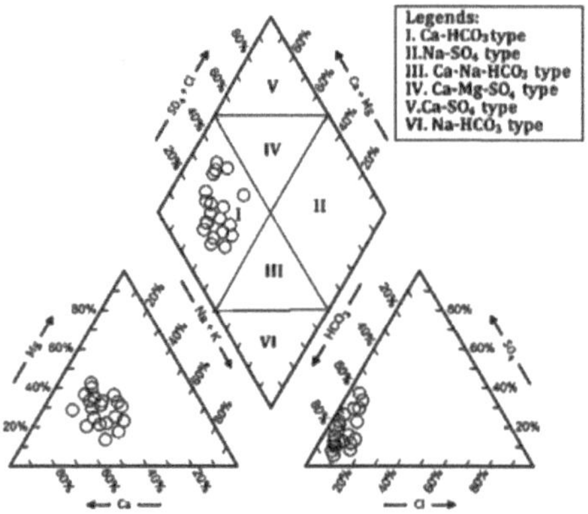

Figure 2. Piper diagram depicting water type of the river island.

3.3 *Accumulation of arsenic in biological samples*

The average As concentration in urine, nail and hair samples of As exposed people is presented in Table 1. Around 95.2% of urine samples showed As concentration over the maximum normal level (50 µg/L. As concentration in urine was still higher in some cases, where 15% of the samples showed As concentrations more than 300 µg/L.

3.4 *Analysis of Variance (ANOVA) test of biological samples of exposed population groups*

The one-way analysis of variance (ANOVA) was applied to examine if any considerable difference exist between the means of biological samples of the different age groups. The single factor ANOVA analysis revealed that urine had significant age group variation (p value <0.05) as shown by higher F calculated values in comparison to the critical F values. This is also supported by the results shown in Table 1 where urine of the children showed much higher arsenic as compared to the adults.

3.5 *Risk characterization by estimating hazard indices (HI)*

Computational results implied that 80% of the studies groups in the area were at probable risk to non-cancer health impacts. The maximum value for the HI exceeded for both adults and children (Goswami *et al.*, 2014), implying that the As exposed population may face severe non-carcinogenic health impacts together with significantly high risk of cancer in near future.

4 CONCLUSIONS

Thr upper Brahmaputra floodplain is enriched with As showing significant correlation with As in biological samples. The accumulation rate was found to vary amongst individuals probably due to the individual's health status, malnutrition level, family socio economic condition, drinking water use and storage conditions. The application of the EHRA framework was found to be helpful for estimation of the health risk due to As exposure via drinking water in a systematic and participatory manner.

REFERENCES

Environmental Health Risk Assessment (EHRA) 2004. Guidelines for Assessing Human Health Risks from Environmental Hazards.

Goswami R. Rahman M.M., Murrill M., Sarma K.P., Thakur R. & Chakraborti D. 2014. Arsenic in the groundwater of Majuli – the largest river island of the Brahmaputra: magnitude of occurrence and human exposure. *J. Hydrol.* 518: 354–362.

Kumar M., Das A., Das N. and Goswami R. 2016. Co-Occurrence perspectives of arsenic and fluoride in the groundwater of Diphu, Northeastern India. *Chemosphere* 150: 227–238.

WHO. 2011. Guidelines for Drinking Water Quality.: World Health Organization, Geneva.

Arsenic in the Environment: Bridging Science to Practice for Sustainable Development –
van der Wal, Ahmad, Petrusevski, Weijma, Savic, van der Wens, Beerendonk, Bhattacharya, Bundschuh & Naidu (Eds)
 ISBN 978-1-032-32928-4

Chronic arsenic exposure dose-dependently increases the risk of hyperglycemia through skeletal muscle mass reduction

K. Hossain[1], V. Mondal[1], Z. Hosen[2], F. Hossen[1], A.E. Siddique[1], S. Hossain[1] & S. Himeno[3]
[1]*Department of Biochemistry and Molecular Biology, University of Rajshahi, Rajshahi, Bangladesh*
[2]*Department of Applied Nutrition and Food Technology, Islamic University, Kushtia, Bangladesh*
[3]*Laboratory of Molecular Nutrition and Toxicology, Faculty of Pharmaceutical Sciences, Tokushima Bunri University, Tokushima, Japan*

ABSTRACT: Although association between high levels of arsenic (As) and diabetes mellitus (DM) has been established, inconsistent results have been reported in the association between low to moderate levels of As exposure and DM. The underlying mechanisms of As-induced DM remain to be clarified. We explored the dose-response relationship between As exposure and hyperglycemia or DM and the underlying mechanism of As exposure-related DM. Aiming these, we have conducted two studies on the same population recruited from high- and low-As exposed areas in Bangladesh. In the first study, we found that chronic As exposure increases the fasting blood glucose (FBG) levels in both male and female subjects, however, females showed greater susceptibility to As-related hyperglycemia than males. In the subsequent study, we found that chronic As exposure is associated with increased insulin resistance with decreasing skeletal muscle mass. Female subjects' insulin resistance showed a greater susceptibility to the muscle mass reduction, possibly causing the greater risk of hyperglycemia than males. The results of the two consecutive studies suggest that chronic As exposure increases the risk of hyperglycemia with greater susceptibility in females than males and the reduction of skeletal muscle mass may be a potential mechanism of As exposure-related insulin resistance.

1 INTRODUCTION

Exposure to arsenic (As) through drinking water and food has made a catastrophic situation in some countries especially Bangladesh where more than half of the total population were exposed to As through drinking water at greater than the maximum permissive limit set by WHO (Kinniburgh & Smedley 2001). Exposure to As causes several chronic diseases including diabetes mellitus (DM). Hyperglycemia or DM is a major global concern because of its increasing prevalence and dire consequence. Association between high level of As exposure and DM has been established, however, inconsistent results have been reported in the association between low to moderate levels of As exposure and diabetes (Maull *et al.*, 2012). Moreover, underlying mechanisms of As-related diabetes remain to be clarified. Therefore, we explored the dose-response relationship between As exposure and hyperglycemia and the underlying mechanism of As exposure-related hyperglycemia or DM.

2 METHODS

Ethical approval for this study was obtained from We recruited study subjects from some high- and low-As exposed villages in the north-west region of Bangladesh (Paul *et al.*, 2019) and conducted two consecutive studies on the same population. Ethical permission was taken from The Institute of Biological Sciences, University of Rajshahi, Bangladesh (661/320/IAMEBBC/IBSc). As exposure levels were assessed by the subjects' drinking water, hair and nail As concentrations through ICP-MS. Subjects with hyperglycemia, impaired glucose tolerance (IGT) or DM were defined according to WHO diagnostic criteria. Pre-diabetic and diabetic subjects were identified by 2-hours blood glucose (2h-BG) levels $\geq$ 140–199 mg/dl and $\geq$ 200 mg/dl, respectively, as determined by oral glucose tolerance test.

3 RESULTS AND DISCUSSION

We conducted two studies on the same population to explore the dose-response relationship between As exposure and hyperglycemia and the underlying mechanism of As exposure-related hyperglycemia or DM. In the first study, we reported that chronic As exposure is associated with the risk of hyperglycemia including IGT and DM (Paul *et al.*, 2019) (Figure 1). Gender was found to be a significant confounder on the risk of hyperglycemia. Intriguingly, in gender specific analyses, female subjects showed clear pattern of dose-response relationship between As exposure and the risk of

DOI: 10.1201/9781003317395-96

hyperglycemia (Figure 1A). Male subjects showed weak and non-significant association between As exposure and the risk of hyperglycemia (Figure 1B). Females also showed stronger association between As exposure metrics and FBG levels (data not shown). Hence the results of the first study clearly demonstrated that As exposure dose-dependently increases the risk of DM or hyperglycemia and females were more susceptible to As exposure-related hyperglycemia or DM than males.

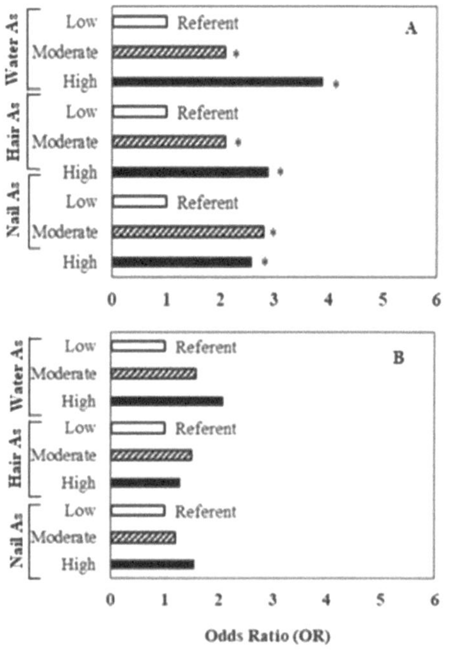

Figure 1. Schematic diagram of associations of As exposure with the risk of hyperglycemia in females (A) and males (B). Schematic diagram were prepared form the results published by Paul *et al.* (2019). *Significant OR.

In a subsequent study (submitted manuscript), we found that subjects' As exposure levels were positively associated with insulin resistance as measured by HOMA-IR (Figure 2). Insulin resistance decreases glucose uptake and utilization by adipocytes and skeletal muscle. Skeletal muscle is a predominating site for insulin-mediated glucose uptake and glycogen synthesis (Ferrannini *et al.*, 1985). Thus decreased muscle mass is one of the risk factor for insulin resistance that ultimately develop DM (Srikanthan *et al.*, 2011). Serum creatinine and lean body mass (LBM) are the important indicators of muscle mass. We found an inverse association between As exposure and muscle mass indicators (Figure 2). These associations provided evidence in support of the sarcopenic effects of prolonged exposure to As. We also found that insulin resistance was more pronounced with the decreasing muscle mass markers in females than males (data not shown) which may be a cause for greater susceptibility of females to As exposure-related DM than the males.

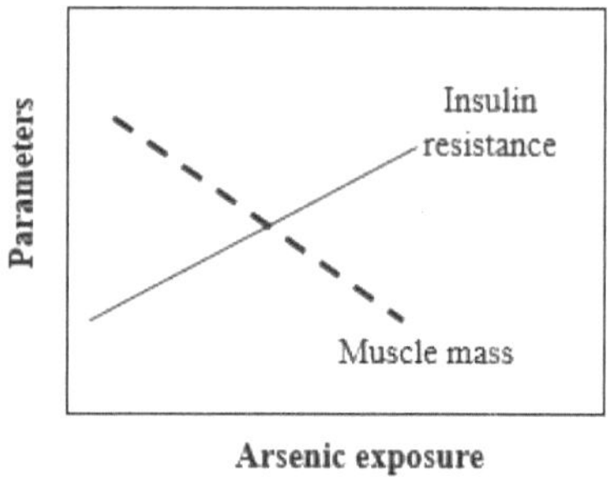

Figure 2. Schematic diagram showing associations of As exposure with insulin resistance and skeletal muscle mass.

4 CONCLUSIONS

Results of our two studies suggest that chronic As exposure increases the risk of DM or hyperglycemia with greater susceptibility in females than males. Decreased muscle mass reduction may be an underlying mechanism of chronic As exposure-related insulin resistance and DM.

ACKNOWLEDGEMENTS

We acknowledge the financial supported given by the University of Peradeniya research grant (RG/2014/31/M) offered to SF and the National Research Council (NRC), Sri Lanka research grant (TO 14-05) offered to RC.

REFERENCES

Ferrannini E., Bjorkman D. & Reichard G.A. 1985. The disposal of an oral glucose load in healthy subjects. a quantitative study. *Diabetes* 34: 580–588.

Kinniburgh D. & Smedley P. 2001. Arsenic contamination of groundwater in Bangladesh. *Br. Geol. Surv.* 1.

Maull E.A., Ahsan H., Edwards J., Longnecker M.P., Navas-Acien A., Pi J., Silbergeld E.K., Styblo M., Tseng C.H., Thayer K.A., Loomis D. 2012. Evaluation of the association between arsenic and diabetes: A national toxicology program workshop review. *Environ. Health Perspect.* 120:1658–1670.

Paul S.K., Islam M.S., Hasibuzzaman M.M., Hossain F., Anjum A., Saud Z.A., Haque M.M., Sultana P., Haque A., Andric K.B. & Rahman A. 2019. Higher risk of hyperglycemia with greater susceptibility in females in chronic arsenic-exposed individuals in Bangladesh. *Sci. Total Environ.* 668: 1004–1012.

Srikanthan P. & Karlamangla A.S. 2011. Relative muscle mass is inversely associated with insulin resistance and prediabetes. Findings from the third national health and nutrition examination survey. *J. Clin. Endocrinol. Metab.* 96: 2898–2903.3.

Arsenic in the Environment: Bridging Science to Practice for Sustainable Development –
van der Wal, Ahmad, Petrusevski, Weijma, Savic, van der Wens, Beerendonk, Bhattacharya, Bundschuh & Naidu (Eds)
© 2023 the Author(s), ISBN 978-1-032-32928-4

Arsenic exposure and endothelial dysfunction: A possible cause of cardiovascular diseases

S. Hossain[1], M.M. Hasibuzzaman[1], E. Hossain[1], S. Himeno[2] & K. Hossain[1]
[1]*Department of Biochemistry and Molecular Biology, University of Rajshahi, Bangladesh*
[2]*Laboratory of Molecular Nutrition and Toxicology, Faculty of Pharmaceutical Sciences, Tokushima Bunri University, Japan*

ABSTRACT: Chronic exposure to arsenic is associated with the increasing risk of morbidity and mortality from cardiovascular diseases (CVDs). However, underlying mechanisms of arsenic-related CVDs has not yet been clearly understood. Endothelial dysfunction plays a central role in the development of atherosclerosis leading to CVDs. We hypothesized that endothelial damage or dysfunction is an early event of arsenic-related CVDs. Soluble thrombomodulin (sTM) in the serum is thought to be a specific marker for endothelial damage or dysfunction. To evaluate the effects of arsenic exposure on endothelial dysfunction, we conducted a population based study recruiting human subjects from arsenic-endemic and non-endemic areas in Bangladesh. We observed that subjects' arsenic exposure levels had significant positive associations with serum sTM levels. Serum sTM levels were higher in the higher exposure gradients of arsenic exposure if we stratified the subjects into three groups (low, medium and high) based on the arsenic concentrations of the subjects' drinking water, hair and nails. In another study conducted almost on the same population, it was observed that chronic exposure to arsenic significantly increased the levels of plasma Big-ET-1, a precursor form of endothelin-1 which has been recognized as a potent vasoconstrictor and a marker of endothelial dysfunction. Interestingly, we observed that sTM and plasma Big-ET-1 levels were significantly higher in the hypertensive groups compared to the normotensive counterpart of the study subjects in arsenic-endemic areas. Taken together our results suggest that chronic exposure to arsenic causes endothelial dysfunction which may be a potential mechanism of arsenic-related hypertension or other forms of CVDs.

1 INTRODUCTION

Arsenic is a major threat to the public health in many countries especially Bangladesh where it has been recognized as the largest mass poisoning in the history of human civilization (Smith *et al.*, 2000). Vascular endothelium plays a central role in vascular homeostasis. However, endothelial dysfunction is the main pathophysiological mechanism of atherosclerosis leading to CVD (Flammer *et al.*, 2012). Thrombomodulin (TM) is an integral membrane glycoprotein and high-affinity receptor for thrombin on the endothelial cell surface, and it has been implicated in the endothelial regulation of fibrinolysis and coagulation (Sadler 1997). After proteolytic cleavage of TM from the injured or damaged endothelial cell surface, soluble form (sTM) is found in blood and its concentration reflects the degree of endothelial damage (Ishii *et al.*, 1991; Seigneur *et al.*, 1993). Additionally, Big-ET-1, the biological precursor of endothelin-1 (ET-1) is a well known vasoconstrictor and a marker of endothelial dysfunction. CVD is a major cause of arsenic-related morbidity and mortality. However, the effect of arsenic exposure on endothelial damage or dysfunction leading to CVDs has not yet been well documented. Therefore, we explore the association between arsenic exposure and serum biomarkers of endothelial dysfunction.

2 METHODS

We recruited study subjects from arsenic-endemic and non-endemic areas in the north-western region of Bangladesh (Karim *et al.*, 2010). Subjects' arsenic exposure levels (i.e., drinking water, hair and nail arsenic concentrations) were measured by Inductively Coupled Plasma Mass Spectroscopy. The subjects' serum sTM levels and plasma Big-ET-1 levels were quantified by immunoassay kit.

3 RESULTS AND DISCUSSION

In our research (Hasibuzzaman *et al.*, 2017), we found that sTM levels were positively and significantly associated with drinking water, hair and nail arsenic concentrations and (data not shown). Next, we checked the dose-dependent associations between arsenic exposure and serum sTM levels (Table 1). All subjects were split into three groups (low, medium and high) based on the subjects' water, hair and nail arsenic concentrations, and we found that sTM levels were higher in the higher exposure gradients based on the water, hair and nail arsenic concentrations. These results were further supported by the findings of another study in which we found that chronic arsenic exposure was significantly

DOI: 10.1201/9781003317395-97

Table 1. Dose-response relationship between arsenic exposure matrices and serum sTM levels. Data are presented as (mean ± SE). p-values are from one way ANOVA.

Exposure markers	sTM (ng/mL)
Water As (μg/l)	
Low	3.15 ± 0.63
Moderate	3.87 ± 0.83 [a*]
High	4.95 ± 0.98 [a***, b***]
Hair As (μg/g)	
Low	3.03 ± 0.59
Moderate	4.10 ± 0.79 [a***]
High	4.89 ± 0.96 [a***, b*]
Nail As (μg/g)	
Low	3.06 ± 0.57
Moderate	4.48 ± 0.84 [a***]
High	4.67 ± 0.86 [a***]

$^{***}p < 0.001$ and $^{*}p < 0.05$. [a]Significantly different from low groups; [b]Significantly different from moderate group

associated with plasma Big-ET-1 (data not shown). Both sTM and Big-ET-1 is the specific marker for endothelial damage or dysfunction (Hossain *et al.*, 2012). Thus, these results indicate that chronic arsenic exposure dose-dependently causes the endothelial dysfunction. Hypertension is a major risk of many forms of CVDs. It has also been sometimes recognized as form CVD. It is the most common form of CVD worldwide. Chronic exposure to arsenic has been reported to be associated with the risk of hypertension. In our study, we also found that chronic arsenic exposure to arsenic was positively associated with blood pressure (data not shown). Intriguingly, we observed that both sTM (Figure 1) and

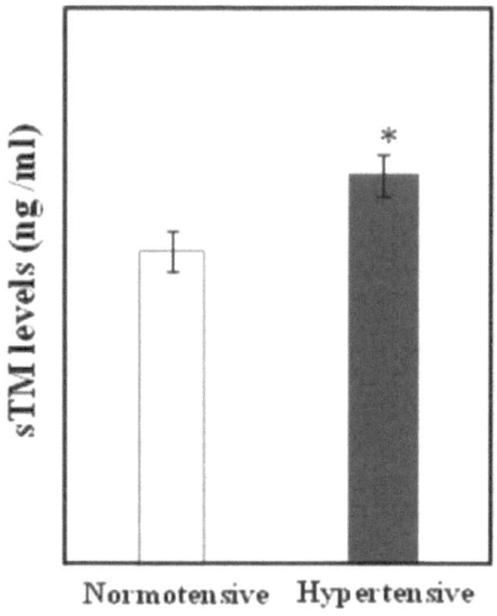

Figure 1. Comparison of serum sTM levels between normotensive and hypertensive subjects. Data were presented as (mean ± SE). $^{*}p < 0.05$.

plasma Big-ET-1 levels (data not shown) were significantly higher in the hypertensive groups compared to the normotensive subjects in arsenic-endemic areas suggesting that endothelial dysfunction may be an underlying mechanism of chronic arsenic exposure-related hypertension or other forms of CVDs.

4 CONCLUSIONS AND RECOMMENDATIONS

Results of our research suggest that chronic arsenic exposure causes endothelial damage or dysfunction, and endothelial dysfunction may be an underlying mechanism of chronic arsenic exposure-related hypertension and other forms of CVDs.

REFERENCES

Flammer A.J., Anderson T., Celermajer D.S., Creager M.A., Deanfield J., Ganz P., Hamburg N.M., Lüscher T.F., Shechter M., Taddei S., Vita J.A. & Lerman A. 2012. The assessment of endothelial function: From research into clinical practice. *Circulation* 126: 753–767.

Hasibuzzaman M.M., Hossain S., Islam M.S., Rahman A., Anjum A., Hossain F., Mohanto N.C., Karim M. R., Hoque M.M., Saud Z.A., Miyataka H., Himeno S. & Hossain K. 2017. Association between arsenic exposure and soluble thrombomodulin: a cross sectional study in Bangladesh. *PloS One.* 12(4): e0175154.

Hossain E., Islam K., Yeasmin F., Karim M.R., Rahman M., Agarwal S., Hossain S., Aziz A., Al Mamun A., Sheikh A., Haque A., Hossain MT., Hossain M., Haris PI., Ikemura N., Inoue K., Miyataka H., Himeno S. & Hossain K. 2012. Elevated levels of plasma big endothelin-1 and its relation to hypertension and skin lesions in individuals exposed to arsenic. *Toxicol Appl Pharmacol.* 259:187–194

Ishii H., Uchiyama H. & Kazama M. 1991. Soluble thrombomodulin antigen in conditioned mediumis increased by damage of endothelial cells. *Thromb Haemost* 65: 618–623.

Karim M.R., Salam K.A., Hossain E., Islam K., Ali N., Haque A., Saud Z.A., Yeasmin T., Hossain M., Miyataka H., Himeno S. & Hossain K. 2010. Interaction between chronic arsenic exposure via drinking water and plasma lactate dehydrogenase activity. *Sci. Total. Environ.* 409, 278–283.

Sadler J.E. 1997. Thrombomodulin structure and function. *Thromb Haemost.*78: 392–395.

Seigneur M., Dufourcq P., Conri C., Constans J., Mercie P., Pruvost A., Amiral J., Midy D., Baste J.C. & Boisseau MR. 1993. Levels of plasma thrombomodulin are increased in atheromatous arterial disease. *Thromb Res.* 71: 423–431.

Smith A.H., Lingas E.O. & Rahman M. 2000. Contamination of drinking-water by arsenic in Bangladesh: a public health emergency. *Bull World Health Organ.* 78: 1093–1103.

Arsenic in the Environment: Bridging Science to Practice for Sustainable Development – van der Wal, Ahmad, Petrusevski, Weijma, Savic, van der Wens, Beerendonk, Bhattacharya, Bundschuh & Naidu (Eds)
© 2023 the Author(s), ISBN 978-1-032-32928-4

Effects of arsenic exposure on the gut microbiome in exposed populations of Bihar, India

D. Mondal[1], D. Paul[2], Y. Shouche[2], S. Suman[1,3], R. Kumar[3], A.K. Ghosh[3], S.K. Singh[4,5] & M.M. Rahman[6]
[1]*School of Science, Engineering and Environment, University of Salford, UK*
[2]*National Centre For Microbial Resource, NCCS, Pune, India*
[3]*Mahavir Cancer Institute and Research Center, Patna, India*
[4]*Department of Geography, A.N. College, Patna, India*
[5]*Department of Environment and Water Management, A.N. College, Patna, India*
[6]*Global Centre for Environmental Remediation, The University of Newcastle, Callaghan, NSW, Australia*

ABSTRACT: Arsenic (As) exposure from drinking water and food might interact with the gut-microbiome. We aim to determine the effects of As exposure on the microbial diversity in an exposed population of Bihar, India.

1 INTRODUCTION

Arsenic (As), affecting hundreds of millions of people worldwide is known to have one of the highest detrimental environmental health impacts in India. Though numerous mechanisms have been proposed for As-induced diseases, no scientific consensus has been reached regarding its mode of action. Recently, ability of As to affect gut microbes has been proposed as a possible mode of action. Experimental studies (largely in vitro and few in vivo using animal models) suggested As-gut microbiome interaction to be potentially double-edged in their consequences: a) microbiome contributing to pre-systemic biotransformation of ingested As, sometimes to more toxic forms (Rubin *et al.*, 2014); and b) As induced compositional changes in the gut microbiome (dysbiosis) and hence perturbing metabolic profiles at the function level (Lu *et al.*, 2014). The aim of this study is to determine the effects of As exposure from both drinking water and diet on the gut microbiome of As exposed populations in Bihar, India.

2 METHODS

2.1 *Sample collection and analysis*

Samples were collected from participants residing in As contaminated areas of Bihar, India as a part of on-going Nutri-SAM project (Figure 1). After taking informed consent, data on dietary intake, socioeconomic and health status were collected using questionnaires. Food consumption pattern and nutrient intake was determined using food frequency questionnaire and 24-hour recall methods, as detailed in Suman *et al.* (2019). Fecal samples were collected in sterile container with RNAlater (Invitrogen, Carlsbad, CA) and stored immediately at -80 °C. Total DNA was extracted from the fecal samples using ‘QIAamp fast DNA stool Mini Kit’ following the protocol as specified by the manufacturer (QI-AGEN, Germany) and all the extracted DNA samples were stored at -20 °C for further analysis. 16S rRNA gene sequencing and bacterial community analysis was performed at National Centre for Microbial Resource, Pune, India following the established protocol as detailed in Bodkhe *et al.* (2019). Food (rice, wheat and potato) and drinking water samples were collected from every household, processed and shipped to University of Newcastle, Australia where food samples were analysed following the established protocol as detailed in Suman *et al.* (2019). Drinking water was analysed using Atomic absorption spectrophotometer.

2.2 *Data analysis*

Microbial diversity and composition analysis were done using the phyloseq (v1.22.3) and microbiome (v1.0.2) packages in R4. Differential abundance of bacterial taxonomic groups between high exposure and low exposure groups were analyzed for each taxonomic level. Using Stata (version 11.2, StataCorp, USA) a generalised linear model (GLM) was used to determine the factors potentially associated with the change in relative abundances of bacterial communities at phyla level with increase of As exposure.

3 RESULTS AND DISCUSSION

Sampling was done from 10 out of 19 As effected districts of Bihar (Figure 1). We present with the available results of this on-going study.

3.1 *Arsenic, diet and health*

In the paired samples collected from the households (Table 1), we found a significant correlation between

DOI: 10.1201/9781003317395-98

Figure 1. Sampling locations in Bihar, India.

drinking water and cooked rice (Spearman's rho= 0.4541 ($p < 0.05$)); and drinking water and wheat flour (Spearman's rho= 0.3174 ($p < 0.05$)), indicating that with the increase in As in the drinking water, if food is cooked or processed with the same contaminated water it might result in an increased As exposure.

We estimated the overall inorganic As exposure from drinking water, cooked rice, potato and wheat intake and obtained a mean exposure of 1.99 μg/kg bw/day (n=153, min 0.03 and max 11.17 μg/kg bw/day) in the surveyed population. 24% of the participants (n=37) were found to have overall As exposure above the recommended upper limit of 3 μg/kg bw/day of inorganic As proposed by the Joint FAO/WHO Expert Committee on Food Additives (Mondal *et al.*, 2019). European Food Safety Authority recommended a range between 0.3–8 μg/kg bw/day of inorganic As exposure based on 1% increased incidence in lung, skin and bladder cancer and skin lesions (Mondal *et al.*, 2019) and we found 72% of the surveyed participants to have an exposure within the range, while 23.5% below the range and 4.5% above this proposed range of As exposure.

Table 1. Summary of arsenic concentrations.

Sample (N, unit)	Mean ± SD	Range
Water (77, ug/l)	29.35±37.88	0.05–156.13
Raw rice (76. jug kg)	96.60±31.21	15.37–230.55
Cooked rice (57. |ng/kg)	200.57±248.27	17.38–1128.28
Whole wheat (72, jig/ kg)	43.64±48.19	0.96–234.30
Wheat flour (58, ug kg)	49.80±74.08	3.59–448.25
Potato (72, ug kg)	43.78±34.49	5.65–176.84

3.2 *Microbiota composition*

Currently, microbial data collection and analysis have been performed for 38 participants with average inorganic As exposure of 1.24±1.45 μg/kg bw/day. Among these 38 participants, 14 had overall inorganic As exposure <0.3 μg/kgbw/day (mean ± SD: 0.19±0.06) considered as low exposed group and 24 had exposure ≥ 0.3 μg/kgbw/day (mean ± SD: 1.85±1.52) considered as the high exposed group. While inorganic As exposure was significantly (one-way ANOVA, $p<0.05$) different between these two groups, overall gut microbial compositions in the high and low exposure groups (n=38) were similar with most prevalent phyla being Firmicutes and Bacteriodetes, followed by Proteobacteria and Actinobacteria with relative abundances of 37%, 34%, 14% and 11% respectively (Figure 2). Total microbial frequency was higher in the high exposed group.

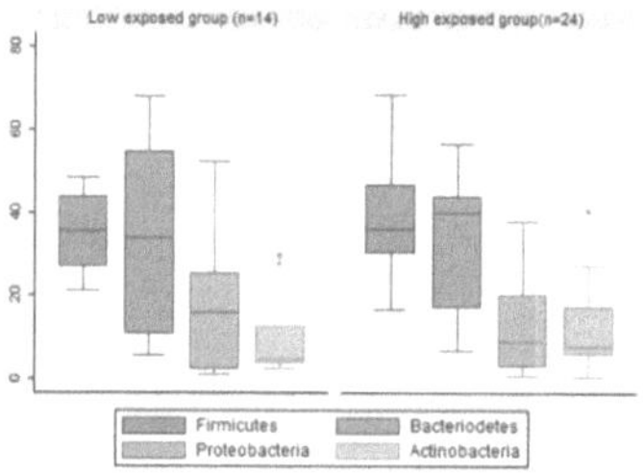

Figure 2. Microbial composition in high and low As exposure groups based on 16S rRNA sequencing.

Though not significantly different, a greater number of participants consuming milk and yogurt (regularly or frequently) were found to be in the higher exposed group. We found an increased (but not statistically significant) relative abundances of Firmicutes and Bacteriodates in high exposed group and decreased abundances of Proteobacteria and Actinobacteria after adjusting for gender, age milk and yogurt intake.

4 CONCLUSIONS

Arsenic exposure in the studied population might have an affect on the microbial diversity. Even at the observed environmentally relevant concentrations of As exposure the gut microbiota might get altered. This is a part of an on-going study and further results will be presented at the conference.

ACKNOWLEDGEMENTS

This study is funded by the DST-UKIERI Thematic partnership project. We thank all the participants.

REFERENCES

Bodkhe R., Shetty S.A., Dhotre D.P. *et al.* 2019. Comparison of small gut and whole gut microbiota of first-degree relatives with adult Celiac disease patients and controls. *Front. Microbiol.* 10: 164.

Lu K., Abo R.P., Schlieper K.A. *et al.* 2014. Arsenic exposure perturbs the gut microbiome and its metabolic profile in mice: An integrated metagenomics and metabolomics analysis. *Environ. Health Perspect.* 122 (3): 284–291.

Mondal D., Mwale T., Xu L. *et al.* 2019. Risk perception of arsenic exposure from rice intake in a UK population. *Palgrave Commun.* 5: 89.

Rubin S.S.C.DC., Alava P. *et al.* 2014. Arsenic thiolation and the role of sulfate-reducing bacteria from the human intestinal tract. *Environ. Health Perspect.* 122 (8): 817–822.

Suman S., Sharma P.K., Siddique A.B. *et al.* 2019. Wheat is an emerging exposure route for arsenic in Bihar, India. *Sci. Total Environ.* 703: 134774.

Arsenic in the Environment: Bridging Science to Practice for Sustainable Development – van der Wal, Ahmad, Petrusevski, Weijma, Savic, van der Wens, Beerendonk, Bhattacharya, Bundschuh & Naidu (Eds)
© 2023 the Author(s), ISBN 978-1-032-32928-4

Health risk assessment of arsenic from natural origin in groundwater sources in Mexico

X. Gutiérrez-Aviña[1] & I. Navarro-González[2]
[1]*Programa de Maestría en Ingeniería Ambiental, Universidad Nacional Autónoma de México (UNAM), Mexico City, Mexico*
[2]*Instituto de Ingeniería, Universidad Nacional Autónoma de México (UNAM), Mexico City, Mexico*

ABSTRACT: The occurrence of arsenic (As) in groundwater sources is worldwide problematic. In Mexico, the As concentrations exceed the acceptable limit of 10 µg/L established by the WHO (World Health Organization), and by the Mexican regulations of 25 µg/L, causing adverse health effects in the population. The aim of this study was to conduct a health risk assessment for two populations (children and adults, for the occurrence of As in groundwater sources with the historical data of the states of Chihuahua, Coahuila, Durango, San Luis Potosí, Sonora and Zacatecass.

1 INTRODUCTION

In Mexico, the arsenic (As) concentrations in groundwater sources exceed the acceptable limit established by the Mexican regulations of 25 µg/L (NOM-127), (Salud Ambiental 1994) and the guideline value of 10 µg/L established by the WHO. This problem of As occur mainly in the northern part of the country, in the states of Baja California Sur, Coahuila, Chihuahua, Durango, Hidalgo, Michoacán, San Luis Potosí, and Sonora (Vega 2002), due to the mineral composition of the soil. It should be pointed out that the 58.5% of the water supply in Mexico is through the groundwater (CONAGUA 2017), for this reason the primary route of exposure to this contaminant is through drinking water, that can cause adverse effects to health like skin cancer, cardiovascular and dermal diseases (keratosis and pigmentation). The probabilistic health risk assessment is a useful tool to estimate the risk to a given population for the exposssure to a chemical or hazardous material and the probability of any adverse health effects.

2 MATERIALS AND METHODS

2.1 *Study area*

An analysis data was conduct from the review of the data collected of the arsenic concentrations in groundwater in different states of the Mexican Republic (Bocanegra 2006; Espino-Valdés *et al.*, 2009; Rosas *et al.*, 1999; Vega 2002; Wyatt *et al.*, 1998). Five states of them were selected (Chihuahua, Coahuila, Durango, San Luis Potosí and Sonora) according to available data and the frequency of the concentrations higher than 25µg/L (Table 1).

2.2 *Sample preparation and analysis*

The probabilistic health risk assessment was conduct for two populations: children (<15 years old) and adults, for the carcinogenic risk (CR) and the non-carcinogenic risk (HQ) for drinking water ingestion contaminated with arsenic, using the software Crystal Ball. The daily doses ingestion (DI) was calculate using Equation (1) where the values for the expossure variables: ingestion rate (IR), expossure duration (ED) and body weigth (BW), expressed as distribution functions for children and adults (Table 2),

Table 1. Arsenic concentration in groundwater (µg/L).

State	n	Range	Mean	Distribution Function*
Chihuahua	21	40–800	195±178	µ = 0.20 σ = 0.20
Coahuila	16	26–740	180±216	µ = 0.40 σ = 3.82
Durango	67	26–590	81±94	µ = 0.09 σ = 0.20
San Luis Potosí	28	26–151	54±31	µ = 0.06 σ = 0.04
Sonora	12	32–117	62±29	µ = 0.06 σ = 0.04

Table 2. Data for daily dose exposure variables.

	IR (L/d)	BW(kg)	ED (years)
Children (< 15 years)			
n	135	135	135
Range	0.375-2.75	8-75	0.6-14
Mean	1.39±0.558	33.06±14.4	8.1±3.8
Distribution function	Gamma α=0.2 β=7.18	Beta α=1.8 β=3.0	Weibull α=16.6 β=4.6
Adults			
n	110	101	108
Range	0.4-3.5	40-95	1-82
Mean	1.889±0.634	61.50±11.8	21.7±16.6
Distribution function	Normal µ=1.89 σ=0.63	Weibull α=31.2 β=2.5	Log-Normal µ=21.7 σ=16.5

DOI: 10.1201/9781003317395-99

correspond to the rural areas in Mexico (Navarro *et al.*, 2007).

$$DI = CXIRXEDXEF/BWXAT \quad (1)$$

The carcinogenic risk (CR) and the non-carcinogenic risk (HQ) for drinking water contaminated were calculate using Equation (2) and (3), and the toxicological reference values (USEPA-IRIS 2017) for the slope factor (SF=1.5 mg/kg-day) and reference dose (RfD = 3×10-4 mg/kg day) for arsenic.

$$HQ = DI/RfD \quad (2)$$

$$CR = DIXSF \quad (3)$$

3 RESULTS AND DISCUSSION

The probabilistic health risk assessment results for HQ in children, showed a probability of 95% (95th percentile) that ingestion of water contaminated with the reported concentrations of arsenic can present an adverse health effect, since the results exceed the reference value, HQ>1 (Table 3).

Table 3. Probabilistic health risk assessment results for children and adults in the study states.

	Children 95th Percentile		Adults 95th Percentile	
State	CR	HQ	CR	HQ
Chihuahua	1.1E-02	23.8	9.8E-03	22.2
Coahuila	1.9E-02	46.1	1.7E-03	36.8
Durango	4.7E-03	11.4	4.3E-03	9.2
San Luis Potosí	2.8E-03	6.0	2.4E-03	5.5
Sonora	2.6E-03	5.8	2.3E-03	5.3

Nevertheless, despite the high concentrations reported for the states of Chihuahua, Coahuila, Durango, San Luis Potosí and Sonora, exists a 7, 28, y 35% of probability for children and a 15, 35 and 45% of probability for adults, respectively, that the arsenic exposure do not represent a health risk (HQ<1). Even so, those probabilities indicated that the risk exposure is higher for children. The carcinogenic risk for the two populations exceed the acceptable limit, 1.0E-04, with 95th percentile values between 1.7E-03 and 1.9E-02. Turned out to be the states of Chihuahua and Coahuila the states with the highest cancer risk probabilities, with average values (50th percentile) of three and six cases of cancer, respectively, in 1,000 people for both population.

Maldonado *et al.* (2019) conducted a health risk assessment for children (6 to 16 years old) in three localities of Sonora, with arsenic concentrations of 6.2 µg/L, 36 µg/L and 108.2 µg/L; their average cancer risk results were, 5.73E-05, 3.53E-04 and 9.37E-04, for the localities of Cocorit, Vicam and Potam, respectively. Regarding to the non-carcinogenic effects (HQ) the values obtained were 0.94, 6.02 and 16.64. Therefore, it is highlighted the similarity in the risks assessed with the reported in this study showing that the exposure variables obtained for the Mexican rural areas were appropriate for this assessment.

4 CONCLUSIONS AND RECOMMENDATIONS

Generally for the population exists a higher health risk in Chihuahua and Coahuila states, where the highest arsenic concentrations were reported, and some of the carcinogenic and non-carcinogenic health effects, skin cancer and hyperpigmentation, keratosis and some of the vascular diseases in the exposed population to arsenic can appeared. According to the results of the probabilistic health risk assessment the children presents a higher risk to both effects and the ingestion water rate is the mainly influence factor, as the adults the mainly influence factor is the exposure duration. Therefore is necessary a water treatment process to remove arsenic from water, and with this, decrease the health risk.

REFERENCES

Bocanegra M. 2006. Evaluación de Riesgo en Salud Por la Exposición a Fluoruro y Arsénico en Agua de Pozo Para Consumo Humano de Las Zonas Altiplano, Centro y Media del es-Tado de San Luis Potosí. S.L.P., México: Universidad Autónoma de San Luis Potosí.

CONAGUA. 2017. *Estadísticas del Agua en México*. Ciudad de México. México: Comisión Nacional del Agua.

Espino-Valdés M.A., Barrera-Prieto Y. & Herrera-Peraza E. 2009. *Presencia de Arsénico en la Sección Norte Del Acuífero Meoqui-Delicias Del estado de Chihuahua, México. Medio Ambiente y Desarrollo Sustentable* 3 (1): 8–18.

IRIS. 2017. Integrated Risk Information System https://cfpub.epa.gov/ncea/iris2/chemicalLanding.cfm?substance_nmbr=278

Maldonado-Escalante J.F., Meza-Figueroa D., Devóra-Figueroa A.G., García-Rico L., Burgess J.L., Lantz R. C., Yañez-Estrada L., Martínez-Cinco M.A., Balderas-Cortes J.J., Mondaca-Fernandez I., Meza-Montenegro M.M. 2019. An integrated health risk assessment of indigenous children exposed to arsenic in Sonora, Mexico. *Hum. Ecol. Risk Assess.* 25(3): 706–721.

Navarro I., Jiménez B., Maya C. & Lucario E.S. 2007. Assessment of potential cancer risk from trihalomethanes in water supply at Mexican rural communities. *International Symposium on New Directions in Urban Water Management*.

Rosas I., Belmont R., Armienta A. & Báez A. 1999. Arsenic concentrations in water, soil, milk and forage in Comarca Lagunera, México. *Water Air Soil Pollut.* 112: 133–149.

Salud Ambiental. 1994. Agua Para uso y Consumo Humano. Límites Permisibles de Calidad y Tratamientos a Que Debe Someterse el Agua Para su Potabilización. NOM-127-SSA1-1994. Diario Oficial de la Federación. 22 noviembre de 2000.

Vega S. 2002. *Riesgo Sanitario Ambiental Por la Presencia de Arsénico y Fluoruros en Los Acuíferos De México*. D. F., México: Comisión Nacional del Agua.

Wyatt C.J., Fimbres C., Romo L., Méndez R.O. & Grijalva M. 1998. Incidence of heavy metal contamination in water supplies in Northern Mexico. *Environ. Res.*76: 114–119.

Arsenic in the Environment: Bridging Science to Practice for Sustainable Development – van der Wal, Ahmad, Petrusevski, Weijma, Savic, van der Wens, Beerendonk, Bhattacharya, Bundschuh & Naidu (Eds)
© 2023 the Author(s), ISBN 978-1-032-32928-4

A medical geology perspective of arsenic as a poison and medicinal agent

J.A. Centeno
Founding Member and Past President, International Medical Geology Association, Sao Paulo, Brazil

ABSTRACT: The public health concerns for environmental exposure to arsenic has been recognized for decades. However, recent human activities have resulted in even greater arsenic exposures and the potential increase for chronic arsenic poisoning on a worldwide basis. The natural sources of arsenic exposure vary from burning of arsenic-rich coal (China) and mining activities (Malaysia, Japan) to the ingestion of arsenic-contaminated drinking water (e.g., Taiwan, Mexico, Chile, Bangladesh). The groundwater arsenic contamination in Bangladesh and the West Bengal Delta of India has received the greatest international attention because of the large number of people potentially exposed and the high prevalence of arsenic-induced disease. This presentation will focus on providing a medical geology perspective on arsenic as an environmental and medicinal agent.

1 INTRODUCTION

Medical geology – the science that deals with the impacts of natural geologic materials and processes on animal and human health is aimed at strengthening the interactions between geoscientists, environmental, biomedical and public health communities, and by stimulating increased research collaboration among these disciplines (Centeno 2008, 2016; Selinus *et al.*, 2011, 2013). Medical geologists are a group of scientists that are primarily interested in outbreaks of disease in which the characteristics of the local geological constituents contribute to the occurrence of various disease states. For the most part, diseases of interest have often included the health and toxicological effects associated with deficiency or toxicity of a variety of metallic and non-metallic elements on various systemic organs.

2 BACKGROUND

Medical geologists are scientists (geoscientists, biomedical/public health scientists, chemists, toxicologists, epidemiologists, hydrologists, geographers, etc.) who generally collaborate on a wide range of environmental health problems seeking causes and solutions. Among these problems are the health impacts of geogenic (natural) dusts, naturally occurring elements in surface water, ground water and soil, geologic processes such as volcanoes, erosions, earthquakes, tsunamis, etc., occupational exposure to natural materials and natural radiation, and long-term effects of exposure to oncogenic elements, the most prominent example being arsenic (Centeno *et al.*, 2016; Shand *et al.*, 2007; Zikovsky & Chah 1990).

Medical geologists try to determine the sources, transport and fate of potentially harmful trace elements such as arsenic, fluorine, selenium, copper and other metals. They try to determine the pathways of exposure and produce maps that illustrate local, regional and/or global geologic and geochemical factors and their relationship to existing or potential health problems. A good example of collaborative research on medical geology is the arsenic issue in Bangladesh and West Bengal, India. In this region, medical geologists are working together to determine the source of the high arsenic levels in well water that put at risk the health of thousand of people in this region (Smith *et al.*, 2000).

This presentation explores problems inferring risk and disease causation from natural exposures to arsenic, particularly for chronic outcomes, and will argue for the importance of the ecological perspective in assessing pathogenesis. Additionally, the potential beneficial aspects of arsenic as a medicinal agent will be explored.

REFERENCES

Centeno J.A. 2008. Editorial—introducing a special theme issue on medical geology. *J. Environ. Monit.* 10:1391–1393.

Centeno J.A., Finkelman R.B. & Selinus O. 2016. Medical geology: impacts of the natural environment on public health *Geosciences* 6(1): 8.

Selinus O., Finkelman R.B. & Centeno J.A. 2011. Principles of medical geology. *In*: J.O. Nriagu (ed.) *Encyclopedia of Environmental Health*; Elsevier: New York, NY, USA, Volume 2, pp. 669–676.

Selinus O., Alloway B., Centeno J.A., Finkelman R.B., Fuge R., Lindh U. & Smedley P. (eds.) 2013. *Essential in Medical Geology*. Elsevier: New York, NY, USA, 805p.

Shand P., Edmunds W M., Lawrence AR., Smedley P L. & Burke S. 2007. The natural (baseline) quality of groundwater in england and wales. *British Geological Survey Research Report No. RR/07/06*. ISBN 978 085272595, 84pp.

Smith A.H., Lingas E.O. & Rahman M. 2000. Contamination of drinking-water by arsenic in Bangladesh: a public health emergency. *Bull World Health Organ.* 78: 1093–1103.

Zikovsky L. & Chah B. 1990. The log-normal distribution of radon concentration in ground water. *Groundwater* 28: 673–676.

DOI: 10.1201/9781003317395-100

Arsenic in the Environment: Bridging Science to Practice for Sustainable Development –
van der Wal, Ahmad, Petrusevski, Weijma, Savic, van der Wens, Beerendonk, Bhattacharya, Bundschuh & Naidu (Eds)
© 2023 the Author(s), ISBN 978-1-032-32928-4

Linear or non-linear – association of low-level inorganic arsenic exposure from rice with age-standardized mortality risk of cardiovascular disease

L. Xu[1], D.A. Polya[1], Q. Li[1] & D. Mondal[2]
[1]*Department of Earth and Environmental Sciences, School of Natural Sciences and Williamson Research Center for Molecular Environmental Science, University of Manchester, Manchester, UK*
[2]*School of Science, Engineering & Environment, University of Salford, Salford, UK*

ABSTRACT: To quantify and compare the linear and non-linear associations between local authority level daily per capita inorganic arsenic (iAs) exposure from ingestion of rice ($E\text{-}iAs_{ing,rice}$) and the age-standardised mortality rates (ASMR) of cardiovascular disease (CVD), an ecological study has been conducted with local authority level iAs exposure and health information obtained from public domain sources for England and Wales. This study reveals that compared with linear models, a non-linear model might be a better fit. In the linear model, $E\text{-}iAs_{ing,rice}$ was observed to be associated with a slight but significant rise in the local authority CVD ASMR both categorically and continuously. In the preferred non-linear model, higher risks of CVD could be observed for local authorities with higher exposure levels but, in contrast to the linear model, CVD ASMR decreased before increasing from 0.3 μg/person/day. Given the otherwise nutritional benefits of rice consumption, this study suggests high rice consumers should eat rice in a smarter way.

1 INTRODUCTION

People may be exposed by inorganic arsenic (iAs), a widespread human carcinogen, via several pathways (Chen *et al.*, 2011; Polya & Middleton 2017). Drinking water is a well-recognized source of iAs exposure and the association between iAs exposure from drinking water and cardiovascular diseases (CVD) have already been reported (Chen *et al.*, 2013). However, growing evidence has suggested that everyday foodstuffs may also be important exposure pathways (European Food Safety Authority 2009, 2014; Mondal & Polya 2008). It is especially the case for rice as it contains substantially more As than other major staples and provides more than 20% of the daily calorie intake for over 3 billion people in the world (Food and Agriculture Organization of the United Nations 2008; Torres-Escribano *et al.*, 2008). Unfortunately, still limited epidemiological evidence, to our knowledge, exists on CVD risks arising from iAs exposure from rice intake. Meanwhile, though the linear associations between As exposure and CVD risks have already been confirmed epidemiologically and biologically (Mateen *et al.*, 2017; Moon *et al.*, 2017; Tsinovoi *et al.*, 2018), exploration of non-linear associations is still lacking (D'Ippoliti *et al.*, 2015). Therefore, it is necessary to further explore the effects of iAs exposure from rice intake on CVD risks both linearly and non-linearly.

Our aim was to quantify and compare the linear and non-linear associations between local authority level daily per capita iAs exposure from ingestion of rice ($E\text{-}iAs_{ing,rice}$) (μg/person/day) and CVD mortality, specifically in England and Wales. After discussing our exploratory findings and indicating the importance of non-linear assessment, we also made recommendations both for future work and for smarter ways to eat rice, reducing potential health burden arising from avoidable exposure to iAs.

2 MATERIALS AND METHODS

In this study, we modelled, for England and Wales for the year 2012, the potential linear and non-linear associations between iAs exposure from rice and CVD age-standardised mortality rates (ASMR) through an ecological design with a series of linear and non-linear regression models. We obtained local authority level iAs exposure and health information from public domain sources for England and Wales. On one hand, CVD ASMR were calculated from crude local-authority level CVD mortality rates and age distributions via standardisation using European Standard Population 2013 (Glickman 2013). On the other hand, local authority level $E\text{-}iAs_{ing,rice}$ was modelled using Office for National Statistic reported ethnicity as a proxy for rice consumption level. The fitness of a series of linear and non-linear regression models was compared through several objective model comparison criteria, notably Akaike's Information Criterion (AIC).

3 RESULTS AND DISCUSSION

For the linear model, when corrected for some significant known confounders which are reported to be not only associated with high $E\text{-}iAs_{ing,rice}$ but also

DOI: 10.1201/9781003317395-101

protective against CVD, E-$iAs_{ing,rice}$ was observed to be related with a slight but significant increase in the local authority CVD ASMR both categorically and continuously. In the best-fitted linear model (AIC value 3287.4), there was a significant (p-value for trend lower than 0.001) and positive steady rise of CVD ASMR with E-$iAs_{ing,rice}$. Also, a 6% increase of CVD ASMR was observed for local authorities in the highest exposure quartile (0.375–2.71 μg/person/day) when compared with that in the lowest quartile (< 0.265 μg/person/day) in this model. However, the non-linear model was regarded as a better model when compared with the linear model, partly because of a lower (though not significantly lower) AIC (3286.9) but also due to the fact that it provided a more realistic dose-response extrapolation for higher rice consumption rates. In the preferred non-linear model, higher risks of CVD could be observed for local authorities with higher exposure levels but, in contrast to the best-fitted linear model, CVD ASMR CVD risks decreased up to an exposure of 0.3 μg/person/day followed by an increase.

4 CONCLUSIONS AND RECOMMENDATIONS

This study reveals that a non-linear model may provide a better fit than a linear model, of the association between iAs exposure from rice consumption and local authority level CVD ASMR in England and Wales. The linear model conservatively estimated that about 1.21% (95% confidence interval: −1.38–3.88%) of CVD ASMR is associated with iAs exposure from rice consumption in England and Wales. Similarly, application of the preferred best-fitted non-linear model indicated that around 2.0% (95% confidence interval: 0.0–11.5%) of CVD ASMR is related with E-$iAs_{ing,rice}$. However, in this study there are important uncertainties, including not only those common to the ecological design, such as inability to adequately capture the underlying confounding information especially contrasting nutritional status, dietary habit and genetics indicators, but also those specific to this study, such as the use of ethnicity as a proxy for rice consumption level. Notwithstanding the above-mentioned uncertainties, this study was the first quantifying the association between iAs exposure from rice intake and CVD risks, signifying the importance of non-linear analysis of As effects and calling for future individual-level studies with sufficient data of As and adverse health effects reporting non-linear association. Lastly, given the nutritional benefits of rice consumption, this study also suggests high rice consumers to eat rice in a smarter way – including but not limited to, consuming polished rice from non-endemic areas with low iAs rice (European Food Safety Authority 2014) and using better cooking methods, for example cooking with a large excess of water (Mwale *et al.*, 2018), to remove part of iAs from the final food for consumption.

ACKNOWLEDGEMENTS

We are grateful to Nick Warren and Jackie Morton (Health and Safety Executive's Health and Safety Laboratory, Buxton) and Martie van Tongeren (University of Manchester) for discussions. LX acknowledges receipt of a University of Manchester President's Doctoral Award. This abstract includes excerpts and other material from our recent article (Xu *et al.*, 2020, Science of the Total Environment, https://doi.org/10.1016/j.scitotenv.2020.140534) reproduced here in accordance with the terms and conditions of the Creative Commons Attribution (CC BY) license: http://creativecommons.org/licenses/by/4.0/).

REFERENCES

Chen C., Qian Y., Chen Q. & Li C. 2011. Assessment of daily intake of toxic elements due to consumption of vegetables, fruits, meat, and seafood by inhabitants of Xiamen, China. *J. Food Sci.* 76(8): T181–T188.

Chen Y., Wu F., Liu M., Parvez F., Slavkovich V. *et al.* 2013. A prospective study of arsenic exposure, arsenic methylation capacity, and risk of cardiovascular disease in Bangladesh. *Environ. Health Perspect.* 121(7): 832–838.

D'Ippoliti D., Santelli E., De Sario M., Scortichini M., Davoli M. & Michelozzi P. 2015. Arsenic in drinking water and mortality for cancer and chronic diseases in Central Italy, 1990–2010. *Plos One* 10(9): e0138182.

European Food Safety Authority 2009. *Scientific Opinion on Arsenic in Food.* (Vol. 7, pp. 1351–1550). European Food Safety Authority. Rome, Italy.

European Food Safety Authority 2014. *Dietary Exposure to Inorganic Arsenic in the European Population.* (Vol. 12, pp. 3597–3664). Rome, Italy: EFSA.

Food and Agriculture Organization of the United Nations 2008. *Rice Market Monitor.* (Vol. XI, pp. 1–44). Rome, Italy: United Nations.

Glickman M. 2013. *Revised European Standard Population* 2013 (2013 ESP).

Mateen F.J., Grau-Perez M., Pollak J.S., Moon K.A., Howard B.V., Umans J.G. *et al.* 2017. Chronic arsenic exposure and risk of carotid artery disease: the strong heart study. *Environ. Res.* 157: 127–134.

Mondal D. & Polya D.A. 2008. Rice is a major exposure route for arsenic in chakdaha block, nadia district, West Bengal, India: a probabilistic risk assessment. *Appl. Geochem.* 23(11): 2987–2998.

Moon K.A., Navas-Acien A., Grau-Perez M., Francesconi K.A., Goessler W., Guallar E. *et al.* 2017. Low-moderate urine arsenic and biomarkers of thrombosis and inflammation in the strong heart study. *Plos One* 12(8): e0182435.

Mwale T., Rahman M.M. & Mondal D. 2018. Risk and benefit of different cooking methods on essential elements and arsenic in rice. *Int. J. Environ. Res. Public Health* 15 (6): 1056–1066.

Polya D.A. & Middleton D.R.S. 2017. Arsenic in Drinking Water: Sources & Human Exposure Routes. In: P. Bhattacharya, D.A. Polya & D. Jovanovic (eds.), *Best Practice Guide on the Control of Arsenic in Drinking Water.* IWA Publishing. UK. pp. 1–20.

Torres-Escribano S., Leal M., Vélez D. & Montoro R. 2008. Total and inorganic arsenic concentrations in rice sold in Spain, effect of cooking, and risk assessments. *Environ. Sci. Technol.* 42(10): 3867–3872.

Tsinovoi C. L., Xun P. C., McClure L. A., Carioni V. M. O., Brockman J.D., Cai J.W. *et al.* 2018. Arsenic exposure in relation to ischemic stroke the reasons for geographic and racial differences in stroke study. *Stroke* 49(1): 19–26.

Arsenic in the Environment: Bridging Science to Practice for Sustainable Development –
van der Wal, Ahmad, Petrusevski, Weijma, Savic, van der Wens, Beerendonk, Bhattacharya, Bundschuh & Naidu (Eds)
© 2023 the Author(s), ISBN 978-1-032-32928-4

Arsenic in keratinized matrices from patients with chronic kidney disease of uncertain etiology (CKDu) in Sri Lanka

S. Diyabalanage[1], S. Fonseka[2] & R. Chandrajith[3]
[1]*Instrument Centre, Office of the Dean, Faculty of Applied Sciences, University of Sri Jayewardenepura, Nugegoda, Sri Lanka*
[2]*Department of Pharmacology, Faculty of Medicine, University of Peradeniya, Peradeniya, Sri Lanka*
[3]*Department of Geology, Faculty of Science, University of Peradeniya, Peradeniya, Sri Lanka*

ABSTRACT: Chronic Kidney Disease of uncertain etiology (CKDu) prevails among the rural dry zone farming communities in Sri Lanka and now consider as an endemic disease that occurred due to chronic exposures to toxic trace elements. Agro-genic Arsenic (As) have been widely suspected as a risk a factor for the pathogenesis of CKDu. Arsenic content in keratinized matrices such as hair and nail from CKDu patients and age matched healthy controls from an urban area were measured with Inductively Coupled Plasma Mass Spectrometer (ICP-MS) after microwave aided acid digestion. The hair As levels varied from 0.007 to 0.165 µg/g and 0.005 to 0.143 µg/g with mean values of 0.037 µg/g and 0.054 µg/g in CKDu subjects and controls. In nails, the As level varied from 0.015 to 0.250 µg/g in CKDu subjects and 0.013 to 0.204 µg/g in controls. No significant differences were observed between cases and controls, thus indicate that the patients were not exposed to toxic levels of As.

1 INTRODUCTION

Trace elements play a significant role in the human physiological system according to their essentiality to the body. In contrast, the human body can tolerate toxic elements up to a certain dosage although these elements are not involved in any biochemical processes (Dissanayake & Chandrajith 1999). The available concentrations of such elements in biological samples reflect the nutritional status, environmental and occupational exposure and possible causes for existing adverse health effects. Therefore monitoring of trace element intake and its accumulation in the human body is extremely important for exposure assessments and bio-monitoring. In recent years, monitoring of trace elements in environmental samples and biological specimens has gained wider attention with the emergence of certain chronic diseases associated with trace element exposures such as kidney diseases, neurological disorders and cancer.

In the last two decades, there has been a wider interest in the alarming increase of chronic kidney disease (CKD) in several tropical regions such as Sri Lanka, India, El Salvador and Nicaragua (Athuraliya *et al.*, 2011; Reddy & Gunasekar 2013; Torres *et al.*, 2010). The etiology of these emerging nephropathies is not related to any known risk factors hence the disease is described as Chronic Kidney Disease of Unknown etiology (CKDu), which is most common among low socioeconomic farming communities. In Sri Lanka, the prevalence of CKDu is shifting towards the younger generation creating a massive negative impact on the sociological wellbeing of the country. Since the disease is confined to a specific geographical region of Sri Lanka, most early studies have hypothesized environmental exposures to agro-genic arsenic (As) may have been the risk factors for the pathogenesis of CKDu (Jayatilake *et al.*, 2013; Jayasumana *et al.*, 2013). It is well known that part of inhaled or ingested As accumulates in keratinized materials such as hair and nails. We hypothesized that hair and nail analysis will perhaps provide better information on variable environmental exposures. The levels of As content in hair and nails from both CKDu patients and healthy controls were investigated in this study. Simultaneously, the cutaneous manifestations of CKDu patients and any changes on the skin due to chronic As exposure were also investigated.

2 MATERIALS AND METHODS

2.1 *Sample collection*

Ethical approval for this study was obtained from the Ethical Review Committee of the Faculty of Medicine, University of Peradeniya while written consent was obtained from all participants. Samples of hair (77) and nails (76) were collected from biopsy proven CKDu patients who attended the Renal Care Clinic of the Girandurukotte Base Hospital (7.46684°N; 81.0507°E) from which 41 samples were obtained from male patients. Control samples were collected from age matched volunteer individuals from the Kandy region (7.29067°N; 80.6340°E) where CKDu is considered as non-endemic. All patients and controls were carefully examined by a consultant dermatologist for any skin manifestations. About 10 g of hair were clipped close to the scalp from the occipital area using a pair of sterilized ceramic scissors while the fingernails were collected using a stainless steel nail clipper.

DOI: 10.1201/9781003317395-102

2.2 *Sample preparation and analysis*

Hair and nail samples were washed few times with deionized water and acetone. The cleaned samples were dried at 55°C overnight and stored in a desiccator. 0.10 g of subsamples was digested with 4.5 mL of HNO_3 and 0.5 mL of H_2O_2 using Mars-6 microwave digester equipped with EasyPrep-Plus high pressure vessels. Final solutions were diluted to 25 mL. Concentration of trace elements including As were determined by Thermo ICapQ Inductively Coupled Plasma Mass Spectrometry (ICP-MS).

3 RESULTS AND DISCUSSION

3.1 *Dermatological inspections*

The age of the subjects of patients ranged from 16 to 87 years with the average age being 56 years. Among them, 52% were males and almost all of them were paddy farmers. Nearly 61% of the selected CKDu subject population use agrochemicals and most of them do not practice safety precautions to avoid exposure during the application of agrochemicals. Out of 77 observed patient subjects, 74 showed some kind of skin change. All patient subjects exhibited ichthyosis, while some had developed eczema. Most of the changes were noted in the palms but none of the patients exhibited gross As poisoning features as described in literature (Figure 1).

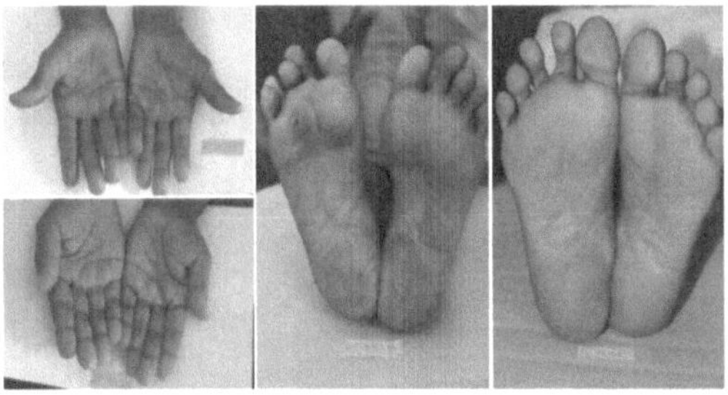

Figure 1. Examination for skin manifestations.

3.2 *Arsenic in hair and nail*

According to the results, hair As levels varied from 0.007 – 0.165 µg/g in samples from CKDu patients while it was 0.005 – 0.143 µg/g in controls. Arsenic levels in nail samples varied from 0.015 – 0.250 µg/g in patients while it varied from 0.013 – 0.204 $\mu g\ g^{-1}$ in controls. Jayatilake *et al.* (2013) reported higher As levels in hair of CKDu patients ranging from <0.001 µg/g to 0.452 µg/g (mean: 0.144 µg/g; median: 0.139 µg/g) compared to controls (mean = 0.125 µg/g, median = 0.103 µg/g) from the endemic area. Jayasumana *et al.*, 2013, have found hair As level of 1.27 – 7.03 µg/g in CKDu patients while it was between 0.22 – 4.58 µg/g in control samples. The values obtained in our study are roughly comparable to values of Jayatilake *et al.* (2013), but significantly lower than the values reported by Jayasumana *et al.* (2013). The As levels obtained in this study is lower than the hair As level in ordinary population prescribed by the World Health Organization (<1 µg/g) and some other Asian countries such as India (3.43 µg/g), Malaysia (0.28 µg/g), Japan (0.05 µg/g), Iran (0.07 µg/g), Cambodia (1.77 µg/g) and Vietnam (0.62 µg/g). According to Mann-Whitney U test, As shows a statistically significant difference between patient and control samples of hair ($p = 0.039$) while it is not significantly different in nail samples ($p = 0.186$).

4 CONCLUSIONS AND RECOMMENDATIONS

The present work reveals the distinct variation in As concentrations in scalp hair and nails of CKDu patients and healthy controls. Significantly lower levels of As were reported from CKDu patients indicating that patient populations were not exposed to any form of toxic components with this elements. Although cutaneous manifestations in CKDu patient subjects were observed, all such symptoms are associated with kidney failures. This study clearly indicates that CKDu patients are not exposed to chronic As as suggested by previous studies.

ACKNOWLEDGEMENTS

We acknowledge the financial supported given by the University of Peradeniya research grant (RG/2014/31/M) offered to SF and the National Research Council (NRC), Sri Lanka research grant (TO 14-05) offered to RC.

REFERENCES

Athuraliya N.T.C., Abeysekera T.D.J., Amerasinghe P.H., Kumarasiri R., Bandara P., Karunaratne U., Milton A.H. & Jones A.L. 2011. Uncertain etiologies of proteinuric-chronic kidney disease in rural Sri Lanka. *Kidney Int.* 80 (11): 1212–1221.

Dissanayake C.B. & Chandrajith R. 1999. Medical geochemistry of tropical environments. *Earth Sci. Rev.* 47 (3–4): 219–258.

Jayasumana M., Paranagama P., Amarasinghe M., Wijewardane K., Dahanayake K., Fonseka S., Rajakaruna M.A., Samarasinghe K.D.L.M.P.U. & Senanayake V. 2013. Possible link of chronic arsenic toxicity with chronic kidney disease of unknown etiology in Sri Lanka. *J. Nat. Sci. Res.* 3 (1): 64–73.

Jayatilake N., Mendis S., Maheepala P. & Mehta F.R. 2013. Chronic kidney disease of uncertain aetiology: prevalence and causative factors in a developing country. *BMC Nephrol.* 14 (1): 180.

Reddy D. & Gunasekar A. 2013, Chronic kidney disease in two coastal districts of Andhra Pradesh, India: role of drinking water. *Environ. Geochem. Health* 35(4): 439–454.

Torres C., Aragón A., González M., Jakobsson K., Elinder C.G., Lundberg I. & Wesseling C. 2010. Decreased kidney function of unknown cause in Nicaragua: a community-based survey. *Am. J. Kidney Dis.* 55 (3): 485–496.

Arsenic in the Environment: Bridging Science to Practice for Sustainable Development –
van der Wal, Ahmad, Petrusevski, Weijma, Savic, van der Wens, Beerendonk, Bhattacharya, Bundschuh & Naidu (Eds)
© 2023 the Author(s), ISBN 978-1-032-32928-4

Urinary arsenic speciation in a Brazilian population serving as a baseline for future environmental biomonitoring studies

J.C. Ng[1], D. Campolina[2], X. Gutiérrez Aviña[1,3], Md. N. Islam[1] & V.S.T. Ciminelli[4]
[1]*Queensland Alliance for Environmental Health Sciences, The University of Queensland, Brisbane, QLD, Australia*
[2]*Campolina Toxicologia Comp. Nucl. e Audio ME, Belo Horizonte, Brazil*
[3]*Instituto de Ingeniería, Universidad Nacional Autónoma de México (UNAM), Mexico City, Mexico*
[4]*INCT-Aqua, Universidade Federal de Minas Gerais, Belo Horizonte, Brazil*

ABSTRACT: Minas Gerais (MG) is a mineral-rich State in Brazil hosting mining activities since the late 17th century, when large gold deposits were discovered. Arsenic contamination is often associated with precious and base-metal extraction. It is sometimes a challenge to establish what might be a representative background level of arsenic in a population biomonitoring program. This study aimed to establish such a background with a representative sample amongst the population in Belo Horizonte, a modern urban city without mining areas. Urine samples were collected for total (n = 506) and arsenic speciation (n = 138) measurements. The selection of study subjects across Belo Horizonte approximated the demographics of the city in terms of sex and age proportions. The geometric mean total As was 9.68 µg/g creatinine (9.17–10.22, n = 506) and the sum of inorganic and methylated As was 5.21 µg/g creatinine (4.32-6.28, n=138) with no significant differences were observed between males and females. The results of this study may serve as a background As-exposure and methylation profile in future environmental exposure studies.

1 INTRODUCTION

Minas Gerais State (MG) is the world's largest producer of niobium and tantalum and Brazil's largest producer of iron ore, gold, zinc, phosphate rock, and lithium products, in addition to gemstones. The mineral province of approximately 7,000 km^2, the so-called Iron Quadrangle, is centrally located in Minas Gerais and hosts large deposits of iron, and gold ore bodies. This region together with other mineral-rich areas in Brazil are known for natural arsenic anomalies associated with gold deposits. The potential health risk of environmental arsenic contamination has drawn increasing attention from the regulators, the extractive industry, and the communities in general (Teixeira *et al.*, 2020). However, there is a scarcity of human biomonitoring program available for a more detailed assessment of arsenic exposure and health effects in Brazil. Specific background level assessment in the population is needed so that anthropogenic exposure can be compared. It is the aim of this study to obtain a background arsenic level of arsenic in MG using urinary total arsenic and its metabolic profile as a biomarker of arsenic exposure. This data can be utilised for future biomonitoring studies in Brazil.

2 MATERIALS AND METHODS

2.1 *Study area and urine collection*

Belo Horizonte, a modern city, is the capital of MG and the sixth-largest municipality in Brazil with an estimated population of 2.4 million people (IBGE 2010). The city is also a service hub of many mining and extractive industries accommodating head offices of these companies. It is considered a suitable study site for obtaining representative background for the mineral-rich region of MG.

Nine regions of Belo Horizonte including Barreiro, Centro-South, East, Northeast, Northwest, North, West, Pampulha, and Venda Nova were selected for the urine sample collection (Figure 1). A total of 506 urine samples were collected *ad hoc* from study subjects with consents ranging from infants to 60+ of age in 2018–19. Each urine sample was collected into a clean plastic container with a lid. An aliquot of 99 mL urine, was acidified with 1 mL high purity HNO_3 then transferred into 2 aliquots of 10 mL polypropylene plastic tubes with screw-on lids for total arsenic and speciation analyses. An aliquot of 1.5 mL urine without the addition of acid was transferred into a 2 mL Eppendorf tube for creatinine analysis. All tubes were paper-labelled and given a unique sample i.d. and stored at –80°C before analysis.

Figure 1. Map of Brazil showing Belo Horizonte city in Minas Gerais and the nine sampling regions.

 DOI: 10.1201/9781003317395-103

2.2 *Total urinary arsenic and speciation*

Total urinary arsenic concentrations (n=506) were measured by ICP-MS following nitric acid digestion in a local laboratory in Belo Horizonte. Selected samples (n=138) were sent to Australia for total arsenic and speciation analyses using ICP-MS (Agilent 7900 ICP-MS) and HPLC-ICP-MS (HP 1000 HPLC coupled to Agilent 7500 ICP-MS), respectively, at the University of Queensland. A Hamilton PRP-X100 ion exchange column (250 mm×4.1 mm i.d., 10 μm) was used for the arsenic speciation of "toxic-As" (AsIII, AsV, MMA, and DMA) and non-toxic arsenobetaine. No other organic arsenicals were measured. The mobile phase consisted of 10 mM $NH_4H_3PO_4$, 3 mM $NaNO_3$, and 0.2 mM EDTA-Na_2. Representative blanks, spiked recovery samples, and CRMS NIES-18 and ClinChek-Control I & II were included for QA/QC. Urinary creatinine was determined by the Jaffe's method using Beckman Coulter urine calibrator (Lot #10234) and Biorad Liquicheck urine biochemistry control (Lot #68540). Geometric mean and 95% confidence interval were calculated after post-stratification and population factor correction applied to the statistical survey design.

3 RESULTS AND DISCUSSION

3.1 *Demographics*

Table 1 shows the demographics of the subjects tested for total arsenic (TAs) and toxic-As (AsIII, AsV, MMA, and DMA) that approximated the population distribution pattern as closely as possible to the census data in terms of sex and age, and by regions of the city (IBGE 2010) and considered a good representation of the local population.

Table 1. Demographic data of subjects in this study compared to 2.4 M people in Belo Horizonte city.

	BH City (%)	TAs (%)	Toxic-As %
Female/Male	53.1/46.9	56.5/43.4	60.1/39.9
Age (years): 0-9	11.7	3.8	3.6
10-19	14.9	7.9	5.8
20-59	60.8	66.8	63.0
60+	12.6	21.5	27.5
Region: Barreiro	11.9	8.7	8.0
Centro-South	11.9	17.0	18.8
East	10.0	9.9	10.9
Northeast	12.2	12.8	11.6
Northwest	11.3	12.5	16.7
North	8.9	8.3	9.4
West	13.0	12.6	10.9
Pampulha	9.5	6.5	4.3
Venda Nova	11.2	11.7	9.4

3.2 *Total arsenic and speciation profile*

Excellent QA/QC results were obtained when compared to CRMs and spiked recoveries (details not shown). Results for TAs and the "toxic-As" adjusted for creatinine are shown in Table 2. The geometric mean of TAs is within the acceptable range although is slightly higher by about 2 μg/g creatinine compared to that of the NHANES survey (NHANES 2015-2016) whereas the toxic-As is very similar to the US population 4.94 μg/g creatinine (4.70-5.20, n=3091), males 4.42 (4.12-4.75, n=1537), and females 5.49 (5.23-5.77, n=1554).

Table 2. Geometric mean (GM) and 95% confidence interval (CI) of total urinary arsenic (TAs) and the sum of Toxic-As (μg As/g creatinine).

Category	n	TAs	n	Toxic-As
All	506	9.68 (9.17-10.22)	138	5.21 (4.32-6.28)
Females	286	9.70 (9.05-10.40)	83	4.77 (3.79-5.99)
Males	220	9.66 (8.87-10.52)	55	5.74 (4.32-7.63)
Age 0-9 y	19	10.91 (8.23-14.46)	5	5.71 (3.73-8.73)
10-19	40	9.73 (8.29-11.42)	8	7.22 (2.92-17.89)
20-59	338	9.38 (8.82-9.98)	87	5.14 (4.11-6.43)
60+	109	10.22 (9.29-11.25)	38	5.11 (4.11-6.35)
Barreiro	44	10.46 (9.14-11.98)	11	5.44 (4.52-6.55)
Centro-South	86	10.06 (8.51-11.88)	26	4.62 (3.15-6.77)
East	50	9.90 (8.72-11.23)	15	5.20 (3.42-7.89)
Northeast	65	8.93 (7.72-10.33)	16	5.96 (2.61-13.60)
Northwest	63	10.98 (9.08-13.29)	23	5.24 (3.82-7.18)
North	42	9.31 (7.83-11.07)	13	4.45 (3.67-5.40)
West	64	9.72 (8.33-11.35)	15	6.96 (3.21-15.09)
Pampulha	33	9.27 (7.93-10.85)	6	5.75 (3.49-9.47)
Venda Nova	59	12.39 (9.58-16.02)	13	3.64 (2.95-4.49)

Our previous studies on food and environmental matrices suggest that food is a significant source of arsenic intake (Ng *et al.*, 2019), and rice and bean are the major contributors to the total dietary As intake in the Brazilian population (Ciminelli *et al.*, 2017).

4 CONCLUSIONS AND RECOMMENDATIONS

The TAs in the BH population is generally below 10 μg/g creatinine but higher than "toxic-As". The speciation technique is capable of separating the arsenic species of interest. This data can serve as a population baseline for future biomonitoring studies in mining towns or around extractive industries.

ACKNOWLEDGEMENTS

We thank the support of participants and Kinross Brasil Mineração (KBM).

REFERENCES

Ciminelli V.S.T., Gasparon M., Ng J.C., Silva G.C. & Caldeira C.L. 2017. Dietary arsenic exposure in Brazil: the contribution of rice and beans. *Chemosphere* 168: 996–1003.

IBGE 2010. Instituto Brasileiro de Geografia e Estatistica (ibge.org.br). Accessed 8 Dec 2020.

Ng J.C. Ciminelli V. Gasparon M. & Caldeira C. 2019. Health risk apportionment of arsenic from multiple exposure pathways in Paracatu, a gold mining town in Brazil. *Sci. Total Environ.* 673: 36–43.

NHANES. 2015-2016. National Health and Nutrition Examination Survey 2015-2016 Data Documentation, https://wwwn.cdc.gov/Nchs/Nhanes/2015-2016/UTAS_I.htm Accessed 8 Dec 2020.

Teixeira M.C. Santos A.C. Fernandes C.S. & Ng J.C. 2020. Arsenic contamination assessment in Brazil – past, present and future concerns – an historical and critical review. *Sci. Total Environ.* 730: 138217–138237.

3.2 Genetic predisposition of chronic arsenic poisoning

Arsenic in the Environment: Bridging Science to Practice for Sustainable Development – van der Wal, Ahmad, Petrusevski, Weijma, Savic, van der Wens, Beerendonk, Bhattacharya, Bundschuh & Naidu (Eds)

MicroRNAs may have some important contribution in arsenic induced skin lesions and other non-dermatological health effects in human

N. Banerjee[1,2] & A.K. Giri[1]
[1]*Molecular Genetics Division, CSIR-Indian Institute of Chemical Biology, Kolkata, India*
[2]*Biomedical Research and Diagnostic Centre, Manovikas Kendra Rehabilitation and Research Institute for the Handicapped, Kolkata, India*

ABSTRACT: Recent studies suggest that epigenetic alterations contribute to arsenic induced adverse health effects in arsenic exposed populations, but such evidences involving microRNAs are lacking. So, here we did global plasma microRNA (miRNA) profiling in 12 arsenic exposed individuals with skin lesions and 12 and arsenic unexposed individuals of West Bengal. We found 221 miRNAs which were differentially regulated with chronic arsenic exposure. The results were validated by qRT-PCR for some selected miRNAs in 60 individuals. Pathways in cancer were found to be the most enriched pathways having maximum targets. miRNA expressions were associated with the commonly occurring non-dermatological health effects: respiratory problems and conjunctival irritation of the eyes also. Thus, our results indicate that miRNA expressions are altered, which leads to differential effects in their target pathways causing the development of skin lesions and other adverse health effects in the arsenic exposed individuals with skin lesions.

1 INTRODUCTION

Ground water arsenic contamination is a curse which affects millions of people in the world. In India, West Bengal is the worst affected state where more than 26 million people are exposed to arsenic by drinking arsenic-laden water (Chakraborti *et al.*, 2009). This leads to the development of arsenic induced skin lesions which are the hall marks of chronic arsenic toxicity. Other detrimental health effects include cancers of the internal organs, peripheral neuropathy, respiratory distress, conjunctivitis of the eyes, cardiovascular disease, diabetes mellitus and liver diseases (Guha Mazumder 2008). Altered miRNA expression profiles have been associated with various diseases including cancers (Calin & Croeci 2006). So, in this present study we wanted to identify whether microRNAs have any role to play in development of arsenic induced skin lesions and other health effects in the chronically exposed individuals of West Bengal.

2 MATERIALS AND METHODS

2.1 *Study site and participants*

The study participants consisted of 32 arsenic exposed individuals with skin lesions from the arsenic affected Murshidabad district and 28 unexposed individuals from the unaffected Midanpur district of West Bengal. The arsenic content in the drinking water of the exposed population was much above the permissible limit of 10 μg/L, while the unexposed individuals consumed drinking water having arsenic content well within this limit. The details of the study sample collection are described previously (Banerjee *et al.*, 2019). This study was conducted in accordance with the Helsinki II Declaration and approved by the Institutional Human Ethical Committee of CSIR-Indian Institute of Chemical Biology.

2.2 *Arsenic estimation in water and urine samples*

Drinking water and first morning voids of urine samples were collected from the study participants following standard protocols and processed as described before (Banerjee *et al.*, 2019). Arsenic measurement was done employing the atomic absorption spectrometer (Shimadzu AA-7000) coupled to a Graphite furnace atomizer (Shimadzu) using the AA Wizard software and arsenic lamp (lamp current 380 mA).

2.3 *Collection of blood and isolation of plasma*

Venous blood was drawn by vein-puncture method, collected in EDTA-vacutainer tubes, and immediately put on ice. Blood samples were brought to the laboratory on ice within 2 hours after collection and the subsequent works were done. Plasma (about 500 μl) was collected from by centrifugation of blood samples for 5 minutes at 1500 rpm at room temperature.

DOI: 10.1201/9781003317395-104

2.4 *miRNA profiling*

Plasma RNA was isolated from 24 individuals (12 from each group of the following groups: arsenic exposed individuals with skin lesions and arsenic unexposed individuals). OD was measured in a nanodrop analyzer, the RNA was quantified and the samples were then sent to the *Life Technologies*, Bangalore for microarray analysis using the Affymetrix Gene Chip miRNA 2.0 platform.

2.5 *cDNA conversion and quantitative real time PCR*

Total RNA was converted to cDNA using Biobharati Super Reverse Transcriptase (India) cDNA synthesis kit following manufacturer's instructions. The primers used were similar as used in our previous work (Banerjee *et al.*, 2019). The miRNA expression level was determined by quantitative (SYBR green low ROX kit) real time PCR kit (Eurogentec) following the manufacturer's instructions. Another 16 control and 20 exposed subjects in addition to the 24 samples profiled were included for realtime PCR study. The data were normalized using the $2^{-\Delta\Delta CT}$ method relative to the U6 small nuclear RNA.

2.6 *Identification of non-dermatological health effects in the study population*

Identification of the non-dermatological health effects like, respiratory problems, peripheral neuropathy and conjunctival irritations of the eyes was done by expert physicians in the relevant field as explained previously (Banerjee *et al.*, 2019).

3 RESULTS AND DISCUSSION

Micro RNAs play important roles in various cellular processes such as development, cell growth, differentiation, angiogenesis, invasion and cellular stress responses. Recent evidences show that epigenetic modifications play important roles in arsenic induced susceptibility and carcinogenesis (Paul *et al.*, 2014), so we hypothesized that altered miRNA expressions might contribute to arsenic toxicity as well. Thus, we did plasma miRNA microarray analysis in the arsenic exposed individuals with skin lesions and unexposed individuals followed by downstream validation by qRT-PCR.

3.1 *Demographic characteristics of the study participants*

The study participants were age–gender matched. Arsenic content in drinking water and urine samples was significantly higher ($p<0.001$) in the exposed group compared to the unexposed group (Table 1).

Table 1. Demographic characteristic of the study participants.

Parameters	Unexposed	Exposed individuals with skin lesions
Total Subjects (n)	28	32
Age(Mean± SD)	44.67 ± 12.23	44.93 ± 9.95
Male (N)	16	18
Female (N)	12	14
Water (μg/L)	6.09± 1.39	168.24±99.28*
Urine (μg/L)	19.28±7.48	230.72±97.68*

$p<0.001$, Students unpaired T test.

3.2 *MicroRNA profiling of the study participants*

The results are shown in as Heat map in Figure 1 which shows that among 221 miRNAs, 218 were upregulated and 3 were down regulated in the exposed group with skin lesions compared to the unexposed group.

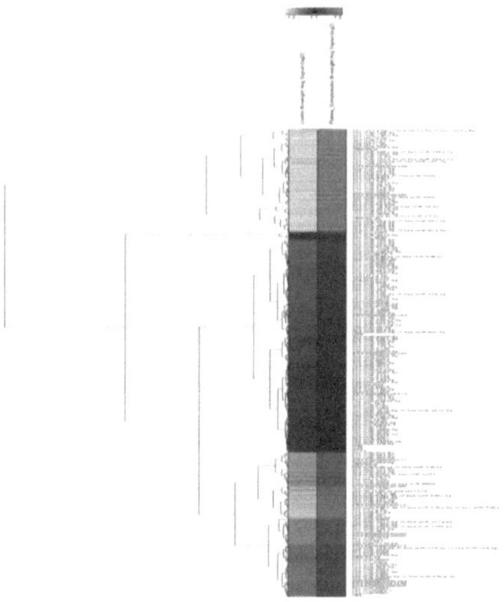

Figure 1. Heatmap showing global miRNA expressions in the control vs exposed individuals.

3.3 *Downstream validation by quantitative real-time PCR*

From all the differentially expressed microRNAs, based on whether they were significantly up/down regulated and found to be closely associated with different cancers including that of the skin we did quantitative real-time PCR analysis for miR21, miR23a, miR124, miR126, miR619, miR 3613-3p, miR1282 and miR4530. We found that miR21, miR23a, miR124, miR126, 3613-3p were significantly up regulated (fold change>2) and miR1282 and miR4530 were significantly

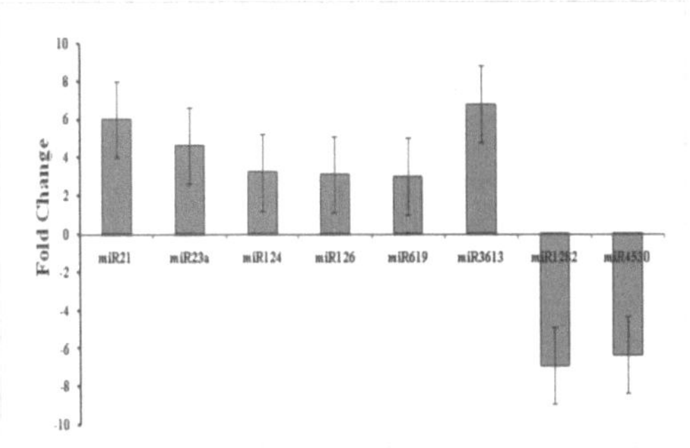

Figure 2. Fold change in the exposed vs controls (real time).

downregulated in the exposed individuals with skin lesions compared to the unexposed group (Figure 2). The upregulated miRNAs are established oncomers and their increased expressions have been previously associated with different types of cancers and inflammatory conditions. Their up regulation in the individuals with skin lesions explains the occurrence of precancerous and cancerous skin lesions, and other forms of cancers in them (García-Sancha *et al.*, 2019). TARGETSCAN was used for pathway prediction of the differentially regulated miRNAs. *Pathways in cancer* were found to be most enriched with 111 potential targets.

3.4 *Association of miRNA expression patterns with non-dermatological health effects*

We found that miRNAs were differentially associated with the commonly occurring non-dermatological health effects also. miR21, miR 3613, miR23a, miR124 and miR126 expressions were more than 2 fold higher in individuals with respiratory distress and conjunctival irritation of the eyes or both as is shown in the Table 2. The differentially expressed miRNAs might contribute to increased inflammatory conditions which contribute to the respiratory diseases and conjunctival irritation of the eyes in the study subjects.

Table 2. miRNA expressions affecting non-dermatological health effects.

miRNA	Diseaseconditions	Fold change of miRNAs (*fold change>2)
miR 21	Respiratory problems	2.98*
miR 3613	Respiratory problems	2.86*
	Conjunctival irritation of eyes	2.74*
miR 23a	Respiratory Problems	2.84*
miR124	Conjunctival irritation of eyes	2.73*
miR126	Respiratory problems	2.89*

4 CONCLUSION

Thus we can conclude that, chronic arsenic exposure leads to alteration in the micro RNA expressions in the exposed individuals which affects different functional pathways. This leads to the development of skin lesions and other non-dermatological health effects in them.

REFERENCES

Banerjee N., Das S., Tripathy S., Bandyopadhyay A.K., Sarma N., Bandyopadhyay A. & Giri A.K. 2019. MicroRNAs play an important role in contributing to arsenic susceptibility in the chronically exposed individuals of West Bengal, India. *Environ. Sci. Pollut. Res. Int.* 26: 28052–28061.

Calin G.A. & Croce C.M. 2006. MicroRNA signatures in human cancers. *Nat. Rev. Cancer* 6: 857–866.

Chakraborti D., Das B., Rahman M.M., Chowdhury U. K., Biswas B., Goswami A.B., Nayak B., Pal A., Sengupta M.K., Ahamed S., Hossain A., Basu G., Roychowdhury T. & Das D. 2009. Status of groundwater arsenic contamination in the State of West Bengal, India: a 20-year Study Report. *Mol. Nutr. Food Res.* 3: 542–551.

García-Sancha N., Corchado-Cobos R., Pérez-Losada J. & Cañueto J. 2019. Microrna dysregulation in cutaneous squamous cell carcinoma. *Int. J. Mol. Sci.* 2:20

Guha Mazumder D.N. 2008. Chronic arsenic toxicity & human health. *Ind. J. Med. Res.* 128: 436–447.

Paul S., Banerjee N., Chatterjee A., Sau, T.J., Das J.K., Mishra P.K., Chakrabarti P., Bandyopadhyay A. & Giri A.K. 2014. Arsenic-induced promoter hypomethylation and over-expression of ERCC2 reduces DNA repair capacity in humans by non-disjunction of the ERCC2-Cdk7 Complex. *Metallomics* 6: 864–873.

Arsenic in the Environment: Bridging Science to Practice for Sustainable Development –
van der Wal, Ahmad, Petrusevski, Weijma, Savic, van der Wens, Beerendonk, Bhattacharya, Bundschuh & Naidu (Eds)
© 2023 the Author(s), ISBN 978-1-032-32928-4

Arsenic exposure in population of Bihar and increasing cancer incidences: A correlative study

A. Kumar, M. Ali & A.K. Ghosh
Mahavir Cancer Sansthan & Research Centre, Patna, Bihar, India

ABSTRACT: In the entire world, about 300 million populations are exposed to arsenic poisoning in groundwater. In Bihar, India about 10 million people are drinking arsenic contaminated water. Study on 1500 cancer patients was carried out to know the correlation between cancer and arsenic contamination. The study showed significant correlation between arsenic toxicity and cancer incidences in the exposed population. High blood arsenic concentration was observed in the samples of cancer patients. The number of incidences is very high in the Gangetic zone in comparison to non-Gangetic zone. Unfortunately, the arsenic contamination in the ground water has also been maximum reported in the Gangetic zone. Hence, urgent strategy to control the disease burden is required.

1 INTRODUCTION

Arsenic menace in the ground water in entire Ganga Meghna Brahmaputra has caused serious health hazards in the exposed population. In Bihar state, an estimated 10 million population is exposed to arsenic. The exposed population exhibit typical symptoms of arsenicosis like keratosis, melanosis, rain drop pigmentation, Mee's lines (Kumar *et al.*, 2015, 2016, 2019).

In the recent times the cancer incidences in the state has drastically increased many folds. An estimated 70,000 new cancer patients are reported annually in the state. The disease like gall bladder cancer, liver cancer, renal cancer, breast cancer, prostate cancer and skin cancer is prevalent in the entire Gangetic zone of the state. The etiology of the disease burden is still a mystery. The patients with symptoms of arsenicosis and cancer disease are however reported with few numbers but their number have increased many folds in the recent times.

The present study deals with the evaluation of arsenic concentration in biological samples of cancer patients and its relationship with cancer.

2 MATERIALS AND METHODS

2.1 *Location of the study area*

The study was carried out in the state of Bihar, India.

2.2 *Sample collection and survey*

Ethical approval was obtained from the Institutional Ethics Committee (IEC) of the institute before collection of the biological samples of the studied subjects. Altogether, 150 control subjects and 1500 cancer patient's biological samples blood and hair were collected for the arsenic estimation by Atomic Absorption Spectrophotometer (Pinnacle 900T, Perkin Elmer, Singapore).

All the results were statistically analysed through Graphpad Prism 5 software, USA. For the epidemio-logical study, the cancer patients were interrogated through questionnaire.

3 RESULTS AND DISCUSSION

3.1 *Results of the studied subjects*

- About n=57 cancer patients had significantly very high concentration i.e more than 400 μg/L.
- Most of the cancer patients were from the Gangetic zone in comparison to the non Gangetic zone.

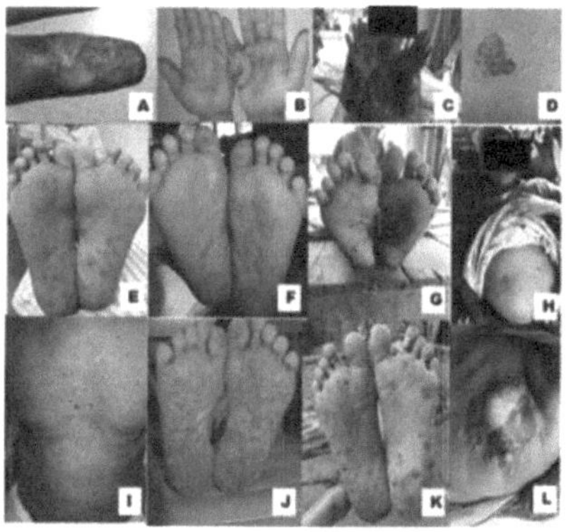

Figure 1. Cancer patients with arsenicosis symptoms.

DOI: 10.1201/9781003317395-105

- The number of female cancer patients were n = 912 while male was n = 588 (Figure 2).
- The control subjects (n = 150) had no arsenic concentration levels.
- In the present study the patients with arsenicosis symptoms were very few in number (n = 21).
- Out of 1500 cancer patients n = 747 cancer patients had blood arsenic concentration < 10 μg/L (normal ranges).
- The rest n = 753 had arsenic concentration in their blood more than the normal levels (Figures 3, 4).
- The district wise map of the cancer patients also shows that the disease burden is very high in the Gangetic zone in comparison to the non-Gangetic zone (Figure 5).

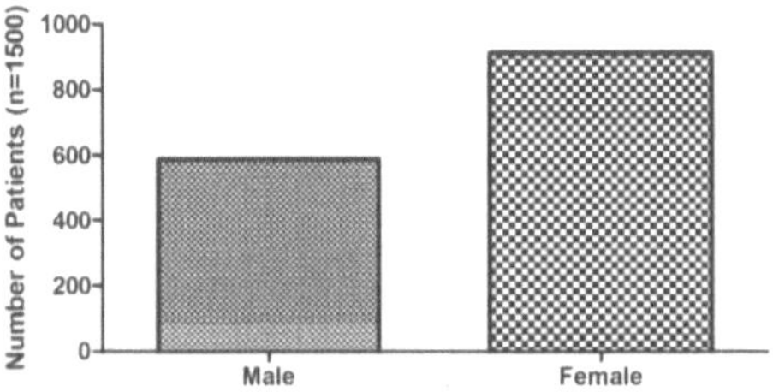

Figure 2. Gender-wise numbers of cancer patients.

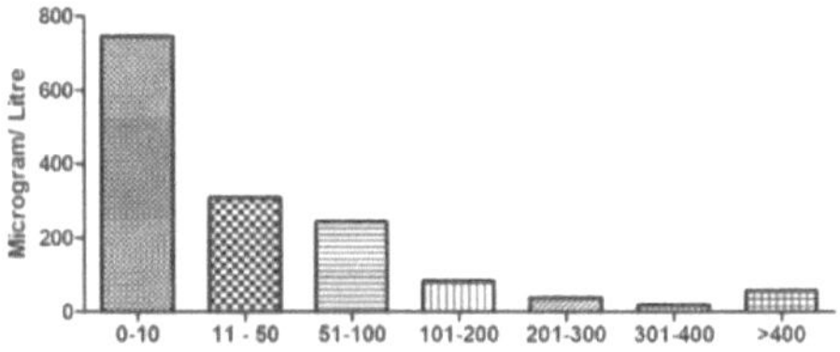

Figure 3. Arsenic concentration in blood samples of cancer patients (n=1500) samples assayed through AAS.

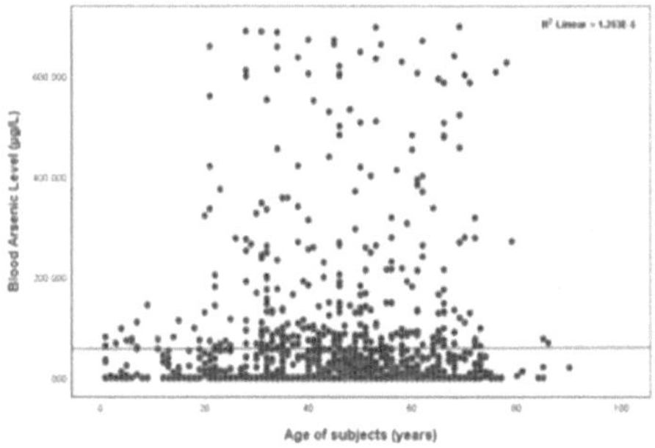

Figure 4. The correlation coefficient between blood arsenic levels and age of the subject (in years) (r= 1.253 & P<0.05).

- In the present study, there were significant high arsenic concentration in the blood samples of cancer patients especially in the females.

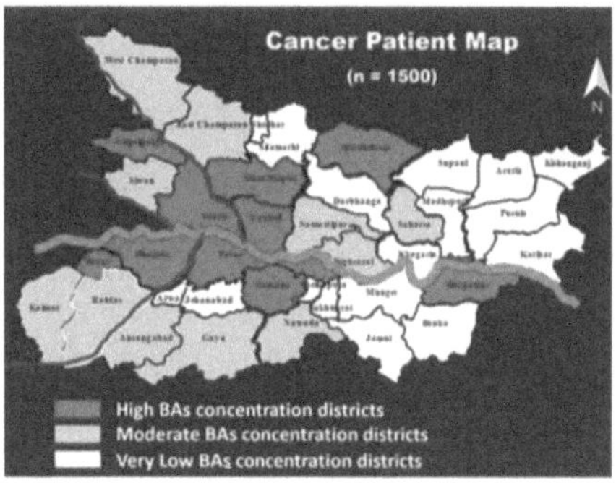

Figure 5. Cancer patient map showing blood arsenic concentration district wise.

- In extradiol dioxygenase family enzymes, a conserved hist This denotes that there is some pathway, which in the cancer patients of the state Bihar, is first making them diseased and finally leading to cancer.
- The significant levels of arsenic concentration have somehow strong correlation with the incidences of cancer patients.

4 CONCLUSIONS

The geogenic arsenic exposure to the population of Bihar has become major problem in the present time. The exposure has caused serious health hazards in them. The population is becoming diseased after drinking arsenic contaminated water for longer duration. The blood arsenic concentration in the cancer patients shows strong correlation in more than 50% of the patients. Unfortunately, the incidences of the cancer incidences is also increasing many folds in these population. Hence, specific strategy is immediately required to combat the problem.

REFERENCES

Kumar A., Ali M., Rahman M.S., Iqubal M.A., Anand G., Niraj P.K. *et al.,* 2015. Ground water arsenic poisoning in "Tilak Rai Ka Hatta" Village of Buxar district, Bihar, India causing severe health hazards and hormonal imbalance. *J. Environ. Anal. Toxicol.* 5: 290.

Kumar A., Kumar R., Rahman M.S, Iqubal M.A., Ali, M., Niraj P.K., Anand G., Prabhat K., Abhinav. & Ghosh A.K. 2016. Ground water arsenic contamination: a local survey in India. *Int. J. Prev. Med.* 7: 100.

Kumar A., Ali M., Kumar R., Rahman M.S., Srivastava A., Chayal N.K., Sagar V., Kumari R., Parween S., Kumar R., Niraj P.K., Anand G., Singh S.K. & Ghosh A.K. 2020. High arsenic concentration in blood samples of people of village Gyaspur Mahaji, Patna, Bihar drinking arsenic-contaminated water. *Expos. Health* 12: 131–140.

Arsenic in the Environment: Bridging Science to Practice for Sustainable Development –
van der Wal, Ahmad, Petrusevski, Weijma, Savic, van der Wens, Beerendonk, Bhattacharya, Bundschuh & Naidu (Eds)
 ISBN 978-1-032-32928-4

As3MT polymorphisms and Vitamin D: Effects in arsenic elimination and genotoxic damage in women of Poopó Lake – Bolivia

J. Mamani & N. Tirado
Genotoxicity Unit, Genetic Institute, Medicine School, Universidad Mayor de San Andrés, La Paz, Bolivia

ABSTRACT: The variability in the metabolic capacity of arsenic elimination has been associated with a large number of factors, such as gender, age, exposure level, ethnicity, nutritional status, the presence of polymorphisms in the AS3MT gene related to methylation of inorganic arsenic (iAs) to its organic monomethylarsenic (MMA) and dimethylarsenic (DMA) forms. In this study arsenic exposure was assess in the communities around Lake Poopó and evaluate the influence of two AS3MT polymorphisms and the concentration of vitamin D on the arsenic elimination metabolism and genetic damage. The results showed a higher frequency of the protective variants of the SNPs of the AS3MT (rs11191439: T = 0.91 and rs3740393: C = 0.93) that are associated with a high methylation capacity of As and less % of genotoxic damage. On the other hand, the median plasma level of Vitamin D was 19.14 ng/mL, women with vitamin D sufficiency status have 2% less MMA, and lower% genetic damage compared to deficit states.

1 INTRODUCTION

The Aymara-Quechua and Uru communities around Lake Poopó in the department of Oruro-Bolivia, have human populations with chronic arsenic (As) exposure through drinking water up to 250 µg/L; (Ormachea Muñoz *et al.*, 2013). Initial biomonitor- ing studies of these communities conducted by the Genotoxicity Unit - Genetic Institute in cooperation with the IRD, France and the Karolinska Institutet, Sweden found that women in these populations have an efficient inorganic As metabolism present in drinking water (De Loma *et al.*, 2019). The present study aims to assess the role of As3MT polymorphisms and the influence of vitamin D concentration on the metabolism of iAs and genetic damage caused by chronic exposure to water contaminated with As in women from communities around Lake Poopó.

2 MATERIALS AND METHODS

2.1 *Study area and population*

Rural communities near Lake Poopó were studied in the northern and southern region: Llapallapani, Puñaca, Poopó, Bengal Vinto, Crucero Belén, Sevaruyo, Vilañeque and San Felipe de Chaytavi.

Drinking water samples were taken in order to confirm the concentrations of As observed previously. Multistage sampling was performed with the selection of women by simple random sampling. The inclusion of the participants was voluntary and according to selection criteria and signed informed consent. Demographic characteristics, family history, sun exposure, work and domestic activities, previous illnesses, lifestyle habits, food preferences, and risk factors such as alcohol intake and smoking were obtained through an interview, medical exam and individual application of questionnaires; peripheral venous blood samples were collected in two EDTA tubes (Vacuette®): From one tube with 3ml blood, plasma was separated by sedimentation within 12 hours after sampling of whole blood. 250-500 ul of plasma was aliquoted in Eppendorf tubes, for Vitamin D determination subsequently protected from light and stored at -20 ° C, until processed. In the second tube 3ml whole blood was taken for comet assay and AS3MT polymorphisms, these samples were kept at 4°C and transported to the Genetic Institute UMSA for analysis.

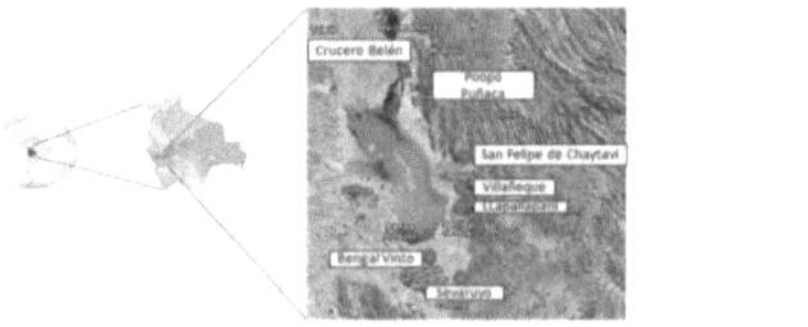

Figure 1. Study area and sampling locations.

2.2 *Assessment of arsenic exposure*

Spot midstream urine samples were obtained for assessment of As exposure, the samples were frozen and transported to the trace element laboratory at Karolinska Institutet, Sweden, for determination of As concentration and metabolites. Exposure to As was determined based on the sum of concentration of As of iAs and its metabolites (iAs+MMA+DMA) in urine (U-As) by high-performance liquid chromatography, coupled with hydride generation (HG) and inductively coupled plasma mass spectrometry (Agilent 7500ce, Agilent Technologies, Tokyo). Arsenic concentrations were adjusted to the mean specific gravity (SG).

2.3 *AS3MT (SNP rs3740393 and rs11191439) polymorphisms analysis*

DNA extraction was performed using QIAamp DNA Blood Mini Kit (250) QiAmp®. Genotyping was

 DOI: 10.1201/9781003317395-106

performed by allelic discrimination on the Rotor Gen (Quiamp®) device with Rotor-Gen Q software – Pure Detection v.2.1.0. The amplification was performed using TaqMan SNP probes (Apply Biosystem, USA) and with TaqMan Genotyping Master Mix (Apply Biosystem, USA).

2.4 *Genotoxic damage: Comet assay*

Genotoxic damage was determined in peripheral blood lymphocytes according to protocol (Larrea Poma *et al.*, 2010). The measurement of genotoxic damage was evaluated by quantification of DNA present in tail using Comet Score TM Freeware v1.5 software.

2.5. Plasma concentration of vitamin D 25 OH

Plasma concentration of Vitamin D2/D3 was measured by ELISA using the 25-OH Vitamin D3/D2 ORGENTEC Diagnostika GmbH kit (catalog #ORG 270).

3 RESULTS AND DISCUSSION

3.1 *Assessment of arsenic exposure*

The U-As values, expressed in µg/L, were between 11.78 and 288.70 µg/L, with a mean of 93.30 µg /L and a median. Descriptive analysis of urine metabolites showed that the study population has an efficient As removal metabolism, since the percentage of DMA is greater than 80% and the percentage of MMA% was low.

3.2 *AS3MT (SNP rs3740393 and rs11191439) polymorphisms analysis*

Respect rs3740393 the C allele that is considered the least frequent allele in different populations, in our population it presented a MFA of 0.93, similar to that observed in the population of San Antonio de los Cobres (MFA: 0.70), but not so in the population of Bangladesh where it is less frequent (MFA: 0.18). While, in the case of rs11191439, the C allele presented a frequency similar to that observed in populations of Bangladesh or San Antonio de los Cobres.

3.3 *Comet assay*

An average of 10.55% and a median of 8.7% of DNA in tail were observed, in terms of Tail moment values, an average of 3.54 and a median of 2.01. These values are similar to those observed in other studies such as the population in Southern Assam in India (Roy *et al.*, 2016). It was also observed that genotypes related to greater methylation capacity were also related to lower genetic damage, since the TT genotype of rs111191439 and the CC genotype showed a difference of 4.7% and 0.5%, respectively.

3.4 *Plasma concentration of vitamin D 25 OH*

The median plasma concentration of vitamin D in the study women was 16.11 ng / mL. In the categor ical analysis of vitamin D status, it was observed that the majority of women have a state of sufficiency (43.2%) followed by a state of insufficiency (41.9%) and finally a state of deficiency (14.9%).

Regarding the relationship of VitD status with the methylation coefficients of first it was shown that women with a sufficient level of VitD have a lower MMA/iAs index (0.48) in comparison with the states of insufficiency (0.65) or deficiency (0.65) ($p = 0.311$). Besides that, women with a level of VitD sufficiency have a higher DMA/MMA index (14.89), while deficiency and deficiency states had values of 10.99 and 10.40, respectively ($p = 0.013$). This could suggest a beneficial effect of vitamin D for proper elimination of iAs.

4 CONCLUSIONS AND RECOMMENDATIONS

The median plasma concentration of vitamin D in the study women was 16.11 ng/mL. In the categorical analysis of vitamin D status, it was observed that the majority of women have a state of sufficiency (43.2%) followed by a state of insufficiency (41.9%) and finally a state of deficiency (14.9%).

Regarding the relationship of VitD status with the methylation coefficients of first it was shown that women with a sufficient level of VitD have a lower MMA/iAs index (0.48) in comparison with the states of insufficiency (0.65) or deficiency (0.65) ($p = 0.311$). Besides that, women with a level of VitD sufficiency have a higher DMA/MMA index (14.89), while deficiency and deficiency states had values of 10.99 and 10.40, respectively ($p = 0.013$). This could suggest a beneficial effect of vitamin D for proper elimination of iAs.

ACKNOWLEDGEMENTS

We acknowledge the women in the communities for participation in the study. To Professor Karin Broberg for the scientific support. The Swedish research Council trough "Toxbol" project and Swedish Internatonal Development Cooperation Agency (Sida) for funding the research.

REFERENCES

De Loma J., Tirado N., Ascui F., Levi M., Vahter M., Broberg K. & Gardon J. 2019. Elevated arsenic exposure and efficient arsenic metabolism in indigenous women around lake Poopó Bolivia. *Sci. Total Environ.* 657: 179–186.

Larrea Poma M., Tirado Bustillos N. & Ascarrunz M.E. 2010. Daño genotóxico por exposición a plaguicidas en agricultores del municipio de Luribay. *Biofarbo* 18:31.

Ormachea Muñoz M., Wern H., Johnsson F., Bhattacharya P., Sracek O., Thunvik R., Quintanilla J. & Bundschuh J. 2013. Geogenic arsenic and other trace elements in the shallow hydrogeologic system of Southern Poopó basin, Bolivian Altiplano. *J. Hazard. Mater.* 262, 924–940.

Roy P., Mukherjee A. & Giri S. 2016. Ecotoxicology and environmental safety evaluation of genetic damage in tobacco and arsenic exposed population of Southern Assam, India using buccal cytome assay and comet assay. *Ecotoxicol. Environ. Saf.* 124: 169–176.

Arsenic in the Environment: Bridging Science to Practice for Sustainable Development –
van der Wal, Ahmad, Petrusevski, Weijma, Savic, van der Wens, Beerendonk, Bhattacharya, Bundschuh & Naidu (Eds)
© 2023 the Author(s), ISBN 978-1-032-32928-4

Structure of the ArsI C-As lyase with bound substrate roxarsone and mutational studies of active site residues: Elucidating the catalytic mechanism of degradation of organoarsenicals

V.S. Nadar, M. Yoshinaga & B.P. Rosen
Department of Cellular Biology and Pharmacology, Herbert Wertheim College of Medicine, Florida International University, Miami, FL, USA

ABSTRACT: Organoarsenicals such as monosodium methyl arsenate (MSMA or MAs(V)) and roxarsone (4-hydroxyl-3-nitro-benzenenarsenate or Rox(V)) have been extensively used as an herbicide and growth enhancers for poultry, respectively. The degradation of organoarsenicals to inorganic acid As(III) contaminates crops and drinking water. The bacterial enzyme ArsI is responsible for aerobic degradation of methylarsenite (MAs(III)) and trivalent roxarsone (Rox(III)) by C-As bond cleavage. The previous crystal structure of ArsI suggests that a loop gating mechanism controls the catalytic reaction. However, the full catalytic mechanism of ArsI such as the mode of substrate binding and activation of dioxygen was unclear because of lack of a substrate binding structure. Here we report the crystal structure of ArsI with bound Rox(III) and studies of site directed mutagenesis of active site residues. These studies help to elucidate the catalytic mechanism of ArsI and will enhance understanding of the recycling of environmental organoarsenicals.

1 INTRODUCTION

Organoarsenicals such as MAs(V) and Rox (V) have been used for many years as herbicides and antimicrobial growth promoters for farm animals (Bednar *et al.*, 2002; Garbarrino *et al.*, 2003). Microbial communities degrade these organoarsenicals by cleavage of the C-As bond by the *arsI* gene product (Yoshinaga *et al.*, 2011, 2015). The ArsI C-As lyase belongs to the family of non-heme iron dependent extradiol dioxygenases. Our previous crystallographic and biochemical studies reveal that, ArsI residues Gln8, His65 and Glu117 are critical for metal binding, and Cys98 and Cys99 form the substrate binding site (Nadar *et al.*, 2016; Pawitwar *et al.*, 2017). Lys105 and Gln103 are in the second coordination sphere, and Tyr38 interacts with the second coordination sphere via a water molecule (Nadar *et al.*, 2016). To determine the role of those residues in catalysis, we mutated each residue. The ArsI catalytic mechanism is unclear because of the lack of structures of ArsI with bound arsenical substrates. We co-crystallized the inactive mutant protein K105A with Rox(III) and Fe. This structure shed light on the mechanism of ArsI catalysis.

2 MATERIALS AND METHODS

2.1 *Structure of ArsI_K105A with bound Rox (III)*

Protein purification and crystallization experiments were done as previously (Nadar *et al.*, 2014). Crystallization was done in an anaerobic tent filled with argon gas to avoid protein oxidation. The yellow crystals with Rox(III) present were harvested and sent to the Advanced Light Source (ALS), Berkeley for diffraction experiments. The crystal diffracted to 2.23 Å resolution with a tetragonal space group. The difference map at 15.0 σ and the anomalous map at 5.0 σ level confirms the presence of Rox(III). The structure was refined and deposited in protein data bank (PDB ID: 5V0F).

2.2 *In vitro activity*

In vitro activity of each mutant was analyzed by high pressure liquid chromatography coupled with inductively coupled plasma mass spectrometry (HPLC-ICP-MS) as reported (Yoshinaga *et al.*, 2015).

2.3 *Isothermal Titration Calorimetry (ITC)*

The binding constants of ArsI-Fe(II) interactions were determined by ITC using a MicroCal iTC200 (GE Healthcare Bio Sciences). Fe(II) and protein were prepared in degassed distilled water. Spectra were analyzed using Origin 7.0, and binding constants were calculated with a one-site binding model.

2.4 *Fluorescence assays*

The intrinsic tryptophan fluorescence quenching assays were done by using a Quanta-Master UV-vis QM-4 steady state spectrofluorometer (PTI, Birmingham, NJ). Fluorescence of 1 µM ArsI was assayed in 50 mM MOPS and 0.5 mM NaCl (pH 7.5) at excitation/emission wavelengths of 295/340 nm.

3 RESULTS AND DISCUSSION

Residues Y38, K105 and Q103 were mutated to Y38F, K105A/E/R and Q103A/H by site-directed

DOI: 10.1201/9781003317395-107

mutagenesis. MAs(III) demethylation activity of wild type and mutants were determined using HPLC-ICP-MS. Y38F and Q103H retained activity, and K105A/E/R and Q103A lost demethylation activity in cells. Purified Y38F exhibited lower activity than wild type. Q103H retained full activity. K105A/E/R and Q103A were inactive.

Metal binding affinity was assayed by ITC with purified enzymes and Fe(II). The dissociation constant (K_d) enzyme with Fe(II) is similar to BmArsI and non-heme Fe(II) dependent ring-cleaving dioxygenases (Pawitwar *et al.*, 2017). Fe(II) binds to K105R, Y38F and Q103H with affinity similar to that of wild type. K105A and K105E showed 10-fold lower affinity with Fe(II) compare with wild type enzyme. Binding of arsenical substrates was determined by intrinsic tryptophan fluorescence quenching. There were no significant differences between the affinity of wild type ArsI and its mutants. This confirms that the mutations do not affect substrate binding. Co-crystallization with wild type of ArsI and mutants with the substrate Rox(III) and co-substrate Fe(II) was attempted. Only ArsI_K105A co-crystallized with both substrate and metal. The arsenic atom of Rox(III) was directly coordinated with Cys98 and Cys99. The distance between the sulfur and arsenic atom is approximately 2.0 Å (Figure 1). The aromatic ring of Rox(III) stacked with the five membered ring of His65 on one side and the six membered ring of Trp107. The hydroxyl group of Rox(III) forms a hydrogen bond with the side chain of Asp64. Hydrophobic interactions between Ile53 and the aromatic ring of Rox(III) help to stabilize substrate binding. The arsenic atom of Rox(III) has a three-coordinate trigonal pyramidal geometry. Cys98 and Cys99 bring the substrate to the metal center. They also hold the arsenical substrate in position for dioxygen attack of the C-As bond with a distance between the metal and sulfur atom of Cys99 of 4.0 Å.

In extradiol dioxygenase family enzymes, a conserved histidine residue involves acid catalysis in which a proton is transferred to the metal-bound oxygen atom (Lipscomp 2008). Based on this concept, we hypothesized that the conserved residue Lys105 in ArsI may involve a proton donor to the metal-bound dioxygen molecule and promoting attack of the initial adduct on the substrate. A combination of metal and substrate binding studies and activity analysis both *in vitro* and *in vivo* demonstrate that the conserved residue Lys105 is critical for catalysis. Tyr38 interacts with the metal

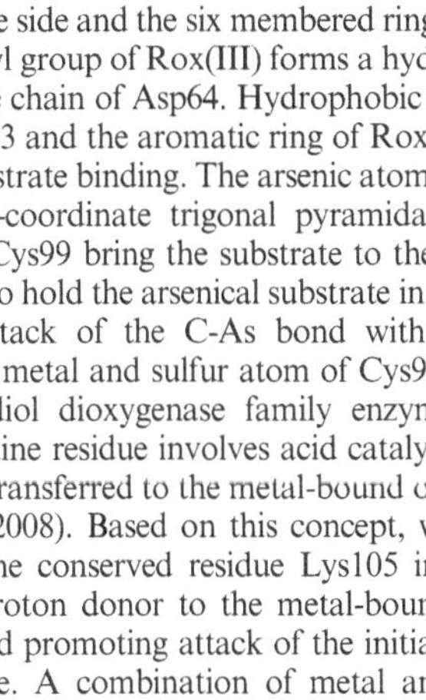

Figure 1. Binding of Rox(III) with ArsI. The metal binding residues are shown in cyan. The brown and red spheresare Fe and water molecules, respectively. Distances of 3.5 Å or less, shown in dotted lines.

through a water molecule at the same time it interacts the sulfur atom of Cys99 through a water molecule. We hypothesized that Tyr38 is involved in formation of a stabilizing intermediate during C-As bond cleavage, similar to how the conserved tyrosine residue stabilizes the alkly peroxo intermediate by hydrogen bonding in catechol dioxygenase. Partially conserved residue Gln103 interacts with the metal through a water molecule. When it is changed to Ala, it loses catalytic activity, but when changed to His, it retains activity. Gln103 stabilizes the active site loop when it enters the catalytic site. In the Q103A mutant, that residues can no longer form the hydrogen bond and does not stabilize the active site loop at the catalytic site.

4 CONCLUSIONS

The structure of ArsI with bound Rox(III) shows that the binding mode of the arsenical substrate with the metal center is different from that of other extradiol dioxygenases and is a unique feature of this C-As lyase even though their structures are similar. The results of mutational studies reveal that Lys105 is critical for catalytic activity and may activate the dioxygen molecule.

ACKNOWLEDGEMENTS

Supported by grant NSF BIO/MCB grant 1817962 and a pilot project grant from the Herbert Wertheim College of Medicine (project#80000843) to MY and NIH grants GM55425 and ES023779 to BPR.

REFERENCES

Bednar A.J., Garbarino J.R., Ranville J.F. & Wildman T.R. 2002. Preserving the distribution of inorganic arsenic species in groundwater and acid mine drainage samples. *Environ. Sci. Technol.* 36: 2213–2218.

Garbarrino J.R., Bednar A.J., Rutherford D.W., Beyer R.S. & Wershaw R.L. 2003. Environmental fate of roxarsone in poultry litter. I. degradation of roxarsone during composting. *Environ. Sci. Technol.* 37: 1509–1514.

Lipscomp J.D. (2008) Mechanism of extradiol aromatic ring-cleaving dioxygenases. *Curr. Opin. Struct. Biol.* 18: 644–649.

Nadar S.V., Yoshinaga M., Kandavelu P., Sankaran B. & Rosen B.P. 2014. Crystallization and preliminary X-ray crystallographic studies of the ArsI C-As lyase from Thermomonospora Curvata. *Acta Cryst. F* 70: 761–764.

Nadar S.V., Yoshinaga M., Pawitwar S.S., Kandavelu P., Sankaran B. & Rosen B.P. 2016. Structure of the ArsI C-As lyase: insights into the mechanism of degradation of organoarsenical herbicides and growth promoters. *J. Mol. Biol.* 428: 2462–2473.

Pawitwar S.S., Nadar S.V., Kandegedara A., Stemmler T.L., Rosen B.P. & Yoshinaga M. (2017) Biochemical characterization of ArsI: a novel C-As lyase for degradation of environmental organoarsenicals. *Environ. Sci. Technol.* 51: 11115–11125.

Yoshinaga M. & Rosen B.P. 2014. A C-As lyase for degradation of environmental organoarsenical herbicides and animal husbandry growth promoters. *Proc. Natl. Acad. Sci. USA* 111: 7701–7706.

Yoshinaga M., Cai Y. & Rosen B.P. 2011. Demethylation of methylarsonic acid by a microbial community. *Environ. Microbiol.* 13: 1205–1215.

Arsenic in the Environment: Bridging Science to Practice for Sustainable Development –
van der Wal, Ahmad, Petrusevski, Weijma, Savic, van der Wens, Beerendonk, Bhattacharya, Bundschuh & Naidu (Eds)
 ISBN 978-1-032-32928-4

Organoarsenicals inhibit bacterial peptidoglycan biosynthesis by targeting MurA

L.D. Garbinski, B.P. Rosen & M. Yoshinaga
Department of Cellular Biology and Pharmacology, Herbert Wertheim College of Medicine, Florida International University, Miami, USA

ABSTRACT: Trivalent organoarsenicals such as methylarsenite (MAs(III)) act as antibiotics that provide a competitive advantage for producers in microbial communities. While mechanisms of resistance to trivalent organoarsenicals have been identified and characterized, how they exert antibiotic-like properties is largely unknown. To identify targets of MAs(III), a genomic library of the highly versatile gram-negative bacterium, *Shewanella putrefaciens* 200, was expressed in *Escherichia coli* and selected for MAs(III) resistance. One clone contained the *S. putrefaciens murA* (*SpmurA*), the gene for the first committed step in peptidoglycan biosynthesis. Overexpression of *SpmurA* confers MAs(III) resistance in *E. coli*. Purified SpMurA was inhibited by MAs(III), phenylarsenite (PhAs(III)) or the phosphonate antibiotic fosfomycin, but not by inorganic arsenite (As(III)). Fosfomycin inhibits MurA by binding to a conserved catalytic cysteine residue. A SpMurA with a mutation in the cysteine became resistant to fosfomycin but remained sensitive to MAs(III), suggesting that the two compounds have different mechanisms of action. Our results demonstrate that organoarsenicals have the potential to be a new class of peptidoglycan biosynthesis inhibitors that can address the global threat of antibiotic resistance.

1 INTRODUCTION

Arsenic is a ubiquitous toxic metalloid that has been present in the environment since the origin of life. Microbes inhabiting arsenic-rich environments for billions of years have evolved mechanisms to detoxify arsenic and even to harness it to produce organoarsenical antimicrobial compounds such as methylarsenite (MAs (III)), which are significantly more toxic than inorganic arsenite (As(III)) (Zhu *et al.*, 2014). The strong antibiotic properties of MAs(III) give a competitive advantage to producers in microbial communities, which in turn leads other members to develop resistance mechanisms that transform it to less toxic species, forming a part of the arsenic biogeochemical cycle (Chen *et al.*, 2019). A number of MAs(III) resistance mechanisms have been identified and characterized, while the mode of action of MAs(III) was largely unknown. Here we show that MurA, which catalyzes the first committed step in peptidoglycan biosynthesis, is a target of MAs(III) inhibition in bacteria. Furthermore, we demonstrate that the mode of action of MAs(III) is different from that of fosfomycin, the well-characterized MurA inhibitor. Our results shed light on the antibiotic properties of MAs(III) and showcase the potential of organoarsenicals as a new class of bacterial cell wall biosynthesis inhibitors.

2 MATERIALS AND METHODS

2.1 *Identifying targets of MAs(III) inhibition*

A genomic library of *Shewanella putrefaciens* 200 was constructed using pUC118 and transformed into *E. coli* TOP10. Transformants were selected for MAs (III) resistance. The plasmid was extracted from a resistant isolate, and the sequence of the insert fragment was determined, yielding two putative genes. Stop codons were introduced into each gene on the plasmid, producing two single-mutant plasmids. Each plasmid was transformed into *E. coli* TOP10, and MAs(III) resistance of each transformant was compared with *E. coli* cells carrying a control vector.

2.2 *MurA activity and inhibition*

Wild-type *S. putrefaciens* 200 MurA (SpMurA), a cysteine-to-aspartate mutant SpMurA derivative ($SpMurA_{C117D}$), and wild-type *Staphylococcus aureus* MurA (SaMurA) were overexpressed in *E. coli* BL21 and purified using affinity chromatography. The activity of each purified MurA enzyme was quantified as described previously (Webb 1992) in the absence or presence of As(III), MAs(III), phenylarsenite (PhAs (III)), fosfomycin (Fos) or *N*-ethylmaleimide (NEM).

3 RESULTS AND DISCUSSION

3.1 *Identification of MurA as a target of MAs (III)*

Overexpression of targets can confer resistance to antibiotics (Palmer & Kishony 2014). To isolate potential targets of MAs(III) inhibition, a genomic library of *S. putrefaciens* 200, a well-characterized, versatile environmental bacterium, was constructed and expressed in *E. coli*. Selection for resistance to MAs(III) yielded a colony that was confirmed to

 DOI: 10.1201/9781003317395-108

confer MAs(III) resistance. Sequencing and mutagenesis analyses demonstrated that *S. putrefaciens 200 murA* (*SpmurA*), which encodes the first committed enzyme in peptidoglycan biosynthesis, is responsible for the MAs(III)-resistance phenotype. Purified SpMurA was selectively inhibited by MAs (III) but not inorganic As(III) (Figure 1), confirming that MurA is a target molecule of the MAs(III) antibiotic. The trivalent aromatic organoarsenical phenylarsenite (PhAs(III)) inhibits MurA more strongly than MAs(III).

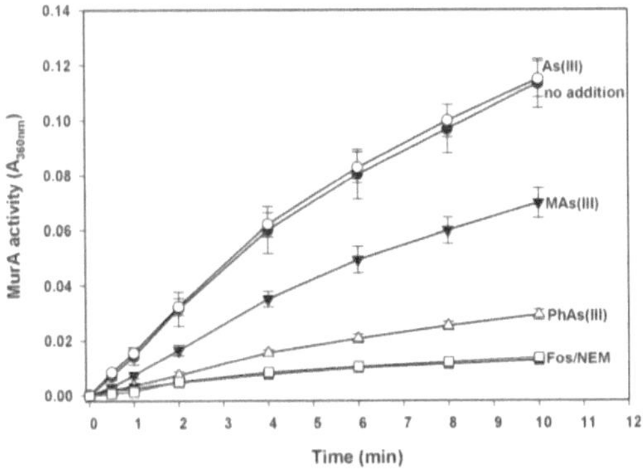

Figure 1. Activity of wild-type SpMurA. See the text for detail.

3.2 *The mode of action of MAs(III)*

MurA is inhibited by the broad-spectrum antibiotic fosfomycin (Fos), which covalently binds to a highly conserved cysteine residue that corresponds to Cys117 in SpMurA. However, some bacteria, such as the global pathogen *Mycobacterium tuberculosis,* encode MurA with the cysteine replaced by aspartate and are resistant to Fos (De Smet *et al.*, 1996). Wild type SpMurA was sensitive to Fos (Figure 1), whereas the C117D derivative ($SpMurA_{C117D}$) was Fos-resistant (Figure 2). In contrast, both wild-type SpMurA (Figure 1) and $SpMurA_{C117D}$ (Figure 2) remained sensitive to either MAs(III) or PhAs(III).

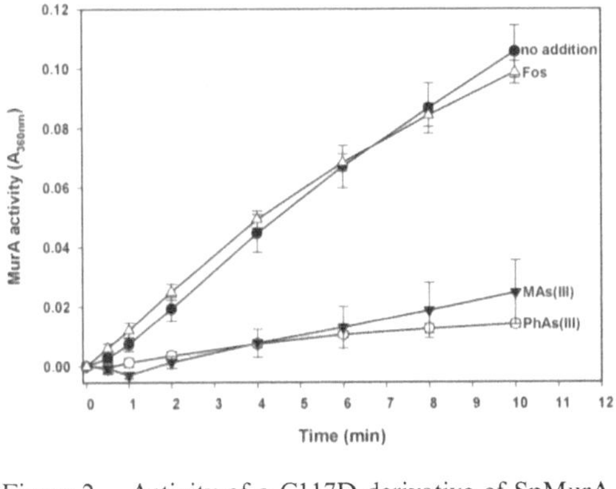

Figure 2. Activity of a C117D derivative of SpMurA. See the text for detail.

Furthermore, MurA from *Staphylococcus aureus,* which contains only the single conserved catalytic cysteine, was sensitive to Fos but relatively resistant to trivalent organoarsenicals compared with SpMurA, which has eight other cysteine residues in addition to Cys117. The cysteine modifying reagent *N*-ethylmaleimide (NEM) completely inhibited SpMurA (Figure 1), suggesting that inhibition of MurA by MAs (III) may involve one of the other cysteine residues.

4 CONCLUSIONS

MAs(III) has antibiotic-like properties and is produced by a number of bacterial species in microbial communities as a way to gain a competitive advantage. However, MAs(III) targets have not been identified. Here we demonstrate that trivalent organoarsenicals exert antimicrobial activity in part through its ability to inhibit MurA, the first committed enzyme in peptidoglycan biosynthesis. Clinically used MurA inhibitors including Fos are ineffective against pathogens with MurA lacking a conserved cysteine residue, while MAs(III) still inhibits. Thus, the mechanism of action of trivalent organoarsenicals may be different from other inhibitors of cell way biosynthesis. Our studies demonstrate that MAs(III) and derivatives may lead to the design of new antimicrobial drugs that target bacterial cell wall biosynthesis.

ACKNOWLEDGEMENTS

This work was supported by NIH R01 grants GM055425 and ES023779 to BPR, NSF BIO/MCB Grant 1817962 and a pilot project grant from the Herbert Wertheim College of Medicine (Project #800008403) to MY. LDG was supported by the Florida Education Fund's McKnight Doctoral Fellowship program.

REFERENCES

Chen J., Yoshinaga M. & Rosen B.P. 2019. The antibiotic action of methylarsenite is an emergent property of microbial communities. *Mol. Microbiol.* 111(2):487–494.

De Smet K.A.L., Kempsell K.E., Gallagher A., Duncan K. & Young D.B. 1999. Alteration of a single amino acid residue reverses fosfomycin resistance of recombinant MurA from *Mycobacterium Tuberculosis. Microbiology* 145: 3177–3184.

Palmer A.C. & Kishony R. 2014. Opposing effects of target overexpression reveal drug mechanisms. *Nat. Commun.* 5: 4296

Webb M.R. 1992. A continuous spectrophotometric assay for inorganic phosphate and for measuring phosphate release kinetics in biological systems. *Proc. Natl. Acad. Sci. U S A.* 89(11):4884–4887.

Zhu Y.-G., Yoshinaga M., Zhao F.-J. & Rosen B.P. 2014. Earth abides arsenic biotransformations. *Annu. Rev. Earth Planet. Sci.* 42: 443–446

Arsenic in the Environment: Bridging Science to Practice for Sustainable Development –
van der Wal, Ahmad, Petrusevski, Weijma, Savic, van der Wens, Beerendonk, Bhattacharya, Bundschuh & Naidu (Eds)
© 2023 the Author(s), ISBN 978-1-032-32928-4

Identification of the biosynthetic gene cluster for the organoarsenical antibiotic arsinothricin

A.E. Galván[1], S.M. Utturkar[2], B.P. Rosen[1] & M. Yoshinaga[1]
[1]*Department of Cellular Biology and Pharmacology, Florida International University, Herbert Wertheim College of Medicine, Miami, USA*
[2]*Purdue University Center for Cancer Research, Purdue University, West Lafayette, USA*

ABSTRACT: Antimicrobial resistance is an emerging global public health crisis, calling for urgent development of novel potent antibiotics. We recently demonstrated that arsinothricin (AST), the novel organoarsenical produced from arsenite (As(III)) by a soil bacterium, is effective against both Gram-positive and Gram-negative bacteria, suggesting that microbes utilize environmental arsenic to synthesize potent broad-spectrum antimicrobials for dominance in the community. AST is effective against resistant pathogens, demonstrating the potential to address the global threat of drug resistance. Here we report identification of the biosynthetic gene cluster (BGC) of AST from the producer, *Burkholderia gladioli* GSRB05. We sequenced the *B. gladioli* genome and found a seven-gene BGC containing *arsM* for As(III) *S*-adenosylmethionine (SAM) methyltransferase. Heterologous expression of the BGC confers AST production in *Escherichia coli*. Mutagenesis analysis suggests that only two genes, *arsM* and an adjacent gene encoding a radical SAM protein, termed *arsL*, are required for AST biosynthesis. The role of the additional genes is under investigation. Our results demonstrate that the biosynthetic pathway of AST is relatively straightforward.

1 INTRODUCTION

Arsenic is the most persistent environmental toxin and carcinogen, contaminating our water and food sources. Humans have harnessed this poison against pathogens since the eras of ancient Greece and China. In the early 20th century, arsphenamine or Salvarsán, the organoarsenical "magic bullet", was the first chemotherapeutic drug that was most effective against syphilis until the discovery of penicillin.

New potent antibiotics are urgently needed due to the spread of bacterial resistance. Recently, the rice rhizosphere bacterium *Burkholderia gladioli* GSRB05 was shown to produce a novel potent organoarsenical antibiotic termed arsinothricin (AST) (Figure 1A) (Kuramata *et al.*, 2016), suggesting that bacteria have harnessed arsenic to synthesize antimicrobials billions of years before the dawn of mankind's "antibitoic era". AST effectively works against resistant pathogens, whereas it exhibits low cytotoxicity on human monocytes (Nadar *et al.*, 2019), demonstrating the potential to be a progenitor of new magic bullets. The biosynthetic pathway of AST was unknown until this study. Here we identified the biosynthetic gene cluster (BGC) of AST from the genome of *B. gladioli* GSRB05, and further demonstrated that only two genes, *ars*M and arsL, encoding different SAM-dependent enzymes, are necessary for AST production.

2 MATERIALS AND METHODS

2.1 *Genome sequencing and analysis*

Genome sequencing of *B. gladioli* GSRB05 was outsourced to Center for Genome Technology, University of Miami, Miller School of Medicine (Miami, FL). Using BLAST, *ars*M orthologous genes were searched against predicted proteins of the *B. gladioli* GSRB05, resulting in identification of an *ars*M-containing cluster of seven genes (Figure 1C).

2.2 *Cloning and mutagenesis*

The BGC with *ars*M was entirely or partially PCR-amplified from the *B. gladioli* GSRB05 genomic DNA and cloned into pUC118, generating plasmids carrying the entire BGC (cluster 1), the cluster lacking *gnt*T (cluster 1a) or both *gnt*T and *ars*M (cluster 1b). Stop codons were introduced into each gene in cluster 1b, producing five single-mutant plasmids. Each plasmid was transformed into *E. coli* TOP10.

2.3 *Arsenic speciation*

E. coli cells carrying the plasmids were grown at 37 °C to an A600 of 1 in LB medium. The cells were

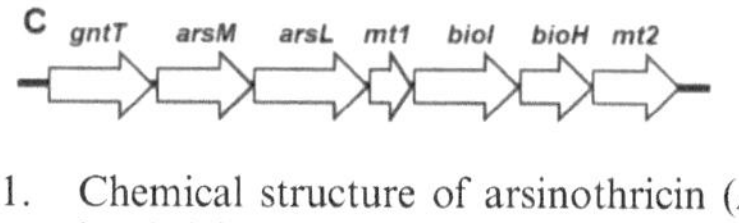

Figure 1. Chemical structure of arsinothricin (A) and hydroxy-arsinothricin (B) and the gene cluster from *B. gladioli* GSRB05 genome (C). Gene names and annotations: *gntT*, gluconate transporter; *arsM*, As(III) SAM methyltransferase; *arsL*, radical SAM protein; *mt1* and *mt2*, SAM-dependent methyltransferases; *bioI* and *bioH*, biotin biosynthesis proteins.

DOI: 10.1201/9781003317395-109

transferred to ST 10^{-1} medium supplemented with 0.5% glycerol and 1 μM As(III) at a density of A600= 3 and cultured for 24 h. Arsenic species in the cultures were analyzed by HPLC-ICP-MS.

3 RESULTS AND DISCUSSION

3.1 *Assessment of arsenic exposure*

When cultured with As(III), *B. gladioli* GSRB05 produces hydroxyarsinothricin (AST-OH) (Figure 1B), the unmethylated form of AST, followed by conversion to AST, suggesting that AST-OH is a precursor of AST, and that the final step of AST biosynthesis is methylation of AST-OH to AST (Kuramata *et al.*, 2016). Microbial arsenic methylation is catalyzed by the enzyme As(III) *S*-adenosylmethionine (SAM) methyltransferase, or ArsM (Qin *et al.*, 2006). Predicting that an *arsM* genes in the genome and found a seven-gene cluster (cluster 1) including *ars*M and an adjacent gene encoding a radical SAM enzyme, termed *ars*L (Figure 1C). Recently an ArsM and a different radical SAM enzyme, ArsS, have been shown to function together to catalyze the initial steps in arsenosugar biosynthesis in the cyanobacterium *Synechocystis* sp. PCC 6803 (Xue *et al.*, 2019). It is reasonable to consider that ArsM and ArsL function together in AST bio-synthesis. In the cluster 1, there are five other genes in addition to *ars*M and *ars*L (Figure 1C).

To examine whether the BGC is involved in AST production, we first cloned cluster 1 and transformed it into E. coli. However, the transformants poorly grew. We next constructed E. coli cells carrying cluster 1a lacking gntT, which grew normally and produced AST and smaller amount of AST-OH from As (III) (Figure 2, line A). In contrast, *E. coli* cells carrying cluster 1b lacking *gnt*T and *ars*M produced only AST-OH (Figure 2, line B). These results suggest that at least several of the five genes downstream of *ars*M are involved in AST-OH biosynthesis, and that ArsM catalyzes methylation of AST-OH to AST.

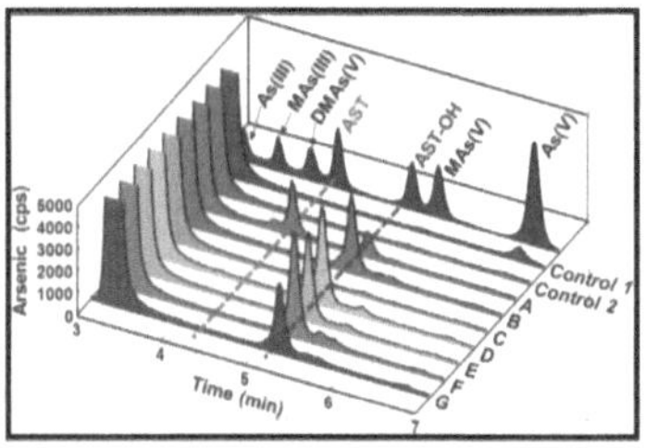

Figure 2. Production of AST by heterologous expression of the arsM-containing BGC in *E. coli*. See the text for detail. Control 1: no cells; Control 2: *E. coli* cells carrying an empty vector.

3.2 *Two genes are required for AST biosynthesis*

To determine which genes are required for AST-OH production, we introduced stop codons into the sequences of each of the five genes. Mutation in arsL resulted in loss of the ability to produce AST-OH (Figure 2, line C), while mutation in the other genesshowed little effect (Figure 2, lines D-G), suggesting that only arsL is required for AST-OH production.

Our results strongly suggest that the biosynthetic pathway of AST is composed of two steps: 1) conversion of As(III) to AST-OH catalyzed by the new radical SAM protein ArsL, and 2) methylation of AST-OH to AST catalyzed by ArsM. Phosphinothricin (PT), the bioactive moiety of the prodrug PT tripeptide (PTT), is a phosphonate mimetic of AST (Blodgett *et al.*, 2016). Given that the BGC of PTT is composed of 24 genes, it is striking that only two genes are necessary for AST biosynthesis.

4 CONCLUSIONS

Arsinothricin is a novel organoarsonate antibiotic that has the potential to provide a countermeasure against the global threat of resistant pathogens. Here we demonstrate that a seven-gene cluster identified in B. gladioli GSRB05 genome encodes the AST biosynthetic pathway. Strikingly, only two genes in the cluster are required to produce AST from As(III). The role of the additional five genes is not known but could be involved in AST modification or transport. Our discovery illustrates the previously unknown complexity of the arsenic biogeochemical cycle and provides insight for the development of more of effective organoarsenical antibioticss.

ACKNOWLEDGEMENTS

This work was supported by NSF BIO/MCB grant 1817962 and a pilot project grant from the Herbert Wertheim College of Medicine (project # 800008403) to M.Y. and NIH R01 grants GM55425 and ES023779 to B.P.R

REFERENCES

Blodgett J., Zhang J., Yu X. & Metcalf W. 2016. Conserved biosynthetic pathways for phosalacine, bialaphos and newly discovered phosphonic acid natural products. *J. Antibiot.* 69(1): 15–25.

Kuramata M., Sakakibara F., Kataoka R., Yamazaki K., Baba K., Ishizaka M., Hiradate S., Kamo T. & Ishikaw S. 2016. Arsinothricin, a novel organoarsenic species produced by a rice rhizosphere bacterium. *Environ. Chem.* 13 (4): 723–731.

Nadar V.S., Chen J., Dheeman D., Galván A.E., Sakurai K., Kandavelu P., Sankaran B., Kuramata M., Ishikawa S., Rosen B.P. & Yoshinaga M. 2019. Arsinothricin, an arsenic-containing non-proteinogenic amino acid analog of glutamate, is a broad-spectrum antibiotic. *Commun. Biol.* 2: 131.

Qin J., Rosen B.P., Zhang Y., Wang G., Franke S. & Rensing C. 2006. Arsenic detoxification and evolution of trimethylarsine gas by a microbial arsenite S-adenosylmethionine meThyltransferase. *Proc Natl Acad Sci USA*. 103(7): 2075–80.

Xue X., Ye J., Raber G., Rosen B.P., Francesconi K., Xiong C., Zhu Z., Rensing C. & Zhu Y. 2019. Identification of steps in the pathway of arsenosugar biosynthesis. *Environ. Sci. Technol.* 53: 634–641.

3.3 Reliable biomarkers for arsenic exposure

Arsenic in the Environment: Bridging Science to Practice for Sustainable Development –
van der Wal, Ahmad, Petrusevski, Weijma, Savic, van der Wens, Beerendonk, Bhattacharya, Bundschuh & Naidu (Eds)
 ISBN 978-1-032-32928-4

Arsenic exposure and cancer-related biomarkers in indigenous populations in Bolivia – modification by arsenic metabolism efficiency

J. De Loma[1], N. Tirado[2], M. Levi[1], J. Gardon[3] & K. Broberg[1]
[1]*Institute of Environmental Medicine, Karolinska Institutet, Stockholm, Sweden*
[2]*Genetics Institute, Genotoxicology Unit, Universidad Mayor de San Andrés, La Paz, Bolivia*
[3]*Hydrosciences Montpellier, Institut de Recherche pour le Développement, CNRS, University of Montpellier, France*

ABSTRACT: Inorganic arsenic is a known carcinogen. Telomere length (TL) and mitochondrial DNA copy number (mtDNAcn) in peripheral blood may serve as biomarkers for genotoxicity and cancer risk. Our aim is to assess if arsenic exposure influences TL and mtDNAcn in women living around Lake Poopó (Bolivia) from two ethnicities (Aymara-Quechua and Uru). Arsenic exposure was evaluated as the sum of arsenic metabolite concentrations in urine (U-As) measured by high-performance liquid chromatography online with hydride generation and inductively coupled plasma-mass spectrometry (HPLC-HG-ICP-MS), and as total arsenic in blood (B-As) measured by ICP-MS. Efficiency of arsenic metabolism was evaluated using the relative fractions of urinary metabolites, and arsenic methylating genetics (*AS3MT* rs3740393 and rs1046778) measured by TaqMan allelic discrimination or chip-based genotyping. TL and mtDNAcn were determined in blood by real-time PCR. Our results show that arsenic exposure (assessed as U-As and B-As) was associated with longer TL and higher mtDNAcn in this study population, and the associations were modified by arsenic metabolism capacity and *AS3MT* genotype.

1 INTRODUCTION

Elevated arsenic concentrations in water are widespread in Latin American countries (Bundschuh *et al.*, 2012). We recently presented the most comprehensive study so far on human arsenic exposure in Bolivia. The study took place around Lake Poopó, in the Bolivian Altiplano, where indigenous women presented a wide variation in inorganic arsenic exposure and an efficient metabolism (De Loma *et al.*, 2019). Yet, there is still a need to understand the health consequences of such exposure since no toxicological data is currently available from this region.

Arsenic is a potent carcinogen, and a key susceptibility factor for arsenic toxicity is the individual's capacity to metabolize arsenic (Vahter 2002). Arsenic is methylated into monomethylarsonic acid (MMA) and further into dimethylarsinic acid (DMA), which are both excreted in urine in combination with unmethylated inorganic arsenic (iAs). AS3MT is the main arsenic methylating protein in humans.

Longer telomere length (TL), and lower and higher (depending on cancer type) mitochondrial DNA copy number (mtDNAcn) in peripheral blood are associated with increased cancer risk. Recently, a study with individuals from the Argentinean Andes found associations between arsenic exposure and increased TL and mtDNAcn (Ameer *et al.*, 2016) suggesting arsenic-related carcinogenic changes in this study group. Our current aim is to evaluate if arsenic exposure also is associated with TL and mtDNAcn in the Bolivian Andes, and to study how arsenic metabolism efficiency may modify the associations.

2 MATERIALS AND METHODS

2.1 *Study group*

We recruited women (N=201) from ten villages around Lake Poopó, in the Bolivian Altiplano, as described in De Loma *et al.* (2019). For this study, we selected women for which we had urinary and blood arsenic, genotype, TL and mtDNAcn data (n=183).

2.2 *Arsenic determination*

We determined arsenic exposure as the sum concentration of arsenic metabolites (iAs, MMA and DMA) in urine (U-As), measured by high-performance liquid chromatography online with hydride generation and inductively coupled plasma-mass spectrometry (HPLC-HG-ICP-MS). Efficiency of arsenic metabolism was evaluated using the relative fractions of urinary metabolites. Variations in urinary dilutions were compensated by adjusting to the mean urinary osmolality (726 mOsm/kg; range 129-1161). Arsenic was also measured in whole blood using ICP-MS.

2.3 *TL and mtDNAcn*

We extracted DNA from whole peripheral blood with EZNA Blood DNA Mini kit (Omega, Bio-teck, USA). TL and mtDNAcn were determined by real-time PCR. In each run, a standard curve and control samples were included. All samples were run in triplicates. Relative TL and mtDNAcn were calculated as the ratio with hemoglobin beta (*HBB*), as a single copy gene, and normalized for the interplate variation by dividing by the ratio of a control sample.

DOI: 10.1201/9781003317395-110

2.4 *Statistical analyses*

We performed Spearman's rank tests to evaluate the correlation between TL, mtDNAcn, arsenic exposure markers (U-As and B-As), and potential covariates. We log2-transformed TL and mtDNAcn to fulfil the assumptions of linear regression. Linear regression analyses were used to evaluate the association between arsenic exposure (as U-As or B-As) and genotoxicity biomarkers (as TL or mtDNAcn). The crude model did not include any adjustments, while the adjusted model included ethnicity (for mtDNAcn), or ethnicity, age, and hemoglobin (for TL). We also studied how arsenic metabolism efficiency influenced the association between arsenic exposure and genotoxicity biomarkers by including interactions terms (between U-As or B-As, and a variable defined as below or above median %iAs or % MMA) and by stratifying the data by median values of %iAs or %MMA. The same approach was used to assess the influence of SNPs associated with arsenic metabolism efficiency (*AS3MT* rs3740393 and rs1046778) on the association between arsenic exposure and the studied cancer-related biomarkers.

3 RESULTS AND DISCUSSION

In the adjusted linear regression models, increasing arsenic exposure was associated with longer TL (U-As: B=0.001, p=0.01; B-As: B=0.018, p=0.11) and non-significantly with higher mtDNAcn (U-As: B=0.001, p=0.073; B-As: B=0.023, p=0.29).

The efficiency of arsenic metabolism influenced the association between arsenic exposure and TL and mtDNAcn. For TL, the interaction between %iAs and U-As was p=0.066. When stratifying the data by the median values of metabolite fractions, only individuals with higher %iAs or %MMA (less efficient metabolizers) presented a significant association between arsenic exposure and longer TL (%iAs: for U-As p=0.014, for B-As p=0.020; %MMA: for U-As p=0.042, for B-As p=0.055). No significant interactions were found between arsenic exposure and metabolite fractions on mtDNAcn. However, for individuals with higher %MMA (less efficient metabolizers), there was a significant positive association between U-As and mtDNAcn (B=0.002, p=0.034). A less clear pattern was seen for the modification by % iAs. This is similar to results by Ameer *et al.* (2016), where individuals from the Argentinean Andes with a less efficient arsenic metabolism had significantly longer TL and higher mtDNAcn with increasing U-As.

Furthermore, we evaluated how *AS3MT* rs3740393 and rs1046778, known to predict arsenic metabolism efficiency, modify the associations between arsenic exposure and the studied genotoxicity biomarkers. Arsenic exposure (as U-As and B-As) was significantly associated with longer TL and higher mtDNAcn (see Figure 1) in those individuals with genotypes associated with more %MMA (less efficient metabolizers).

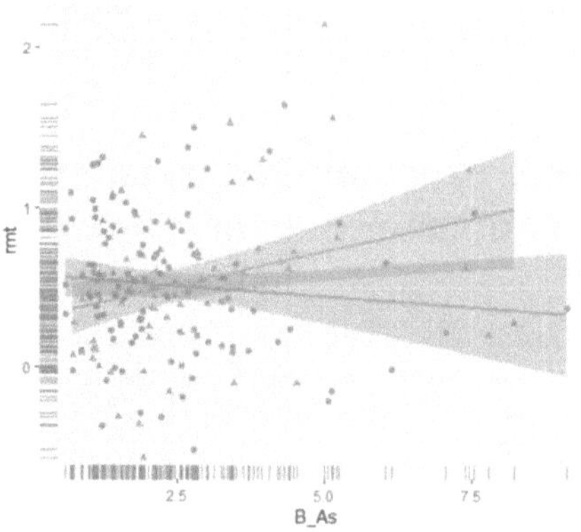

Figure 1. Linear fit between mtDNAcn and B-As (ng/g) stratified by *AS3MT* rs1046778 genotype (Blue: TC + TT; Red: CC, associated with lower %MMA) showing different associations.

4 CONCLUSIONS

Arsenic exposure is associated with longer TL and higher mtDNAcn. These associations were modified by arsenic metabolism efficiency, supporting the role of arsenic metabolism as a susceptibility factor for arsenic-related toxicity.

ACKNOWLEDGEMENTS

We acknowledge the Swedish Research Council, Eric Philip Sörensens Foundation, and Karolinska Institutet for supporting the project.

We thank all the participants, the Servicio Departamental de Salud de Oruro (SEDES), the Salud Familiar Comunitaria e Intercultural (SAFCI) program, and all collaborators that assisted the recruitment.

REFERENCES

Ameer S.S., Xu Y., Engström K., Li H., Tallving P., Nermell B., Boemo A., Parada L.A., Peñaloza L.G., Concha G., Harari F., Vahter M. & Broberg K. 2016. Exposure to inorganic arsenic is associated with increased mitochondrial DNA copy number and longer telomere length in peripheral blood. *Front. Cell Dev. Biol.* 4: 87.

Bundschuh J., Litter M.I., Parvez F., Román-Ross G., Nicolli H.B., Jean J.-S., Liu C.-W., López D., Armienta M.A., Guilherme L.R., Cuevas A.G., Cornejo L., Cumbal L. & Toujaguez R. 2012. One century of arsenic exposure in Latin America: a review of history and occurrence from 14 countries. *Sci. Total Environ.* 429: 2–35.

De Loma J., Tirado N., Ascui F., Levi M., Vahter M., Broberg K. & Gardon J., 2019. Elevated arsenic exposure and efficient arsenic metabolism in indigenous women around Lake Poopó, Bolivia. *Sci. Total Environ.* 657: 179–186.

Vahter M., 2002. Mechanisms of Arsenic Biotransformation. *Toxicology* 181–182: 211–217.

Arsenic in the Environment: Bridging Science to Practice for Sustainable Development –
van der Wal, Ahmad, Petrusevski, Weijma, Savic, van der Wens, Beerendonk, Bhattacharya, Bundschuh & Naidu (Eds)

Severe arsenic exposure in the population of Chapar village of Samasatipur district of Bihar

A. Kumar[1], M. Ali[1], R. Kumar[1], A.K. Ghosh[1], P. Salaun[2] & A.C.G. Gourain[2]
[1]*Mahavir Cancer Sansthan & Research Centre, Patna, Bihar, India*
[2]*School of Ocean Sciences, University of Liverpool, UK*

ABSTRACT: In Bihar an estimated 10 million population are exposed to arsenic. This has caused severe health hazards in the population. The study was carried out in n = 100 subjects of Chapar village of Mohiuddinagar district of Samastipur district of Bihar. In the present study, severe arsenic contamination in ground water was observed. The subjects exhibited typical symptoms of arsenicosis like hyperkeatosis, melanosis, rain drop pigmentation etc. The arsenic contamination in hair samples was also very high in the exposed population. Presently, village population have no options of safe drinking water, hence immediate intervention is required.

1 INTRODUCTION

Arsenic is considered to be major environmental health disaster. It is estimated that more than 200 million people are exposed to high level (<10 μg/L) of arsenic toxicity through drinking water and about 13 regions of world are greatly affected by high arsenic level in aquifers. Drinking water is considered to be one of the primary routes of exposure of inorganic arsenic. In Asia India and Bangladesh is the most affected country due to arsenic groundwater contamination. Arsenic poisoning causes a great health hazard for the people residing in the Ganga-Brahmaputra fluvial plains. Bihar is the second most arsenic contaminated state in India. Bihar is divided into 38 districts out of which 18 districts are affected with groundwater arsenic contamination. It is assumed that a population of about 10 million are consuming arsenic contaminated water with arsenic concentration more than 50 μg/L (Singh *et al.*, 2014). This has caused severe health hazards in the exposed population of the state (Kumar *et al.*, 2015, 2016, 2019). The present study deals with the evaluation of arsenic contamination in groundwater and health evaluation of exposed population of Chapar village of Samastipur district of Bihar.

2 METHODS/EXPERIMENTAL

2.1 *Location*

The study was carried out in Chapar village (25°32‘56.4“N 85°39’58.8”E) of Harail Panchayat of Mohiuddinagar block of Samastipur district of Bihar, India (Figure 1).

Figure 1. Map of the study area around Chapar village of Samastipur district of Bihar.

2.2 *Sample collection & survey*

Ethical approval was obtained from the Institutional Ethics Committee (IEC) of the institute before collection of the blood samples of the subject from the targeted site.

Altogether, 100 water samples from Chapar village while 100 hair samples from same household subjects were collected for arsenic assessment through atomic absorption spectrophotometer (Pinnacle 900T, Perkin Elmer, Singapore). All the results were statistically analysed through Graphpad Prism 5 software, USA. Arsenic related health problems among village people were also studied. A survey in the entire area was conducted to estimate the per capita consumption of drinking water through hand pumps by the villagers. For determining the exact location of the hand pump, hand held global positioning system (GPS) receivers (Garmin etrex10, of USA) with an estimated accuracy of ≈ 10 m were utilised. All the results were statistically analysed through Graphpad Prism 5 software, USA. For the epidemiological study, the cancer patients were interrogated through questionnaire.

DOI: 10.1201/9781003317395-111

3 RESULTS AND DISCUSSION

3.1 *Arsenic assessment*

The study shows novel findings ever explored in this area. The maximum arsenic concentration in the groundwater was found as 655 μg/L. The maximum arsenic concentration in hair sample of subjects was 46404 μg/kg.

3.2 *Health assessment*

The rural population exhibited the typical symptoms of arsenicosis like hyperkeratosis in palm and sole, melanosis in palm and sole, skin irritation, gastritis, constipation, loss of appetite, bronchitis and cough etc. The typical symptoms were also observed in children of age 8 year old (Figure 2).

The population were unaware about the arsenic exposure. The population had no option of safe drinking water.

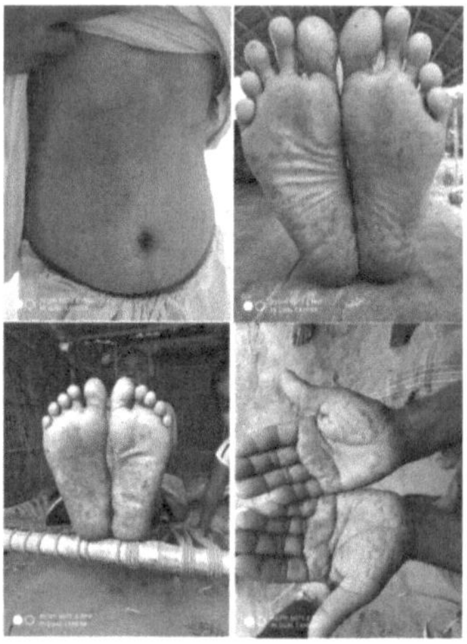

Figure 2. Arsenicosis symptoms in exposed population of Chapar village.

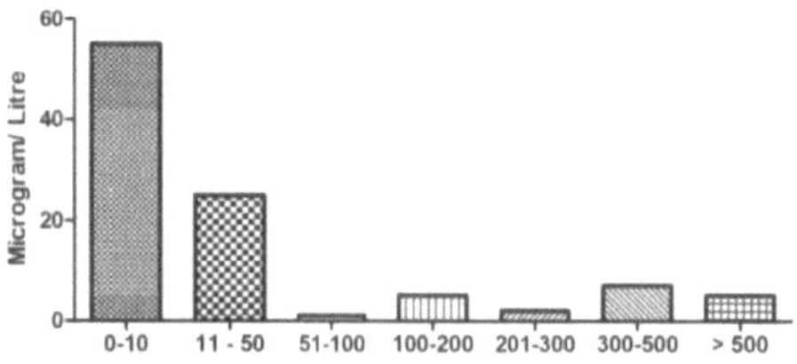

Figure 3. Arsenic concentration in handpumps in Chapar village of Mohiuddinagar district (n = 100 samples assayed through AAS).

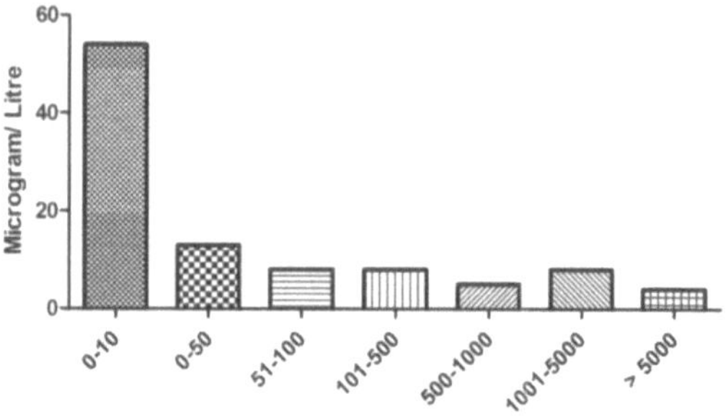

Figure 4. Arsenic concentration in hair samples of subjects of Chapar village of Mohiuddinagar district (n = 100 samples assayed through AAS).

4 CONCLUSIONS

The present study reveals the arsenic contamination ground water in Chapar village. The subjects exhibited typical symptoms of arsenicosis. They also had very high arsenic in their hair samples denotes the severity of the toxicity. The village people were unaware about the arsenic exposure. Presently, they have no safe drinking water access hence, require urgent interventions.

REFERENCES

Kumar A., Ali M., Rahman M.S., Iqubal M.A., Anand G., Niraj P.K. *et al.* 2015. Ground water arsenic poisoning in "Tilak Rai Ka Hatta" village of Buxar district, Bihar, India causing severe health hazards and hormonal imbalance. *J. Environ. Anal. Toxicol.* 5: 290.

Kumar A., Kumar R., Rahman MS, Iqubal MA., Ali, M., Niraj P.K., Anand G., Prabhat K., Abhinav & Ghosh A.K. 2016. Ground water arsenic contamination: a local survey in India. *Int. J. Prev. Med.* 7: 100.

Kumar A., Ali M., Kumar R., Rahman M.S., Srivastava A., Chayal N.K., Sagar V., Kumari R., Parween S., Kumar R., Niraj P.K., Anand G., Singh S.K. & Ghosh A.K. 2020. High arsenic concentration in blood samples of people of village Gyaspur Mahaji, Patna, Bihar drinking arsenic-contaminated water. *Expos. Health* 12: 131–140.

Singh S.K., Ghosh A.K., Kumar A., Kislay K., Kumar C., Tiwari R.R., Parwez R., Kumar N. & Imam M.D. 2014. Groundwater arsenic contamination and associated health risks in Bihar, India. *Int. J. Environ. Res.* 8(1):49–60.

3.4 Neurophysiological and IQ impacts of arsenic

Arsenic in the Environment: Bridging Science to Practice for Sustainable Development – van der Wal, Ahmad, Petrusevski, Weijma, Savic, van der Wens, Beerendonk, Bhattacharya, Bundschuh & Naidu (Eds)
 ISBN 978-1-032-32928-4

Cognitive impairment and its relation to serum brain-derived neurotropic factor in arsenic-exposed adult individuals in Bangladesh

M.S. Islam[1], A.E. Siddique[2], Y. Karim[2], F. Hossen[2], V. Mondal[2], Z. Hosen[1], S. Himeno[3] & K. Hossain[2]

[1]*Department of Applied Nutrition and Food Technology, Islamic University, Bangladesh*
[2]*Department of Biochemistry and Molecular Biology, University of Rajshahi, Bangladesh*
[3]*Laboratory of Molecular Nutrition and Toxicology, Faculty of Pharmaceutical Sciences, Tokushima Bunri University, Japan*

ABSTRACT: Arsenic (As) toxicity is one of the emerging public health concerns worldwide. A limited number of studies have been conducted to explore the effects of As exposure on adult cognition. Moreover, biochemical basis of As-related cognitive impairment is almost unknown. The objectives of our study was to explore the association between As exposure and cognitive impairment by Mini-Mental State Examination (MMSE) especially with regard to the serum brain-derived neurotrophic factor (sBDNF) levels in adult human subjects recruited from As-endemic and non-endemic areas in Bangladesh. In this study, we found that MMSE score of the study subjects were 0.6, 0.48 and 0.41 times, and sBDNF levels were 0.7, 0.62 and 0.4 times lower for 1 unit increase in water, hair and nail As concentrations, respectively. Additionally, we found that As exposure levels were increased and sBDNF levels were decreased with greater severity of cognitive impairment. Thus the decreased BDNF level may be implicated in As exposure-related cognitive impairment.

1 INTRODUCTION

Arsenic (As) is a potent environmental pollutant that is naturally present at high levels in the groundwater in many countries including Bangladesh. According to WHO, at least 140 million people in 50 countries have been drinking water containing As at levels above the WHO provisional guideline value of 10 µg/L. Inorganic As is a well-established carcinogen, however, many recent studies have reported that chronic As exposure causes several non-malignant diseases like cardiovascular diseases, peripheral neuropathy and diabetes mellitus. Chronic exposure to As has also been found to be associated with cognitive impairment but most of the previous studies in this regard have conducted on the cognition of children. Relatively, little attention has been given on the association between chronic As exposure and the risk of cognitive impairment in adults. Most of the previous studies in this regard have limitations in exposure metrics selection. Moreover, biochemical basis of the effects of As exposure on adult cognition is almost unknown. Brain-derived neurotrophic factor (BDNF) is deeply implicated in the growth and survival of neuronal cells in the central and peripheral nervous system, and reduced levels of BDNF are associated with neurological disorders including cognitive dysfunction. However, effect of As exposure on BDNF levels in human is unknown. Therefore, this study was designed to examine the association between As exposure and cognitive impairment especially with regard to sBDNF levels in adult individuals in Bangladesh who were exposed to varying concentrations of As through drinking water.

2 MATERIALS AND METHODS

Adult (18–60 years old) individuals from high and low As exposure areas in Bangladesh were recruited as study subjects as we described previously (Karim *et al.*, 2019). Ethical permission was taken from Bangladesh Medical Research Council and Institute of Biological Science, Rajshahi University, Bangladesh. Measurement of the As levels in drinking water, hair and nails of the study subjects were performed by Inductively Coupled Plasma Mass Spectroscopy (ICP-MS). The Bangla version of the MMSE was used as a cognitive assessment tool (Ghose *et al.*, 2015). Based on the MMSE scores; we classified the subjects into normal (24–30), mild (18–23), and severe (0–17) cognitive impairment groups (Tombaugh *et al.*, 1992). Statistical analyses were performed by SPSS software, ver. 21.0.

3 RESULTS AND DISCUSSION

The total number of subjects recruited for this study was 693. Subjects' drinking water, hair and nail As concentrations were used as exposure metrics. Subjects' MMSE scores were 0.6, 0.48 and 0.41

DOI: 10.1201/9781003317395-112

times, and sBDNF levels were 0.7, 0.62 and 0.4 times lower for 1 unit increase in water, hair and nail As concentrations, respectively. (Figures 1A and 1B) suggesting that subjects' As exposure levels were inversely associated with MMSE scores and serum BDNF levels in a dose-dependent manner.

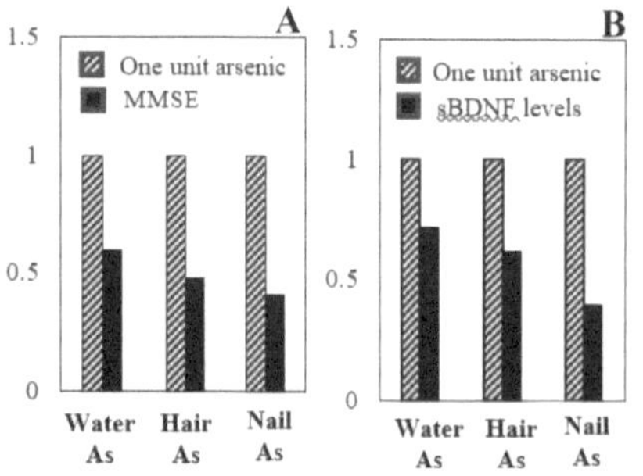

Figure 1. Association of arsenic exposure with MMSE scores (A), and serum BDNF levels (B). As: arsenic.

Next we divided our study subjects' into three cognitive impairment groups such as normal (24–30), mild (18–23) and severe (0–17) based on the MMSE scores, and we found that As exposure levels of the subjects were increased and sBDNF levels were decreased with greater severity of cognitive impairment (Table 1).

Table 1. Arsenic exposure and sBDNF levels in the three cognitive status groups.

Parameters	n	Normal (no cognitive impairment) (24–30)§	n	Mild cognitive impairment (18–23)§	n	Severe cognitive impairment (0–17)§
Water As (0.03–1799 μg/L)	280	96.7 ± 151.2	255	147.9 ± 228.2 a†	158	200.6 ± 220.4 a*, b‡
Hair As (0.02–62.0 μg/g)	280	2.75 ± 4.59	255	3.33 ± 4.5	158	5.69 ± 7.87 a*, b*
Nail As (0.11–47.8 μg/g)	280	5.2 ± 6.48	255	6.24 ± 7.11	158	9.65 ± 8.66 a*, b*
sBDNF (0.93–80.6 ng/ml)	188	25.6 ± 11.4	181	20.9 ± 10.2 a¶	121	14.9 ± 9.62 a*, b*

Data are presented as mean ± SD. n: number of subjects. The As exposure and serum BDNF levels among the three cognitive status groups in the one-way ANOVA followed by a Bonferroni multiple comparison tests between each cognitive status group. §MMSE score; a, bSignificantly different from the normal and mild cognitive impaired groups, respectively. *p < 0.001; †p<0.01; ‡p< 0.05

It has been reported that BDNF level in blood is changed in a similar manner as observed in brain (Klein *et al.*, 2011). Reduced BDNF level or the inhibition of BDNF-signaling pathway in brain (especially hippocampus) is associated with cognitive impairment (Pandey *et al.*, 2017). Thus the inverse association of As exposure with MMSE scores and serum BDNF levels observed in our study suggest that chronic exposure to As decreases cognitive function in adults possibly through the reduction of BDNF.

4 CONCLUSIONS

Results of our study suggest that chronic exposure to As causes cognitive impairment in adults and decreased BDNF level may be an important biochemical basis of As exposure-related cognitive impairment.

REFERENCES

Ghose S.K., Chowdhury A.H., Hasan A.T.M.H., Khan M.Z.R., Karim A.S.M.R., Saha K., Habib M., Hoque M.A., Alam M.B. & Mohammad Q.D. 2015. Comparison of mini-mental state examination (MMSE) and Bangla mini-mental state examination (MMSE-B) among healthy elderly in Bangladesh. *J. Dhaka Med. Coll.* 24(1):30–35.

Karim Y., Siddique A.E., Hossen F., Rahman M., Mondal V., Banna H.U., Hasibuzzaman M.M., Hosen Z., Islam M.S., Sarker M.K., Nikkon F., Saud Z.A., Xin L., Himeno S. & Hossain K. 2019. Dose-dependent relationships between chronic arsenic exposure and cognitive impairment and serum brain-derived neurotrophic factor. *Environ. Int.* 131: 105029.

Klein A.B., Williamson R., Santini M.A., Clemmensen C., Ettrup A., Rios M., Knudsen G.M. & Aznar S. 2011. Blood BDNF concentrations reflect brain-tissue BDNF Levels Across Species. *Int. J. Neuropsychopharmacol.* 14 (3): 347–353.

Pandey R., Rai V., Mishra J, Mandrah K, Roy S, Kumar & Bandyopadhyay S. 2017. Arsenic induces hippocampal neuronal apoptosis and cognitive impairments via an up-regulated BMP2/smad-dependent reduced BDNF/TrkB signaling in rats. *Toxicol. Sci.* 159 (1):137–158.

Paul S.K., Islam M.S., Hasibuzzaman M.M., Hossain F., Anjum A., Saud Z.A., Haque M.M., Sultana P., Haque A., Andric K.B., Rahman A., Karim M.R., Siddique A.E., Karim Y., Rahman M., Miyataka H., Xin L., Himeno S. & Hossain K. (2019). Higher risk of hyperglycemia with greater susceptibility in females in chronic arsenic-exposed individuals in Bangladesh. *Sci. Total Environ.* 668: 1004–1012.

Tombaugh T.N. & McIntyre N.J. 1992. The mini-mental state examination: a comprehensive Review. *J. Am. Geriatr. Soc.* 40(9): 922–935.

3.5 Risk assessment of chronic ingestion

Arsenic in the Environment: Bridging Science to Practice for Sustainable Development – van der Wal, Ahmad, Petrusevski, Weijma, Savic, van der Wens, Beerendonk, Bhattacharya, Bundschuh & Naidu (Eds)
© 2023 the Author(s), ISBN 978-1-032-32928-4

Updated problem formulation and protocol for the inorganic arsenic (iAs) IRIS assessment

J.S. Lee, J.A. Davis, J.S. Gift, I. Druwe & K. Thayer
U.S. Environmental Protection Agency, Center for Public Health and Environmental Assessments, Research Triangle, Park, USA

ABSTRACT: In May 2019, the U.S. Environmental Protection Agency (EPA) released the updated problem formulation and protocol for the iAs assessment. The updated problem formulation includes the refined scope that specifies which health outcomes are prioritized for dose-response analyses and toxicity value derivation. The protocol includes the methods and approaches proposed for use in developing the assessment. Epidemiology studies will be the focus of the assessment, consistent with prior National Research Council (NRC) of the National Academy of Science (NAS) input. EPA considered strength of the epidemiological evidence for hazard by relying on conclusions from assessments conducted by other health agencies or by conducting new systematic reviews of the existing literature. Based on qualitative hazard analyses of the iAs literature, the following health outcomes were identified for potential dose-response analyses based on sufficient human evidence: cancers of the bladder, lung, kidney, liver, and skin; and noncancer effects of iAs on the circulatory system (ischemic heart disease, hypertension, and stroke), reproductive system (including pregnancy and birth outcomes), developmental outcomes (including neurodevelopmental toxicity), endocrine system (including diabetes), immune system, respiratory system, and skin. The NRC released their review of these materials, supporting EPA's refined scope and methods, in October 2019.

1 INTRODUCTION

The U.S. Environmental Protection Agency (EPA) is developing an updated Toxicological Review of Inorganic Arsenic (iAs) that considers the substantial body of new data and refined methods for hazard assessment and dose-response analysis that have emerged since the previous assessment was published in 1995. Given the size and complexity of the evidence base for iAs, input on the scope of this assessment has been sought from the National Academy of Sciences (NAS), EPA program and regional offices, other federal agencies, and public stakeholders to help focus the scope and objectives of the assessment and ensure it is transparently conducted using the best available scientific data and methods, including systematic review methodology.

2 METHODS

A literature search was originally conducted in three online scientific databases (PubMed, Web of Science, and Toxline) in 2012 with an update in 2019. Human health effect studies and PBPK models were identified based on defined PECO (Population, Exposure, Comparator, Outcome) criteria. Outcomes were selected for full systematic review based on recommendations from the NAS and lack of conclusions from other agencies and previous health assessments. Risk of bias (ROB) for each epidemiology study was evaluated using questions across seven evaluation domains (i.e., selection, confounding, performance, attrition, detection, selective reporting bias, and other) adapted from the Office of Health Assessment and Translation (OHAT) (NTP 2013). ROB conclusions were combined with conclusions about sensitivity for each study to arrive at a conclusion about study confidence (high, medium, low, uninformative). Evidence was synthesized by outcome and potential human health hazards were evaluated. Details on methods can be found in the *Updated Problem Formulation and Protocol for the Inorganic Arsenic IRIS Assessment* (US EPA 2019).

3 RESULTS

Literature search and screening results are summarized in Figure 1 and can be found in the Health Effects Research Online (HERO) database.
https://heronet.epa.gov/heronet/index.cfm/project/page/project_id/2211

DOI: 10.1201/9781003317395-113

Table 1. PECO criteria for inorganic arsenic.

PECO element	Evidence
Populations	Human: Any population and life-stage (occupational or general population, including children and other sensitive life stages or populations).
Exposures	Human: Subchronic- or chronic-duration studies of interest provide quantitative estimates of exposure with measurements based on biomonitoring data (e.g., hair, nails, urine, or blood), inhalation (air exposures [μg/m^3]), drinking water exposures (μg/L), cumulative exposures (μg/m^3-yr; μg/L/-yr), and doses expressed as μg/d and μg/kg-d. Studies with episodic or acute exposures will be excluded (i.e., poisonings or other short-term exposures that last up to 30 d). Studies using arsenicals, primarily arsenic trioxide and Fowler's solution will be excluded because chemotherapeutic agents are not within the scope of this review. Studies using arsenide (As^{3-}), an inorganic form of arsenic, also will be excluded. Exposures usually occur via the gas arsine and result in a different, distinctive toxicological profile based on binding to hemoglobin and red blood cell lysis.
Comparators	Human: A comparison or referent population with no detectable exposure or exposure to lower levels of inorganic arsenic (at or below background levels in the US population). Exposure-response quantitative results are presented in sufficient detail (e.g., odds ratios or relative risks with associated confidence intervals, numbers of cases/controls, etc.).
Outcomes	All health outcomes (both cancer and noncancer) as identified by the NRC (2013).
Physiologically based pharmacokinetic (PBPK) models	Studies describing PBPK models for inorganic arsenic will be included. Studies describing quantitative models or data for understanding kinetics in biological media will be tracked as "potentially relevant supplemental material."

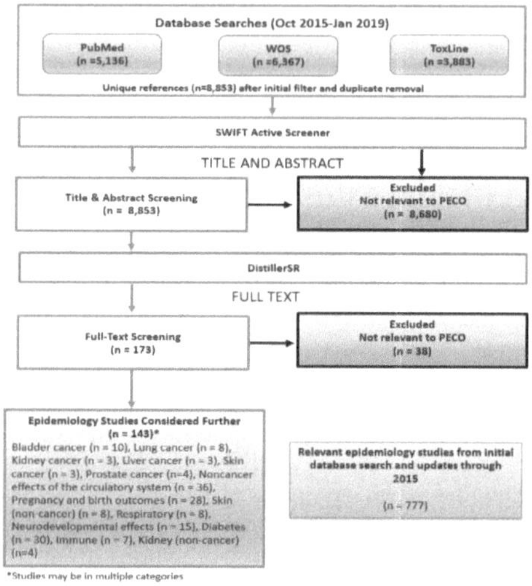

Figure 1. Literature search and screening.

Skin, bladder and lung cancer and skin lesions are accepted hazard outcomes for inorganic arsenic based on conclusions from other health agencies (ATSDR 2007; Health Canada 2006; IARC 2004, IARC 2012; NRC 2013) and considered to have sufficient human evidence for hazard. Other health outcomes (cancers of kidney, liver, prostate, and pancreas; or noncancer effects of the circulatory system, pregnancy and birth outcomes, neurodevelopmental effects, diabetes, immune system, respiratory disease (non-malignant), kidney disease) were determined to have sufficient human evidence to determine hazard with prostate and pancreatic cancer and kidney disease having the weakest evidence. Evidence synthesis was based primarily on studies of *high* and *medium* confidence based on study evaluation.

4 CONCLUSIONS

U.S. EPA has recently updated the scope of the inorganic arsenic IRIS assessment. Cancers of the bladder, lung, kidney, liver, and skin; and non-cancer effects of iAs on the circulatory system (ischemic heart disease, hypertension, and stroke), reproductive system (including pregnancy and birth outcomes), developmental outcomes (including neurodevelopmental toxicity), endocrine system (including diabetes), immune system, respiratory system, and skin will be considered for further dose-response analysis. For non-cancer outcomes, EPA will develop RfD and RfC values. These values will preferably be derived using Bayesian meta-regression methods, but if not possible due to data quality or poor model fits, a traditional BMD approach will be used. For cancer health outcomes, EPA will derive upper-bound U.S. population-specific risk estimates with confidence intervals (that account for identified sources of variability and uncertainty to the extent the data can support such analyses) from epidemiological data over a broad range of iAs intake doses (μg/kg-day) above U.S. background levels. The upper-bound linear relationships will be analogous to oral slope factor (OSF) and inhalation unit risk (IUR) estimates that EPA has historically provided for cancer risks. The NRC reviewed these materials and supports EPA's refined scope and methods (NAS 2019).

ACKNOWLEDGEMENTS

We thank Ila Cote, Ellen Kirrane, Tom Luben, Ryan Jones, Andrew Kraft, Dave Thomas, Hisham El-Masri, and others at the U.S. EPA. We thank Ali Goldstone, Katie Duke, Audrey Turley, Robyn Blain, Bill Mendez, Sorina Eftim, Cara Henning, and others at ICF International.

REFERENCES

ATSDR (Agency for Toxic Substances and Disease Registry). 2007. *Toxicological Profile for Arsenic (Update)*. Atlanta, GA: U.S. Department of Health and Human Services.

Health Canada. 2006. *Guidelines for Canadian Drinking Water Quality: Guideline Technical Document-Arsenic*. Ottawa, Ontario: Water Quality and Health Bureau, Health Environments and Consumer Safety Branch, Health Canada.

IARC (International Agency for Research on Cancer). 2004. *Some Drinking Water Disinfectants and Contaminants, Including Arsenic*. Lyon, France: World Health Organization.

IARC (International Agency for Research on Cancer). 2012. *A Review of Human Carcinogens. Part C: Arsenic, Metals, Fibres, and Dusts*. Lyon, France: World Health Organization.

NAS (National Academy of Sciences). 2019. *Review of EPA's Updated Problem Formulation and Protocol for the Inorganic Arsenic IRIS Assessment*. Washington, D.C.: the National Academies Press.

NRC (National Research Council). 2013. *Critical Aspects of EPA's IRIS Assessment of Inorganic Arsenic: Interim Report*. Washington, D.C.: The National Academies Press.

NTP (National Toxicology Program). 2013. *Draft OHAT Approach for Systematic Review and Evidence Integration for Literature-Based Health Assessments – February 2013*. National Institute of Environmental Health Sciences, National Institutes of Health. http://ntp.niehs.nih.gov/ntp/ohat/evaluationprocess/draftohatapproach_february2013.pdf.

US EPA (U.S. Environmental Protection Agency) 2019. *Updated Problem Formulation and Protocol for the Inorganic Arsenic IRIS Assessment*. Office of Research and Development, Washington, DC.

Arsenic in the Environment: Bridging Science to Practice for Sustainable Development –
van der Wal, Ahmad, Petrusevski, Weijma, Savic, van der Wens, Beerendonk, Bhattacharya, Bundschuh & Naidu (Eds)
© 2023 the Author(s), ISBN 978-1-032-32928-4

Health risk of inorganic arsenic from rice-based diets

S. Islam[1,2], M.M. Rahman[1] & R. Naidu[1]
[1]*Global Centre for Environmental Remediation (GCER), Faculty of Science and Information Technology, The University of Newcastle, University Drive, Callaghan NSW, Australia and Cooperative Research Centre for Contamination Assessment and Remediation of the Environment (CRC CARE), Australia*
[2]*Department of Soil Science, Bangladesh Agricultural University, Mymensingh, Bangladesh*

ABSTRACT: Elevated concentrations of arsenic (As) have been reported in rice-based diets from many countries and the level of concentration is above the international guideline. Total and inorganic As in rice-based diets collected from Australian supermarkets were studied to investigate the contamination levels and how consumption of these products poses potential health risks to young children and adults. About 52.5% of the samples exceeded the European Union (EU) recommended value (100 μg/kg) of As for young children and 22% samples had higher than maximum levels of 200 μg/kg recommended for adults. These also contributes a significant amount of other heavy metals as well. Therefore, the risk cannot be excluded from these rice-based products and the manufacturer's recommended serving could deliver a significant level of inorganic As into different age group of baby, children and or even adults.

1 INTRODUCTION

A number of studies have reported that the contamination of arsenic (As) in rice and their exposure could be considerable a significant risk factor (Islam *et al.*, 2016). The young children are the most vulnerable consumer groups to As toxicity and they have a higher As body burden than adults. The wholegrain and polished rice, and bran are used in a variety of foods such as crisped rice, puffed rice, rice crackers, cereal bars, rice flour, rice noodles, rice malt, etc. and these rice-based products are very popular worldwide. There have been few studies on health risk of As to children from their rice-based food products in Europe and U.S.A. (Signes-Pastor *et al.*, 2016; Sun *et al.*, 2009) and little is known about the health risk to Australian population from the consumption of rice-based food products available in the local markets. A study on total and inorganic As in Australian rice showed that As content substantially higher, notably the brown rice and it can pose a serious health risk to adult immigrants (Rahman *et al.*, 2014). Thus, it is crucial to investigate the total and inorganic As present in rice-based product available in Australia.

2 MATERIALS AND METHODS

2.1 *Sample collection*

Samples of rice crackers (n = 10), rice cakes (n = 11), puffed rice (n = 6), baby rice (n = 4), ready to eat rice (n = 22) and rice-based other snacks (n = 6) of 25 different popular commercial brands were purchased from supermarkets in South Australia.

2.2 *Sample preparation, digestion and analysis*

All rice-based products except ready-to-eat rice were ground with a glass mortar before digestion for analysis. For total As analysis, a microwave digestion system were used with 2 ml MQ, 3 mL trace analytical grade HNO_3 and 2 mL H_2O_2. The ready to eat rice was directly weighed and followed the same digestion procedure as described above. An Agilent 7500c (Agilent Technologies, Tokyo, Japan) inductively coupled plasma mass spectrometry (ICP-MS) was used for the determination of As in digested samples. For quality control a standard reference material (SRM 1568b rice flour) was used to verify the analytical results. The nature of As speciation analysis digestion of the samples followed by the extraction procedure of Signes-Pastor *et al.* (2016).

2.3 *Calculation of health risk*

In this study, the total diet study approach was used to determine the dietary intake of inorganic As from rice-based products.

Inorganic As intake per serving (μg) from rice-based products:

$$\text{EIPS} = (\text{CiAs} \times \text{RS rice} - \text{based products}) \quad (1)$$

Where, EIPS is estimated intake per serving (μg); CiAs is inorganic As concentration in rice-based products (μg/kg); RS: Average recommended serving (g/person/day).

3 RESULTS AND DISCUSSION

Total average As content in rice-based products ranged from 30 to 273 μg/kg. The variation

DOI: 10.1201/9781003317395-114

represents the source and type of rice. For all rice products investigated in this study, the predominant As species were inorganic (34.7 to 100%), with the remainder being DMA and it varies between different products, probably due to the variability of total As concentration and speciation inherently found in rice and also processing of raw rice (Figure 1). This study also represents the higher As in brown rice-based products compared to white or basmati rice. The average total As in rice crackers from Australian rice was 239 µg/kg, but the product from Thailand contents 171.2 µg/kg total As. Speciation of As in rice crackers samples (n = 10) showed that inorganic As species comprised the major fraction ranging from 41.5% to 82.6% compared to organic species.

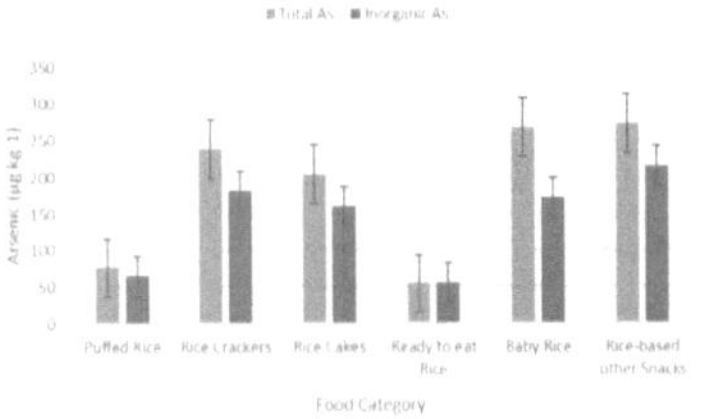

Figure 1. Total and inorganic As contents varies in different rice-based diets.

The average total As contents found in the puffed rice imported from New Zealand (213.3 µg/kg) was 3 times higher than that in Australian puffed rice (76.6 µg/kg). Speciation showed inorganic As the major species (31.9% – 73%). Though the New Zealand rice contains total As 3 times higher than Australian rice but the average inorganic As contents in Australian origin was lower (45.1 µg/kg) compared to New Zealand origin (54.1 µg/kg). Like other rice-based products, rice cake is used for light dinner or lunch, and in Australia they are made of 100% Australian-grown whole grain brown rice and the concentrations of total As ranged from 58.1 to 384.4 µg/kg (Table 1).

Total As concentration in baby rice samples ranged from 58.6 to 356.9 µg/kg. The inorganic As concentration from four different commercial brands

Table 1. Total and inorganic As concentration in different rice-based diets.

Food Items	tAs (µg/kg)	iAs (µg/kg)	Weekly intake (µg/kg bw)
Puffed rice	76.6	64.8	0.0871
Rice crackers	239.0	181.7	0.3157
Rice cakes	204.3	160.7	0.2314
Ready-to-eat rice	54.0	45.3	0.0929
Baby rice	268.2	172.6	0.1900
Rice-based other snacks	273.0	215.3	0.2629

ranged from 45.5 to 108.1 µg/kg that represent 22% to 100% of the speciated As. Total As concentration in ready to eat rice samples of five different commercial brands ranged from 13.8 to 90.1 µg/kg.

Intake of inorganic As and health risk for young children: The results showed that 53% rice-based products exceeded the EU recommended value (100 µg/kg) of As for young children and 22% samples had higher than maximum levels of 200 µg/kg recommended for adults. Out of six categories rice-based products, except ready-to-eat rice, all other products exceeded the EU recommended value for young children. Even manufactured recommended serving deliver significant amount (0.56 to 6.87 µg) of inorganic As. These results within the range of $BMDL_{01}$ values indicated by the European Food Safety Authority (EFSA), which represent the risk cannot be avoided for young children and adults considering the levels of total and inorganic As in rice-based products. On average, young children under 5 years of age eat three times more foods than adults on a body weight basis. Therefore, this sub-population is much vulnerable than the adults.

4 CONCLUSIONS

The daily consumption of rice-based diets therefore constitutes an important exposure pathway for inorganic As to young children and even adults. It is crucial to minimize the inorganic As exposure from rice-based diets. For this, rice varieties those accumulate less inorganic As should be used for the preparation of the rice-based diet to abate the inorganic As exposure.

ACKNOWLEDGEMENTS

The author acknowledge the financial support from CRC CARE and The University of Newcastle for the laboratory support and post-graduate scholarship.

REFERENCES

Islam S., Rahman M.M., Islam M. & Naidu R. 2016. Arsenic accumulation in rice: consequences of rice genotypes and management practices to reduce human health risk. *Environ. Int.* 96: 139–155.

Rahman M.A., Rahman M.M., Reichman S.M., Lim R.P. & Naidu R. (2014). Arsenic speciation in Australian-grown and imported rice on sale in Australia: implications for human health risk. *J. Agric. Food Chem.* 62(25): 6016–6024.

Signes-Pastor A. J., Carey M. & Meharg A.A. 2016. Inorganic arsenic in rice-based products for infants and young children. *Food Chem.* 191: 128–134.

Sun G.-X., Williams P.N., Zhu Y.-G., Deacon C., Carey A.-M., Raab A., Feldmann J. & Meharg A.A. 2009. Survey of arsenic and its speciation in rice products such as breakfast cereals, rice crackers and Japanese rice condiments. *Environ. Int.* 35(3): 473–475.

 ISBN 978-1-032-32928-4

Health exposure due to arsenic toxicity: A risk assessment study in West Bengal, India

M. Joardar, A. Das, N. Roy Chowdhury, A. De & T. Roychowdhury
School of Environmental Studies, Jadavpur University, Kolkata, India

ABSTRACT: Risk of suffering through consumption of arsenic (As) contaminated drinking water and rice grain is an area of global concern for the populations residing in arsenic-affected areas of rural Bengal. In this study, the population is presently not consuming arsenic contaminated drinking water but they are exposed to chronic arsenic toxicity. The cancer risk for the affected population highlights that the male individuals are more prone to severe health risk compared to the female individuals. The non-cancer risk assessment (through inorganic As) for both male and female individuals (HQ value) is very much higher than 1. The population not only needs to avoid the vital use of contaminated groundwater for the household (drinking, cooking) purposes but also during agricultural season to prevent the health exposure due to arsenic.

1 INTRODUCTION

The risk of suffering through natural groundwater arsenic contamination is globally well-known as a crucial water quality problem (Chakraborti *et al.*, 2009). Arsenic (As) in drinking water and rice grain is the main source of inorganic form, commonly known as potent carcinogen and toxicant (IARC 2012; NRC 2001). Rice is considered the staple crop which is the main reason for arsenical health problems in major arsenic endemic regions of the world. Exposure to chronic and acute toxicity due to the arsenic contaminated water is a major threat to human health. It has been reported that the arsenic concentration in the body tissues increases with the rate of exposure to arsenic contamination (Roychowdhury 2010).

The present study aims to survey i) scenario of drinking water and rice grain of the studied population, ii) health exposure through As contamination, iii) probable risk analysis due to consumption of inorganic As in drinking water and rice grain.

2 MATERIALS AND METHODS

2.1 *Study area*

The studied population belongs from Gobordanga Municipality (Gaighata block) situated in North 24 Parganas district in West Bengal. Gaighata is a distinct As affected area in West Bengal (Roychowdhury 2010).

2.2 *Sample collection, preparation & preservation*

A detailed sampling was performed on the families of the Gobordanga Municipality. Domestic tubewell water samples were collected in sterile polyethylene containers by adding 2 drops of concentrated HNO_3 acid and spot urine samples were collected without any chemical for preservation, stored at 4°C prior to analysis. The water and urine samples were filtered using Whatman 42 to remove the colloidal particles followed by As analysis. The rice grain samples were collected in zip packets and then subjected to digestion for As estimation. Hair and nail samples were collected from the scalp and the tip of the finger from the exposed population respectively. Both the hair and nail samples were washed thoroughly using distilled water and soap in a sonicator to remove the externally adsorbed As from its surface. The washed hair and nail samples were dried by adding acetone and placing it in the hot air oven for 6–8 h at 50°C. The dried hair and nail samples were preserved in the zip lock and further digested for As analysis. Details of sample collection, preservation, storage and preparation have been described earlier (Roychowdhury 2010).

2.3 *Analysis*

Analysis of total As of all the digested solid and liquid samples were performed by using Flow

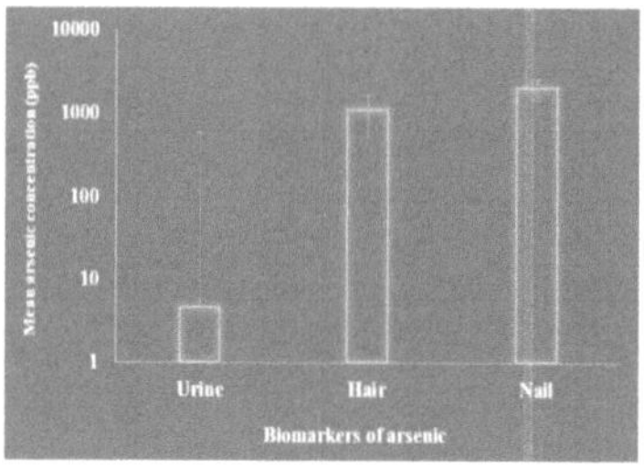

Figure 1. Mean arsenic concentration in different biomarkers.

DOI: 10.1201/9781003317395-115

Injection-Hydride Generation-Atomic Absorption Spectrophotometric (FI-HG-AAS) method. An Atomic Absorption Spectrophotometry [Varian model AA140 (USA)] was used for estimation of total As.

3 RESULTS AND DISCUSSION

3.1 *Scenario of drinking water and rice grain of the studied population*

The domestic tube-well water samples collected from the families residing in Gobordanga (Gaighata block) showed that the population usually consumes higher As than the permissible limit. The As concentration in water samples ranged 0.02–0.163 mg/L with an average concentration of 0.07 ± 0.05 mg/L. Apart from drinking water, the affected population consumes As through their daily dietary foodstuff which is being cultivated using As-contaminated groundwater. The rice grain samples collected from each family (n = 16), showed a considerable amount of As concentration, which is higher than the WHO recommended permissible limit in rice i.e. 0.1 mg/kg. The rice grain samples showed As with an average concentration of 0.249 ± 0.146 mg/kg ranging 0.07–0.680 mg/kg.

3.2 *Health exposure through arsenic contamination*

Biomarkers of As (urine, hair, and nail) were analyzed for health exposure study (Figure 1). Urine and body tissue samples determine the acute and chronic effect of As consumption in the internal body system. Urine samples collected (n = 52) showed an average concentration of 0.005 ± 0.003 mg/L, whereas the permissible limit of As concentration in urine is 0.003–0.026 mg/L. Due to high content of keratin protein in the tissues of hair and nail, the rate of deposition is higher than in other body parts of human. The exposed hair samples (n = 55) showed As concentration in the range of 0.08 to 4.06 μg/g with an average concentration of 1.09 ± 1.02 μg/g, where the permissible limit of As in hair is 0.08–0.25 μg/g. Similarly, nail samples (n = 55) showed an average As concentration of 2.03 ± 1.73 μg/g ranging from 0.13 to 6.72 μg/g, where the permissible limit of As in nail is 0.43–1.08 μg/g.

3.3 *Probable risk analysis due to consumption of inorganic As in drinking water and rice grain*

The health risk analysis was evaluated using the following as mentioned in (Shakoor *et al.*, 2015)

$$ADD = (C * IR * ED * EF)/(BW * AT) \quad (1)$$

$$CR = (ADD * CSF) \quad (2)$$

CSF (cancer slope factor) = 1.5 per mg/kg per day for direct ingestion of As.

Cancer risk through inorganic arsenic (drinking water and rice grain) for male and female has been shown in Figure 2.

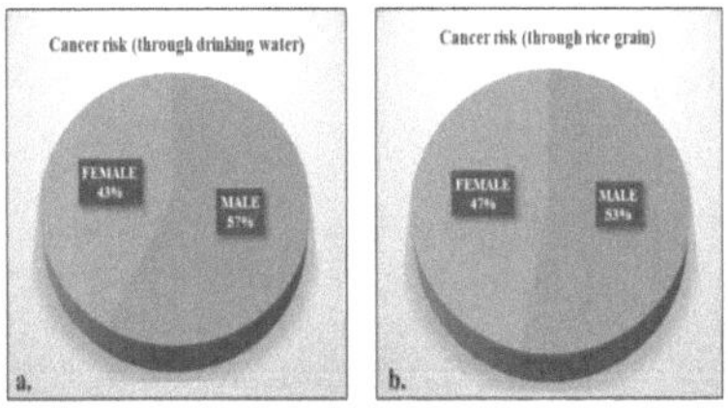

Figure 2. Cancer risk analysis in male and female individuals.

4 CONCLUSIONS AND RECOMMENDATIONS

Our study revealed that the affected population the male individuals are prone to higher cancer risk compared to the female individuals. So, the population are prone to several health hazards not only through contaminated drinking water but also through contaminated foodstuffs found in the affected areas. So, the study recommends the vital use of arsenic-free water for drinking purpose among the exposed population. To prevent the severity of arsenic exposure awareness programs, the use of arsenic-free water (treated surface water) needs to be implemented among the populations of the arsenic exposed areas.

ACKNOWLEDGEMENTS

We acknowledge the Department of Science & Technology, Govt. of West Bengal for financial support towards and the villagers for their cooperation during my work.

REFERENCES

Chakraborti D., Das B., Rahman M.M. *et al.* 2009. Status of groundwater arsenic contamination in the state of West Bengal, India: a 20 year studyreport. *Mol. Nutr. Food Res.* 53(5):542–551.

IARC 2012. *A Review of Human Carcinogens: Arsenic, Metals, Fibres, and Dusts.* World Health Organization, Lyon, France.

NRC (2001) Arsenic in Drinking Water: 2001 Update.

Roychowdhury T. 2010. Groundwater arsenic contamination in one of the 107 arsenic-affected blocks in West Bengal, India: status, distribution, health effects and factors responsible for arsenic poisoning. *Int. J. Hyg. Environ. Health* 213(6), 414–427.

Shakoor M., Niazi N.K. *et al.* 2015. Unraveling health risk and speciation of arsenic from groundwater in rural areas of Punjab, Pakistan. *Int. J. Environ. Res. Public Health* 12(10): 12371–12390.

Arsenic in the Environment: Bridging Science to Practice for Sustainable Development –
van der Wal, Ahmad, Petrusevski, Weijma, Savic, van der Wens, Beerendonk, Bhattacharya, Bundschuh & Naidu (Eds)
© 2023 the Author(s), ISBN 978-1-032-32928-4

Comparative study to evaluate the changes in arsenic dietary intake caused by washing and cooking rice with groundwater from the Bengal Delta, India

A. Shrivastava[1,4], M. Jaafar[2,3], S. Bose[1], M. Felipe-Sotelo[2] & N.I. Ward[2]
[1]*Department of Earth Sciences, Indian Institute of Science Education and Research Kolkata, India*
[2]*ICP-MS Facility, Chemistry Department, University of Surrey, Guildford, Surrey, UK*
[3]*Universiti Malaysia Terengganu, Kuala Terengganu, Terengganu, Malaysia*
[4]*Amity University Uttar Pradesh, Noida, India*

ABSTRACT: Contamination of groundwater with arsenic(As) from natural sources is endemic in the West Bengal region of India. Despite increased awareness regarding the consumption of safe water for drinking, the use of groundwater for agricultural purposes, cooking and other domestic tasks still prevails. This study evaluates the effect on the contents of the three metals induced by washing and cooking the rice with groundwater, which provides a more realistic estimation of the dietary intake of As, than the concentration in the raw foodstuff alone. There is a marked increase on the As contents in the cooked rice (up to 232%), however, the accumulation is highly dependent on the type of rice, with processed commercial samples (many of them parboiled) presenting a lower capacity to retain As during washing and cooking that the local varieties available among the villagers.

1 INTRODUCTION

Arsenic (As) contamination of groundwater in different parts of the world is an outcome of natural and/or anthropogenic sources, leading to adverse effects on human health and ecosystem. Millions of people from different countries are heavily dependent on groundwater containing elevated level of As for drinking purposes (Shankar *et al.*, 2014). In West Bengal, India and Bangladesh, natural release of As from aquifer rocks has been reported to contaminate the groundwater (Fendorf *et al.*, 2010) and prolonged use of As-contaminated groundwater for irrigating crops has led to an increase of As content in agricultural soils of the Bengal Delta (Shrivastava *et al.*, 2014).

Most studies have assessed the risk of As exposure via rice consumption considering only the total and inorganic As contents in the raw (uncooked) rice. Although, recent investigations have reported that the concentration and speciation of As in the cooked rice can be different from that in the corresponding raw rice; this mainly depends on the As concentration in cooking water and the processes of rice cooking (Halder *et al.*, 2014). Human intake of As from consumption of rice can be substantial, particularly for populations on a subsistence rice diet (Bhattacharya *et al.*, 2009). It is therefore crucial to undertake the comparative study of the content of As in different rice varieties and the effect of washing and cooking on the As content of those rice grains. Thus, this study focussed to investigate the effect of washing and cooking on the As content of a range of rice varieties.

2 MATERIAL AND METHODS

2.1 *Study area*

The field study site was the Sarapur village (latitude 23°01′15.31″ N and longitude 88°38′36.86″ E) in Chakdaha, West Bengal, India (Figure 1). Agriculture is the primary occupation of the area and farmers in the village use the As contaminated groundwater for various purposes especially for irrigation of the traditional continuously flooded rice paddies.

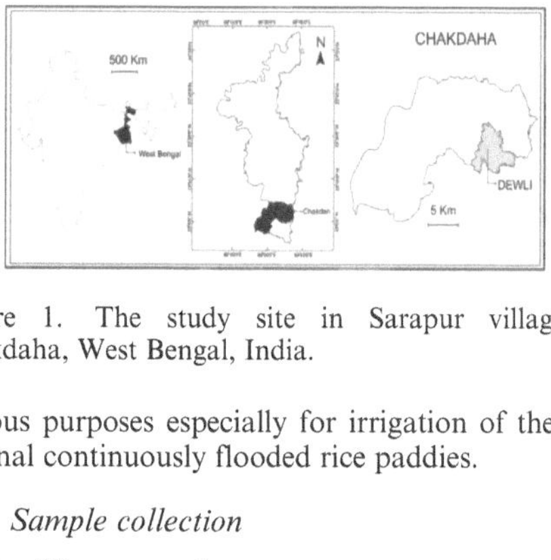

Figure 1. The study site in Sarapur village in Chakdaha, West Bengal, India.

2.2 *Sample collection*

2.2.1 *Water samples*

A total of 16 samples were collected from different sources: contaminated groundwater, ponds, and tap water. For the determination of total trace elements, about 15 mL of filtered water samples were collected at each sampling location into pre-acidified) polypropylene tubes.

DOI: 10.1201/9781003317395-116

2.2.2 *Rice samples*

Local variety of rice from contaminated and control fields were collected for the study. Additionally, rice grains from other sources; namely, local market (2 varieties of short and bold grained rice) and supermarkets (2 commercial brands of Basmati long and slender grained rice) were also taken. The samples were taken randomly from the plots and stores, stored in sterile sealed polypropylene zipped bags and transferred to the laboratory for further treatments.

2.3 *Quantification of arsenic in water and rice grains samples*

All samples were analysed using ICP-MS (Agilent 7700x) at the University of Surrey, UK. Optimisation and tuning of the ICP-MS were performed daily and the instrument was operated in collision-cell mode.

2.4 *Assessment of arsenic exposure from rice consumption*

The oral intake of As from rice was estimated following the formula as shown in Equation 1:

$$\begin{aligned} DIA(\mu g/day) = & [CGrain(\mu g/kg) \\ & \times Dailyriceconsumption(kg/day)] \\ & + [CWater(\mu g/L) \\ & \times Dailywaterconsumption(L/day) \end{aligned} \quad (1)$$

where, DIA is the daily intake of As (μg/day), CGrain corresponds to the mean As concentrations (μg/kg d. w.) accumulated in grains, and CWater is the mean As concentrations (μg/L) in drinking water.

3 RESULTS AND DISCUSSION

3.1 *Analysis of water samples*

The data show significant contamination of As (>100 ppb), exceeding the national and international limits for drinking water and irrigation, while both pond and tap waters were below these recommendations (10 ppb). It was observed that in the case of oxic water (pond and tap), As is mainly present as arsenate, i-AsV over the range 50.30% and 73.91% of the total As forms in pond and tap water respectively. Whereas, in the groundwater (anoxic environment), As is mainly found as i-AsIII (69.57% – 94.37%) of the total As.

3.2 *Analysis of raw unwashed, washed and cooked rice samples*

Results showed that the concentration of total As in raw unwashed rice ranged from 0.03 to 0.24 mg/kg d. w., with the samples of locally grown unpolished rice showing higher concentrations (0.14–0.24 mg/kg d. w.) than polished rice purchased in the local market (0.08 – 0.10 mg/kg d.w.) and supermarkets (0.05 – 0.07 mg/kg d.w.).

Moreover, overall washing of all rice varieties with tap water (2.05 μg/L As) led to a decrease in the As content from 7 to 32%. Furthermore the effect of cooking on the concentration of As in rice showed two clearly different trends; general decrease of As levels was found in grains cooked with As-safe water (2.05 < 10 As μg/L), while levels increased for rice boiled with groundwater containing ca. 94 μg/L As from 90% to 230% for most varieties of rice.

3.3 *Estimated human exposure to arsenic through consumption of water and rice*

The daily intake (DI) obtained from consuming groundwater and rice cooked in water from the same source ranged from 444 to 950 μg As/day (equivalent to 377 – 808 μg i-As/day) for adults and 262 – 600 μg As/day (223 – 510 μg i-As/day) for children, which are well above the recommended tolerable daily intake (TDI) (124 μg/day for adults and 43 μg/day for children).

4 CONCLUSIONS

The study showed that the effects of washing the rice with groundwater on the accumulation of As was much lower than the increase of concentration observed during cooking, although it was not negligible. Moreover, the effect of the food preparation varied widely depending on the variety of rice consumed, with much higher rates of accumulations in the locally grown varieties of rice than for the commercial types, most of which had been parboiled. The estimations of the daily intakes of As in the area reived from the consumption of water and rice, showed that cooking foodstuff with contaminated water can contribute very significantly to the total consumption of As (from 21% to 67% of the As total intake coming from rice cooked in groundwater), which highlights the importance of assessing the method of preparing and cooking the food in order to be able to provide realist estimation of elemental dietary intakes in contaminated areas.

ACKNOWLEDGEMENT

We acknowledge Department of Science & Technology (GoI), British Council (UK) and Ministry of Education (Malaysia) for funding this work.

REFERENCES

Bhattacharya P., Samal A.C., Majumdar J. & Santra S.C. 2009. Transfer of arsenic from groundwater and paddy soil to rice plant (Oryza sativa L.): a micro level study in West Bengal, India, *World J. Agric. Sci.* 5: 425–431.

Fendorf S., Michael H.A. & van Geen A. 2010. Spatial and temporal variations of groundwater arsenic in south and Southeast Asia. *Science* 328: 1123.

Halder D., Biswas A., Šlejkovec Z., Chatterjee D., Nriagu J., Jacks G. & Bhattacharya P. 2014. Arsenic species in raw and cooked rice: implications for human health in rural Bengal. *Sci. Total Environ.* 497: 200–208.

Shankar S., Shanker U. & Shikha. 2014. Arsenic contamination of groundwater: a review of sources, prevalence, health risks, and strategies for mitigation. *ScientificWorldJournal* 2014:304524.

Shrivastava A., Barla A., Yadav H. & Bose S. 2014. Arsenic contamination in shallowground water and agricultural soil of Nadia block, West Bengal, india. *Front. Environ. Sci.* 2: 50.

Health risk assessment associated with arsenic contamination in Gomti River Basin: Impacts of pre- and post-COVID-19 lockdown

R. Khan[1], A. Saxena[1], S. Shukla[1] & P. Bhattacharya[2]
[1]*Faculty of Civil Engineering, Institute of Technology, Shri Ramswaroop Memorial University, Lucknow, India*
[2]*KTH-International Groundwater Arsenic Research Group, Department of Sustainable Development, Environmental Science and Engineering, KTH Royal Institute of Technology, Stockholm, Sweden*

ABSTRACT: The primary aim of the current study was to determine the arsenic (As) concentration across an approximate streath of 61 km of River Gomti at Lucknow district. We also attempted to evaluate assess the impact of lockdown on the As concentration as compared the pre-COVID-19 lockdown. The results suggested ~28.5% reduction in As concentration after the lockdown, with only two downstream sites having As >10 μg/L. The human health risks associated with consumption of As contaminated drinking water in children and adults were also evaluated. It was found that ~40% sites for children were safe for drinking purpose for childrenduring both the sampling periods. Whereas, in case of adults, 40% sites in 2019, and 50% sites in 2020 were found to be safe. Children were more affected by the presence of As in drinking water due to their lower body weight and developing immunity.

1 INTRODUCTION

Rapid pace of urbanization and industrialization has led to depleted water quality and elevated levels of heavy metals (HMs). The non degrable, noxious nature of HMs with their associated health risks have emerged as a huge problem. River Gomti is a groundwater fed river and also a primary source of drinking and domestic purposes in the city of Lucknowin northern India. The novel coronavirus (COVID-19) affected nearly 145 million people across the world and was declared a global pandemic on 11th March 2021. A nationwide lockdown was imposed in India to prevent the outbreak of COVID-19 amongst of a population of ~1.35 billion. The lockdown was considered as a 'ventilator' for reinstatement of environment with various reports on improved water quality of rivers. Hence, this study was conducted to evaluate the health impacts of As contamination in river water.

2 MATERIALS AND METHODS

2.1 *Study area*

The Gomti River Basin (GRB) experiences a of humid subtropical climate with dry summers (March–May), heavy monsoons (June–September) and winters (December–February). Mostly, rainfalls occur in the monsoon season with a total precipitation of approximately 1025 mm. The GRB reportedly stores 75% of the precipitation as groundwater and evaporated into the atmosphere. River Gomti is an major alluvial tributary of River Ganga and traversing anapproximate stretch of ~980 kms from Madho-Tanda to the Himalayan foothills (Khan *et al.*, 2021).

2.2 *Sample collection and analysis*

In the present study, thirty samples (three samples at each sampling station) were collected across a total stretch of ~61 km from River Gomti in the month of June 2020, after the COVID-19 unlocking. The sampling stations have been illustrated in Figure 1. The collection, storage, and analysis of water samples was done in strict accordance with APHA. The levels of of As in all the water samples after digestion were analysed through Atomic Absorption Spectrophotometer (ECI: Model 1441 and 4341).

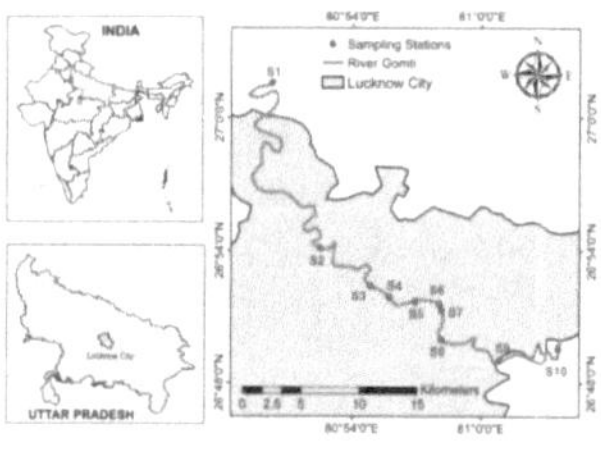

Figure 1. Sampling stations along the stretch of River Gomti in Lucknow city.

2.3 *Arsenic and human health risk assessment*

Arsenic is one of the most critical element which affect the human health. Human health risk assessment is considered to be an efficient tool which assess the probable risk of exposure towards these contaminants present in water. The potential non-carcinogenic health risks through oral intake of As is done through estimation of Hazard Quotients (HQ) in this study. In the first step, the chronic daily intake (CDI) of As through drinking water was computed using Equation (1), proposed by Environmental

DOI: 10.1201/9781003317395-117

Protection Agency (Shukla & Saxena 2020; USEPA 2005).

$$CDI = \frac{C_W \times DI \times EF \times EP}{BW \times AT} \quad (1)$$

The values of DI, EP, and BW vary according to the climatic conditions and residents' behaviour in the study area. In this study the values of these parameters were taken considering the previous studies (Rishi *et al.*, 2020) for the arid-climatic conditions which prevails in northern India. Further, the Hazard quotient (HQ) was computed through Equation (2):

$$HQ = \frac{CDD}{RfD} \quad (2)$$

where, the oral toxicity reference dose values (RfD) was be taken as 0.3 μg/kg/day for As as per the previous studies (Etikala *et al.*, 2019). When HQ<1, water is assumed to be safe for drinking purpose for the residents (Adimalla *et al.*, 2020).

3 RESULTS AND DISCUSSION

3.1 *Variation of arsenic in River Gomti*

The maximum concentration of As was reported at site S10, which is the downstream site at River Gomti. It can be observed that due to the imposed lockdown the mean As concentration showed a ~28.5% reduction in River Gomti. Only two downstream sites (S9 and S10) had As concentration >10 μg/L.

3.2 *Human Health Risk Assessment (HHRA)*

The variation of HQ values in adults and children during 2019 and 2020 are presented in Figure 2. It can be observed that the COVID-19 lockdown resulted in a reduction in the health hazards in 2020 compared to 2019. Before the COVID-19 lockdown (October 2019), the HQ values for children varied between 0 to 5.02, whereas in adults it was found to be varying between 0 to 3.72. In June 2020, 40% of the sites for children and 50% sites for adults had HQ<1, suggesting a slight improvement in conditions due to the lockdown. The study also suggested that adults have

Figure 2. Site-wise variation of HQ in adults and children during 2019 and 2020.

lower health risks as compared to children. This is attributed towards lower body weight and developing immunity in children.

4 CONCLUSIONS AND RECOMMENDATIONS

The mean As concentration suggested a slight improvement in conditions due to the nationwide lockdown. Closure of these anthropogenic factors can be attributed as the reason behind lower As levels during June 2020. Children were found to be more susceptible towards the exposure of As through drinking water as compred to adults due to their lower body weight and immunity. It can be concluded that monitoring of the status of As and other heavy metal contamination is very important at both regional and national level for reinstatement to its natural state.

ACKNOWLEDGEMENTS

Authors are thankful to Dr. A. K. Singh, Vice-Chancellor, Shri Ramswaroop Memorial University for providing lab facilities to conduct this study.

REFERENCES

Adimalla N., Qian H. & Li P. 2020. Entropy water quality index and probabilistic health risk assessment from geochemistry of groundwaters in hard rock terrain of Nanganur county, South India. *Chem. Erde.* 80(4): 125544.

APHA. 2012. *Standard Methods for the Examination of Water and Wastewater* (2nd ed.). American Public Health Association.

Etikala B., Golla V., Adimalla N. & Marapatla S. 2019. Factors controlling groundwater chemistry of Renigunta area, Chittoor district, Andhra Pradesh, south india: a multivariate statistical approach. *HydroResearch* 1: 57–62.

Khan R., Saxena A., Shukla S., Sekar S. & Goel P. 2021. Effect of COVID-19 lockdown on the water quality index of river Gomti, India, with Potential Hazard of Faecal-Oral Transmission. *Environ. Sci. Pollut. Res.* 28: 33021–33029.

Rishi M. S., Kaur L. & Sharma S. 2020. Groundwater quality appraisal for non-carcinogenic human health risks and irrigation purposes in a part of Yamuna sub-basin, India. *Hum. Ecol. Risk Assess.* 26(10): 2716–2736.

Shukla S., & Saxena A. 2020. Appraisal of groundwater quality with human health risk assessment in parts of indo-gangetic Alluvial plain, North India. *Arch. Environ. Contam. Toxicol.* 80: 55–73.

USEPA. 2005. *Guidelines for Carcinogen Risk Assessment.* United States Environmental Protection Agency. URL: https://www3.epa.gov/airtoxics/cancer_guidelines_final_3-25-05.pdf

Arsenic in the Environment: Bridging Science to Practice for Sustainable Development –
van der Wal, Ahmad, Petrusevski, Weijma, Savic, van der Wens, Beerendonk, Bhattacharya, Bundschuh & Naidu (Eds)
© 2023 the Author(s), ISBN 978-1-032-32928-4

Arsenic in cooked and uncooked rice: A field scale study in rural West Bengal, India

U. Mandal, D. Chatterjee, M. Mazumder & P. Ghosh
Department of Chemistry, University of Kalyani, Kalyani, Nadia, West Bengal, India

ABSTRACT: Cooked rice consumption is now been established as a well-recognized exposure route of arsenic (As) for many subpopulations. The aim is to investigate the various factors that might exert control on the final As level in cooked rice following the indigenous cooking practice pursued by the rural villagers of West Bengal. The study highlights the As sourcing among the rural population of West Bengal particularly by rice consumption. Results also show that for the cooking method employed, genotype and grain size, background As concentration in raw rice, cooking water As concentration, top soil and irrigating water As levels and local cultivation practice are important predisposing factors that direct the accumulation of As in cooked form. Finally, the risk analysis indicates that genotype and grain size are the most vulnerable to the potential health threat of dietary As exposurey.

1 INTRODUCTION

Globally, several articles has been published which shows that millions of people worldwide are exposed to high levels of arsenic (As) from drinking As contaminated water. Currently, rice is the staple food for 2/3 of the world population, particularly in developing Asian countries, where an average of 75% of the daily calorie intake per capita is provided from rice. Cooked rice has the potential to accumulate more As than other food crops. As a result, rice intake in rural Bengal has provided a facile route for As exposure. The risk posed from rice depends upon both the amount of rice consumed and the concentration of As in rice grains (notably inorganic), moderated by gut bioavailability. In the present study, rice samples are cooked in individual households in rural Bengal affected by endemic arsenicism, following the traditional cooking method adopted by the villagers. The several factors (genotype, grain size, up take, cooking water, land use pattern) are considered to study their effect on As content in cooked rice and explored so as to evaluate the reduction in As exposure for the population.

2 MATERIALS AND METHODS

2.1 *Study area*

The study area is located in Chakudanga village of Chakdaha block in Nadia district, West Bengal. The agricultural system is highly dependent upon groundwater to meet the water requirement for irrigation of crop.

2.2 *Rice cooking and sample collection*

For this study, rice has been cooked in individual household following the procedure adopted by rural villagers. Generally, rice was washed 3 – 4 times with water and the washed rice soaked in excess water (10–12 times the weight of raw rice) for 15–20 minutes and finally cooked. After cooking, the excess starch water (gruel) has been discarded by tilting the pan against the lid. The entire cooking procedure as described above has been done in rural villagers house.

2.3 *Sample analysis*

Cooking water, raw rice, starch water, and cooked rice are analyzed for As by Hydride Generation Atomic Absorption Spectroscopy (HG-AAS) (Varian AA240).

2.4 *Statistical analysis*

SPSS statistical software, version 17.0 by IBM has been used for data analysis. Independent variables such as genotype, grain size, As concentration in raw rice and cooking water are tested for multiple linear regression analysis with As content in cooked rice. Statistical significance was indicated by values of $p < 0.05$.

3 RESULTS AND DISCUSSION

The door to door survey revealed that Ratna, Satabdi, Pratik, Parijat, Ranjit, Swarna are the most common rice varieties that are used by the

DOI: 10.1201/9781003317395-118

villagers in the study area. The median As concentration of raw rice (uncooked) for the six rice varieties investigated in this study is 228 μg/kg with a range of 105–510 μg/kg. Genotype categorizing has been done for the collected raw rice. The results indicate that SB (Swarna) had the highest grain As concentration (median: 380 μg As/kg; range: 270 – 510 μg As/kg) followed by another SB (Ranjit) (median: 221 μg As/kg; range: 206 – 360 μg As/kg) including SB (Parijat) (median: 205 μg As/kg; range: 140 – 370 μg As/kg).MS (Pratik) (median: 194 μg As/kg; range: 155 – 358 μg As/kg) is low in As levels. Both LS (Ratna, median: 179 μg As/kg; range: 130 – 235 μg As/kg) along with LS (Satabdi, median: 163 μg As/kg; range: 105 – 356 μg As/kg) had the lowest As concentration.

Studies have shown that there can be severe variation in the final As concentration in cooked rice from the initial As concentration in raw rice. It has been found in this study, low land cultivated genotype (short bold) showed the maximum As concentration compared to the upland rice varieties. However, the discrepancies in As concentration with top soil (upto 200 mg/m^2) and irrigating water (200 μg/L) is somewhat perturbed when plotted for the cooked form indicating that the As concentration in cooked rice may vary greatly from the raw rice. Moreover, depending upon cooking water and pattern, genotype are also important. Results indicate that As in cooking water can be a very important determinant for As in cooked rice. Elevated concentration of As in uncooked rice can result in the influx of As in cooked rice, thereby causing an increase in As content in the cooked form. Moreover, genotype and grain size are also influenced the amount of As influx in the cooked rice samples, thereby being a principal factor controlling the cooked rice As concentration. The risk analysis reveals that the variation of TDI-iAs (Total Daily Intake) has been observed for genotype and grain size. The rice consumption pattern and BMI are also contributing for risk factor. Moreover, the leach-out As from rice grain into the gruel and thereby ultimately creating concern for the As exposure from gruel intake. The results confirm the risk which have shown consumption in As content in cooked rice following the use of high consumption rate.

4 CONCLUSIONS AND RECOMMENDATIONS

This study further suggests that, provided As-safe cooking water are used, the traditional cooking practice can actually be beneficial for the rural people of Bengal. The cooking procedure and the large volume of washing/cooking water enable the contaminated rice with sufficient contact time to leach As from the rice grains into the starch water,thereby food grain highly elevated with As may be made suitable for human consumption. However, gruel intake will be high risk associated. Therefore, although strategies on decreasing the As content in uncooked rice grains along with low As uptake genotype verity should be promoted. Finally, As in cooked rice is the actual point of convergence for the dietary exposure for the population and attempts to limit the As in cooked rice must be taken into consideration to successfully combat the menace of As.

ACKNOWLEDGEMENTS

We would also like to thank UGC–SAP program and DST-FIST program to the Department of Chemistry, University of Kalyani, West Bengal, India.

REFERENCES

Chatterjee D., Halder D., Majumder S., Biswas A., Nath B., Bhattacharya P., Bhowmick S., Mukherjee-Goswami A., Saha D., Maity P.B., Chatterjee D., Mukherjee A. & Bundschuh J. 2010. Assessment of arsenic exposure from groundwater and rice in Bengal delta region, West Bengal, India. *Water Res.* 44: 5803–5812.

Meharg A.A., Williams P.N., Adomako E., Lawgali Y. Y., Deacon C., Villada A., Cambell R.C., Sun G., Zhu Y.G., Feldmann J., Raab A., Zhao F.J., Islam R., Hossain. S. & Yanai J. 2009. Geographical variation in total and inorganic arsenic content of polished (white) rice. *Environ. Sci. Technol.* 43: 1612–1617.

Nriagu J., Bhattacharya P., Mukherjee A., Bundschuh J., Zevenhoven R. & Loeppert R., 2007. Arsenic in soil and roundwater: an overview. In: P. Bhattacharya, A.B. Mukherjee, J. Bundschuh, R. Zevenhoven & R.H. Loeppert (eds.), *Arsenic in Soil and Groundwater Environment*. Elsevier, Amsterdam, pp. 3–60.

Arsenic in the Environment: Bridging Science to Practice for Sustainable Development –
van der Wal, Ahmad, Petrusevski, Weijma, Savic, van der Wens, Beerendonk, Bhattacharya, Bundschuh & Naidu (Eds)
© 2023 the Author(s), ISBN 978-1-032-32928-4

Poverty dynamics, arsenic exposure and adolescent mortality: A prospective finding

M. Rahman[1], N. Sohel[2] & M. Yunus[3]
[1]*Pure Earth, Dhaka, Bangladesh*
[2]*Department of Health Research Methods, Evidence, and Impact, McMaster University, Hamilton, Ontario, Canada*
[3]*icddr, b Dhaka, Bangladesh*

ABSTRACT: Earlier we reported in 2019 that higher mortalities in young adults who were exposed to higher arsenic through drinking water. Widespread arsenic contamination in underground water is a well-documented public health concern that threatens millions of lives worldwide. In other word, Bangladesh belongs to low-middle income country (LIMC) and poverty is still high in Bangladesh. One in five individuals are living below the national poverty line. In other word, Poverty, often linked to chronic diseases, is a multi-dimensional phenomenon. Although it is evident that high levels (> 300 µg/L) of arsenic exposure from drinking water are related to adverse health outcomes, health effects of arsenic exposure at low-to-moderate levels (10–300 µg/L) are not well understood. Since poorer people living in the rural areas have limited access to safe drinking water and had to rely on the tubewell water which is the point of arsenic commination in humans. Therefore, research is warranted to understand the additional sufferings of the poor segment of the society due to chronic high level arsenic exposure. Given the complex scenario, we investigated if poverty dynamic plays a role in young adult deaths in Bangladesh.

1 INTRODUCTION

Arsenic (As) is abundant in the earth's crust and can be released into groundwater (Kinniburgh & Smedley 2001). Nearly 200 million people worldwide, including ~57 million in Bangladesh and ~17 million in the United States (US), are at increased risk of cancers, cardiovascular diseases (CVD) and diabetes mellitus (DM) due to chronic exposure to inorganic As. Long-term exposure to high levels of As exposure (>300 µg/L) in drinking water has been linked to increased risk of bladder, lung, skin, kidney, and liver cancers, as well as diabetes, cardiovascular disease (CVD), adverse pregnancy outcomes, and a decrease in children's intellectual function. Although As has been classified as a Class I human carcinogen by the International Agency for Research on Cancer, evidence on the risk of internal cancers, as well as CVD and DM associated with low-to-moderate levels. Arsenic exposure through drinking water has made a catastrophic situation in Bangladesh where more than half of the total population were exposed to As at greater than the maximum permissive limit set by WHO (Maull *et al.*, 2012).

2 METHOD

A prospective cohort study of 58,406 individuals were enrolled who were 5–18 years at baseline on January 01 2003. Participants were included from active Matlab HDSS (Health and Demographic Surveillance System) which started in 1960. Baseline data were collected during January through December in 2003 and participants were followed-up every two months till December 31, 2015. Each individual's As exposure was calculated at cumulative As exposure throughout the study period and poverty indices were categorized based on SES and education. Ultra poor population was defined who has asset score below 40% percentiles with no education. Non-poor was defined the rest. Cox proportional hazards models were run to estimate the mortality risk due to cumulative As exposure after adjusting age, sex, poverty indices. Verbal autopsy (VA) using International Classification of Diseases (ICD-10) were used to identify the causes of young adults' death.

3 RESULTS AND DISCUSSION

Of the 58,406 young-adults, a total of 411 died during follow-up. Deaths due to cerebro-vascular

DOI: 10.1201/9781003317395-119

Table 1. Hazard ratio (HR) for mortality due to cancer in childhood participants in relation to cumulative arsenic exposure and ultra poor population.

			Ultra poor		Non-poor	
Exposure	Participant	PersonYear	Deaths (N = 14)	Adjusted HR* (95%CI)	Deaths (N = 36)	Adjusted HR* (95%CI)
≤ 1827	29,205	271751	8	1.0	20	1.0
>1827	29,201	271751	6	7.18 (1.6–32.7)	16	1.77 (0.7–4.1)

disease, cardio-vascular disease, and respiratory disease (N = 40). Girls had higher values of cumulative As exposure than boys (median: 1858.5 µg/year/L vs. 1798.8 µg/year/L). Ultra poor population with higher level of As exposure (>1827.0 µg/year/L) was found 7.1 times higher chance of deaths due to cerebro-vascular disease, cardio-vascular disease, and respiratory disease compare to those who are exposed with lower As level after adjusting covariates. Additionally, mortality risk of non-poor segment of the population still remain high (HR 1.77) due to chronic high level of As accumulation (>1827.0 µg/year/L) in the body.

4 CONCLUSION

This paper shed light on the poverty dynamics to human demography at individual level. Higher concentration of, and chronic exposure to As in drinking water, increases the mortality risk among the young adults, who lives below poverty line. This study indicated additional burden of higher mortality risk along with existing risk of communicable and non-communicable diseases among the poorer segment of the rural Bangladesh. Future prospective analyses of clinical endpoints and related host susceptibility will enhance our knowledge on the health effects of low-to-moderate levels of As exposure, elucidate disease mechanisms, and give directions for prevention.

REFERENCES

Ferrannini E., Bjorkman D. & Reichard G.A. 1985. The Disposal of an Oral Glucose Load in Healthy Subjects. A Quantitative Study. *Diabetes* 34: 580–588.

Kinniburgh D. & Smedley P. 2001. Arsenic contamination of groundwater in Bangladesh. *Br. Geol. Surv.* 1.

Maull E.A., Ahsan H., Edwards J., Longnecker M.P., Navas- Acien A., Pi J., Silbergeld E.K., Styblo M., Tseng C.H., Thayer K.A. & Loomis D., 2012. Evaluation of the association between arsenic and diabetes: a national toxicology program workshop review. *Environ. Health Perspect.* 120: 1658–1670.

Paul S.K., Islam M.S., Hasibuzzaman M.M., Hossain F., Anjum A., Saud Z.A., Haque M.M., Sultana P., Haque A., Andric K.B. & Rahman A. 2019. Higher risk of hyperglycemia with greater susceptibility in females in chronic arsenic-exposed individuals in Bangladesh. *Sci. Total Environ.* 668: 1004–1012.

Srikanthan P. & Karlamangla A.S. 2011. Relative muscle mass is inversely associated with insulin resistance and prediabetes. findings from the third national health and nutrition examination survey. *J. Clin. Endocrinol. Metab.* 96: 2898–2903.

Arsenic in the Environment: Bridging Science to Practice for Sustainable Development –
van der Wal, Ahmad, Petrusevski, Weijma, Savic, van der Wens, Beerendonk, Bhattacharya, Bundschuh & Naidu (Eds)
© 2023 the Author(s), ISBN 978-1-032-32928-4

Arsenic contamination of drinking water and health risk assessment in Dagestan region, Russia

T.O. Abdulmutalimova
Institute of Geothermal and Renewable Energy, Joint Institute of High Temperature, Russian Academy of Sciences, Makhachkala, Russia

ABSTRACT: Due to naturally occurring arsenic in the sediments the groundwater in the south part of Russia content high level of arsenic contamination and more than 500 thousands of people might be at a considerable risk of chronic arsenic poisoning. The evaluation of cancer risks from oral exposure to As were found to be above the acceptable U.S. EPA and WHO cancer health risk range of 1×10^{-6} to 1×10^{-4}. A relatively high content of arsenic in the hair of the people surveyed was found. The results of this study allow us to reveal areas in region with high levels of arsenic in drinking water, determine the exposed population to the implementation of risk mitigation measures and also direct correlation between the arsenic content in drinking water and its content in the hair of the exposed people.

1 INTRODUCTION

About 200 million people worldwide are exposed to arsenic with concentrations exceeding the recommended limit of 10 μg/L water (WHO 2012). Previous studies (Abdulmutalimova & Revich 2017; Kurbanova *et al.*, 2013) had revealed that groundwater, used by population for drinking, in some areas of Russia also contained arsenic above international guidelines. But the extent of those risks was unknown. The aim of this study is to show spatial distribution of arsenic in the region with the high As concentration, it's accumulation in hair of exposed people and human health risk assessment of chronic oral exposure.

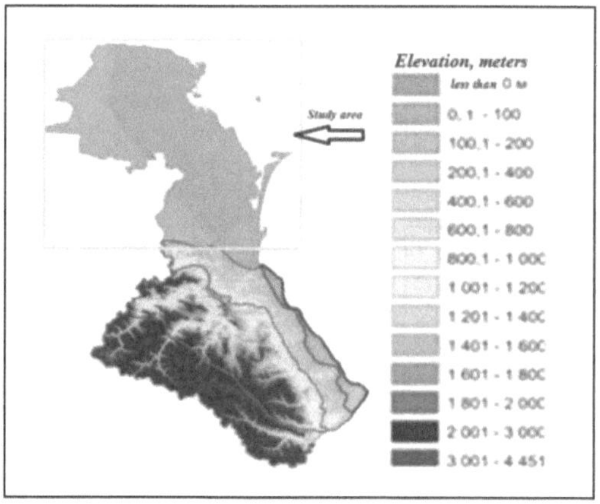

Figure 1. Daghestan region and study area.

2 MATERIALS AND METHODS

2.1 *Study area*

Daghestan district is the south Federal subject of Russia located in the North Caucasus region with the total area of 50,300 sq.km and a population of nearly 3.06 million.

The study area (Figure 1) is located in the north part of Daghestan within longitude 45°10′–47°50′ W and latitude 43° 20′ – 44° 70′ N with an estimated total area about 30 thousand sq. km.

The territory is characterized by arid and semi-arid climate with high temperatures (maximum daily values averages around 30–40°C). Average estimated annual evapotranspiration (800–1000 mm/year) far exceeds annual rainfall values (200 – 400 mm/year). The geology of the area consists of Pliocene and Pleistocene rocks. The rural population of over 500 thousand people depends on groundwater resources for drinking.

2.2 *Water and hair sampling, arsenic analysis*

A total 910 water samples were collected from the arsenic-contaminated groundwater, used by population for drinking in different villages of Daghestan (Figure 2). During sampling, the plastic containers were rinsed several times with the sample before the final sample was taken. The water samples were acidified by adding 1 mL of 10% analytical grade nitric acid and then analyzed for

 DOI: 10.1201/9781003317395-120

As by flow-injection hydride generation-atomic absorption spectrophotometry (FI-HG-AAS).

Hair analysis has potential merit as a screening procedure for long-term arsenic exposure. Hair samples of exposed people from villages with different As concentrations were collected. Approximately 100 strands of hair were collected at the neck nape and stored in a labeled Ziploc bag. Arsenic in scalp hair was determined by the graphite furnace technique with an atomic absorption spectrometer.

2.3 *Human health risk assessment and carcinogenic health risk calculations*

Cancer health risk for population was evaluated in this study using human health risk assessment method. The exposure route evaluated is oral. Carcinogenic risks for oral exposure to As were calculated by using following equation:

$$\text{Cancer health risk}(CR_{oral}) = (ADD) \times CSF \quad (1)$$

where ADD is the average daily dose of As in water via oral exposure route in the study area by population; CSF – cancer slope factor for oral exposure to As which is 1.5 mg/kg/day (USEPA 2005).

2.4 *Data analysis and creation of spatial database*

For spatial As distribution in study area was used ArcGIS 9 ArcMap 9.3 software (ESRI, USA) with the location and description of water sample. For statistical calculations of results of human health risk assessment and biomonitoring were perfomed using Microsoft Excel 2007.

3 RESULTS AND DISCUSSION

3.1 *Spatial distribution of arsenic in drinking water*

Arsenic concentration in nearly 97% water samples revealed a significant variation in concentrations ranging from 10 to 500 µg/L (mean: 190 µg/L (Figure 2). Most of the water samples revealed As concentrations higher than WHO guideline value of 10 µg/L (Table 1). 4.8% water samples contained As at the levels 400–500 µg/L. 53,9% of population use drinking water with the level of As 10–40 µg/L. Groundwater with the high concentration of As (400–500 µg/L) used by 3% of the northern districts population.

The level of As in the water is detected 20 times or more in 12 villages with the total population of about 16 thousands of people (Table 1).

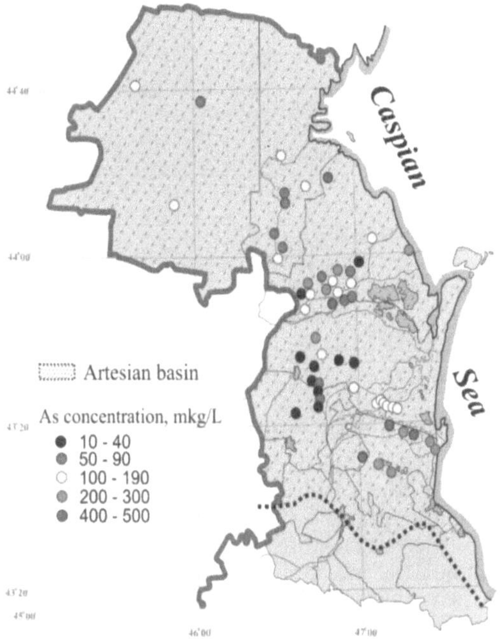

Figure 2. Spatial distribution of arsenic.

Table 1. Association between exposed population and As level in drinking water.

S. No.	As levels in drinking water µg/L	Water samples %	Exposed population	% of total Population in North Daghestan
1	10–40	15.8	167134	53.9
2	50–90	24.7	16985	5.5
3	100–190	36.8	108147	34.9
4	200–300	17.9	9023	2.9
5	400–500	4.8	8444	2.8
	10–500	100	309733	100

3.2 *Cancer risk results*

In the study area, ADD for oral exposure ranged from 0.0003 mg/kg/day to mg/kg/day. It has been shown that the lifetime individual cancer risks were at the minimum As concentration (10 µg/L) – 4.3E-4; at maximum As concentration (500 µg/L) – 2.1E-2, respectively, with a mean of 190 µg/L – 8.1E-3. The lifetime individual cancer health risks results were found to be high and not permissible for population (Table 2). The annual cancer health risks for population were from 1 to 95 additional cases of possible occurrence of cancer.

Table 2. Cancer risk for exposed population in the Daghestan Republic.

The level of arsenic in drinking water, µg/L	Individual cancer health risk
10 – 40	$4.29 \times 10^{-4} - 1.71 \times 10^{-3}$
50 – 90	$2.14 \times 10^{-3} - 3.86 \times 10^{-3}$
100 – 190	$4.29 \times 10^{-3} - 8.14 \times 10^{-3}$
200 – 300	$8.57 \times 10^{-3} - 1.29 \times 10^{-2}$
400 – 500	$1{,}71 \times 10^{-2} - 2{,}14 \times 10^{-2}$

3.3 *Results of hair analysis*

In the region with the highest arsenic content in drinking water (400–500 µg/L) in 10% of the examined people – its content exceeds the threshold value in the hair, at which clinical symptoms of arsenoses can be observed; in 58% of the examined individuals, the arsenic content in the hair is at the background level; in 32% the arsenic content was found to be 0.5–0.9 µg/g, which indicates accumulation, but does not reach the threshold value.

4 CONCLUSIONS AND RECOMMENDATIONS

High arsenic concentrations in drinking water and its long-term exposure is causally related to increased risks for human health. The results of this study prove the high lifetime individual cancer health risks that not permissible for population. Preliminary studies of the hair arsenic confirm the assumption of arsenic in the body due to chronic oral exposure in the studied population group.

REFERENCES

Abdulmutalimova T.O. and Revich B.A. 2017. Assessment of carcinogenic risk to population health due to high arsenic content in drinking artesian water of the North Dagestan. *Rus. J. Hygiene & Sanitation* 96(8): 743–746.

Kurbanova L.M., Samedov Sh.G., Gazaliev I.M. and Abdulmutalimova T.O. 2013. Arsenic in the groundwaters of the North Dagestan Artesian basin. *Geochem. Int.* 51(3): 237–239.

WHO. 2012. *Arsenic Fact Sheet No 372*. World Health Organization; Geneva. Available from: http://www.wwho.int/mediacentre/factsheets/fs372/en/ [Accessed: 2014 May 16].

Arsenic in the Environment: Bridging Science to Practice for Sustainable Development –
van der Wal, Ahmad, Petrusevski, Weijma, Savic, van der Wens, Beerendonk, Bhattacharya, Bundschuh & Naidu (Eds)
© 2023 the Author(s), ISBN 978-1-032-32928-4

Report about advances and challenges during the first steps of the project: "Arsenic in Uruguayan groundwater and associated health risk"

K. Pamoukaghlián[1], P. Collazo[1], N. Mañay[2] & E. Alvareda[3]
[1]*Instituto de Ciencias Geológicas, Facultad de Ciencias – UDELAR, Montevideo, Uruguay*
[2]*Centro Especializado en Química Toxicológica (CEQUIMTOX) Área Toxicología, DEC, Facultad de Química, UDELAR, Montevideo, Uruguay*
[3]*Water Department, Centro Universitario Regional Litoral Norte (CENUR), Universidad de la República, Salto, Uruguay*

ABSTRACT: In Uruguay there is a great concern about the arsenic (As) geogenic contamination in groundwater as an environmental health problem, after rather high levels of As in groundwater in some areas (above 20 μg L^{-1}) were reported and, considering that the main part of the rural population consumes drinking water from private aquifers wells. On the other side Uruguay has indeed well quantity and quality groundwater and this is why there is a must beware of the Uruguayan aquifers as important resources. In this respect we proposed and start the execution of a multidisciplinary and interinstitutional program: AsURU Project ("Arsenic in Uruguayan groundwater and associated health risk" (website: facebook.com/AsURUArsénico).

1 INTRODUCTION

According to our national drinking water regulations, there are no systematic studies on the quality of groundwater or on incidence of health effects, regarding arsenic (As) exposure. However, a first medical geology approach is being performed, trying to assess health risks restrospectively to correlate melanoma cancer with As higher ground water levels, distributed by different localities (Mañay *et al.*, 2019). To look for integral solutions to this environmental problem, we have presented and are currently coordinating the project "Arsenic in groundwater and associated health risks" (AsURU). (Collazo *et al.*, 2019) with the incorporation of new researchers and the parti cipation of delegates of several public bodies in the regular meetings.

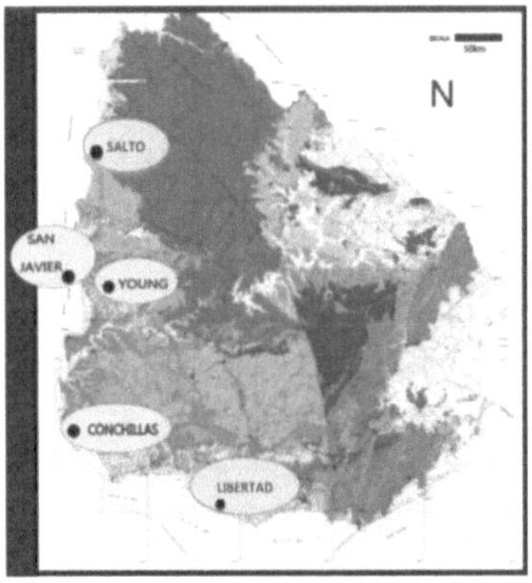

Figure 1. Pilot areas on the geological map of Uruguay (modified from Bossi *et al.*, 2001).

2 MATERIALS AND METHODS

2.1 *Study area*

Study areas the AsURU Project considers *a priori* all the country, but main aquifers which represent most important water resources and which are affected by the problem of high levels of arsenic (As) are specially taken into account as the southwest area where Raigon Aquifer is placed; west littoral area where Mercedes Aquifer is placed and North Littoral where Salto Aquifer is an important water resource. For each of these aquifers a pilot area has been selected (Figure 1).

2.2 *Work group strategy*

AsURU Project has stated a working plan that consists in: (a) identification of main Uruguayan aquifers with relervant As data, (b) subdivision in study areas, (c) pilot areas selection following specific criteria: i) areas which have drinking water supplied by aquifers with As levels above WHO guidelines (WHO 2010); ii) they should represent a significative extension of the contaminated aquifer; iii) have enough information about the aquifer hydraulic parameters; and iv) quantitative data and population's exposure timeline to groundwater information.and evidences about health adverse effects available; v) hydrogeologic, hydrodynamic and hydrogeochemical study as to identify the origin of As and space-time movility, vi) identification and selection of affected population and evaluation about As exposition through groundwater in the pilot areas; and vii) propose solutions like well's construction and remediation actions.

DOI: 10.1201/9781003317395-121

3 RESULTS AND DISCUSSION

As concentration values in groundwater samples were anomalous in several cases, and lightly superior to the admissible maximum values, according to the existing regulations in Uruguay (0,02 mg/L) (UNIT 2008). This difference is in fact rather bigger when we compare Uruguayan aquifers As concentration values with the OMS norms (0,01 mg/L). According to this, there were defined five pilot zones: (1) Conchillas with average values of As>0,03 mg/L in groundwater samples; (2) Young with As>0,01 mg/L in groundwater samples; (3) San Javier with As>0,01 mg/L, where the drinking water supplied, the State-owned company (O.S.E) is implementing remediation using osmosis; (4) Libertad with As between 0,01 and 0,03 mg/L; (5) Salto (North Littoral) with As between 0.004 mg/L and 0.05 mg/L values. To solve this problem about the admissible rates in groundwater is not only a must for the country but also a necessity in order to warranty the health and life quality of its population.

4 CONCLUSIONS AND RECOMMENDATIONS

AsURU Project has begun with advances and difficulties. Main advances have been made according to the proposed activities program: (1) the identification of pilot areas with As concentration levels in groundwater above 20 μg L^{-1} (corresponding limit according to uruguayan norms); and even with As level concentrations above 10 μg L^{-1} (corresponding to international norms of the O.M.S.); (2) the formation of three work groups that are just developing different related projects: (a) Hydric Groundwater Resource Group, with Dr. Paula Collazo and Dr. Karina Pamoukaghlián as responsibles; Chemistry and Medical Geology Group, leadered by Dr. Nelly Mañay; (b) North Littoral Group with Elena Alvareda as responsible; (3) there have been held several meetings, like the First Meeting of AsURU Project in CeReGAS (Regional Center for Groundwater Management); the I Symposium of Arsenic in Uruguayan Groundwater, with the participation of various State-owned companies (the Sanitary Management company O.S.E., the Environmental State Department DINAMA, The Water State Department DINAGUA, The Mining and Geology Department DiNaMiGe), a mixed Private-State company for Rural Housing MEVIR) and the UdelaR (Chemistry, Science, Medicine and Engineering Faculties).

However several difficulties and challenges were found in the way of performance. Despite State companies have participated and shown their concern about the problem and interest in the project, it is not so easy that the responsible State companies have a real involvement in order to fully develop this or "other" project. Moreover, the fact that there are really scarce resources is a great difficulty and we will have to priories the problem to give solutions.

Considering that there are already a lot of cases where As levels are above the admissible maximum values according the local guidelines for drinking water, and in fact the difference will rise if Uruguay changes this to the World Health Organization guidelines, as it is expected for the year 2021, the most important recommendation is that this problem needs to be treated by the point of view of Medical Geology, having just started with some researches by the Chemistry and Medical Geology Group (Desai *et al.*, 2020; Falchi *et al.*, 2018; Kordas *et al.*, 2016). However geo-stadistic studies about the incidence in health, risk maps for a real definition of the applicable limit levels are still very little in Uruguay.

ACKNOWLEDGEMENTS

We aknowledge to the Littoral North data from Research and Innovation National Agency (ANII) for supporting the ANII FMV_1_2017_1_135656 Project. We also acknowledge to the State-owned company O.S.E. for the provided information.

REFERENCES

Bossi J., Ferrando L., Montaña J., Campal N., Morales N., Gancio F., Schipilov A., Piñeiro P. & Sprechman P. 1998. *Carta Geológica del Uruguay 1/500.000*, Cátedra de Geología – Facultad de Agronomía, Geoeditores.

Collazo P., Mañay N. & Pamoukaghlián K. 2019. Proyecto AsURU: Arsénico en Agua Subterránea de Uruguay y Riesgo a la Salud asociado. Avances y Dificultades en Su gestión. *Actas IX Congreso Uruguayo de Geología*, Flores, Uruguay.

Desai G., Barg G., Vahter M., Queirolo E., Peregalli F., Mañay N., Millau A.E., Yu J., Browne R.W. & Kordas, K. 2020. Low level arsenic exposure, B-vitamins, and achivement amorg Uruguayan school children. *Int. J. Hyg. Environ. Health* 223: 124–131.

Falchi L., Pizzorno P., Iaquinta F. & Cousillas A. 2018. Relevamiento de la concentración de arsénico total en agua proveniente de varias fuentes en una zona arrocera de Uruguay. *Revista del Laboratorio Tecnológico de Uruguay* 17: 10–17.

Kordas K., Queirolo E.I., Mañay N., Peregalli F., Hsiao P. Y., Lu Y. & Vahter M., 2016. Low-level arsenic exposure: nutritional and dietary predictors in first-grade Uruguayan children. *Environ. Res.* 147: 16–23.

Mañay N., Pistón M., Cáceres M., Pizzorno P. & Bühl V. 2019. An overview of environmental arsenic issues and exposure risks in Uruguay. *Sci. Total Environ.* 686: 590–598.

UNIT, *Agua Potable-Requisitos*, 2010. Recovered from: http://www.aiqu.org.uy/documentos/medioambiente/unit.pdf

WHO. 2010. *Exposure to Arsenic: A Major Public Health Concern*. Recovered from: World Health Organization: http://www.who.int/ipcs/features/arsenic.pdf?ua=1

Arsenic in the Environment: Bridging Science to Practice for Sustainable Development – van der Wal, Ahmad, Petrusevski, Weijma, Savic, van der Wens, Beerendonk, Bhattacharya, Bundschuh & Naidu (Eds)
© 2023 the Author(s), ISBN 978-1-032-32928-4

Private wells in Uruguay: Evaluating groundwater arsenic levels and finding new areas for population's health risks assessment

V. Bühl[1], P. Pizzorno[2], I. Machado[1], E. Alvareda[3] & N. Mañay[2]
[1]*Analytical Chemistry, Faculty of Chemistry, DEC, Universidad de la República (UdelaR), Montevideo, Uruguay*
[2]*Toxicology, Faculty of Chemistry, DEC, Universidad de la República (UdelaR), Montevideo, Uruguay*
[3]*Water Department, Centro Universitario Regional Litoral Norte, Universidad de la República, Salto, Uruguay*

ABSTRACT: Recent studies on arsenic (As) occurrence particularly in private wells in Uruguay show that several sources of drinking water have elevated concentrations above national and international guidelines. In this study, As results have been used to find possible study zones for risk assessment studies. Acquisition of relevant information about water consumption through lifetime, nutritional aspects as personal and family data is being collected for develop health risk assessment for arsenic-associate diseases.

1 INTRODUCTION

Uruguay is one of the smallest South American countries and is located between Argentina and Brazil. It has a surface area of 176,215 km^2 (68,037 sq mi). The country has a population of approximately 3.3 million people, with almost 40% residing in the capital city, Montevideo. Drinking water is supplied to 94% of the population by the state company Obras Sanitarias del Estado (OSE 2018). Most of the water supplied to Montevideo is from surface water. Until 2010 the maximum allowed arsenic (As) concentration level in drinking water was 50 μg/L. From 2010 to now, Uruguayan sanitary authorities adopted 20 μg/L as a provisional limit with the compromise of lowering it in 2021, to reach the target value of 10 μg/L recommended by the World Health Organization (WHO) guidelines (UNIT 2010).

However, some rural homes and small towns do not have water supply from OSE. In these cases, individual drilled wells, are used to waterworks using available aquifers. Since Uruguay economy is mainly based in livestock and agriculture, there are numerous private perforations that are located mostly in rural areas. This could be shown in Figure 1. Water from these perforations is used by people who work and live in those areas as drinking water as well as to feed the cattle and watering crops. There are 39 drilling companies registered by the Ministry of Housing Territorial Planning and Environment (MVOTMA) and according to information collected by DINAMIGE (National Directorate of Mining and Geology of the Ministry of Industry, Energy and Mining) there are 41,957 wells declared (2019). As far as it is known, drinking water from those private wells are not properly controlled for As levels as they are out of the scope of OSE, although they should accomplish drilling wells regulations.

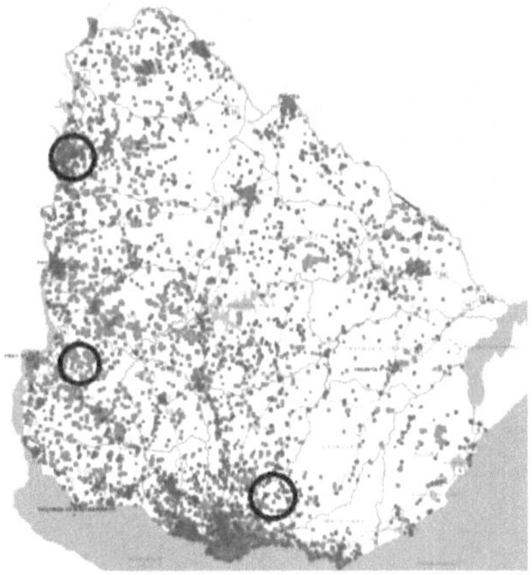

Figure 1. Uruguayan map with reported wells from map viewer of Ministry of Industry, Energy and Mining (MIEM). (green: private wells, blue: irrigation wells and red: DINAMIGE wells. Black circles: selected areas.

The main objectives of this work are to gather information from arsenic levels in private wells in order to identify areas for population's health risk assessment studies.

2 MATERIALS AND METHODS

2.1 *Water sampling*

A total of 74 groundwater samples were collected from different sites of Uruguay at depths between 30 and 50 m. The sampling points were randomly selected considering their spatial distribution in the country. Each sample has the corresponding geo-referencing data. Samples were stored in polypropylene bottles at 4–5°C until analysis.

2.2 *Laboratory analysis*

For analytical determinations of As by ET-AAS, an atomic absorption spectrometer (iCE 3500, Thermo Scientific) equipped with a graphite furnace atomizer and Zeeman-based correction was employed. Argon (Ar 99.99%, Linde) was used as purge and protective gas. The analytical line used was 193.7 nm and the signal used for quantification was integrated absorbance (peak-area). Palladium matrix modifier (Merck) was used (5 μg). For all the determinations, pyrolytically coated graphite tubes (Thermo Scientific) were used. ET-AAS temperature program is shown on Table 1.

Table 1. ET-AAS temperature program for As determination.

Parameter	Temperature (°C)	Ramp rate (°C s^{-1})	Hold time (s)
Drying	100	10	50
Pyrolysis	1200	150	20
Atomization	2200	0	3
Cleaning	2600	0	3

Ar flow: 0.2 mL/min, except during atomization step. Injection volume: 30 μL.

3 RESULTS AND DISCUSSION

Considering As results from those 74 samples and their coordinates, we defined different areas with As levels above international and national regulations These results are shown in Figure 1 (black circles). i) The northern selected area is located in Salto department and is known as "Thermal Corridor". We found As values between 20.7 and 58.9 μg/L. This is a livestock area, with production of vegetables and fruits and also with several thermal resorts and hotels; ii) the center west zone is located in Soriano department. We found As values between 35.9 and 53.6 μg/L. This zone has mainly farms and dairy farms; iii) the southern area is located in Canelones department. This area has an important agriculture production as well as dairy farm and vineyards. We found As values between 63.2 and 120.4 μg/L.

4 ONGOING RESEARCH

We are starting a pilot study in a subarea (within selected areas) with a location less than 500 inhabitants.

Subsequently, we plan to sample and analyze the groundwater in the pilot area and conduct a survey that will take into account those necessary aspects and variables to apply the stages of the recommended process in the WHO (2017) tool for health risk assessment for arsenic-associated diseases, such as water consumption through lifetime, certain food intake, personal and family data, etc.

ACKNOWLEDGEMENTS

We acknowledge the support from Comisión Sectorial de Investigación Científica (CSIC), PEDECIBA-Química and Agencia Nacional de Investigación e Innovación (ANII), Uruguay.

REFERENCES

Map viewer (Geological and Mining), *DINAMIGE (National Directorate of Mining and Geology of the Ministry of Industry, Energy and Mining* http://visualizadorgeominero.dinamige.gub.uy/DINAMIGE_mvc2/.

MVOTMA, *Ministry of Housing Territorial Planning and Environment*. 2019. https://app.mvotma.gub.uy/SIH-JSF/tramites/consulta/MVOTMA_ConsultaEmpresa Perforadoras.xhtml

OSE, *Obras Sanitarias del Estado*, 2018. Water. Supply. http://www.ose.com.uy/agua

UNIT, 2010. http://www.ose.com.uy/descargas/Clientes/Reglamentos/unit_833_2008_.pdf

WHO, 2017. *Human Health Risk Assessment Toolkit: Chemical Hazards*. ISBN: 978-92-4-354807-4 https://www.who.int/ipcs/publications/ra_toolkit/es/

WHO, 2018. *Arsenic. Key Facts*. Geneve, Switzerland. http://www.who.int/news-room/fact-sheets/detail/arsenic

Arsenic in the Environment: Bridging Science to Practice for Sustainable Development –
van der Wal, Ahmad, Petrusevski, Weijma, Savic, van der Wens, Beerendonk, Bhattacharya, Bundschuh & Naidu (Eds)

Arsenic bioavailability and bioaccessibility in mining tailings from the Brazilian Iron Quadrangle

A. Santos[1] & M.C. Teixeira[1,2]
[1]*Graduating Program in Environmental Engineering (PROAMB), Pharmacy Department, School of Pharmacy, Federal University of Ouro Preto (UFOP), Minas Gerais, Brazil*
[2]*Pharmacy Department, School of Pharmacy, Federal University of Ouro Preto (UFOP), Minas Gerais, Brazil*

ABSTRACT: The mobilization of the non-essential elements from the mineral matrix to the environment as a consequence of mining activities represents a potential risk to human health since some of those elements are toxic and bioaccumulate into the trophic chain. This study aimed the investigation of the bioavailability and bioaccessibility of non-essential elements detected in some soil samples collected into and close to an ancient gold mine. Samples were sieved and classified according their particle sizes, prior to the chemical analysis. Thirty-one elements were found, including arsenic (As). The highest As content (10,739.2 mg/kg) was found in the smallest particles (<0.037 mm). The gastric (IVBA) and lung bioacessibilities of samples were assessed through *in vitro* tests. As bioavailabilty was directly related to the total As and Mn content and inversely proportional to the Fe concentration. Fortunately, our results pointed to a low carcinogenic risk to the exposed population.

1 INTRODUCTION

Ouro Preto is located in the middle of the Iron Quadrangle (IQ), one of the more important mining region in Brazil. In such region, arsenic (As) enriched gold minerals as arsenopyrite are usually found (Matschullat *et al.*, 2007) and gold had been explored since the 18th century. As a consequence, high levels of As have been reported in soil (21,000 mg/kg), sediment (3,300 mg/kg) (Matschullat *et al.*, 2007), water (1,700,000 μg/L) (Borba *et al.*, 2004) and groundwater (224 μg/L) (Gonçalves *et al.*, 2007). Therefore, As bioavailability and bioaccessibility deserves attention, since humans could be exposed to this environmental contaminants which is largely distributed through different environmental compartments. The present study aims: (i) to investigate the distribution of As in solid particles taking into consideration their sizes; (ii) assess As bioavailability and bioaccessibility, mainly for the fine particles (which could be inhaled or ingested) and, (iii) to estimate the population carcinogenic risk.

2 MATERIALS AND METHODS

2.1 *Study area and sampling characterization.*

Samples were collected at Ouro Preto, Minas Gerais, at Chico Rei Mine (CR) and at a backyard of a residence nearby (CS) (Figure 1). Samples were collected at 15 cm depth, homogenized, dried and sieved. The following parameters were analyzed: particle size distribution, pH, TOC, zeta potential, polydispersity index and chemical composition after *aqua regia* extraction. Samples were named according their particle sizes, being the CR01 and CS01 the smallest (<0.037 mm) and CR08 and CS08 (0.177 < Dp < 0.250), the largest.

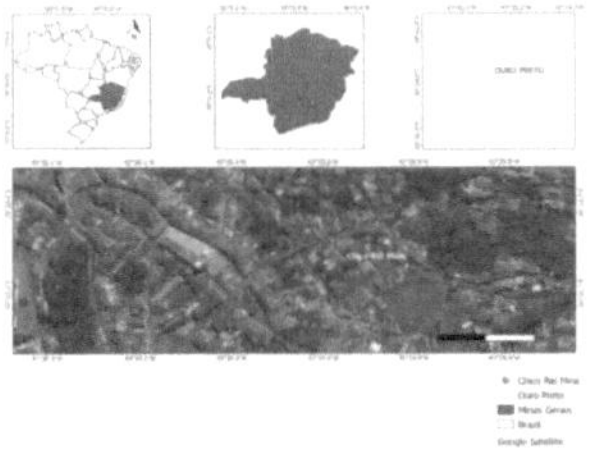

Figure 1. Study area and soil sampling location.

2.2 *Laboratory analysis*

Bioavailability was determined by BCR (Community Bureau of Reference) protocol (Ure *et al.*, 1993). Oral bioaccessibility was estimated for CR01, CR06, CS01 and CS06 samples (USEPA 2017). The pulmonary bioaccessibility was carried out according protocols adapted from Pelfrenê *et al.* (2017). Elements were determined by ICP-OES and ICP-MS.

2.3 *Carcinogenic risk*

The risk to human health were estimated only considering the residents of urban areas following Brazilian Environmental Agencies (CETESB) recommendations and models (CETESB 2013).

3 RESULTS AND DISCUSSION

3.1 *Chemical analysis*

Thirty-one chemical elements were identified. CS has an alkaline pH (8.0) while CR was acidic (3.0). Particle size analysis demonstrated the highest As concentration were found in the smallest particles, 10,739 mg/kg for CR01 (Figure 2). An opposite

DOI: 10.1201/9781003317395-123

feature was observed for samples from the CS group (Figure 3). Organic carbon values were below the limit of quantification by TOC method. Elements mobility, including As, increased with decreasing pH. An opposite behavior was observed for alkaline samples.

According to BCR extraction, the As content of CR samples was considered exchangeable (i.e. bioavailable) and reducible (i.e. bioavailable under reducing conditions). CR01 and CR08 presented As concentration about 23 and 12 times higher than the guiding value adopted in Brazil, i.e. 55 mg/kg (BRASIL 2009), respectively. As bioavailability was directly related to the total As and Mn content and inversely proportional to the Fe concentration. Regarding As bioavailability and bioaccessibility it was considered limited even for CR01 (Figure 4).

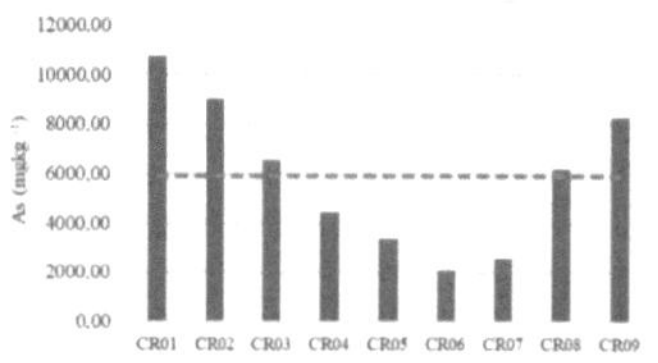

Figure 2. Arsenic concentration in CR samples according to particle size distribution, with CR01 being the thinnest. The dashed line refers to the average concentration value of the element.

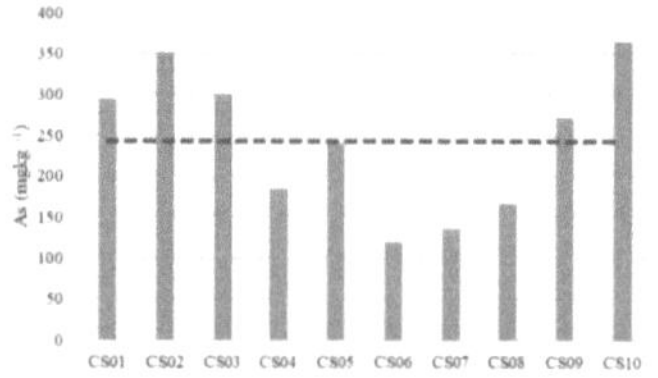

Figure 3. Arsenic concentration in CS samples according to particle size distribution, with CS01 being the thinnest. The dashed line refers to the average concentration value of the element.

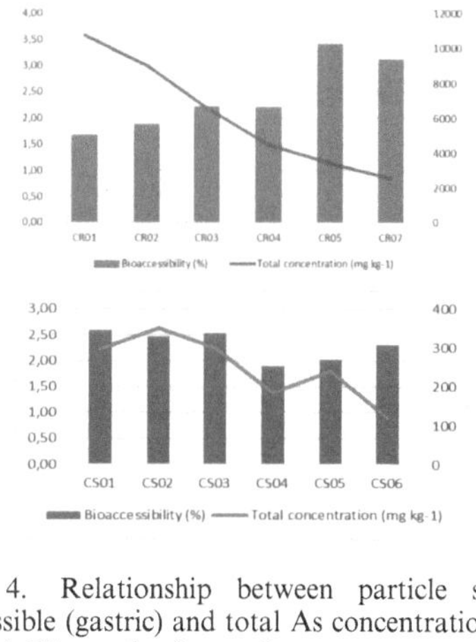

Figure 4. Relationship between particle size, As bioaccessible (gastric) and total As concentration in CR (top) and CS samples (bottom).

3.2 *Carcinogenic risk*

The cumulative carcinogenic risk for CR soil (As intake, inhalation and dermal contact) was 3.1×10^{-3} and 1.8×10^{-3} for children and adults, respectively. The values for CS sample were 5.18×10^{-4} for children and 3.08×10^{-4} for adults. However, the tolerable level of risk to human health is 1×10^{-5} (BRASIL 2009), therefore, considering only the bioaccessible As (ingestion), the carcinogenic risks were acceptable: 1.04×10^{-5} and 5.57×10^{-6}, for children and adults, respectively.

4 CONCLUSIONS AND RECOMMENDATIONS

The results obtained after BCR sequential extraction revealed a relatively high As bioavailability. Contrarily, oral and pulmonary bioaccessibility tests proved As mobility under physiologically simulated conditions is limited. Taking this information into consideration, the estimation of the carcinogenic risks of the studied material, using a conventional approaching, i.e. considering the total As content, could produce a misleading and overestimated result. Considering the bioaccessible As only, our results demonstrated a low carcinogenic risk to humans.

ACKNOWLEDGEMENTS

CNPq, Capes, FAPEMIG and the UFOP for support is acknowledge.

REFERENCES

Borba R.P., Figueiredo B.R. & Cavalcanti J.A. 2004. Arsênio na água subterrânea em ouro preto e mariana, quadrilátero ferrífero (MG). *Revista Escola de Minas* 57(1): 45–51.

BRASIL. 2009. Resolução CONAMA N° 420.

CETESB 2013. *Spreadsheets for Evaluation*. Available in https://cetesb.sp.gov.br/areas-contaminadas/planilhas-para-avaliacao/

Gonçalves J.A.C., de Lena J.C., Paiva J.F. *et al.* 2007. Arsenic in the groundwater of Ouro Preto (Brazil): its temporal behavior as influenced by the hydric regime and hydrogeology. *Environ. Earth Sci.* 53(4): 785–793.

Matschullat J., Birman K. *et al.* 2007. Long-term environmental impact of arsenic-dispersion in Minas Gerais, Brazil. *Trace Metals and Other Contaminants in the Environment*, 9(6): 365–382.

Pelfrêne A. Pelfrêne A., Cave M.R. *et al.* 2017. In vitro investigations of human bioaccessibility from reference materials using simulated lung fluids. *Int. J. Environ. Res. Public Health* 14(2): 112.

Ure A.M., Quevauviller P., Muntau H. & Griepink B. 1993. Speciation of heavy metals in soils and sediments. *Int. J. Environ. Anal. Chem.* 51(1–4): 135–151.

USEPA. 2017. *Sop for in Vitro Bioaccessibility for Lead and Arsenic in Soil.*

Section 4: Technologies for arsenic removal from water

4.1 Adsorptive processes

Arsenic in the Environment: Bridging Science to Practice for Sustainable Development – van der Wal, Ahmad, Petrusevski, Weijma, Savic, van der Wens, Beerendonk, Bhattacharya, Bundschuh & Naidu (Eds)
© 2023 the Author(s), ISBN 978-1-032-32928-4

Arsenate removal from drinking water by application of pelletized iron (hydr)oxides

L. de Waal[1], A. Ahmad[2] & C.H.M. Hofman-Caris[1]
[1]*KWR Water Research Institute, Nieuwegein, The Netherlands*
[2]*SCR-Sibelco N.V., Antwerp, Belgium*

ABSTRACT: In drinking water production iron sludge is formed, either by aeration of groundwater, or as a result of the addition of coagulant (e.g. $FeCl_3$) to surface water. For the Netherlands this means a yearly production of 80,000 tons of iron sludge. Iron(hydr)oxide is known for its good adsorption capacity for arsenate, phosphate and sulfide. However, large scale application is hindered by the high water content of iron sludge (70–90 weight %), which causes problems in transport and handling both in the supply chain and by the customer.

1 INTRODUCTION

In the Netherlands drinking water is produced from either groundwater, surface water or river bank filtrate. Most groundwater sources contain ferrous iron, which upon aeration is turned into iron (hydr)oxides. In case surface water is used as a source for drinking water, Fe(III)-salts are added as a coagulant. In both cases a sludge is formed, with a high iron content (up to about 50% Fe in dry matter) In total this amounts to a yearly production of 80,000 tons of iron rich sludge in the Netherlands.

Freshly produced iron(hydr)oxides are known to have good adsorption capacity for arsenate (Nagar *et al.*, 2010; Ociński *et al.*, 2016) and other anions like phosphate (Ippolito *et al.*, 2011; Makris *et al.*, 2005). The presence of arsenic in groundwater is unwanted if it is used as a source for drinking water production and high phosphate concentrations are a problem in surface or wastewater.

Large scale application of wet iron(hydr)oxide sludges as adsorbents is limited by high transportation costs and environmental impact due to their low the solids content, which varies between 10–30% (Hofman-Caris *et al.*, 2017).

By turning the iron (hydr)oxides into stable pellets with high solids content, storage and transport will become much more practical and new applications as filter material for drinking- and wastewater polishing become available. In this project we have studied methods to produce stable pellets for the removal of arsenate from drinking water.

2 METHODS

Different types of iron(hydr)oxides were studied, both from a groundwater and surface water plants. As a binder material 1 or 2 wt% bentonite (Cebogel IPR-012) or carboxymethyl cellulose (CMC, Gabrose P300D) was used. Drying occurred at 105, 500 or 800°C. Pellet sizes were 0.6–1.4 mm.

The iron (hydr)oxides were characterized by means of their specific surface area (S_{BET}), elemental composition (X-ray fluorescence), crystallinity (X-ray diffraction), and maximum compression strength (robotic compression tester).

Arsenate adsorption performance of (prewashed) iron(hydr)oxide pellets was measured in column experiments carried out at KWR water research institute.

3 RESULTS AND DISCUSSION

Pelletization experiments showed that a binder material will be required. Carboxymethyl cellulose appeared to give good results. This binder has the advantage that it can also be obtained in "food grade". Heating of the pellets significantly increased the pellet strength, but also enhanced crystallization to goethite (Hofman-Caris *et al.*, 2015). This results in a huge decrease in both specific surface area and adsorption capacity. A similar effect can be observed as a result of aging of the iron sludge. Another problem was caused by the pelletization technique used. Applying high

DOI: 10.1201/9781003317395-124

pressure, e.g. by using an extruder, results in stable pellets with low porosity, and as a result low adsorption capacity.

During column experiments (Figure 1), it was found that iron sludge pellets are suitable adsorbents for arsenate removal, but that desorption of other compounds (manganese and dissolved organic carbon) may occur. By pre-washing of pellets at different pH conditions, desorption behaviour of these components could be improved.

Arsenate adsorption performance also appears to be influenced by washing the pellets before application, as shown in Figure 2.

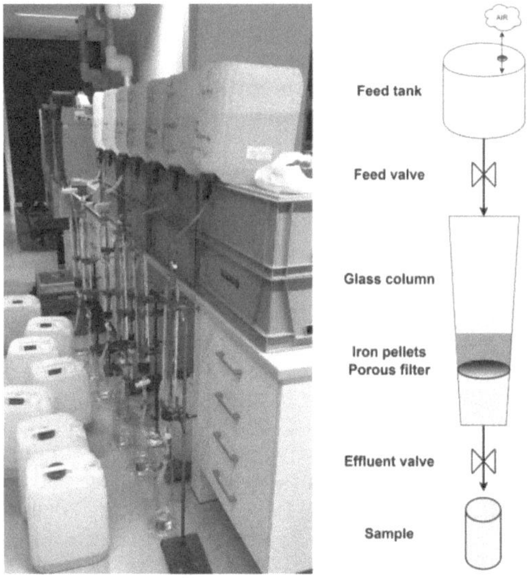

Figure 1. Column experiment for arsenate adsorption experiments.

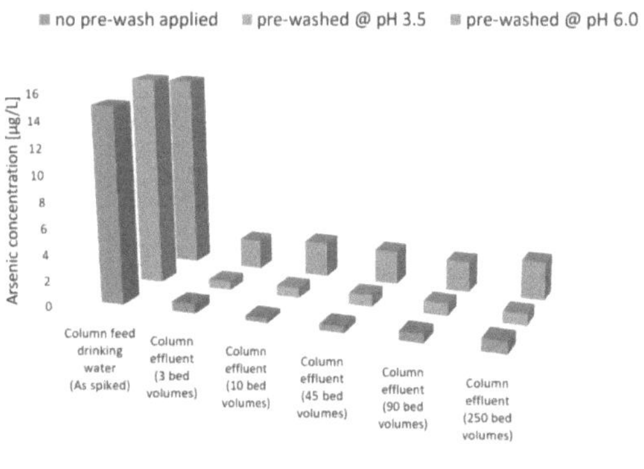

Figure 2. Arsenate adsorption on (pre-washed) pellets in first 250 bed volumes.

At the moment pellet production techniques are being optimized, to obtain stable pellets with high adsorption capacity, and to prevent leakage of unwanted compounds from the pellets.

4 CONCLUSIONS

It can be concluded that iron sludge can be used to produce effective adsorbent for the removal of arsenate from drinking water. However, for large scale application it is important that a stable pellets can be produced, without, however, losing their adsorption capacity. A rotating disk pelletization technique and binder materials are being optimized for the production of such pellets.

Furthermore, attention should be paid to preventing leakage of other compounds from the pellets, especially when applied as polishing step in the drinking water production process. This is realized by selection of suitable iron sludge compositions, and optimizing a washing procedure for the pellets.

REFERENCES

Hofman-Caris R., Ahmad A., Siegers W., Rahn S., Voort J.-W. & van der Kolk O. 2017. *Pelletized Drinking Water Treatment Residuals for the Removal of Arsenic and Phosphate (KWR 2017.036).* Retrieved from http://api.kwrwater.nl//uploads/2017/12/KWR-2017.036-Pelletized-drinking-water-treatment-residuals-for-the-removal-of-arsenic-and-phosphate-from-water-(PUBLIC).pdf

Hofman-Caris R., Siegers W., Hofman J., Elings J., Kolk O. van der & Jong A. de. 2015. *Granular Iron (hydr)oxide for Phosphate Removal From Water (KWR 2015.008).*

Ippolito J.A., Barbarick K.A. & Elliott H.A. 2011. Drinking water treatment residuals: a review of recent uses. *J. Environ. Qual.* 40(1): 1–12.

Makris K.C., Harris W.G., O'Connor G.A., Obreza T. A. & Elliott H.A. 2005. Physicochemical properties related to long-term phosphorus retention by drinking-water treatment residuals. *Environ. Sci. Technol.* 39(11): 4280–4289.

Nagar R., Sarkar D., Makris K.C. & Datta R. (2010). Effect of solution chemistry on arsenic sorption by Fe- and Al-based drinking-water treatment residuals. *Chemosphere* 78(8): 1028–1035.

Ociński D., Jacukowicz-Sobala I., Mazur P., Raczyk J. & Kociołek-Balawejder E. 2016. Water treatment residuals containing iron and manganese oxides for arsenic removal from water – characterization of physicochemical properties and adsorption studies. *Chem. Eng. J.* 294: 210–221.

*Arsenic in the Environment: Bridging Science to Practice for Sustainable Development –
van der Wal, Ahmad, Petrusevski, Weijma, Savic, van der Wens, Beerendonk, Bhattacharya, Bundschuh & Naidu (Eds)*

Arsenic removal from drinking water with Granular Ferric Hydroxide (GEH): Effect of vanadium and phosphorous on the adsorption capacity

C. Bahr
GEH Wasserchemie GmbH & Co. KG, Osnabrück, Germany

ABSTRACT: Groundwater contaminated with geogenic arsenic of volcanic origin typically also contains elevated concentrations of vanadium and phosphorus. Due to the chemical similarity of these three oxyanions, adsorption competition occurs, which leads to a reduction in the efficiency of drinking water treatment plants. Laboratory tests have shown that the competition between the three substances is equally high when adsorbed on Granular Ferric Hydroxide (GEH). The analysis of loaded adsorbents from existing waterworks confirmed this strong competition between arsenic and vanadium. In contrast, a lower competition effect was found for phosphorus, although the total effect of phosphorus is of course considerable due to the higher concentration in the water.

1 INTRODUCTION

Various technologies are available for removing arsenic from drinking water (Ahmad *et al.*, 2017). Besides iron flocculation, ion exchangers or membrane processes, the use of adsorption filters is the technically simplest and most reliable process. In practice, adsorbents based on iron hydroxide show the greatest performance in terms of selectivity and capacity (Amy *et al.*, 2005). Groundwater contaminated with geogenic arsenic of volcanic origin can be found for example in France (Massif Central) or in Italy (e.g. Lazio region). In addition to arsenic, these waters typically also contain elevated concentrations of phosphorus and vanadium, which lead to a reduction in the arsenic adsorption capacity and thus in practice decreases the service life of the adsorption filters. In earlier laboratory investigations, the competitive adsorption was already investigated in small columns tests (Nguyen *et al.*, 2011). The objective of the present study was to determine the competitive adsorption of the three oxyanions arsenate, vanadate and phosphate and to compare the results with the loading profile in full-scale adsorbers in existing waterworks.

2 MATERIALS AND METHODS

2.1 *Granular Ferric Hydroxide (GEH)*

Granular Ferric Hydroxide (GEH) is a synthetically produced, pure iron hydroxide, which is mineralogically composed of akaganeite (β-FeOOH) and ferrihydrite ($Fe(OH)_3$). It was developed in the 1990s at the Technical University of Berlin, Germany, and was the first commercially available iron-based adsorbent (Driehaus *et al.*, 1998). The product quality complies with the European Standard EN 15029 and the product is certified according to NSF/ANSI Standard 61 for the treatment of drinking water.

2.2 *Laboratory batch tests*

In laboratory experiments the adsorption competition of arsenic vs. vanadium (As/V), arsenic vs. phosphorus (As/P) and vanadium vs. phosphorus (V/P) was investigated. For this purpose, DI water was mixed with different molar ratios of the oxyanions, the sum always being 50 µmol/L. The GEH was dried, ground (<63 µm) and added to the individual batches (200 mL) using a fixed dosage of 74 mg/L (dry solid). The bottles were shaken for 48 h and the pH value was kept constant at pH 7.5. After filtration (0.45 µm) the remaining concentration of oxyanions in the solution was analyzed. As and V were determined by ICP-MS (ISO 17294-2), P was analyzed photometrically (ISO 6878). The loading q was calculated from the concentration difference and the dosed amount of GEH.

2.3 *Full-scale GEH water treatment plant*

The waterworks, located in the Auvergne region, France, uses 4 GEH adsorbers, each with 9.6 t GEH in parallel operation, to treat more than 3 million cubic meters of drinking water per year. The water composition is typical for this volcanic region and is characterized by increased values for As, V and P (Table 1). At this site, the GEH adsorber achieves a specific treatment capacity of approx. 50,000 bed volumes (BV) (<10 µg/L As). This is comparatively low in relation to the low arsenic concentration and the low pH value. Therefore, a competitive adsorption is indicated.

DOI: 10.1201/9781003317395-125

Table 1. Water composition (inlet to GEH filter).

pH	EC	Hardness	As	P	V	SiO_2
–	μS/cm	mmol/L	μg/L	μg/L	μg/L	mg/L
6.5	164	0.5	24	300	12	30

3 RESULTS AND DISCUSSION

3.1 *As, V, P adsorption competition in batch tests*

The laboratory experiments show that all three oxyanions adsorb similarly well on GEH (Figure 1). Irrespective of the molar ratio of the compounds (always 50 μmol/L in total), an approximately equally high sum loading of 540 μmol/g GEH (dry solid) was achieved in all batches. This behavior was found to be the same for all 3 model systems As/V, As/P and V/P, which means that the competition between each oxyanion does not differ significantly.

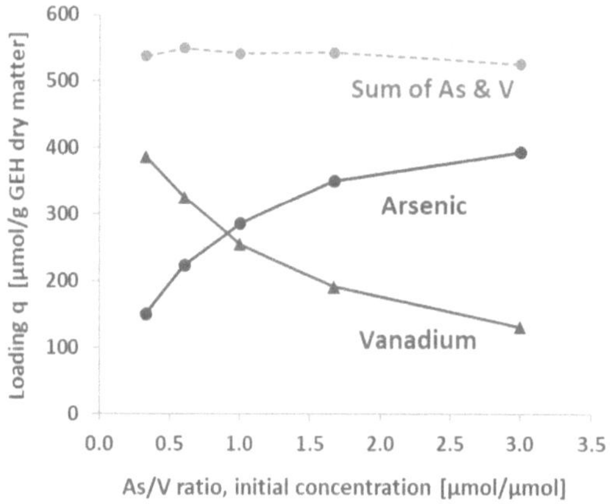

Figure 1. Arsenic and vanadium loading for different molar As/V ratios in the water.

3.2 *As, V, P adsorption competition in full-scale treatment plant*

The GEH adsorber in the waterworks was sampled vertically over the entire bed height of 135 cm at the time of arsenic breakthrough (i.e. 10 μg/L As in the effluent). The analysis of the loaded GEH was performed by ICP-MS (ISO 17294-2) after previous acid digestion.

As expected, the highest loadings were obtained at the bed surface. With increasing bed depth, the loading decreases almost linearly.

For vanadium, which is present in a molar concentration comparable to that of arsenic, the same loadings were achieved as for arsenic, thus confirming the competitive As/V behavior determined in laboratory tests. For phosphorus, on the other hand, which was present in water in a higher concentration by a factor of 10, high loads were also achieved, but these were only about 2–3 times higher than for arsenic and vanadium and thus lower than expected (Figure 2).

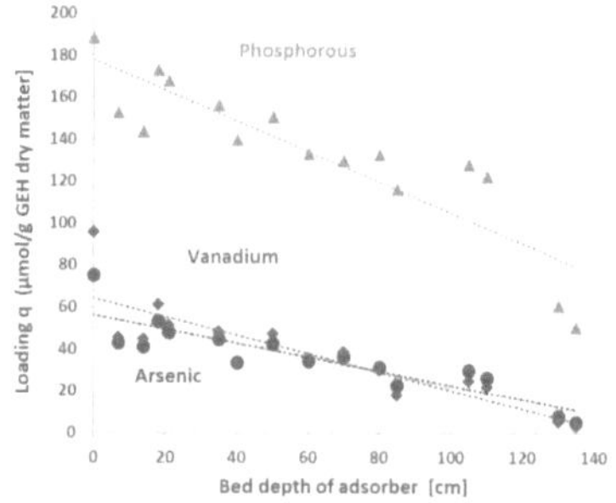

Figure 2. Vertical profile of As, P, V loading in GEH bed. Raw water concentrations: 0.32 μmol/L As; 3.3 μmol/L P; 0.24 μmol/L V.

4 CONCLUSIONS

Under laboratory conditions it could be shown that the adsorption competition between arsenic, vanadium and phosphorous is approximately the same. This was confirmed by GEH samples from waterworks for arsenic and vanadium. The phosphorous loadings, on the other hand, were lower than expected, so that in practice competition is expected to be less intense than for vanadium. When designing arsenic removal plants, the strong competitive effect of vanadium should not be overlooked. Furthermore, vanadium itself is considered a new problem substance in drinking water and a future limit value is being discussed. The first plants for the selective removal of vanadium with GEH have already gone into operation in Germany (Bahr *et al.*, 2019).

REFERENCES

Ahmad A., Richards L.A. & Bhattacharya P. 2017. Arsenic remediation of drinking water: an overview. (chapter 7). In: P. Bhattacharya D.A. Polya & D. Jovanovic (eds.) *Best Practice Guide on the Control of Arsenic in Drinking Water*. IWA Publishing, UK. pp. 79–89.

Amy G., Chen H., Drizo A., von Gunten U., Brandhuber P., Hund R., Chowdhury Z., Kommineni S., Sinha S., Jekel M. & Banerjee K. 2005. *Adsorbent Treatment Technologies for Arsenic Removal*, AWWA Research Foundation.

Bahr C., Payer H. & Simon A. 2019. Entfernung von vanadium aus trinkwasser mit granuliertem eisenhydroxid: ergebnisse eines pilotversuchs im Saarland. *GWF Wasser Abwasser* 160(6): 79–88.

Driehaus W., Jekel M. & Hildebrandt U. 1998. Granular ferric hydroxide—a new adsorbent for the removal of arsenic from natural water. *Aqua* 47: 30–35.

Nguyen V.L., Chen W.H., Young T. & Darby J. 2011. Effect of interferences on the breakthrough of arsenic: rapid small scale column tests. *Water Res.* 45(14): 4069–4080.

Arsenic in the Environment: Bridging Science to Practice for Sustainable Development –
van der Wal, Ahmad, Petrusevski, Weijma, Savic, van der Wens, Beerendonk, Bhattacharya, Bundschuh & Naidu (Eds)
© 2023 the Author(s), ISBN 978-1-032-32928-4

Dynamic membrane pre-coated with micro-sized iron oxyhydroxide for arsenic removal: Application study and mathematical modelling

M. Usman, A.I. Belkasmi & M. Ernst
Institute for Water Resources and Water Supply, Hamburg University of Technology, Hamburg, Germany

ABSTRACT: Arsenic is among the major drinking water contaminants affecting populations in many countries because it causes serious health problems on long-term exposure. In this study, the application of dynamic membranes (DM's) pre-coated with micro-sized fraction of granular ferric hydroxide (μGFH) in microfiltration process is investigated to remove As(V) from contaminated water. DM shows As(V) removal efficiency of 90% in continuous flow experiments at constant water flux. The performance of the DM's obtained depended on the water flux and the amount of μGFH particles deposited. The linear driving force (LDF) model is used to predict the As(V) removal rates and the simulation results show that LDF model could describe the As(V) removal rates correctly. The results of this study exhibit that DM's can be a viable option in removing arsenic for drinking water production.

1 INTRODUCTION

The use of low pressure membranes alone or in combination with other physiological treatment methods has become popular and appealing in drinking water treatment, because porous membranes serve as a nearly complete barrier to particulates and colloidal matter including pathogenic microorganisms from natural water, however they are not able to remove mono and polyvalent ions, i.e., arsenate. A dynamic membrane (DM), which may also called secondary membrane, is formed on an underlying support material, e.g., a porous membrane or a filter cloth, when the filtered solution contains suspended solid particles such as microbial cells, flocs, and adsorbent particles. Formation of dynamic cake layer can determine removal efficiency of the system since the deposited cake layer so-called DM will act as a "secondary" membrane prior the "real" membrane (Ersahin *et al.*, 2012). DM's potentially offer easy formation and replacement of dynamic layer by simple backwashing.

In our formed study, we have reported that μGFH is a very effective adsorbent for arsenic removal (Usman *et al.*, 2018, 2020). This fine fraction of adsorbent media may be used as dynamic layer forming material in UF membrane processes. Accordingly the cake layer of μGFH formed over the support material (e.g.,) MF membrane) will adsorb arsenic from contaminated water.

The primary objective of study is to explore the potential of μGFH, a low cost iron oxyhydroxide, as dynamic cake layer forming material in a low-pressure driven membrane applications. A simplified mathematical model based on adsorption equilibrium and mass transport is used to predict the As(V) removal under different experimental conditions.

2 LDF MODEL

A mass balance over an infinitesimal element of the UF filter leads to eq. 1. A detailed description of the overall mass balance was reported elsewhere (Sperlich *et al.*, 2008).

$$\epsilon_{CL}\frac{\partial C}{\partial t} + v_F \quad \frac{\partial C}{\partial z} + 3(1-\epsilon_{CL})\frac{k_f}{r_p}(C_p - C_s) = 0 \tag{1}$$

where, ϵ_{CL} is the porosity of the dynamic cake layer, k_f is mass transfer rate due to film diffusion. Intraparticle transport (D_s, which characterizes the diffusion rate of adsorbate along the surface of the adsorbent particle) is described by linear driving force (LDF) model, which was derived by (Sperlich *et al.*, 2008). This model assumes that the rate of mass transfer is directly proportional to the concentration gradient developed between μGFH surface and feed water. According to LDF model, the rate of adsorption of an adsorbate in the adsorbent particle is given by:

$$\frac{dq_{avg}}{dt} = \frac{15D_s}{r_p^2}\left(q_s(t) - q_{avg}(t)\right) \tag{2}$$

with q_{avg} is the average adsorbed phase concentration in the adsorbent particle, q_s is the adsorbed phase concentration at the external surface of the adsorbent particle or at solid-liquid interface, and r_p is the particle radius.

$$q_s = K_F(C_s)^n \quad (3)$$

C_s is the liquid phase at the solid-liquid interface. C_s and q_s are correlated based on the Freundlich isotherms expression of eq. (3). LDF model calculations are performed using the software fixed bed simulation tool (FAST 2.1) developed by (Sperlich *et al.*, 2008).

3 METHOD AND MATERIALS

Flat sheet PES-based MF membranes with a nominal size of 200 nm used a support material. The mean diameter (d_{50}) of μGFH (GEH Wasserchemie GmbH & Co, Osnabrück, Germany) is 3.5 μm (original μGFH with particle size < 250 μm passed through 230 mesh sieve) and total surface area (BET) of 283 m^2/g. Pre-coated DM is formed on the MF support material by filtration of μGFH suspension (either 300 or 400 mg with 150 mL water). DM is then applied in dead-end filtration set-up (Figure 1).

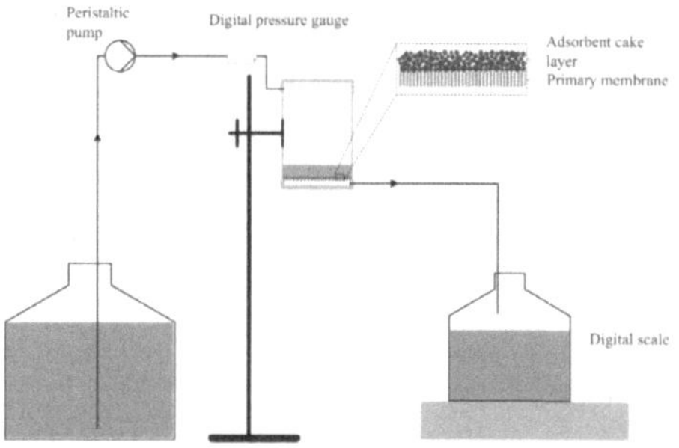

Figure 1. Schematic diagram of the dead-end filtration setup.

4 RESULTS AND DISCUSSION

4.1 *Adsorption equilibrium isotherms*

As(V) adsorption isotherms are determined to obtain the isotherm equilibrium parameters at an equilibration time of 7 days and T = 20°C. As(V) adsorption isotherms are shown in Figure 2. The values of Freundlich parameters (K_F = 4.5 μg/mg and n = 0.268) are computed for batch adsorption isotherm data and subsequently used as inputs for the LDF model.

4.2 *As(V) removal using μGFH DM*

Figure 2 depicts the As(V) permeate concentration as a function of the operation time for two different membrane water fluxes and amount of μGFH pre-coated on supporting membrane. The results show that DM pre-coated with μGFH results in an

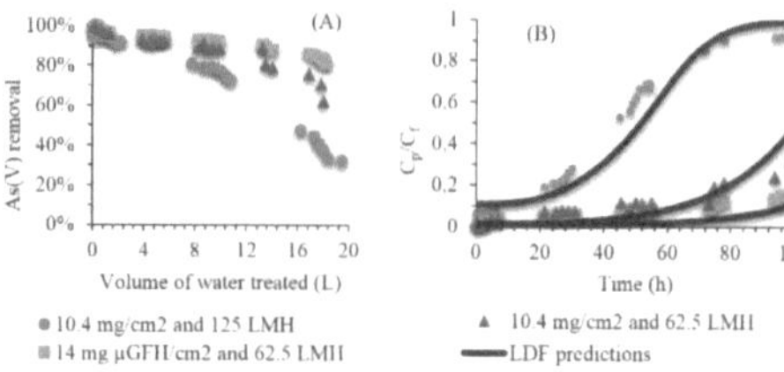

Figure 2. As(V) removal under different experimental conditions with As(V) feed concentration of 380 ug/L (A) and C_p/C_f vs. time profiles of As(V) permeate concentration and corresponding model simulations at D_s = 1.1 × 10^{-18} m^2/s (B). Solid circles reflects the experimental data, while solid lines represent values predicted by model.

immediate decrease in As(V) permeate concentration with the As(V) concentration reaching a minimum value after which the As(V) permeate concentration starts to increase with time (Figure 2B). At constant water flux of 62.5 LMH and varying amount of adsorbent particles deposited, larger than 12 L of volume of water can be treated with As(V) removal efficiency of more than 80% (Figure 2A).

It can also be seen in Figure 2B, that the C_p/C_f profiles can be affected by the amount of μGFH deposited at the surface of support membrane and water flux. The higher the amount of μGFH coated at supporting membrane, the lower the normalized As(V) permeate concentration especially after a operation time of 72h (Figure 2B). From the results shown in the Figure 2, lower flux favors the As(V) removal. At higher water flux, the As(V) contaminated water has short contact time (~2.9 s) with μGFH deposited at the surface of MF, while at lower water flux the contact time of As(V) with μGFH is ~5.8 s.

4.3 *Mathematical modelling of results of μGFH DM*

LDF model is applied to predict As(V) removal rates and this model has been used to predict the removal rates of fixed-bed filters as well as submerged hollow fiber membrane filtration hybrid system (*Sperlich et al.*, 2008; Vigneswaran *et al.*, 2003). Figure 3B shows experimental data and values predicted by the model over a range of conditions. It is evident that the model predicts the data trend effectively and only tends to slightly over predict the As(V) removal.

5 CONCLUSIONS AND RECOMMENDATIONS

The DM's tests exhibit that μGFH has significant potential to be utilized as dynamic cake layer forming material in UF processes. Moreover, performance of DM's can be controlled by changing the water flux and amount of particles deposited at the supporting material. The results presented demonstrates that use of these inexpensive materials would enhance the sustainability of the industrial production process of conventional granular adsorbents by utilizing the wastes created during the process of adsorbent production.

ACKNOWLEDGEMENTS

We acknowledge the German Academic Exchange Service (DAAD) and Hamburg University of technology for supporting the project.

REFERENCES

Ersahin M.E. *et al.* 2012. A review on dynamic membrane filtration: materials, applications and future perspectives. *Bioresour. Technol.* 122: 196–206.

Sperlich A. *et al.* 2008. Predicting anion breakthrough in granular ferric hydroxide (GFH) adsorption filters. *Water Res.* 42(8–9): 2073–2082.

Usman M. *et al.* 2018. Performance evaluation of small sized powdered ferric hydroxide as arsenic adsorbent. *Water* 10(7): 957.

Usman M. *et al.* 2020. Mathematical modeling of arsenic(V) adsorption onto iron oxyhydroxides in an adsorption-submerged membrane hybrid system. *J. Hazard. Mater.* 400: 123221.

Vigneswaran S. *et al.* 2003. Application of a PAC-membrane hybrid system for removal of organics from secondary sewage effluent: experiments and modelling. *Sep. Sci. Technol.* 38(10): 2183–2199.

Arsenic in the Environment: Bridging Science to Practice for Sustainable Development –
van der Wal, Ahmad, Petrusevski, Weijma, Savic, van der Wens, Beerendonk, Bhattacharya, Bundschuh & Naidu (Eds)
 ISBN 978-1-032-32928-4

Removal of As(III) and As(V) from water samples using metallurgical slags sourced from the steel and iron industries

A. Chicken & J. Gonzálex
Faculty of Chemistry, National Autonomous University of Mexico, Mexico City, Mexico

ABSTRACT: Water is an essential resource, and as a result of demographic and industrial growth, large masses of water around the world are contaminated. Arsenic is a common and hazardous contaminant, and even though there are various removal techniques, these are expensive and provide low yields. Therefore, there is a need to find new adsorbents with better yields and lower costs. In this work 5 slags sourced from metallurgical processes were characterized and used as potential adsorbents of As (III) and As (V). Some variables such as stirring time, slag amount, arsenic concentration and their effect on arsenic removal were studied. Arsenic determination was performed by pulse differential polarography and the reduction reaction from As (V) to As (III) was optimized. Results show that the Corsa slag can remove as much as the 90% of both contaminants.

1 INTRODUCTION

Because large masses of water around the world are contaminated, this is a health-related issue, hence, the reduction of water contaminants is of paramount importance (Briones & Razo 2008). Arsenic even at ppb levels it represents a health risk for living beings, some examples of arsenic removal techniques are precipitation, electrocoagulation, osmosis, adsorption, filtration, and ion exchange, although usually, they are expensive and inefficient; however, adsorption is a simple and easy to operate method, and activated charcoal is, despite its cost, the most widely used material as an adsorbent. Nevertheless, researchers are currently looking for cheaper materials with similar or even better yields than those for activated charcoal. A material requires certain characteristics like availability in large quantities, not being hazardous, a high carbon and oxygen content, thermal stability, a small pore diameter and a high surface area in order to be considered a good adsorbent. When obtaining and recovering metals by metallurgical processes, residues or slags are produced, whose chemical composition depends on the mineral from which they are extracted, on the fluxes used and on the furnace temperatures (Ros-Latienda & Fernández 2013). Slags are mainly composed of silicates, metals and calcium, iron, manganese and magnesium oxides. It is considered that their presence in large quantities makes them potential adsorbents in water treatment (Shanbazi *et al.*, 2017).

2 MATERIALS AND METHODS

2.1 *Characterization*

Surface area and pore diameter were determined with a BET equipment, in which a previous cleaning and activation treatment was carried out with N2 for 24 hours. Subsequently, N2 was adsorbed at 77 K using an accurately weighed amount of slag for the determination of its surface area. The composition of slags and whether they had a crystalline or amorphous character were determined by fluorescence and X-ray diffraction. A Z meter device was used to determine the surface charge of slags, and their pH was adjusted from 3 to 11. In order to assess if these slags release some other contaminant into the environment, leachings were made at the pH obtained as a result of the mix between slag and water. For this analysis, the slag was weighed and mixed with distilled water (proportion 1:20), then samples were stirred for 18 hours on an axial shaker and subsequently the liquid phase was analyzed by ICP-OES.

2.2 *Arsenic quantification*

Arsenic(III) was determined by differential pulse polarography, and HCl 1 M was used as a support electrolyte because in this technique, the material is suspended and the other slag components do not interfere with the quantification. An area vs. concentration calibration curve was plotted from 0 to 10 ppm in order to confirm its linearity. As (V) is not electroactive but because of the suitability of the electrochemical method to quantify arsenic without interferences, the reduction from As(V) to As(III) was optimized. Sodium metabisulfite was used in the presence of KI as a catalyst and acidic medium, and the reaction was carried out at three different temperatures (25, 50 and 70°C).

2.3 *Removal of As (III) and As (V)*

The As(III) and As(V) removal tests were done with the five slags using a 2^k experimental design

 DOI: 10.1201/9781003317395-127

with the following variables and levels: amount of slag (50 and 250 mg), concentration of As (3 and 10 ppm) and stirring time (15 and 60 min).

3 RESULTS AND DISCUSSION

3.1 *Characterization*

Slags have a surface area ranging from 1 to 10 m^2/g (Table 1), which is considerably smaller than that of activated charcoal, however, this is not the only factor affecting the ability to remove contaminants from a material. The pore size range from 6–17 nm, therefore, the material can be classified as mesoporous.

Table 1. Surface area and pore diameter of the five slags.

Slag	Surface area (m^2/g)	Pore diameter (nm)
Ahmsa	9.40	8.93
Corsa	5.53	7.57
Nardo	2.57	16.09
Mittal	1.60	16.86
Matehuala	10.08	6.56

With X-ray fluorescence we observe that elements found in slags are calcium, iron, silicon, magnesium, aluminum and manganese (Table 2), and these elements are frequently found in materials used as adsorbents such as clays, zeolites, among others.

Table 2. Main components percentage of the five slags.

Comp	Ahmsa	Corsa	Mateh	Mittal	Nardo
MgO	7.7	3.6	0	8.17	4.4
Al_2O_3	11.4	13	5.5	8.5	4.5
SiO2	34.8	50.1	36.8	15.8	14
CaO	40.2	3.8	22.7	22.9	44.4
Fe_2O_3	1.4	24.2	27.5	42.8	24.7

With X-ray diffraction we observed that Corsa and Nardo have a greater amorphous character due to the background noise in the diffractogram. In general, it is known that amorphous materials are better adsorbents, hence, we expect good results from these two slags used as adsorbents (Figures 1a & b).

The zero load point of the slags was determined in order to determine the surface charge as a function of pH and to assess if this could explain the mechanism by which arsenic removal is performed. It is observed that the five slags have a zero charge point close to 5.5, except Mittal with 4.5, therefore, this means that below this pH the surface charge of the slags is positive and negative above this value (Figure 2). When the slag is mixed with water in natural conditions, the resulting pH

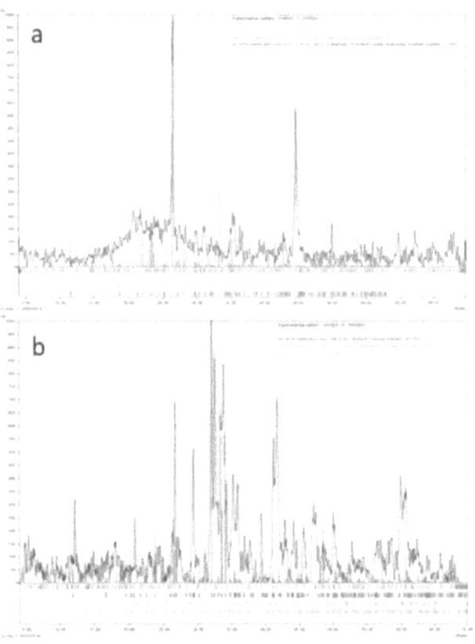

Figure 1. XRD patterns for a) Corsa slag; b) Nardo slag.

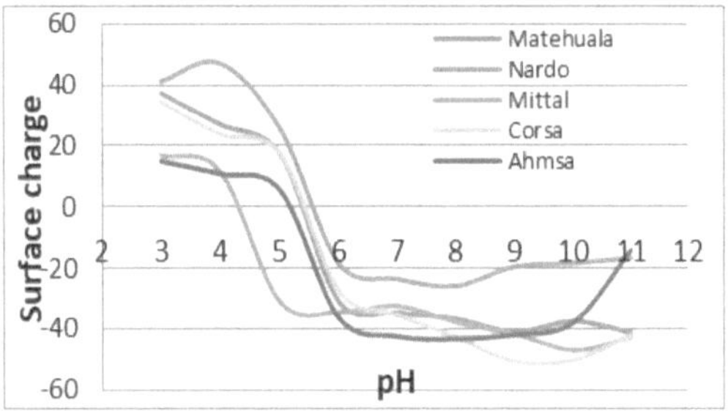

Figure 2. Surface charge of the five slags vs pH.

is near 10, therefore, the surface charge for the material is negative. As (III) is in the $H_2AsO_3^-$ form and As (V) as $HAsO_4^{2-}$, hence, we rule out any electrostatic interaction.

Figure 3 shows elemental composition in the slags leachates, where red lines indicate the

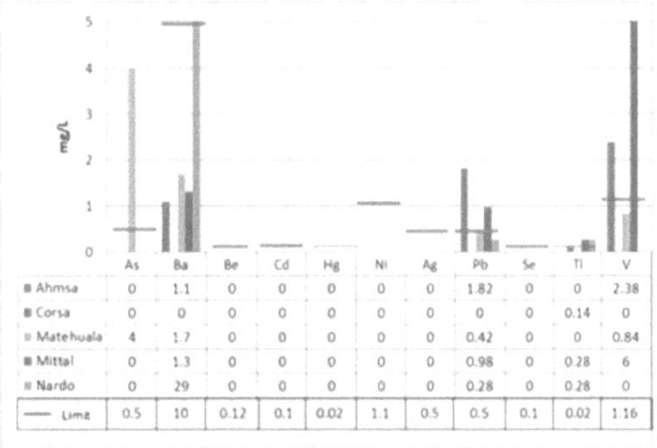

Figure 3. Main contaminants concentrations (mg/L) in the leachate at the natural pH of the five slags.

maximum permissible limit established by the NOM-147SEMARNAT/SSA1 (2004), and Corsa does not release any contaminant above these limits.

3.2 *Arsenic quantification*

The polarogram of As (III) shows two signals, one at −0.48 and the other at −0.88 V vs Ag/AgCl (Figure 4), which corresponds to the reduction from As^{3+} to As^{0} and from As^{0} to As^{3-}, respectively.

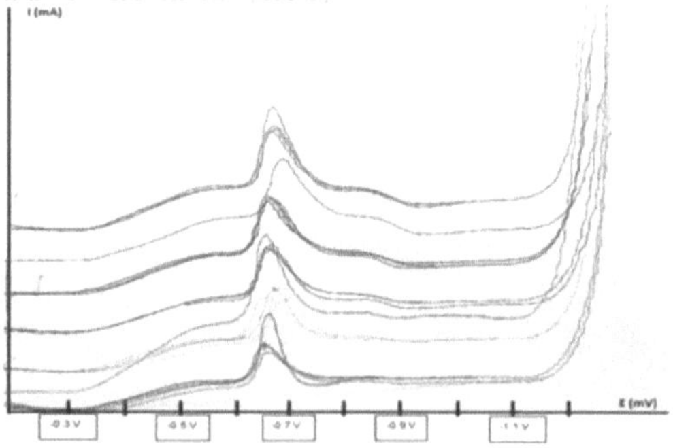

Figure 4. Polarogram of As (III) in HCl 1M.

The kinetic study showed that As(V) reduction was carried out completely with a shorter stirring time at 70°C in acidic medium (Figure 5).

3.3 *Removal of As(III) and As(V)*

Figure 6 presents the effects of different variables on the As removal process. It is worth noting that the slag with the highest percentage of As(III) and As(V) removal was Corsa with almost 90% and Matehuala, Ahmsa and Mittal are the ones with less contaminant removal. For both contaminants, the removal percentage increases as the amount of slag and stirring time increases.

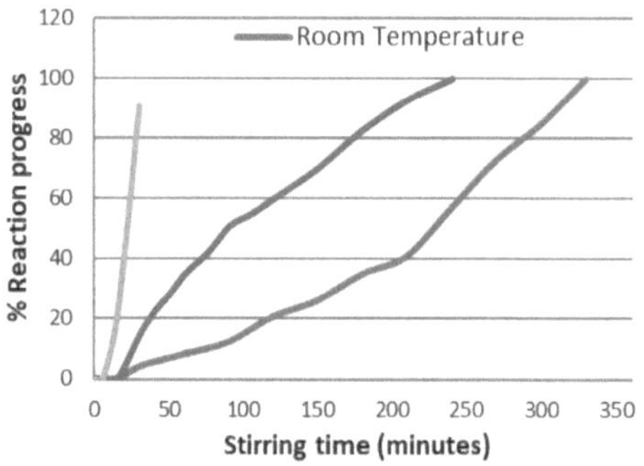

Figure 5. Reduction kinetics from As (V) to As (III).

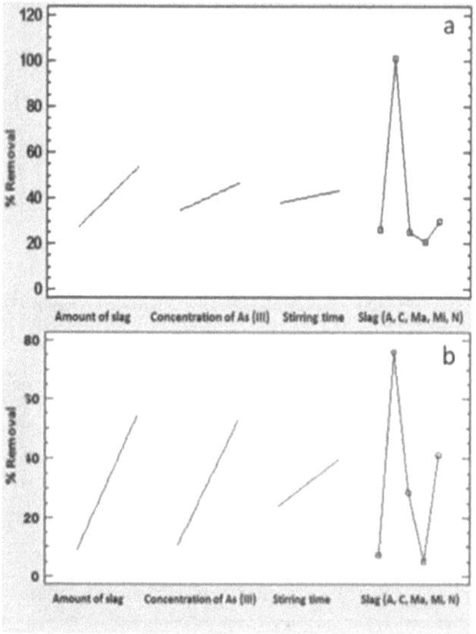

Figure 6. Effect of different variables on the process of removal for: a) As(III) b) As(V).

4 CONCLUSION

As(III) and As(V) were removed with slags sourced from the steel industry. Corsa removes approximately 90% of the contaminant. This slag showed amorphous characteristics and a high content of iron, which is believed to be the responsible for arsenic removal. Despite having a low surface area, the Corsa slag has produced satisfactory results without releasing any other contaminant into the environment.

REFERENCES

Briones R. & Razo I. 2008. Contaminación de Aguas Naturales Por Arsénico Asociado a la Actividad Minero-Metalúrgica: *Caso de Estudio en San Luis Potosí, México*. 181–188.

Mexican Official Norm. NOM-147SEMARNAT/SSA1-2004. *Secretaria de Medio Ambiente y Recursos Naturales (200)*. Que Establece Criterios Para Determinar las Concentraciones de Remediación de Suelos Contaminados por Arsénico, Bario, Berilio, Cadmio, Cromo Hexavalente, Mercurio, Níquel, Plata, Plomo, Selenio, Talio y/o Vanadio. Mexico: Author.

Parkhurst D.L. 1995. Users guide to PHREEQC-A computer program for speciation, reaction-path, advective-transport, and inverse geochemical calculations. *U.S. Geol. Surv. Water Resour. Invest. Rep.* 95–4227.

Parkhurst D.L. & Appelo C.A.J. 1999. *User's Guide to PHREEQC (version 2)*. U.S. Geol. Surv. *Water Resour. Invest.*, 99–4529, 312 pp.

Ros L. & Fernández J. 2013. Caracterización de escorias metalúrgicas procedentes de yacimientos arqueológicos de Navarra (Siglos II A. C. – IV d. C.). *Revista de Metalurgia* 49(6). 438–448.

Shahbazi M., Rowshanzamir M., Mahdi S. & Mahdi H. 2017. Optimization of carpet waste fibers and steel slag particles to reinforce expansive soil using response surface methodology. *Appl. Clay Sci.* 142: 185–192.

Arsenic in the Environment: Bridging Science to Practice for Sustainable Development –
van der Wal, Ahmad, Petrusevski, Weijma, Savic, van der Wens, Beerendonk, Bhattacharya, Bundschuh & Naidu (Eds)
© 2023 the Author(s), ISBN 978-1-032-32928-4

Selective adsorption of arsenate and antimonate from phosphate-rich waters

B. Dousova[1], M. Lhotka[1], K. Kremenic[1], D. Kolousek[1] & A. Cechova[2]
[1]*University of Chemistry and Technology Prague, Prague, Czech Republic*
[2]*GLAZURA, s.r.o., Dobříň, Czech Republic*

ABSTRACT: The similarity of arsenic (As), antimony (Sb) and phosphorus (P) originates in their identical periodicity, which results in the formation of the same inorganic species, analogical binding properties etc. While As and currently also Sb denote permanent environmental risk originating in their toxicity and carcinogenic properties, P as the biogenic element is obviously desired in water for its fertilizing effects. As, Sb and P oxyanions, which are abundant in natural systems, mostly behave as the competitors to binding positions in surface complexation. Three manganese-based adsorbents (birnessite, Mn^{2+} modified zeolite and Mn^{2+} modified biochar) were tested for selective adsorption of arsenate (As^V) and antimonate (Sb^V) in presence of phosphate (P^V). Birnessite and Mn-modified biochar were excellent for the removal of individual oxyanions (80–95% of P^V, 93–100% of As^V, and ~85% of Sb^V was removed), while a preferential As^V/Sb^V adsorption only for birnessite was observed (86% of As^V with 32% of P^V, and 68% of Sb^V with 7% of P^V were removed).

1 INTRODUCTION

Phosphorus (P), arsenic (As) and antimony (Sb) are the elements with identical periodicity, similar chemistry and binding properties, but quite different environmental impact and main sources. Phosphorus belongs to essential nutrient for plants, animals and humans and its potential environmental risk (eutrophication) originates mostly in the overproduction of detergents, fertilizers and commonly used products (Nedjah *et al.*, 2015). Arsenic and antimony are extremely toxic with carcinogenic, mutagenic and teratogenic effects. While As risk comes from both anthropogenic and geogenic sources, Sb results mostly from anthropogenic activities and its anthropogenic mobilization factor AMF (mass ratio of Sb produced by humans normalized by naturally released Sb) is high (Klee & Graedel 2004). A co-occurrence of favorable phosphates beside inadmissible arsenates or antimonates in waters can complicate decontamination and/or purification processes. Arsenate, antimonate and phosphate show similar sorption properties competing for adsorption sites on the sorbent surface. According to the position in periodic table (ionic radius), a decreasing adsorption affinity of P > As > Sb can be expected. The studies of e.g. Violante & Pigna (2002) describe the manganese (Mn)-like adsorbents having a better affinity to As than to P. The aim of this work was to prepare and verify three Mn-based sorbents for selective adsorption of P^V, As^V and Sb^V oxyanions from contaminated water. The usability of Mn-based sorbents for preferential As^V/Sb^V adsorption beside competing P^V was also studied.

2 MATERIALS AND METHODS

2.1 *Mn-based sorbents*

For birnessite preparation concentrated hydrochloric acid (HCl) was added dropwise to a boiling solution of potassium permanganite ($KMnO_4$) in 2.5 L of water (H_2O), with vigorous stirring. After boiling for ten minutes, the precipitate was filtered and washed.

Natural zeolite (clinoptilolite from Nižný Hrabovec, Slovakia) and biochar (Novo Carbo GmbH, Germany) were modified with Mn^{2+} ions, which involved the alkalization of 0,8M $MnSO_4$. H_2O with 4M NaOH to pH ≈ 8.0 – 8.3, stirring 1 L of alkalized solution with raw sorbent (20 g) in sealed polyethylene bottle at laboratory temperature for 24 h, filtration, washing of solid product with distilled water, drying at 40°C and homogenizing. The raw and modified sorbents were characterized by X-ray fluorescence (XRF), specific surface area (S_{BET}) and scanning electron microscopy (SEM) to determine the changes related to surface modification (Table 1).

2.2 *Model solutions*

Model solutions of PO_4^{3-}, AsO_4^{3-} and SbO_4^{3-} were prepared from inorganic salts (KH_2PO_4, KH_2AsO_4 and $NaSbO_3{\cdot}3H_2O$, respectively) of analytical grade and distilled water, at the concentration of 5×10^{-4} mol/L and their natural pH values (≈3.5). The mixed solutions of AsO_4^{3-}/PO_4^{3-} and SbO_4^{3-}/PO_4^{3-} in the molar ratio of 1:1 were prepared in the same manner at the concentration of 5×10^{-4} mol/L for each oxyanion.

DOI: 10.1201/9781003317395-128

2.3 *Adsorption process*

A suspension of PO_4^{3-}, AsO_4^{3-} or SbO_4^{3-} model solution (50 mL) and defined dosage (1–20 g/L) of appropriate sorbent was shaken using a batch procedure at laboratory temperature (20°C) for 24 hours (Doušová *et al.*, 2006). The product was filtered off and the filtrate was analyzed for P, As or Sb content, while the residual saturated sorbent was tested by S_{BET} measurement, XRF and SEM. The adsorption data were fitted to the Langmuir model, the kinetic data of P, As or Sb adsorption were calculated according to first- and second-order kinetics. The adsorption from mixed AsO_4^{3-}/PO_4^{3-} and SbO_4^{3-}/PO_4^{3-} solutions ran in the same manner.

3 RESULTS AND DISCUSSION

3.1 *Adsorption of PO_4^{3-}, AsO_4^{3-} and SbO_4^{3-} from separate solutions*

Adsorption efficiencies of separate oxyanions adsorption (Figure 1) illustrated a high selectivity of all Mn-based sorbents (Mn-biochar ≈ birnessite > Mn-zeolite) to PO_4^{3-}, AsO_4^{3-} and SbO_4^{3-}.

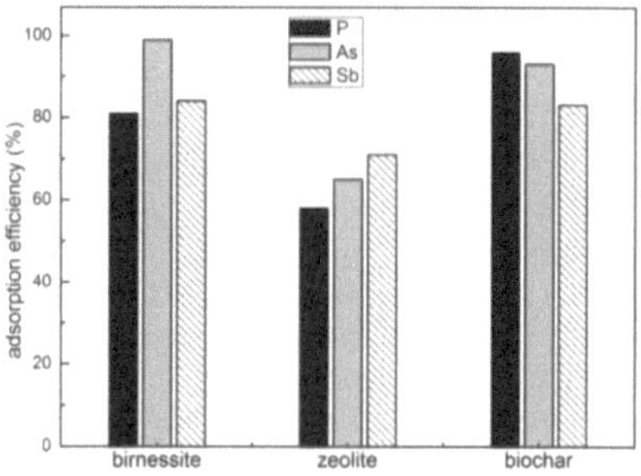

Figure 1. Adsorption efficiency of separate PO_4^{3-}, AsO_4^{3-} and SbO_4^{3-} adsorption.

In terms of perspective adsorption, birnessite showed the best properties (Table 1), while Mn-biochar was the most balanced sorbent of separate oxyanions, with the expected selectivity order: $PO_4^{3-} > AsO_4^{3-} \geq SbO_4^{3-}$.

Table 1. Mn content and S_{BET} of Mn-based sorbents.

Adsorbent	Mn(% *wt.*)	Mn valence	S_{BET} (m^2/g)
Birnessite	92	2.9	42.4
Mn-zeolite	40	2.3	35.5
Mn-biochar	35	2.7	24.7

3.2 *Adsorption of P^V, As^V and Sb^V from mixed PO_4^{3-}/AsO_4^{3-} and PO_4^{3-}/SbO_4^{3-} solutions*

Adsorption efficiencies of simultaneous adsorption of 2 oxyanions (PO_4^{3-}/AsO_4^{3-} or PO_4^{3-}/SbO_4^{3-}) in the molar ratio of 1:1 (Figure 2) indicated a different sorbent preferences over competing ions. The adsorption on Mn- zeolite and Mn-biochar proceeded according to the expected manner: (i) in PO_4^{3-}/AsO_4^{3-} system Mn phase slightly inhibited P^V adsorption in favor of As^V. (ii) in PO_4^{3-}/SbO_4^{3-} system the effect ionic radius prevailed and P^V was preferably adsorbed on both adsorbents.

Conversely, birnessite clearly declared preferential As^V and Sb^V over P^V adsorption. The theoretical adsorption capacities of simultaneous adsorptions (Table 1) corresponded to mentioned trend.

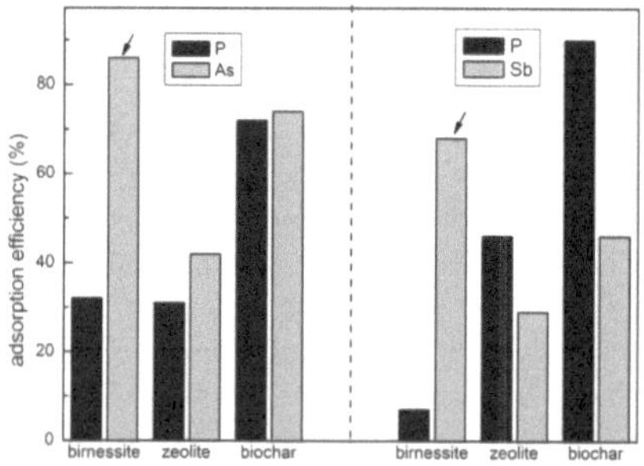

Figure 2. Adsorption efficiency of simultaneous PO_4^{3-}/AsO_4^{3-} and PO_4^{3-}/SbO_4^{3-} adsorption.

4 CONCLUSIONS

Mn-based sorbents showed a high selectivity to separate PO_4^{3-}, AsO_4^{3-} and SbO_4^{3-} in the order: Mn-modified biochar ≈ birnessite > Mn-modified zeolite. In the case of simultaneous PO_4^{3-}/AsO_4^{3-} or PO_4^{3-}/SbO_4^{3-} adsorption only birnessite was appropriate to prior AsO_4^{3-}/SbO_4^{3-} adsorption with considerable inhibiting PO_4^{3-} adsorption.

ACKNOWLEDGEMENTS

This research was founded by Czech Science Foundation under the project 19-04682S.

REFERENCES

Doušová B., Grygar T. Martaus A., Fuitová L. Koloušek D. & Machovič V. 2006. Sorption of As(V) onaluminosilicates treated with FE(II) nanoparticles. *J. Coll. Interface Sci.* 302: 424–431.

Klee R.J. & Graedel T.E. 2004. Elemental cycles: a status report on human or natural dominance. *Ann. Rev. Environ. Res.* 29: 69–107.

Nedjah N., Hamdaoui O. & Laskri N. 2015. Phosphorus removal of urban wastewater by physico-chemical treatment: waterways euthrophication prevention. *Int. J. Environ. Sci. Dev.* 6(6): 435–438

Violante A. & Pigna M. 2002. Competitive sorption of arsenate and phosphate on different clay minerals and soils. *Soil Sci. Soc. Am. J.* 66 (6): 1788–1796.

Arsenic in the Environment: Bridging Science to Practice for Sustainable Development –
van der Wal, Ahmad, Petrusevski, Weijma, Savic, van der Wens, Beerendonk, Bhattacharya, Bundschuh & Naidu (Eds)
© 2023 the Author(s), ISBN 978-1-032-32928-4

Arsenite removal through an adsorbent developed from industrial waste

N. Jain & A. Maiti
Polymer & Process Engineering Department, IIT Roorkee Saharanpur Campus, Saharanpur, U.P., India

ABSTRACT: Arsenic toxicity in groundwater have the severe impact on the health of the living beings. There are many countries affected by the elevated level of arsenic in the groundwater and the recommended value of arsenic in drinking water given by WHO (<10 μg/L). In the current study, we have developed the adsorbent consist of oxides of manganese and aluminum. Tetravalent manganese oxidizes the arsenite into arsenate and simultaneously forms the co-precipitate with trivalent manganese and the trivalent aluminum. Ultimately, the concentration of elements in the treated water meets the standard of the drinking water. The prepared adsorbent shows the good adsorption capacity for the arsenite (21.56 mg/g) species. XANES and XPS techniques have been used to confirm the oxidation state of arsenic and manganese in the pristine as well as the spent adsorbent.

1 INTRODUCTION

Arsenic mobility and its toxicity is the major challenge for the scientific community to develop the economical as well as efficient technology (Maiti *et al.*, 2012). High level of arsenic in the drinking water leads to various diseases like skin cancer, kidney malfunction, organ imperfection in the mankind. Drinking water especially the groundwaters are contaminated worldwide by inorganic arsenic species i.e arsenite and arsenate. Arsenite species in groundwater are more lethal and carcinogenic compared to arsenate (Maiti *et al.*, 2009). This acute toxicity of arsenite appears because of its neutral form in the water, on the other hand the arsenate is present in the anionic form (Singh & Pant 2006). Hence, the removal of arsenate compared to arsenite is quite effective. So, the main focus of the study is to eliminate the arsenite toxicity in the groundwater by adopting the effective and economical technology. Although various technologies such as adsorption, oxidation, co-precipitation, phytoremediation, electro-oxidation and much more have been developed to mitigate the arsenite toxicity (Maiti *et al.*, 2012; Villalobos *et al.*, 2014). Adsorbents consist of various iron oxides, aluminum oxides, manganese oxides, copper oxides; titania, and ceria have shown the good remediation of arsenite by first fully or partially oxidizing arsenite into arsenate and then finally decontaminate it (Gude *et al.*, 2017; Sherman 2005; Villalobos *et al.*, 2014). Manganese based adsorbents have shown the better arsenite removal due its inherent property to oxidize the arsenite in even in presence of visible light sources (Gude *et al.*, 2017). In the current study, we have developed the adsorbent from the ferromanganese slag (waste from the steel industry) through thermo-chemical treatment and called it as treated ferromanganese slag (TFS). Now, the TFS has shown more than 99% removal of arsenite through oxidation followed by co-precipitation. The developed adsorbent works well even in the dark environment and the removal rate will increase by providing any kind of light sources. So, the study proves the TFS has the potential to be used as the arsenite filter media for the household water filter as well as commercial applications.

2 MATERIALS AND METHODS

2.1 *Materials*

Ferromanganese slag (FS) was obtained from the steel industry, then it was crushed and passed through the sieve screen of 90-micron before introducing it for the thermo-chemical treatment. Treatment includes the leaching of elements with the hydrochloric acid, and alkaline treatment with sodium hydroxide solution to precipitate the adsorbent as the mixed metal oxides/hydroxides, named as treated ferromanganese slag (TFS). The black precipitate of TFS is obtained due to the formation of manganese(IV) oxide/hydroxides along with aluminum and iron oxides/ hydroxides. The precipitate was dried at 70°C and pulverized by pestle and mortar for adsorption study of both arsenic species in isotherm and kinetic modes. Double distilled water (DDW) was used to prepare working solutions. Arsenite and arsenate solutions were prepared in DDW by dissolving the AR grade 98% $NaAsO_2$ of Loba Chemie Pvt. Ltd, Mumbai, India, and by dissolving the required amount of AR grade 98.5% $Na_2HAsO_4.7H_2O$ of Himedia, Mumbai, India,

DOI: 10.1201/9781003317395-129

respectively. The standard solutions were prepared using 5% nitric acid by diluting the 500 mg/L of arsenic standard (Reagecon REICPCAL29A; Ireland). AR grade NaOH with 98% purity from Himedia and HCl of AR grade from the Rankem, India were used in the process of adsorbent preparation and for pH adjustment.

2.2 *Analytical techniques*

The pH was measured by the digital pH meter (Hanna Instrument Pvt. Ltd, Mumbai, India). Concentrations of metal ions like Mn, Al, Fe, Ca, Si, As in the filtrate were measured by inductively coupled plasma mass spectrometer (ICP-MS) (Perkin Elmer SCIEX The Elan Series, USA). Characterize the TFS in the wavelength range of 200 to 800 nm with the solid state UV/Visible spectrophotometer (Shimadzu 2450 Germany). The X-ray absorption near edge structure (XANES) analysis has been carried out at the Extended X-ray absorption fine structure (EXAFS) beamline (BL-9) at the Indus-2 Synchrotron Source (2.5 GeV, 200 mA) at Raja Ramanna Centre for Advanced Technology (RRCAT), Indore, India. This beamline of XANES operates in the energy range of 4 KeV to 25 KeV. Brunauer–Emmett-Teller (BET) surface analyzer, (ASIQ Quantachrome instrument, USA) was used to characterize the FS and TFS fresh/spent adsorbents. Before performing the liquid N2 sorption/desorption in surface area analyzer, the samples were degassed before the surface area analysis at 150°C for 3 h under vacuum to achieve complete purging.

2.3 *Adsorption kinetics and equilibrium assay*

Arsenic adsorption experiments were investigated at 298 ± 0.5 K with an adsorbent dose of 0.5 g/L in 100 mL solutions of As(III) and As(V) separately in the concentration range of 0.2 to 20 mg/L. The initial pH of the experimental solutions was adjusted to 7.0 ± 0.2 with 1.0 N HCl and 1.0 N NaOH solutions. The solutions were agitated using an orbital shaker at 200 rpm for each equilibrium adsorption study. The kinetic study was performed identically as of equilibrium study, and samples were collected at a particular interval of time. The kinetic adsorption data were fitted into kinetic models of pseudo first order and pseudo-second-order. In the equilibrium study, the isotherm data were fitted into the adsorption isotherm models, namely Langmuir isotherm and Freundlich isotherm. The effect of light sources (visible, sunlight and dark environment) on the arsenic adsorption behavior of TFS was investigated by performing experiments with 100 mL of 2.0 mg/L arsenic solution at TFS dose 0.5 g/L. The solutions were agitated at 200 rpm for 24 h.

3 RESULTS AND DISCUSSION

3.1 *Characterization*

Total specular reflection was measured from the sample in the wavelength region of UV and visible radiations of 200 nm–700 nm by the Solid state UV-visible spectrophotometer. For the materials like pure manganese dioxide, the bandgap is found in the range from 0.8 eV to 2.0 eV depending upon the type of mineral phases of manganese dioxide (Sherman 2005). Tangent on the curves of the first peak was drawn to find the energy bandgap of the material through the intercept on the x-axis. The energy band gap is obtained as 1.33 eV for the commercial MnO2 and 1.24 eV for the TFS (shown in Figure 1a). The lower bandgap of TFS might favor the oxidation of arsenite under indoor as well as in the outdoor environment.

The total specific surface area of the FS and the TFS is estimated by using nitrogen adsorption-desorption isotherm data fitted into the BET isotherm model. Figure 1b reveals that the TFS follows the type IV isotherm along with H3 type hysteresis according to the IUPAC nomenclature. Such type of isotherm indicates the presence of mesoporous structure. The BET specific surface area of the TFS is obtained as 119.3 ± 2.0 m^2/g,

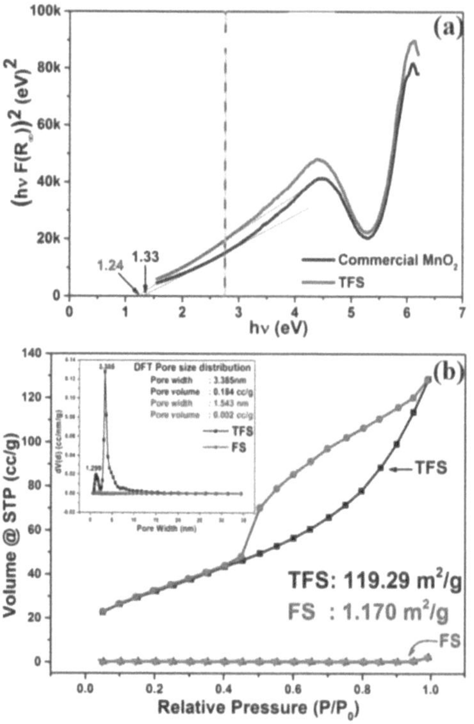

Figure 1. Characterization of TFS by a) solid state UV-Visible spectroscopy; b) BET surface area analyzer.

whereas the specific surface area for raw FS is 1.17 m^2/g. Further, the BJH isotherm model has confirmed the mesoporous nature of the TFS with an average pore diameter of 3.8 nm (inset in Figure 1b), which is much larger compared to FS, i.e., 1.4 nm.

3.2 *Adsorption study*

The isotherm data of both arsenic species is best fitted to the Langmuir isotherm and reveals the monolayer adsorption of arsenic on active surface sites of TFS with equal energies. On the other hand, Freundlich isotherm predicts the multilayer adsorption of adsorbate. In presence of 100% initial As(III) species, first As(III) oxidizes to As(V) on the TFS surface and probably new active sites may be generated during that oxidation step. Hence, the overall removal of As(III) species on TFS has improved due to the oxidation of the As(III) which enables the surface passivation (Figure 2a).

3.3 *Adsorption in the presence of different light sources*

The impact of different light sources on the oxidation kinetics of arsenite species as well as adsorption kinetics of arsenic species (as total) by TFS is examined to assess the oxidative power of it under the indoor/outdoor environment. Samples were collected at predetermined time intervals and measured the residual arsenic concentration of supernatants after centrifuging the samples at 10000 rpm. Figure 2b shows the adsorption of arsenic at different time intervals in the presence of different light sources, and it is observed that the arsenic concentration reaches below the maximum permissible limit (<10 μg/L) of the drinking water standard within 15 minutes of contact time under both indoor lighting (visible light) and outdoor conditions (in presence of sunlight). The concentration of arsenic reached well below the maximum permissible limit of arsenic in the drinking water within 60 minutes when adsorption experiments performed in the dark environment. Thus TFS could be used as an arsenic scavenger in a household filtration system. In the presence of visible light sources, more amount of electron-hole pairs generate due to the favorable redox potential of manganese dioxide. The electron-hole pair oxidizes the arsenite and enhances the total arsenic removal efficiency. In the presence of a visible light source, the arsenic removal rate is found to be much faster. Thus, in the real household filters using TFS, visible light can act as the catalyst to increase the arsenic removal rate.

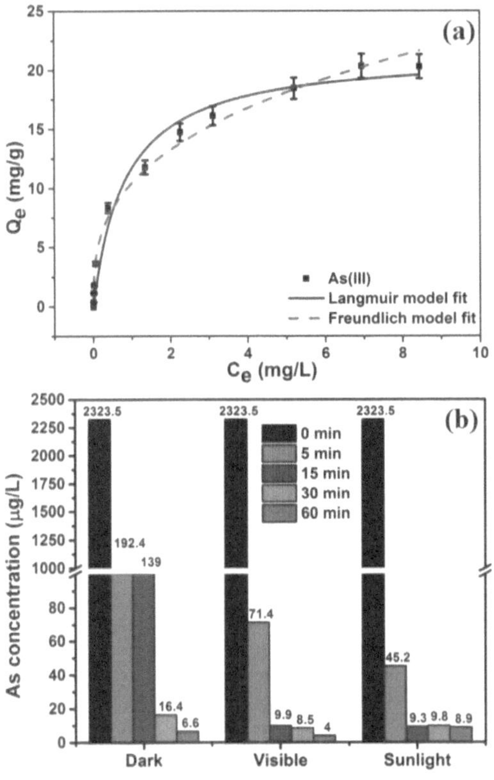

Figure 2. Arsenite (a) adsorption isotherm; (b) removal in presence of different light sources.

4 CONCLUSIONS

This study has demonstrated an easy and sustainable chemical treatment method to produce valuable product from a steel industrial waste namely ferromanganese slag. The chemically treated ferromanganese slag (TFS) has shown significantly higher arsenite oxidation with simultaneous arsenic adsorption from the arsenic-contaminated groundwater compared to ferromanganese slag (FS). The freshly prepared TFS containing Mn (IV) oxide were mainly responsible for the oxidation of arsenite into arsenate and simultaneously co-precipitation of Mn(II) with arsenic species in the aqueous medium. TFS has shown maximum arsenic adsorption capacity of 21.6 ± 0.4 mg/g for initial 100% arsenite. Arsenic adsorption kinetic data for TFS has shown faster adsorption even at the low arsenic concentration (<1000 μg/L) in arsenic-contaminated groundwater. The arsenite adsorption by TFS is found to be faster compared to arsenate species because; the oxidation step of arsenite increases the surface active sites of TFS for adsorption of more arsenic species. TFS in presence of visible light source is capable to bring the final arsenic concentration below the maximum permissible limit of drinking water (10 μg/L) from initial concentration of 2000 μg/L. The process of preparation method of TFS is simple and efficient adsorbent has been prepared from low cost industrial waste namely ferromanganese slag. The preparation method of TFS is easy to scale-up at industrial scale and provides sustainable approaches to minimize the industrial waste as well as to produce drinking water from arsenic contaminated groundwater.

ACKNOWLEDGEMENTS

We acknowledge the IIT Roorkee for providing the facility to research under IMPRINT 2A project IMP/2018/001228. We heartily acknowledge Mr. Ashok Yadav and Dr. S.N. Jha of RRCAT, Indore, for giving us the valuable time to access the Beamline BL-9 for the experimental purpose.

REFERENCES

Bonnin D. 2000. Method of removing arsenic species from an aqueous medium using modified zeolite minerals.-. *U.S. Patent No.* 6,042,731.

Dousova B., Grygar T., Martaus A., Fuitova L., Kolousek D. & Machovic V. 2006. Sorption of AsV on aluminosilicates treated with FeII nanoparticles. *J. Colloid Interface Sci.* 302: 424–431.

Gude J.C.J., Rietveld L.C. & van Halem D. 2017. As(III) oxidation by MnO_2 during groundwater treatment. *Water Res.* 111, 41–51.

Herzogova, L. 2014. *The Study of Surface Changes on Treated Sorbents Based on Aluminosilicates.* PhD Thesis, UCT Prague, Prague, Czech Republic.

Izumi Y., Masih D., Aika K. & Seida Y. 2005. Characterization of intercalated Iron(III) nanoparticles and oxidative adsorption of arsenite on them monitored by X-ray absorption fine structure combined with fluorescence spectrometry. *J. Phys. Chem. B* 109: 3227–3232.

Maiti A., Basu J.K. & De S. 2012. Experimental and kinetic modeling of As(V) and As(III) adsorption on treated laterite using synthetic and contaminated groundwater: effects of phosphate, silicate and carbonate ions. *Chem. Eng. J.* 191, 1–12.

Maiti A., Sharma H., Basu J.K. & De S. 2009. Modeling of arsenic adsorption kinetics of synthetic and contaminated groundwater on natural laterite. *J. Hazard. Mater.* 172: 928–934.

Ouvrard S., Simonnot M.O. & Sardin M. 2002. Reactive behavior of natural manganese oxides toward the adsorption of phosphate and arsenate. *Ind. Eng. Chem. Res.* 41: 2785–2791.

Sherman D.M. 2005. Electronic structures of Iron(III) and Manganese(IV) (hydr)oxide minerals: thermodynamics of photochemical reductive dissolution in aquatic environments. *Geochim. Cosmochim. Acta* 69: 3249–3255.

Singh T.S. & Pant K.K., 2006. Kinetics and mass transfer studies on the adsorption of arsenic onto activated alumina and iron oxide impregnated activated alumina. *Water Qual. Res. J. Canada* 41, 147–156.

Villalobos M., Escobar-Quiroz I.N. & Salazar-Camacho C. 2014. The influence of particle size and structure on the sorption and oxidation behavior of birnessite: I. adsorption of As(V) and oxidation of As(III). *Geochim. Cosmochim. Acta* 125, 564–581.

Arsenic in the Environment: Bridging Science to Practice for Sustainable Development –
van der Wal, Ahmad, Petrusevski, Weijma, Savic, van der Wens, Beerendonk, Bhattacharya, Bundschuh & Naidu (Eds)
© 2023 the Author(s), ISBN 978-1-032-32928-4

Fe/Mn – modified adsorbents to arsenic and antimony removal from contaminated water

M. Lhotka & B. Dousova
University of Chemistry and Technology Prague, Prague, Czech Republic

ABSTRACT: Natural solids such as aluminosilicates, metal oxides, lignin, biowaste belong to low-cost and environmentally friendly materials with favorable surface properties. Their adsorption selectivity in water primarily depends on the actual pH value to the zero point of charge (pH_{ZPC}) of the solid. The surface modification of cation-selective (aluminosilicates) or uncharged (biochar) sorbents with Fe/Mn ions can change the surface charge (pH_{ZPC}) causing a strong selectivity to anions. Three different aluminosilicates and two biomaterials were treated with Fe^{2+}, Fe^{3+} and Mn^{2+} ions. The raw and modified sorbents were characterized by XRD, XRF, SEM, S_{BET}, pH_{ZPC}, particle size and IR spectroscopy. The adsorption properties and selectivity of prepared sorbents were tested in solutions containing arsenate (As^{V}), arsenite (As^{III}) and antimonate (Sb^{V}) ions . The mechanism and kinetic of surface-mediated processes relied on the structure of raw material, the applied treating method and water quality. The adsorption parameters were calculated by Langmuir and Freundlich adsorption models and by the formal kinetic models.

1 INTRODUCTION

Aluminosilicates represent a challenging material due to several aspects, such as its chemical stability, availability, environmental safety issues, fineness, and appropriate surface properties. Most of them belong to low-cost and environmentally acceptable materials. In water systems at pH $\approx$ 5 and above, aluminosilicates are typical cation-selective sorbents thanks to low pH of zero point of charge (pH_{ZPC}). Recently described method of their surface modification with Fe/Mn ions has expanded their possible use in adsorption processes due to the change of surface charge (pH_{ZPC}) causing a strong selectivity to anions (Dousova *et al.*, 2006).

Three pre-treating methods including Fe^{II} (Bonnin 2000), Fe^{III} (Izumi *et al.*, 2005), and Mn^{II} (Herzogova 2014) were applied to different types of aluminosilicates, i.e. natural kaolin from the Merkur quarry, Czech Republic, calcined at 550°C (metakaolin), raw bentonite from the same region and natural commercially available clinoptilolite from the Nizne Hrabovce deposit (Zeocem, Inc., Slovak Republic), respectively. As organic materials biochar (Novo Carbo GmbH, Germany) and waste lignin from southern Bohemia, Czech Republic were used. The present study aims to prepare stable powdered sorbents with new physical chemical and surface properties using interaction of raw material and a metal salt solution. Reactive ion-exchangeable sites and poorly crystalline hydrated oxides in stable oxidation state (Fe^{III}, $Mn^{III/IV}$) arise on the solid surface.

2 MATERIALS AND METHODS

2.1 *Preparation of modified sorbents*

For this study caolinites from West Bohemia were calcined at 550°C for 3 h and converted to metakaolinites. Clinoptilolite from Nižný Hrabovec deposit, Zeocem, Ltd. (Slovakia), Montmorillonite from Arizona (USA) and raw bentonite from Kada⊠, (West Bohemia) were also used for preparation of sorbents modified by Fe^{II}, Fe^{III} and Mn^{II} ions. The modifications were as follows: (i) Fe^{II} (0.6 M $FeSO_4.7H_2O$; solid-liquid 1:50; agitation at 20°C for 24 hours; filtration; washing with distilled water; drying at 50°C; homogenization). (ii) Fe^{III} (0.025 M $Fe(NO_3)_3.9H_2O$ and 0.05 M NaOH (partly hydrolyzed); solid-liquid 1:50; agitation at 20°C overnight; filtration; washing with distilled water; drying at 50°C; homogenization). (iii) Mn^{II} (alkalization of 0,8M $MnSO_4.H_2O$ with 4M NaOH to pH $\approx$ 8.0 – 8.3; solid-liquid 1:50; agitation at 20°C 24 h; filtration; washing with distilled water; drying at 40°C; homogenization.

2.2 *Analytical methods*

The information on the potential surface activity of raw and treated adsorbents was determined by adsorption of nitrogen. Equilibrium adsorption isotherms of nitrogen were measured at 77 K using static volumetric adsorption systems (TriFlex analyzer, Micromeritics). The adsorption isotherms were fitted in the BET specific surface area method. Fourier transform infrared (FTIR) spectra were collected on a Nicolet 6700 FTIR (Thermo Nicolet Instruments Co., Madison, USA) with a N_2 purging system.

The model solutions of oxyanions As^{III}, As^{V} and Sb^{V} with a concentration of 2.1-4M were used. The suspension of model solution and sorbent (6 g/L) was shaken in sealed polyethylene bottle at room temperature for 24 hours. The product was filtered off; the filtrate was analyzed for residual As and Sb concentrations and pH value. The concentration of As and Sb in aqueous solutions was determined by HG-AFS using PSA 10.055 Millennium Excalibur.

DOI: 10.1201/9781003317395-130

3 RESULTS AND DISCUSSION

3.1 *Characterization of modified sorbents*

The samples of modified sorbents were prepared from different raw clays. The specific surface area S_{BET} (Figure 1) differed depending on the type of clay (1:1, 2:1) and the character of modifying particle (Fe^{II}, Fe^{III}, Mn^{II}). The FTIR spectra of modified sorbents indicated new metal phases on the solid surface (e.g. FeOOH or MnOOH aggregates).

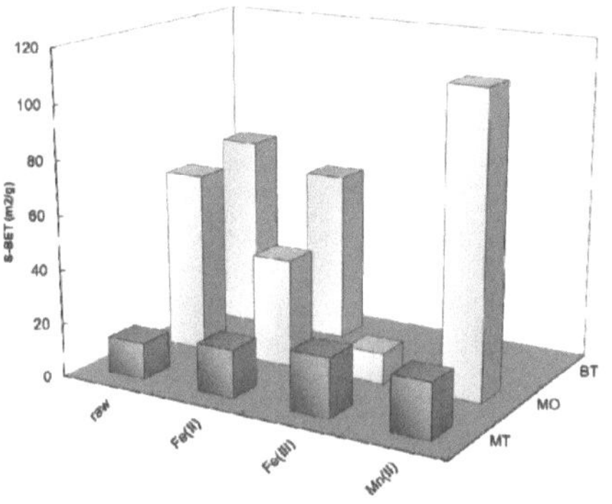

Figure 1. The specific surface area of raw and modified alumino-silicates. MT – metakaolins, BT – bentonite, MO – Montmorillonite.

3.2 *Sorption of arsenic and antimony*

The adsorption properties and selectivity of pre-pared sorbents were tested on toxic oxyanions, such as arsenate (As^{V}), arsenite (As^{III}) and antimonate (Sb^{V}) (Figure 2). The results showed, that all investi-gated methods were proper for bentonite treatment, where the sorption efficiency of $As^{III/V}$ varied from 92 to > 99%. In the case of metakaolin, Fe^{III} – treatment was less appropriate thanks to the steric effect of Fe^{III} hydrolyzed particles binding to the surface during the process (FeOOH aggregates are much larger compared to isolated Fe^{III} particles arising from the Fe^{II} – treatment). The clinoptilolite showed to be a suitable carrier for both the Fe^{II} and Fe^{III} modifications. A less effective Mn^{II} method can be successfully applied in several specific adsorption processes, such as in the competing removal of two similar anions (Ouvrard *et al.*, 2002).

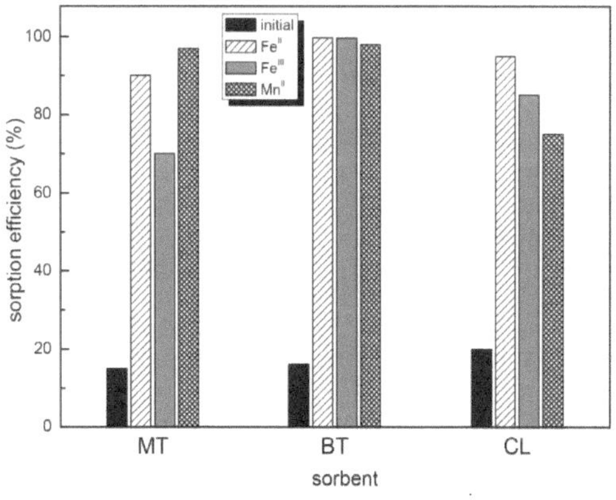

Figure 2. Adsorption efficiencies of raw and modified alumino-silicates. MT – metakaolin, BT – bentonite, CL – clinoptilolite.

4 CONCLUSIONS AND RECOMMENDATIONS

The mechanism and kinetics of surface processes depended on the structure of raw material (steric properties, active surface sites), applied treating method and water quality (pH, inicial concentration of anion, competitive ions). The adsorption parameters were calculated by Langmuir and Freundlich adsorption models and by the formal kinetic models. The raw and modified sorbents were characterised by XRD, XRF, SEM, S_{BET}, pH_{ZPC}, particle size distribution, voltametry of microparticles and IR spectroscopy. The characterization of solid phase with above mentioned methods will permit to identify the actual anion-sorbing phases and to tailor the sorbents with the optimal sorption properties.

ACKNOWLEDGEMENTS

This research was founded by Czech Science Foundation under the projects 19-04682S and 19-11027S.

REFERENCES

Bonnin D. (2000): Method of removing arsenic species from an aqueous medium using modified zeolite minerals.-. *U.S. Patent No.* 6,042,731.

Dousova B., Grygar T., Martaus A., Fuitova L., Kolousek D. & Machovic V. 2006. Sorption of AsV on aluminosilicates treated with FeII nanoparticles. *J. Colloid Interface Sci.* 302: 424–431.

Herzogova L. (2014): *The Study of Surface Changes on Treated Sorbents Based on Aluminosilicates.* PhD Thesis, UCT Prague, Prague, Czech Republic.

Izumi Y., Masih D., Aika K. & Seida Y. 2005. Characterization of intercalated Iron(III) nanoparticles and oxidative adsorption of arsenite on them monitored by X-ray absorption fine structure combined with fluorescence spectrometry. *J. Phys. Chem. B* 109: 3227–3232.

Ouvrard S., Simonnot M.O. & Sardin M. 2002. Reactive behavior of natural manganese oxides toward the adsorption of phosphate and arsenate. *Ind. Eng. Chem. Res.* 41: 2785–2791.

Arsenic in the Environment: Bridging Science to Practice for Sustainable Development –
van der Wal, Ahmad, Petrusevski, Weijma, Savic, van der Wens, Beerendonk, Bhattacharya, Bundschuh & Naidu (Eds)
© 2023 the Author(s), ISBN 978-1-032-32928-4

Sequential ferrous iron oxidation for arsenic removal from phosphate-containing groundwater

M. Annaduzzaman[1], L.C. Rietveld[1], B.A. Hoque[2] & D. van Halem[1]
[1] *Water Management Department, Faculty of Civil Engineering and Geosciences, Delft University of Technology, Delft, The Netherlands*
[2] *Environment and Population Research Centre, New DOHS. Mohakhali, Dhaka, Bangladesh*

ABSTRACT: Groundwater arsenic (As) contamination is a severe drinking water quality problem and a threat to human health in Bangladesh and other countries. Appropriate arsenic removal technology has been a global challenge. Biological As(III) oxidation to As(V) and co-precipitation with groundwater-native Fe^{3+}-precipitates is a promising approach, as it does not require adsorbents or chemicals. In order to optimize this process for application in Bangladesh, it is critical to prevent competition of phosphate with As(V) for removal with Fe^{3+}-precipitates in groundwater. The study aimed to investigate delayed Fe^{2+} oxidation in solution towards more efficient As(III) and As(V) removal in phosphate-containing waters (2 mg/L). The research was conducted with 280 μg/L As(III) and 2.5 mg/L of Fe^{2+}, with an initial pH of 7.0, being representative for the concentrations in the targeted village in rural Bangladesh. The experiments consisted of aerated single and multiple-step jar tests, with the multiple-steps jar test being the simulation of delayed Fe^{2+} oxidation through sequential dosing of 1.25 mg/L Fe^{2+}. In the single-step jar test, the As(III) containing system with and without phosphate resulted in 24% and 47% As removal respectively, whereas the multiple-steps jar test, enhanced total As removal to 57% and 94% correspondingly. These results show that delayed Fe^{2+} aeration is beneficial to achieve more efficient As removal in the presence of phosphate.

1 INTRODUCTION

In Bangladesh, over 20 million people are exposed to As concentrations that exceed the Bangladesh Drinking Water Standard (BDWS) of 50 μg/L, while the World Health Organization (WHO) guideline value for As is 10 μg/L. This makes it one of the country's prime drinking water problems (WHO & UNICEF 2018). In anaerobic, reducing groundwater at circumneutral pH, As(III) is a thermodynamically stable species as $H_3AsO_3^0$, which is challenging to remove by standard treatment technologies compared to oxidized and negatively charged As(V). Recent research has demonstrated the potential of biological As(III) oxidation in conventional rapid sand filtration combined with aeration (Katsoyiannis *et al.*, 2015). In these aerated sand filters, As(III) oxidizes to As (V) by indigenous arsenic oxidizing bacteria (AsOB) natively present in groundwater. Subsequently, oxidized As(V) is partially removed by hydrous ferric oxides (HFO) flocs generated by groundwater ferrous iron oxidation (Gude *et al.*, 2018). However, the efficiency of As removal with HFO flocs is sensitive to the water composition (Roberts *et al.*, 2004). The commonly present oxyanions such as silicate (i.e., H_4SiO_4) and phosphate (i.e., $H_2PO_4^-$ or HPO_4^{2-}) have been identified to change the HFO flocs structure, size, and composition. Moreover, they can compete with As(V) for adsorption sites on HFO flocs (Kanematsu *et al.*, 2013). In order to optimize aeration-filtration for application in high phosphate and silicate waters in Bangladesh, it is critical to overcoming the competing effect of these oxyanions. Therefore, in this study, different sequences of Fe^{2+} and As(III) oxidation were investigated, aiming at achieving higher As removal efficiencies per milligram of Fe^{2+}, naturally available in groundwater.

2 MATERIALS AND METHODS

The experiments consisted of aerated single and multiple-steps jar tests, with the multiple-steps jar test being the simulation of sequential Fe^{2+} oxidation through repetitive dosing of 1.25 mg/L Fe^{2+} at the first and third steps after a sodium hypochlorite oxidation (as a means to accelerate As(III) oxidation for simulation purposes) at second step. The laboratory experiments were conducted using tap water from the TU Delft WaterLab. The jar tests were conducted with 280 μg/L of As(III/V), 2 mg/L of phosphate, and a total of 2.5 mg/L of Fe^{2+}, with an initial pH of 7.0; being representative for the concentrations in the targeted village in rural Bangladesh (Rajshahi). The phosphate, and As removal efficiencies were evaluated by collecting filtered (0.45 μm filter) and unfiltered (15 ml) acidified water samples. The elemental concentration was analyzed with Inductively Coupled Plasma Mass Spectrometry (ICP-MS) in the WaterLab, TU Delft, the Netherlands. Arsenic speciation was done using ion-exchange Amberlite® IRA-400 chlorite resin.

3 RESULTS AND DISCUSSION

Figure 1 represents the arsenic oxidation and removal, dissolved iron, and phosphate levels in different jar tests.

DOI: 10.1201/9781003317395-131

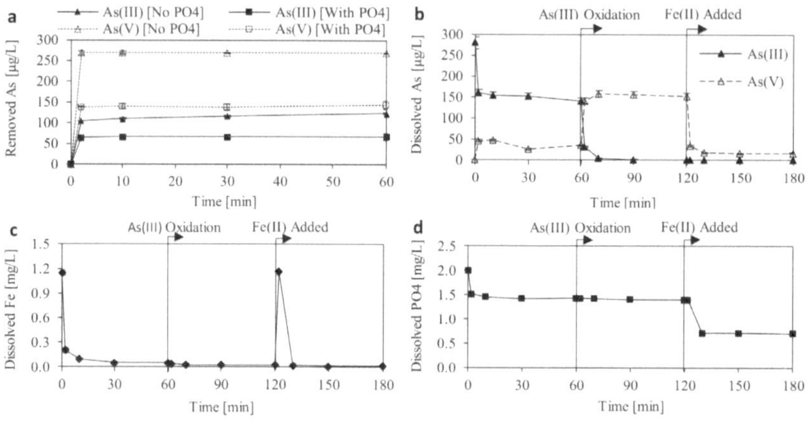

Figure 1. Single-step (a) and multiple-step (b-d) jar tests results for As species, iron and phosphate.

In the single-step jar test, once only As(V) used as initial As species with and without phosphate, the total As removal was 51% (C_e=137 µg/L) and 96% (C_e=11 µg/L), respectively (Figure 1a). In the case of As(III) species used, the As removals were 24% (C_e=212 µg/L) and 44% (C_e=166 µg/L) with and without phosphate, correspondingly (Figure 1a).

In the multiple-steps jar tests, where ferrousiron was dosed in sequence (1.25 mg/L at first and third step), the total As removal was enhanced and resulted in 57% (C_e=120 µg/L) and 94% (C_e=17 µg/L) with and without phosphate correspondingly (Figure 1b). The Fe^{2+} oxidation occurred within the first 2 minutes after introducing into the system in the respective steps (Figure 1c). Importantly, the initial (first-step) 1.25 mg/L Fe^{2+} in the multiple-step test was enough for a similar amount of As removal in the single-step. In the single-step jar test, the phosphate removal was 75% (C_e=0.5 mg/L) whereas, in multiple-step tests, phosphate removal after the first step was 30% (C_e=1.4 mg/L) and further removal occurred to attain a final removal of 65% (C_e = 0.7 mg/L) once Fe^{2+} introduced in the third step (Figure 1d). The results reveal that delaying Fe^{2+} oxidation and allowing As (III) oxidation enhanced total As removal by double compared to instant Fe^{2+} oxidation (Kanematsu *et al.*, 2013; Roberts *et al.*, 2004). The results also show that the sequential Fe^{2+} oxidation helps to reduce competitional effects among As(V) and phosphate for HFO sites by partial phosphate removal and facilitated more As^{III} oxidation and removal (30 µgAs/mgFe versus 160 µgAs/mgFe in the presence of phosphate).

4 CONCLUSIONS AND RECOMMENDATIONS

This study was conducted to investigate the effects of phosphate on As(III/V) removal with HFO flocs generated during sequential Fe^{2+} oxidation.

- Sequential Fe^{2+} oxidation is beneficial to achieve more efficient As removal (nearly double).
- Reconfirms that As removal with Fe based co-precipitation is affected by phosphate and thus sensitive to the water matrix.
- Preference of stepwise phosphate removal with partially Fe^{2+} oxidized HFO flocs can improve As^{III} oxidation and removal.

ACKNOWLEDGMENTS

This study was supported by the DELTAP project, funded by the NWO-WOTRO research program UDW. The authors also thankful to Erik Kraaijeveld for his support during laboratory studies in the Water Lab, Department of Water Management, TU Delft.

REFERENCES

Gude J.C.J., Rietveld L.C. & van Halem. D. 2018. Biological As(III) oxidation in rapid sand filters. *J. Water Process. Eng.* 21: 107–115.

Kanematsu M., Young T.M., Fukushi K., Green P.G. & Darby. J.L. 2013. Arsenic(III, V) adsorption on a goethite-based adsorbent in the presence of major co-existing ions: modeling competitive adsorption consistent with spectroscopic and molecular evidence. *Geochim. Cosmochim. Acta* 106: 404–28.

Katsoyiannis I.A., Voegelin A., Zouboulis A.I. & Hug S.J. 2015. "Enhanced As(III) oxidation and removal by combined use of zero valent iron and hydrogen peroxide in aerated waters at neutral pH values. *J. Hazard. Mater.* 297.

Roberts L.C., Hug S.J., Ruettimann T., Billah M., Khan A.W. & Rahman M.T. 2004. Arsenic removal with Iron(II) and Iron (III) in waters with high silicate and phosphate concentrations. *Environ. Sci. Technol.* 38(1): 307–315.

WHO & UNICEF. 2018. Arsenic primer guidance on the investigation and mitigation of arsenic contamination. http://www.unicef.org/wes.

Arsenic in the Environment: Bridging Science to Practice for Sustainable Development –
van der Wal, Ahmad, Petrusevski, Weijma, Savic, van der Wens, Beerendonk, Bhattacharya, Bundschuh & Naidu (Eds)
© 2023 the Author(s), ISBN 978-1-032-32928-4

Arsenic co-precipitation with iron oxidation products and retention during precipitate aging: Effects of phosphate, silicate and calcium

A. Voegelin, A.-C. Senn, R. Kaegi & S.J. Hug
Eawag, Swiss Federal Institute of Aquatic Science and Technology, Dübendorf, Switzerland

ABSTRACT: Arsenic (As) co-precipitates with ferric iron (Fe(III)) precipitates formed by the oxidation of ferrous iron (Fe(II)) in water. This process is of great importance in the removal of As from drinking water and for the fate of As in environmental systems. We studied the effects of phosphate (P), silicate (Si) and calcium (Ca) on the co-precipitation of arsenate (As(V)) with fresh Fe(II)-derived Fe(III)-precipitates as well as the retention or release of As(V) during precipitate aging. The structure of the Fe oxidation products and their transformation over time are greatly affected by P, Si and Ca, with substantial consequences for the fate of co-transformed As(V). Both Si and Ca favorably impact on As(V) removal and retention: Si inhibits the formation of crystalline Fe(III)-solids with lower sorption capacity, and Ca enhances P removal in more stable Ca-Fe(III)-phosphate and thereby reduces the adverse effect of P on As(V) removal. Our results emphasize that the structural diversity of Fe oxidation products caused by variations in water chemistry and resulting variations in precipitate reactivity must be taken into account when assessing the effects of Fe(III)-precipitate formation and transformation on As.

1 INTRODUCTION

The oxidation of dissolved ferrous Fe(II) by O_2 in aerated water leads to the formation of poorly soluble ferric Fe(III)-precipitates. Because of their high sorption capacity, these Fe(III)-precipitates critically affect the fate of As and other trace elements in environmental and engineered systems. Methods based on the oxidation and precipitation of Fe are widely employed to remove As from drinking water. To form As-removing Fe(III)-precipitates, these methods make use of the Fe (II) that is naturally present in anoxic groundwaters, of Fe(II) added as salt, or of Fe(II) produced by zerovalent iron corrosion or electrocoagulation (Kaegi *et al.*, 2010; Neumann *et al.*, 2013; Roberts *et al.*, 2004; van Genuchten *et al.*, 2012; 2014; Voegelin *et al.*, 2010).

In natural waters, Fe(III)-precipitates form in the presence of co-precipitating solutes such as Ca, P, Si that affect precipitate structure and reactivity. The overarching goals of our recent studies on Fe(III)-precipitates was to gain a more detailed understanding of individual and interdependent effects of P, Si and Ca on the structure and reactivity of fresh Fe(III)-precipitates and their transformation over time, and to assess implications for As(V) removal and retention by Fe(III)-precipitates (Senn *et al.*, 2015, 2017, 2018; Voegelin 2019).

2 MATERIALS AND METHODS

Fresh Fe(III)-precipitates were synthesized by the oxidation of 0.5 mM Fe(II) in bicarbonate-buffered synthetic groundwater at pH 7.0. Six electrolytes (Na, Ca, low Ca, Mg, Na + Si and Ca + Si) were prepared by dissolving 8 mM $NaHCO_3$ (Na), 4 mM $CaCO_3$ (Ca), 0.5 mM $CaCO_3$ and 7 mM $NaHCO_3$ (low Ca), or 4 mM MgO (Mg) in CO_2-purged water, followed by Si addition ($(Si/Fe)_{init}$ = 1) where needed (Na + Si; Ca + Si). After raising the pH to 7.0 by purging with air, 7 μM (525 μg/L) As(V) (0.014 $(As/Fe)_{init}$) and 0 to 1 mM P were added (12 levels; $(P/Fe)_{init}$ from 0 to 2). Fe oxidation and precipitation were initiated by addition of 0.5 mM Fe(II). After 4 h, the fresh suspensions were filtered to collect Fe(III)-precipitates and solutions for analysis. A second set of fresh suspensions was aged for 30 days at 40°C. During aging, suspension pH increased to pH 7.9 ± 0.3 due to CO_2 outgassing.

All fresh and aged precipitates were analyzed by X-ray absorption spectroscopy (XAS) at the Fe K-edge, selected solids by Ca K-edge XAS, transmission electron microscopy (TEM), X-ray diffraction (XRD), or Fourier-transform infrared spectroscopy (FTIR). Dissolved element concentrations were measured using inductively coupled plasma mass spectrometry (ICP-MS).

3 RESULTS AND DISCUSSION

3.1 *Structure of fresh Fe(III)-precipitates*

The structural characterization of the fresh Fe(III)-precipitates by Fe K-edge XAS revealed that fresh Fe (III)-precipitates can be described in terms of three endmember phases.

(i) Amorphous (Ca-)Fe(III)-phosphate with precipitate P/Fe ratios ($(P/Fe)_{ppt}$) of >0.52 (Na) to >0.86 (Ca) forming in P-containing solutions. In Ca-Fe(III)-phosphate, Ca causes a higher $(P/Fe)_{ppt}$ and more extensive Fe polymerization.
(ii) Nanocrystalline Si-ferrihydrite with an $(Si/Fe)_{ppt}$ of ~0.1 forming in P-free solutions at $(Si/Fe)_{init}$ of 1.0.

DOI: 10.1201/9781003317395-132

(iii) Poorly-crystalline lepidocrocite forming in P- and Si-free solutions.

In solutions whose $(P/Fe)_{init}$ ratios were lower than the ratios required for exclusive (Ca-)Fe(III)-phosphate formation, (Ca-)Fe(III)-phosphate forms until dissolved P is depleted, followed by the formation of Si-ferrihydrite (with Si) or lepidocrocite (without Si). Consequently, Fe(III)-precipitates formed under such conditions can be rationalized as mixtures of the end-member phases listed above.

3.2 *Structural transformations during aging*

During aging, fresh Fe(III)-precipitates can undergo structural changes due to various interdependent processes: (i) Outgassing of CO_2 and increase in pH. (ii) pH-driven continuing Fe(III)-polymerization and crystallization. (iii) pH-driven precipitation of Ca-phosphate and Ca-carbonate. The extent of transformation depends on initial precipitate structure and solution chemistry. Structural transformation is most evident for Fe(III)-phosphate in solutions with low dissolved P, and rather subtle for precipitates formed in the presence of Ca and Si. In the latter case, Ca stabilizes Ca-Fe(III)-phosphate against further Fe (III) polymerization and Si prevents ferrihydrite transformation into a more crystalline Fe(III)-phase with lower oxyanion sorption capacity.

3.3 *As(V) co-precipitation and retention*

Residual dissolved As(V) concentrations after Fe precipitation varied by nearly 3 orders of magnitude. The least As(V) was removed by (Ca-)Fe(III)-phosphates formed at high $(P/Fe)_{init}$, the most As(V) was removed by Si-ferrihydrite formed at low $(P/Fe)_{ini}$. At intermediate $(P/Fe)_{init}$, substantially more As(V) was removed in Ca-containing than Ca-free electrolytes. Summarily, both Si and Ca enhanced and P inhibited As(V) removal by fresh Fe(III)-precipitates.

During aging, As(V) was released from all precipitates, but to variable degrees. Release of As(V) was highest for Fe(III)-phosphates formed at intermediate $(P/Fe)_{init}$. In general, both Si and Ca reduced As(V) resolubilitsation, Si via its inhibiting effect on Fe(III)-polymerization, Ca by stabilizing Ca-Fe(III)-phosphate and by removing P via Ca-phosphate and Ca-carbonate precipitation.

4 CONCLUSIONS AND RECOMMENDATIONS

Our studies highlight the intricate interdependent effects of Si, P and Ca on Fe(III)-precipitate structure and aging and consequences for As(V) removal and retention. The water chemistry should therefore be considered when developing and evaluating Fe-based water treatment systems for As removal.

The present studies were focusing on the removal and retention of As(V). Since As(III) prevails in anoxic groundwater, future studies should also address the confounding impacts of As(III) co-oxidation during Fe(II) oxidation, precipitate formation and aging. With respect to the impact of Fe(III)-precipitates on As cycling in natural systems, also the effects of organic compounds in combination with inorganic solutes warrant further study.

ACKNOWLEDGEMENTS

We thank Irene Brunner, Numa Pfenninger, and Thomas Rüttimann (Eawag) for their support in the laboratory, and acknowledge the Swiss National Science Foundation for the funding of the PhD student project of Anna-Caterina Senn (contracts No. 200021-132123 and 200020-152993).

REFERENCES

Kaegi R., Voegelin A., Folini D. & Hug S.J. 2010. Effect of phosphate, silicate, and Ca on the morphology, structure and elemental composition of Fe(III)-precipitates formed in aerated Fe(II) and As(III) containing water. *Geochim. Cosmochim. Acta* 74: 5798–5816.

Neumann A., Kaegi R., Voegelin A., Hussam A., Munir A. K.M. & Hug S.J. 2013. Arsenic removal with composite iron matrix filters in Bangladesh: a field and laboratory study. *Environ. Sci. Technol.* 47: 4544–4554.

Roberts L.C., Hug S.J., Ruettimann T., Billah M.M., Khan A.W. & Rahman M.T. 2004. Arsenic removal with Iron(II) and Iron(III) in waters with high silicate and phosphate concentrations. *Environ. Sci. Technol.* 38: 307–315.

Senn A.C., Kaegi R., Hug S.J., Hering J.G. & Voegelin A. (2018) Arsenate co-precipitation with Fe(II) oxidation products and retention or release during precipitate aging. *Wat. Res.* 131: 334–345.

Senn A.C., Kaegi R., Hug S.J., Hering J.G., Mangold S. & Voegelin A. (2015) Composition and structure of Fe(III)-precipitates formed by Fe(II) oxidation in near-neutral water: interdependent effects of phosphate, silicate and ca. *Geochim. Cosmochim. Acta* 162: 220–246.

Senn A.C., Kaegi R., Hug S.J., Hering J.G., Mangold S. & Voegelin A. (2017) Effect of aging on the structure and phosphate retention of Fe(III)-precipitates formed by Fe(II) oxidation in water. *Geochim. Cosmochim. Acta* 202: 341–360.

van Genuchten C.M., Addy S.E.A., Pena J. & Gadgil A.J. 2012. Removing arsenic from synthetic groundwater with iron electrocoagulation: an Fe and As K-edge EXAFS study. *Environ. Sci. Technol.* 46: 986–994.

van Genuchten C.M., Peña J., Amrose S.E. & Gadgil A.J. (2014) Structure of Fe(III) precipitates generated by the electrolytic dissolution of Fe(0) in the presence of groundwater ions. *Geochim. Cosmochim. Acta* 127: 285–304.

Voegelin A., Kaegi R., Frommer J., Vantelon D. & Hug S.J. 2010. Effect of phosphate, silicate, and Ca on Fe(III)-precipitates formed in aerated Fe(II)- and As(III)-containing water studied by X-ray absorption spectroscopy. *Geochim. Cosmochim. Acta* 74: 164–186.

Voegelin A., Senn A.-C., Kaegi R. & Hug S.J. 2019. Reductive dissolution of As(V)-bearing Fe(III)-precipitates formed by Fe(II) oxidation in aqueous solutions. *Geochem Trans.* 20(1): 2.

Arsenic in the Environment: Bridging Science to Practice for Sustainable Development –
van der Wal, Ahmad, Petrusevski, Weijma, Savic, van der Wens, Beerendonk, Bhattacharya, Bundschuh & Naidu (Eds)
 ISBN 978-1-032-32928-4

Molecular-scale insights into Fe(II), As(III) and Mn(II) co-oxidation by weak and strong oxidants: Pros and cons of O_2, NaOCl and $KMnO_4$

C.M. van Genuchten[1] & A. Ahmad[2,3]
[1]*Geochemistry Department, Geological Survey of Denmark and Greenland, Copenhagen, Denmark*
[2]*KWR Water Cycle Research Institute, Nieuwegein, The Netherlands*
[3]*KTH-International Groundwater Arsenic Research Group, Department of Sustainable Development, Environmental Science and Engineering, KTH Royal Institute of Technology, Stockholm, Sweden*

ABSTRACT: The co-removal of Fe(II), As(III) and Mn(II) during oxidation by O_2, NaOCl and $KMnO_4$ was investigated by combining ICP-MS measurements of ion uptake and solid phase characterization by X-ray absorption spectroscopy. We studied O_2, NaOCl and $KMnO_4$ as oxidants because they have different reaction rates with reduced species (O_2 < NaOCl < $KMnO_4$) and all have potential use in groundwater treatment. Our results indicated that the oxidant identity largely controlled the solid phase speciation and efficiency of Fe(II), As(III) and Mn(II) removal. In the O_2 system, Fe(II) oxidation was complete (0.5 h reaction time), but As(III) and Mn(II) were only partially oxidized and removed, producing Mn(III)-incorporated lepidocrocite with sorbed As(III,V). Co-oxidation was more effective with NaOCl, but a significant fraction of sorbed Mn(II) was still present in the reaction products, which consisted of As(V)-sorbed hydrous ferric oxide (HFO). $KMnO_4$ was the most effective oxidant, producing solids containing a mixture of As(V)-sorbed HFO and MnO_2. However, the solids produced in the $KMnO_4$ system were colloidally stable in solutions free of bivalent cations (i.e. Ca) and difficult to remove from the treated water, which is explained by the highly negative charge of MnO_2 at neutral pH preventing particle aggregation. Our results suggest a potential tradeoff between effective Fe(II), As(III) and Mn(II) co-oxidation with $KMnO_4$ in soft waters and reduced particle trapping in rapid sand filters. This diversity in particle behavior must be considered when enhancing As(III) oxidative removal in water treatment, especially with $KMnO_4$.

1 INTRODUCTION

One of the most simple and inexpensive methods of treating groundwater contaminated by arsenic (As) is aeration. This method is based on using atmospheric O_2 to oxidize natural Fe(II) present in As-rich groundwater to form Fe(III) precipitates that bind As. Although this approach requires very little infrastructure, it can be ineffective at lowering As levels to below target limits. For example, aeration alone often cannot decrease the high initial As content common in South Asian groundwater to below the WHO's 10 μg/L limit (Roberts *et al.*, 2004) nor can it reliably remove the lower initial As levels in Dutch groundwater to below the new 1 μg/L Dutch target (Gude *et al.*, 2016). To enhance co-oxidative removal of Fe(II) and As(III), strong chemical oxidants, such as permanganate (MnO_4^-), can be used (Ahmad *et al.*, 2018). Strong oxidants can also have the added benefit of co-oxidizing Mn(II), an ion of growing concern that is often present in As-bearing groundwater. However, little is known about how different oxidants impact the pathways of Fe(II), As(III) and Mn(II) co-oxidation. Here we investigate Fe(II), As(III) and Mn(II) co-oxidation by weak, intermediate and strong oxidants (O_2, NaOCl and $KMnO_4$). Our results can be used to optimize As removal in decentralized and conventional systems.

2 MATERIALS AND METHODS

2.1 *Co-precipitation experiments*

Co-precipitation experiments were performed in $N_{2(g)}$-purged, bicarbonate-buffered, pH 7.5 solutions initially containing 90 μM Fe(II), 9 μM Mn(II) and 0.7 μM As(III) (50 μg/L As). The impact of solution composition was studied by performing experiments in the absence or presence of systematically varied P, Si and Ca levels. Experiments were initiated by either aerating the solution (O_2 set to 9 mg/L) or adding stoichiometric amounts of NaOCl or $KMnO_4$. The reaction proceeded for 0.5 h under vigorous mixing. The removal of Fe, As and Mn was determined by passing the suspension through 0.45 μm filters and measuring the Fe, As and Mn concentration in the filtered solution by ICP-MS.

DOI: 10.1201/9781003317395-133

2.2 *X-ray absorption spectroscopy*

Structural analysis of the filtered solids was performed with X-ray absorption spectroscopy (XANES and EXAFS). Spectra were recorded at room temperature at the DUBBLE beam line of the European Synchrotron Radiation Facility. The X-ray energy was calibrated with Fe(0) (7112 eV) and Mn(0) (6539 eV) foils. The Fe K-edge EXAFS spectra were analyzed by linear combination fits (LCFs), with shell-by-shell fits performed on a subset of samples. The Mn K-edge XANES and EXAFS data were analyzed by LCFs and shell-by-shell fits, respectively. Sixpack software was used for all spectral analyses (Webb 2005).

3 RESULTS AND DISCUSSION

3.1 *Ion uptake and solid phase composition*

The removal of Fe(II), As(III) and Mn(II) depended on the oxidant identity, with increased removal in order of O_2 < NaOCl < $KMnO_4$. In O_2 experiments, Fe(II) was completely oxidized, but only a small fraction of Mn(II) was removed, leading to particles with a Mn/Fe mol ratio of 0.02. Furthermore, <50% As(III) removal was measured in O_2 experiments. When NaOCl was the oxidant, over 80 and 95% removal of Mn(II) and As(III) was observed, respectively, producing solids with a Mn/Fe mol ratio near 0.08. In the $KMnO_4$ system, nearly complete removal of all reduced species was measured, which equated to a solid phase Mn/Fe mol ratio of 0.5, given the Mn added in the form of $KMnO_4$ in these experiments. The impact of oxyanions and bivalent cations on Fe, As and Mn uptake was generally similar, regardless of oxidant. The removal of Mn(II) increased in the presence P and Si, but decreased in the presence of Ca, whereas As(III) removal decreased in the presence of P and Si, but increased in the presence of Ca (Figure 1).

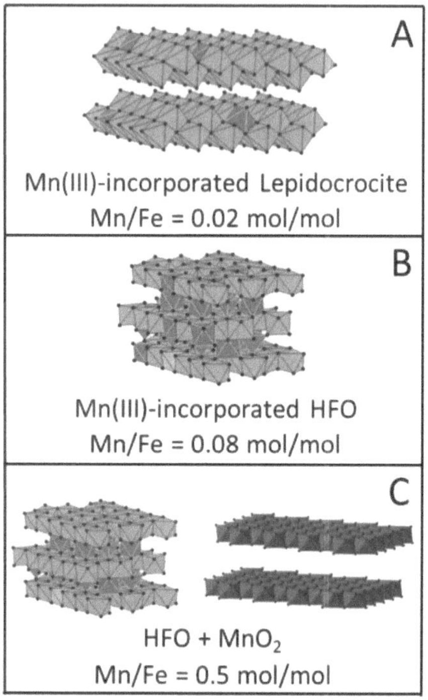

Figure 1. Solids ratios and structural models derived from X-ray absorption spectra of Fe and Mn-bearing precipitates produced in the A) O_2, B) NaOCl and C) $KMnO_4$ experiments.

3.2 *Solid phase speciation*

Similar to ion uptake, the solid phase speciation was governed by the oxidant identity. In the O_2 system, the XANES and EXAFS data indicated that Fe(II) oxidation formed lepidocrocite and Mn(II) was removed via oxidation to Mn(III) and structural incorporation into the Fe(III) phase. The removal of As in the O_2 system occurred by partial As(III) oxidation and sorption of As(III) and As(V) to the Fe (III) precipitate. In the NaOCl system, poorly ordered hydrous ferric oxide (HFO) formed and Mn(II) was oxidized to Mn(III), which incorporated into the HFO structure. However, a fraction of sorbed Mn(II) was also present, particularly in experiments with high initial P and Si concentrations. More effective As(III) oxidation was observed in the NaOCl system than the O_2 system, which is consistent with the lower residual aqueous arsenic levels. In the $KMnO_4$ system, complete oxidation of reduced species was observed, with the solids consisting of a mixture of As(V)-sorbed HFO and MnO_2. Notably, nearly all suspensions produced in $KMnO_4$ experiments were colloidally stable and were difficult to remove by vacuum filtration, which contrasts suspension behavior in the O_2 and NaOCl systems. This observation is explained by the highly negative charge of MnO_2 minerals at neutral pH, which prevents particle aggregation (Post 1999). Our results suggest a tradeoff between effective Fe(II), As(III) and Mn(II) co-oxidation using $KMnO_4$ and reduced particle trapping in rapid sand filters due to poor suspension aggregation. This diversity in structure and aggregation behavior must be considered when adjusting water treatment processes to enhance As(III) oxidation and removal, particularly when $KMnO_4$ is used.

4 CONCLUSIONS AND RECOMMENDATIONS

The oxidation of As(III) and Mn(II) can be a critical step to enhance groundwater treatment. Powerful oxidants, such as $KMnO_4$, can oxidize reduced species completely. However, strong oxidants can negatively impact rapid sand filtration due to the decreased aggregation of solids containing MnO_2 at neutral pH especially in the absence of Ca and Mg, which aid aggregation. Oxidants with reactivity intermediate between O_2 and $KMnO_4$, such as NaOCl or H_2O_2, might be ideal to enhance co-oxidative removal of Fe (II), As(III) and Mn(II) in soft waters (i.e. waters with low Ca and Mg) because they are less likely to produce MnO_2. Our results are a critical addition to other important information, including chemical and infrastructure costs, plant size, operator skill level and potential for toxic by-product formation (e.g. chlorinated organics, bromate), to inform the selection of the best oxidant for groundwater treatment.

ACKNOWLEDGEMENTS

CMvG acknowledges funding from a Veni grant from the Dutch Organization for Scientific Research.

REFERENCES

Ahmad A., Cornelissen E., van de Wetering S., van Dijk T., van Genuchten C., Bundschuh J., van der Wal A. & Bhattacharya P. 2018. Arsenite removal in groundwater treatment plants by sequential permanganate—ferric treatment. *J. Water Process. Eng.* 26: 9.

Gude J.C.J.; Rietveld L.C.; van Halem D. 2016. Fate of low arsenic concentrations during full-scale aeration and rapid filtration. *Water Res.* 88: 566–574.

Post J. 1999. Manganese oxide minerals: crystal structures and economic and environmental significance. *Proc. Natl. Acad. Sci. U.S.A* 96(7): 3447–3454.

Roberts L.C., Hug S.J., Ruettimann T., Billah M., Khan A.W. & Rahman M.T. 2004. Arsenic removal with Iron(II) and Iron(III) in waters with high silicate and phosphate concentrations. *Environ. Sci. Technol.* 38 (1): 307–315.

Webb S. 2005. SIXPACK: a graphical user interface for XAS analysis using IFEFFIT. *Physica Scripta* T115: 1011–1014.

Arsenic in the Environment: Bridging Science to Practice for Sustainable Development – van der Wal, Ahmad, Petrusevski, Weijma, Savic, van der Wens, Beerendonk, Bhattacharya, Bundschuh & Naidu (Eds)

Arsenic precipitation using arsenic-bearing hematite residues

V.S.T. Ciminelli[1,2], H.L. Mendes[3] & C.L. Caldeira[1,2]
[1]*Department of Metallurgical and Materials Engineering, Universidade Federal de Minas Gerais-UFMG, Minas Gerais, Brazil*
[2]*National Institute of Science and Technology on Minerals Resources, Water and Biodiversity, INCT-Acqua, Brazil*
[3]*Anglo Gold Ashanti, Minas Gerais, Brazil*

ABSTRACT: The need to treat large volumes of concentrated arsenic (As) industrial effluents is challenging. The increasing costs of effluent treatment and disposal, and the risks of contamination related to the long-term stability of the residues are some of the challenges worth to be mentioned. The use of secondary sources of reagents becomes attractive, in particular when there are restrictions in the application, as in the case of As-bearing residues. The present study reports the results of As removal from an industrial solution using and As-bearing hematite as the iron source. Arsenic concentration dropped from 1500 to 5 mg/L by the addition of ferric sulfate and lime in two stages (at pH 4.5 and 8.5-9) in a pilot plant. Ferric sulfate was produced *in situ* by sulfation of As-bearing hematite, followed by dissolution in water. In the final sludge, As(V) is mainly associated (77-88%) with amorphous and nano crystalline Fe(Al) (hydr)oxides, as as Fe (Al)-As coprecipitates, Ca and Al arsenates or strongly adsorbed (17%) onto these (hydro)oxides.

1 INTRODUCTION

The removal of arsenic (As) from relatively concentrated (e.g., 0.1-1 g/L) industrial effluents usually comprises an oxidation step followed by precipitation or co-precipitation with iron salts, adsorption onto suitable surfaces and lime treatment. Ion-exchange resins or membrane processes may be applied when polishing is required (Cortina *et al.*, 2016). It is well established that the extent of As removal by precipitation, as well as the stability of As residues, is related to the Fe/As ratio. The increase of the Fe/As ratio increases the As removal and the stability of As residues (Pantuzzo & Ciminelli 2010). The treatment of large volumes of industrial solution implies costs related to the consumption of iron salts, and therefore less costly alternatives have been evaluated (Araujo *et al.*, 2019). The disposal of large amounts of toxic wastes with low As content (e.g., 3–8%) results in additional costs and environmental liabilities.

Arsenic-bearing, iron-rich materials, and residues, available in some industrial operations, are now investigated for effluent treatment. The use of these materials reduces operational costs in effluent treatment and creates an application for toxic wastes.

2 MATERIALS AND METHODS

The iron-rich hematite is produced in the extraction of a refractory gold ore, where gold is found finely disseminated mostly in pyrite (FeS_2), and rarely in arsenopyrite (FeAsS). The process requires the roasting of gold-bearing sulfides in order to produce a porous iron oxide matrix (calcine), from which gold is leached. Following gold extraction, the iron-rich oxide tailings (As-HMT) are disposed of in tailings dam.

During roasting, volatilized As is absorbed in water and then removed by addition of ferric sulfate and lime. The final residue (As-sludge) is disposed of in lined tailings dams.

X-ray diffraction (XRD) analyses of solid samples were performed using a PANalytical Empyrean Cu anode diffractometer (1.54 Å). The samples were also analyzed by scanning electron microscopy-based automated image analysis (Mineral Liberation Analyzer – MLA FEI Quanta 650 F and Bruker Quantax X-Flash 5030 EDS detectors), and transmission electron microscopy (TEM) (LaB6-TEM Tecnai G2-20 (FEI), 200 kV, equipped with a 30 mm^2 window Si(Li) EDS detector and electron energy loss spectroscopy (EELS) (Gatan Image Filter – GIF Quantum SE), as described by Ciminelli *et al.* (2018). Raman spectroscopy was performed on Horiba Jobin Yvon Labram-HR 800, equipped with a 633 mm He-Ne laser.

The sequential extraction protocol (SEP) followed Pantuzzo & Ciminelli (2010): (i) pH 5, 1 mol/L NaOAc/HOAc, 25°C, 1 h to dissolve Ca-compounds and others; (ii) pH 5, 0.1 mol/L Na_2HPO_4/HNO_3, 70°C, 1 h, to extract As-strongly adsorbed onto Fe (Al) (hydr)oxides and Fe-As coprecipitates; (iii) 4.2 mol/L HCl/Ascorbic acid 0.66%/Na-citrate 0.88%, 80°C, 2.5 h, to dissolve amorphous and crystalline Fe(Al) (hydr)oxides; (iv) digestion in aqua regia ($HCl:HNO_3$ 3:1 v/v). The solid samples were submitted to digestion in aqua regia. Elemental composition was quantified by ICP-OES (Perkin Elmer, Optima 7300DV) for digested and aqueous samples.

DOI: 10.1201/9781003317395-134

3 RESULTS AND DISCUSSION

According to XRD analyses, the iron-rich calcines contains hematite (Fe_2O_3) and quartz (SiO_2), with minor anhydrite ($CaSO_4$), and trace amounts of As trioxide (As_2O_3), and silicates.

Analyses by MLA indicated As associated with iron (hydr)oxides and in complex compounds containing Fe, Ca, Al, Si, S, and other elements. The calcine was mixed with concentrated sulfuric acid (sulfation) at 200°C for 10 min., and then with process water to produce a ferric sulfate solution used for As precipitation. The solution to be treated feed (Table 1) contains typically 1500 mg/L aqueous As(III) species during the contact of roaster off-gas with water in the washing towers.

The ferric sulfate solution is added to the feed solution, and pH is raised to 4.5 and 8.5-9 with lime, in two tanks, for As and Mn removal. The effluent (5.3 mg/L As) is recirculated into the industrial circuit.

The As sludge contains gypsum ($CaSO_4.2H_2O$) and bassanite ($CaSO_{4}.{}_{1/2}H_2O$), with minor amounts of hematite, and quartz, according to XRD analyses. Complex compounds (As, Fe, Ca, Al and others) were detected. Raman spectra indicated poorly crystalline Fe (hydr)oxides, such as ferrihydrite, typical bands of As(III)/As(V) coprecipitated/adsorbed, with incipient formation of goethite. The Fe/As molar ratio is 2.4. The chemical compositions of iron-rich calcines and As-sludge, shown in Table 2, support the mineralogical findings.

To further characterize the partitioning of As in the sludges, gypsum and other calcium sulfates were leached with water. The XRD pattern of the leached residues exhibited the two typical broad peaks of poorly crystalline Fe-As coprecipitates, As-ferrihydrite, and ferrihydrite. Crystalline Ca-Fe (hydr)arsenates, quartz, hematite, and other trace constituents were also identified. Clusters (50-200 nm) of amorphous nanoparticles containing Fe, As, and O predominate in the samples. However, elements such as Al, Ca, Zn, Si, P, S, Mn, Mg and K were detected in trace amounts, by TEM/EDS. Pentavalent As predominates in the As-sludge, according to EELS, and also in the treated aqueous effluent. Figure 1 summarizes As partition in the As-HMT and As-sludge. In the latter, As is mainly associated (77-88%) with Fe(Al) (hydr)oxides, as Fe (Al)-As coprecipitates, Ca and Al arsenates and strongly adsorbed As (17%, exchange reaction with phosphate) was found. Weakly-bound As is low (5%), thus indicating low mobility in aqueous solutions (Mendes *et al.*, 2021).

Table 1. Chemical composition (mg/L) of industrial solutions before and after arsenic removal.

Element	Al	As	Ca	Cu	Fe	K
Feed	152.1	1503	472	19.6	166.5	52.9
Effluent	<2.5	5.3	491	<0.1	<2.5	100.7
Element	Mg	Mn	Ni	Pb	S	Zn
Feed	506	47.2	1.8	2.2	2211	137.3
Effluent	16.6	<0.2	<0.1	<0.1	436.4	<0.1

Table 2. Chemical composition (% weight or *mg/kg).

Element	Fe	As	S	Si	Ca	Al
As-HMT	39.4	2.0	2.7	10.0	2.2	2.6
As-Sludge	7.6	4.2	13.8	0.6	17.2	1.1
Element	Mn*	Ni*	Cu*	Pb*	Mg	Zn
As-HMT	1300	797.7	1051	215.3	0.9	0.43
As-Sludge	1465	250.9	782	169.3	1.7	0.48

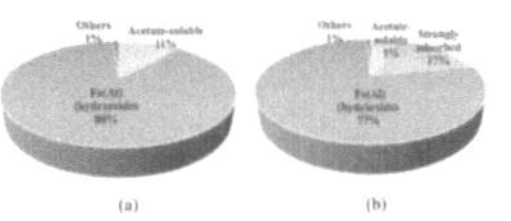

Figure 1. Arsenic partition according to sequential extraction procedure in the As-hematite (a) and in the As-sludge (b).

4 CONCLUSIONS

Arsenic was efficiently removed (1503 to 5.3 mg/L As) from an industrial effluent using As-bearing hematite tailings as an iron source. The final As-sludge was enriched with As, thus implying in reducing disposal area and costs, with potential up to 60% decrease in the effluent operating costs. Further investigation should address the long-term stability of the As residues.

ACKNOWLEDGEMENTS

We acknowledge Anglo Gold Ashanti for funding this research, and the Center of Microscopy - UFMG for the use of microscopy facilities. We would like to thank Filipe A. T. Alves and Hellen A. Andrade for performing the chemical analyses, and Dr. Erico Freitas and Dr. Itamar D. Delbem for their support on microscopy analyses. V. Ciminelli acknowledges the support from CNPq and CAPES.

REFERENCES

Araujo S.F., Caldeira C.L., Ciminelli V.S.T., Silva A. & Amorim C.C. 2019. Versatility of iron-rich steel waste for the removal of high arsenic and sulfate concentrations in water. *Environ Sci. and Poll. Res.* 26(5): 4266–4276.

Ciminelli V.S.T., Antônio D.C, Caldeira C.L., Freitas E.T.F., Delbem I.D., Fernandes M.M., Gasparon M. & Ng J.C, 2018. Low arsenic bioaccessibility by fixation in nanostructured iron (hydr)oxides: quantitative identification of As-bearing phases *J. Hazard Mater.* 353: 261–270.

Cortina J.L., Litter M.I., Gibert O., Valderrama C., Sancha A. M.S.G. & Ciminelli V.S.T. 2016. Latin american experiences in arsenic removal from drinking water and mining effluents. In: M. Bryjak, N. Kabay, B.L. Rivas & J. Bundschuh (eds). *Innovative Materials and Methods for Water Treatment*, 1st ed. Taylor & Francis- pp. 391– 416.

Mendes H.L., Caldeira C.L. & Ciminelli V.S.T. 2021. Arsenic Removal From Industrial Effluent: In-situ Ferric Sulfate Production and Arsenic Partitioning in the Residues Min. Eng. 169: 106945.

Pantuzzo F.L. & Ciminelli V.S.T. 2010. Arsenic association and stability in long-term disposed arsenic residues. *Water Res.* 44: 5631–5640.

Arsenic in the Environment: Bridging Science to Practice for Sustainable Development – van der Wal, Ahmad, Petrusevski, Weijma, Savic, van der Wens, Beerendonk, Bhattacharya, Bundschuh & Naidu (Eds)
© 2023 the Author(s), ISBN 978-1-032-32928-4

Tailored Metal-Organic Frameworks (MOFs) for arsenic-free drinking water

S. Ramanayaka & M. Vithanage
Ecosphere Resilience Research Center, Faculty of Applied Sciences, University of Sri Jayewardenepura, Nugegoda, Sri Lanka

ABSTRACT: Arsenic in drinking water as a contaminant has become a worldwide environmental concern due to the health effects. Therefore, As removal technologies have been received a primary attention whereas the adsorption technology is considered to be a prevailing method, as of its high removal efficiency and the easy operation. Metal-Organic frameworks (MOFs) have become a primary material for adsorptive removal of contaminants in water, due to their high surface area, adjustable porosity, and recyclability. As MOFs are unstable in water, pre and post synthetic modification and functionalization is common. Various MOFs have been used for arsenic removal with different synthesis techniques, composition and modifications. This review addresses the performance and mechanisms of As removal in various modified MOFs in detail. In order to compare the performance of MOFs, here we used partition coefficient (PC) instead of maximum adsorption capacity, which is sensitively influenced by initial loading concentrations. Furthermore, it discusses the scale-up issues and forthcoming pathway for research and development needs of MOFs for effective As removal. This review further elucidates the main removal mechanisms of As by MOFs.

1 INTRODUCTION

High concentrations of arsenic (As) have been reported in many countries in various continents and the source, distribution and mobilization of As in aqueous environment are associated mostly with geogenic processes. Limited anthropogenic activities such as mining, combustion of fossil fuels, petroleum refining and wood and agrochemicals, release arsenic into the environment (Wang *et al.*, 2018). Most common species of As in the environment are inorganic arsenic oxides, such as arsenite (AsIII), arsenate (AsV) and methylated forms (Sharma *et al.*, 2014). The World Health Organization (WHO) reduced the guideline value for As in drinking water from 50 to 10 μg/L in 1993 whereas many countries are now focusing it to be reduced further to 1 μg/L due to the toxicity effects (Ahmad & Bhattacharya 2019).

Various methods have been tested in order to remove As from water. Among many techniques, adsorption received a substantial focus due to easiness and less cost compared to other technologies. Although many adsorbents were experimented for As removal, due to the distinct chemistry of As(III) and As(V), not many promising sorbents exist for the removal of both species simultaneously. Midst of various adsorbents such as activated carbon, clays and nanomaterials, Metal-Organic Frameworks (MOFs) have become a recent interest for As removal in aqueous media (Ramanayaka *et al.*, 2019). MOFs are highly ordered, crystalline, and porous materials, known as for their extraordinary performances in remediating various contaminants (Dhaka *et al.*, 2019). The present review aims to investigate: i) capacities of MOFs in removing various As species in water; ii) effect of different modifications of MOFs on As removal; iii) and future needs for further developments.

2 METAL-ORGANIC FRAMEWORKS AS ADSORBENTS

2.1 *Synthesis*

Both conventional methods as well as alternative methods are used to prepare MOFs. Conventional slow evaporation synthesis is carried out in the room temperature; therefore external energy is not required, however, this method is time consuming. Microwave-assisted synthesis methods are extensively used to produce nanoscale metal oxides and very rapid method to synthesize MOFs. Furthermore, electrochemical, sonochemical and mechanochemical synthesis have been applied for MOF synthesis.

2.2 *Capacity*

The diverse nature, high surface area, tunable pore size, and high porosity of MOFs make them more attractive for water treatment than other materials, (Dhaka *et al.*, 2019). MOFs demonstrate high partition coefficient, which indicate high capacity of removal, which is independent of the initial conditions (Ramanayaka *et al.*, 2019). It is evident from the literature that the maximum adsorption capacity does not provide a good base to select the best adsorbent due to the different initial conditions applied in various experiments, and hence, Partition Coefficient (PC) is calculated (Vikrant & Kim 2019).

DOI: 10.1201/9781003317395-135

2.3 *Modifications*

MOFs are further investigated for various modifications to improve the adsorption capacity. Modifications can be introduced through inner and outer surfaces of MOFs by solvent molecules, which can easily penetrate to the interior channels of a MOF due to its high porosity.

3 ADSORPTIVE REMOVAL OF ARSENIC

As(V) is the predominant form of arsenic in surface water and hence, MOFs have been synthesized with As (V) adsorption capabilities. The calculated adsorption performance of MOF Fe-BTC under different pH conditions demonstrated As(V) removal efficiency higher than 96% for an initial concentration of As(V) 5 mg/L at pH of 4. It showed 37 times effectiveness than Fe_2O_3 nanoparticles for As(V) removal and the PC value of 0.17 L/g.

Arsenic uptake by Zeolitic Imidazolate Framework-8 (ZIF-8), has achieved 49.5 mg/g for As(V) and 60 mg/g for As(III) at neutral pH (Drout *et al.*, 2018). Although the same amount of As(III) and As(V) were used for the study in mg/g, it is not the same concentration in mol/L. Therefore, As(III) depicts a lower PC value of 0.85 L/g, while As(V) exhibits 1.88 L/g value of PC, which specifies that the ZIF-8 shows more than twice higher performance for As(III) than that of the As(V). Arsenic was adsorbed on to the framework material surface as the As (V) and As(III) ions were more substantial than the ZIF-8 pore spaces. ZIF-8 has different morphologies, however it was found that the surface area does not correlate with the adsorption capacity of the MOF.

The arsenic uptake by MOF was independent of other anions in the aqueous media than phosphate as expected. Arsenate species interacted with the MOF node while thiolated ligands were bound to arsenite species with an uptake of arsenite and arsenate from UiO-66-$(SH)_2$ (Prum *et al.*, 2018). However, no information was given regarding the possibility of forming thiolated arsenic species during the study, which is a cytotoxic to biological systems.

MIL-53(Fe) is another type of MOF that has a variable adsorption capacity for As(V) as the ion in the metal node changes. Lewis acid-base interactions of anionic $H_2AsO_4^-$ and the metal ion of MIL-53 resulted in an adsorption capacity of 21 mg/g with a PC value of 1.06 L/g. MIL-53 (Al), which is very similar to MOF-53 (Fe), with a different ion at the node where its maximum adsorption capacity was reported as 106 mg/g for As(V) in the form of $HASO_4^{2-}$ at a pH of 8 (Vu *et al.*, 2015).

However, studies are limited on adsorption of As (III) and As(V) by MOFs, and the data provided are insufficient for evaluating performance. Therefore, thorough studies on adsorption of As by MOFs are essential to be conducted. Among the reported details ZIF-8 demonstrated the best performance for As removal and specifically for As(V). Mostly, the studies on arsenic adsorption by MOFs indicate ion exchange as the prominent mechanism. Data on reduction of As (V) to As(III) during adsorption and thermodynamics are lacking.

Although the As removal were promising and tested in laboratory scale, the number of commercial-scale products is limited because of scale-up issues such as production capacity (space-time yield), expenditure, properties, purity, and stability. Due to these limitations and challenges, despite the remarkable academic interest which produced thousands of new MOFs with diverse applications, only a few of them are being translated from laboratory to large scale to be used in real-world applications.

4 CONCLUSIONS

MOFs have consistently shown enhanced adsorption capacity and performance for As than for conventional adsorbents. Overall, MOFs will inevitably be considered as an effective alternative to conventional adsorbents provided that their water stability, selectivity and reusability can be ensured.

REFERENCES

Ahmad A. & Bhattacharya P. (2019). Arsenic in drinking water: is 10 µg/L a safe limit? *Curr. Pollut. Rep.* 5(1): 1–3.

Dhaka S., Kumar R., Deep A., Kurade M.B., Ji S.W. & Jeon B. H. 2019. Metal–organic frameworks (MOFs) for the removal of emerging contaminants from aquatic environments. *Coord. Chem. Rev.* 380: 330–352.

Drout R.J., Robison L., Hanna S. & Farha O.K. 2018. Can metal–organic framework composites contain the water contamination crisis? *ACS Cent. Sci.* 4(3): 321–323.

Prum C., Dolphen R. & Thiravetyan P. 2018. Enhancing arsenic removal from arsenic-contaminated water by echinodorus cordifolius–endophytic arthrobacter creatinolyticus interactions. *J. Environ. Manage.* 213: 11–19.

Ramanayaka S., Vithanage M., Sarmah A., An T., Kim K.-H. & Ok Y.S. 2019. Performance of metal–organic frameworks for the adsorptive removal of potentially toxic elements in a water system: a critical review. *RSC Adv.* 9(59): 34359–34376.

Sharma A.K., Tjell J.C., Sloth J.J. & Holm P.E. 2014. Review of arsenic contamination, exposure through water and food and low cost mitigation options for rural areas. *Appl. Geochem.* 41: 11–33.

Vikrant K & Kim K.H. 2019. Nanomaterials for the adsorptive treatment of Hg(II) ions from water. *Chem. Eng. J.* 358: 264–282.

Vu T.A., Le G.H., Dao C.D., Dang L.Q., Nguyen K.T., Nguyen Q.K., Dang P.T., Tran H.T.K., Duong Q.T., Nguyen T.V. & Lee G.D. 2015. Arsenic removal from aqueous solutions by adsorption using novel MIL-53(Fe) as a highly efficient adsorbent. *RSC Adv.* 5(7): 5261–5268.

Wang J., Zhang T., Li M., Yang Y., Lu P., Ning P. & Wang Q. 2018. Arsenic removal from water/wastewater using layered double hydroxide derived adsorbents, a critical review. *RSC Adv.* 8(40): 22694–22709.

Arsenic in the Environment: Bridging Science to Practice for Sustainable Development –
van der Wal, Ahmad, Petrusevski, Weijma, Savic, van der Wens, Beerendonk, Bhattacharya, Bundschuh & Naidu (Eds)
 ISBN 978-1-032-32928-4

Visual MINTEQ simulation for prediction of the adsorption of arsenic on ferrihydrite

R. Irunde[1,2], P. Bhattacharya[2,3], J. Ijumulana[1,2], F.J. Ligate[1,2], A. Ahmad[2,3,4], F. Mtalo[1] & J. Mtamba[1]
[1]*DAFWAT Research Group, Department of Water Resources Engineering, College of Engineering and Technology, University of Dar es Salaam, Dar es Salaam, Tanzania*
[2]*KTH-International Groundwater Arsenic Research Group, Department of Sustainable Development, Environmental Science and Engineering, KTH Royal Institute of Technology, Stockholm, Sweden*
[3]*KWR Watercycle Research Institute, Nieuwegein, The Netherlands*
[4]*Department of Environmental Technology, Wageningen University and Research (WUR), Wageningen, The Netherlands*

ABSTRACT: The surface of ferrihydrite adsorbs arsenic effectively. In this investigation, the arsenic (As) laced water samples collected from Geita and Mara regions within the Lake Victoria Basin (LVB) under DAFWAT project were simulated on Visual MINTEQ 3.1 software to determine the amount of ferrihydrite required to adsorb a given amount of As from water. Model simulations show that As concentration of $\leq$ 1 mM can be completely adsorbed by 4 g/L ferrihydrite. Previous studies show that the lower pH 4 to 4.5 influences adsorption, while it decreases as pH increases as well as when As concentration increases. The increase of adsorbent dose to 4 g/L has shown to improve As(V) adsorption on pH 5 to 8 at 100%. The amount of adsorbent can now be used for laboratory adsorption experiments by using iron-based materials or commercial ferrihydrite.

1 INTRODUCTION

High levels of arsenic (As) have been reported both in surface water and groundwater in several African countries (Ahoulé *et al.*, 2015). Elevated concentrations of As has recently reported in the northern part of Tanzania, particularly around the Lake Victoria Basin (LVB) and the Lake Victoria Goldfields (LVGF) (Ijumulana *et al.*, 2016, 2017; Lucca 2017; Mnali 2001).

Geochemical modeling tools such as PHREEQC (Parkhurst 1995; Parkhurst & Appelo 1999) and Visual MINTEQ (Gustafsson 2009) are useful to model the mobility and fate of arsenic in water sources. The present study aims to use Visual MINTEQ for prediction of amount of iron oxide that can be applied on arsenic laced water samples for adsorptive removal. The simulation results of groundwater samples collected from Geita and Mara are presented.

2 MATERIALS AND METHODS

2.1 *Study area and water sampling*

Lake Victoria Basin, Tanzanian part, is one of the 9 river basins in Tanzania mainland covering area of 119,442 sq. km. The region has little seasonal variation but the eastern section where the study area lies (Geita and Mara), average only 750 – 1000 mm of rain. Favourable climatic conditions for agriculture and livestock and the abundance of natural resources have supported the livlihood of the rural population of over 35 million people (Lucca 2017). The geology of the Tanzanian LVB consists of Archean granitoids-greenstone belts hosted in the Tanzanian craton. More than 80% of rural population depends on groundwater resources for various uses.

Water sampling was carried out at the end of dry season during October 2016 from Geita and Mara (Figure 1). A total of 29 water samples collected from Mara and 18 samples from Geita were simulated using the Visual MINTEQ.

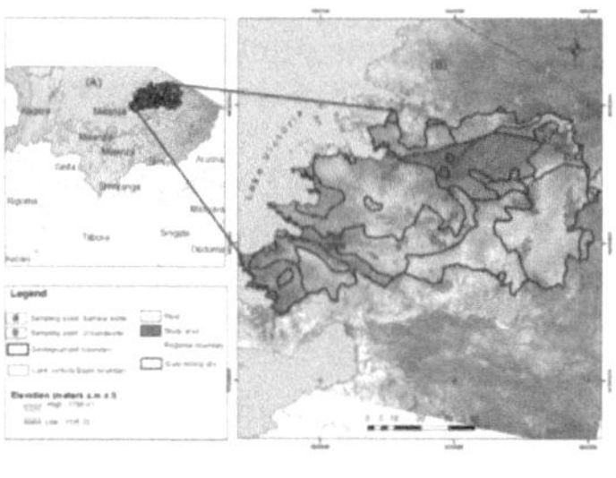

Figure 1. Map of the study area around the LVB region (A) and water sampling area locations (B).

2.2 *Visual MINTEQ parameter setting*

Groundwater quality data for Geita and Mara such as pH and the concentration of As(V) and As(III) (Lucca 2017) were imported in the Visual MINTEQ. The following parameters were fixed in

 DOI: 10.1201/9781003317395-136

the model for Geita sample; NaCl = 0.01 M, temperature = 25°C, pH = 6.3 and ferrihydrite = 4 g/L; and for Mara sample; NaCl = 0.01 M, temperature = 26°C, pH = 7.3 and ferrihydrite = 4 g/L.

3 RESULTS AND DISCUSSION

The results of the geochemical modelling were concentrations of As adsorbed onto ferrihydrite. The percentage of As sorbed was calculated using the following equation:

$$As_{sorbed} = (\text{Final } As_{conc}/\text{Initial}As_{conc}) \times 100\% \quad (1)$$

The percentage of adsorbed As(V) and As(III) on ferrihydrite were plotted against pH to visualize the strength of Visual MINTEQ 3.1 on arsenic adsorption (Figure 2).

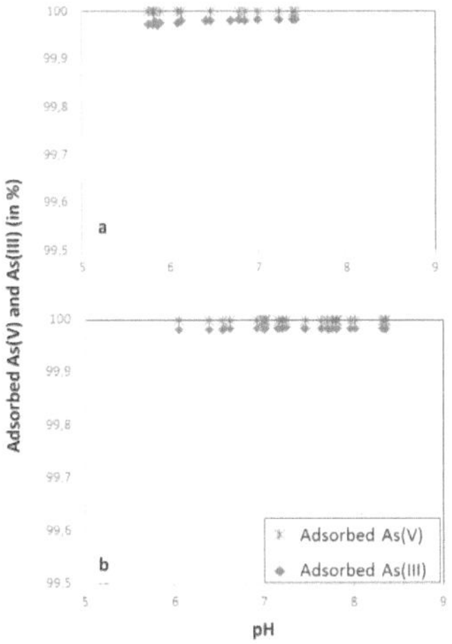

Figure 2. Adsorption of arsenic onto ferrihydrite from water sample from a) Geita and b) Mara sites.

The As(V) concentrations at Geita and Mara ranged from 10 μg/L to 300 μg/L which is equivalent to ≤ 1 mM and hence only 4 g/L ferrihydrite added in the Visual MINTEQ was adequate to remove As from water samples in the pH range of 5 to 8.5. The model results from Geita (Figure 2a) and Mara (Figure 2b) show maximum adsorption of As(V) leading to 100% removal, while As(III) is removed at 99%. This is because As(V) a negatively charged species, is strongly bound to surface of the ferrihydrite and is thus easily removable from water upto a pH close to 8.5 (Ahmad *et al.*, 2014). The groundwater of Geita and Mara have higher concentrations of As(V) because of an oxidizing environment where As(III) can be oxidized to As(V) as well as Fe^{2+} being oxidized to Fe^{3+}.

4 CONCLUSION

The Visual MINTEQ software is a useful tool to gain insight on the adsorption processes. The software is recommended for prediction of adsorbent amount that is required for As removal from water samples. Simulation results predict 100% of As(V) removal from Geita and Mara in the pH range of 5 to 8.5 when 4 g/L ferrihydrite was dosed.

ACKNOWLEDGEMENTS

We acknowledge the Swedish International Development Cooperation Agency (Sida) for supporting the DAFWAT program (Contribution: 51170072).

REFERENCES

Ahmad A., van de Wetering S., Groenendijk M. & Bhattacharya P. 2014. Advanced oxidation-coagulation-filtration (AOCF)—an innovative treatment technology for targeting drinking water with <1 μg/L of arsenic. In: M.I. Litter, H.B. Nicolli, M. Meichtry, N. Quici, J. Bundschuh, P. Bhattacharya & R. Naidu (eds.) *One Century of the Discovery of Arsenicosis in Latin America (1914–2014) (As2014)*. CRC Press/Taylor and Francis, pp. 817–819.

Ahoulé D.G., Lalanne F., Mendret J., Brosillon S. & Maïga A.H. 2015. Arsenic in African waters: a review. *Wat. Air Soil Poll.* 226(9): 302.

Gustafsson J.P 2009. Visual-MINTEQ, Version 2.61: A Windows Version of MINTEQA 2 (version 4.0) http://hem.bredband.net/b108693/

Ijumulana J., Bhattacharya P. & Mtalo F. 2016. Arsenic occurrence in groundwater sources of Lake Victoria Basin in Tanzania. In: P. Bhattacharya, M. Vahter, J. Jarsjö, J. Kumpiene, A. Ahmad, C. Sparrenbom, G. Jacks, M.E. Donselaar, J. Bundschuh, & R. Naidu (eds.) *Arsenic Research and Global Sustainability (As2016)*. CRC Press/Taylor and Francis (ISBN 978-1-138-02941-5), pp. 86–87.

Ijumulana J., Lucca E., Bhattacharya P. & Mtalo F. (2017) Mineral solubility controls on drinking water quality in the areas of gold mining in Geita and Mara regions of Northern Tanzania. *Geol. Soc. Amer. Abstracts with Programs*, 49(6): doi: 10.1130/abs/2017AM-308005.

Kassenga G.R. & Mato R.R. 2008. Arsenic contamination levels in drinking water sources in mining areas in Lake Victoria Basin, Tanzania, and its removal using stabilized ferralsols. *Int. J. Biol. Chem. Sci.* 2(4): 389–400.

Lucca E. 2017. *Geochemical Investigation of Arsenic in Drinking Water Sources in Proximity of Gold Mining Areas in the Lake Victoria Basin, in Tanzania.* MSc Thesis. KTH Royal Institute of Technology.

Mnali, S. 2001. Assessment of heavy metal pollution in the lupa gold field, SW Tanzania. *Tanzania J. Science* 27(2): 15–22.

Parkhurst D.L. 1995. Users Guide to PHREEQC-A Computer Program for Speciation, Reaction-path, Advective-Transport, and Inverse Geochemical Calculations. *U.S.Geol. Surv. Water Resour. Invest. Rep.* 95–4227.

Parkhurst D.L. & Appelo C.A.J. 1999. User's Guide to PHREEQC (version 2), U.S. *Geol. Surv. Water Resour. Invest.* 99–4529, 312 pp.

Arsenic in the Environment: Bridging Science to Practice for Sustainable Development –
van der Wal, Ahmad, Petrusevski, Weijma, Savic, van der Wens, Beerendonk, Bhattacharya, Bundschuh & Naidu (Eds)
© 2023 the Author(s), ISBN 978-1-032-32928-4

Application of a novel mesoporous MFT/SBA-15 composite materials on arsenic elimination from aqueous solutions

L. Tao[1] & N.Y. Yu[2]
[1]*Guangdong Key Laboratory of Integrated Agro-environmental Pollution Control and Management, Guangdong Institute of Eco-environmental Science & Technology, Guangzhou, P.R. China*
[2]*Key Laboratory of Sustainable Resources Processing and Advanced Materials of Hunan Province, National & Local Joint Engineering Laboratory for New Petrochemical Materials and Fine Utilization of Resources, Hunan Normal University, Changsha, P.R. China*

ABSTRACT: A novel chelating resin/mesoporous silica composite material (MFT/SBA-15) was synthesized via a co-condensation method, and applied in the treatment of elimination of arsenic (As) contaminated water. Kinetics results illustrated that As adsorption onto MFT/SBA-15 obeyed the pseudo second-order model and the chemical adsorption was the rate limiting step in the adsorption process. Thermodynamics results presented that As adsorbed onto MFT/SBA-15 obeyed the Langmuir isothermal adsorption model and carried the characteristic of single molecular layer adsorption. The formation of coordination bond (C = S→As) between the lone electron pairs of S atom and As during the adsorption processes led to the fixation processes of As, further illustrated that the novel mesoporous MFT/SBA-15 materials combine both the advantages of chelating resins and mesoporous silica together, and have excellent application potentials in the As wastewater treatment.

1 INTRODUCTION

Arsenic (As), the 33rd element on the periodic chart, carrying an average atomic weight of 75, and approximately 1.5 mg/kg in the Earth's crust, is one of the priority contaminants of major concern in the environment. Among conventional techniques for removing As from water, adsorption methods are preferred because of their advantages in removal effectiveness, cost and equipment handling (Babel & Kurniawan 2003). Nevertheless, the absorbent is the critical factor in deciding the As removal efficiency. Chelating resins can selectively chelate specific metal ions from solutions. Mesoporous silica take the advantage of high specific surface area and easily modified pore surface. Hence, the present study aims to investigate: i) the synthesis of a functional composite material combined the advantages of both chelating resins and mesoporous silica together; ii) the extent and rate of As elimination by chelating resin/mesoporous silica composite materials; and iii) the functional groups in the composite materials and its mechanism on the stabilization of As.

2 MATERIALS AND METHODS

2.1 *Materials*

The synthesis processes of MFT/SBA-15 are detailed presented in Wu's MSc Thesis (Wu 2016). The organic components in materials were analysis by combustion method, as well as analysis by the high temperature calcination method. The surface area analysis as well as the pore size measurements of these samples was analyzed by Surface Area & Pore Size by Gas Sorption surface area analysis and pore size measurements. Thermo gravimetric (TG) analysis was conducted at TG-FTIR (Netzsh TG 209C & Bruker Vector^{TM} 22).

2.2 *Analytical techniques*

Typical suspensions prepared for the reactor contained 0.4 mM As(V) and 25.0 mg MFT/SBA-15 at 298 K, placed on a rotator at 200 rpm. Triplicate samples were destructively sampled at each specific time point of incubation. Batch studies were conducted to access the influence of pH as well as ionic strength of solution on As(V) adsorption onto MFT/SBA-15. At given time intervals, the serum bottles were taken from the shaker and passed through a 0.45-μm (pore size) membrane filter. As(V) concentrations in the filtrate were determined by a hydrogen generation – atomic fluorescence spectrometer. As(III) was determined by liquid chromatography–atomic fluorescence spectrometry (Wang *et al.*, 2018).

3 RESULTS AND DISCUSSION

3.1 *Typical physical-chemical characteristics of MFT/SBA-15*

The organic contents in materials obtained after calcination was approximately 91.70 ± 0.20 wt.%, which

DOI: 10.1201/9781003317395-137

were equal to the theoretical calculation re-sults of 91.8 wt.%. Furthermore, the nitrogen adsorption-desorption isotherm of the material presents the adsorption isotherm of IV type, carrying the characteristics of surface areas (149 m^2/g), pore size (7.7 nm) and pore volume (0.57 cm^3/g).

3.2 *As adsorption kinetics onto MFT/SBA-15*

It was very interested to find that the adsorption of As(V) adsorbed onto MFT/SBA-15 exhibited increase trends with the increase of solution pH from 2 to 9, whereas the amount of As(V) adsorption onto MFT/SBA-15 sharply decreased with the increase of ionic strength. Besides, it should be noted that the adsorption of As(V) adsorbed onto MFT/SBA-15 exhibited time-dependent patterns with faster adsorption processes occurring in the beginning (Figure 1A). Furthermore, As(V) adsorption onto MFT/SBA-15 obeyed the pseudo second-order model and chemical adsorption was the rate limiting step in the adsorption process (Figure 1B & 1C).

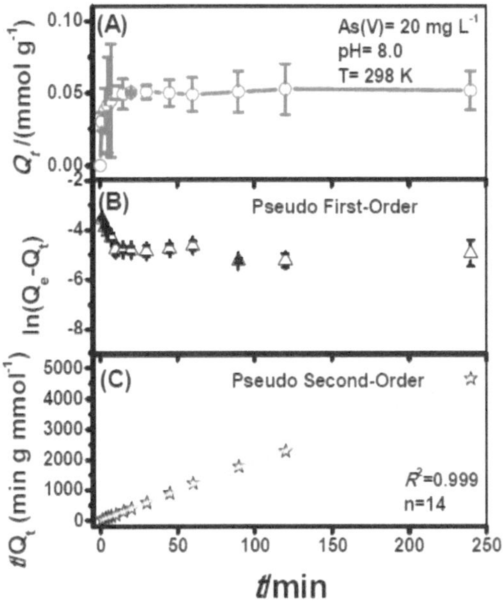

Figure 1. Effect of contact time on As(V) adsorption (A), kinetic plots by pseudo first-order model (B) and pseudo second-order model (C).

3.3 *Sorption of arsenic and antimony*

Figure 2 present the adsorption isotherms of As(V) adsorbed onto MFT/SBA-15 under different temperatures. It should be noted that the quantity of adsorption (Q) increase with an increase in the reaction temperature (Figure 2A), indicating that adsorption of As(V) onto MFT/SBA-15 was an endothermic reaction. The obtained Q value increase with the rise concentration of soluble As (V) (Ce), and them reach its saturated state and keep at adsorption equilibrium. The fitting results in Figures 2B and C clearly illustrate that As(V) adsorbed onto MFT/SBA-15 have the characteristic of single molecular layer adsorption.

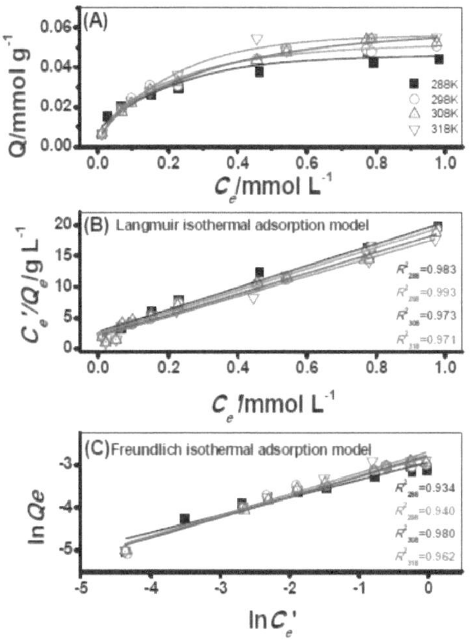

Figure 2. Adsorption isotherms (A), fitting plots of the adsorption data to Langmuir (B), and Freundlich (C).

3.4 *Critical role of S in the stabilization of As*

X-Ray Photoelectron Spectroscopy (XPS) analysis were applied to analyze the chemical bond characteristics of these elements including Si, C, N, S and O during the adsorption reaction of As(V), and the results were presented in Figure 3. For these elements including Si, C, N, and O, any fresh peak or conjunction energy shifting after As(V)

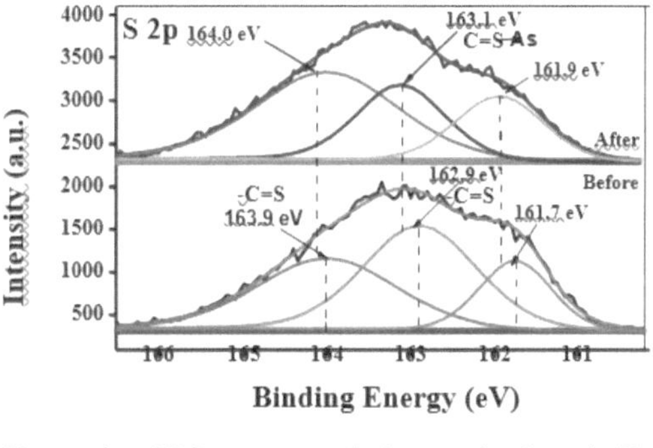

Figure 3. XPS spectrum before and after As(V) adsorption.

adsorption, further illustrated that no conduction electron pairs existed between As atom and Si/C/N/O atom after adsorption. For comparison, the formation of coordination bond (C = S→As) between the lone electron pairs of S atom and As decreases the electron cloud density around S and increases the conjunction energy of S, further leading to the positive shift (approximately 0.2 eV) of peaks in S 2p.

4 CONCLUSIONS AND RECOMMENDATIONS

A novel mesoporous MFT/SBA-15 materials combine both the advantages of chelating resins and mesoporous silica together were synthesized and have excellent application potentials in the Arsenic wastewater treatment. The organic components within MFT/SBA-15 were the core function groups on As(V) adsorption, whereas the formation of coordination bond between the lone electron pairs of S atom and As during the adsorption processes led to the synchronous fixation processes of As(V).

ACKNOWLEDGEMENTS

We acknowledge the National Natural Science Foundation of China (No. 41877038), Guangdong Natural Science Funds for Distinguished Young Scholars (No. 2016A030306019), the Local Innovative and Research Teams Project of Guangdong Pearl River Talents Program (No. 2017BT01Z176) and the Natural Science Foundation of Guangdong Province (No. 2018A030313385).

REFERENCES

Babel S. & Kurniawan T.A. 2003. Low-cost adsorbents for heavy metals uptake from contaminated water: A review. *J. Hazard. Mater.* 97(1): 219–243.

Wang X.Q., Liu T.X., Li F.B., Li B. & Liu C.P. 2018. Effects of simultaneous application of ferrous iron and nitrate on arsenic accumulation in rice grown in contaminated paddy soil. *ACS Earth Space Chem.* 2 (2): 103–111.

Wu K. 2016. *Preparation of Chelating Resin/SBA-15 Composite Materials for Adsorption of As(V) Species.* Hunan Normal University, MSc Thesis.

 ISBN 978-1-032-32928-4

Removal of arsenic from groundwater in Tanzania using locally available magnesite

R. Irunde[1,2,3], F.J. Ligate[4], J.P. Maity[4], A. Ahmad[5], J. Ijumulana[1,2], P. Bhattacharya[1] & F. Mtalo[2]
[1]*KTH-International Groundwater Arsenic Research Group, Department of Sustainable Development, Environmental Science and Engineering, KTH Royal Institute of Technology, Stockholm, Sweden*
[2]*DAFWAT Research Group, Department of Water Resources Engineering, College of Engineering and Technology, University of Dar es Salaam, Dar es Salaam, Tanzania*
[3]*Department of Chemistry, College of Natural and Applied Sciences, University of Dar es Salaam, Tanzania*
[4]*Department of Earth and Environmental Sciences, National Chung Cheng University, Min-Hsiung, Chiayi, Taiwan*
[5]*SIBELCO Ankerpoort NV, Maastricht, The Netherlands*

ABSTRACT: Arsenic removal study was carried at laboratory level through batch experiment to investigate efficient of raw magnesite from Tanzania on arsenic removal. Parameters such as initial concentration, dosage, contact time and pH were investigated. Adsorption efficient was high about 98% when high dose magnesite was used. Optimization of magnesite through calcination at 500°C increases surface area for about 4 times of the raw magnesite and adsorption was increased to 99% at 20 minutes contact time at higher doses. The kinetic adsorption fitted the pseudo second order on raw and calcined magnesite with R^2 0.99 and 1 respectively. Adsorption isotherm fitted strongly with Freundlich isotherm with R^2 0.98, n of 1.8. Application of calcined magnesite on real water from Masinki stream, Tanzania shows arsenic reduction from 117 µg/L AsV and 5.32 µg/L AsIII to below 0.1 µg/L. The high dose of 10 g magnesite removes arsenic up to 10 mg/L in a short time of 10 min.

1 INTRODUCTION

Arsenic (As) contamination to drinking water sources has been a global challenge since 1990's (Bhattacharya *et al.*, 2004). In recent years, As has been widely detected in groundwaters around gold mining areas in Tanzania (Almås & Manoko 2012; Ahoulé *et al.*, 2015; Ijumulana *et al.*, 2016; Kassenga & Mato 2008; Nyanza *et al.*, 2014; Taylor *et al.*, 2005) in which severe health effects are predicted to occur in future. The WHO guideline of As in drinking water is 10 µg/L. Thus, supply of drinking water exceeding the WHO limit may lead to adverse impacts on human health (Kapaj *et al.*, 2006). The impacts include skin lesions, hyperkeratosis and hyperpigmentation, circulatory disorders, diabetes as well as cancers of the bladder, lung, kidney, and liver (Centeno *et al.*, 2002; Smith *et al.*, 1992). Thus, there is a need to investigate proper methods that remove arsenic from water to secure society from As exposure. The previous study on arsenic removal using magnesite in South Africa was tested on wastewater (Masindi & Gitari 2016). However, there is no study reported to apply magnesite from Tanzania for arsenic removal from water and especially drinking water from Tanzania.

The current study aims to evaluate the adsorption efficient of magnesite minerals collected from Tanzania for arsenic removal from water as well as drinking water sources in Tanzania. The laboratory scale focus on kinetic studies and adsorption isotherms.

2 MATERIALS AND METHODS

2.1 *Study area*

The gold mining area around Geita and Mara were reported to have arsenic levels above 10 µg/L of WHO guideline. Water samples were collected from Geita and Mara. The water sample collected from Masinki in Tarime Mara was used for adsorption batch experiment.

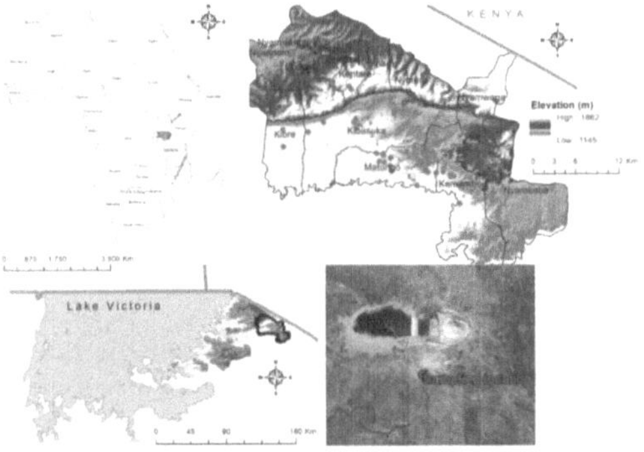

Figure 1. Study area and water sampling locations.

DOI: 10.1201/9781003317395-138

2.2 *Sampling and laboratory experiment*

The raw magnesite rocks were collected from Chambogo in Same district, Moshi, Tanzania. Magnesite rocks were brushed clean, washed, grinded, and sieved to below 0.3 mm and used for arsenic adsorption process. The natural water sample fetched from Masinki stream in Tarime, Mara, Tanzania was used for arsenic adsorption in real world. The synthetic arsenic water was prepared in the laboratory to investigate removal capacity 1 g of the adsorbent at high arsenic concentrations.

Arsenic solutions were prepared by dilution of the stock arsenic solution of 100 mg/L $Na_2HAsO_4.7H_2O$. The initial As concentrations prepared at near neutral pH was varied from 0.1, 0.5, 1, 2, 3, 5 and 10 mg/L for kinetic studies to determine equilibrium time, concentrations, equilibrium capacity, and the effect of initial concentration on arsenic removal. To the volume of 10 or 20 ml As solution, 1 g magnesite was introduced and shaked vigorously at different times such as 20 min, 40 min, 60 min up to 4 hours while samples were filtered and kept in the fridge for analysis. The kinetic reactions such as pseudo second order and Freundlich adsorption isotherm were used to evaluate batch results.

3 RESULTS AND DISCUSSION

Arsenic speciation of water sample natural water from Tanzania (Masinki stream) shows 117 µg/L As V, 5.32 µg/L AsIII, <0.2 µg/L monomethyarsonate (MMA) and 0.832 µg/L dimethylarsnate (DMA). Upon treatment with the calcined magnesite for 2 hours, the AsIII and AsV were removed to <0.1 µg/L while MMA was <0.2 µg/L and DMA was 0.364 µg/L. The observation shows the adsorption of organic arsenic was less favourable compared to inorganic arsenic.

The efficient of calcined magnesite was extremely increased at higher magnesite dosage when it shows concentration below 10 µg/L of about 100% removal after shaking for 4 hours. The raw and calcined magnesite fitted strongly in the pseudo second order (Figure 2). From raw magnesite, equilibrium adsorption capacity was 0.03 mg/g at 3 h (Figure 2) and its rate of reaction K_2 was 111 g/mg min.

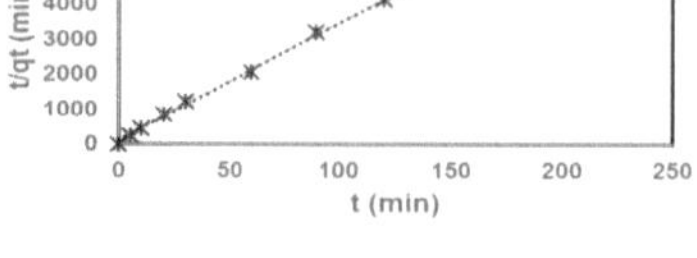

Figure 2. Pseudo second order kinetic for interaction of arsenic with 1g magnesite at 10 mL, 3.8mg/L AsV.

Adsorption isotherm such as Freundlich indicate presence of enough sites for arsenic adsorption in which 1 g magnesite provide required surface for As adsorption. The Freundlich equation was fitted well with R^2 of 0.9789 and n value of 1.8.

4 CONCLUSIONS

In this study, arsenic removal increases with increasing magnesite dosage, the ratio magnesite dosage to arsenic concentration was varied to 1 g Mg^{2+}:0.01 g As. The process obeys pseudo second order with R^2 of 0.999 on 1 g raw magnesite and 1 on calcined at high dosage. The adsorption process fitted strongly in Freundlich to ensure multilayer provide multisite for arsenic adsorption.

ACKNOWLEDGEMENTS

We acknowledge the Swedish International Development Cooperation Agency (Sida) for supporting the DAFWAT program (Contribution: 51170071).

REFERENCES

Ahoulé D.G., Lalanne F., Mendret J., Brosillon S. & Maïga A.H. 2015. Arsenic in african waters: a review. *Wat. Air, Soil Poll.* 226: 302.

Almås Å.R. & Manoko M.L.K. 2012. Trace element concentrations in soil, sediments, and waters in the vicinity of Geita gold mines and North Mara gold mines in Northwest Tanzania. *Soil Sed. Contam.* 21(2): 135–159.

Bhattacharya P., Welch A.H., Ahmed K.M., Jacks G. & Naidu R. 2004. Arsenic in groundwater of sedimentary aquifers. *Appl. Geochem.* 19(2): 163–167.

Gbaruko B.C., Ana G.R.E.E. & Nwachukwu J.K. 2008. Ecotoxicology of arsenic in the hydrosphere: implications for public health. *Afr. J. Biotechnol.* 7(25): 4737–4742.

Ijumulana J., Mtalo F., Bhattacharya P. & Bundschuh J. 2016. Arsenic occurrence in groundwater sources of lake victoria basin in tanzania. In: P. Bhattacharya *et al.* (eds.) *Arsenic Research and Global Sustainability (As2016)*. CRC Press/Taylor and Francis (ISBN 978-1-138-02941-5), pp. 86–87.

Kapaj S., Peterson H., Liber K. & Bhattacharya P. 2006. Human health effects from chronic arsenic poisoning – a review. *J. Environ. Sci. Health A* 41(10): 2399–2428

Kassenga G. & Mato R. 2008. Arsenic contamination levels in drinking water sources in mining areas in Lake Victoria Basin, Tanzania, and its removal using stabilized ferralsols. *Int. J. Biol. Chem. Sci.* 2(4): 389–400.

Masindi V. & Gitari W.M. 2016. Removal of arsenic from wastewaters by cryptocrystalline magnesite: complimenting experimental results with modelling. *J. Cleaner Prod.* 113: 318–324.

Nyanza E.C., Dewey D., Thomas D.S.K., Davey M. & Ngallaba S.E. 2014. Spatial distribution of mercury and arsenic levels in water, soil and Cassava plants in a community with long history of gold mining in Tanzania. *Bull. Environ. Contamin. Toxicol.* 93(6): 716–721.

Taylor H., Appleton J.D., Lister R., Smith B., Chitamweba D., Mkumbo O., Machiwa J.F., Tesha A.L. & Beinhoff C. 2005. Environmental assessment of mercury contamination from the Rwamagasa artisanal gold mining centre, Geita District, Tanzania. *Sci. Total Environ.* 343 (1–3): 111–133.

Arsenic in the Environment: Bridging Science to Practice for Sustainable Development –
van der Wal, Ahmad, Petrusevski, Weijma, Savic, van der Wens, Beerendonk, Bhattacharya, Bundschuh & Naidu (Eds)

Effect of the mole ratio of Mn/Fe composites on arsenic(V) adsorption

S.E. Garrido Hoyos[1] & J.L. Álvarez Cruz[2]
[1]*Mexican Institute of Water Technology (IMTA), Postgraduate Subcoordination, Jiutepec, Morelos, Mexico*
[2]*Posgraduate Program in Environmental Engineering UNAM-IMTA, Jiutepec, Morelos, Mexico*

ABSTRACT: Iron and manganese have been studied as a proposal for new materials that could be used for the adsorption of arsenic(V) [As(V)], in order to remove this contaminant. The objective of this work was to study the effect of the molar ratio of three Mn/Fe + Mn composites (X = 0.17, 0.32, 0.47) on the properties of adsorbent media, and to determine their influence on arsenic removal by comparing them with two metallic oxyhydroxides, that are commonly used as adsorbents of As(V) in aqueous solution (goethite and birnessite). These media were synthesized by chemical precipitation while controlling particle size. The surface area (286 m^2/g) of the composite with a molar ratio of X = 0.17 was larger than that of the other media. The adsorption kinetics and isotherms were fitted to the mathematical models, specifically, the pseudo-second order and Langmuir, respectively. The X = 017 composite had an adsorption capacity of 3.28 mg/g and removed 99% of As(V) with an initial concentration of 0.5 and 97% with an initial concentration of 10 mg/L, at 180 min, 25°C, and pH 7. The five adsorbent media were tested with well water with an initial As(V) concentration of 0.075 mg/L, and the best behavior was exhibited with a molar ratio of X = 0.17 at 90 min, resulting in 100% removal of As(V). The results suggest that this material is an effective and viable alternative to remove this contaminant from water.

1 INTRODUCTION

Goethite (α – Fe OOH) is one of the most important iron oxides phases, given its use as an adsorbent of As(V) in active sites on the surface (Rout *et al.*, 2014). Alvarez *et al.* (2006) have reported that incorporating Mn (isomorphic substitution) in the goethite structure modifies the crystallinity and the dehydroxylation of the surface and reported that goethite is the predominant crystal phase for mole ratios (Mn/Fe + Mn) < 0.14, while for mole ratios between 0.14 and 0.45 jacobsite appears as a second phase, along with goethite. And jacobsite and hausmannite phases appear with X = 0.55. Rout *et al.* (2014) doped goethite with manganese (0.32 – 1.87 wt%) using the precipitation method and reported changes in unit cell like as the size and geometric shape of the acicular goethite particles. Birnessite is a manganese oxide (MnO_2). In both its natural and synthesized forms, it is an excellent purifier of natural water containing arsenic. Hou *et al.* (2017) reported on the oxidation of As(III) to As(V), and that the latter forms complexes on the crystal surface of the birnessite and between its layers, since it is a laminar oxide. This oxidation produces changes new active sites on the surface. The objective of this work was to study the effect of the molar ratio of three Mn/Fe + Mn composites (X = 0.17, 0.32, 0.47) on the properties of adsorbent media, and their influence on the removal of arsenic.

2 MATERIALS AND METHODS

2.1 *Synthesis and structural characteristics of the absorbent media*

The adsorbent media were obtained according to the methodology described by Cornell & Schwertmann (2003) and Garrido & Romero (2016). *Surface Area and Porosity.* The samples were analyzed using Micromeritics ASAP 2020 V301 H equipment. *Determination of the Structure.* The crystal structure of the adsorbent media was identified using a Bruker D8 Advance diffractometer. *Determination of the Isoelectric Point.* Solutions were prepared with different solutions pH values (from 3 to 11) in 50 mL tubes, to which 0.1 g of adsorbent medium was added and the potential was then measured with Microtrack Zeta Check® equipment. *Semiquantitative Analysis of the Composition of the Media.* The morphology and elemental chemical compositions of the adsorbent media were studied using scanning electron microscopy with energy dispersive x-ray spectrometry (SEM-EDS), with a Carl Zeiss environmental scanning electron microscope (model EVO LS 10) (Alvarez & Garrido 2019).

2.2 *Adsorption kinetics and isotherms*

The kinetics and isotherms were obtained according to the methology described by Alvarez & Garrido., 2019. The As_i concentration: 0.5 mg/L The arsenic in solution was analyzed with a Varian ICP 730-ES.

3 RESULTS AND DISCUSSION

3.1 *Physicochemical properties*

Table 1 presents a summary of the analysis of the pore surface area, volume, size, IEP and manganese and iron for each adsorbent medium.

Figure 1 shows the XRD spectrum of the adsorbent media. Figure 2 shows as Mn (X = 0.17) was added to the goethite structure, these was predominant phase

DOI: 10.1201/9781003317395-139

Table 1. Texture properties and chemical composition of all adsorbent media.

Sample	SA (m^2/g)	PV (cm^3/g)	PS (Å)	Wt.% Fe	Wt.% Mn	pH IEP
Goethite	135	0.12	37	66	0	7
X = 0.17	286	0.35	49	66	9	4.5
X = 0.32	121	0.26	85	52	19	4.4
X = 0.47	103	0.17	66	37	28	4.4
Birnessite	62	342	828	0	68	2.5

SA = Surface Area; PV = Pore Volume; PS = Pore Size; IEP = Isoelectric point

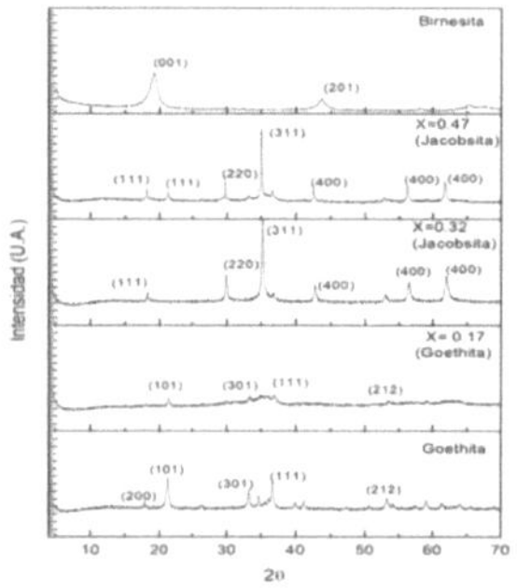

Figure 1. XRD patterns of all adsorbent media.

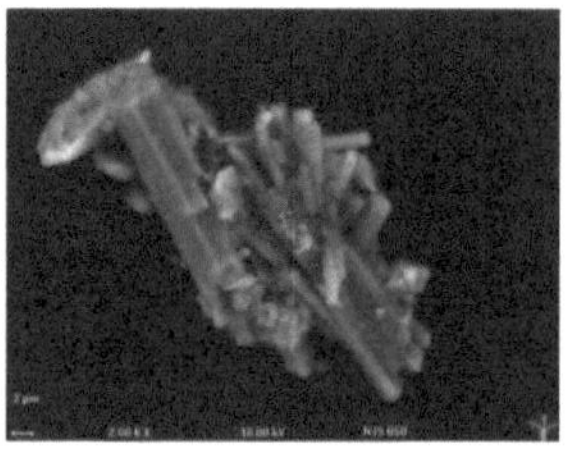

Figure 2. SEM micrograph of the synthesized adsorbent media: X = 0.17. Size 2 μm.

and the acicular structure of the goethite was conserved, while clusters were deposited heterogeneously throughout the surface of the goethite crystal, thereby decreasing crystallinity and changing its properties such as cell and phase parameters.

3.2 *Adsorption kinetics and isotherms*

Table 2 shows the kinetic constants obtained using the mathematical pseudo-second order model. The adsorption capacity constant (qe) of X = 0.17 was larger than all the other adsorbent media, and the initial adsorption velocity required was slower given that the incorporation of that Mn ratio into the goethite structure increases the surface area, producing more active OH sites than goethite.

Table 2. Kinetic constants with 0.5 mg/L of As(V).

Sample	q_e (mg/g)	K_{ad} (g/mg min)	R^2	h (mg/g min)
Goethite	0.2403	8.7831	0.9999	0.5071
X = 0.17	0.2460	4.8724	0.9998	0.2899
X = 0.32	0.2404	3.3108	0.9997	0.1912
X = 0.47	0.2384	5.4453	0.9999	0.3094
Birnesite	0.2386	5.0923	0.9998	0.2899

Table 3 presents the model parameters that were obtained from the linear regression Langmuir isotherm model. The As(V) Q_m of X = 0.17 (3.2873 mg/g) was larger than other media even reported in the literature. The factor R_L for this study was between 0 and, confirming that the adsorption of arsenic on the composites was favorable. Figure 3 shows the arsenic removal from the contaminated well (Tepetzingo, Mor.), with each of the five synthesizes adsorbent media, compared with commercial media, the best result was for X = 0.17, which remove 100% of the As(V) after 20 min of adsorption.

Table 3. Adsorption isotherm constants for As(V).

Sample	K_L (L/mg)	Q_m (mg/g)	R^2	R_L
Goethite	7.4390	1.2825	0.9348	0.1007
X = 0.17	6.6196	3.2873	0.8265	0.111
X = 0.32	4.0108	1.6705	0.8431	0.172
X = 0.47	2.1046	0.6996	0.917	0.283
Birnessite	2.8364	2.2246	0.8782	0.227

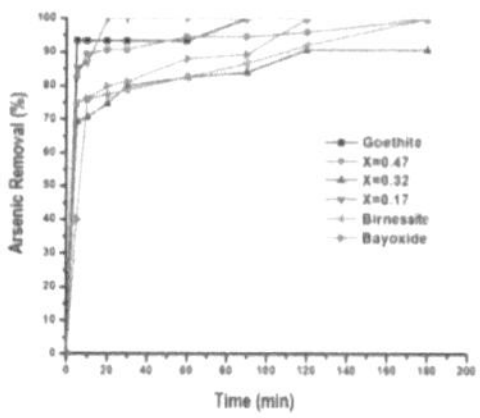

Figure 3. Arsenic removal kinetic with respect to time. In well water contaminated with As_i 0.075 mg/L, at 25°C.

4 CONCLUSIONS

Changes in pH_{IEP} were observed as the manganese in the mol ratios, this is because manganese is incorporated in the OH sites in the goethite structure, causing dihydroxylation and decrease in pH_{IEP} which results in changes in the crystal structure, unit cell parameters and pore surface area, size and volume.

REFERENCES

Alvarez Cruz J.L. & Garrido Hoyos S.E. 2019. Effect of the mole ratio of Mn/Fe composites on arsenic (V) adsorption *Sci. Total Environ.* 668: 47–55.

Alvarez M., Rueda E.H. & Sileo E.E. 2006. Structural characterization and chemical reactivity of synthetic Mn-goethites and hematites. *Chem. Geol.* 231(4): 288–299.

Cornell R.M. & Schwertmann U. 2003. *The Iron Oxides* (2nd ed.). Willey VCH GmbH & KGaA.

Garrido Hoyos S. & Romero Velazquez L. 2016. Synthesis of minerals with iron oxide and hydroxide contents as a sorption medium to remove arsenic from water for human consumption. *IJERPH*,13(69).

Rout K., Dash A., Mohapatra M. & Anand S. 2014. Manganese doped goethite: structural, optical and adsorption properties. *J. Environ. Chem. Eng.* 2(1), 434–443.

Arsenic in the Environment: Bridging Science to Practice for Sustainable Development –
van der Wal, Ahmad, Petrusevski, Weijma, Savic, van der Wens, Beerendonk, Bhattacharya, Bundschuh & Naidu (Eds)
© 2023 the Author(s), ISBN 978-1-032-32928-4

Optimization of IOCP coating to improve arsenic adsorption capacity

Y.M. Slokar[1], N.S. Nkiriti[1], K. Huysman[2] & B. Petrusevski[1]
[1]*IHE Delft Institute for Water Education, Delft, The Netherlands*
[2]*PIDPA, Antwerp, Belgium*

ABSTRACT: One of the most attractive technologies for the removal of arsenic from groundwater used for drinking water production is adsorptive filtration with iron (hydr)oxide rich media. In our previous research we showed that such media can be produced by filtration of Fe(II) containing water through porous carrier, *e.g.*, pumice. However, the produced material, Iron Oxide Coated Pumice (IOCP), showed poorer adsorption capacity than Iron Oxide Coated Sand (IOCS) from iron removal plants. This research focused on further optimization of the IOCP coating procedure by varying mode of iron addition and duration of coating. The efficiency of coating was evaluated by batch adsorption experiments and compared with IOCS from an iron removal plant based on aeration and rapid sand filtration. Under studied conditions, the results showed higher adsorption capacity of IOCP than IOCS. It was also determined that prolonged pumice coating can result in blocking of meso pores, responsible for the adsorption of arsenic, and as such reduce it removal capacity.

1 INTRODUCTION

Arsenic (As) presence in groundwater intended for the production of drinking water is a worldwide problem. Among the existing technologies for As removal, adsorptive filtration on iron oxide based composites seem to be one of the more effective and simple ones. Due to commercial As adsorbents being rather costly, research efforts are aimed at finding cheaper alternatives.

One of such alternatives is Iron Oxide Coated Sand (IOCS). It is effective in As removal and since it is a by-product of adsorptive iron removal plants, it is also very cheap. However, its quantities are limited and quality variable, which instigated search for another low cost As adsorbent.

In our previous research we have demonstrated the potential of Iron Oxide Coated Pumice (IOCP) as another potential low-cost adsorbent (Slokar & Petrusevski 2016). Objective of the study presented here was to improve the efficiency of in-situ coating of pumice and thus improve arsenic adsorption capacity of IOCP. The studied parameters were iron (Fe) concentration and its mode of introduction into the feed, and coating time. The efficiency of coating was assessed by adsorption experiments and compared with results of IOCS coated under field conditions in a facility with conventional groundwater treatment.

2 METHODS

Material to be coated was commercially available pumice with a size fraction 0.3–1 mm. Coating was carried out under laboratory conditions in a shallow filter column with 80 cm bed depth. The filter was operated in an up-flow mode, with filter bed expansion of approximately 40%. The coating cycle consisted of:

- 15 min dosing of acidified Fe(II) solution pumped into feed (tap) water; and
- 30 min oxidation of adsorbed Fe by dissolved oxygen (DO) present in tap water.

Two modes of Fe addition were applied: (i) constant dosage and (ii) variable (increasing) dosage. Coating cycles were running continuously for cumulative 15 and 5 days of coating with constant and variable c(Fe), respectively. Coating was paused overnight for practical reasons. For both modes, pH of the feed was kept at 6.7 ± 0.3.

The extent of coating was quantified by coating extraction to determine the amount of Fe present on the pumice. The adsorption capacity of As was assessed in short and long (isotherm) batch adsorption experiments. Results were compared with IOCS (size fraction 1–3 mm) produced under field conditions in a full-scale iron removal plant based on aeration-rapid sand filtration. IOCS samples were taken after the first 121 days of operation with virgin sand, at filtration rate of 14 m/h.

Specific surface area, pore size distribution and pore volume were analyzed on a Micromeritics TriStar 3000 gas adsorption analyzer based on N2 isotherms at 77 K and t-plot method (Groen *et al.*, 2008).

3 RESULTS AND DISCUSSION

Results of Fe extraction and short batch adsorption experiments are given in Figure 1. The samples depicted are:

- IOCP-con media produced with constant c(Fe) in 15 days;
- IOCP-var media produced with variable c(Fe) in 5 days;
- IOCS media produced in industrial set-up in 121 days.

Based on the As affinity for iron oxides it was expected the As adsorption would improve with

DOI: 10.1201/9781003317395-140

increased Fe content in the coating. However, from Figure 1 it can be seen, this was not the case. In fact, the As adsorption capacity was inversely proportional to the amount of Fe in the coating. Short term adsorption efficiency was confirmed also with isotherms for all three tested materials (Figure 2).

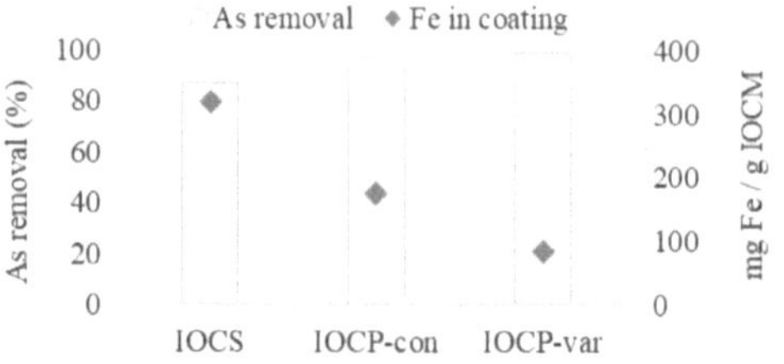

Figure 1. Removal of As(III) with Iron Oxide Coated Media (IOCM) and amount of Fe in its coating. Model water: pH 7, c(As) ≈ 200 μg/L. IOCM dosage 1 g/L, contact time 48 h.

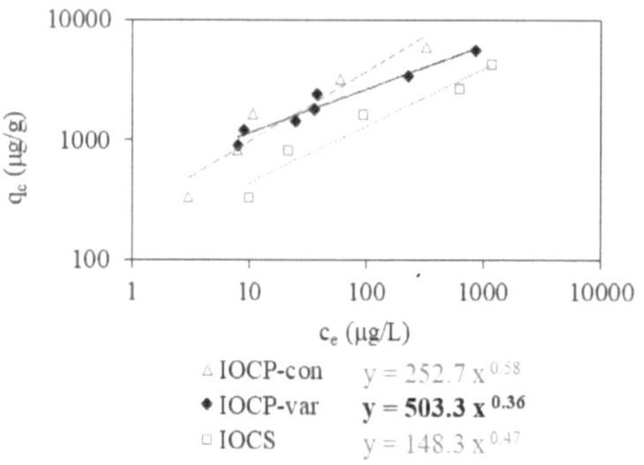

Figure 2. Comparison of Freundlich isotherms for As(III) adsorption on IOCM. Model water pH 7.

Freundlich isotherms (Figure 2) confirmed that among the tested materials arsenic adsorption capacity was best for IOCP with the lowest Fe content – IOCP_var, coated for 5 days only. Results implied that with prolonged coating time part of the Fe (hydr) oxides start blocking the pores and create multiple stacked Fe layers, which becomes inaccessible for arsenic. To further investigate this, the surface area and pore size and distribution measurements were carried out for two samples taken from the same coating experiment, but from different heights of the filter column, with different extent of coating (Figure 3).

As it can be seen from Figure 3, increased Fe in coating resulted in increased BET surface area (S BET) and the volume of *micro* pores (sized <2 nm). However, the volume of *meso* pores (2–50 nm) decreased. *I.e.*, continuous addition of Fe layers in the coating increased overall porosity of the adsorbent, but by blocking the larger meso pores, where arsenic can

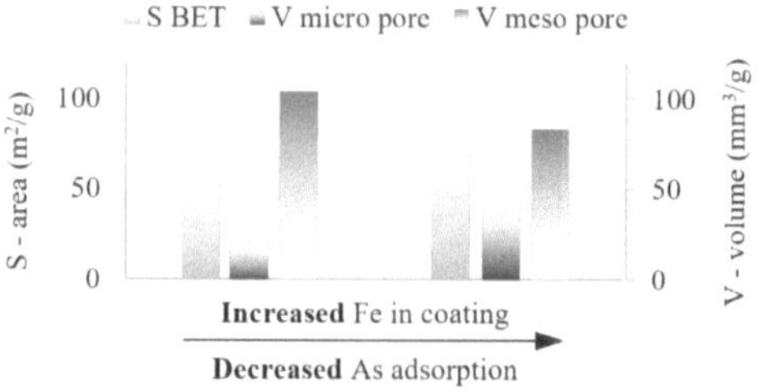

Figure 3. Effect of Fe coating extent on pore size of IOCP.

attach during the adsorption process, the adsorption capacity reduced. The same phenomenon likely contributed to lower As adsorption capacity of IOCS, even though it had higher Fe content.

4 CONCLUSIONS

Replicating conventional Fe removal by filtration through filter containing porous carrier can produce iron (hydro)oxide rich coated media that has high affinity for arsenic. Introducing coating Fe solution with variable (increasing) dosage increases efficiency of coating by creating less active adsorption sites at the start of the coating and catalytically increasing them with the increased amount of Fe (hydro)oxides on the pumice and feed c(Fe). In addition, this reduces cost of the process, as less Fe is lost at the beginning of the coating.

Prolonged coating increases surface area of the adsorbent, however too thick coating can start blocking meso pores which are large enough to provide active sites for the adsorption of As. If care is taken as to not over-coat the material, the produced IOCP has potential to be an efficient alternative to IOCS.

ACKNOWLEDGEMENTS

The authors would like to acknowledge the efforts of IHE Delft MSc participants who carried out the research presented in this paper.

REFERENCES

Groen J.C., Abelló S., Villaescusa L.A. & Pérez-Ramírez J. 2008. Mesoporous beta zeolite obtained by desilication. *Microporous Mesoporous Mater.* 114(1): 93–102.

Slokar Y.M. & Petrusevski B. 2016. Iron oxide coated pumice: promising low cost arsenic adsorbent. In: P. Bhattacharya, M. Vahter, J. Jarsjö, J. Kumpiene *et al.* (eds.) *Arsenic Research and Global Sustainability (As2016)*. CRC Press/Taylor and Francis (ISBN 978-1-138-02941-5). pp. 549–550.

Arsenic in the Environment: Bridging Science to Practice for Sustainable Development – van der Wal, Ahmad, Petrusevski, Weijma, Savic, van der Wens, Beerendonk, Bhattacharya, Bundschuh & Naidu (Eds)
© 2023 the Author(s), ISBN 978-1-032-32928-4

Arsenic remediation onto photocatalytic synergistic Mn-Al-Fe impregnated rGO hybrid adsorbent (MAF-rGO)

Y.K. Penke, J. Ramkumar & K.K. Kar
Advanced Nano-engineering Materials Laboratory (ANML), Materials Science Programme (MSP) and Department of Mechanical Engineering, Indian Institute of Technology Kanpur, Kanpur, India

ABSTRACT: A novel Mn-Al-Fe impregnated RGO (Mn-RGO) hybrid adsorbent is synthesized, and verified for possible arsenic (As(III), As(V)) remediation, and mitigation. A maximum arsenic adsorption loading ability is observed around 402, and 339 mg g^{-1} for As(III), and As(V), respectively (C_i ~ 0.1 – 100 mg/L). Adsorption isotherms experiments are observed with Freundlich isotherm plot-based multilayer sorption. The kinetics study on arsenic adsorption [combined As(III), and As(V)) in competing for ion environment] also confirmed a better removal ability. The influence of solar irradiation onto adsorption kinetics is verified with in-situ photo-current response experiments. The photo-catalytic activity (PCA), and photocatalytic oxidation (PCO) of Mn-RGO is understood in terms of zero-bias photo-current of 17 μA. XPS study on As 3d, and detailed analysis onto individual metal elements (Mn 2p, Fe 2p, Al 2p), functional groups (O 1s, C 1s) supports redox, and surface ligand exchange phenomenon.

1 INTRODUCTION

Arsenic (As), which is mostly a geogenic contaminant poisoning underground aliquot systems mainly in South Asian countries, and other continents. Even though the problem is being well addressed in the past few years, the scientific community is not successful in suggesting a full time or permanent solution to address the issue on a global scale. Various kinds of arsenic remediation methodologies like ion exchange (IX), reverse osmosis (RO), coagulation-precipitation (C-P), adsorption, photo-remediation, and bio-remediation are in usage in recent times (Rahman & Singh 2019). But the economic drawbacks of various demographic societies and supply to demand ratio at mass scale were few reasons which have put a limitation on a few of these proposed techniques. Considering remote level applications adsorption technique is encouraged due to ease in installation, operation, and maintenance. Manganese based systems (e.g., MnO_2, Mn-Fe, and Mn-Fe loaded zeolite) were reported in huge proportions due to their synergistic redox capability [i.e., oxidation state flipping behavior] for mitigating highly toxic As(III) species (Zhu *et al.*, 2009). In the present study, we synthesized first of its kind novel hybrid adsorbent (i.e., Mn-Al-Fe impregnated RGO, here after Mn-RGO), and verified for arsenic removal in batch scale laboratory studies. The photocatalytic active behavior of the Mn-RGO hybrid is also verified for arsenic remediation using various in-situ experiments.

2 MATERIALS AND METHODS

2.1 *Materials*

Chemicals such as $MnCl_2.4H_2O$, $Al(NO_3)_3.9H_2O$, Fe $(NO_3)_3.9H_2O$, and NaOH pellets [Loba Chemie, India] of A.R grade were procured and used as received. Chemicals such as natural flake graphite [Sigma Aldrich], perchloric acid [$HClO_4$] [Rankem], acetic anhydride [$(CH_3CO)_2O$] [Qualigens, India], Potassium Paramagnet [$KMnO_4$] [Qualigens, India], and hydrazine hydrate [Loba Chemie, India]. Sodium arsenite ($NaAsO_2$), sodium arsenate heptahydrate ($NaH_2AsO_4{\cdot}7H_2O$) (S.D fine – chemicals Ltd, India) were used for arsenic-based studies. HCl and NaOH [Loba Chemie, India] solutions of 1 M were used for pH titration. All synthesis processes and adsorption experiments were performed at room temperature conditions in the normal atmosphere using double-distilled (DD) water.

2.2 *Methods*

MAF-rGO hybrid was synthesized as observed in our recent study (*Penke et al.*, 2020). Adsorption kinetics studies were performed in dark and light irradiated conditions. In adsorption kinetics 100 mL aliquots of As(III)/As(V) (C_i ~ 7–8 mg/L) combined with individual SO_4^{2-}, PO_4^{3}, NO_3^{-}, and CO_3^{2-} solutions (5 mg/L) in pH 7.0 ± 0.1 by dispersing MAF-rGO (50 mg). The arsenic aliquots were kept on orbital shaker, and samples were collected in different intervals (1 – 90 min). In light exposed in-situ kinetics experiments, the aliquot systems were saturated for 5 min in Light-off condition before irradiating simulated Light (i.e., solar simulator: Newport device, specs: Sol3A 94023A, 450 W Xenon lamp, 100 mW cm^{-2}). Adsorption isotherms study was performed for As(III) and As(V) solutions in 0.1 – 100 mg/L by dispersing 10 mg of Mn-RGO hybrid adsorbent.

3 RESULTS AND DISCUSSION

3.1 *Adsorption studies*

The adsorption kinetics study is observed with a Pseudo Second Order (PSO) model supported

DOI: 10.1201/9781003317395-141

adsorption behavior. The corresponding plots and data are shown in the Figure 1 (Table 1). This PSO behavior conforms both physisorption and chemisorption behaviors.

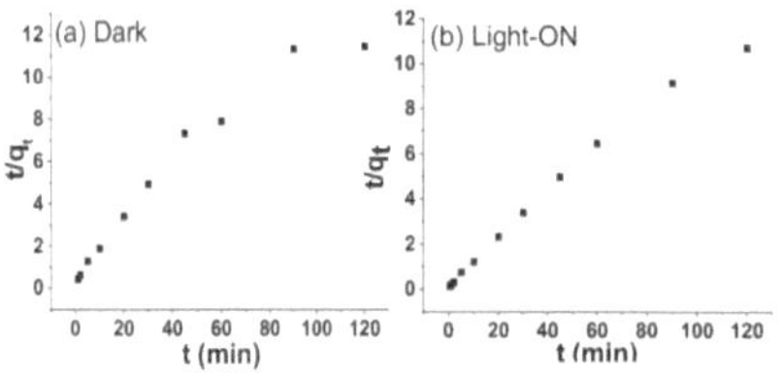

Figure 1. Adsorption kinetics (PSO) plots of MAF-rGO for (a) DARK, and (b) LIGHT-ON systems.

Table 1. Adsorption kinetics data of MAF-rGO hybrid (PSO Model).

Arsenic	K_2k_2 (g/mg/min)	q (mg/g)	R^2	$q_{e,exp}$ (mg/g)
DARK	0.01	10.47	0.93	10
LIGHT	0.023	10.50	0.98	10.76

The maximum loading abilities for As(III) and As (V) in higher concentrations ($C_i \sim 100$ mg/L) was observed around 402, and 339 mg/g. The present adsorption pattern resembles Freundlich isotherm based multi-layer adsorption phenomenon (Data not shown). A better mesoporous surface might influenced this phenomenon by means of both surface and bulk diffusion. In low arsenic loadings (100 – 500 μg/L) low equilibrium concentration ($Ce \sim 10 - 20$ μg/L) were also observed in pH 7.

3.2 *In-situ photocurrent (PC) measurement:*

A detailed in-situ PC response (I-V) adsorption kinetics experiments under solar simulator were performed to understand the influence of white light on arsenic remediation as shown in Figure 2. This study is performed for rGO, and metal-oxide impregnated rGO for a comparative analysis.

In dark, significant variations were not observed in respective I-V plots over the stipulated duration, where all plots followed a similar path. Even though the pattern is similar in white light irradiated systems, in few individual intervals (30 – 60 min) were observed with variations in path and increment in the current values by few μA in + 0.6 to + 0.9 V potential band. This might be due to the possible redox reactions in that particular band. A detailed study on photo electro chemical behavior of Mn-rGO is detailed in our recent study (*Penke et al.*, 2020). After white light irradiation the increment in the generated current was observed where it went up to 30 μA for 90 min study. This study confirmed the intrinsic photo-catalytic activity of Mn-RGO is the plausible reason for better arsenic remediation in light irradiation samples over dark mode experiments (Table 1). The 2-Dimensional layered rGO has provided a possible conduction pathway for e^-: h^+ pair in the conduction band.

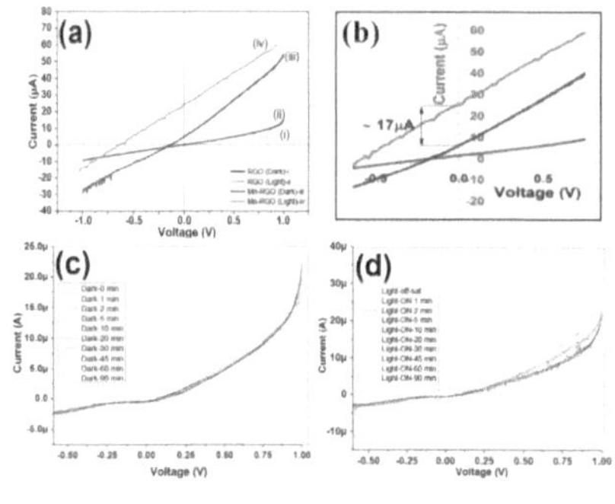

Figure 2. PC generation behavior (I-V plot) of Mn-RGO in dark, and white light irradiation conditions (a-b), in-situ I-V plot of kinetics experiment in dark (c), in light (d) condition.

4 CONCLUSIONS AND RECOMMENDATIONS

A novel ternary metal oxide (Mn-Al-Fe) impregnated RGO hybrid was synthesized, and verified as an efficient arsenic remediating agent by means of photo-redox catalytic synergism. Photo-catalytic synergism revealed the intrinsic photo-current generation ability of the MAF-RGO hybrid due to the schottky junction behavior of metal oxide and graphene combination. The low equilibrium concentrations and better photo-catalytic impact suggests the possible utilization of MAF-rGO for arsenic treatment applications.

ACKNOWLEDGEMENTS

The authors would like to thank IMPRINT-Proposal-6134 [Ministry of Textiles, and Ministry of Human Resource Development (Government of India)] for their support to carry out this research.

REFERENCES

Penke Y.K., Yadav A.K., Sinha P., Malik I., Ramkumar J. & Kar K.K, 2020. Arsenic remediation onto redox and photocatalytic/electrocatalytic Mn-Al-Fe impregnated RGO: sustainable aspects of sludge as supercapacitor. *Chem. Eng. J.* 124000.

Rahman Z. & Singh V.P. 2019. The relative impact of toxic heavy metals (THMs) (arsenic (As), cadmium (Cd), chromium (Cr)(VI), mercury (Hg), and lead (Pb)) on the total environment: an overview. *Environ. Monit. Assess.* 191(7): 419.

Zhu M., Paul K.W., Kubicki, J.D & Sparks D.L. 2009. Quantum chemical study of arsenic (III, V) adsorption on Mn-oxides: implications for arsenic(III) oxidation. *Environ. Sci. Technol.* 43: 6655–6661.

Arsenic in the Environment: Bridging Science to Practice for Sustainable Development –
van der Wal, Ahmad, Petrusevski, Weijma, Savic, van der Wens, Beerendonk, Bhattacharya, Bundschuh & Naidu (Eds)
© 2023 the Author(s), ISBN 978-1-032-32928-4

Synthesis and characterization of β-Cyclodextrin fortified hydrous iron-zirconium hybrid oxide for dearsenification

I. Saha[1], K. Gupta[2], S. Ahmed[1], D. Chatterjee[3] & U.C. Ghosh[2]
[1]*Department of Chemistry, Sripat Singh College, Jiaganj, Murshidabad, India*
[2]*Department of Chemistry, Presidency University, Kolkata, India*
[3]*Department of Chemistry (Analytical), University of Kalyani, Nadia, India*

ABSTRACT: Groundwater is a vital resource which is now sharply on the decline both in developed and developing countries. It is now the principle source of drinking water supply notably in rural areas and becoming a challenge due to rapid decline of per capita availability and supply. Much of global research is now geared towards employing modified adsorbent materials for arsenic removal. β-cyclodextrin (β-CD) which is a non-toxic, largely inexpensive oligosaccharide, is one such material which can be effectively deployed for the fortification of hydrous metal oxides conventionally used for groundwater arsenic remediation. The aim of this contribution is to provide a detailed insight regarding the synthesis, characterization and exploration of de-arsenification potential of synthetic iron-zirconium hybrid oxide (9:1 mole ratio) (CHIZO), surface functionalized in an optimized fashion by β-CD vis-à-vis the unmodified oxide.

1 INTRODUCTION

Groundwater is a vital resource which is now sharply declined both in developed and developing countries. The resource plays a key role in routine water supply due to its easy availability and user friendliness. Groundwater is now the principle source of drinking water supply notably in rural areas and becoming a challenge due to rapid decline of per capita availability and supply. Nevertheless, several countries are now seriously suffering from arsenic exposure from groundwater when extracted from shallow aquifers (<50 m). Presently, the scenario is developing into a serious threat to human health due to detrimental effects of arsenic throughout South and SE Asia where a vast majority of rural population is reliant on groundwater for their drinking and domestic purposes (Ritchie 1961; Smedley & Kinniburgh 2002). In Bengal Delta Plain (BDP), the widespread occurrence of arsenic in shallow alluvial aquifers and its tentacles have now led to an alarming state of affairs with regard to the supply of safe drinking water (Biswas *et al.*, 1998; Vaughan 2006). Widespread efforts are afoot by national and international agencies as well as state/federal governments to supply safe water among affected populations. However, the quality of treated water is often questionable with regard to WHO guideline ($\leq$ 10 µg/L As). The reasons are several such as technologies, problems in O&M, cost, user friendliness and environmental friendliness. Furthermore, several treatment plants are found defunct/ or malfunctioning after a period of time which have been implemented by local/state governments due to the lackadaisical approach with regard to their maintenance and sustainability. In West Bengal alone, a population of more than 6 million people residing in 12 districts covering 111 blocks are forced to drink groundwater laced with As concentration often breaching the WHO recommended values (Guha Mazumder *et al.*, 2010). At this critical juncture, the urgent need is to develop new methodologies for arsenic remediation which could be easily implemented in affected areas. Much of global research is now geared towards employing modified adsorbent materials for arsenic removal. β-cyclodextrin (β-CD) which is a non-toxic, largely inexpensive oligosaccharide, can be effectively deployed for the fortification of hydrous metal oxides conventionally used for groundwater arsenic remediation. The proposed adsorbent has seven α-(1,4)-linked glycosyl units and a host of –OH groups which can provide several binding sites for metal oxides. Due to the presence of surface –OH groups, these oligosaccharides possess appreciable adsorption capacity through either surface complexation or hydrogen bond formation. The aim of this contribution is to provide a detailed insight regarding the synthesis, characterization and exploration of de-arsenification potential of synthetic iron-zirconium hybrid oxide (9:1 mole ratio), CHIZO, surface functionalized in an optimized fashion by β-CD vis-à-vis the unmodified oxide. The proposed adsorbent has been tested to remove As(III) which is more toxic than As(V). Moreover, the conventional techniques often used hazardous oxidants to convert As(III) to As(V) for optimal removal efficiency. The developed adsorbent has unique property to remove solely As(III) from groundwater along with fluoride which is often encountered in groundwater contamination.

2 MATERIALS AND METHODS

Co-precipitation method was employed for the synthesis of HIZO which is a well-known synthetic process which is mentioned in the pre reported work by Saha *et al.* Aiming for the functionalization of as prepared hybrid oxide, aqueous solutions of β-cyclodextrin in a pre-optimized ratio were injected in to the acidic solutions of $FeCl_3$ and $ZrOCl_2$ and co precipitation performed by dilute base for an extended period of time. After filtration of the brown mass the residue washed with distilled water

DOI: 10.1201/9781003317395-142

and kept steady for 48 hours followed by drying at 100°C. The dried mass was grounded to a particle size ranging from 140–290 μm (52–100 mesh size).

3 RESULTS AND DISCUSSION

The synthesized hybrid mixed oxide and its functionalized version were characterized very precisely by various sophisticated instrumental techniques like XRD, FE SEM –EDX, AFM, BET, FTIR, TGA, UV-Vis and cross polarized 1H and C13 solid state NMR (ssNMR) spectra for the conformation of the surface area, morphology, composition and the metal organic structural pattern. The comparative morphological study between HIZO and CHIZO has been executed by using scanning electron microscope (SEM) and atomic force microscope (AFM). The elemental configuration of both materials also examined by Electron dispersive X-Ray photometry (EDX) and shown in Figure 1. HIZO surface exhibits non-continuous rough nature comparative to CHIZO. A clear observation in both low and high magnified microscopic images is after modification of the HIZO surface it smoothened enough and homogeneity in surface morphology increase. There is a clear indication of the presence of carbon content in the EDX spectra of CHIZO rather than HIZO. The metal organic structural pattern examined in details by using cross polarized 1H and C13 solid state NMR (ssNMR) spectra and the resulted outcome has been examined. Observable fact in 1H NMR pattern of pure β-Cyclodextrin there is sharp peak with some small peaks appears. This strong signal is due to the hydroxyl proton (C6) which is present in a separate highly electronegative environment in the molecule and attached with other similar type of protons, cumulatively exhibits high field intensity. In case of pure HIZO a broad peak appears which is responsible for all the hydroxyl proton present in the identical environment on the mixed metal oxide surface. Interestingly in the composite CHIZO the broad peak which was representing the hydroxylic proton of the oxide surface fragmented in to many sharp peaks indicating the presence of high electronegative environment. And certainly that is coming from the hydroxyl proton (C6) of β-cyclodextrin which clearly supporting the surface modification. In the C13 solid state NMR (ssNMR) spectral data there is a comparative study performed between empty rotor and CHIZO packed rotor and the difference spectra also extracted. The evident fact that there is a clear peak splitting at chemical shift (δ) = 50 to 110 ppm region indicating the presence of carbon originated from β-cyclodextrin which also supported by the C13 spectrum of pure β-cyclodextrin. From the spectral data a clear outcome is the obvious existence of β-cyclodextrin on the mixed metal oxide surface and the bonding with the hydroxyl group of HIZO can be predicted by the ether bond which have a supporting spectral data from Infrared Spectrum of CHIZO.

4 CONCLUSIONS AND RECOMMENDATIONS

A directed synthesized tailored adsorbent material (CHIZO) was characterized rigorously and used for

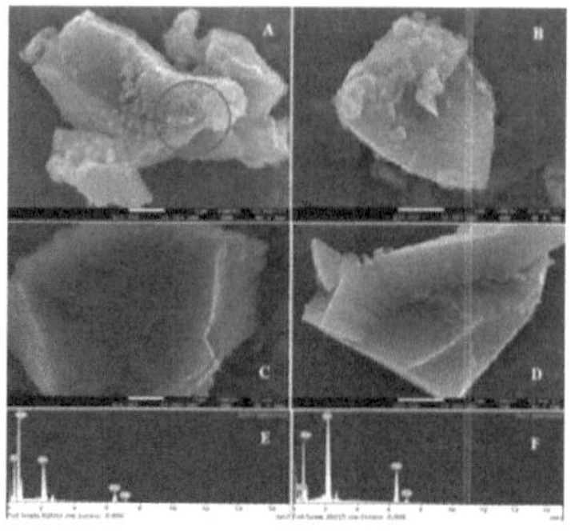

Figure 1. Field emission scanning electronic microscopic image of HIZO (A = low magnification, B = High magnification) and CHIZO (C = low magnification, D = High magnification). E = EDX spectra from the selected area of C and F = EDX spectra from the selected area of A.

As(III) adsorption. The smooth surface morphology obtained and the maximum thermal tolerance limit obtained 230°C. Cross polarized 1H and C13 solid state NMR (ssNMR) spectra confirms the presence of β-cyclodextrin on the surface. The adsorption kinetics following pseudo 2nd order kinetic model and the maximum monolayer capacity obtained 32.012 $mg.g^{-1}$. The arsenic adsorption mechanism proposed via intramolecular hydrogen bonding which instrumentally supported by IR spectroscopic study.

ACKNOWLEDGEMENTS

The authors are thankful to Sripat Singh College for providing laboratory facilities and WBDST for partial funding.

REFERENCES

Biswas B.K., Dhar R.K., Samanta G., Mandal B.K., Faruk I., Islam K.S., Chowdhury M.M., Islam A., Roy S. & Chakraborti D. 1998. Detailed study report of Samata, one of the arsenic affected villages of Jessore district, Bangladesh. *Curr. Sci.* 74(2): 134–145.
Guha Mazumder D.N., Ghosh A., Majumdar K.K., Ghosh N., Saha C. & Guha Mazumder R.N. 2010. Arsenic contamination of ground water and its health impact on population of district of Nadia, West Bengal, India. *Indian J Community Med.* 35(2): 331–338.
Ritchie J.A. 1961. Arsenic and antimony in some New Zealand thermal waters, *NZ J Sci.* 4: 218–229.
Smedley P.L. & Kinniburgh D.G. 2002. A review of the source, behaviour and distribution of arsenic in natural waters, *Appl. Geochem.* 17: 517–568.
Vaughan D. J. 2006. Arsenic, *Elements* 2:71–75.
Wang L., Chen A.S.C., Tong N. & Coonfare C.T. 2007. Arsenic removal from drinking water by ion exchange, U.S. EPA demonstration project at Fruitland, ID, Six month Evaluation Report. EPA/600/R-07/017. *United States Environmental Protection Agency, Water Supply and Water Resources Division, National Risk Management Research Laboratory, Cincinnati, OH.*

*Arsenic in the Environment: Bridging Science to Practice for Sustainable Development –
van der Wal, Ahmad, Petrusevski, Weijma, Savic, van der Wens, Beerendonk, Bhattacharya, Bundschuh & Naidu (Eds)
© 2023 the Author(s), ISBN 978-1-032-32928-4*

Biosurfactant mediated arsenic mobilization

L.S.S. Araújo[1] & M.C. Teixeira[1,2]
[1]*Environmental Engineering Graduating Program-ProAmb, Federal University of Ouro Preto, Ouro Preto, Brazil*
[2]*Department of Pharmacy, Federal University of Ouro Preto, Ouro Preto, MG, Brazil*

ABSTRACT: Arsenic (As) is naturally found in association with gold minerals widely exploited in Minas Gerais, Brazil since the 17th Century, mainly at the region named Iron Quadrangle (IQ). As a consequence of this economic activity, large amounts of As containing mine tailings were, and still are, discharged. Biological approaches are being tested to decontaminate As containing wastes. Soil-flushing technique, using biosurfactants, was tested to promote As mobilization from contaminated tailing samples collected at an abandoned gold mine. The optimum condition for biosurfactant production by the chosen microbial consortium was determined. The efficiency of the biosurfactant containing supernatant was tested against other extractants: distilled water, $CaCl_2$, SDS and Saponin in sequential extractions. The cumulative arsenic removal was approximately 25% in the column soil-flushing tests, running at room temperature, pH 11.0.

1 INTRODUCTION

Arsenic (As) is a natural element widely distributed in the environment by natural and anthropogenic factors. As is naturally released in the environment by oxidation of arsenopyrite in the presence of oxygen and as a consequence of bacterial activities (Natarajan 2008). The Iron Quadrangle (IQ), located in the state of Minas Gerais, Brazil, has an endured gold and iron ore mining historic dated back the Seventeenth Century. After more than 300 years of exploitation, it is estimated that more than 390,000 ton of As released in the environment (Figueiredo *et al.*, 2007). The released As leads to the contamination of either surface and groundwater. Biosurfactants (BS) biomolecules are metabolic products of bacteria, fungi and, plants that have been largely used for industrial purposes and, more recently, for environmental applications, such as soil remediation, as well (Arab & Mulligan 2018; Sarubbo *et al.*, 2015; Wang & Mulligan 2009). Compared to the synthetic surfactants, the biosurfactants possess interesting properties as low toxicity, high biodegradability and, high tolerance to temperature, pH and salinity (Santos *et al.*, 2016).

Therefore, the hypotheses that As could be removed from solid wastes by soil leaching techniques using biosurfactant solutions was tested. The present study aims: (i) to investigate the performance of BS produced by mixed microbial cultures for As removal from a natural contaminated soil and (ii) to compare the efficiency of the bioproduct against other well-known chemical extractants.

2 MATERIALS AND METHODS

2.1 *Soil sample*

As contaminated surficial tailing sample was collected (20 cm, depth) at an abandoned gold mine named Chico Rei, in Ouro Preto, MG, Brazil (20°23′09.6″S 43°29′57.5″W). Material was dried (air), homogenized and sieved ($0.105 \leq x \leq 0.25$ mm). The pH, Eh and chemical composition of the sample was assessed.

2.2 *Consortium enrichment*

The initial microbial samples were collected at Lagoa do Gambá, Ouro Preto, Brazil and were initially cultivated in Postgate C Medium under low oxygen pressure. After few cultivation batches, the consortium was slowly adapted to aerobiosis. The consortium cells where then cultivated in Postgate B medium and selected for these studies. For the biosurfactant production, the consortium cells (1 g) were inoculated in 500 mL of salt medium containing (g/L): $MgSO_4.7H_2O$ (0.50), KCl (0.10), $CaCl_2$ (0.01), $NaNO_3$ (7.00), $FeSO_4.7H_2O$ (0.01), K_2HPO_4 (1.00), yeast extract (0.01), 10% (w/v) glucose and glycerol, 1.5% (w/v), pH 7.0. The glucose/glycerol (glu/gly) ratios of 0.5, 1.0, 1.5, 2.0 and 2.5 were evaluated in a univariate analysis, pH (5.5-9.5), temperature (25, 30, 35°C) and incubation time (1-7 days) to determine the optimum production condition (OPC). The BS production was assessed by emulsification index (EI24) calculated by the ratio of the height of column emulsion and the column, after transferring 2 mL of paraffin and 2 mL of cell-free supernatant onto a test tube and stirring vigorously in vortex for 1 min.

2.3 *Sequential extraction*

The extraction efficiency of SDS, 1%; Saponin, 0.1%; $CaCl_2$, 0.1 mol/L and, distilled water had been compared to the BS cell-free supernatants (0.5, 1.0, 1.5, 2.5 glu/gly ratios and OPC). Polypropylene columns (13.8×3.8 cm) filled with 5 g of soil (Figure 1a) and percolated with the selected extractant (10 mL). The extractions were conducted in one first cycle of 30 day followed by three 7 day cycles.

DOI: 10.1201/9781003317395-143

In a further assay, a 18.5×4.8 cm column was packed with 75 g of soil. The sequential extraction was performed in four cycles of 24 h using the BS cell-free supernatant obtained under optimum production condition (OPC).

3 RESULTS AND DISCUSSION

The soil sample was classified as sandy soil (ABNT 1995) with low organic matter content (<5 mg/kg), pH 3.49. The chemical characterization had indicated Fe (52903.13 mg/kg), As (2760.49 mg/kg), Al (2365.68 mg/kg) and Mn (3485.87 mg/kg) as major elements. The higher EI24, 55.1% and 58%, were obtained using a medium glu/gly ratio of 0.5 (pH 5.5, 30°C, 7 days) and glu/gly ratio 2.5 (pH 9.5, 25°C, 3 days), respectively. The higher glycerol concentration and the acidic pH of the 0.5 glu/gly ratio medium strongly indicate the suitability of the use of crude glycerol as an alternative carbon source, aiming the reuse of secondary products thus enhancing the economic feasibility of the process. Although similar EI24 were verified for the described conditions, shorter incubation periods of BS production for increased surface activity property is an interesting characteristic regarding environmental applications. Hence, the later condition was established to be the OPC. After the four cycles of extraction, the distilled water, $CaCl_2$ (0.1 mol/L), SDS 1% and Saponin 0.1% removed, respectively, 0.42, 0.03, 0.67 and 0.14% of As. The biosurfactant cell-free supernatants presented the greatest efficiency if compared to the other extractants. The results have also indicated that, the higher glu/gly ratio, the higher the As extraction. OPC cell-free supernatant demonstrated to be also the best extractant, presenting a 10% As accumulated removal and, in this way, it was used for the further soil-flushing tests using bigger column. The pH influenced the biosurfactant extract efficiency. At alkaline conditions, the results were better, following the results reported by Wang & Mulligan (2009b).

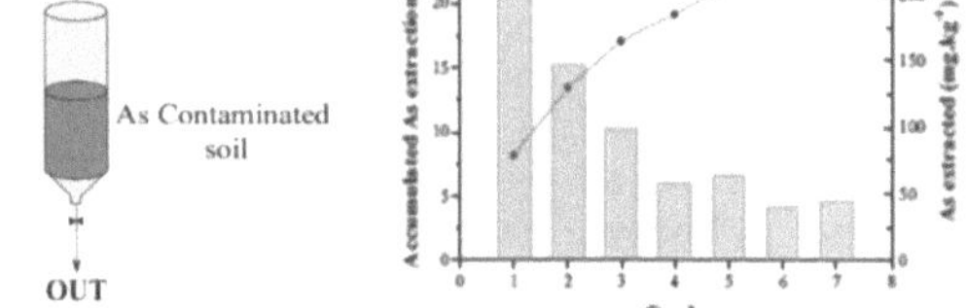

Figure 1. a) Scheme for column extraction, b) arsenic mobilization from the mine tailing by the biosurfactant extract (OPC) at pH 11.

The results of the 75 mL column are depicted in Figure 1b. A removal of approximated 24.6% (678.39 mg/kg) of the initial As content was achieved after seven 24 h cycles. Approximately 3.7 and 6.6% of the Mn and Al original contents were also solubilized, respectively. The removal of those elements was slight if compared to As, thus explaining the selectivity of the biosurfactant for arsenic.

4 CONCLUSION

The polluting potential of the mining tailings was confirmed by comparing the As solubilization obtained using just water as extractant and the guidelines of WHO for potable water. The effectiveness of the BS extract for the soil flushing of the As contaminated material was proved. The results discussed here emphasized the potential use of biosurfactants for the remediation of toxic metals and metalloids contaminated wastes. Biosurfactants has been recognized as promising candidates for the replacement of synthetic extractants during contaminated soil bioremediation treatments.

ACKNOWLEDGEMENTS

This work was supported by CNPQ, CAPES and by the Federal University of Ouro Preto.

REFERENCES

ABNT - Associação Brasileira de Normas Técnicas. 1995. *NBR 6502 Rochas e solos. Abnt.*

Arab F. & Mulligan C.N. 2018. An eco-friendly method for heavy metal removal from mine tailings. *Environ. Sci. Pollut. Res.* 25: 16202–16216.

Figueiredo B.R., Borba R.P. & Angélica R.S. 2007. Arsenic occurrence in Brazil and human exposure. *Environ. Geochem. Heallth* 29: 109–118.

Natarajan K.A., 2008. Microbial aspects of acid mine drainage and its bioremediation. trans. *Nonferrous Met. Soc. China* (English Ed.) 18: 1352–1360.

Santos D.K.F., Rufino R.D., Luna J.M., Santos V.A. & Sarubbo L.A. 2016. Biosurfactants: multifunctional biomolecules of the 21st century. *Int. J. Mol. Sci.* 17: 1–31.

Sarubbo L.A., Rocha R.B., Luna J.M., Rufino R.D., Santos V.A. & Banat I.M., 2015. Some aspects of heavy metals contamination remediation and role of biosurfactants. *Chem. Ecol.* 31: 707–723.

Wang S. & Mulligan C.N. 2009a. Rhamnolipid biosurfactant-enhanced soil flushing for the removal of arsenic and heavy metals from mine tailings. *Process Biochem.* 44: 296–301.

Wang S., Mulligan C.N., 2009b. Arsenic mobilization from mine tailings in the presence of a biosurfactant. *Appl. Geochem.* 24: 928–935.

4.2 Nanomaterials related novel applications in arsenic treatment

Arsenic in the Environment: Bridging Science to Practice for Sustainable Development – van der Wal, Ahmad, Petrusevski, Weijma, Savic, van der Wens, Beerendonk, Bhattacharya, Bundschuh & Naidu (Eds)

Arsenic removal from water using a new generation adsorbent: Titanium dioxide coated with magnetic nanoparticles

J. Nikić, M. Watson, A. Tubić, M. Šolić & J. Agbaba
Environmental Protection Research Group, Department of Chemistry, Biochemistry and Environmental Protection, Faculty of Sciences, University of Novi Sad, Novi Sad, Serbia

ABSTRACT: Arsenic in drinking water resources is a global health problem. Of the As species commonly found in groundwater, As(III) is generally more mobile and toxic than As(V). In this work, commercial TiO_2 was modified with magnetic nanoparticles (TiO_2-MNp), in order to develop an adsorbent which can combine oxidation and adsorption properties and which can be readily separated from water using a magnetic field. The adsorption capacity of TiO_2-MNp and arsenic removal mechanism of this novel material was investigated through kinetic and equilibrium experiments and their corresponding mathematical models. It was found that the arsenic adsorption mechanism on MNp-FeMn involves chemisorption and intraparticle diffusion. The maximal adsorption capacity, q_{max} of TiO_2 for As (III) (21 mg/g) was improved after modification with magnetic nanoparticles (31 mg/g), while q_{max} for As(V) were 16 and 20 mg/g, respectively. The high oxidation and adsorption capacity of TiO_2-MNp and magnetic properties, suggest this material is a highly promising adsorbent for treatment of arsenic contaminated groundwater.

1 INTRODUCTION

Arsenic in drinking water resources is a global health problem. Millions of people, mostly in rural and developing countries, are still exposed to high levels of arsenic via the intake of arsenic rich groundwater. It is well know that long term consumption of arsenic contaminated drinking water can cause various health problems with cancer being the most serious consequence of chronic arsenic exposure (Abdul *et al.*, 2015). In order to reduce health risks, the World Health Organization (WHO) recommends 10 μg/L as the maximum allowed concentration for arsenic in drinking water (WHO 2011). Several physicochemical techniques have been developed and applied for arsenic removal from water (Nicomel *et al.*, 2016). However, adsorption is considered as one of the most promising, due to its cost-effectiveness, simple operation, and minimal waste generation (Kalan *et al.*, 2016). Different synthetic and natural materials have been used as adsorbents for arsenic removal. However the majority of them show a greater affinity for As(V) in comparison with the much more mobile and toxic As(III). TiO_2 as an adsorbent for arsenic removal has been investigated by many authors since this material has great oxidation perfomance for conversion of As(III) to As(V) (Pirilä *et al.*, 2011). However, its low adsorption capacity and the problem of separating TiO_2 powder from the aqueous solution have usually limited its application. Therefore, many efforts have been taken to improve the performance of TiO_2 and to facilitate the application of TiO_2 in real practice (Kalan *et al.*, 2016). One approach to overcome these hurdles is to modify TiO_2 with magnetic nanoparticles. However, most researchers have investigated the use of these materials as photocatalysts. To date, very few studies are available into their application as a new generation adsorbent for arsenic removal.

The main objectives of this work were therefore to i) produce TiO_2 coated magnetic nanoparticles (TiO_2-MNp) and ii) investigate sorption behaviour and mechanism of both As(III) and As(V) on TiO_2-MNp in batch systems, through kinetic and isotherm experiments and their corresponding mathematical models.

2 MATERIALS AND METHODS

2.1 *Synthesis of TiO_2 coated magnetic nanoparticles (TiO_2-MNp)*

Magnetic nanoparticles (MNp) were prepared in a single step process by conventional co-precipitation method, using $Fe^{3+}$and Fe^{2+} salts at a molar ratio of 2:1. The synthesized magnetic nanoparticles (3 g) were then dispersed in 100 ml of water and suspension was sonicated for 10 min. Similarly, TiO_2 (10 g) was dispersed in 100 ml of water, after which the resulting suspension was gradually added to the magnetite nanoparticles. The mixture was stirred on an ultrasonic bath for 1h. The synthesized TiO_2-MNp particles was then removed from the suspension by external magnet, washed several times with destilled water to pH 7, dried in an oven at 105°C for 2 h and then annealed at 450°C for 3h.

2.2 *Characterisation of TiO_2-MNp*

The specific surface area was obtained by nitrogen adsorption using the Brunauer–Emmett–Teller (BET) method and the mesopore and micropore volumes determined using the BJH and t-test methods, using a Quantachrome autosorb TMiQ surface area analyzer.

DOI: 10.1201/9781003317395-144

2.3 *Adsorption experiments*

Adsorption kinetics was performed in a synthetic matrix with 10 mg of adsorbent and 1 mg/L As(III) or As(V) solution at pH 7.0 ± 0.1. Adsorption isotherms were obtained using 20 mL As(III)/As(V) solution in concentration ranges from 0.1–20 mg/L, and with 10 mg of the adsorbent, over 24 h. After adsorption, the supernatants were separated from the adsorbents using an external magnet. Residual arsenic concentrations in the supernatant were analyzed by inductively coupled plasma mass spectrometry (ICP/MS).

3 RESULTS AND DISCUSSION

The BET surface areas of MNp, TiO_2, TiO_2-MNp were found to be 37.4, 54.6, 56.6 m^2/g, respectively. Mesopore volumes of TiO_2 (0.958 cm^3/g) decreased after the modification with MNp (0.483 cm^3/g) The point of zero charge (pH_{pzc}) of the MNp, TiO_2 and TiO_2-MNp were 5.8, 6.3 and 5.2, respectively.

3.1 *Adsorption kinetics*

Kinetic experiments show that adsorption equilibrium for both As(III) and As(V) on TiO_2-MNp was reached after 8 h, indicating specific adsorption occuring between the arsenic species and the surface of the adsorbent. Different models (pseudo-first, pseudo-second and Weber-Morris) were employed to model the kinetic data (Table 1). The pseudo-second order model best fit the experimental data for both As(III) and As(V) suggesting that the adsorption of As on TiO_2-MNp occurs via chemisorption (Pirilä *et al.*, 2011). The Weber-Morris model shows that intraparticle diffusion is not the only limiting step of the adsorption process and that other interaction mechanisms must be acting simultaneously.

Table 1. Kinetic parameters for arsenic adsorption on TiO_2-MNp.

Model	Parameters	As(III)	As(V)
Pseudo-first order	k_1(1/h)	0.536	1.50
	q_e (mg/g)	1.49	1.12
	R^2	0.9678	0.9780
Pseudo-second order	k_2 (g/mg h)	0.673	1.78
	q_e (mg/g)	1.59	1.26
	$q_e$exp(mg/g)	1.66	1.56
	R^2	0.9724	0.9890
Weber-Morris	k_i(mg/g $h^{-0.5}$)	0.755	0.432
	C_i	0.112	0.548
	R^2	0.8643	0.7669
	k_i (mg/g $h^{0.5}$)	0.031	0.0389
	Ci	1.47	1.37
	R^2	0.2149	0.7824

3.2 *Adsorption isotherms*

In order to evaluate the sorption capacities of the TiO_2-MNp for As(III) and As(V), equilibrium data were fitted with the Freundlich and Langmuir models. The Freundlich isotherm (Figure 1) was the more suitable for describing the adsorption behaviour of both arsenic species on TiO_2-MNp indicating that the adsorption of As preferably followed multilayer and heterogeneous adsorption processes. The maximal adsorption capacity q_{max} of TiO_2 for As(III) (21 mg/g) was improved after modification with magnetic nanoparticles (31 mg/g), while q_{max} for As(V) were 16 and 20 mg/g, respectively.

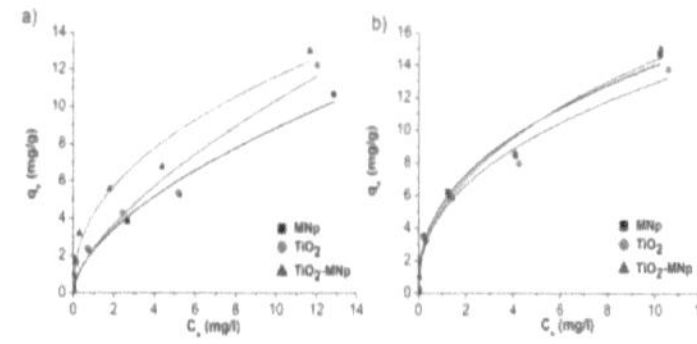

Figure 1. Freundlich isotherms of a) As(III) and b) As(V) on MNp, TiO_2 and TiO_2-MNp.

4 CONCLUSIONS AND RECOMMENDATIONS

TiO_2 was successfully modified with magnetic nanoparticles in a single step process, for enhanced arsenic removal from drinking water. The adsorption mechanism of arsenic species on TiO_2-MNp involved chemisorption with intraparticle diffusion contributing to the overall adsorption process. Modification of TiO_2 with MNp increased its adsorption capacity for As(III) while capacities for As(V) were similar. The magnetic properties of this adsorbent, which allow for its easy separation from water, eliminating the need for more complex and expensive separation techniques, along with their high potential for the simultaneous removal of both As(III) and As(V) without a pre-oxidation step, makes this material a very promising candidate for treatment of arsenic contaminted waters.

ACKNOWLEDGEMENTS

The authors acknowledge financial support of the Ministry of Education, Science and Technological Development of the Republic of Serbia (Grant No. 451-03-68/2020-14/200125).

REFERENCES

Abdul M.K.S., Jayasingheb S.S., Chandanaa E.P.S. *et al.* 2015. Arsenic and human Health effects: *A review*. *Environ. Toxicol. Pharmacol.* 40: 828.

Kalan R.E., Yaparatne B.C. *et al.* 2016. P25 titanium dioxide coated magnetic particles: preparation, characterization and photocatalytic activity. *Appl. Catal. B Environ.* 187: 249.

Nicomel N.N., Leus K., Folens K. *et al.* 2016. Technologies for arsenic removal from water: Current status and future perspectives. *Int. J. Environ. Res. Public Health* 13(1):1.

Pirilä M., Martikainen M., Ainassaari K. *et al.* 2011. Removal of aqueous As(III) and As(V) by hydrous titanium dioxide. *J. Coloid. Interface Sci.* 353(1):257.

World Health Organisation (WHO) *Guidelines Drinking-Water Quality*. 4th ed.; WHO: Geneva, Switzerland, 2011.

Arsenic in the Environment: Bridging Science to Practice for Sustainable Development –
van der Wal, Ahmad, Petrusevski, Weijma, Savic, van der Wens, Beerendonk, Bhattacharya, Bundschuh & Naidu (Eds)
 ISBN 978-1-032-32928-4

Effect of temperature on adsorptive removal of arsenic from water by hollow polyaniline microsphere/Fe_3O_4 nanocomposite

S. Dutta[1], A.K. Gupta[1], S.K. Srivastava[2] & M.K. Yadav[3]
[1]*School of Water Resources, Indian Institute of Technology Kharagpur, Kharagpur, India*
[2]*Department of Chemistry, Indian Institute of Technology Kharagpur, Kharagpur, India*
[3]*School of Environmental Science and Engineering, Indian Institute of Technology Kharagpur, Kharagpur, India*

ABSTRACT: Arsenic (As) pollution in water has been considered as a severe public health concern throughout the world. In this study, prepared polyaniline hollow microsphere (PNHM)/Fe_3O_4 magnetic nanocomposites were applied to the effective removal of toxic As(III) and As(V) from aqueous solutions at different temperatures. Batch adsorption study shows that arsenic adsorption capacity of PNHM/Fe_3O_4-40 (Fe_3O_4 content 40%) can be achieved up to 24–61 mg/g at neutral pH. Additionally, adsorption data fitted well with the Freundlich isotherm model for both the arsenic species. However, the present study emphasizes the effect of temperature on the adsorption of arsenic from contaminated water and the analysis of the thermodynamic parameters. Results show that the adsorption of As(III) and As(V) on the surface of PNHM/Fe_3O_4-40 nanocomposites increase with the increase in temperatures from 293 K to 313 K indicating the endothermic nature of the process.

1 INTRODUCTION

Arsenic is introduced into the water from several natural and anthropogenic sources. Contamination of drinking water by arsenic beyond the permissible limit (10 µg/L) of the World Health Organization could result in severe toxicological and carcinogenic effects on human health (Hao *et al.*, 2018). Different adsorbents are available for the remediation of arsenic from water. However, recent days, nanostructure materials are receiving considerable attention due to its large effective surface area and high adsorption capacity. Additionally, the magnetic behavior of Fe_3O_4 is beneficial for the arsenic adsorption and separation of the adsorbent after its use. The purpose of the present work is to establish the effect of temperature on the adsorptive removal of As(III) and As(V) from contaminated water using a polyaniline hollow microsphere (PNHM)/Fe_3O_4 (40 wt.%) magnetic nanocomposites (termed as: PNHM/Fe_3O_4-40).

2 MATERIALS AND METHODS

2.1 *Preparation of PNHM/Fe_3O_4-40 nanocomposite*

The polyaniline hollow microsphereis synthesized by using polystyrene core as a templet. Followed by decorating the surface of PNHM with in-situ deposition of magnetic Fe_3O_4 (40 wt.%) nanoparticles at room temperature by co-precipitation method.

2.2 *Sorption study*

Effect of temperature in adsorption of arsenic onto PNHM/Fe_3O_4-40 was performed at three different temperatures (293 K, 303 K, and 313 K) by varying initial arsenic concentrations (0.1 to 1000 mg/L). 1 g/L of the adsorbent dose was added into the arsenic solution by keeping the pH at ~7 for the adsorption of arsenic. Finally, arsenic concentrations were determined in Atomic adsorption Spectroscopy. The schematic representation of the arsenic removal process is described in Figure 1.

3 RESULTS AND DISCUSSION

3.1 *Effect of temperature*

Effect of temperature on adsorption of As(III) and As (V) onto PNHM/Fe_3O_4-40 was studied by varying the temperature in between 293K to 313K at different initial concentrations of arsenic. The experimental result displayed in Figures 2a,b, shows that arsenic removal capacity increased with the rise of temperature for all initial concentrations.

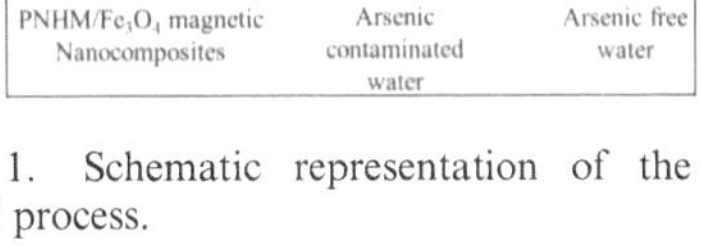

Figure 1. Schematic representation of the arsenic removal process.

3.2 *Thermodynamics study*

Experimental data were fitted in different isotherm models, and results show that the Freundlich model better described the adsorption of the As(III) and As (V) onto PNHM/Fe_3O_4-40 at all three temperatures.

DOI: 10.1201/9781003317395-145

The thermodynamics study is conducted at three different temperatures (293 K, 303 K and, 313 K) to understand the feasibility and spontaneity of the adsorption process. Thermodynamics parameters are calculated based on the following equations:

$$\Delta G^0 = \Delta H^0 - T\Delta S^0 \qquad (1)$$

$$lnk_d = \frac{-\Delta H^0}{RT} + \frac{\Delta S^0}{R} \qquad (2)$$

$$k_d = K_f M_w(T)\left[\frac{1}{\frac{M_w(T)}{1000}}\right]^{\left(1-\frac{1}{n}\right)} \qquad (3)$$

$$S^* = (1-\theta)exp\left(\frac{-E_a}{RT}\right) \qquad (4)$$

$$\theta = \left(1 - \frac{C_e}{C_0}\right) \qquad (5)$$

where ΔG^0 indicates the change of standard Gibbs free surface energy (kJ/mol), R is universal gas constant (8.314×10^{-3} kJ/mol/K), ΔH^0 is the change in standard enthalpy (kJ/mol), ΔS^0 is changed in standard entropy (kJ/mol/K). k_d has been calculated from different methods, Ghosal and Gupta (2015) have analyzed thoroughly various approaches to determine the value of k_d. In the current study, the Freundlich isotherm was followed. Subsequently, the k_d value is calculated using the Freundlich constant K_f. $M_w(T)$ indicates the mass of water per liter in gram at a temperature (T) in Kelvin. S^* indicates the sticking probability, θ is surface coverage, E_a is activation energy. C_e and C_0 show the equilibrium and initial arsenic concentrations. Corresponding graphs are shown in Figure 2.

The calculated values of thermodynamic parameters are tabulated in Table 1. The value of ΔH^0 is obtained from Figures 2c,d and the positive value of ΔH^0 (Table 1) suggests that the adsorption process is endothermic. Whereas, the positive value of ΔS^0 indicates the increase in randomness at the adsorbate-adsorbent interface at the time of arsenic adsorption. The negative value of ΔG^0 at all three temperatures suggests that the adsorption process is spontaneous.

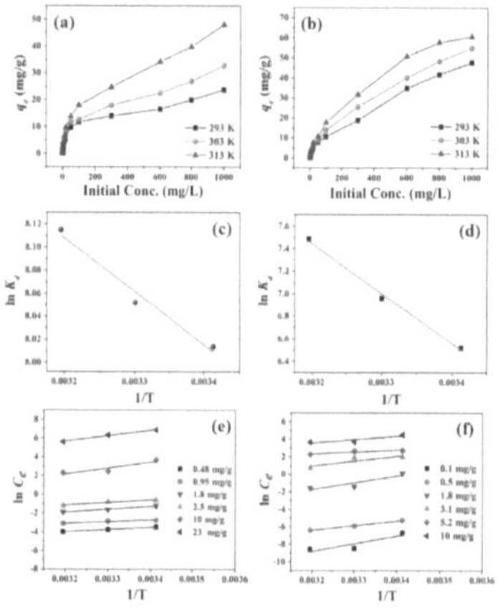

Figure 2. Effect of temperature on removal capacity (a) As(III) and (b) As(V) adsorption onto PNHM/Fe_3O_4-40 at different initial concentrations; Van't Hoff plot to determine the thermodynamic parameters of adsorption of (c) As(III) and (d) As(V) onto PNHM/Fe_3O_4-40 nanocomposites; Plots of ln*Ce* for adsorption of (e) As(III) and (f) As(V) at constant amounts adsorbed as a function of 1/T.

Table 1. Thermodynamics parameters for As(III) and As(V) adsorption onto PNHM/Fe_3O_4-40.

Temp (K)	ΔG^0 (kJ/mol)	ΔH^0 (kJ/mol)	ΔS^0 (kJ/mol/K)	E_a (kJ/mol)	S^*
			As(III)		
293	−19.50	3.84	0.08	19.47	
303	−20.30				5.84×10^{-5}
313	−21.09				
			As(V)		
293	−15.83	36.76	0.18	34.54	
303	−17.62				2.72×10^{-8}
313	−19.43				

Isosteric heat of adsorption (ΔH_x) is calculated from the slope of the plot lnC_e vs. 1/T shown in Figures 2e,f. The corresponding values of ΔH_x are 14–52 kJ/mol for As(III) and 17–72 kJ/mol for As(V), indicated that the physisorption is prevailed ($\Delta H_x < 80$ kJ/mol)(Ghosal & Gupta 2015) on the heterogenous surface of Fe_3O_4 coated PNHM composites.

The values of $0<S^*<1$ indicates the favorable sticking of arsenic species onto the surface of PNHM/Fe_3O_4-40. Again, the positive value E_a suggests the adsorption process is endothermic in nature. However, the values of 5 kJ/mol$<E_a<$40 kJ/mol (Table 1) implies that physisorption is predominated for the adsorption of arsenic onto PNHM/Fe_3O_4-40 (Ghosal & Gupta 2015).

4 CONCLUSIONS

Experimental results suggested that PMHM/Fe_3O_4-40 nanocomposite effectively adsorbed As(III) and As(V) at a temperature range of 293 K to 313 K, and the removal efficiency was increased with an increase in the temperature. The thermodynamic study result shows that the arsenic adsorption process is endothermic and spontaneous in nature. The physisorption process predominates the adsorption of As (III) and As(V) onto PMHM/Fe_3O_4-40.

REFERENCES

Ghosal P.S. & Gupta A.K. 2015. An insight into thermodynamics of adsorptive removal of fluoride by calcined Ca–Al–(NO_3) layered double hydroxide. *RSC Adv.* 5: 105889–105900.

Hao L., Liu M., Wang N. & Li G. 2018. A critical review on arsenic removal from water using iron-based adsorbents. *RSC Adv.* 8: 39545–39560.

Arsenic adsorption using immobilized magnetic iron oxide nanoparticles in electrospun PVA

N. Torasso[1], J. Palatnik[2], A. Vergara-Rubio[3], A. Londonio[2,3], P. Smichowski[2,4] & S. Goyanes[1,4]
[1]*Universidad de Buenos Aires. Facultad de Ciencias Exactas y Naturales, Departamento de Física, Laboratorio de Polímeros y Materiales Compuestos (LP&MC), Instituto de Física de Buenos Aires (IFIBA-CONICET), Buenos Aires, Argentina*
[2]*Comisión Nacional de Energía Atómica (CNEA), Gerencia Química, Buenos Aires, Argentina*
[3]*Universidad de San Martín (UNSAM), 3iA, Argentina*
[4]*Consejo Nacional de Investigaciones Científicas y Técnicas (CONICET), Buenos Aires, Argentina*

ABSTRACT: In line with the scientific community effort to create efficient and accessible ways to remove arsenic from water, this work shows the ability of immobilized magnetic iron oxide nanoparticles (MIONs) into electrospun poly(vinyl alcohol) (PVA) membranes to adsorb As(V) from water. Electrospinning solutions were prepared using two different approaches: suspending dry MIONs in a PVA solution and in situ synthesis of the MIONs. Results show a high adsorption capacity of 68 mg/g and well dispersed particles in the nanofibers. These straightforward and scalable methods are promising alternatives for water remediation.

1 INTRODUCTION

Arsenic presence in drinking water leads to acute or chronic diseases. Removal of arsenic from water by adsorption onto nanomaterials is being studied as a promising alternative to conventional methods (reverse osmosis, nanofiltration, etc.). In this sense, nanoscaled iron oxides and iron containing complexes such as magnetite and maghemite have been extensively studied as arsenic adsorbents (Hao *et al.*, 2018). In order to maximize available adsorption sites and increase efficiency, agglomeration must be avoided. Moreover, the use of nanoparticles alone may have fast adsorption kinetics but must be filtered afterwards before consumption. These issues can be overrun by the immobilization of the particles in a water permeable matrix. Electrospinning is a well-known, versatile and scalable technique to produce polymeric membranes. Its porous structure made up of crisscrossed nanofibers allows the water to flow while exposing a large surface area, ideal for interchange and adsorption processes.

Poly(vinyl alcohol) (PVA) is a hydrophilic, biodegradable and biocompatible polymer that has been electrospun to produce nanoparticle containing fibers. Different methods reported in literature can be used for this purpose, but none of them has tested the resulting material as an As adsorbent (Shi *et al.*, 2020).

Herein we report As adsorption onto magnetite immobilized in electrospun PVA fibers using two approaches: suspension of previously synthesized nanoparticles in the electrospinning solution and in situ synthesis of the nanoparticles.

2 MATERIALS AND METHODS

2.1 *Elecrospinning solutions*

Electrospinning solution was prepared using PVA (Sigma-Aldrich Mowiol 98 MW 68000 g/mol 98% degree of hydrolysis). Magnetic iron oxide nanoparticles (MIONs) were synthesized by the coprecipitation method using $FeCl_3$ and $FeSO_4*7H_2O$ as described elsewhere (Yazdani *et al.*, 2016) and according to the following reaction:

$$2Fe^{3+} + Fe^{2+} + 8OH^- \rightarrow Fe_3O_4 + 4H_2O \quad (1)$$

First approach: in order to properly disperse the particles, dried MIONs were added to a 1% PVA solution and probe-sonicated at 300 W until no precipitate was observed. Afterwards, the rest of PVA was added and dissolved in the dispersion for 1 h at 85°C. Total weight of MIONs was 0.1% of PVA. A higher ratio of MIONs could not be suspended using this approach.

Second approach: 12% PVA solution was dissolved along with precursor salts. Then, a solution containing NaOH in 10% excess was dropwise added at room temperature and under a sonication bath. The stabilized suspension was further vortexed to homogenize the electrospinning solution. Quantities were chosen in such a way that total weight of MIONs was 1% of PVA.

Both solutions were electrospun using 30 kV, an injection flux of 0.4 mL/h and a distance between collector and injector of 10 cm. In order to give the PVA-nFe3O4 membrane water stability, it was heated at 190°C for 10 minutes. Membranes were washed in deionized water before batch experiments to remove possible remaining salts and loose MIONs.

2.2 *Batch experiments*

Adsorption isotherm experiments were performed at room temperature, adjusted pH at 5 with HCl, fixed amount of adsorbent (6 mg/L for 1st approach and 30 mg/L for the 2nd approach) and varying initial arsenate concentration. Measurements were made with a HG-AAS.

DOI: 10.1201/9781003317395-146

3 RESULTS AND DISCUSSION

3.1 *Electrospinning solutions*

Figure 1 shows the electrospinning solutions. In situ approach allowed a higher loading of MIONs in the suspension, longer stability and better dispersion at the expense of containing reaction by-products, which may compromise spinnability. In fact, a 5% loading was prepared but could not be electrospun.

Figure 1. Stable magnetite suspensions in the presence of PVA for electrospinning.

3.2 *Morphology of the membranes*

Figure 2 show electron micrographs of the as-obtained nanofibers. MIONs can be observed in both cases, with diameters between 5 and 20 nm. For the *in situ* synthesis, there are ramifications growing from the fibers, possibly related to the growth of nanoparticles or crystallization of by-products along PVA chains.

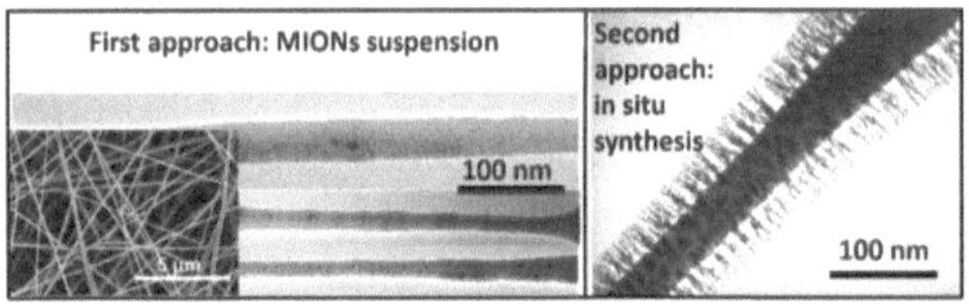

Figure 2. TEM images of the immobilized nanoparticles. SEM image of the 0.1% MIONs membrane in the lower left corner.

As(V) adsorption of the 0.1% MIONs membranes showed a linear isotherm for low equilibrium concentrations, with a maximum adsorption capacity of 87 mg_{As}/g. This value is higher than most of those reported in literature for MIONs, suggesting well dispersed MIONs along the fibers (Su 2017). Preliminary tests on 1% membranes using 2nd approach suggest similar adsorption capacities. It is important to remark the ability of adsorption even though the MIONs are located inside the polymeric nanofibers. This evidence that, after swelling of the hydrophilic polymer, arsenic ions in solution diffuse into the fibers and get adsorbed by the MIONs. This statement is supported by adsorption kinetic studies, which show a delay of approximately 1 hour before a significant drop in As(V) concentration. That is the time required for full swelling of the membrane.

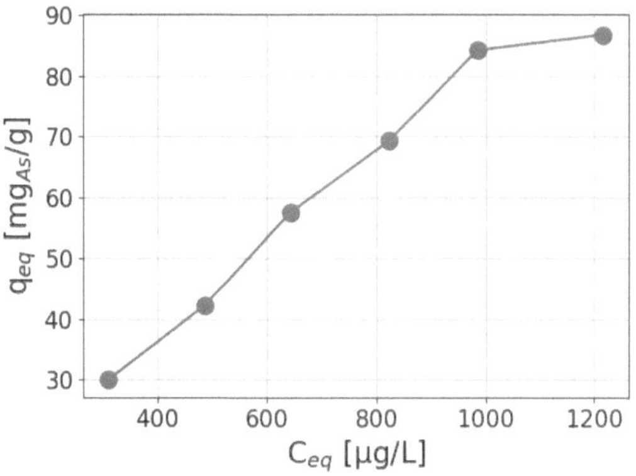

Figure 3. Adsorption isotherm for first approach membrane containing 0.1% MIONs.

4 CONCLUSIONS

Magnetic iron oxide nanoparticles have been successfully suspended in a PVA solution and then electrospun to give arsenic adsorbent membranes. Suspension of dried MIONs results in high adsorption capacity (87 mg/g) at very low adsorbent doses (6 mg/L) but the amount of MIONs that can be loaded into the fibers is limited because of particles precipitation. The approach consisting of *in situ* synthesis gives evenly distributed particles along the fibers and a greater potential because of higher loads and an easier production process.

ACKNOWLEDGEMENTS

We acknowledge the support of UBA (UBACYT 2018-2020 No 20020170100381BA), ANPCyT (PICT 2017-2362, PICT 3590) and PIP 00078. We also thank Juan Troncoso and Alicia Fernández Cirelli from CETA – UBA.

REFERENCES

Hao L., Liu M., Wang N. & Li G. 2018. A critical review on arsenic removal from water using iron-based adsorbents, *RSC Advances* 8: 39545–39560.

Shi S., Xu C., Wang X., Xie Y., Wang Y., Dong Q., Zhu L., Zhang G. & Xu D. 2020. Electrospinning fabrication of flexible Fe3O4 fibers by sol-gel method with high saturation magnetization for heavy metal adsorption. *Mater. Des.* 186: 108298.

Su C. 2017. Environmental implications and applications of engineered nanoscale magnetite and its hybrid nanocomposites: a review of recent literature. *J. Hazard. Mater.* 322: 48–84.

Yazdani F. & Seddigh M. 2016. Magnetite nanoparticles synthesized by co-precipitation method: the effects of various iron anions on specifications. *Mater. Chem. Phys.* 184: 318–323.

Arsenic in the Environment: Bridging Science to Practice for Sustainable Development – van der Wal, Ahmad, Petrusevski, Weijma, Savic, van der Wens, Beerendonk, Bhattacharya, Bundschuh & Naidu (Eds)
 ISBN 978-1-032-32928-4

As(III) removal from aqueous solution by calcium titanate nanoparticles prepared by the sol gel method

R. Tamayo[1,4], R. Espinoza-González[1], M. Flores[2] & E. Sacari[3]
[1]*LabMAM, Departamento de Ingeniería Química Biotecnología y Materiales, FCFM, Universidad de Chile, Santiago, Chile*
[2]*Laboratorio de Superficies, Deptartamento de Física, FCFM, Universidad de Chile, Santiago, Chile*
[3]*Laboratorio de Nanotecnología, Facultad de Ingeniería, Universidad Nacional Jorge Basadre Grohmann, Tacna, Perú*
[4]*Departamento de Ingeniería de Materiales, Facultad de Ingeniería de Procesos, Universidad Nacional San Agustín de Arequipa, Arequipa, Perú*

ABSTRACT: In this work, an innovative adsorbent based on $CaTiO_3$ nanoparticles (CTO) was prepared by the sol-gel technique for the removal of As(III) from aqueous solution. X-ray diffraction of the CTO nanoparticles powders confirmed the CTO phase. Transmission electron microscopy observations indicated an average particle size of 27 nm, while energy dispersive X-ray spectroscopy analysis showed the presence of Ca, Ti, and O in the expected stoichiometric amounts. The surface specific area measured by (BET) isotherm was 43.9 m^2/g. Batch adsorption experiments were used to study the effect of pH on the equilibrium adsorption of As(III), using an arsenite solution with 15 mg/L as initial concentration. The highest removal was achieved at pH 3, reaching an efficiency of up to 73%, determined by X-ray fluorescence from the residual As(III) in the solution. Time dependent adsorption experiments at different pHs exhibited a pseudo-second order kinetics with an equilibrium adsorption capacity of 11.12 mg/g at pH 3. Moreover, CTO nanoparticles were regenerated and evaluated for four cycles, decreasing their arsenic removal efficiency by 10% without affecting their chemical structure. X-ray photoelectron spectroscopy analysis of the CTO surface after removal experiments, showed that arsenic was present as As(III) and partially oxidized to As(V).

1 INTRODUCTION

It is known that in Latin America that the problem of As pollution is present in more than 14 countries (Bundschuh *et al.*, 2012). Waters contaminated with As are a hazard to humans, and chronic exposure to contaminated water intake causes different diseases. Some studies indicate that As could be associated with skin cancer and internal organs (Mandal & Suzuki 2002). As is mainly found as compounds of As(V) and As(III). The former is present in surface waters and the second in groundwater. It is known that As(III) is more toxic and difficult to remove from water than As(V), so a strategy for removal is first to oxidize the As(III) to As(V) to achieve a more effective removal (Corkhill & Vaughan 2009).

This work proposes the use of nanoparticles of perovskite type $CaTiO_3$ obtained by the sol-gel methods, for the adsorbent function of As(III), which has been little studied with this type of perovskite. In addition, it will allow us to look for the generation of a new system that facilitates the removal of As(III) using a simple ceramic prepared from earth abundant materials.

2 MATERIALS AND METHODS

2.1 *Synthesis of $CaTiO_3$ nanopowder*

Pure $CaTiO_3$ (CTO) perovskite nanoparticles were prepared by the sol-gel method. In this synthesis, titanium (IV) isopropoxide $Ti(OC_3H_7)_4$, calcium (II) nitrate tetrahydrate $Ca(NO_3)_2{\cdot}4H_2O$, 2-propanol and citric acid monohydrate $C_6H_8O_7{\cdot}\ H_2O$, were used as starting materials. The synthesis steps are presented in the scheme below (Figure 1).

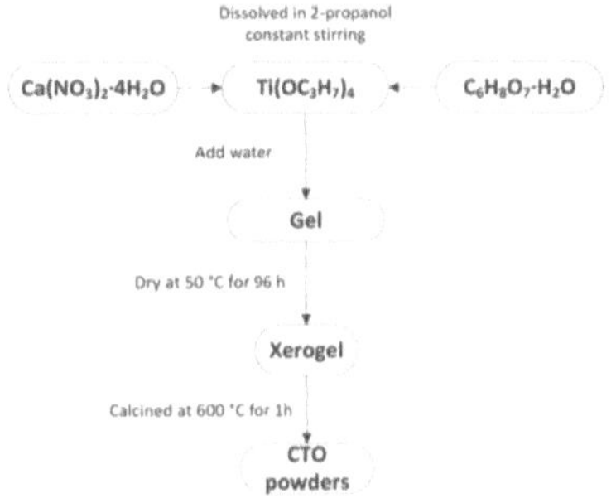

Figure 1. Flowchart for $CaTiO_3$ nanoparticles preparation procedure.

2.2 *Characterization*

The obtained powders were characterized by X-ray diffraction (XRD) in a D8 Bruker diffractometer. The morphology and microstructure of the samples were studied by transmission electron microscopy (TEM) in a FEI Tecnai. The specific surface area of the powders was measured by nitrogen adsorption isotherm BET equation in a Micromeritics ASAP

DOI: 10.1201/9781003317395-147

2010 apparatus. The point of zero charge was measured employing a ZSNano Zetasizer instrument.

2.3 *Batch adsorption experiments*

Different sets of 100 mL solutions containing As(III) were prepared using sodium arsenite ($NaAsO_2$) with an initial concentration of 15 mg/L and different pHs (1 to 11), and 100 mg CTO was added. This mixture was stirred and 2 mL aliquots were taken at 6, 12, 18, 24, 30, 40, 60, and 80 min. The total As concentration in the aliquots was determined by using X-ray fluorescence (XRF) analysis using a benchmark MiniPal4 (PANalytical). On the other hand, the final solutions after 80 min. of the adsorption experiments were centrifuged of at 5000 rpm on 10 min to recover the solids of CTO nanoparticles. The solids were dried at 115°C over 3 h to study their superficial chemical composition after adsorption, which was analyzed by means of the X-ray photoelectron spectroscopy technique (XPS).

2.4 *Ion effect and adsorption/desorption cycles*

The tests were conducted using the coexisting amounts of anions in natural waters using the batch experiment methodology described in Section 2.3. To prepare the solutions, different sodium salts ($NaNO_3$, NaCl, $NaCO_3$, and $NaSO_4$) were used. Each one was added in concentrations of 100, 250, and 500 mg/L to solutions of As (III) with 15 mg/L initial concentration and 1 g/L of CTO at pH 3. Additionally, adsorption/desorption cycles were performed using the same batch adsorption conditions.

3 RESULTS AND DISCUSSION

The XRD pattern of calcined powders, which shows the formation of the single-phase CTO in the powders after the calcination step. The observed peaks in the pattern can be indexed as CTO according to the JCPDS card N° 78-1013. The highest As(III) adsorption is reached at pH 3 compared to those at pH 1, 4, and 7. These results are coherent with those obtained at different pHs (Figure 2). The morphology and microstructure of calcined CTO powders were analyzed by TEM. The CTO particles exhibited an irregular shape, which size ranges between 10 and 40 nm, and an average diameter of 27 nm. The surface specific area of CTO powders obtained by BET isotherm was 43.9 ± 0.1 m^2/g, that is higher than the 1.6 m^2/g of commercially available CTO used by Jia *et al.* (2013). Zhuang *et al.* (2014) reported surface areas between 28.34 and 108.14 m^2/g for CTO prepared by the hydrothermal method using different titanium precursors.

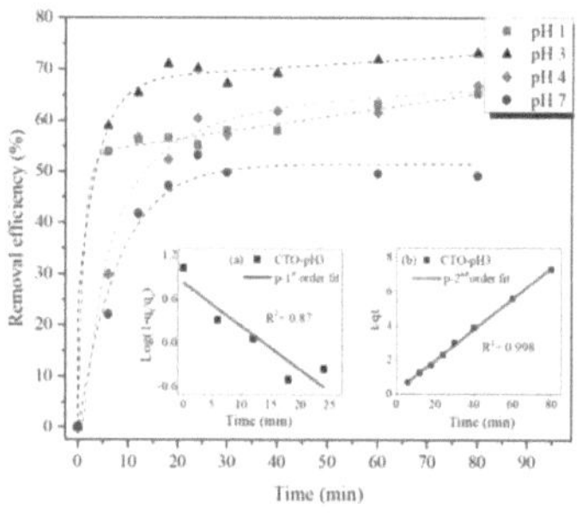

Figure 2. Time dependent adsorption behavior of As(III) by CTO na-noparticles at different pHs. Inset: Linear fit of the kinetic adsorp-tion at pH 3 by a) pseudo-first and b) pseudo-second order models.

4 CONCLUSIONS AND RECOMMENDATIONS

The CTO nanoparticles prepared by the sol-gel technique were demonstrated to be a good adsorbent of As (III) with a maximum adsorption capacity of 11.12 mg/g at room temperature, which is higher than other values reported in similar studies. It was shown that the adsorption capacity of As varies with the pH of the solution, and that the maximum adsorption capacity was given at pH 3 which coincides with the isoelectric point of the CTO nanoparticles. This is related to an inner-sphere surface complex formation of As(III)–O–Ti(CTO) bonding. The study on adsorption kinetics shows that this system follows a pseudo-second order model. The X-ray photoelectron spectroscopy analysis of the CTO surface after removal experiments, showed that arsenic was present as As(III) and partially oxidized to As(V). Under these conditions of removal, the addition of interfering ions (carbonates, nitrates, sulfates and chlorides) did not cause a significant decrease of As(III) adsorption using CTO. Additionally, the regeneration of the CTO is possible and confirms that the adsorption is reversible.

ACKNOWLEDGEMENTS

This research was funded by CONICYT-Chile, grant number FONDECYT Regular 1150652. We also acknowledge MINECON-Chile for the support through project Millennium Nucleus MULTIMAT-ICM/MINECON. RT-C thanks the scholarship support N° 078-FINCYT-BDE-2014, Perú.

REFERENCES

Bundschuh J., Litter M.I., Parvez M.F. *et al.* 2012. One century of arsenic exposure in Latin America: a review of history and occurrence from 14 countries. *Sci. Total Environ.* 429: 2–35.

Corkhill C.L. & Vaughan D.J. 2009. Arsenopyrite oxidation – a review. *Appl. Geochem.* 24: 2342–2361.

Jia C., Gao J., Li J., Gu F., Xu G.W. *et al.* 2013. Nickel catalysts supported on calcium titanate for enhanced CO methanation. *Catal. Sci. Technol.* 3(2): 490–499.

Mandal B. & Suzuki K. 2002. Arsenic round the world: a review. *Talanta* 58(1): 201–235.

Zhuang J., Tian Q., Lin S. *et al.* 2014. Precursor morphology-controlled formation of perovskites cation and their photo-activity for As(III) removal. *Appl. Catal. B* 156–157: 108–115.

Arsenic in the Environment: Bridging Science to Practice for Sustainable Development –
van der Wal, Ahmad, Petrusevski, Weijma, Savic, van der Wens, Beerendonk, Bhattacharya, Bundschuh & Naidu (Eds)
 ISBN 978-1-032-32928-4

Sulphur modified nano zerovalent iron (S-nZVI) for enhanced removal of arsenic from aqueous solution

P. Singh[1], S. Bhowmick[1], P. Mondal[2] & D. Chatterjee[3]
[1]*Kolkata Zonal Center, CSIR – National Environmental Engineering Research Institute (NEERI), Kolkata, West Bengal, India*
[2]*Ceramic Membrane Division, CSIR-Central Glass and Ceramic Research Institute, Kolkata, India*
[3]*Department of Chemistry, University of Kalyani, Kalyani, West Bengal, India*

ABSTRACT: Sulphur modified nano zerovalent iron (S-nZVI) was synthesized and used for removal of arsenic (As) from aqueous solution under anaerobic condition. The synthesized material showed a core Fe^0 structure with an outer rough and flake like distorted oxide shell structure comprising of iron oxide/hydroxide and FeS as evidenced from TEM and XRD analysis. S-nZVI showed higher rate of As removal from aqueous solution as compared to nZVI with ~99% of As removed within two hours of interaction. The removal rate was dependent upon the dosing of sulphur with maximum removal reported for Fe/S molar ratio of 0.1. XPS analysis of the As-reacted S-nZVI showed As(III) and As (V) primarily removed through adsorption of As(III) and As(V) species respectively, along with surface precipitation of As_2S_3. This study therefore illustrates great potential of S-nZVI in reducing elevated levels of As in contaminated water.

1 INTRODUCTION

Arsenic (As) poisoning has now become a globally challenging issue. Continuous exposure to arsenic (As) affects the health greatly. The major source of As exposure is from drinking As contaminated drinking water. Therefore, As remediation from contaminated water has now become very important. For this purpose, several techniques like anion exchange, reverse osmosis, microbial transformation, adsorption etc are being applied and several researches on further improvement of these techniques are being carried out. Among these techniques, adsorption is the most popular technique owing to its low cost and ease of operation (Bhowmick *et al.*, 2014). Recently, there are reports that show doping sulfur onto nZVI (S-nZVI) yields an effective material for removal of inorganic from contaminated water. The objective of this study is therefore to evaluate the efficacy of S-nZVI towards removal of inorganic As from aqueous solution and study the effect of various factors in rate of As removal. Finally, a probable mechanism of As removal by S-nZVI has been proposed.

2 EXPERIMENTAL

2.1 *Synthesis of S-nZVI*

S-nZVI particle was synthesized by sodium dithionite – borohydride reduction method where 4M of $NaBH_4$ with required amount of $Na_2S_2O_4$ (so as to maintain the desired S/Fe mole ratio) were added drop-wise to N_2-purged 0.5M $FeCl_3$ solution followed by 30–40 minutes of additional mechanical stirring. The solution was finally centrifuged, washed with water, filtered and vacuum dried overnight at 80°C and stored under anaerobic condition.

2.2 *Batch experiment*

All the batch experiments were performed in anaerobic condition. S-nZVI dose of 0.5 mg/L and an aqueous solution of either As(III) and As(V) of 10 mg/L was taken for the kinetic experiment. pH was adjusted using HCl or NaOH. All the experiments were performed in triplicates.

3 RESULTS AND DISCUSSION

3.1 *Characterization*

TEM images of the synthesized S-nZVI showed flake like shell structure with a Fe^0 core (Figure 1a). Unlike nZVI where the shell is more intact, S-nZVI has a disturbed outer shell and such flake structures increases the surface area of the S-nZVI.

Powdered XRD analysis of the S-nZVI shows presence of Fe^0 as well as magnetite/maghemite and FeS (Figure 1b). Therefore from the above analysis, it can be concluded that the outer shell consist of iron oxide/hydroxides and FeS, with inner core of Fe(0).

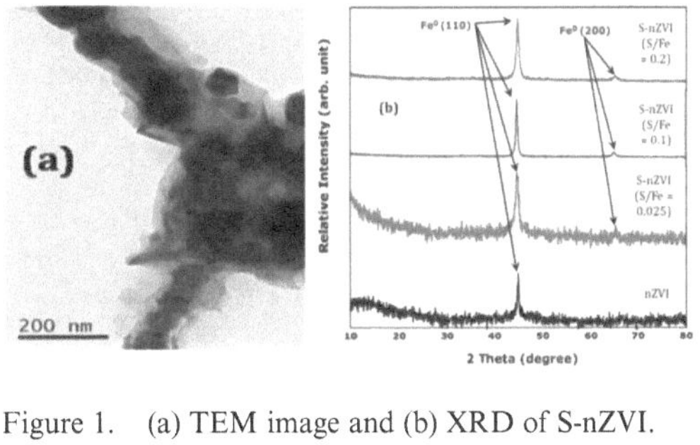

Figure 1. (a) TEM image and (b) XRD of S-nZVI.

DOI: 10.1201/9781003317395-148

3.2 *Effect of S/Fe molar ratio*

The effect of different S/Fe molar ratio was evaluated for the removal of As(III) and As(V). For both As(III) and As(V), it was observed that on adding a small amount of sulphur to nZVI, initially the adsorption decreases to S/Fe molar ratio of 0.025 and then the rate of adsorption increases from S/Fe ratio 0.025 to 0.1. At low sulphur concentration, formation of dense layer of FeS inhibit Fe corrosion, and thereby adsorption and co-precipitation of As is inhibited. Further increasing the S/Fe molar ratio from 0.025 to 0.1, corrosion of Fe^0 increases, thereby increasing the As adsorption efficiency. However, on further increasing the sulphur concentration from S/Fe ratio of 0.1, formation of FeS is increased and this reduced the number of open pores and also decreased the corrosion. Therefore, with the smaller number of open sites and decreasing corrosion, finally the removal of As from aqueous solution is decreased.

3.3 *Adsorption kinetics experiment*

Since S-nZVI with S/Fe molar ratio of 0.1 showed the highest arsenic adsorption, all the other experiments were conducted with this molar ratio. The data obtained from adsorption kinetics were fitted well with Pseudo – second order kinetic model:

$$t/q_t = 1/k_2 q_e^2 + t/q_e \quad (1)$$

where, q_e and q_t are the amount of As adsorbed (mg/g) at equilibrium and at time t (h) respectively and k_2 (g/mg/h) represents the pseudo-second-order rate constant.

The pH experiments were carried out at pH 5.0 ± 0.5, 7.0 ± 0.5 and 9.0 ± 0.5. For both As(III) and As(V), the adsorption was slightly better at acidic condition but decreased sharply at basic pH. The pHzpc of S-nZVI is usually around near-neutral pH and thus S-nZVI surface tends to remain positively charged in acidic pH and negatively charged in alkaline pH. The higher rate of Fe corrosion and the higher electrostatic attraction between negatively charged oxyanion of As(V) and the positive surface of S-nZVI at acidic condition, increases the adsorption of As(V). However, at basic pH, both the corrosion of Fe decreases as well as there is electrostatic repulsion between negatively charged oxyanions of As(V) and negatively charged S-nZVI surface, so adsorption decreases. As(III) exists as neutral H_3AsO_3 at acidic and neutral condition and as $H_2AsO_4^-$ at pH > 9. Therefore, higher rate of adsorption of As(III) in acidic pH was due to higher rate of Fe corrosion in acidic medium, while the decrease in adsorption in basic pH was due to reduced Fe corrosion and electrostatic repulsion between negatively charged As oxyanions and negatively charged S-nZVI surface.

3.4 *Removal mechanism*

To investigate the removal mechanism, XPS analysis of the As reacted S-nZVI was performed (Figure 2). It

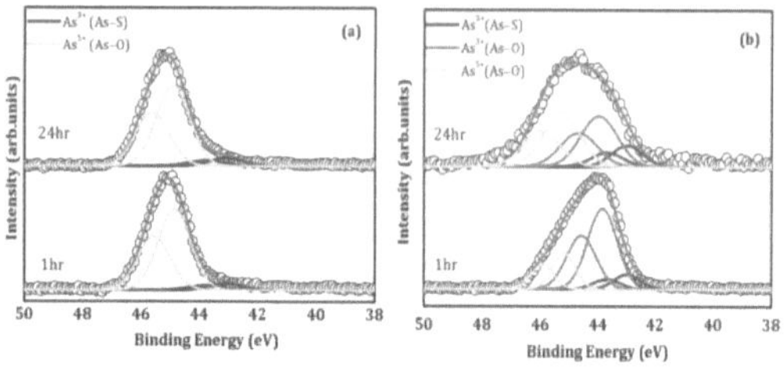

Figure 2. XPS images of As3d spectra of (a) As(V) and (b) As(III) reacted S-nZVI.

was found that for As(V)-reacted S-nZVI, a major doublet peak of As(V)-O was found along with small doublet peak of As(III)-S. The peak pattern did not change with progress in reaction time to 24 hr. In case of As(III), two doublets peaks of As(III)-O and As(V)-O was found along with a doublet peak of As(III)-S. However with increase in reaction time from 1hr to 24hr, the As(III)-O peak decreased and the As(III)-S peak increased. Thus for both As(III) and As(V), there was no reduction of As species to its lower valence state, including As(0) upon uptake by S-nZVI surface. Based on the results, it was hypothesized that As removal by S-nZVI involved adsorption as As(III) and As(V) oxyanion respectively along with As_2S_3 surface precipitation. The partial oxidation of As(III) to As(V) in anaerobic condition is through reactive Fe(II) – Fe(III) species (Amstaetter *et al.*, 2012).

4 CONCLUSIONS

Based on this study, it can be concluded that, doping a suitable amount of sulphur in nZVI enhanced the removal efficiency of As compared to nZVI. The S/Fe molar ratio is one of the most important factors dictating the rate of As removal with S/Fe molar ratio of 0.1 showing the highest rate of As removal. As uptake by S-nZVI involve adsorption as As(III) and As(V) oxyanion without reduction to lower valent species along with formation of As_2S_3.

REFERENCES

Amstaetter K., Borch T., Larese-Casanova P. & Kappler A. 2010. Redox transformation of arsenic by Fe(II)-activated goethite (α-FeOOH). *Environ. Sci. Technol.* 44: 102–108.

Bhowmick S., Chakraborty S., Mondal P., Renterghem W. V., van der Bergh S., Roman-Ross G., Chatterjee D. & Iglesias M. 2014. Montmorillonite – supported nanoscale zero – valent iron for removal of arsenic from aqueous solution: kinetics and mechanism. *Chem. Eng. Jour.* 243: 14–23.

4.3 Biochar and Chitosan-based novel technologies for removal of arsenic

Arsenic in the Environment: Bridging Science to Practice for Sustainable Development –
van der Wal, Ahmad, Petrusevski, Weijma, Savic, van der Wens, Beerendonk, Bhattacharya, Bundschuh & Naidu (Eds)
© 2023 the Author(s), ISBN 978-1-032-32928-4

Enhanced As(V) removal from aqueous solution by Zr/Zr-Fe modified biochar

M.A. Rahman[1,2], D. Lamb[1], M.M. Rahman[1], M.M. Bahar[1], P. Sanderson[1], S. Abbasi[1], A.S.M.F. Bari[1] & R. Naidu[1]
[1]*Global Centre for Environmental Remediation (GCER), The University of Newcastle, Callaghan, NSW, Australia*
[2]*Department of Public Health Engineering (DPHE), Zonal Laboratory, Khulna, Bangladesh*

ABSTRACT: In this investigation, we synthesize and characterize zirconium (Zr) and Zr-Fe coated biosolid biochar (BSBC) to determine the sorption capacity of arsenate, As(V). The combined results of specific surface area (BET-N_2), FTIR, XRD, SEM, and TEM confirmed that Zr(IV) and Zr-Fe were successfully coated onto BSBC. The effects of reaction time, pH, initial As(V) concentration, adsorbate doses, temperature, ionic strength, at different concentrations on the adsorption of As(V) were investigated. At equilibrium, the experimental sorption data were successfully modelled using Langmuir (R^2 = 0.98) and Freundlich (R^2 = 0.99) isotherms where Freundlich model fitted well. Kinetic data were fitted with pseudo first order and second order model where best fitted with second order kinetics (R^2 = 0.99). The maximum sorption capacity of Zr-BSBC and Zr-FeBSBC composites were 18.97 and 27.73 mg/g, respectively at pH 6. The thermodynamic parameters (*viz.* changes in Gibbs free energy, enthalpy, and entropy) suggested that the adsorption process is spontaneously endothermic and feasible in nature. The results suggest that Zr-O and Zr-O-Fe coated BSBC could serve as promising adsorbents for remediating As(V) contaminated water.

1 INTRODUCTION

Arsenic (As) is a class I toxic metal(loid) and has been recognized as a human carcinogen (IARC 2004). In the aquatic environment, As exists in different forms such as arsenious (H_3AsO_3, $H_2AsO_3^-$, $HAsO_3^{2-}$), arsenic (H_3AsO_4, $H_2AsO_4^-$,$HAsO_4^{2-}$), arsenites [As(III)], arsenates [As(V)], methylarsenic, dimethylarsinic acid, and arsine (Mohan and Pittman Jr 2007). Both As(III)) and As(V) are the predominant species in water, where As(V) is stable and present under oxidizing conditions; and exists as oxyanions of arsenic acid while As(III) prevails under mildly reducing conditions (Smedley & Kinniburgh 2002). Arsenic contamination occurs by both natural process as well as anthropogenic activities (Niazi *et al.*, 2018). Biochar is an emerging and low-cost potential sorbent for removing environmental contaminants, and can be produced from a wide variety of agricultural waste, plant, urban, and industrial residues (Agrafioti *et al.*, 2014) which has attracted significant attention for beneficial use in waste-water treatment (Shimabuku *et al.*, 2016). Biochar is an effective material in removing contaminants due to high specific surface areas and the presence of different surface functional groups. In this study, a zirconium (Zr) and/or iron biosolid biochar (Zr-BSBC and Zr-Fe-BSBC) composite have been synthesized and used to evaluate the sorption efficiency of (As(V)) from aqueous solution. The effects of various experimental conditions, such as solution pH, contact time, initial concentration, adsorbate dosage, ionic strength, competitive co-existing ions, effect of temperature on As(V) adsorption were investigated.

2 MATERIALS AND METHODS

2.1 *Synthesis of Zr-BSBC and Zr-FeBSBC*

Biochar was produced from biosolid biomass by slow pyrolysis at 300°C for 30 min under a nitrogen atmosphere (Vithanage *et al.*, 2015). Biosolid biochar was treated with a) 0.1 M Zr(IV) solution and the pH was adjusted to 12.5 and; b) Zr(IV) + $FeCl_3$ (pH was adjusted to 6.5) by dropwise addition of NaOH by continuous stirring for 12 hours. Finally, the biochar composites were washed several times with MQ followed by centrifuged at 5000 rpm for 15 min, filtered (0.22 μm) and dried at 80°C for 12 h. The modified Zr-BSBC and Zr-FeBSBC were kept in a desiccator for further experiments.

2.2 *Biochar characterization*

The synthesized biochars were characterized using XRD, SEM-EDS, TEM for mineralogy and surface morphology. The functional groups and surface area were determined by FTIR and Surface Area Analyzer, respectively.

2.3 *Batch sorption experiment*

Kinetic experiments were performed to assess As(V) sorption by the adsorbents as a function of reaction time. To evaluate the maximum adsorption capacity of Zr-BSBC and Zr-FeBSBC, isotherms were determined with As(V) concentration between 1.0 and 200 mg/L. The effect of pH on As(V) adsorption was determined at pH range 2–10. To assess the impact of different adsorbent dosage on the As(V) sorption, different

DOI: 10.1201/9781003317395-149

dosages of biochars (solid: solution) (1:100, 1:250, 1:500 and 1:1000) were used. In this study, $NaNO_3$ was used to stimulate as various ionic strengths 0.01, 0.1, 0.5 and 1.0 M respectively. The adsorption thermodynamics were also examined at varying temperatures of 4, 15, 20, 25, 30 and 37°C. All batch sorption experiments were conducted in sealed 50 mL tubes at room temperature (22 ± 0.5°C).

3 RESULTS AND DISCUSSION

The surface texture and morphology of modified biochars were different from one another, due to covering by high densities of fine Zr, or Zr-Fe particles, indicating the successful coating of Zr and/or Zr-Fe on biochars surface (Figure 1). Moreover, some fractions of Zr(IV) was located randomly onto the biochar surface which indicates heterogeneous coating during precipitation The chemical modification by Zr and/or Zr-Fe of the adsorbent was attributed to the formation of porous structure that leads to an increase in internal surface as well as sorption capacity of As(V).

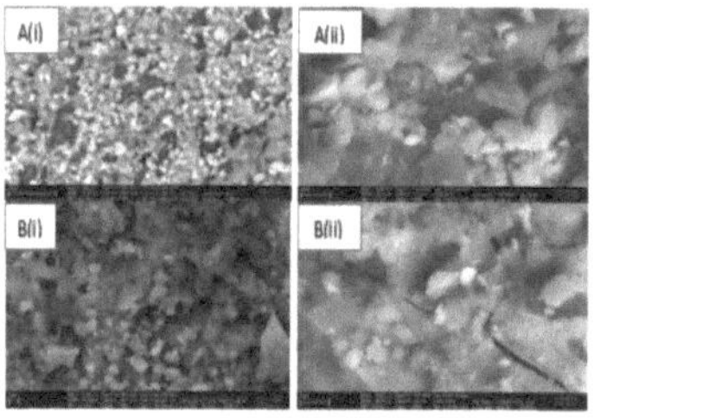

Figure 1. SEM of A(i) and B(i) before and A(ii) and B (ii) after As(V) sorption of ZrBSBC and Zr-$FeCl_3$BSBC, respectively.

The EDX spectrum of ZrBSBC and Zr-$FeCl_3$BSBC confirmed the presence of Zr into the Zr and Zr-Fe treated biochars at 2.1 keV. After sorption of As(V) the presence of sorbed As(V) (at 10.25 keV) along with other elements such as C, O, N, Fe, P, and S was also confirmed.

The sorption isotherms and pH edges indicated that the affinity of As(V) toward the Zr-BSBC and Zr-$FeCl_3$BSBC were strongly dependent on solution pH and biochar characteristics. Sorption of As(V) was favored at pH 6.0 (Figure 2) as a consequence of the higher affinity between the positively charged biochar surface and the predominant As species ($H_2AsO_4^-$ and H_3AsO_4). The kinetics of As sorption of both composites were well fitted the second order model. The Freundlich isotherm model described the As(V) experimental data well for Zr-BSBC (R^2 = 0.99) and Zr-$FeCl_3$BSBC (R^2 = 0.98), whereas the Langmuir isotherm model fitted for Zr-BSBC (R^2 = 0.97) and for Zr-$FeCl_3$BSBC (R^2 = 0.98). The maximum sorption capacity of Zr-BSBC and Zr-$FeCl_3$BSBC were 18.97 mg/g (at pH 6.0) and 27.73 mg/g (at pH 6.0), respectively.

A higher ionic strength significantly decreased the sorption capacity at 72 h when $NaNO_3$ concentrations increased from 0.01–1.0 M. At higher concentrations,

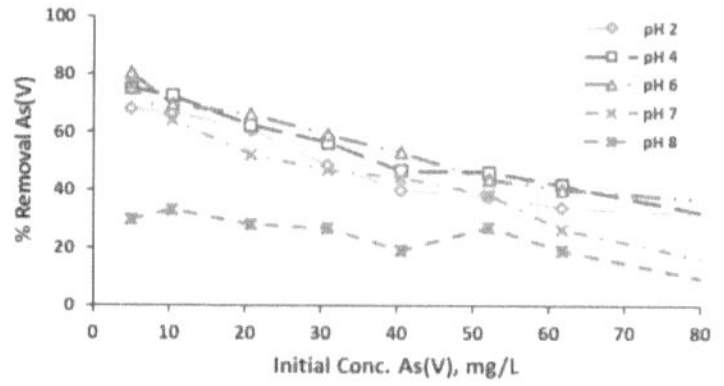

Figure 2. Removal efficiency of As(V) onto Zr-BSBCas a function of pH.

the NO_3^- anion might compete with AsO_3^- for the available sites on the biochar surface, and prevent As (V) approaching to the biochar. The calculated free energy (ΔG) was found negative, indicating the feasibility and spontaneous nature of the adsorption process in both Zr-BSBC and Zr-$FeCl_3$BSBC. Positive values of ΔH° suggested that As(V) sorption was endothermic and increased at higher temperatures.

4 CONCLUSIONS AND RECOMMENDATIONS

Zr-BSBC and Zr-$FeCl_3$BSBC composite were successfully synthesized, characterized and utilized for effective As(V) removal from aqueous solutions. Zr-$FeCl_3$BSBC revealed improved sorption capacity of As(V) when compared to Zr-BSBC. In, conclusion, Zr-O and Zr-O-Fe treated BSBC could serve as promising adsorbents for decontamination of As(V) from contaminated water.

ACKNOWLEDGEMENTS

Md. Aminur Rahman is thankful to The University of Newcastle for providing UNIPRS and UNRSC central scholarship.

REFERENCES

Agrafioti E., Kalderis D. & Diamadopoulos E. 2014. Ca and Fe modified biochars as adsorbents of arsenic and chromium in aqueous solutions *J. Environ. Manage.* 146: 444–450

Mohan D. & Pittman Jr. C.U. 2007. Arsenic removal from water/wastewater using adsorbents—a critical review *J. Hazard. Mater.* 142: 1–53

Niazi NK *et al.* (2018) Arsenic removal by perilla leaf biochar in aqueous solutions and groundwater: an integrated spectroscopic and microscopic examination *Environ. Pollut.* 232: 31–41

Shimabuku K.K., Kearns J.P., Martinez J.E *et al.* 2016. Biochar sorbents for sulfamethoxazole removal from surface water, stormwater, and wastewater effluent *Wat. Res.* 96:236–245

Smedley P.L. & Kinniburgh D. 2002. A review of the source, behaviour and distribution of arsenic in natural waters *Appl. Geochem.* 17: 517–568

Vithanage M., Rajapaksha A.U., Ahmad M. *et al.* 2015. Mechanisms of antimony adsorption onto soybean stover-derived biochar in aqueous solutions *J. Environ. Manage.* 151: 443–449.

Arsenic in the Environment: Bridging Science to Practice for Sustainable Development – van der Wal, Ahmad, Petrusevski, Weijma, Savic, van der Wens, Beerendonk, Bhattacharya, Bundschuh & Naidu (Eds)
© 2023 the Author(s), ISBN 978-1-032-32928-4

Biochar as a potential material for removal of arsenic from water

N.K. Niazi[1], I. Bibi[1] & M. Shahid[2]
[1]*University of Agriculture Faisalabad, Faisalabad, Pakistan*
[2]*COMSATS University Islamabad, Vehari Campus, Vehari, Pakistan*

ABSTRACT: Biochar is considered to be a low-cost, eco-friendly and sustainable material for various environmental implications, including water treatment. Here, we evaluated arsenic (As) removal efficiency of various biochars produced at 300, 500, and 700°C in aqueous solutions and drinking water. Both arsenite (As(III)) and arsenate (As(V)) species were successfully removed from aqueous solutions, with slightly higher sorption affinity, Q_L, for As(V) than As(III). The As removal spanned 81% to 99% for different biochars, with the maximum sorption obtained for perilla leaf biochar at pH 7 and 6. Surface functional moieties contributed to As sequestration by the biochars examined here. Arsenic K-edge XANES spectroscopy demonstrated oxidation of As(III) to As(V) which could be an additional benefit of biochar in sorption process and As purification. The study highlights some important aspects of As-biochar interactions before its application in treatment of and As-contaminated drinking well water.

1 INTRODUCTION

Arsenic (As) contamination of groundwater represents a global, health and environmental issue due to its toxic nature. Although many types of sorbents (e.g., activated carbon, Fe-coated granular AC, Fe-containing fly ash, nano-adsorbents) have been used for toxic metal/metalloid removal from water, biochar is considered to be the low-cost, environmentally-friendly sustainable tool for treatment of water, including As-contaminated well water. In this study, As removal efficiency of different biochars, derived from perilla leaf, Oak wood, maize residues (produced at low 350, 500, 700°C temperature), was explored in aqueous solutions, as well as their potential application for removal of As from groundwater which is used for drinking in Pakistan.

2 METHODS/EXPERIMENTAL

2.1 *Production and analysis of biochars*

The perilla leaf, oak wood chipping and maize residues wastes were collected, washed with tab water followed by rinsing with deionized water to remove impurities. After washing, wood chippings were air-dried and ground (<2 mm) to yield uniform size particles prior to pyrolysis process. Various physicochemical properties of biochars were determined.

2.2 *Laboratory-scale sorption experiments*

Batch sorption experiments were performed in 50 mL polyethylene (reaction) vials using the biochars in a background electrolyte solution of 0.01 M NaCl. All experiments were performed at a constant temperature of 20 ± 1°C with an equilibration time of 2 h. Effect of pH (pH 3–10) was evaluated at an initial As(V) and As (III) concentrations of 4 mg/L and constant biochar suspension density of 1 g/L Suspension pH was maintained by adding 0.01 M HCl or NaOH solutions.

Sorption isotherm experiments were performed at initial As(V) and As(III) concentrations of 0.05–7.0 mg/L. Arsenic was analysed using a hydride generation-atomic absorption spectrometer (HG-AAS; Agilent AA240 with VGA 77). Fourier transform infrared spectroscopy, SEM-EDX and X-ray absorption near edge structure (XANES) spectroscopy were employed to determine surface functional groups, morphology and speciation of As in As(III)- and As(V)-loaded biochar samples.

3 RESULTS AND DISCUSSION

3.1 *Laboratory-scale sorption experiments*

Arsenic sorption isotherm experimental data indicated that As removal by biochars increased the increasing initial As(V) and As(III) concentrations. Relatively greater sorption was obtained for As(V) (0.056–2.02 mg/g) than that of As(III) (0.031–1.86 mg/g). Among the four isotherm models, Langmuir model achieved the best fit for As(III) and As(V) sorption to all biochars, with a slightly better fit for As(V) than that of As(III). This indicates that monolayer (homogeneous) As sorption is a dominant mechanism for As species on the biochars examined here (Abid *et al.*, 2016).

DOI: 10.1201/9781003317395-150

The effect of solution pH on As(III) and As(V) removal percentage by biochars indicated an increase in As(III) removal (72–81%) at pH spanning 3.1–7.2, with the highest As(III) removal at pH ~7. Whilst, above pH 7 a decline was observed in the As sorption (74–68%) in As(III)-biochar sorption systems. Conversely, As(V) removal percentage was the highest at pH 6 (89%) in As(V)-biochars experiments (*Niazi et al.*, 2018).

Different surface functional groups, mainly –OH, –COOH, –C–O, $-CH_3$, are thought to be responsible in As removal by these biochars, suggesting the surface complexation, precipitation and/or electrostatic interaction of As on biochar surface. The scanning electro microscopy combined with energy dispersive X-ray spectroscopy (SEM-EDX) indicated relatively more rough, irregular and porous micro-structure of As-unloaded perilla leaf biochars, which is due to its high specific surface area (476 m^2/g). The elemental dot maps from SEM-EDX analysis indicated that As was mainly sorbed via interaction with carbon and oxygen bearing groups and/or calcium-bearing mineral species, notably in perilla leaf biochar. Arsenic binding, as As(V) could occur with calcium (present as $CaCO_3$-like species in biochars), thus making

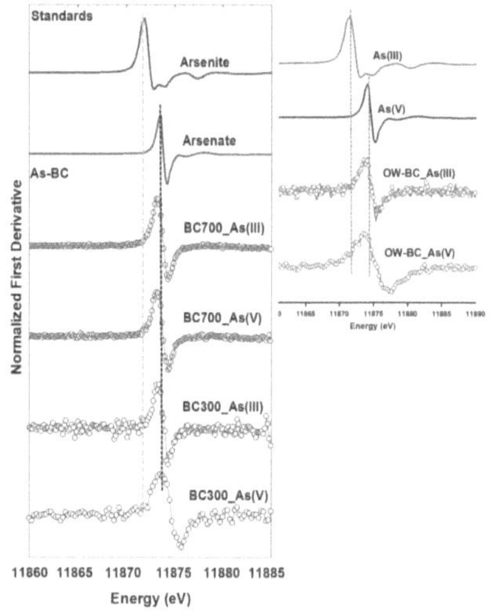

Figure 1. First derivative As K-edge XANES spectra for As(III) and As(V) under aquatic conditions compared to reference compounds (absorption edge energy (eV) of references compounds is 11869.4 (arsenite) and 11873.4 (arsenate)). As(V) and As(III), standards were included in the linear combination fitting (LCF) (circles show the experimental data and solid black color lines represent the LCF to experimental data). BC300 and BC700: perilla leaf biochar at 300 and 700; OW-BC: oakwood biochar.

relatively insoluble Ca-As precipitates to immobilize As in biomass.

Arsenic K-edge X-ray absorption near edge structure (XANES) spectroscopy revealed that As(III) was partially oxidized to As(V) (38%) in As(III)-biochars system, indicating that biochar could be efficient in As(III) oxidation and subsequently facilitating As removal on sorbent surface (Figure 1).

Results showed that 87 to 100% of As was removed in As-contaminated groundwater which is used for drinking (total As: 21–149 μg/L; total samples = 11), showing a great potential for remediation of As-rich drinking well water.

4 CONCLUSIONS

The biochars used in this study removed both As(III) and As(V) in aqueous solutions, as well as As from As-contaminated drinking water. Langmuir model demonstrated the best fit for As(III) and As(V) sorption to biochars, which was also supported by our FTIR spectroscopy and SEM-EDX elemental dot mapping data. The XANES data revealed complex redox transformation of As(III) $\Leftrightarrow$ As(V) in the As (III)- and As(V)-biochars systems. Application of biochar, especially perilla leaf and oak wood biochars, to As-contaminated drinking water showed up to 100% As removal despite in the presence of a broad suite of anions and cations co-occurring in water. However, drinking water after biochar treatment was saline in nature due to dissolution of high amount of soluble salts from biochar, especially in maize residue biochar.

ACKNOWLEDGEMENTS

The authors thank the members of the Pohang Accelerator Laboratory (PAL), Korea for providing synchrotron user facilities for XAFS spectroscopy experimentation.

REFERENCES

Abid M., Niazi N.K., Bibi I., Farooqi A., Ok Y.S., Kunhikrishnan A., Ali F., Ali S., Igalavithana A.D. & Arshad M. 2016. Arsenic(V) biosorption by charred orange peel in aqueous environments. *Int. J. Phytoremediation* 18(5): 442–449.

Niazi N.K., Bibi I., Shahid M., Ok Y.S., Shaheen S.M., Rinklebe J., Wang H., Murtaza B., Islam E., Nawaz M.F. & Lüttge A. 2018. Arsenic removal by Japanese oak wood biochar in aqueous solutions and well water: investigating arsenic fate using integrated spectroscopic and microscopic techniques. *Sci. Total Environ.* 621: 1642–1651.

Arsenic in the Environment: Bridging Science to Practice for Sustainable Development – van der Wal, Ahmad, Petrusevski, Weijma, Savic, van der Wens, Beerendonk, Bhattacharya, Bundschuh & Naidu (Eds)
© 2023 the Author(s), ISBN 978-1-032-32928-4

Arsenic adsorption using synthesized iron-oxide biochar nanocomposites: Mechanistic study and application over field samples

P. Singh[1,2] & D. Mohan[1]
[1]*School of Environmental Sciences, Jawaharlal Nehru University, New Delhi, India*
[2]*Shaheed Rajguru College of Applied Sciences for Women, Vasundhara Enclave, Delhi, India*

ABSTRACT: Arsenic (As) in groundwater is a greatest threat to the human health. This paper investigates the role of biochar nanocomposites derived from locally available rice and wheat husk followed by impregnation with iron-oxide to remove As from water. The batch adsorption studies showed high Langmuir adsorption capacity for As(III) and As(V) at pH 7.5 and 5.5, respectively. The biochar composites were characterized for their porosity, surface area, and crystallinity. Kinetic and isotherm data suggested removal of As due to electrostatic force of attraction between As oxyanions and iron hydroxyl surface via dissociative/associative pathways. Fixed-bed column adsorption studies were conducted at flow rate 5.0 ml/min. Regeneration and multicomponent studies revealed that RHIOB and WHIOB have high potential to remove As from contaminated drinking water.

1 INTRODUCTION

Arsenic (As) is one of the most toxic contaminant in water. Because of its carcinogenic, mutagenic impacts on living systems, WHO has reported As contamination in drinking water to have caused greatest number of deaths as compared to other heavy metals (Ahmed 2000). WHO's permissible limit for As in drinking water is 0.01 mg/L. In India, states of West Bengal, Jharkhand, Bihar, Uttar Pradesh, Assam, Manipur and Chhattisgarh are reported to be most affected (Bhattacharya *et al.*, 2008). In response to the large number of people affected by As and fluoride in water in India, government has launched National Water Quality Sub-mission scheme in 2017 (DDWS 2017). It aims at providing clean and safe drinking water in already identified 28000 As and fluoride affected areas. Hence, exploration of highly efficient, cost-effective and easy to use removal techniques for As is a need of an hour. Adsorption is one such method (Mohan *et al.*, 2011). Because developing countries are mostly affected by As toxicity, locally available, affordable adsorbents that can be reused and regenerated needs to be developed.

2 METHODS/EXPERIMENTAL

2.1 *Synthesis of RHIOB and WHIOB*

Iron-oxide composite biochars were prepared by modified co-pyrolysis method using rice husk and wheat husk precursor biomass (filed in Indian Patent Application No. 20181010032) (Singh *et al.*, 2020). The biomass were thoroughly dried. Rice husk and wheat husk were soaked separately in 2.8 M iron salt ($FeCl_3.6H_2O$) solution in distilled water for 24h. The iron soaked biomass(s) were oven dried at 80°C for 2 h followed by slow pyrolysis at 600°C for 1 h using muffle furnace in N_2 supply. The pyrolysed biomass(s) were washed, oven dried, sieved into 30–50, 50–100, 100–200 B.S.S. mesh size and kept in airtight containers. 50–100 mesh sized composite biochars used for studies and experimentations. The synthesized biochar(s) were coded as rice husk iron-oxide biochar (RHIOB) and wheat husk iron-oxide biochar (WHIOB) composites.

3 RESULTS AND DISCUSSION

3.1 *Characterization of adsorbents*

RHIOB and WHIOB were characterized for their surface area and porosity. High surface area and porosity was reported. C, H, N, S, ash content, metallic composition, surface chemistry and functional group identification was carried out using different instruments in order to establish the mechanism involved in the adsorption of As(III) from water.

3.2 *Batch sorption studies*

Sorption studies were performed in batch mode to optimize the equilibrium and kinetic constants and pH impact over As adsorption from water over RHIOB/WHIOB. The maximum As(III) and As(V) removal occurred at pH 7.5 and 5.5, respectively. Kinetic data fitted pseudo-second-kinetic model for As(III)/(V) adsorption, indicating chemisorption as the main mechanism during the process. Diffusion model suggested particle diffusion at lower concentration and film diffusion at higher concentration as the rate limiting steps.

DOI: 10.1201/9781003317395-151

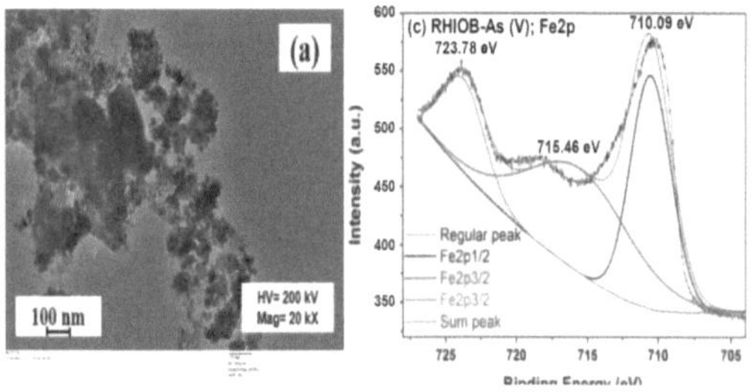

Figure 1. (a) TEM image of RHIOB 20KX (b) XPS analysis deconvoluted peaks of RHIOB-As(V) Fe2p peak.

RHIOB and WHIOB reported high isotherm adsorption capacities ~96.2 µg/g and ~110.7 µg/g for As(III) and ~386.9 µg/g and ~421.4 µg/g for As(V), respectively (Figures 2a,b). Thermodynamic studies suggested process to be an endothermic process.

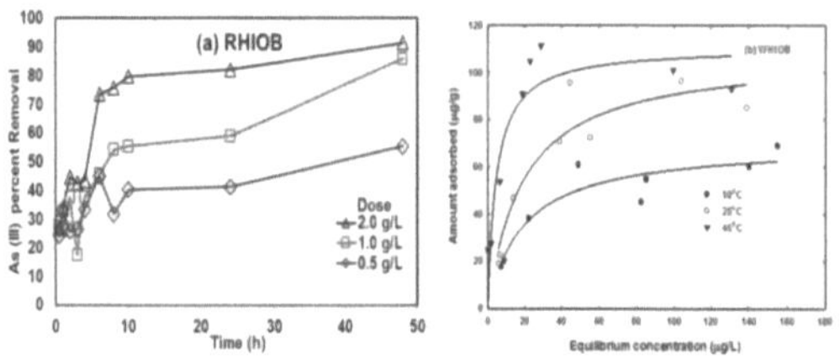

Figure 2. (a) Dose studies of As(III) removal over RHIOB (b) Isotherm studies, WHIOB over different temperatures.

3.3 *Fixed-bed column studies*

Adsorption studies in continuous mode was done through column studies at a constant flow rate of 5.0 ml/min (Figure 3a). The contact time for the empty bed were 1.6 min and 1.8 min respectively for RHIOB and WHIOB. High breakthrough was obtained for WHIOB as compared to RHIOB for As adsorption (Figure 3b). Adsorption capacities were reported by applying Thomson's model.

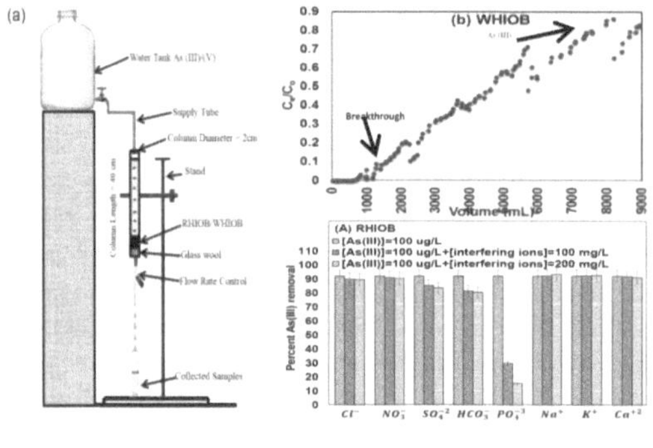

Figure 3. (a) Fixed-bed column set up; (b) Breakthrough curve for WHIOB over As(III).

3.4 *Regeneration and multicomponent studies*

Single and multicomponent As sorption on RHIOB and WHIOB was examined in presence and absence of chloride, nitrate, sulphate, bicarbonate, phosphate, sodium, potassium, and calcium. Negligible chloride and nitrate influence on As adsorption was observed for RHIOB and WHIOB. Regeneration studies were conducted using 0.1N NaOH and 0.1N NaCl. High adsorption efficiency was reported even after four consecutive adsorption-desorption cycle. Studies were conducted over field samples collected from Balia and Kushinagar, Uttar Prades, India. ~99% removal of As was reported using RHIOB and WHIOB.

4 CONCLUSIONS

Iron oxide modification of widely available rice husk and wheat husk biochars was achieved by pyrolysis of the $FeCl_3$-pre-impregnated husks at 600°C (1h). These hybrid adsorbents successfully remediated As from low concentration. Reactions at the Fe-OH surface sites to form Fe-O-As(III) linkages, which chemisorbed the As. RHIOB and WHIOB exhibited high Langmuir maximum capacities with excellent adsorption over the pH range of 3.0–10. Uptake rates were fast both in batch and column studies and followed pseudo second order kinetics. High breakthrough,regeneration capacity and high adsorption of As in presence of multiple ions suggest RHIOB and WHIOB as highly potential, cost effective and sustainable material for As adsorption from water.

ACKNOWLEDGEMENTS

One of the authors (PS) thanks to DST-INSPIRE for providing financial support to this work.

REFERENCES

Bhattacharya A.K., Naiya T.K., Mandal S.N. & Das S. K. 2008. Adsorption, kinetics and equilibrium studies on removal of Cr(VI) from aqueous solutions using different low-cost adsorbents. *Chem. Eng. J.* 137: 529.

Department of Drinking Water and Sanitation. 2017. *National Water Quality Sub-Mission Scheme*.

Mohan D., Sarswat A., Singh V.K., Alexandre-Franco M. & Pittman C.U. 2011. Development of magnetic activated carbon from almond shells for trinitrophenol removal from water. *Chem. Eng. J.* 172(2): 1111–1125.

Singh P., Sarswat A., Pittman C.U., Mlsna T. & Mohan D. 2020. Sustainable low-concentration arsenite [As (III)] removal in single and multicomponent systems using hybrid iron oxide–biochar nanocomposite adsorbents—a mechanistic study. *ACS Omega* 5(6), 2575–2593.

Arsenic in the Environment: Bridging Science to Practice for Sustainable Development –
van der Wal, Ahmad, Petrusevski, Weijma, Savic, van der Wens, Beerendonk, Bhattacharya, Bundschuh & Naidu (Eds)
 ISBN 978-1-032-32928-4

Removal of roxarsone and its metabolites by a sludge-based, biochar supported zerovalent iron nanocomposite: Adsorption and redox transformation

Ming Lei, Bingyu Li, Baiqing Tie & Huihui Du
College of Resource & Environment, Hunan Agricultural University, Changsha, P.R. China

ABSTRACT: Roxarsone is a phenyl-substituted arsonic acid comprising both arsenate and benzene rings. Few adsorbents are designed for the effective capture of both the organic and inorganic moieties of ROX molecules. Herein, nano zerovalent iron (nZVI) particles were incorporated on the surface of sludge-based biochar (SBC) to fabricate a dual-affinity sorbent that attracts both the arsenate and benzene rings of ROX. The incorporation of nZVI particles significantly increased the binding affinity and sorption capacity for ROX molecules compared to pristine SBC. The enhanced elimination of ROX molecules was ascribed to synergetic adsorption and degradation reactions, through π-π^* electron donor/acceptor interactions, H-bonding, and As-O-Fe coordination. Among these, the predominate adsorption force was As-O-Fe coordination. During the sorption process, some ROX molecules were decomposed into inorganic arsenic and organic metabolites by the reactive oxygen species (ROS) generated during the early stages of the reaction. The degradation pathways of ROX were proposed according to the oxidation intermediates. This work provides a theoretical and experimental basis for the design of adsorbents according to the structure of the target pollutant.

1 INTRODUCTION

Phenyl-substituted arsonic acids (PSAAs) are a class of veterinary drugs that have been widely used in the breeding/husbandry industry because of their various functions, including the control of intestinal parasite, promotion of feed efficiency, and meat pigmentation (Xu *et al.*, 2007). ROX can be transformed into highly toxic inorganic arsenic (arsenate and arsenite) and other phenyl derivatives via abiotic and biotic reactions in soil or the leachates of feces. These metabolites are highly mobile and can easily accumulate in rice and vegetables or be transported into groundwater, posing a significant threat to the environment. Thus, practical measures must be considered to reduce the hazards posed by ROX molecules in soils and surface water/groundwater. In the present study, nano zerovalent iron on sludge-based biochar (nZVI@SBC) was designed as a multifunctional sorbent and fabricated via the in-situ liquid reduction and growth of nZVI on the surface of SBC. The main objectives of this study were to (i) demonstrate the successful incorporation of nZVI on the SBC matrix, (ii) evaluate the performance of nZVI@SBC for the removal of ROX molecules from aqueous solution in comparison to pristine SBC, and (iii) uncover the mechanism by which nZVI@SBC eliminates ROX and its metabolites using a series of analytical techniques.

2 MATERIALS AND METHODS

2.1 *Synthesis and characterization of SBC and nZVI@SBC*

Municipal sludge was collected from a local wastewater treatment plant in Changsha city, the obtained sludge was pyrolyzed in a muffle furnace at 300°C under oxygen-deficient conditions. The nZVI@SBC nanocomposite was fabricated via the in-situ liquid reduction and growth of iron nanoparticles on the surfaces of sludge biochar (Wang *et al.*, 2017; Wei *et al.*, 2018).

2.2 *Batch experiments and analysis*

Kinetics and isotherms studies were used here to evaluate the performance of SBC and nZVI@SBC on ROX removal. ROX concentrations were determined by HPLC. The primary degradation intermediates of ROX (arsenate, arsenite) were identified by HPLC coupled with atomic fluorescence spectrometer (HPLC-AFS-9530, Haiguang Instrument, China).

2.3 *Mechanistic investigations*

Subtle changes in the organic moieties during the kinetic reaction were monitored by ultraviolet-visible (UV-VIS) spectrophotometry (Nicolet Evolution-300; Thermo Scientific) over the spectral

DOI: 10.1201/9781003317395-152

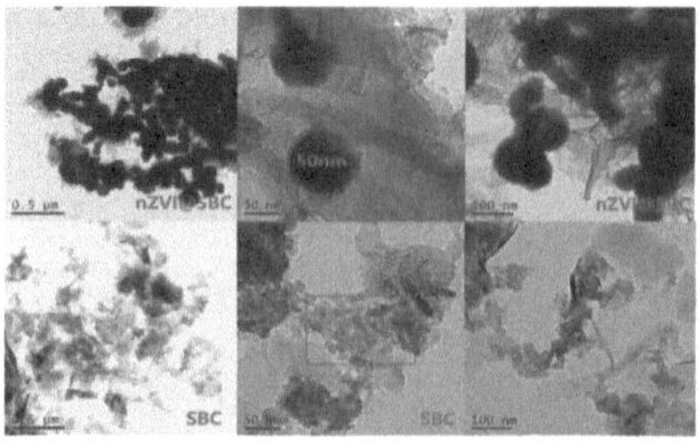

Figure 1. TEM micrographs of SBC (bottom row) and nZVI@SBC (top row).

range of 190–600 nm at regular time intervals. The changes in elemental composition and chemical bonds after reaction were identified by XPS and FTIR. The degradation pathway was explored by ultraperformance liquid chromatography quadrupole time-of-flight mass spectrometry (UPLC-Q-TOF-MS; Agilent Technologies).

3 RESULTS AND DISCUSSION

3.1 *Characterization of SBC and nZVI@SBC*

The synthesized SBC and nZVI@SBC samples were characterized by FE-SEM, FE-TEM, XRD, BET, and XPS, respectively, to verify that the zerovalent iron nanoparticles have been successfully soldered on the surface of SBC.

The magnified TEM images clearly showed that the nZVI nanoparticles were uniformly dispersed on the biochar matrix. The results of XPS and XRD proved the presence of zero valent iron on the surface of SBC.

3.2 *Batch adsorption experiments*

Batch adsorption experiments showed that the nZVI greatly enhanced the sorption affinity of SBC towards ROX. Kinetics model simulation results indicated that adsorption of ROX occurred via chemisorption, which involved valence forces through the sharing or exchange of electrons between nZVI@SBC and ROX. The overall rate-limiting step of the adsorption process was controlled by intraparticle diffusion and film diffusion.

3.3 *Mechanisms of the enhanced removal of ROX by nZVI@SBC*

The sorption of ionizable and less polar organic compounds (e.g., arsanilic acid and ROX) on sorbents such as activated carbon and iron oxides is usually controlled by noncovalent interactions (π–π stacking and hydrogen bonding), or surface complexation/coordination, and/or Lewis acid/base interactions. (Chen *et al.*, 2012). Our results revealed that the sorption of ROX by nZVI@SBC was mainly dominated by As-Fe coordination between arsenate moieties of ROX and nZVI and π–π interactions between benzene rings and aromatic C = C sites of SBC. Besides, the redox transformation of ROX during the reaction process was also contributed to the removal of ROX, the results of HPLC/ESI-QTOF-MS indicated that a portion of ROX molecules were transformed into 2-nitro-hydroquinone, hydroxyquinol and hydroquinone through hydroxylation and/or electrophilic addition reactions.

4 CONCLUSIONS

The transformation of PSAAs in aqueous solutions or manure leachates can lead to widespread contamination by toxic inorganic arsenic species. Therefore, the strategies used to eliminate PSAAs should focus on reducing/preventing arsenic leaching. In this study, an nZVI@SBC nanocomposite was designed to simultaneously capture arsenate and the benzene rings of ROX molecules from aqueous solution. The dual-affinity sorbent exhibited superior adsorption performance and affinity for ROX molecules compared to pristine SBC, demonstrating the potential of dual-affinity sorbents to reduce hazards related to arsenic leaching. The coordination and degradation abilities of the nanocomposite toward ROX molecules simultaneously attracted degraded inorganic arsenic species and ROX molecules. The mechanism of ROX removal by nZVI@SBC was proposed based on hydrogen bonding, As–O–Fe coordination, and π–π* EDA interactions. The results shed new light on the design of adsorbents for organic/inorganic molecules.

ACKNOWLEDGEMENTS

The authors are grateful for the financial support provided by the National Natural Science Foundation of China (41671475), the Ministry of Science and Technology of China (2018YFD0800700) and the Excellent Doctoral Dissertation Cultivation Fund of Hunan Agricultural University (YB2018004).

REFERENCES

Chen W.R. & Huang C.H. 2012. Surface adsorption of organoarsenic roxarsone and arsanilic acid on iron and aluminum oxides. *J. Hazard. Mater.* 227–228: 378–385.

Wang S., Gao B., Li Y., Creamer A.E. & He F. 2017. Adsorptive removal of arsenate from aqueous solutions by biochar supported zero-valent iron nanocomposite: batch and continuous flow tests. *J. Hazard. Mater.* 322: 172–181.

Wei D., Li B., Huang H., Luo L., Zhang J., Yang Y., Guo J., Tang L., Zeng G. & Zhou Y. 2018. Biochar-based functional materials in the purification of agricultural wastewater: fabrication, application and future research needs. *Chemosphere* 197: 165–180.

Xu T., Kamat P.V., Joshi S., Mebel A.M., Cai Y. & O'Shea K.E. 2007. Hydroxyl radical mediated degradation of phenylarsonic acid. *J. Phys. Chem. A* 111 (32): 7819–7824.

Arsenic in the Environment: Bridging Science to Practice for Sustainable Development –
van der Wal, Ahmad, Petrusevski, Weijma, Savic, van der Wens, Beerendonk, Bhattacharya, Bundschuh & Naidu (Eds)
© 2023 the Author(s), ISBN 978-1-032-32928-4

A promising stable chitosan-Fe (III) adsorbent for arsenate removal from drinking water

S. Todd-Supuy[1], R. González-Rodríguez[2], A. Caballero-Chavarria[2], O. Rojas-Carrillo[3] & L.G. Romero-Esquivel[2]

[1]*Licenciature student, Faculty of Engineering, School of Chemical Engineering, University of Costa Rica, San José, Costa Rica*
[2]*Environmental Protection Research Center (CIPA), School of Chemistry, Instituto Tecnológico de Costa Rica, Cartago, Costa Rica*
[3]*School of Chemistry, Universidad Nacional de Costa Rica, Heredia, Costa Rica*

ABSTRACT: Chitosan-Fe (III) beads have been reported as good adsorbents for arsenic. NaOH is normally reported as iron hydroxide precipitating agent and high amounts of washing water is needed to stabilize the pH. In this study, NH_4OH was used, as it evaporates easily, a quick rinsing was enough to get neutral pH. Batch kinetic and isotherms tests with 1 mg/L As (V) solutions were carried out. The adsorption process was slow (5d for 80% removal) and followed a pseudo-second order kinetic model, meaning that chemical sorption was the rate-limiting step. However, the adsorption in the kinetic study and the Langmuir adsorption capacity were 8.69 mg/g and 16.22 mg/g respectively. Furthermore, the calculation of adsorption capacity in equilibrium with 10 µg/L (q_{10}) using Freundlich parameters gave 3.15 mg/g. This means that the material has a similar As (V) adsorption capacity (mg/g) to the ones reported for iron-chitosan beads and others adsorbents. Therefore, the material is promising, and further fixed bed studies are recommended.

1 INTRODUCTION

Removal of arsenic from water by adsorption is widely applied. Among the adsorbents, chitosan-iron (oxyhydr)oxide beads have been reported as effective due to the formation of a very stable Fe:As complex (De Marques Neto *et al.*, 2013). Normally, beads are produced by dropping a chitosan-iron solution into aqueous sodium hydroxide (He *et al.*, 2016). However, the beads produced need high amounts of washing water to remove the residual alkaline reagent to obtain the beads with neutral pH. Besides, depending on the procedure, swollen and unstable beads can be obtained. In this study, chitosan-Fe(III) material, stable and with neutral pH, were obtained by using NH_4OH as precipitating solution. Besides the improvement in the method, the aim of the study was to evaluate the adsorption capacity of the material by batch adsorption and kinetic experiments.

2 MATERIALS AND METHODS

2.1 *Chitosan-Fe(III) beads synthesis*

Chitosan powder (2.5% w/v) was dissolved in acetic acid (0.5%) and commercial $FeCl_3$ was added (0.12 M). The spherical shape beads were obtained by dropping the solution into NH_4OH (1.75% v/v) under continuous stirring using a magnetic stirrer. After 3 h, the beads were collected, and rinsed with a small amount of water since the evaporation of the ammonia result in beads with almost neutral pH. Finally, the beads were oven dried at 45°C overnight.

2.2 *Kinetic adsorption experiments*

Duplicate adsorption experiments were conducted in an orbital shaker at 20°C and 130 rpm. Tap water was spiked with 1 mg/L As(V) and placed in contact with 0.1 g/L of the chitosan-Fe(III) beads for 8 days. The pH was adjusted to 7.0 using HCl or NaOH. Samples were taken after 4 h and every 24 h thereafter. The data was fitted using the pseudo first and pseudo second order models (Ngah *et al.*, 2005), respectively (Equations 1 and 2).

$$\log(q_e - q_t) = \log q_e - \frac{k_1 t}{2.303} \tag{1}$$

$$\frac{t}{q_t} = \frac{1}{k_2 q_e^2} + \frac{t}{q_e} \tag{2}$$

where q_e and q_t are the amount of As(V) adsorbed on adsorbent (mg/g) at equilibrium and at time t, respectively. k_1 and k_2 are the rate constants for pseudo-first order adsorption (/min) and pseudo-second order adsorption (g/mg/min) respectively.

2.3 *Batch adsorption studies*

Adsorption experiments were carried out by shaking 0.5 L of 1 mg/L As(V) and pH 7.0 solutions in an orbital shaker at 130 rpm and 25°C for 7 days. The dose of adsorbent was varied from 0.02 g/L up to 0.1 g/L. All adsorption experiments were conducted in duplicate. The data were fitted using the linearized Freundlich and Langmuir isotherm models according to equations (4) and (5), respectively.

$$\log(q_e) = \log(K_F) + \left(\frac{1}{n}\right)\log(C_e) \quad (4)$$

$$\frac{1}{q_e} = \frac{1}{Q_m} + \frac{1}{Q_m \quad b}\left(\frac{1}{C_e}\right) \quad (5)$$

where q_e (mg/g) and C_e (mg/L) are the adsorption capacity and the concentration of As(V) in the equilibrium. K_F and 1/n are the Freundlich parameters related to adsorption capacity and adsorption intensity, respectively. The parameter b is the Langmuir sorption equilibrium constant (L/mg) and Q_m is the maximum adsorption capacity (mg/g), which corresponds to the monolayer adsorption capacity.

2.4 *Analytical methods*

The arsenic measurement was performed following the 3113A method (APHA, AWWA, & WEF, 2005) with a Perkin Elmer AAnalyst 800 (DL: 2.5 µg/L and QL: 5 µg/L).

3 RESULTS AND DISCUSSION

3.1 *Synthesis of the chitosan-Fe(III) beads*

In contrast to others beads procedures (Gupta *et al.*, 2009; He *et al.*, 2016; Padilla-Rodríguez *et al.*, 2014), in this case, NH_4OH quickly evaporates, therefore, excess base is removed easily. Although the formulated material swelled around 85% in contact with water, it maintained an appropriate mechanical stength for use in filtration systems.

The formulated beads also presented a uniform rounded shape with a mean grain size of 1.2 mm, making them suitable for column packaging.

3.2 *Adsorption kinetics*

The arsenic adsorption by the beads was relatively slow (Figure 1), only around 10% was adsorbed in the first 4h of the experiment. Moreover, the equilibrium and maximum adsorption capacity were reached after around 120 h (5 d) with 80% removal. Contrary, Marques Neto *et al.* (2013) and He *et al.* (2016) reported similar chitosan iron materials, with and without glutaraldehyde, more than 90% removal in 2h and 24h, respectively. Even though the differences are significant, it is important to consider that both were performed at higher temperature (25°C instead of 20°C) and using higher initial concentration, 75 mgAs/L and 4.9 mgAs/L versus 1 mgAs/L.

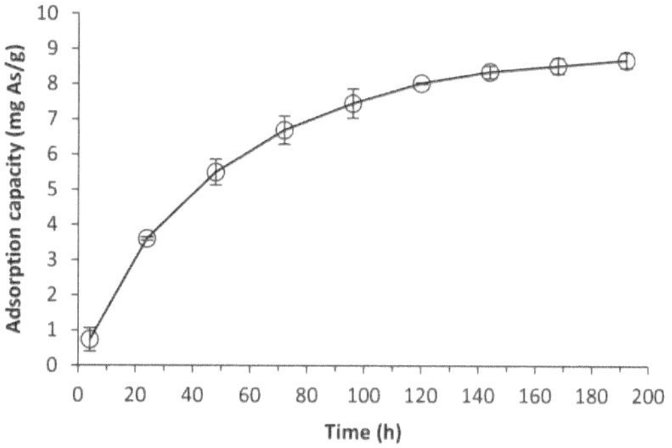

Figure 1. Kinetics of As(V) removal (initial As(V) concentration 1 mg/L, adsorbent dose 0.1 g/L, shaking for 8 days at 20°C).

Regarding the kinetic modelling the data adjusted better to the pseudo second order (R^2 0.997) than the pseudo first (R^2 0.922) suggesting that the rate-limiting step may be chemical adsorption (Ngah *et al.*, 2005). The experimental adsorption capacity (q_e = 8.69 mg/g) was close to the calculated one (q_e = 11.25 mg/g), confirming the pseudo second order model. The capacity obtained was higher than 2.05 mg/g and lower than 13.24 mg/g reported by He *et al.* (2016) and Marques Neto *et al.* (2013) respectively. The rate constant ($k_2 = 2.86 \times 10^{-5}$ g/(mg·min), which was two orders of magnitude lower than the studies of Marques Neto *et al.* (2013) and He *et al.* (2016), 7.09×10^{-3} g/(mg·min) and 5.84×10^{-3} g/(mg·min), respectively, confirms that the adsorption process is slower. However, the high adsorption capacity of this experiment and the isotherms (below) confirms that this material has potential and column studies are suggested.

3.3 *Equilibrium isotherms*

The data fitted well both Langmuir (R^2 0.962) and Freundlich (R^2 0.966) models (Figure 2). The Langmuir monolayer maximum capacity, Q_m of 16.22 mg/g, was about eight times higher than previous studies for similar iron chitosan granular materials. Gupta *et al.* (2009) and Padilla-Rodríguez *et al.* (2014) reported 2.24 mg/g and 2.72 mg/g, respectively.

The Freundlich isotherm, with K_F of 15.38 $(mg/g)(L/mg)^{1/n}$ and 1/n of 0.345, indicates good

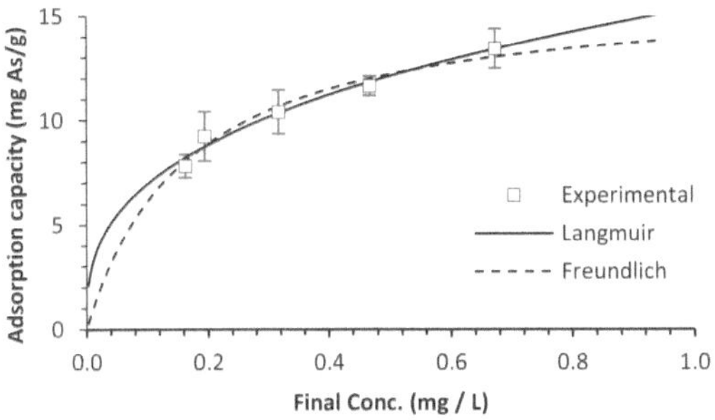

Figure 2. Adsorption isotherms (initial As (V) concentration 1 mg/L, adsorbent dose 0.02–0.1 g/L, 130 rpm, 7 day, 20°C).

adsorption strength and relatively high adsorbent loading at low concentrations, showing favorable isotherms (Worch 2012). Besides, using the Freundlich parameter, a q_{10} adsorption capacity in equilibrium with 10 μgAs /L was calculated. The q_{10} of 3.15 mg/g obtained is higher to 0.42 mg/g calculated by using the data of Padilla *et al.* (2014) for iron chitosan beads. Accordingly, with Sinha *et al.* (2011) the high capacity obtained in the present study is comparable to magnetic ion exchange resins MIEX and granular ferric hydroxide GFH with 3.16 mg/g and 2.51 mg/g at pH 5.5 and 7.5 respectively.

4 CONCLUSIONS AND RECOMMENDATIONS

The synthesized beads were stable, and a small amount of washing water was needed to have a constant pH. The adsorption process was rather slow needing up to 5 days for 80% removal. However, both the pseudo second order kinetic model, and the Langmuir and Freundlich isotherms results showed very high adsorption capacity, superior or similar to previous chitosan beads and other types of arsenic adsorbents. Therefore, the results indicate a promising material and fixed bed filtration evaluation is recommended.

ACKNOWLEDGEMENTS

This research was funded by the Research and Extension Council of ITCR, Costa Rica project no. 1460-044.

REFERENCES

De Marques Neto J.O., Bellato C.R., Milagres J.L., Pessoa K.D. & De Alvarenga E.S. 2013. Preparation and evaluation of chitosan beads immobilized with Iron(III) for the removal of As(III) and As(V) from water. *J. Braz. Chem. Soc.* 24: 121–132.

Gupta A., Chauhan V.S. & Sankararamakrishnan N. 2009. Preparation and evaluation of iron–chitosan composites for removal of As(III) and As(V) from arsenic contaminated real life groundwater. *Water Res.* 43: 3862–3870.

He J., Bardelli F., Gehin A., Silvester E. & Charlet L. 2016. Novel chitosan goethite bionanocomposite beads for arsenic remediation. *Wat. Res.* 101: 1–9.

Ngah W.S.W., Ab Ghani S. & Kamari A. 2005. Adsorption behaviour of Fe(II) and Fe(III) ions in aqueous solution on chitosan and cross-linked chitosan beads. *Bioresour. Technol.* 96: 443–450.

Padilla-Rodríguez A., Perales-Pérez O. & Román-Velázquez F.R. 2014. Removal of As(III) and As(V) oxyanions from aqueous solutions by using chitosan beads with immobilized iron(iii). *Int. J. Hazard. Mater.* 2: 7–17.

Sinha S., Amy G., Yoon Y., Her N., 2011. Arsenic removal from water using various adsorbents: magnetic ion exchange resins, hydrous ion oxide particles, granular ferric hydroxide, activated alumina, sulfur modified iron, and iron oxide-coated microsand. *Environ. Eng. Res.* 16(3): 165–173.

Worch E. 2012. *Adsorption Technology in Water Treatment. Fundamentals, Processes, and Modeling.* Walter de Gruyter GmbH & Co., Göttingen, Germany.

4.4 Membrane technologies and applications

Arsenic in the Environment: Bridging Science to Practice for Sustainable Development –
van der Wal, Ahmad, Petrusevski, Weijma, Savic, van der Wens, Beerendonk, Bhattacharya, Bundschuh & Naidu (Eds)
© 2023 the Author(s), ISBN 978-1-032-32928-4

Chitosan/PVA electrospun nanofiber membranes for the adsorption of As(V) from water

J. Cimadoro[1], J. Palatnik[2], N. Torasso[1], A. Londonio[2,3], S. Cerveny[4], P. Smichowski[2,5] & S. Goyanes[1,5]
[1]*Universidad de Buenos Aires. Facultad de Ciencias Exactas y Naturales, Departamento de Física, Laboratorio de Polímeros y Materiales Compuestos (LP&MC), Instituto de Física de Buenos Aires (IFIBA-CONICET), Buenos Aires, Argentina*
[2]*Comisión Nacional de Energía Atómica (CNEA), Gerencia Química, Buenos Aires, Argentina*
[3]*Universidad de San Martín (UNSAM), 3iA, Argentina*
[4]*Centro de Física de Materiales (CSIC, UPV/EHU)-Materials Physics Center (MPC), San Sebastián, Spain*
[5]*Consejo Nacional de Investigaciones Científicas y Técnicas (CONICET), Buenos Aires, Argentina*

ABSTRACT: In this study, chitosan/poly(vinyl alcohol) (CS/PVA) 50/50 nanofiber membranes were fabricated by the electrospinning technique. These membranes were subjected to a thermal treatment to achieve water stability. Membrane morphology was characterized by FE-SEM and its thermal properties by modulated DSC. Preliminary tests of AS(V) adsorption for concentrations between 0.1 and 5 mg/L were evaluated using HG-AAS. These membranes adsorbed more than 80% of the As(V) in aqueous solution, regardless of the starting concentration and in a less than 30 min. The results obtained agree with previous studies reported in literature for chitosan-based adsorbents.

1 INTRODUCTION

In Latin America, high levels of arsenic (As) have been found in drinking water in 14 of its 20 countries. It is the main contaminant of water in Argentina, where concentrations of As(V) above the WHO limit have been detected in 16 of its 23 provinces.

One way to remove As(V) from water is through its retention with amino groups, such as those present in chitosan. Chitosan (CS) has been studied for water treatment due to its biodegradability, non-toxicity, biocompatibility and the existence of hydroxyl in its structure, making it a low-cost and recyclable sorbent for organic dyes or heavy metal ions through electrostatic interaction or chelation.

In the last decades, electrospinning technique attracted attention in the field of water treatment because it produces nanofiber membranes with a large surface-to-volume ratio, pore sizes within the nano range, a flexible structure and high porosity (Cimadoro *et al.*, 2020). In particular, it is possible to manufacture membranes from mixtures of chitosan with other polymers and to retain As. In order to improve the spinnability of CS in dilute acetic acid, blending with poly(vinyl alcohol) (PVA) is an efficacious way. PVA is a non-toxic, chemically stable and water-soluble biodegradable polymer (Mei *et al.*, 2019). In the present study, CS/PVA 50/50 nanofiber membranes were fabricated via the electrospinning method using 10% w/w acetic acid in water as solvent and then tested as an As(V) adsorbent.

2 MATERIALS AND METHODS

2.1 *Materials*

CS (Chitosan from shrimp shells DDA $\geq$ 75%), PVA (Mowiol® 10-98) and Glacial Acetic Acid ($\geq$ 99% ReagentPlus®) were obtained from Merk (Argentina).

2.2 *Preparation of CS/PVA membranes*

Hydrolysis was performed to increase the degree of deacetylation and decrease the molecular weight of chitosan (Chitosan from shrimp shells DDA $\geq$ 75%). First, 40 g of NaOH was added to 80 g of distilled water. Subsequently, 2.5% w/w of chitosan was added to the solution. The solution was stirred for 24 h at 90°C. Afterward, the samples were filtered and washed by distilled water. Finally, the filtered chitosan powders were dried in an oven at 60°C for 12 h.

Then, a 3% w/w CS solution was prepared by dissolving CS in aqueous acetic acid solution at room temperature under a magnetic stir for 12 h and a 3% w/w PVA solution was prepared by dissolving PVA in water at 90°C under stirring for 2 h. Then, PVA solution was mixed with CS solution in a weight ratio of 50/50 and stirred for 3 h at room temperature.

The electrospinning equipment was reported in previous work (Cimadoro *et al.*, 2018). Solutions contained in 10 mL syringes were pushed using a syringe pump with 0.5 mL/h feed rate through a six 21gauge needles (0.9 mm inner diameter) multi-needle device. An electrical voltage of 30 kV was applied to the needle using a

DOI: 10.1201/9781003317395-154

high-voltage supply unit (max 20 mA). The polymeric solution ejected from the needles tips falls on a rotating drum of 6 cm diameter covered by aluminium foil, located at 10 cm from the tips. After 3 h, a mat 40–50 μm thick was obtained. These membranes then undergo a thermal treatment at 190°C for 10 min to achieve water stability (Cimadoro *et al.*, 2020).

2.3 *Membranes characterization*

Morphology of CS/PVA nanofibrous membrane, was determined using a field emission scanning electron microscope. Thermal characterization of the CS/PVA membranes was measured with a modulated DSC heating at 3 K/min with ± 0.5 K modulation every minute.

2.4 *Adsorption studies*

Batch experiments were performed to test the adsorption capacity (Q) of the membranes. The initial concentration of As(V) was varied between 0.1 and 5 mg/L at an initial pH of 3.5. The adsorbent dose was 0.5 g/L and final concentration was measured after 30 minutes by HG-AAS. Adsorption capacity was calculated as:

$$Q = V(C_0 - C)/m \tag{1}$$

where C_0 and C are the initial and final concentrations of As(V) in solution, V is the solution volume and m is the adsorbent mass.

3 RESULTS AND DISCUSSION

Microscopy studies (Figure 1) reveal that nanofibers surface is smooth, without beads and with a mean diameter of (82 ± 10) nm. This value is lower than (110 ± 13) nm reported in the literature for CS/PVA 50/50 membranes (Mei *et al.*, 2019) and lower than (125 ± 25) nm for 100% PVA nanofibers (Cimadoro *et al.*, 2020).

In DSC studies, reversing heat flow indicated a glass transition temperature (Tg) of CS/PVA membrane at (335 ± 1) K. Reversing hat flow component was necessary because the presence of moisture masks the Tg in the total heat flow curve. Tg for pure PVA electrospun membranes is (349 ± 1) K, which indicates that the measured Tg for CS/PVA membrane is possibly associated to its PVA rich phase.

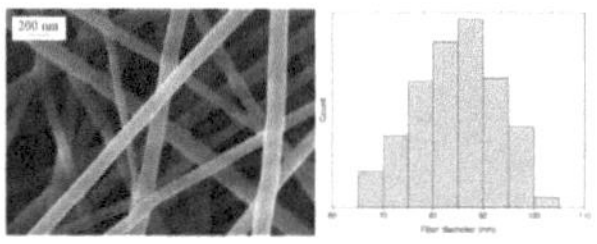

Figure 1. SEM image of nanofiber membrane and nanofiber diameter distribution.

During adsorption studies, the PVA rich zones of the membrane swell because of water absorption. This process lowers the Tg because water acts as a plasticizer and changes the state of the membrane from glassy to rubbery during batch experiments despite being at room temperature. This is confirmed by the flexibility of the membrane in wetting conditions.

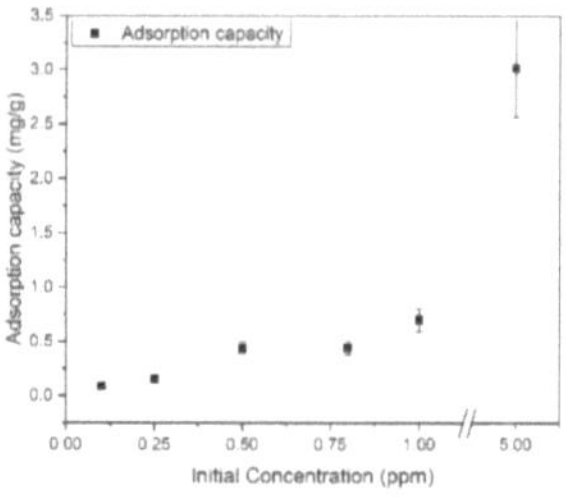

Figure 2. As(V) adsorption capacity vs initial concentration.

Figure 2 shows the adsorption capacity of CS/PVA membrane for initial concentrations between 0.1 and 5 mg/L, giving a maximum value of 3 mg/g, in accordance to other works (Min *et al.*, 2015). As expected from Le Chatelier's principle, adsorption capacity increases with initial concentration of As(V). It is important to remark the fast adsorption kinetics (less than 30 minutes), suggesting the presence of amino groups in the fibers surface. Removal rates varied in between 80 and 100% depending on the initial concentration.

4 CONCLUSIONS

Membranes obtained in this work have relatively low fiber dimeter than those reported so far in literature and with a high CS/PVA ratio. Preliminary results show that their capacity to adsorb As(V) is at least 3 mg/g, with removal rates between 80 and 100%. These results are encouraging and show the possibility of obtaining membranes for As in water removal by a novel method.

ACKNOWLEDGEMENTS

We acknowledge the support of UBA (UBACYT 2018-2020 No 20020170100381BA), ANPCyT (PICT 2017-2362, PICT 3590), PIP 00078 and linkB20012.

REFERENCES

Çay A., Miraftab M. & Kumbasar E.P.A. 2014. Characterization and swelling performance of physically stabilized electrospun poly (vinyl alcohol)/chitosan nanofibres. *Euro. Polym. J.* 61: 253:262.

Cimadoro J. & Goyanes S. 2020. Reversible swelling as a strategy in the development of smart membranes from electrospun polyvinyl alcohol nanofiber mats. *J. Polym. Sci.* 1: 10.

Cimadoro J., Ribba L., Ledesma S. & Goyanes S. 2018. Electrospun mats: From white to transparent with a drop. *Macromol. Mater. Eng.* 303(10): 1800237.

Mei Y., Runjun S., Yan F., Honghong W., Hao D. & Chengkun L. 2019. Preparation, characterization and kinetics study of chitosan/PVA electrospun nanofiber membranes for the adsorption of dye from water. *J. Polym. Eng.* 39(5): 459–471.

Min L.L., Yuan Z.H., Zhong L.B., Liu Q., Wu R.X. & Zheng Y.M. 2015. Preparation of chitosan based electrospun nanofiber membrane and its adsorptive removal of arsenate from aqueous solution. *Chem. Eng. J.* 267: 132–141.

Arsenic in the Environment: Bridging Science to Practice for Sustainable Development –
van der Wal, Ahmad, Petrusevski, Weijma, Savic, van der Wens, Beerendonk, Bhattacharya, Bundschuh & Naidu (Eds)
© 2023 the Author(s), ISBN 978-1-032-32928-4

Removal of arsenic from water by liquid membrane based separation technology

S. Sarkar & P. Saha
Department of Chemical Engineering, Indian Institute of Technology Guwahati, Assam, India

ABSTRACT: It is necessary to remove arsenic from drinking water, as it is very harmful, toxic and carcinogenic in nature. However, there are several separation techniques for removal of arsenic from water; liquid membrane separation technique is being studied for its selective and high separation factor. Two-phase equilibrium study has been done to identify the environmentally benign organic solvent and extractant for the extraction of arsenic ions from its aqueous solution. Extraction of As(III), As(V) and a combination of both As(III) and As(V) in different ratios Have been studied. The effect of extraction on variation of parameters including extractant concentration, pH of feed phase, stirrer speed have been studied. The experiments were designed with the central composite design of response surface methodology to get an optimum range of input variables for maximum extraction of arsenic ions. The objective is to design stable flat sheet supported liquid membrane based technique for arsenic removal for purification of drinking water. The solute-selective nature, ability for uphill transport and high separation factor makes liquid membrane separation technique an efficient way of removing arsenic from drinking water.

1 INTRODUCTION

The toxicological effects of arsenic has led to an exponential rise in scientific research for its removal in the past few decades. There are several separation techniques for removal of arsenic from water including adsorption, oxidation, filtration, etc. Liquid membrane based separation technique (LMST) is being studied for its selective and high separation factor. In LMST, there is an organic phase immiscible with water containing extractant or complexing agent that is selective towards the target element. It enhances the extraction process by temporarily binding with the target element present in the source phase and releasing it to the organic phase or receiving phase (in case of three-phase LMST) due to concentration gradient. Thus, the target element can be extracted and recovered simultaneously in a single step. N. N. Li. has done few studies in the removal or extraction of arsenic with the development of LMST. The use of vegetable oil as liquid membrane in extraction is being explored for its environment friendly nature, having less toxicity in comparison to conventional chemical solvents.

The aim of this research work is to: i) identify environmentally benign solvent preferably vegetable oils that is easily available, non-toxic and low cost for extraction of As(III), As(V) and combination of both As(III) and As(V) from aqueous solution; ii) identify extractant suitable for enhancing the transport of solute from the feed phase (aqueous phase) through the solvent- extractant pair acting as the intermediate organic membrane phase. Extractants or carrier compounds used in this work are methyltrioctyl ammonium chloride (Aliquat 336), di-2-ethylhexyl phosphoric acid (D2EHPA), tributyl phosphate (TBP) and trioctylamine (TOA); and iii) determine the operating conditions for two-phase equilibrium study through variation in parameters.

2 MATERIALS AND METHODS

2.1 *Experimental procedure*

Equal volume of feed phase and organic liquid membrane solutions were taken in a conical flask and placed in the shaking incubator for mixing to reach equilibrium. The feed phase consisted of 150ppb concentration of As(III), As(V) and both As(III) and As(V) in 1:1, 1:2 and 2:1 ratio. The mixture was then kept undisturbed for few hours until the aqueous and organic phases separated into two distinct layers due to their different densities. The samples collected from feed (aqueous) phase were analyzed in atomic absorption spectrophotometer (AAS) to find the concentration of arsenic ions in the feed phase and then calculate the amount extracted in to the organic phase by mass balance. the percentage of extraction is calculated by the following equation.

$$\text{Extraction\%} = (i - c)/i \times 100 \tag{1}$$

where, i is the initial concentration of arsenic ions in the feed phase and c is the concentration of the feed phase at a time t.

2.2 *Chemicals and reagents*

Sodium arsenite and sodium arsenate used for preparing feed phase solution was supplied from Merck® India. For Atomic Absorption Spectrophotometric (AAS) analysis, arsenic standard solution (1000 ppm) supplied from Merck® India was used. Solutions were prepared by dissolving required amount in Milli-Q® deionized water (Millipore, USA). Refined vegetable oils of good quality (Fortune, Adani Wilmer Limited) such as coconut oil,

DOI: 10.1201/9781003317395-155

mustard oil, sunflower oil, soybean oil and sesame oil were used as green solvents. Different extractants like tributyl phosphate (TBP), trioctylamine (TOA), methyltrioctyl ammonium chloride (Aliquat 336) and di-2-ethylhexyl phosphoric acid (D2EHPA) were procured from Sigma Aldrich India. All the chemicals and reagents used are of Guaranteed Reagents (GR) grade.

2.3 *Theory*

In the feed phase, As(III) ions are mostly in the undissociated state as H_3AsO_3 while As(V) ions are in dissociated forms as $H_2AsO_4^-$ and $HAsO_4^{2-}$. These react with aliquot in the feed-membrane interphase to form complexes as given in the following equations.

$$H_3AsO_3 + NR + Cl^- \leftrightarrow NR_4Cl(H_3AsO_3) \quad (2)$$

$$H_2AsO_4^- + NR + Cl^- \leftrightarrow NR_4(H_2AsO_4) + Cl^- \quad (3)$$

$$HAsO_4^{2-} + 2NR_4 + Cl^- \leftrightarrow (NR_4)2(H_2AsO_4) + 2Cl^- \quad (4)$$

2.4 *Design of experiments*

A set of experiments were prepared with central composite design of response surface methodology by the variation of all relevant factors simultaneously. It creates an entity of experiments that work together to explore the region of optimal solutions. This helps in understanding the influence of and interaction between all factors. It estimates the optimum operating conditions through organized approach with fewer experiments and minimization of cost. It consists of the following three steps:

Step 1: Design, which includes the input variables or factors, the minimum and maximum range of these factors. This creates a set of experiments to be performed.
Step 2: Analysis includes evaluation and statistical analysis of the model by ANOVA.
Step 3: Optimization that generates numerical solutions for optimum range of input variables, which would give a maximum output.

3 RESULTS AND DISCUSSIONS

Sesame oil and aliquat 336 was selected as the solvent and the extractant, respectively as given in Figures 1 and 2. The significant input factors predicted by the model for extraction of As(III) and As(V) includes extractant concentration. pH plays a significant role for extraction of arsenic ions when combination of both As(III) and As(V) is present in water. 80–88% extraction of arsenic ions is achieved with feed phase pH in the range of 9–10. As(III) is available in the form of undissociated H3AsO3 and As(V) is in the dissociated form of HAsO42- within this pH range and both of these arsenic species readily form complex with aliquat. The predicted duration of operation for maximum extraction lies within the range of 2–12 hours. The predicted stirring speed for maximum extraction

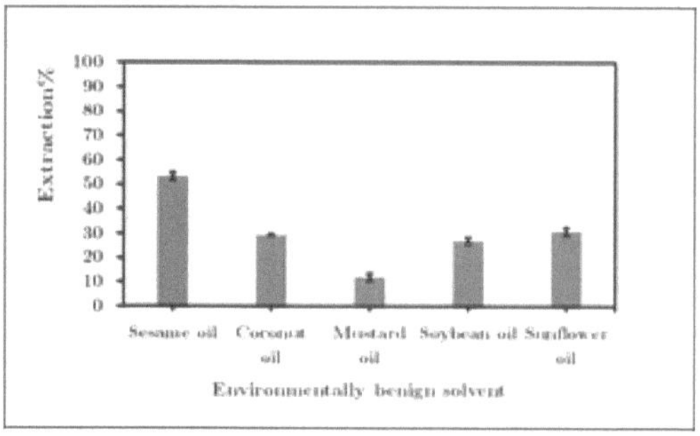

Figure 1. Selection of solvent from different vegetable oil used in two-phase study.

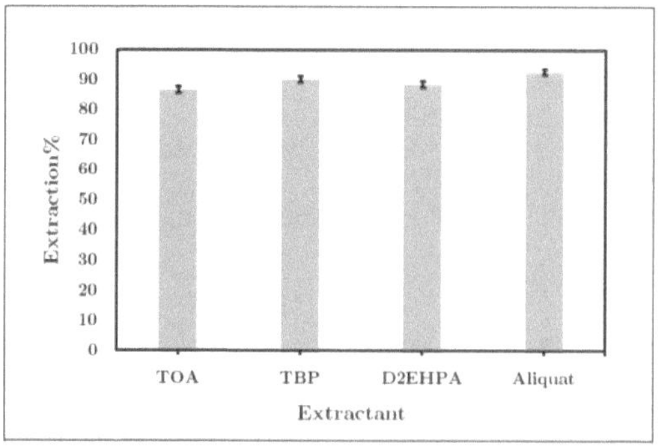

Figure 2. Selection of extractant from two-phase study.

varies from 85–150 rpm that lies within the selected initial range of 50–250 rpm.

4 CONCLUSIONS AND RECOMMENDATIONS

The significant input factors predicted by the model for extraction of As(III) and As(V) includes extractant concentration. pH plays a significant role for extraction of arsenic ions when combination of both As(III) and As(V) is present in water. 80–88% extraction of arsenic ions is achieved with feed phase pH in the range of 9–10. Based on this study, three phase FSSLM has been conducted to improve the removal percentage of arsenic from water.

REFERENCES

Li N.N., Cahn R.P. & Shrier A.L. 1973. Liquid membrane process for the separation of aqueous mixtures. *US. Patent 3.779.907*.

Li N.N. 1968. Liquid separation process. *U.S. Patent 3.410.794*.

Pancharoen U., Poonkum W. & Lothongkum A.W. 2009. Treatment of arsenic ions from produced water through hollow fiber supported liquid membrane. *J. Alloys Compd.* 482(1–2): 328–334.

4.5 Biological processes in arsenic removal

Arsenic in the Environment: Bridging Science to Practice for Sustainable Development – van der Wal, Ahmad, Petrusevski, Weijma, Savic, van der Wens, Beerendonk, Bhattacharya, Bundschuh & Naidu (Eds)
 ISBN 978-1-032-32928-4

Molecular characterization of arsenite oxidizing bacteria for water treatment in rural communities of the state of Guanajuato, Mexico

U.E. Rodríguez Castrejón[1,2], A. Serafín Muñoz[1,2], G. Cruz Jiménez[3], C. Cano Canchola[3] & A. Álvarez Vargas[3]

[1]*Postgraduate Degree in Water Sciences, University of Guanajuato, Guanajuato, Mexico*
[2]*Division of Engineering, University of Guanajuato, Guanajuato, Mexico*
[3]*Division of Natural and Exact Sciences, University of Guanajuato, Guanajuato, Mexico*

ABSTRACT: This work focuses on the mining town of Xichu, Guanajuato, where water has been contaminated by arsenic (As) due to the fact that its river passes by the Aurora mine tailings. In the dry season a high concentration of As (98 µg L-1) has been detected, exceeding the limits established by the WHO (10 µg L^{-1}) and the Mexican norm NOM-127-SSA1 (25 µg L^{-1}). Currently, due to the lack of economic resources in this rural area cannot be apply a conventional water treatment as chemical and physical methods are. Bioremediation of waters using heavy metal resistant bacteria-based technologies may provide a better alternative. It was statistically corroborated the capacity of 3 bacteria strains to interact with arsenite, showing their possible application during the oxidation process of arsenite or as a bioaccumulation process.

1 INTRODUCTION

Arsenic (As) contamination in drinking water represents one of the most challenges to human health. Around the world, it is estimated that more than 100 million people are at risk of drinking water contaminated with As (Nordstrom 2002). The situation in Latin America is serious, due to the few studies and publications that reflect the real problem about arsenic in water. For instance, in Mexico is estimated that there are around 2 million population exposure to arsenic water contamination (Armienta *et al.*, 2008). The situation becomes more complex in rural areas, due to the lack of economic resources, in this context, in Mexico is estimated that there are around of 500,000 population in rural communities drinking contaminated water with arsenic, it which is above the international and Mexican normativity. Despite exist several successfully physicochemical water treatment systems to remove As from water such as, coagulation/filtration, ion exchange, etc. these it cannot be applied in rural communities due to they are very expensive and polluting, for this reason the project have been focused in biological treatments, since they do not produce environmental impacts and they are relatively cheaper. Although the high arsenic toxicity, some microorganisms are able to resist high As levels and interact with it by different forms. Among the possible microbial As-transformations, are the redox reactions involving As(III) oxidation and As(V) reduction the most investigated for bioremediation purposes (Kumari & Jagadevan 2016). Microbial As(III)-oxidation is one of the most promising mechanisms, as a precursor step in As removal from contaminated water, since conventional iron-based treatment are more effective in removing As(V) rather than As (III) (Fazi *et al.*, 2016). Microbial As(III) oxidation represents a detoxification process in heterotrophic microorganisms as *Herminiimonas arsenicoxydans* (Bahar *et al.*, 2012), or an energetic metabolism in chemolithoauthotrophic microorganisms, such as *Thiomonas arsenivorans* (Battaglia-Brunet *et al.*, 2006). Both oxidation mechanisms are carried out by the enzyme arsenite oxidase (Crognale *et al.*, 2019). It should be noted that before use any microorganisms for water treatment, is necessary to realize a full molecular characterization.

2 MATERIALS AND METHODS

2.1 *Study area*

The study area is located within the state of Guanajuato, Mex. Xichu is the municipally name. Here, there was mining activity since the sixteen century, leaving tons of mining waste around the principal body water or the region “the Xichu River”. Which have As concentration above 90 µg L^{-1} [8].

DOI: 10.1201/9781003317395-156

2.2 *Growth kinetics*

In order to observe the growth time of the isolates, 3 growth curves were planned, as described below:

Growth kinetics in M9 medium with 100 μgL^{-1} of As(III). Incubation conditions: temperature 30°C at 120 rpm for 8 days. Each assay was measured every 24 hours.

2.3 *Amplification of aox genes and search for arsenic resistance determinants*

The amplification of aox genes was carried out using the following methodology. PCR was mixed with Water miliQ, Buffer PCR (10X), MgCl2, dNTPs, and the initiators aox A and aox B (in the case of arsenite oxidase), and finally DNAg and Taq Polymerase (Invitrogen). The reaction was carried out in a Thermocycler (LabNet), at an alignment temperature of 60°C, ending with the Extension Phase at 72°C for 1 minute. Finally, a cycle was applied for a final extension of 10 min. at 72°C and finally a maintenance stage was programmed at 4°C. The extraction of plasmid DNA was carried out by the alkaline lysis method described by Birboim & Doly (1979).

2.4 *Relation between production of Polyhydroxybutarate (PHB) and the arsenite resistance*

The determination of PHB production was through Sudan Black Staining, and it was used an optic microscope 100X to observe the PHB production within the cell. Solution I (Sudan Black 0.33% (w/v) in 60% ethanol) and solution II (Safranin 0.5% (w/v) in aqueous solution).

2.5 *Statistical analysis*

The growth kinetics tests were performed in duplicate and a statistical analysis of the data was implemented using an ANOVA analysis of a factor, and the Tukey multiple comparisons method ($P < 0.05$ and level of significance of $\alpha = 0.05$). The above by using the MiniTab 2019 software. This to check the real growth of the strains and identify the differences in their growth in contact with As (III).

3 RESULTS AND DISCUSSION

3.1 *Growth kinetics*

According to the observed in the growth kinetics and with the ANOVA and Tukey methods (Figures 1 & 2), *R. gordoniae* was the strain with major capacity to interact with arsenite growing four times more than other strains. Nevertheless,

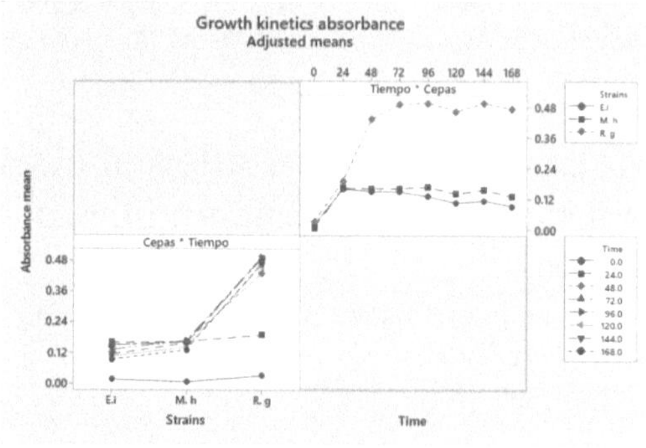

Figure 1. Growth kinetics absorbance with arsenite (adjusted mean). In color blue *E. indic*um, in red *M. hydrocarbonoxydans* and in green *R. gordoniae*.

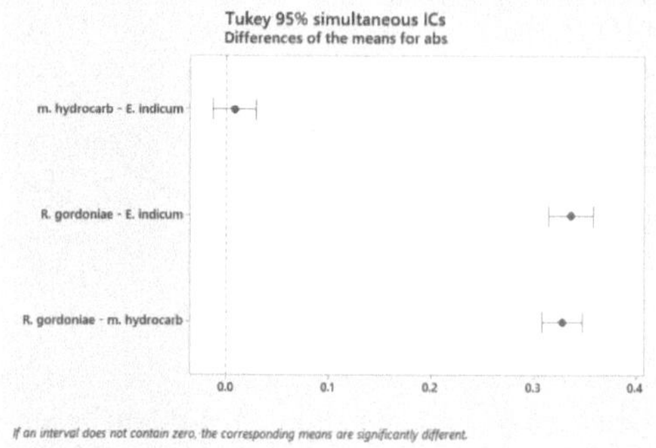

Figure 2. Tukey method of multiple comparisons. If an interval does not contain zero, the corresponding means are significantly different.

as much as *M. hydrocarbonoxydans* and *E. indicum* had an exponential growth in the first 24 hrs.

3.2 *Amplification of aox genes and search for arsenic resistance determinants*

As much as *E. indicum* and *M. hydrocarbonoxydans* were the two strains that expressed the aox genes.

On the other hand, it does not found plasmids in any strain in this study.

3.3 *Relation between production of PHB with the arsenite resistance*

R. gordonia and *E. indicum* were the strains that produced PHB ganules in presences of arsenite. It is important to note that, *R. gordoniae* was the strain more resistant to arsenite, and also the strain that produced major amount of PHB. As a result, this biopolymer is a by-product that can be used to produce a bioplastic.

4 CONCLUSIONS AND RECOMMENDATIONS

So far, 3 bacteria strains were statistically tested in their interaction with arsenite, corroborating their possible application to remove arsenic form water. Being *R. gordoniae* the strain with major growing in presences of arsenite, producing PHB granules as a result of this interaction.

As a recommendation, the application of synchrotron lights techniques, for instance, SEM/EDS, SR-FTIR and Microscope-energy dispersive X-ray spectroscopy would possible to know in which part of cell occurs the interaction with arsenite and what kind of interaction is.

ACKNOWLEDGEMENTS

The authors express their gratitude to the sponsors of the project the University of Guanajuato and CONACYT.

REFERENCES

Armienta M.A., Amat P.D., Larios T. & López D.L. 2008. América central y México. In: Bundschuh J., Pérez-Carrera A., Litter M.I. (eds): *Distribución del Arsénico en las Regiones Ibérica e Iberoamericana. Editorial Programa Iberoamericano de Ciencia y Tecnología Para el Desarrollo*, Buenos Aires, Argentina, 2008, 187–210. Available from: http://www.cnea.gov.ar/xxi/ambiental/Iberoarsen/docs/LIBROdistribucionarsenico.pdf., http://www.cnea.gov.ar.

Bahar M.M., Megharaj M. & Naidu R. 2012. *Arsenic Bioremediation Potential of a New Arsenite-oxidizing Bacterium Stenotrophomonas sp.* MM-7 Isolated from Soil. *Biodegradation* 23: 803–812.

Battaglia-Brunet F., Joulian C., Garrido F., Dictor M., Morin D., Coupland K., Johnson D.B., Hallberg K. B. & Baranger P. 2006. Oxidation of arsenite by Thiomonas strains and characterization of Thiomonas arsenivorans sp. nov. *Antonie Leeuwenhoek* 89: 99–108.

Crognale S., Casetini B., Amalfitano S., Fazi S., Petruccioli M. & Rossetti S. 2019. Biological As(III) oxidation in biofilters by using native groundwater microorganisms. *Sci. Total Environ.* 651: 93–102.

Fazi S., Amalfitano S., Casentini B., Davolos D., Pietrangeli B., Crognale S., Lotti F. & Rossetti S. 2016. Arsenic removal from naturally contaminated waters: a review of methods combining chemical and biological treatments. *Rend. Lincei* 27: 51–58.

Kumari N. & Jagadevan S. 2016. Genetic identification of arsenate reductase and arsenite oxidase in redox transformations carried out by arsenic metabolising prokaryotes. a comprehensive review. *Chemosphere* 163: 400–412.

Nordstrom D.K. 2002. Worldwide occurrences of arsenic in ground water. *Science* 296(5576): 2143–2145.

Arsenic in the Environment: Bridging Science to Practice for Sustainable Development – van der Wal, Ahmad, Petrusevski, Weijma, Savic, van der Wens, Beerendonk, Bhattacharya, Bundschuh & Naidu (Eds)

Extracellular polymeric substances reduce uptake but enhance transformation of arsenic in *Chlamydomonas reinhardtii*

S. Naveed[1], Y. Ge[1], C.H. Li[1], Z.Q. Jiang[1], Q.N. Yu[1], J.Y. Zhang[1] & C.H. Zhang[2]

[1]*Jiangsu Provincial Key Laboratory of Marine Biology, College of Resources and Environmental Sciences, Nanjing Agricultural University, Nanjing, P.R. China*

[2]*Laboratory Centre of Life Science, Nanjing Agricultural University, P.R. China*

ABSTRACT: Arsenic (As) pollution and toxicity in the environment is a serious health concern affecting the millions of individuals health worldwide. Extracellular polymeric substances (EPS) synthesized by microalgae are important metabolites in the "water-algae" interface and have good application prospects in the field of As wastewater treatment. In this study, the bioaccumulation and transformation of As in model green microalgae *Chlamydomonas reinhardtii* with and without EPS on cells were elaborated. The total accumulation and adsorption of As were enhanced and absorption was decreased in intact EPS cells compared to EPS removed cells. Oxidation-reduction of As species were higher in cells with EPS than those of EPS removed, implying that EPS have significant role for the environmental fate and behavior of As and could be potential biosorbent in As remediation by algae-based biotechnologies.

1 INTRODUCTION

Arsenic (As) is a ubiquitous toxic metal(loid) in the environment and microalgae have evolved various As detoxification mechanisms (Wang *et al.*, 2015). One of the strategies is the synthesis of extracellular polymeric substances (EPS), which are a complex mixture of interlaced biological polymers such as polysaccharides and proteins. EPS are enriched with carboxyl, phosphoryl, amine, hydroxyl and sulfhydryl functional groups and form a bio-barrier around the algal cells (Naveed *et al.*, 2019), which may affect the As uptake and transformation processes in microalgae cells and mobility in the environment through various interactions. However, the roles of EPS in the As accumulation and speciation in microalgae remain largely unknown. Therefore, this study was conducted with the objectives (1) to elucidate the As uptake by microalgae cells and (2) to assess the As speciation with and without their EPS under various As concentrations.

2 MATERIALS AND METHODS

2.1 *Microalgae cultivation and EPS extraction*

Strains of *Chlamydomonas reinhardtii* CC-125 wild type (137c) and CC-503 cell wall mutant (cw 92 mt^+) were obtained from Chlamydomonas reinhardtii Resource Center (http://chlamycollection.org) and cultivated in sterilized TAP (tris-acetate phosphate) medium with pH 7.0 ± 0.02 and 12/12 h day-night period at 25 ± 2°C for four days and EPS were removed from algal cells using 20 mM EDTA following the method described by Sheng *et al.* (2005).

2.2 *Arsenic accumulation and speciation in the algal cells with and without EPS*

For the assessment of As bioaccumulation CC-125 and CC-503 with and without EPS were exposed to a series of (0, 10,100, 500 and 1000 μg/L) $As^{(III)}$ and $As^{(V)}$ with an initial optical density of (OD_{680nm}) 0.25 ± 0.03 for 48 h in 100 ml TAP medium. For As speciation, both strains with and without EPS were exposed to 100 pbb $As^{(III)}$ and $As^{(V)}$ for 48 h in 50 ml TAP medium and total accumulated and speciation of As were determined using HPLC-ICP-MS. All the experiments were conducted in triplicates and data were analyzed using SPSS. 20.

3 RESULTS

3.1 *As accumulation and absorption*

CC-125 and CC-503 cells with EPS accumulated up to 18–38% and 32–43% higher As in comparison to EPS-removed cells when exposed to $As^{(III)}$ respectively. Arsenic absorption was 22–44% and 44–59% greater in the EPS removed cells compared to intact EPS cells of CC-125 and CC-503 respectively, adsorption of As was 4.6–6.5 and 2–3 fold higher in CC-503 and CC-125 cells with EPS than those of EPS removed cells respectively (Figure 1a-c). In case of $As^{(V)}$, total accumulated As in the CC-125 and CC-503 cells was 25–32%

DOI: 10.1201/9781003317395-157

and 23–34% greater than the EPS removed cells. Absorption was 23–45% and 32–38% higher after the EPS extraction, both algal cells showed decreased adsorption after removal of EPS (Figure 1d-f).

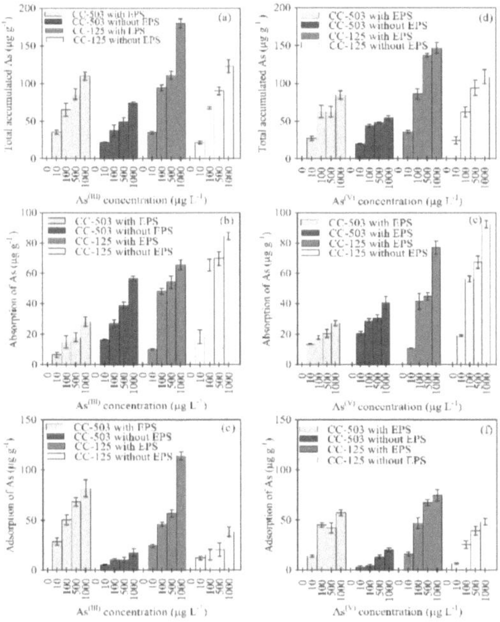

Figure 1. Total As accumulation, absorption and adsorption of CC-503 and CC-125 treated with $As^{(III)}$ (a,b,c, respectively) and $As^{(V)}$(d,e,f, respectively). Data are means of the three replicates (n = 3).

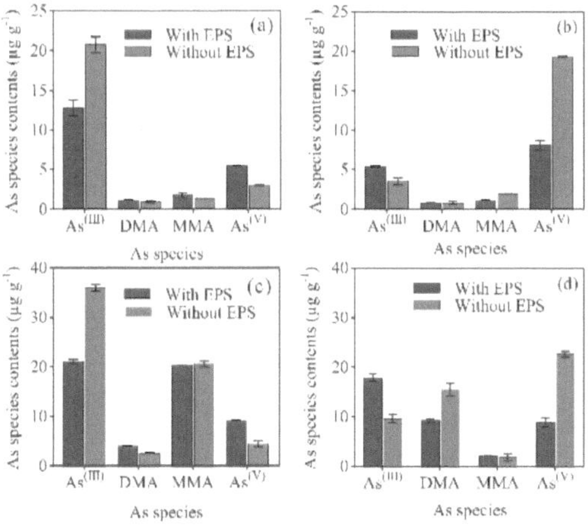

Figure 2. As speciation in CC-503 treated with 100 μg/L $As^{(III)}$ (a) and $As^{(V)}$ (b), As speciation in CC-125 treated with 100 μg/L $As^{(III)}$ (c) and $As^{(V)}$ (d) with and without their EPS for 48 hours. Date are the means of three replicates (n = 3).

3.2 *As speciation*

In case of $As^{(III)}$ treatment, after EPS extraction, CC-125 and CC-503 showed 51% and 32% less As $^{(V)}$ production, while 46% and 32% less $As^{(III)}$ was synthesized respectively, with 5–10% organic species of DMA and MMA with respect to total As detected in algal cells (Figure 2a-d).

4 DISCUSSION

Enhanced accumulation and decreased absorption of As in the microalgae cells with EPS indicated that interactions occurred between functional groups of EPS and As. Further, a greater proportion of $As^{(III)}$ and $As^{(V)}$ oxidation-reduction in cells with EPS implied that the presence of EPS on cell surface might facilitate the transformation of As in microalgae cells. It has been shown that presence of EPS on microalgae cells can reduce the toxicity of metal(loid)s to microalgae cells and enhance As accumulation and transformation (Zhang *et al.*, 2020). The finding of this study confirmed that EPS could reduce the As uptake and intensify the As transformation in microalgae cells.

5 CONCLUSION

The uptake and transformation As of in microalgae cells could be mediated by EPS and the intracellular distribution of As in microalgae was controlled by EPS moieties.

AKNOWLEDGEMENTS

Financial support from the National Natural Science Foundation of China (31770548) and National Key Research and Development Program of China (2016YFD0800306, 2017YFD0800305) are deeply appreciated.

REFERENCES

Naveed S., Li C., Lu X., Chen S., Yin B., Zhang C. & Ge Y. 2019. Microalgal extracellular polymeric substances and their interactions with metal (loid)s: a review. *Crit. Rev. Environ. Sci. Technol.* 49: 1769–1802

Sheng G.-P., Yu H.-Q. & Yu Z. 2005. Extraction of extracellular polymeric substances from the photosynthetic bacterium *rhodopseudomonas acidophila. Appl. Microbiol. Biotechnol.* 67(1): 125–130.

Wang Y., Wang S., Xu P., Liu C., Liu M. & Wang Y. 2015. Review of arsenic speciation, toxicity and metabolism in microalgae. *Rev. Environ. Sci. Biotechnol.* 14(3): 427–451.

Zhang J., Zhou F., Liu Y., Huang F. & Zhang C. 2019. Effect of extracellular polymeric substances on arsenic accumulation in *Chlorella pyrenoidosa. Sci. Total Environ.* 704: 135368.

Arsenic in the Environment: Bridging Science to Practice for Sustainable Development –
van der Wal, Ahmad, Petrusevski, Weijma, Savic, van der Wens, Beerendonk, Bhattacharya, Bundschuh & Naidu (Eds)
 ISBN 978-1-032-32928-4

Integrating biological oxidation of arsenite by arsenic oxidizing bacteria with iron-electrocoagulation: A novel approach for enhanced removal of arsenite from water

M. Roy[1], C.M. van Genuchten[2], L.C. Rietveld[1] & D. van Halem[1]
[1]*Sanitary Engineering, Department of Civil Engineering and Geosciences, Delft University of Technology, Delft, The Netherlands*
[2]*Department of Geochemistry, Geological Survey of Denmark and Greenland, Copenhagen, Denmark*

ABSTRACT: Arsenic (As) is a toxic element and should be removed from water meant for drinking purposes. As per WHO, the As concentration in drinking water should be below 10 µg/L. In water As is present in two forms: arsenite (As(III)) and arsenate (As(V)) and the removal of As(III) by conventional techniques is effective when it is oxidized to the more easily removed As(V) oxyanion. However, the common techniques used for As(III) oxidation and its subsequent removal have various drawbacks. The new concept of biological oxidation of As(III) by arsenic oxidizing bacteria (AsOB) and its removal by adsorption to iron precipitates generated by iron-electrocoagulation (Fe-EC) has shown theoretical advantages over conventional technologies. Hence, this study aimed to combine oxidation of As(III) by groundwater-native AsOB with low charge dosage Fe-EC to remove 150 µg/L of As(III) below 10 µg/L in continuous flow mode. The removal efficiency of As(III) increased from 70% to 99% by applying biological oxidation prior to Fe-EC, making this integrated approach a novel and effective technique to remove As (III) from water.

1 INTRODUCTION

In general, arsenic (As) is present in two forms in water: arsenite (As(III)) and arsenate (As(V)), and its concentration should be below 10 µg/L in drinking water as per the WHO. Under circum-neutral pH, As (III) is neutrally charged and As(V) is negatively charged. Thus, common As removal techniques such as ion exchange or iron hydroxide adsorption, are not very efficient to remove As(III) as these techniques rely on surface charge. Therefore, technologies often aim at As(III) oxidation prior to adsorption or separation. As(III) can be oxidized by air or using O_3, H_2O_2, and $KMnO_4$, but these are either very slow or produce un-wanted by-products. Also, conventional separation techniques have several drawbacks, including chemicals use, high operating cost, pH adjustment (Katsoyiannis *et al.*, 2004).

The novel concept of combined biological oxidation of As(III) by AsOB with Fe-EC has theoretical advantages over conventional techniques. AsOB has shown potential to effectively oxidize different concentrations of As(III) and these bacteria can be easily grown on sand materials by continuous dosing of As (III) spiked water for a few weeks (Gude *et al.*, 2018). Fe-EC is an electro-chemical technique that generates in-situ iron coagulants by passing electric current through iron electrodes in contact with the solution. These iron coagulants act as adsorbents for arsenic (Amrose *et al.*, 2013). Though Fe-EC has shown potential to remove both As(III) and As(V), the removal of initial As(III) requires higher Fe dosage than As(V) (Wan *et al.*, 2011), making it less attractive for application because of more sludge generation, increased costs and power consumed. Therefore, the main objective of this research was to develop continuous down-flow vertical filtration columns able to treat 150 µg/L of As(III) to below the WHO guideline by integrating biological As(III) oxidation with Fe-EC (bio-Fe-EC system).

2 MATERIALS AND METHODS

2.1 *Experimental setup*

To biologically oxidize As(III), AsOB were grown using an anthracite layer (2.0 – 4.0 mm size and 30 cm length) as a bio-carrier at the bottom of two vertical down-flow columns (2 m height and 9 cm diameter). The columns were dosed continuously with tap water spiked with 150 µg/L As(III) ($NaAsO_2$) over a period of 49 days at a rate of 1 ± 0.5 m/hr. After proper growth of AsOB on the bio-carrier (capable of oxidizing 150 µg/L As(III) rapidly) the columns were modified for integration with Fe-EC. In bio-Fe-EC column, the anthracite layer (containing AsOB) was positioned prior to in-line Fe-electrodes (6 cm × 3 cm), whereas in the control column, the Fe-electrodes were placed above the anthracite layer. The dosage of Fe by Fe-EC in both columns was applied at low charge dosage of 10 C/L under 5 C/L/min charge dosage rate resulting in an iron dosage of 2.9 mg/L as per Faraday's law (Amrose *et al.*, 2013). The final filtration columns were run for 6 hours at a flow rate of 1 ± 0.5 m/hr for 3 consecutive days and the

 DOI: 10.1201/9781003317395-158

influent As (III) concentration in both the columns was 150 ± 20 μg/L.

2.2 *Water sampling and laboratory analysis*

All the water samples collected were filtered using 0.45 μm polysthersulfone filters (Macherey-Nagel GmbH & Co. KG) and analyzed for total arsenic, As(III) and As (V) concentrations. As speciation was monitored using ion exchange resin (Amberlite* IRA-400 chlorite form) following Clifford method (Gude *et al.*, 2018). All the analysis were performed using ICP-MS (Alanlytik Jena model PlasmaQuant MS).

3 RESULTS AND DISCUSSION

3.1 *Biological oxidation of As (III) in anthracite layer*

Figure 1 shows the As(III) and As(V) concentration in the water after the anthracite layer of the duplicate columns during ripening with As(III)-containing tap water prior to installing Fe-EC. The influent concentration of As(III) in both columns was 150 μg/L. This figure shows that after approximately 1 month, all the 150 μg/L As(III) in the influent was oxidized to As(V) after passing through the biological anthracite layer. AsOB, native to the tap water, are responsible for oxidizing As(III), similarly to previous work by Gude *et al.* (2018).

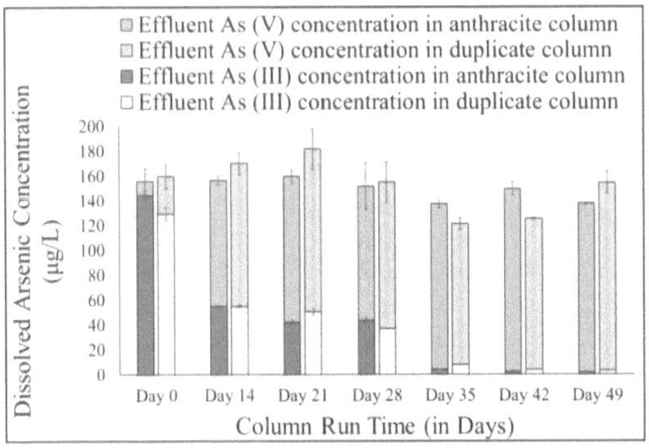

Figure 1. Ripening of biological anthracite layer in duplicate columns during continuous dosing of 150 ± 20 μg/L As(III) prior to integrating Fe-EC.

3.2 *Removal of As (III) in the filtration columns*

Figure 2 shows the removal profile of As in the two filtration columns, one with biological oxidation before Fe-EC, and the other column without the prior oxidation step. In the bio-Fe-EC column, Fe-EC was performed in water containing mainly As(V), whereas Fe-EC was performed in As(III)-rich water in the column without pre-oxidation. For both columns, arsenic removal after Fe-EC was observed, suggesting As adsorption to co-precipitated iron hydroxides generated during Fe-EC. However, in the bio-Fe-EC column, As removal below the WHO standard was achieved, which was not observed in the other column even though identical EC operational parameters were used in both columns. This difference in As removal between the columns clearly demonstrates the major advantage of oxidizing As(III) before Fe-EC because of the higher sorption affinity of As(V) than As(III) for precipitates generated by Fe-EC (Wan *et al.*, 2011).

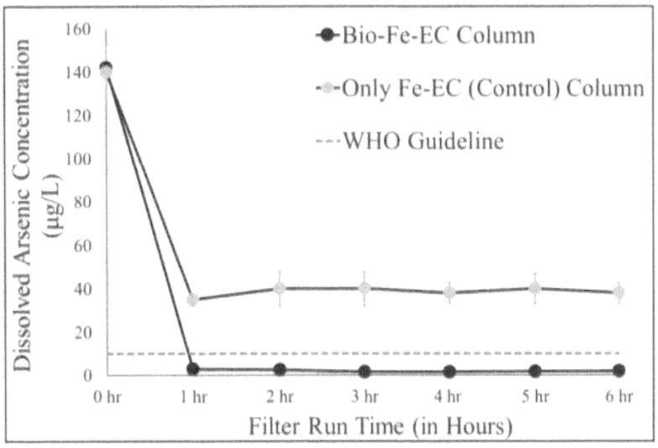

Figure 2. Dissolved As concentration over 6 h in the filtration columns (with and without biological oxidation of As(III) before Fe-EC). Fe-EC was operated with a CD of 10 C/L at CDR of 5 C/L/min.

4 CONCLUSIONS

This study showed successful removal of As(III) by integrating biological oxidation followed by Fe-EC in a continuous flow mode. Performing biological oxidation before Fe-EC increased the removal efficiency of 150 μg/L As(III) from 70% to 99% in tap water. Our results suggest that integrating these biological and electrochemical techniques can be a novel and cost-effective approach to remove As(III) from water.

REFERENCES

Amrose S., Gadgil A., Srinivasan V., Kowolik K., Muller M., Huang J. & Kostecki R. 2013. Arsenic removal from groundwater using iron electrocoagulation: effect of charge dosage rate. *J. Environ. Sci. Health A* 48(9): 1019–1030.

Gude J.C.J., Rietveld L.C. & Van Halem D. 2018. Biological As (III) oxidation in rapid sand filters. *J. Water Process. Eng.* 21: 107–115.

Katsoyiannis I.A. & Zouboulis A.I. 2004. Application of biological processes for the removal of arsenic from groundwaters. *Water Res.* 38(1): 17–26.

Wan W., Pepping T. J., Banerji T., Chaudhari S. & Giammar D. E. 2011. Effects of water chemistry on arsenic removal from drinking water by electrocoagulation. *Water Res.* 45(1): 384–392.

Arsenic in the Environment: Bridging Science to Practice for Sustainable Development – van der Wal, Ahmad, Petrusevski, Weijma, Savic, van der Wens, Beerendonk, Bhattacharya, Bundschuh & Naidu (Eds)
 ISBN 978-1-032-32928-4

Phytoremediation of arsenic—contaminated soil and water through some hyperaccumulator pteridophytic plants

A.C. Samal[1], Piyal Bhattacharya[2], J.P. Maity[3,4], A. Mallick[5] & S.C. Santra[1]
[1]*Department of Environmental Science, University of Kalyani, West Bengal, India*
[2]*Department of Environmental Science, Kanchrapara College, Kanchrapara, West Bengal, India*
[3]*Department of Earth and Environmental Sciences, National Chung Cheng University, Tainan, Taiwan*
[4]*Department of Chemistry, School of Applied Sciences, KIIT Deemed to be University, Bhubaneswar, Odisha, India*
[5]*ENVIS RP on Environmental Biotechnology, Department of Environmental Science, University of Kalyani, West Bengal, India*

ABSTRACT: The arsenic contamination in groundwater-soil-food chain in many parts of the world is in alarming condition causing severe human health effects. To investigate the phytoremediation ability through hyperaccumulation, five commonly available pteridophytic plant species (*Christella dentata, Adiantum caudatum, Pteris* sp., *Marsilea quadrifolia* and *Azolla* sp.) from arsenic—contaminated areas of West Bengal (India) along with a control site were taken for consideration. It was found that all the studied fern plants had significant arsenic hyperaccumulating properties. The probable hyperaccumulating mechanism was examined by studying the polyphenol contents of the selected fern species. A significant correlation was observed between accumulation of arsenic and polyphenol contents. Thus, these pteridophytic plants can be used for sustainable phytoremediation of arsenic from contaminated groundwater and soil in pilot scale through further extensive study.

1 INTRODUCTION

The groundwater and soil have been contaminated by toxic arsenic in many parts of the world through geogenic or anthropogenic sources. A number of researchers had reported the hyperaccumulation and phytoremediation of arsenic from water and soil by the fern *Pteris vittata*, a pteridophytic plant (Chen *et al.*, 2002; Kertulis-Tartar *et al.*, 2006; Ma *et al.*, 2001; Mecwan *et al.*, 2018; Tu *et al.*, 2004).

In Gangetic basin of West Bengal (India) severe arsenic contamination is found in groundwater through geogenic processes. The underground arsenic gets released and contaminates surface soil along with cultivated crops and vegetables due to uncontrolled withdrawal of groundwater for irrigation and household uses. The underground arsenic released to the environment ultimately affects human health (Moulick *et al.*, 2021; Samal *et al.*, 2011, 2013).

Considering this background some common naturally occurring pteridophytic plants from the severely arsenic-contaminated areas of West Bengal were selected to study the hyperaccumulative ability as well as bioremediation efficiency in the present investigation.

2 MATERIALS AND METHODS

The chosen study area belongs to severely arsenic-affected Nadia and North 24 Parganas districts of West Bengal. The pteridophytic plants {*Christella dentata, Adiantum caudatum* and *Pteris* sp. (terrestrial), *Marsilea quadrifolia* and *Azolla* sp. (aquatic fern found in shallow water and marshy paddy fields)} were collected from the agricultural fields along with groundwater and agricultural field soil samples where arsenic-contaminated groundwater is being used for agricultural purposes. Samples were collected from the field with proper preservation techniques and transferred to laboratory for analyzing arsenic concentration. The same plants from an arsenic non-contaminated area (University of Kalyani campus) were collected for comparing the accumulation of arsenic. Soil and plant samples were dried, acid-digested and finally analyzed by FI-HG-AAS (Model: Perkin Elmer AAnalyst 400). Standard reference samples were analyzed to check the reproducibility and were found in good agreement with the certified values. The polyphenol content of the plant-parts were also estimated through standard procedure to know the mechanism of arsenic hyperaccumulation.

3 RESULTS AND DISCUSSION

The Brake fern *Pteris vittata* has been established as a hyperaccumulator for arsenic, and it is a pteridophytic plant. Along with *Pteris vittata* some other abundantly available pteridophytic plants were

 DOI: 10.1201/9781003317395-159

studied for arsenic hyper accumulation in the present study. Different plant parts (shoot, root and whole part) of the selected fern species along with adjacent soils were studied for accumulation of arsenic (Figure 1). It was found that the shoot/frond part of *Adiantum caudatum* accumulated comparatively higher contents of arsenic (23,544.16 µg/kg dry weight) than other species and *Marsillia* sp. (13,103.34 µg/kg dry weight) accumulated lower arsenic. Considering the whole plant accumulation aquatic fern *Azolla* sp. accumulated the highest arsenic (44,069.79 µg/kg dry weight). The average arsenic concentrations in soil and groundwater of the study area were 26,287.92 µg/kg dry weight and 145.5 µg/l, respectively. It was found that the similar plant species collected from the arsenic-free area (the average soil arsenic concentration: 2,961.16 µg/kg and groundwater: below the detection level) accumulated very low amounts of arsenic. The *Marsillia* sp. and *Azolla* sp. in the control area were found to accumulate arsenic below the detection level. This might be due to collecting those plants under water flooding condition. The two plant species *Pteris* sp. and *Marsillia* sp. are consumed as vegetables by the local people. People in these areas should care about consuming these arsenic hyper accumulative plants. The rhizome part of the *Pteris* sp. accumulated the highest concentration of arsenic 64,813.46 µg/kg dry weight). In all cases the root/ rhizome parts were observed to accumulate more arsenic than the shoot/ frond parts.

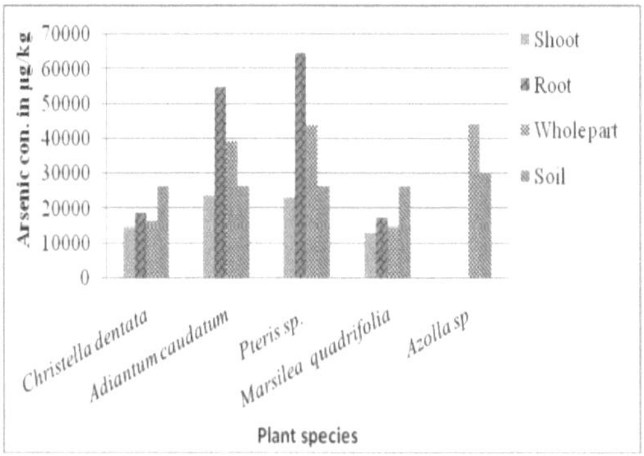

Figure 1. Arsenic hyperaccumulation by the studied fern species.

For understanding the mechanism of hyperaccumulation of arsenic in the ferns species the polyphenol content of the shoot/ frond of selected plants were estimated. It was found that the polyphenol content was higher in the plants collected from arsenic-contaminated area than the control area. The high accumulation of arsenic and the polyphenol contents were observed to have a positive correlation. Some researchers reported that the interaction with phosphate and arsenic is the cause of hyperaccumulation of arsenic by *P. vittata* via the phosphate transporters (Junru *et al.*, 2002; Tu & Ma 2003).

4 CONCLUSIONS

There is great need for remediation of arsenic-rich groundwater and soil from the polluted environment which contaminates the food chain and has severely impacted on human health. Phytoremediation is an ecofriendly cost-effective process for decreasing toxic contaminants from the environment. The studied pteridophytic plants have the properties of hyper-accumulation of arsenic in significant amounts. Thus, these fern species can serve in reducing the arsenic level from the contaminated soil and groundwater through phytoremediation process. Further extensive studies on these types of pteridophytic plants are required for sustainable bioremediation of arsenic from arsenic–contaminated groundwater and soil in pilot scale.

ACKNOWLEDGEMENTS

We acknowledge the University of Kalyani, West Bengal, India for the experimental works.

REFERENCES

Chen T., Wei C., Huang Z., Huang Q., Lu Q. & Fan Z. 2002. Arsenic hyperaccumulator *Pteris vittata* L. and its Arsenic Accumulation. *Chinese Science Bulletin*, 47(11): 902–905.

Kertulis-Tartar G.M., Ma L.Q., Tu C. & Chirenje T. 2006. Phytoremediation of an arsenic-contaminated site using *Pteris vittata* L.: a two-year study. *Int. J. Phytoremediation* 8(4): 311–322.

Ma L.Q., Komar K.M., Tu C., Zhang W., Cai Y. & Kennelley E.D. 2001. A fern that hyperaccumulates arsenic. *Nature* 409(6820): 579.

Mecwan Neha V., Yagnik B.N. & Hitesh A.S. 2018. A review on phytoremediation of arsenic in contaminated soil, *Int. Res. J. Environmental Sci.* 7(4): 27–36.

Moulick D., Samanta S., Sarkar S., Mukherjee A., Pattnaik B.K., Saha S., Awashi J.P., Bhowmick S., Ghosh D., Samal A.C. & Mahanta S. 2021. Arsenic contamination, impact and mitigation strategies in rice agro-environment: an inclusive insight. *Sci. Total Environ.* 800: 149477.

Samal A.C., Kar S., Bhattacharya P. & Santra S.C. 2011. Human exposure to arsenic through foodstuffs cultivated using arsenic contaminated groundwater in areas of West Bengal, India. *J. Environ. Sci. Health A* 46(11): 1259–1265.

Samal A.C., Kar S., Maity J.P. & Santra S.C. 2013. Arsenicosis and its relationship with nutritional status in two arsenic affected areas of West Bengal, India. *J. Asian Earth Sci.* 77: 303–310.

Tu C. & Ma L.Q. 2003. Effects of arsenate and phosphate on their accumulation by an arsenic-hyperaccumulator *Pteris vittata* L. *Plant Soil* 249(2): 373–382.

Tu S., Ma L.Q., Fayiga A.O. & Zillioux E.J. 2004. Phytoremediation of arsenic-contaminated groundwater by the arsenic hyperaccumulating Fern *Pteris vittata* L. *Int. J. Phytoremediation* 6(1): 35–47.

Wang J., Zhao F.J., Meharg A.A., Raab A., Feldmann J. & McGrath S.P. 2002. Mechanisms of arsenic hyper-accumulation in *Pteris vittata*. Uptake kinetics, interactions with phosphate, and arsenic speciation. *Plant Physiol.* 130(3): 1552–1561.

Arsenic in the Environment: Bridging Science to Practice for Sustainable Development –
van der Wal, Ahmad, Petrusevski, Weijma, Savic, van der Wens, Beerendonk, Bhattacharya, Bundschuh & Naidu (Eds)
 ISBN 978-1-032-32928-4

Remediation of arsenic-rich acid mine water in sulfate-reducing bioreactors

F. Battaglia-Brunet[1], C. Joulian[1], H. Tris[1], J. Jacob[1], M. Hery[2] & C. Casiot[2]
[1]*BRGM, Water, Environment, Process and Analyses Division, Orléans, France*
[2]*HydroSciences Montpellier, Univ. Montpellier-CNRS-IRD, Montpellier, France*

ABSTRACT: Arsenic (As) is a toxic element commonly found in mining sites. Removal of As from Acid Mine Drainages (AMD) can be obtained in anaerobic bioreactors thanks to sulfate-reducing bacteria (SRB) through the precipitation of arsenic sulfide. The present study aimed to develop the sulfate-reducing anaerobic step of a bioremediation scheme to treat AMD rich in As, zinc (Zn) and iron (Fe). Column bioreactors worked in series, with differential supply of organic substrates, i.e. yeast extract and glycerol. This strategy allowed to modulate pH and to efficiently precipitate As and Zn in the columns, although it was not possible to selectively recover Zn alone. Most part of the Fe remained in solution, and was treated downstream by oxidation, allowing the co-precipitation of residual As. The SRB biotreatment is currently up-scaled and will be tested on site.

1 INTRODUCTION

Past and present mining activities constitute one of the major sources of arsenic (As) pollution in the environment. Low-cost remediation options for mine water treatment generally combine aerobic and anaerobic steps. Each step must be designed to optimize process efficiency and decrease the cost of the global remediation facility. Sulfate-reduction in acidic conditions was proposed as a bioprocess to precipitate selectively metals from AMD (Ñancucheo *et al.*, 2012). The objective of the present study is to determine the optimal process to remove metals and As from the Carnoulès acid mine water, containing about 1 g/L iron (Fe), 100 mg/L As and 20 mg/L zinc (Zn) as the most concentrated metals and metalloids. In this case, it would be beneficial to precipitate selectively As and Zn as sulfides through the activity of endogenous sulfate-reducing bacteria (SRB), and keep Fe in solution in order to remove it in a downstream oxidation/precipitation step. Thus, the need of organic substrate for SRB would be limited, as Fe is far more concentrated than As and Zn. With this aim, laboratory experiments were performed in two continuously fed columns in series. Then, the SRB bioreactor must be up-scaled from 0.3 L to 100 L to study the bio-treatment on site. Up-scaling implies amplification of the active biomass and modifications of the bioreactor characteristics (design and filler material).

2 METHODS

The small bioreactors consisted of 320 mm high glass columns with an internal diameter of 35 mm. Two columns were disposed in series, with successive glycerol (electron donor) supplies. The exact amount of glycerol necessary for As precipitation was directly added to the main feed entering the first column. Between the two columns, a secondary glycerol supply corresponding to the exact quantity necessary to precipitate Zn was provided as a concentrated solution. The filler material was conceived with two aims: (1) to distribute passively nutrients to the bacteria, and (2) to establish different levels of activities in the two columns. For this last reason, the yeast extract concentration in the filler material was lower in the first column than in the second one. The filler material was prepared with pozzolana pieces impregnated with nutrient agar containing 0.78 g/L H_2PO_4 and 0.2 (column 1) and 2 (column 2) g/L of yeast extract. The columns were first fed (residence time 40 to 80 h in each column) with a synthetic mine water until day 55, then with the real Carnoulès mine water adjusted at pH 3.6 (until day 75), and finally with the real AMD without pH adjustment. The two bioreactors were inoculated simultaneously with a microbial consortium sampled from a previous column experiment (Battaglia-Brunet *et al.*, 2016). For up-scaling, a 0.3 L column was used to inoculate a 5 L column. The filler material of this 5 L column was composed of a mixture of 10% pozzolana embedded in nutrient agar (2 g/L yeast extract) and 90% of clean pozzolana. During up-scaling, the 0.3 L and 5 L columns were fed with the real Carnoules mine water, with no pH modification and enriched with 0.5 g/L glycerol and 1 g/L yeast extract. Fe, As and Zn were quantified by Atomic Absorption Spectrophotometry (AAS), and bacterial communities were identified by cloning-sequencing of 16S genes.

3 RESULTS AND DISCUSSION

3.1 *Small columns experiment*

The sequential supply of electron donors efficiently induced a pH gradient between the feed water, the

 DOI: 10.1201/9781003317395-160

columns 1 outlet and the column 2 outlet (Figure 1). The pH of the column 1 outlet remained stable close to pH 4.4 whereas the pH of the column 2 outlet was more variable (from 4.8 to 6.5).

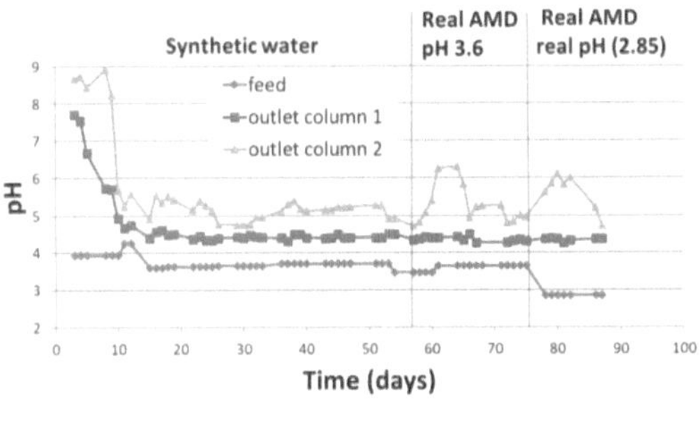

Figure 1. Evolution of pH during experiment with columns in series.

Only a small percentage of As and no Zn was removed in column 1 (Figures 2 & 3). As removal in column 2 was variable, but reached nearly 100% at some periods of the experiment. Zn removal in column 2 also reached 100% during some phases, in particular with the real AMD at pH 3.6 (days 60 to 75). No Fe was precipitated in column 1, however Fe removal reached 20–30% in column 2. At the end of experiment, the most abundant 16S rRNA sequences found in the liquid phase of the bioreactors were affiliated to the genera *Cellulomonas*, *Bacillus*, *Clostridium*, *Burkholderia*, and *Desulfosporosinus* as the only detected sulfate reducer.

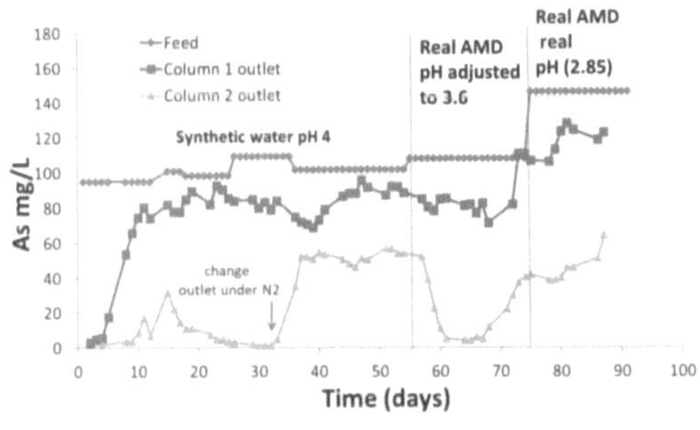

Figure 2. Evolution of As concentration during experiment with 2 columns in series.

3.2 *Up-scaling*

In the 5 L column, pH was rapidly raised from 3.6 (pH of the real AMD with 1g/L yeast extract and 0.5 g/L glycerol added) to pH 7 in the outlet. The objective here was not to treat the water but to develop a volume of inoculum sufficient to start the SRB treatment in a 100 L bioreactor that will work on the site. The total residence time was very high (40 days), however sampling at different levels along the column showed that, in less than 250 h, the SRB removed nearly entirely As, Zn and Fe from the real mine water: the remaining concentration in the mine water sampled at 1/4 of the column total height, the closest sampling port from the feed, was less than 1 mg/L of As, less than 0.5 mg/L Zn and between 0.2 and 36 mg/L of Fe.

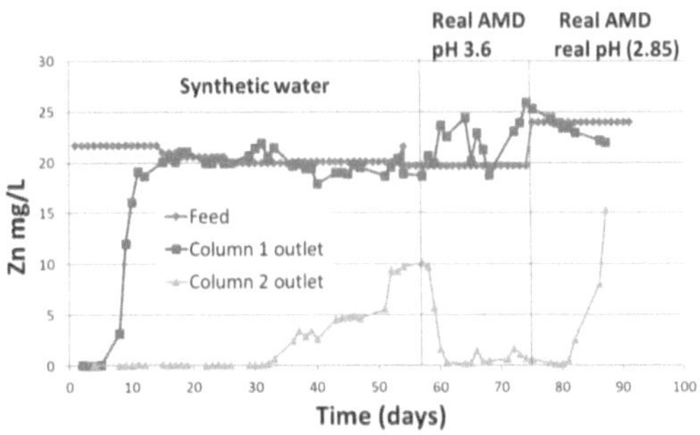

Figure 3. Evolution of Zn concentration during experiment with 2 columns in series.

4 CONCLUSIONS

Results showed that adjusting the availability of organic substrates (yeast extract and glycerol) allows the modulation of pH increase and consequently the type of elements that precipitate as sulfides in the bioreactor. This principle will be applied through up-scaling of SRB bioreactor on site, to remove Zn and partly As (but not Fe) in an anaerobic treatment step. Up-scaling from 0.3 to 5 L, with a 90% decrease of nutrient agar in the filling material, was successfully achieved. The next steps will be performed on site in 25 L then in 100 L bioreactors, accompanied by a decrease of nutrient supply and residence time.

ACKNOWLEDGEMENTS

This work was supported by the French National Research Agency in IngECOST-DMA project (ANR 2013-ECOT-0009) and by ADEME in CompAs project (contract 1872C0004).

REFERENCES

Battaglia-Brunet F., Joulian C. & Casiot C. 2016. Development of a passive bioremediation process based on sulfate-reduction to treat arsenic-containing acidic mine water. In: P. Bhattacharya, M. Vahter, J. Jarsjö, J. Kumpiene, A. Ahmad, C. Sparrenbom, G. Jacks, M.E. Donselaar, J. Bundschuh & R. Naidu (eds.) *Arsenic Research and Global Sustainability (As2016)*. CRC Press/Taylor and Francis (ISBN 978-1-138-02941-5), pp. 129–130.

Ňancucheo I., Hedrich S., Johnson D.B. 2012. New (micro-)biological strategies that enable the selective recovery and recycling of metals from acid mine drainage and mine process waters. *Mineral. Mag.* 76, 2683–2692.

Arsenic in the Environment: Bridging Science to Practice for Sustainable Development – van der Wal, Ahmad, Petrusevski, Weijma, Savic, van der Wens, Beerendonk, Bhattacharya, Bundschuh & Naidu (Eds)

A prospective phytoremedial mechanism for arsenic from contaminated drinking water using *Hygrphila spinose*

N. Roy Chowdhury, D. Sinha, A. Das, M. Joardar & T. Roychowdhury
School of Environmental Studies, Jadavpur University, Kolkata, India

ABSTRACT: Arsenic (As) is a detrimental pollutant of drinking water, causing adverse effects on human health. Various removal technologies have been implemented but now the focus has been shifted on the development of cheap and easy to maintain removal technologies. Plants have a unique ability to remove toxicants from water by its accumulation efficacy. The purpose of the present study is to investigate the remedial capacity of *Hygrophila spinosa* (Starthorn), a phyto-adsorbent of As from contaminated drinking water. A batch study has been performed on different controlling variables, such as adsorbent dose, pH condition, contact time and RPM (Rotation per minute/ agitation speed/ stirring rate). A whole sorption isotherm is considered to be the most appropriate approach for assessing bio-adsorbent capacity of an adsorbent. The results of this study show that phyto-adsorption could remove As (100 μg/L) up to 63.1% by Starthorn leaves following optimum conditions; adsorbent dose: 1 g, pH: 5, contact time: 4 hrs, RPM: 45. Freundlich isotherm further establishes the fact of reduction efficiency.

1 INTRODUCTION

Arsenic (As) is designated as a naturally occurring harmful pollutant of drinking water, causing adverse effects on human health. Since 1993, the World Health Organization has endorsed a maximum contaminant level (MCL) of As is 10 μg/l for drinking water (Zouboulis & Katsoyiannis 2002). In recent scientific history removal of As from drinking water is an important subject that has been involving a lot of attention. The conventional removal technologies used depend upon their efficiency and applicability/ appropriateness of the technologies. Various treatment technologies are available for the removal of As (III) and As (V) ions from water, such as oxidation/ precipitation, coagulation/ co-precipitation, ion-exchange, reverse osmosis and adsorption that have been developed already and have proved themselves to be efficient enough in removing As from contaminated drinking water (Bissen *et al.*, 2003). Bio-sorption as an alternative and effective technology has been widely studied over recent years, because of its wide range of target pollutants, high sorption capacity, excellent performance, eco-friendly nature.

The study of Starthorn leaves aims to decrease the As content from drinking water and it would shed light on the impact of phytoremediation.

2 MATERIALS AND METHODS

2.1 *Collection of adsorbent materials*

The present study has been performed by the leaves of Starthorn (Hygrophila spinosa), an iron enriched plant (Figure 1). Samples were collected from the side of Heritage Institute of Technology, Anandapur, Kolkata.

Figure 1. *Hygrophila spinosa (Starthorn).*

2.2 *Preparation of adsorbent*

The collected plant materials (leaves) were washed to remove all the dirt present on them. Seggregated leaves were again washed with double distilled water and dried in a hot air oven for 3 days at 50–55°C untill they were completely dry. Dried leaves were then ground to a fine powder by grinder (Bajaj rex-500 watt) followed by mortar and pestle. The powder was then sieved through a sieve of pore size 0.15 mm and then treated with 1:2 HCl (12.1 M) solution to bring about desorption and remove impurities. The wet adsorbent was then again dried like before and then stored for future use in an air tight container.

2.3 *Preparation of As stock*

For the adsorption study a 100 μg/l As (V) solution was used, which was prepared from Merck As 1000 mg/l Solution by serial dilution.

DOI: 10.1201/9781003317395-161

2.4 *Preparation for batch adsorption study*

The study was conducted to investigate the effect of controlling parameters like contact time, adsorbent dosage, RPM and pH. For each study, the adsorbent material was weighed and 20 ml of respective As solution was added. The pH was adjusted to the required value using 0.01N HCL and/or 0.01N NaOH. The plastic containers were then placed in a temperature controlled rotary shaker operated at 30°C for the respective number of hours at required RPM, pH using required adsorbent dosages.

2.5 *Estimation of As*

The adsorbent materials were separated from the solution by double filtration (Whatman Ash Less Grade Paper No: 42 and Millipore Filter Paper of 0.45 μm) and then dried in the hot air oven for 3 days at 50–55°C. for future use or reference. The reduction in the concentration of As was studied through the phenomena of adsorption. As concentration was measured by atomic-absorption spectrometry.

2.6 *Calculations for isotherm*

- As Uptake Q_e (mg/g) was determined from the following equation:

$$Q_e = (C_o - C_e).V/M$$

  where C_o and C_e are initial and final As Concentration in aqueous solution (mg/L), V is the Volume of the solution in litre and M (g) is the adsorbent mass, Q_e is the per unit dry weight.
- Freundlich isotherm: Regression analysis ($y = mx + c$) by plotting $\ln(Q_e)$ w.r.t. $\ln(C_e)$
  $n = 1/m$ (n should be preferably 0.7–1)
  $k = e^c$ (where, k = amount (mg) of As is taken up by adsorbent)

3 RESULTS AND DISCUSSION

For this batch adsorption study the adsorbent materials mentioned above had undergone desorption so that the pores became empty and could adsorb the targeted contaminant i.e. As.

3.1 *Effect of adsorbent dosage*

The adsorbent dosage was increased from 0.1 g to 1.0 g with an interval of 0.1 g to check the effect of increase of adsorbent dosage on the reduction of As (pH:7, RPM:30, Time:5 hrs, Temp:30°C).

It was observed that the maximum reduction of As 62.6% has been for the adsorbent dosage of 1.0 g. Also observed from the adsorbent isotherm that 2.09 mg As has been removed by 1 gm of adsorbent.

3.2 *Effect of contact time*

The increase in contact time initially had a positive effect on the increase of As uptake by the adsorbent but later it was observed that after the end of 3 hours there was decrease in the uptake of As by the same adsorbent. In this study, the contact time was differed from 1 hour to 5 hours with the following controlling parameters; pH: 7, RPM: 30, Temp: 30°C, adsorbent dose: 1 g.

It can be seen that maximum reduction of As 60.9% has been for contact time of 4 hours and also observed that 25.7 mg As was removed by 1 g of adsorbent.

3.3 *Effect of pH*

The pH of a solution is a controlling factor in the adsorption process. Thus, the role of hydrogen ion concentration was examined at pH values 4, 5,6,7,8 and 9 (RPM: 30, Temp: 30°C, adsorbent dose: 1 g, contact time: 4 hrs.)

It can be seen that maximum reduction of As 76.6% has been for pH of 5. Simultaneously, from the isotherm it has been observed that 37.8 mg As has been removed by 1 g of adsorbent

3.4 *Effect of RPM*

The agitation speed or RPM has a significant effect on the adsorption capacity of the adsorbent. In this study 30–50 rpm was taken into consideration with an interval of five (Temp: 30°C, adsorbent dose: 1 g, contact time: 4 hrs, pH: 5)

It can be seen that maximum reduction of As 63.1% has been for 45 rpm and from the isotherm it has been observed that 30.1 mg As has been removed by 1 g of adsorbent.

4 CONCLUSIONS AND RECOMMENDATIONS

Arsenic contamination in drinking water is a detrimental problem in today's world. Starthorn leaves showed maximum As removal efficiency of 63.1% from water by following optimum conditions; adsorbent dose: 1 gm, pH: 5, contact time: 4 hrs, RPM: 45. However, there still lies a strong challenge in developing bio-adsorbents for As removal with high uptake efficiency, low cost and a well-elucidated desorption mechanism.

ACKNOWLEDGEMENTS

We acknowledge to WB-DST for supporting (research project grant Memo No.262 (Sanc.)/ST/P/S&T/1G64/2017, dated 25/3/2018).

REFERENCES

Bissen M. & Frimmel F.H. (2003). Arsenic—a review. part II: oxidation of arsenic and its removal in water treatment. *Acta Hydrochim. Hydrobiol.* 31(2): 97–107.

Zouboulis A., & Katsoyiannis I. (2002). Removal of arsenates from contaminated water by coagulation–direct filtration. *Sep. Sci. Technol.* 37(12): 2859–2873.

Arsenic in the Environment: Bridging Science to Practice for Sustainable Development –
van der Wal, Ahmad, Petrusevski, Weijma, Savic, van der Wens, Beerendonk, Bhattacharya, Bundschuh & Naidu (Eds)
© 2023 the Author(s), ISBN 978-1-032-32928-4

Mitigating the impact of irrigation of rice paddies with groundwater containing high levels of arsenic

A. Javed[1,2], R. Ashraf[1] & A. Farooqi[1]
[1]*Environmental Geochemistry Laboratory, Department of Environmental Sciences, Faculty of Biological Sciences, Quaid-i-Azam University, Islamabad*
[2]*Department of Earth and Environmental Sciences, Bahria University, Islamabad, Pakistan*

ABSTRACT: The study builds on assessment of arsenic (As) in a matched set of irrigation water, paddy soil, and rice grain from 60 individual rice fields across the Punjab plains of Pakistan. One rice field with high As in soil was selected along Ravi river flood plain, to see the effects of sprinkler irrigation on translocation of As into rice grain. The rice field was divided into two equal parts; one was irrigated with traditional method (flooded irrigation) whereas other was irrigated with sprinkler irrigation (non-flooded). The As translocation to grain was reduced from 0.005 ± 0.005 mg/kg to 0.002 ± 0.002 by creating aerobic condition through sprinkler irrigation. The total amount of water pumped to grow rice via tube well was 2000 mm compared to 1018 mm used in sprinkler irrigation; clearly suggest the water saving of almost 50%. Despite of less water applied, the yield of rice crop is not affected and was found to be 2000 kg/ha in flooded system and 1950 kg/ha in sprinkler irrigation system.

1 INTRODUCTION

The groundwater used for irrigating rice crop is contaminated with arsenic (As) in northern Punjab which is the country's main rice growing region. (Javed *et al.*, 2019). Rice is the most important summer cereal crop of Pakistan, which earns about US $ 933 annually through its export and is grown on an area of 2.5 million hectares, with an annual production of 6811 tons million tons giving an average yield of 2479 Kg/ha (GOP 2016). Rice is usually irrigated by traditional flooding method, which creates anaerobic conditions favorable for high As availability to rice plant. Hence, irrigation of rice with As contaminated water results in As accumulation in soil and translocate to rice grain, posing a serious health risk to community. Creating an aerobic condition through sprinkler irrigation system appears to be one of the options, to limit As bioavailability to rice plant parts. Therefore, the aim of this study is to see the effect of sprinkler irrigation on translocation of As into rice grain and water conservation by comparing two methods of irrigation (i) traditional method (flooded irrigation) (ii) sprinkler irrigation (non-flooded).

2 METHODS/EXPERIMENTAL

2.1 *Study area and sampling*

In this study, out of total 60 rice fields, spread over three flood plains and two doabs, one rice field which contains high As, both in irrigation well (0.3 mg/L) and in paddy soil (14 mg/kg), was selected in a Khudpur village of Ravi river flood plain. Rice field was divided into two equal parts (630 ft^2), one was irrigated with flooded irrigation system and other was irrigation by sprinkler irrigation systems. Four (4) soil and whole rice plants samples were collected in plastic zipper bags after every two weeks throughout the growing season from both fields and samples were evenly spaced along the diagonal across each field originating from the corner closest to the inlet of the irrigation water to the corner furthest away and were brought to Environmental hydro geo-Chemistry Laboratory in Qauid-i-Azam University, Islamabad, Pakistan for analysis.

2.2 *Analysis*

The soil samples were air-dried, grounded and sieved through 2 mm sieve. Plants samples were washed DI water to remove dust and then air-dried for 72 hours at room temperature 25°C. 5 g of soil was digested with 50 mL 1% HNO_3. Rice plants were slit into separated parts roots, stems, leaf, husk and grain. Rice grains were removed from their husks with the help of pestle and mortar. All rice plants samples were pulverized into powder form using grinder. About 5 g of pulverized plants samples were digested using in 50 mL 1% HNO_3. The digested soil and plant samples were analyzed through AAS (Atomic Absorption Spectroscopy) for As. To verify the accuracy and precision of the method to be with in <5%, a standard of known concentration was included with every run. The total depth of water applied was calculated using the formula; QT = AD (Nawaz *et al.*, 2017), where Q = discharge, T = Time of irrigation, A = Area/Farm size, D = Depth of water.

3 RESULTS AND DISCUSSION

The mean As concentration in different plant parts in the rice field irrigated by flooded irrigation system

DOI: 10.1201/9781003317395-162

and sprinkler irrigation system is presented in Figure 1.

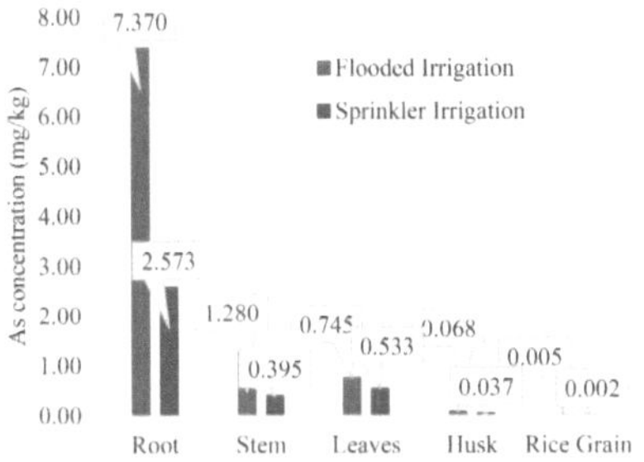

Figure 1. Mean As concentration of As in different plant parts in flooded and sprinkler Irrigation.

Arsenic concentration in plant parts were found in the order of root (7.370 ± 0.87 mg/kg) >stem (1.280 ± 0.93 mg/kg) >leaf (0.745 ± 0.67 mg/kg) >husk (0.068 ± 0.03 mg/kg) >grain (0.005 ± 0.005 mg/kg) in the field irrigated by traditional flooded system. The same order of As concentration was found in the field irrigated by sprinkler irrigation; root (2.573 ± 0.61 mg/kg) > stem(0.395 ± 0.07 mg/kg)>leaves(0.533 ± 0.14 mg/kg) > husk(0.037 ± 0.01 mg/kg) > grain (0.002 ± 0.002 mg/kg). The notable finding is that As translocation to grain was reduced from 0.005 ± 0.005 mg/kg to 0.002 ± 0.005 mg/kg by creating aerobic condition through sprinkler irrigation. In case of soil, however, the As mean concentration of 14.37 ± 0.22 mg/kg in a field irrigated by traditional flooded system was marginally high compared to mean soil As concentration of 13.98 ± 0.15 mg/kg of a field irrigated by sprinkler irrigation.

Table 1 shows the comparison of two systems (Sprinkler irrigation and traditional flooding system). Water was applied for 40 minutes through sprinkler irrigation for the first month and then 40 minutes each consecutive day for the rest of growing season when rice plants have started to grow. In case of traditional flooding system, rice crop was irrigated every after 15 days for 20 minutes each day. It is evident from Table 1. that total amount of water applied via tube well was 2000 mm whereas 1018 mm was used in sprinkler irrigation means 49% of water saved.

Table 1. Depth of water applied for entire growing season.

	Tube well	Sprinkler
Discharge (Q) in ft^3/min	10.6	0.70
Area irrigated (A) in ft^2	630	630
Total Time of irrigation (T) in min	390	3000
Depth of water applied (D) in mm	2000	1018

Table 2 shows the rice yield comparison of both systems. It is evident from Table 2 no significant yield reduction was observed when rice is irrigated with sprinkler irrigation method (only 50 kg per hectare less yield). Our results are supported by Basu *et al.* (2010) that yield is not affected by creating aerobic conditions.

Table 2. Rice yield comparison in both systems.

Irrigation method	Area sampled (m^2)	Dry weight of grain (kg)	Rice yield (kg/ha)
Flooded	10	20	20/10*1000 = 2000
Sprinkler	10	19.5	19.5/ 10*1000 = 1950

4 CONCLUSIONS AND RECOMENDATIONS

Results indicate that the As translocation to grain was reduced by creating aerobic condition through sprinkler irrigation with 50% water saving. Despite of being less water used to grow crop, the rice yield was not affected. The sprinkler irrigation offers one of the good mitigation options. It is, therefore, recommended that sprinkler irrigation can be installed to irrigate rice crop particularly in those areas where As contaminated groundwater is used for irrigation.

ACKNOWLEDGEMENTS

The authors acknowledge US-Pakistan Center for Advanced Studies in Water, Mehran University of Engineering and Technology (MUET), Jamshoro, Pakistan for financial support.

REFERENCES

Basu B., Kundu C., Sarkar S. & Sanyal S. 2010. Deficit irrigation an option to mitigate arsenic load in rice grain. *International Union of Soil Sciences (IUSS)*, pp. 51–53.

GOP, 2016. Chapter agriculture-government of Pakistan. Web http://www.finance.gov.pk/survey/chapters_16/02_Agriculture.pdf.

Javed A., Baig Z., Farooqi A., van Geen A. 2019. Spatial variation of arsenic in irrigation well water from three flood plains (ravi, chenab and jhelum) of Punjab, Pakistan. *Environmental Arsenic in a Changing World: Proceedings of the 7th International Congress and Exhibition on Arsenic in the Environment (As 2018)*, July 1–6, 2018, Beijing, PR China. CRC Press, p. 235.

Nawaz A., Farooq M., Lal R. & Rehman A. 2017. Comparison of conventional and conservation rice-wheat systems in Punjab, Pakistan. *Soil Tillage Res.* 169: 35–43.

4.6 Pilot studies

Arsenic in the Environment: Bridging Science to Practice for Sustainable Development –
van der Wal, Ahmad, Petrusevski, Weijma, Savic, van der Wens, Beerendonk, Bhattacharya, Bundschuh & Naidu (Eds)
 ISBN 978-1-032-32928-4

Small scale setups for arsenic treatment: Examples in Argentina

M.I. Litter
Universidad de San Martín, San Martín, Prov. de Buenos Aires, Argentina

ABSTRACT: Successful examples of setups at small scale recently developed by different groups in Argentina for small rural and periurban communities are presented. The final results indicate that most of the systems can attain the WHO guidelines of 0.01 mg/L for drinking water. Conclusions and recommendations, especially for isolated rural and periurban regions, are includedr.

1 INTRODUCTION

Technologies for As removal from groundwater depend on basic physicochemical processes that can be applied alone, simultaneously or sequentially, including oxidation/reduction, coagulation-filtration, precipitation, adsorption, ionic exchange, membrane technologies, bioremediation, etc. These technologies and other innovative ones can be easily adapted for use at small scale in isolated rural and periurban localities. The literature on this subject can be consulted in various references (e.g., Litter *et al.*, 2008, 2010a, 2010b, 2012, 2019; Litter & Bundschuh 2012).

2 EXAMPLES OF TECHNOLOGIES AT SMALL SCALE IN ARGENTINA

2.1 *First technologies*

The following systems have been applied in Argentina since the 90's: a hydrogel of activated aluminum hydroxide prepared from hydrated aluminum sulfate, calcium hypochlorite and ammonium hydroxide tested in Tucumán, the electrochemical corrosion of a metallic iron fixed bed developed and applied in San Juan, the use of aquatic plants such as *Lemna minor* macrophyte, dry algae, and natural biogenic hydroxyapatite obtained from carbonized cow bones.

An intervention model to provide solutions for the As problem to communities connected to a public water network and to private wells, together with the construction of a simple and low-cost rural device, which uses an optimized coagulation-filtration technology, with a previous oxidation step, and does not require electrical supply, was developed by the Instituto Nacional of Tecnología Industrial. Both the intervention model and the rural device have been applied to groundwaters of Taco Pozo (Chaco province) and of Lobos (Buenos Aires province).

2.2 *Recent results*

A simple and low-cost technology for As elimination from water using geoadsorbents was developed and implemented. The technology uses iron rich mineral species from mineral deposits from nine Argentine provinces. The adsorption process is complemented with a coagulation-sedimentation stage, adding an oxidant (NaClO) and a coagulant (PAC). The process has been scaled at pilot level for application in small localities. The operative cost of the process was 0.10 USD/m^3. Experimental prototypes were installed in rural schools of the Buenos Aires province. The As values in the raw water were in the 150–200 μg/L range and, after the treatment, values lower than 10 μg/L were always obtained. An upscale to treat higher water volumes (20,000 L) is planned.

A water treatment plant for As removal at small scale based on the use of ZVI was designed to treat continuously (500 mL/min) groundwater with high As content and minimal electric energy requirements. Commercial iron wool was the selected iron source due to its low cost and easy accessibility. Laboratory tests carried out with pilot plants showed that the percentage of As removal was higher than 90% for As concentrations up to 300 μg/L. A pilot plant (700 L/day) was constructed to test the efficiency using groundwater with high As levels from a natural source in the Buenos Aires province. The As removal in this test attained 90–95%, with concentrations lower than 10 μg/L in the treated water. Presently, the development of a plant for 4 m^3/day has been concluded and the design and construction of a plant for 20 m^3/day is envisaged.

Treatment systems for As removal based on coprecipitation-filtration through a sand bed containing different iron materials (nails, wires, etc.) were recently developed in Salta. The iron materials have been activated to generate a layer of iron oxyhydroxides. Two treatment plants were scaled in two schools of the Chaco Salteño region, allowing a production of 1,000 L/day of water containing less than 50 μg/L As. Other 10 systems have been installed in Rivadavia, Banda Sur, Joaquín V. González, Que- brachal and El galpón counties.

Low-cost technologies for As removal based on SODIS (solar disinfection), i.e., modified SORAS

DOI: 10.1201/9781003317395-163

(Solar Oxidation for As Removal) and heterogeneousous photocatalysis with TiO_2 have been tested in plastic bottles. In the original SORAS process used in Bangladesh, iron(III) citrate is added to the bottles using lemon juice, and exposed to sunlight; photo- Fenton reactions take place, oxidizing As(III) to As(V), which adsorbs strongly on the iron oxyhydroxides and flocculates. Then, the bottles are placed in a vertical position during the night, and the purified water is decanted from the iron hydroxides or is filtered through textile cloths. The SORAS technology had to be modified for application in Argentina because the waters contained mainly As(V), and the addition of lemon juice was not needed because the water contained important amounts of NOM. As the iron content of the waters was not enough, iron had to be added in the form of different iron materials, the best results being obtained by the addition of packing iron wire, commonly used in the countryside. The method was applied with good results to well waters of Los Pereyra (Tucumán province). Later, zerovalent iron nanoparticles (nZVI) produced by a local company were tested as iron source, and a rapid As(V) removal was obtained using very low concentrations of the iron material (0.005–0.1 g/L), reaching 90% As removal after 150 minutes of contact time at the optimal nZVI concentration. As values lower than 10 μg/L have been obtained.

TiO_2 photocatalytic experiments of As removal with the addition of Fe(III) have been also performed. TiO_2 was previously impregnated on plastic bottles, beads, rods or glass Rashig rings. TiO_2 photocatalytic experiments of As removal with this bottles and Fe(III) addition were performed in waters of six wells of Las Hermanas, Santiago del Estero province.

A diagnosis about access, uses and water quality, together with a study on the ability of some economic materials to remove As in waters was performed recently in two rural localities of Santiago del Estero. One of the zones was close to the capital of the province, in Banda department and the other one was located in the northern Copo department. The study concluded that the Copo zone presented serious problems of the water quality, aggravated by hard socioeconomical conditions, while, in Banda, waters presented neither microbiological contamination nor high As levels. Iron materials such as granular iron oxides and commercial iron nanoparticles were tested for their ability to remove As in samples of Copo wells. Nanoparticles gave very good results after 30 min of treatment, while the granular materials resulted in a low removal, even after 3 h of treatment. In future works, with the most efficient materials, devices such as table jars, under table cartridges, etc. will be designed and installed through an intervention in selected sites of Santiago del Estero to provide safe water to isolated households. This methodology can be easily replicated in other regions of the country.

3 CONCLUSIONS AND RECOMMENDATIONS

In Argentina, there are several examples of successful treatments for As removal, and emergent technologies are continuously investigated in the country by experienced R&D groups. It is important to remark that these methods can be implemented for treating arsenical waters affecting rural or periurban populations in the country.

ACKNOWLEDGEMENTS

To all colleagues from the CONICET Food Safety who have collaborated to collect the data here included. To MINCYT, ANPCYT and CONICET from Argentina for financial support.

REFERENCES

Bundschuh J., Litter M., Ciminelli V., Morgada M.E., Cornejo L., Garrido Hoyos S., Hoinkis J., Alarcón-Herrera M.T., Armienta M.A. & Bhattacharya P. 2010. Emerging mitigation needs and sustainable options for solving the arsenic problems of rural and isolated urban areas in iberoamerica – a critical analysis. *Water Res.* 44: 5828.

Litter M.I. & Bundschuh J. 2012. Emerging options for solving the arsenic problems of rural and periurban areas in latin america. In: J.C. Ng, B. Noller, R. Naidu, J. Bundschuh & P. Bhattacharya (eds.), *Understanding the Geological and Medical Interface of Arsenic (As2012)*, Taylor and Francis Group, London, pp. 267–270.

Litter M.I., Ingallinella A.M., Olmos V., Savio M., Difeo G., Botto L., Farfán Torres E.M., Taylor S., Frangie S. Herkovits J., Schalamuk I., González M.J., Berardozzi E., García Einschlag F.S., Bhattacharya P. & Ahmad A. 2019. Arsenic in argentina: technologies for arsenic removal from groundwater sources, investment costs and waste management practices. *Sci. Total Environ.* 690: 778.

Litter M.I., Alarcón-Herrera M.T., Arenas M.J., Armienta M.A., Avilés M., Cáceres R.E., Cipriani H. N., Cornejo L., Dias L.E., Fernández Cirelli A., Farfán E. M., Garrido S., Lorenzo L., Morgada M.E., Olmos-Márquez M.A. & Pérez-Carrera, A. 2012. Small-scale and household methods to remove arsenic from water for drinking purposes in Latin America. *Sci. Total Environ.* 429: 107.

Litter M.I., Fernandez R.G., Cáceres R.E., Grande Cobián D., Cicerone D. & Fernández Cirelli A. 2008. Tecnologías de bajo costo para el tratamiento de arsénico a pequeña y mediana escala. *Rev. Ing. San. Amb.* 100: 41.

Litter M.I., Morgada M.E. & Bundschuh J. 2010a. Possible treatments for arsenic removal in Latin American waters for human consumption. *Environ. Pollut.* 158: 1105.

Litter M.I., Sancha A.M. & Ingallinella A.M. 2010b. Tecno-Logías Económicas Para el Abatimiento de Arsénico en Aguas, Editorial Programa Iberoamericano de Ciencia y Tecnolo-Gía Para el Desarrollo, Buenos Aires.

Morgada M.E., Levy I.K., Salomone V., Farías S.S. López G. & Litter M.I. 2009. Arsenic (V) removal with nanopar-ticulate zerovalent iron: effect of UV light and humic acids. *Catal. Today* 143: 261.

Arsenic in the Environment: Bridging Science to Practice for Sustainable Development – van der Wal, Ahmad, Petrusevski, Weijma, Savic, van der Wens, Beerendonk, Bhattacharya, Bundschuh & Naidu (Eds)
© 2023 the Author(s), ISBN 978-1-032-32928-4

Modular treatment of arsenic laden saline groundwater in south and southeast Asia – findings of pilot trials in Mekong Delta, Vietnam

J. Hoinkis[1], E.E. Cañas Kurz[1], U. Hellriegel[1], T.V. Luong[2], T. Winkelnkemper[3] & J. Bundschuh[4]
[1]*Center of Applied Research, Karlsruhe University of Applied Sciences, Karlsruhe, Germany*
[2]*Lappeenranta University of Technology, Lappeenranta, Finland*
[3]*Winkelnkemper GmbH – Fermanox Wasseraufbereitung, Wadersloh, Germany*
[4]*Faculty of Health, Engineering and Sciences, University of Southern Queensland, Toowoomba, Australia*

ABSTRACT: High salinity and arsenic (As) concentrations in groundwater are prevalent problems in the tidal delta plains of South and Southeast Asia. Pilot trials for As removal and groundwater desalination were carried out in a private home in Tra Vinh in the Mekong Delta, Vietnam. Desalination was conducted by capacitive deionisation (CDI), a rather new technology showing lower specific energy consumption compared to other desalination technologies. In order to treat anoxic groundwater with As(III), which is usually associated with high Fe, a pre-oxidation step and removal of Fe(oxy)hydroxide is mandatory to avoid clogging of the CDI unit. This was conducted by use of subsurface As removal having the advantage of being a zero waste process. The process was operated by use of photovoltaics (PV) and a small wind turbine.

1 INTRODUCTION

High salinity and arsenic (As) concentrations in groundwater are widespread problems in the tidal delta plains of South and Southeast Asia such as the Ganges-Brahmaputra-Meghna in Bangladesh, the Irrawaddy delta in Myanmar, and the Mekong and Red river delta in Vietnam (Ayers *et al.*, 2016; Buschmann *et al.*, 2008; Luu 2019; Smedley 2005).

Arsenic dissolved in groundwater in South Asia is caused by subsurface sediments, where As is sorbed onto Hydrous Ferric Oxides (HFOs (Hossain *et al.*, 2012; Sankar *et al.*, 2014). Also manganese is often co-associated with iron. Previous studies confirmed that reductive dissolution of HFOs driven by microbial metabolism of organic matter results in increased concentrations of dissolved As in shallow groundwaters in Bangladesh (Nickson *et al.*, 2000).

In addition, groundwater in coastal line aquifers is affected from seawater intrusion due to over-extraction and raising sea level which will be aggravated by climate change in future (Hien *et al.*, 2009; Smedley 2005). The salinity level is subject to seasonal variation during dry and monsoon periods and tidal channels are an important potential source of high salinity water. In southwest Bangladesh electrical conductivity in groundwater ranges from 1 to 10 mS/cm, equivalent to salinities of 0.5–5.2 g/L (Bahar & Rezi 2010). In Mekong delta groundwater, conductivity varies between 1 and >10 g/L and As concentrations ranged from 0.1–1340 µg/L with 37% of the studied wells exceeding the WHO guidelines of 10 µg/L As (Buschmann *et al.*, 2008). Hence sustainable treatment technologies for As, iron (Fe^{2+}) and salinity (TDS) removal will become more important.

The objective of this work is to study a combined subsurface As removal (SAR) with desalination by use of capacitive deionization (CDI). Subsurface treatment aims at removal of As and Fe, upfront of CDI. For this, part of the extracted groundwater water is enriched with oxygen and recycled back into the underground (aquifer). By doing so, dissolved Fe^{2+} and Mn^{2+} ions are oxidized and precipitated as sparingly soluble (hydr)oxides. Around the re-injection zone in the aquifer a zone with water that is free of dissolved Fe^{2+} and manganese (Mn^{2+}) is formed. Under anoxic conditions as it occurs in South Asia, groundwater basically contains arsenite (AsO_3^{3-}) and it will be oxidised to arsenate (AsO_4^{3-}) which is then immobilized in the soil by adsorption onto the iron and manganese(hydr) oxides (Luong *et al.*, 2018).

CDI is a rather new electrosorption process for desalination. In this process, saline water flows between two porous electrodes made of activated carbon, on which a voltage is applied (Al Marzooqi *et al.*, 2014). The positive ions are thereby attracted to the negatively charged electrode and vice versa. After having loaded the electrodes, the salt ions are removed by reversing the polarity of the electrodes. Through a constant change of loading and unloading (regeneration), two streams of water with desalinated water or concentrated salt water are produced. The general advantages of the technology are its low specific energy consumption since no high pressure pumps or heat sources are required. Moreover, it is ideally suited for decentralized applications since it can be directly powered by solar PV energy (Tan *et al.*, 2018).

The study was carried out within the German joint research project WaKap (2019) as part of one of three plants installed in the Mekong Delta for

DOI: 10.1201/9781003317395-164

SAR and/or desalination of brackish water. In contrast to a first SAR plant, this pilot plant was designed for the removal of As and Fe^{2+} only, with the CDI as post-treatment for manganese (Mn^{2+}), ammonium, TDS, etc., and therefore increasing the plant capacity significantly to ca. 10 m^3/L.

2 MATERIALS AND METHODS

2.1 *Groundwater analysis*

The SAR pilot plant in combination with CDI for the removal of Fe, As and desalination from brackish water was installed in June 2018 in the province of Tra Vinh; a coastal region in the Mekong Delta. In addition to initial high Fe (c = 3.1 mg/L) and As (c = 11 μg/L) levels, the groundwater there has an increased salt concentration with max. concentrations TDS >1.5 g/L. The groundwater chemical and physical analysis before trials started is summarized in Table 1 for two samples taken over 7 months.

Table 1. Analysis of raw groundwater in Tra Vinh.

Parameter	Units	Sampling[1] 02/08/2017	Sampling 07/03/ 2018	Vietnamese standard[2]
pH	—	NM[3]	7.0	6.5 – 8.5
EC	mS/ cm	NM[3]	3.3	—
t-As	μg/L	11	2,7	10
Fe	mg/L	3.13	1.8	0.3
Mn	mg/L	0.69	0.26	0.3
Si	mg/L	15.0	—	—
Na^+	mg/L	338	374	200
K^+	mg/L	12.2	10.0	—
Ca^{2+}	mg/L	178	56	—
Mg^{2+}	mg/L	111	110	—
NH_4^+	mg/L	2.4	1.6	3
Cl^-	mg/L	827	955	250
F^-	mg/L	NM[3]	1.5	1.5
HCO_3^-	mg/L	437	NM[3]	300
NO_3^-	mg/L	BDL[4]	NM[3]	50
SO_4^{2-}	mg/L	97.8	138	250
PO_4^{3-}	mg/L	BDL[4]	BDL[4]	—

[1]QUATEST3, HCMC, Vietnam; [2]QCVN 01: 2009/BYT; [3]Not measured; [4]BDL: Below detection limit

2.2 *Pilot plant*

A submersible pump is used to pump the contaminated groundwater through a sand filter into a storage tank (T1). The treatment process (Figure 1) is divided into two steps. Firstly, water in T1 is aerated via an air injector using a booster pump (feed pump 1), stored in the SAR tank and infiltrated back into the aquifer by gravity (Figure 1). Iron oxidation and As-adsorption take place underground and As-free water is obtained around the well. A treatment capacity of up to 10 m^3 As- and Fe-free water can be achieved by infiltrating 1.8 m^3/day.

Secondly, the CDI is used to reduce the high conductivity of the groundwater (3.3 mS/cm) and obtain drinking water. Using a peristaltic pump (feed pump 2) water in T1 is passed through the CDI in alternate cycles (desalination/ regeneration). Product water is stored in T2, which is connected to the water supply whereas the concentrate effluent is discharged directly into the canalization.

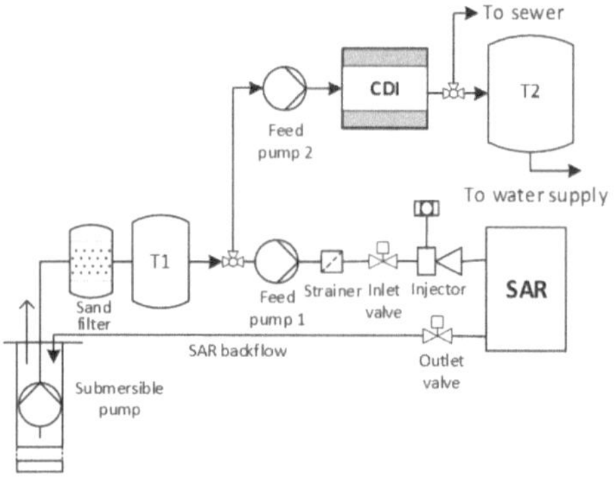

Figure 1. Pilot plant scheme for subsurface arsenic removal (SAR) and capacitive deionization (CDI).

The pilot plant was operated by 3 KW_p solar PV (8 panels SUNERGY – SUN72M-375W) and 2 kW_p wind turbine (Hummer – 2 KW On grid windspot) as well as 4.8 kWh lithium battery storage (Scared Sun).

3 RESULTS

3.1 *Fe, As and Mn-removal*

Results for the first 192 days of operation of the SAR plant are shown in Figure 2. During the first 10 days of operation, water was extracted and analyzed to monitor the development of the concentration in surroundings of the well. While Fe and Mn remained relatively stable (Δc = <5%), the concentrations of As increased by factor 2.5; however, stayed below the WHO limit value of 10 μg/L. This differed to the initial measured value of 11 μg/L, and can be attributed to natural occurring fluctuations.

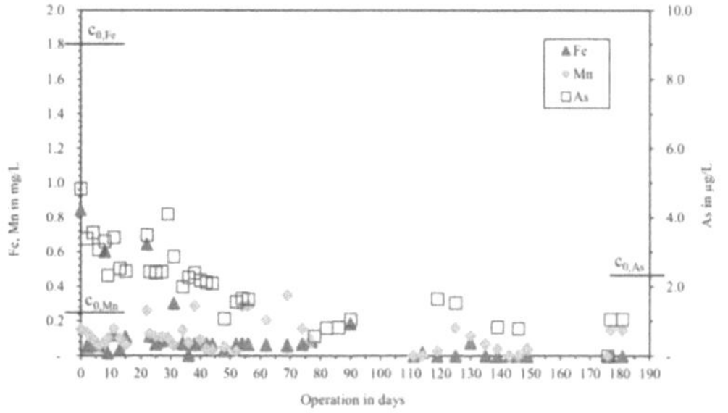

Figure 2. Iron, arsenic and manganese removal with SAR pilot plant in Tra Vinh.

After initiating the infiltration cycles in day 11, results show a decrease in Fe and As concentrations within the first weeks of operation. Values as low as 0.1 mg/L for Fe were measured from the first day, though Fe-concentrations spiked irregularly within the first 40 days until achieving constant values after day 47. Arsenic concentration of 4.8 µg/L at operation start was continuously lowered until reaching values below 0.5 µg/L in day 90 corresponding to elimination rates >90%. Although the SAR process aimed at only Fe and As removal, Mn concentration dropped linearly to below 0.6 mg/L within the first 14 days, indicating a >75% removal. Rises in Mn were observed throughout the first half of operation with concentrations increasing but not exceeding the initial values around 0.26 mg/L.

These rises can be explained by some small technical problems during operation including malfunction of submersible pump, over-extraction (abstraction of more than 10 m^3) and failure of the water meter. In addition, with a total capacity of up to 10 m^3, the quality of the abstracted water can vary. In general, results show a continuous removal trend for all three species, and after more than 190 days of operation no breakthrough was observed.

3.2 *TDS removal*

The desalination step with CDI started on day 122. To assess the overall salt removal and the adsorption efficiency for selected ions, samples were taken before and after the desalination process. Electrical conductivity was reduced from 3.3 to 0.56 mS/cm, corresponding to a desalination of 83%.

Arsenic and Mn^{2+} were completely removed by CDI achieving concentrations around zero (or below detection limit).

3.3 *Renewable energy supply*

During the pilot trials, the solar PV panels produced at a sunny day on average (radiation = 338 W/m^2) ca. 11.2 kWh/d while the wind turbine around 14.5 kWh/d. The average measured wind speed at the pilot location was 4.1 m/s, while the average maximum wind speed was 12.3 m/s. However, the neighborhood of the pilot site complained about high noise of the wind turbine so that it could not be operated continuously. With 50% operation of the wind turbine the daily energy production was estimated to 18.5 kWh/d.

The total average energy consumption for the operation of the SAR and CDI module was 21 kWh per day and a degree of autonomy for the pilot plant of >90% could be achieved.

4 CONCLUSION

- SAR has proven a feasible in-situ remediation technique for Fe, As and Mn in anoxic aquifers.
- Concentrations remained below the initial value even when fluctuations were measured.
- The combination of SAR and CDI show promising results regarding As mitigation and desalination for drinking water production.
- Solar (3 kW_p) and wind energy (2 kW_p) can supply >90% of total required energy, however operational parameters of CDI can still be improved to lower the total energy demand of the plant.

ACKNOWLEDGMENT

This project received funding from the German Ministry of Education and Research under the grant agreement number 02WAV1413.

REFERENCES

Al Marzooqi F.A., Al Ghaferi A.A., Saadat I. & Hilal N. 2014. Application of capacitive deionisation in water desalination: a review. *Desalination.* 342: 3–15

Ayers J.C., Goodbred S. George G., Fry D. Benneyworth L., Hornberger G., Roy K., Karim M.R. & Akter F. 2016. Sources of salinity and arsenic in groundwater in Southwest Bangladesh, *Geochem Trans.*, 17: 4

Bahar M. & Reza M. 2010. Hydrochemical characteristics and quality assessment of shallow groundwater in a coastal area of Southwest Bangladesh. *Environ. Earth Sci.*, 61:1065–1073

Buschmann J., Berg M, Stengel C., Winkel L., Sampson M.L., Pham T., Viet P.H. 2008. Contamination of drinking water resources in the Mekong delta floodplains: arsenic and other trace metals pose serious health risks to population. *Environ Int.*, 34(6): 756–764

Hien L.T., Quy P.N., Viet N.T. 2009. *Assessment of Salinity Intrusion in the Red River Under Effect of Climate Change.* ICEC 2009

Hossain M., Williams P.N., Mestrot A., Norton G.J. *et al.* 2012. Spatial heterogeneity and kinetic regulation of arsenic dynamics in mangrove sediments: the Sundarbans, Bangladesh. *Environ. Sci. Technol.* 46: 8645–8652.

Luong T.V., Canas Kurz E., Hellriegel U. Luu T.L., Hoinkis J. & Bundschuh J. 2018. Iron-based subsurface arsenic removal technologies by aeration: a review of the current state and future prospects. *Water Res.* 133: 110–122.

Luu T.L., 2019. Remarks on the current quality of groundwater in vietnam. *Environ. Sci. Pollut. Res. Int.* 26: 1163–1169.

Nickson R.T., McArthur J.M., Ravenscroft P., Burgess W. G. & Ahmed K.M. 2000. Mechanism of arsenic release to groundwater, Bangladesh and West Bengal. *Appl Geochem.* 15:403–413.

Sankar M.S., Vega M.A., Defoe P.P., Kibria M.G., Ford S., Telfeyan K., Neal A., Mohajerin T.J., Hettiarachchi G.M., Barua S., Hobson C., Johannesson K. & Datta S. 2014. Elevated arsenic and manganese in groundwaters of Murshidabad, West Bengal, India. *Sci. Total Environ.* 488–489: 570–579.

Smedley P. 2005. Arsenic Occurrence in Groundwater in South and East Asia. In: *Towards a More Effective Operational Response: Arsenic Contamination of Groundwater in South and East Asian Countries.* Report No. 31303, Volume II Technical Report, The World Bank.

Tan C., He C., Tang W., Kovalsky P., Fletcher J. & Waite T.D. 2018. Integration of photovoltaic energy supply with membrane capacitive deionization (MCDI) for salt removal from brackish waters. *Water Res.* 147, 276–286.

WaKap. 2019, www.wakap.de. Accessed on 23 January 2020

 ISBN 978-1-032-32928-4

Subsurface arsenic removal (SAR): Lessons learned from a long-term pilot application

E.E. Cañas Kurz[1], U. Hellriegel[1], V.T. Luong[2], J. Bundschuh[3], T. Winkelnkemper[4] & J. Hoinkis[1]
[1]*Center of Applied Research, Karlsruhe University of Applied Sciences, Karlsruhe, Germany*
[2]*Department of Mechatronics and Sensor Systems Technology, Vietnamese-German University, Vietnam*
[3]*Faculty of Health, Engineering and Sciences, University of Southern Queensland, Toowoomba, QLD, Australia*
[4]*Winkelnkemper GmbH, Wadersloh, Germany*

ABSTRACT: Reductive dissolution of arsenic (As) is identified as the principle cause for elevated concentrations in confined aquifers of Southeast Asia. In order to use groundwater resources as an alternative for the production of drinking water, oxidative iron-based, subsurface As removal (SAR) offers a sustainable solution. In this study, the results of two SAR pilot plants installed in the Mekong-Delta, Vietnam are presented. For the first time, both the removal efficiency and the evaluation of operational and technical aspects were investigated in order to assess the long-term applicability of the technology. Initial concentrations of As (81 and 10 µg/L) and iron (8.2 and 2.0 mg/L) were successfully lowered to below the WHO drinking water standard. The long-term study of the oxidation/adsorption process showed there was no As breakthrough after more than 28 months operation proving SAR as a suitable mitigation technique with the salient advantage of no toxic waste production and low-energy requirements.

1 INTRODUCTION

The subsurface arsenic removal (SAR) by aeration is a treatment option based on a well-established, in-situ deironing and demanganisation process called subsurface iron removal (SIR), a method that has been proven technically and economically feasible with long-term experience in Europe (Grischek *et al.*, 2016; Rott & Friedle 2000). However, using SIR for arsenic (As) removal is still yet to be verified as an all-round solution, also for developing countries.

In addition to its low operating and maintenance costs, the salient advantage of SAR is primarily its negligible waste production, since As is bound to the subsurface matrix where it naturally occurs (Karakish 2005; Luong *et al.*, 2018).

For the technical evaluation of SAR, the vulnerability of the process to diverse geochemical conditions such as pH, Fe:As ratio, and the presence of co-ions has to be investigated. In this study, the As removal from groundwater via SAR, based on the oxidation of dissolved iron (Fe) and the adsorption and co-precipitation of As onto Fe-(hydr)oxides was demonstrated.

2 MATERIALS AND METHODS

2.1 *Pilot location*

The first SAR pilot plant (PL1) for As removal was installed inland in the province of An Giang (10.719308, 105.331354). The second SAR plant (PL2) was set up near the coast in the province of Tra Vinh (9.940106, 106.334515) in the Mekong Delta for the treatment of As-laden brackish groundwater.

2.2 *Plant operation*

A delivery pump with a pressure vessel was used to abstract water from the well and aerate it using an injection nozzle (Figure 1, *FERMANOX®-Wasseraufbereitung* type BV 45). The working pressure was controlled with a pressure switch at $3 < p < 4.5$ bar. The oxygenated water was infiltrated back into the aquifer via gravity flow for the formation of oxidation/adsorption zones around the well (*infiltration cycle*). Subsequently, fresh water could be abstracted from the well (*extraction cycle*).

The treatment capacity of each plant was given by the infiltration to abstraction ratio Q_E (Table 1), which gives the volume of aerated water

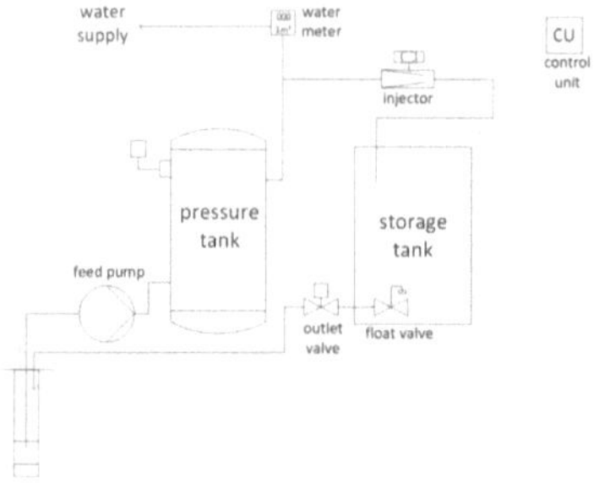

Figure 1. Simplified scheme of a SAR treatment plant with feed pump, air injector and control unit.

DOI: 10.1201/9781003317395-165

Table 1. Treatment capacity of pilot plants.

Plant No.	Capacity (m^3)	Infiltration volume (m^3)	Infiltration/abstraction ratio Q_E
PL1	1.9	2.78	1.46
PL2	8.3	1.80	0.22

needed to produce 1 m^3 of treated water. This ratio is given by the raw water quality at each site. A control unit was used to monitor the *infiltration* and *extraction cycles* based on water demand using a water meter. To avoid abstraction of more water than the daily limit (over-extraction), an alarm was triggered if water consumption exceeded Q_E.

3 RESULTS

This study comprises the results obtained during the operation of two SAR pilot plants (PL1 and PL2) and a summary of technical challenges faced within a period of 28 (PL1) and 16 (PL2) months.

3.1 *Arsenic mitigation*

Initial concentrations of 81 and 10 μg As /L and 8.2 and 2.0 mg Fe /L for PL1 and PL2, respectively, were successfully lowered to below the WHO drinking water standards of 10 μg As/L and 0.5 mg Fe/L. Removals of 91.10% and 99.99% for PL1 and PL2, respectively indicated higher As removal with higher Fe:As ratios (104 and 275). However, the effect of initial As concentrations, the presence of co-ions and hydrogeological conditions must be included.

3.2 *Operational challenges*

Manganese (Mn) deposits: The formation of Mn-oxides was observed within the first month of operation (blackish coloring of pipes). This caused no technical problems, however the prolonged extraction of groundwater containing Mn-levels > 0.5 mg/L posed a risk of blocking.

Pump malfunction: Technical problems with the feed pump included pump breakdown due to corrosion of non-stainless-steel parts and motor overload due to broken pressure switch and a faulty overvoltage protection.

Over-extraction: In several occasions, more water was extracted than the capacity limit, mainly due to consumer's behavior or pump breakdown. Yet, some other reasons were: Faulty water meter, obstruction of water meter due to particles (unsealed well), as well as incomplete *infiltration* cycles due to a leakage between pump and pressure vessel caused by corrosion, which hindered the self-suction property of the pump and the daily automatic infiltrations. If no *infiltration* cycle is carried out, the risk of prolonged over-extraction may result in the rise of concentrations and incrustations around the well.

Well blockage: Yet, well blockage was encountered only once after a period of several weeks of over-extraction due to a pump malfunction and the formation of a biofilm rather than Mn-/Fe-deposits.

Biomass film: Due to high methane concentrations in the groundwater and the growth of methanogenic bacteria a biomass film layer was observed in the well after several months of operation. This was mitigated by chemical and mechanical cleaning, and can be prevented by avoiding over-extraction. Another preventive measure was a periodic cleaning with H_2O_2 to protect the well against biomass deposition.

Well construction: After some weeks operation (PL2), fine particles were discharged from the well and sludge accumulated in the system, starting to block the process. This was mainly due to the unsealed well and unsecure well zone, which is a problem seen in the whole region. A sand filter was additionally installed as a pre-filter to eliminate this problem.

4 CONCLUSIONS AND RECOMMENDATIONS

Long-term results proved SAR as a feasible mitigation technique. Although several technical problems were encountered, all problems were successfully eased and drinking water standard was achieved.

The success of the As removal was attributed to:

- Fixed infiltration-to-abstraction ratios (Q_E) and controlled water demand (water quality safety)
- Automated infiltration cycles (ease of use)
- Sufficient water oxygenation (technical safety)
- Making sure local parts and components are of sufficient quality (stainless steel when needed) (operational safety)

ACKNOWLEDGEMENTS

This project received funding from the German Ministry of Education and Research (02WAV1413)

REFERENCES

Grischek T., Feistel U., Ebermann J., Musche F., Bruntsch S. & Uhlmann W. 2016. Field experiments on subsurface iron removal in the lusatian mining region, in: IMWA 2016, Freiberg, pp. 292–297.

Karakish A.A.K. 2005. Subsurface removal of iron and manganese from groundwater – case study, In: *Ninth International Water Technology Conference*, IWTC9. Cairo, pp. 415 429.

Luong V.T., Cañas Kurz E.E., Hellriegel U., Luu T.L., Hoinkis J. & Bundschuh J., 2018. Iron-based subsurface arsenic removal technologies by aeration: a review of the current state and future prospects. *Water Res.* 133: 110–122.

Rott U. & Friedle M., 2000. 25 Jahre Unterirdische Wasseraufbereitung in Deutschland. *Wasser Spec.* 141: 99–107.

Arsenic in the Environment: Bridging Science to Practice for Sustainable Development –
van der Wal, Ahmad, Petrusevski, Weijma, Savic, van der Wens, Beerendonk, Bhattacharya, Bundschuh & Naidu (Eds)
© 2023 the Author(s), ISBN 978-1-032-32928-4

Introduction of low-cost sustainable solution for household arsenic groundwater treatment in Kudpur, Punjab Pakistan

S. Batool, N. Ali & A. Farooqi
Department of Environmental Sciences, Quaid-i-Azam University, Islamabad Capital Territory, Pakistan

ABSTRACT: Pakistan is among the 4th most affected countries in the globe with regards to Arsenic pollution. Although the extent of arsenic groundwater pollution is not much but there are places which contain one of the highest levls of arsenic contamination in the region (3.7mg/L). Kudpur village in district Lahore, Punjab Pakistan is one of these areas. Most of the population affected by arsenic groundwater pollution in the Province are the low-income rural communities, who neither have the resources to afford sophisticated technologies for water treatment nor the wisdom to do so. Therefore, a low cost easy to maintain and sustainable solution for removal od arsenic in groundwater was desperately needed. This paper presents one such solution. A low cost, solution with minimum operational requirements is introduced to the arsenic groundwater affected community of Kudpur. The paper presents the process of development as well as introduction of the treatment system in the arsenic affected area of Kudpur.

1 INTRODUCTION

Arsenic (As) is a pervasive element found in all environmental matrices. However, its most harmful effects have been observed when it is consumed through potable water (surface and subsurface). It has been reported that out of all groundwater contaminants the most severe health effects on global scale are caused by As (Mudgal *et al.*, 2010). Among the countries most affected by arsenic groundwater pollution Pakistan ranks 4th globally (Jadhav *et al.*, 2015). As accumulation in human body results in arsenicosis. It starts as pigmentation of skin that in later stages develops into various cancers. Its acute poising affects gastro-intestinal tract, nervous system and nervous system (Chouhan & Flora 2010). The most affected As areas in Pakistan are rural population of Punjab and Sindh Provinces(van Geen *et al.*, 2019). Since majority of inhabitants of these areas are low income farmers who rely on groundwater for both agriculture and potable water supply. High levels of As pollution in the groundwater of these areas has put the poor population of these villages at risk (Farooqi *et al.*, 2007). There are visible signs of arsenicoisis among the villagers. Therefore, a low cost, sustainable As removing technology is needed in the area. The paper presents a pilot project by Department of Environmental Sciences Quaid i Azam University to combat As pollution in these low income communities.

2 MATERIALS AND METHODS

2.1 *Study area*

Kudpur is a small village/town in Lahore district, located about 26 km from Lahore, Punjab. Kudpur was selected as the project site because of high As concentrations recorded in blanket testing. The village is not densely populated therefore, a viable option for pilot study.

A total of 228 samples were taken from the study area during blanket sampling. The analysis depicted 58% of the samples with high As (i.e. above 0.01 mg/l). Figure 1 shows satellite image of Kud Pur obtained from Google Earth, locations for some of the notable samples are marked on the image along with the results of blanket testing.

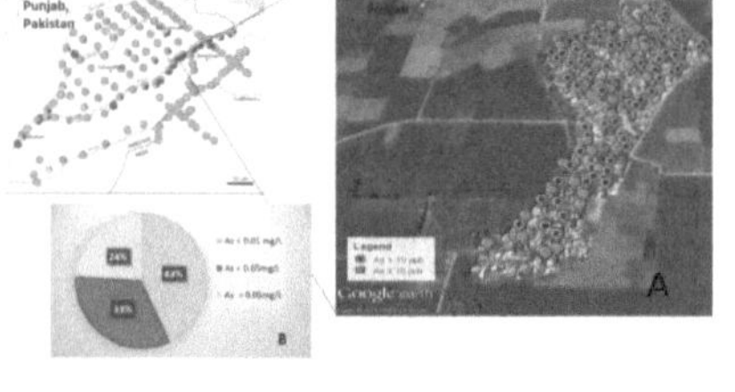

Figure 1. Study Area (A), Arsenic in groundwater as reported in the blanket survey.

DOI: 10.1201/9781003317395-166

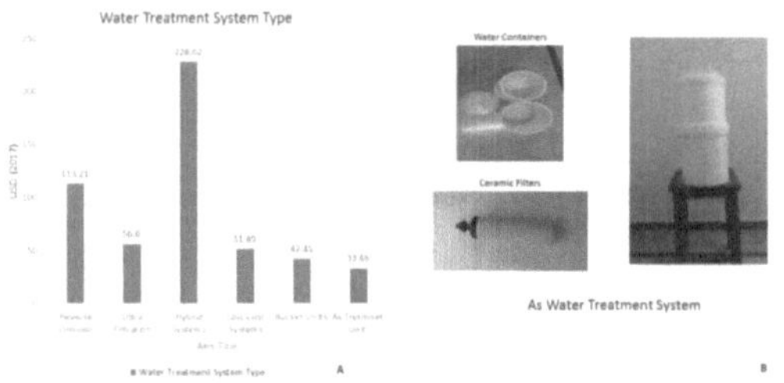

Figure 2. Market validation survey results (A), As treatment system, and the filtration system (B).

2.2 *Development of low-cost treatment system using precipitation, adsorption and filtration techniques*

Highest As concentration observed from the blanket testing (3.7mg/L) was from one of the households in Kudpur. Therefore, groundwater was sampled and collected from this house during the development of the treatment system.

Precipitation was used as the main technique for As removal while adsorption and filtration were used to obtain the optimum dosage for As removal (since this technology is being patented, further information cannot be disclosed). The filtration system was developed in collaboration with a corporate partner ALPA Services (Pvt.) Ltd. The ceramic filters used in the filtration system were also modified by ALPA Services. Figure 2B shows the filtration unit along with the modified ceramic filters.

In order to check the efficiency of removal regular analysis of both raw and treated water were verifed by Pakistan Council of Research on Water Resources (PCRWR). PCRWR is an ISO certified government institution responsible for water resource monitoring in the country.

2.3 *Market validation for the treatment system*

In order to introduce the As treatment system to the affected village communities of Kudpur, market validation surveys were conducted. The first survey was based on water market i.e. technologies/systems available for water purification (especially As removal). The lowest prices of the existing systems were compared with the cost price of As treatment system. The graph in Figure 2A represents the results of the market validation. Furthermore, willingness to pay was also determined through a survey in Kudpur.

2.4 *Introduction of the treatment system to the village communities*

Upon finalization of system production, the next step was introduction of the system to the village communities. In the first phase a short training program was conducted in which the residents were briefed about the harmful affects of As polluted groundwater. The system was introduced at the end of the training.

Contrary to the views expressed in the previous surveys the villagers were reluctant to invest in the treatment system since they belong to the low-income class. Therefore, in order to introduce this technology, these systems were distributed free of charge to the 40 households in Kudpur. This was done under the umbrella Social Integration and Out-reach Program (SOIP) funded jointly by Higher Education Commission and the Quaid i Azam university.

3 CONCLUSIONS AND RECOMMENDATIONS

Arsenic contamination in Punjab had been observed/reported since more than a decade. There are people suffering from various forms of arsinocosis and cancers are also prevalent especially in Kudpur. However, lack of education and conservativeness are a major hinderance in adopting any new technology. Therefore, it is imperative to train the rural communities for water management and conservation.

ACKNOWLEDGEMENTS

We acknowledge the Hydrogeochemistry Laboratory, Department of Environmental Sciences, Quaid-i-Azam University for support in research and analysis required for the successful completion of the project.

REFERENCES

Chouhan S. & Flora S.J.S. 2010. Arsenic and fluoride: two major ground water pollutants. *Indian J. Exp. Biol.* 48(7): 666–678.

Farooqi A., Masuda H. & Firdous N. 2007. Toxic fluoride and arsenic contaminated groundwater in the Lahore and Kasur districts, Punjab, Pakistan and possible contaminant sources. *Environ. Pollut.* 145(3): 839–849.

Jadhav S.V., Bringas E., Yadav G.D., Rathod V.K., Ortiz I., & Marathe K.V. 2015. Arsenic and fluoride contaminated groundwaters: a review of current technologies for contaminants removal. *J Environ Manage.* 162:306–325.

Mudgal V., Madaan N., Singh R.B. & Mishra S. 2010. Effect of toxic metals on human health. *Open Nutraceuticals J.* 3(1): 94–99.

van Geen A., Farooqi A., Kumar A., Khattak J.A., Mushtaq N., Hussain I., Ellis T. & Singh C.K. 2019. Field testing of over 30,000 wells for arsenic across 400 villages of the Punjab plains of Pakistan and India: implications for grprioritizing mitigation. *Sci. Total Environ.* 654: 1358–1363.

Arsenic in the Environment: Bridging Science to Practice for Sustainable Development –
van der Wal, Ahmad, Petrusevski, Weijma, Savic, van der Wens, Beerendonk, Bhattacharya, Bundschuh & Naidu (Eds)
© 2023 the Author(s), ISBN 978-1-032-32928-4

Arsenic and disinfection byproducts concentrations in Osijek's water distribution system, Eastern Croatia

L. Kurajica[1], M. Ujević Bošnjak[1], J. Štiglić[1], Ž. Romić[2], J. Zima[2], F. Dako[2] & V. Santo[3]
[1]*Department for Water Safety and Water Supply, Croatian Institute of Public Health, Zagreb, Croatia*
[2]*Osijek Water Supply Company, Poljski put 1, Osijek, Croatia*
[3]*Department for Water Safety and Water Supply, Institute of Public Health of Osijek-Baranja County, Osijek, Croatia*

ABSTRACT: Studies on arsenic (As) occurrence in groundwaters in Osijek-Baranja county have been reported. Even after water treatment of groundwater, As concentrations are above 10 μg/L, which is maximum contaminant level (MCL). To reduce As concentration, groundwater is mixed with surface water from the river Drava. In this study As concentrations were monitored from February to October of 2019, along with concentrations of disinfection byproducts (DBPs) to see if there is some correlation between these parameters. GIS was used to create a map with a sampling points, and software Statistica was used for descriptive statistics and to calculate Spearman's correlation coefficients. Average As concentration in the water distribution system (WDS) was 22.1 μg/L, with a range from 5.3 to 34.6 μg/L. Average concentrations for THM and HAA were 8.4 μg/L and 4.2 μg/L. As concentrations statistically significantly correlate with THMs; BDCM ($r = -0.28$, $p<0.05$), DBCM ($r = -0.29$, $p<0.05$) and TBM ($r = -0.34$, $p<0.05$) and with one HAA, DBAA ($r = -0.25$, $p<0.05$).

1 INTRODUCTION

High groundwater levels of arsenic (As) in Eastern Croatia, especially in Osijek-Baranja county have been reported (Cavar *et al.*, 2005; Habuda-Stanić *et al.*, 2007; Santo *et al.*, 2002; Ujević *et al.*, 2010). Maximum contaminant level (MCL) for As in tap water is 10 μg/L (European Commission 2018), however the European Commission approved a derogation from the prescribed values for water supply zone Osijek at 30 μg/L by the end of 2020. Corrosion deposits within DWDS can accumulate As and other inorganic contaminants (Gerke *et al.*, 2009) and these accumulations could be released to the user's tap water (Lytle *et al.*, 2004). One of the causes of release of inorganic contaminants from pipe scales is disinfectant change (Boyd *et al.*, 2008). Disinfectant like chlorine could react with natural organic matter (NOM) and cause disinfection byproducts (DBPs) formation (Richardson 2011). Recent laboratory research showed that the simultaneous presence of trihalomethanes (THM) and haloacetic acids (HAA), most common DBPs promoted a release of As from scale-only and scale biofilm conglomerates in finished water (Andra *et al.*, 2014). The aim of this study was to i) determine As concentrations in tap water in the city of Osijek, ii) determine concentrations of DBPs (THM and HAA) and to characterize organic matter (OM), iii) determine temporal variability of measured parameters and iv) determine Spearman's correlation coefficient for measured parameters.

2 MATERIALS AND METHODS

2.1 *Study area*

The study was conducted on water distribution system of Osijek that serves approximately 130 000 inhabitants. The area is situated within the alluvial basins of the Drava river. Groundwater from 18 deep wells situated at the wellfield Vinogradi is used for water supply. Raw water has higher arsenic, iron and manganese concentration and organic matter so before going to system and suppliers it needs treatment. The treatment consists of aeration, sedimentation, filtration and disinfection with chlorine. In order to produce drinking water that contains As in concentrations below 10 μg/L ground water was mixed with surface water from the Drava river. During the sampling campaign groundwater was mixed with surface water in different ratios in May, July and October.

2.2 *Selection of the sampling points*

In order to evaluate the quality of raw water one well (No. 18) was chosen as sampling point, two sampling points were chosen at the WTP, and in order to determine the quality of tap water, sampling points located all over the city were chosen (Figure 1).

DOI: 10.1201/9781003317395-167

Figure 1. City of Osijek with 32 sampling locations.

2.3 *Water sampling and laboratory analysis*

Water sampling was carried out from March till October 2019. A total of 92 water samples were collected, of which 3 of raw groundwater (well No. 18), 4 of raw mixed water after aeration, 4 after disinfection and 81 of tap water (Figure 1). The physio-chemical parameters such as, pH, free residual chlorine, temperature (T), electrical conductivity (EC), oxidation-reduction potential (ORP), pH value and turbidity (TU) were measured in the field. Other parameters were determined in the laboratory. Metals were measured on ICP-MS, disinfection byproduct were determined by gas chromatography coupled with electron capture detector and total organic carbon (TOC) on TOC analyzer. Specific UV absorbance was calculated from TOC value and absorbance at 254 nm.

2.4 *Data analysis*

GIS software was used to create a map comprising the location where water was sampled. Statistica V. 9.1. was used for data analysis (minimum, maximum, average and standard deviation) and for calculating Spearman's correlation coefficient.

3 RESULTS AND DISCUSSION

3.1 *Arsenic in raw and treated water in Osijek*

Average As concentration at well 18 is 117 μg/L, with the range from 74 to 175 μg/L, while average As concentration of raw mixed water after aeration is 224 μg/L, with the range from 181 to 264 μg/L. Average As concentration after disinfection at the exit of the WTP is 30 μg/L, with the range from 25 to 34 μg/L, while the range of As concentration at tap water is from 5.3 to 34.6 μg/L, with average concentration 22.1 μg/L. The lowest As concentrations (<10 μg/L) were obtained by mixing 25% of groundwater and 75% surface water in October.

3.2 *Disinfection byproducts and organic matter (OM) in Osijek*

Total trihalomethanes (TTHM) concentrations ranged from 2.2 μg/L to 51.8 μg/L, with average concentration 8.4 μg/L. Prevalent THM is chloroform (TCM) with average concentration 8.4 μg/L. Total haloacetic acid (THAA) ranged from below LOQ to 13.8 μg/L, with average concentration 4.2 μg/L. Prevalent HAA are dichloroacetic acid (DCAA) and trichloroacetic acid (TCAA) with average concentrations 2.0 μg/L and 1.8 μg/L. Average TOC values at wellfield 18 were 2.5 mg/L, at WTP in mixed raw water after aeration 3.2 mg/L, and in the network 2.5 mg/L. SUVA was calculated to characterize OM. Average SUVA value at wellfield 18 was 6.9, average SUVA value of mixed raw water after aeration was 4.9, and in the network 2.6, meaning that the structure of organic matter is changing. Raw water contains hydrophobic aquatic humics with high molecular weight (MW), while water after treatment in the network contains more hydrophobic, low MW non-humics and hydrophilic humics.

3.3 *Temporal variations of As, OM and DBPs*

Since the two water types were used in distribution system, one with higher As concentrations, and one without, there is a high temporal variability of As (Figure 2). When comparing As and THM concentrations over sampling period one can notice that every time the waters were mixed to reduce As concentrations, THM concentrations rises comparing to previous month. TOC and SUVA values has the similar trend of values like As probably because of often changes in type of raw water. As, TOC, SUVA and THM values are the highest in August, and HAA values in September. SUVA and TOC values are lower in months where groundwater and surface water were mixed.

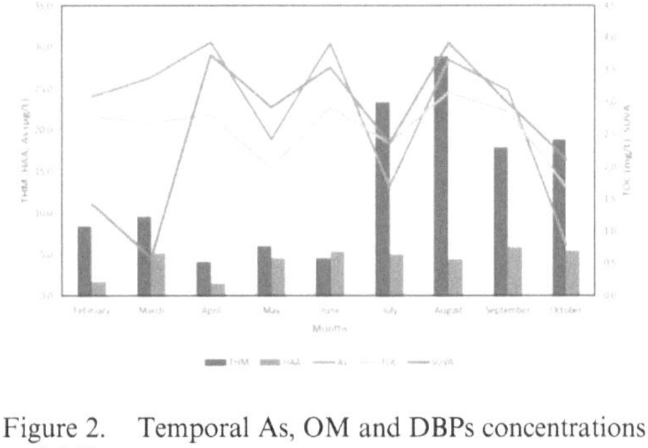

Figure 2. Temporal As, OM and DBPs concentrations variability.

3.4 *Spearman's correlation coefficients*

Spearman correlation coefficients were calculated to determine statistical relationship between concentrations of parameters in distribution system. As concentrations statistically significantly correlate with THMs; bromdichloromethane (BDCM), dibromchloromethane (DBCM) and tribromomethane (TBM) ($r = -0.23$, $p<0.05$) and with one HAA, dibromacetic acid (DBAA) ($r = -0.24$, $p<0.05$). As concentrations also correlate with TOC ($r = 0.22$, $p<0.05$), SUVA ($r = 0.44$, $p<0.05$), chlorate ions ($r = -0.44$, $p<0.05$), turbidity ($r = 0.25$, $p<0.05$), pH value ($r = 0.48$, $p<0.05$), free chlorine ($r = -0.24$, $p<0.05$). There was a significant correlation between As and other metals like copper ($r = -0.19$, $p<0.05$), strontium ($r = 0.67$, $p<0.05$) and barium ($r = 0.64$, $p<0.05$)

4 CONCLUSIONS AND RECOMMENDATIONS

Water treatment plant reduces As concentrations to about 30 μg/L. To reduce it to even lower concentrations, mixing of groundwater with surface water is applied and As concentrations <10 μg/L were found in the distribution network in Osijek. However, mixing of water could cause changes in the water quality and formation of disinfection byproducts. Average TOC values in raw-groundwater and at the tap in the months where only groundwater was in the system are similar, meaning that the amount of OM doesn't change much after water treatment. On the other hand, SUVA values are quite different in raw- groundwater and in the tap water, meaning that the structure of OM is changing. In months where there is surface water in the system, As concentrations are lower, TOC and SUVA values are also lower, and THMs concentrations rises, with the exception of August. Statistical relationship between As and DBPs/OM concentrations was confirmed by calculating Spearman's correlation coefficient. Further investigations are necessary to characterize organic matter and to characterize deposits in DWDS. After that laboratory experiments should be conducted to understand processes that occur in DWDS.

ACKNOWLEDGEMENTS

This work has been supported by the Croatian Science Foundation under the project number [UIP-2017-05-3088].The authors would like to thank their colleagues from the Croatian Institute of Public Health, Department for Water Safety and Water Supply and Unit for Metals and Metalloids for their help with water analyses.

REFERENCES

Andra S.S., Makris K.C., Botsaris G., Charisiadis P., Kalyvas H. & Costa C.N. 2014. Evidence of arsenic release promoted by disinfection by-products within drinking-water distribution systems. *Sci. Total Environ.* 472: 1145–1151.

Boyd G.R., Dewis K.M., Korshin G.V., Reiber S.H., Schock M.R., Sandvig A.M. *et al.* 2008. Effects of changing disinfectants on lead and copper release: Part 1—literature review. *J. Am. Water Works Assoc.* 100: 1–44.

Ćavar S., Klapec T., Jurišić Grubešić R. & Valek M. 2005. High exposure to arsenic from drinking water from several localities in Eastern Croatia. *Sci. Total Environ.* 339: 277–282.

European Commission 2018. Annexes to the Proposal for a Directive of the European Parliament and of the Council on the Quality of Water Intended for Human Consumption (recast). EUR-Lex. Brussels.

Gerke T.L., Scheckel K.G. & Schock M.R. 2009. Identification and distribution of vanadinite ($Pb_5(V_5O_4)_3Cl$) in lead pipe corrosion by-products. *Environ. Sci. Technol.* 43: 4412–4418.

Habuda-Stanić M., Kuleš M., Kalajdžić B., Romić Ž. 2007. Quality of groundwater in Eastern Croatia. The problem of arsenic pollution. *Desalination* 210: 157–162

Lytle D.A., Sorg T.J. & Frietch C. 2004. Accumulation of arsenic in drinking water distribution system. *Environ. Sci. Technol.* 38: 5365–5372.

Richardson S.D. 2011. Disinfection By-Products: Formation and Occurrence in Drinking Water, In: Nriagu J.O. (ed.) *Encyclopedia of Environmental Health*. pp. 110–136.

Santo V., Grgić J., Laslavić B., Dario M. & Valek M. 2002. Concentration of Arsenic, Manganese and Iron in Drinking Water of Osijek-Baranja County. In: E. Lovrić (ed.) *Proceedings of the 6th Symposium Water and Public Water Supply*, 16–19 October 2002, Mlini, Zupa Dubrovacka, Croatia. Croatian Institute of Public Health, Zagreb, pp. 69–77 (in Croatian).

Ujević M., Duić Ž., Casiot C., Sipos L., Santo V., Dadić Ž. & Halamić J. 2010. Occurrence and geochemistry of arsenic in the groundwater of Eastern Croatia. *Appl. Geochem.* 25: 1017–1029.

Arsenic in the Environment: Bridging Science to Practice for Sustainable Development –
van der Wal, Ahmad, Petrusevski, Weijma, Savic, van der Wens, Beerendonk, Bhattacharya, Bundschuh & Naidu (Eds)
 ISBN 978-1-032-32928-4

Arsenic contaminated groundwater in Serbia and the application of pilot scale investigations in the search for sustainable water supply solutions

M. Watson, A. Tubić, J. Nikić, J. Agbaba & B. Dalmacija
Environmental Protection Research Group, Department for Chemistry, Biochemistry and Environmental Protection, Faculty of Sciences, University of Novi Sad, Novi Sad, Republic of Serbia

ABSTRACT: In Vojvodina, the northern province of the Republic of Serbia, around 600,000 people are still supplied with drinking water with elevated arsenic concentrations. The different groundwaters contain up to 270 μg/L of arsenic, and various other natural contaminants such as natural organic matter, sodium and phosphates. These other constituents can make the water difficult to treat efficiently. This work provides three examples of pilot scale investigations conducted in Vojvodina to identify and optimise technologically sound and economically feasible water supply solutions which take into account not only the quality of the source water, but also the levels of human and financial resources available for implementation of water supply infrastructure in the region.

1 INTRODUCTION

Arsenic contamination of groundwater is a wide spread public health problem in many countries, including Serbia. Vojvodina, the northern province of Serbia, lies within the Pannonian basin, the former location of the Pannonian sea which existed during the Miocene and Pliocene epochs, extending into what is now Serbia, Hungary, Romania and Croatia. Aquifers in these sediments may have arsenic concentrations as high as 270 μg/L (Figure 1).

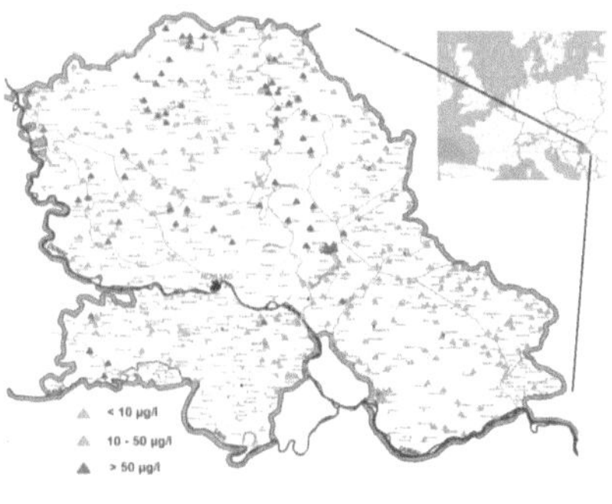

Figure 1. Arsenic contaminated groundwater used for water supply in Vojvodina, Republic of Serbia (adapted from Official Gazette APV 2010).

Unfortunately, water treatment infrastructures in the region are often lacking any sort of arsenic removal technology. As a result, in Vojvodina alone, around 600,000 people are currently still supplied with drinking water which contains more than the 10 μg/L maximum allowable concentration of arsenic (Agbaba & Dalmacija 2013).

In 2018, the GDP per capita in Serbia was 5,190 EUR, five times less than the EU27 average of 27,620 EUR (Eurostat 2020). Given the resulting limitations on the resources of local government, just as much emphasis must be placed on improving the economic feasibility, as well as the technical feasibility, of any arsenic removal technologies required.

2 PILOT INVESTIGATION CASE STUDIES

In addition to our fundamental research into the interactions between arsenic and other water constituents and developing new arsenic treatment technologies in the laboratory, the Environmental Protection Research Group also carries out on-site pilot scale investigations into optimal water treatment solutions. In this work, we present recent examples of pilots applying various technological solutions for three locations in Vojvodina with very different water source qualities.

2.1 *Enhanced coagulation, ozonation and filtration in Novi Bečej*

Novi Bečej (population < 24,000) extracts water from 6 different wells, with arsenic concentrations in the range 160 to 270 μg/L. The water has a high natural organic matter (NOM) content (average 6.7 mg DOC/L) and also has elevated sodium and ammonia contents. This water is very difficult to treat due to the complex nature of the NOM. After preliminary lab-scale and pilot investigations, water from the first aquifer layer was also investigated, adding the potential to apply river bank filtration from the nearby Tisa to the existing sources. Different preliminary treatments were applied to both sources: the current arsenic loaded water was subject to aeration, preozonation, iron chloride coagulation and sedimentation and dual media sand/anthracite filtration. Water from the shallow aquifer was subject to

DOI: 10.1201/9781003317395-168

aeration, preozonation and filtration. These waters were then combined prior to final ozonation and GAC filtration steps. By continuing to utilise the existing source and combining it with better quality water, this second variant requires significantly smaller capital investments for implementation than using the existing water source alone.

2.2 *Adsorption on cerium oxide in Temerin*

The water supplied to the town of Temerin (population <30,000) is of considerably better quality, but has an arsenic concentration of around 40 µg/L, and a relatively high pH of 8.3. In a pilot study designed to provide the data required for the dimensioning and design of a new process line at the waterworks, a simple single column adsorber was operated at around 250 L/h for 4 months of continuous operation. The adsorbent investigated was commercially available granulated cerium oxide. Arsenic breakthrough was eventually observed after 24,000 bed volumes of water were treated, indicating that the adsorbent represents a technologically sound solution to the elevated arsenic concentrations in the water.

2.3 *In-line coagulation and adsorption in Višnjićevo*

Operating a pilot in small rural settlements presents additional challenges in addition to the water quality itself – in the case of Višnjićevo (population < 1000), supporting infrastructure was limited and power cuts to the entire village, frequent. The water from the two wells (only one operates at a time) contains around 130 µg/L arsenic, a somewhat elevated NOM content of 2.45 mg DOC/L, and a phosphate content of 1.33 mg P/L. Although this phosphate level satisfies the drinking water quality regulations, it creates difficult conditions for the removal of arsenic by adsorption, as the phosphate anion has a very similar structure the arsenate anion, leading to the potential for competitive adsorption. Initial investigations using a single adsorber operated at 250 L/h and packed with commercial granular iron oxide adsorbent quickly confirmed the challenge of removing arsenic from phosphate rich waters, reaching breakthrough after only 2000 BV.

It is not feasible in such small settlements to find the financial resources to install, operate and maintain complex treatment lines. However, it was clear that a single adsorption step was not going to provide adequate arsenic removal. After further laboratory experiments, the pilot was extended to include initial pH correction with HCl, in-line $FeCl_3$ coagulation and a sand/anthracite filter, followed by the final iron oxide filter. In this case, by removing most of the phosphates and approximately 50% of the arsenic during the in-line coagulation step, we were able to extend the life of the iron oxide filter media by a factor of 8 (Figure 2).

Maintaining an acceptable iron residual after the inline coagulation step was challenging, such that we believe further optimisation in the operation of a full scale sand/anthracite filter will allow treatment of closer to 25,000 BV before it is necessary to regenerate or replace the arsenic adsorption media.

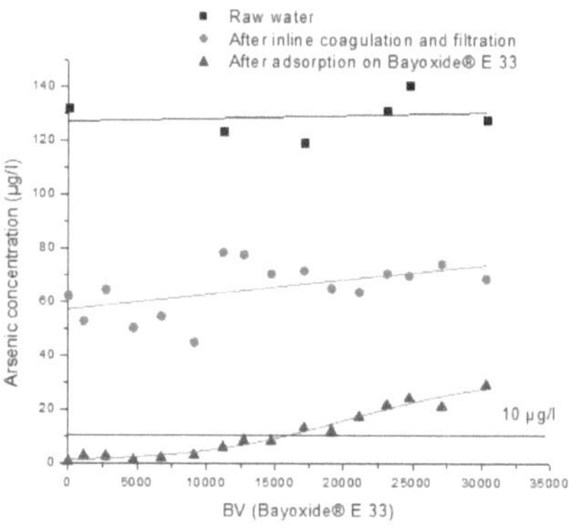

Figure 2. Arsenic concentration at the Višnjićevo pilot after inline coagulation, filtration and adsorption.

3 CONCLUSIONS AND RECOMMENDATIONS

Hundreds of settlements in Vojvodina require arsenic removal technologies to be installed at the treatment plant to provide safe drinking water. These settlements vary in size from less than 1,000 to above 75,000 inhabitants, and have wells which present diverse challenges in terms of source water quality. Serious capital, operation and maintenance investments are required to solve these problems. In order to avoid expensive mistakes during the planning and design of these waterworks, vigorous and robust pilot scale investigations are necessary, to ensure the solutions suggested are both economically and technologically feasible.

ACKNOWLEDGEMENTS

The authors acknowledge financial support of the Ministry of Education, Science and Technological Development of the Republic of Serbia (Grant No. 451-03-68/2020-14/200125), and the assistance of company S-Metalltech 98 KFT, Hungary.

REFERENCES

Agbaba J. & Dalmacija B. (eds.) 2013. *Possible Solutions for Water Supply in Bačka and Northern and Central Banat Based on Micro- and Macro-Regional Systems*, University of Novi Sad Faculty of Sciences, Novi Sad, Serbia.

Eurostat. 2020. *Eurostat Data*, https://ec.europa.eu/eurostat/databrowser/view/sdg_08_10/default/table?lang=en (accessed on 28 02 2020).

Official Gazette APV, No. 1/2010, *Strategy of Water Supply and Water Protection in Vojvodina*.

 ISBN 978-1-032-32928-4

Influence of softening on arsenic adsorption at water treatment plant Ouddorp (The Netherlands)

D. de Ridder[1], S. Abdoel Gafour[1], P. Vollaard[1], B. Hofs[1], B. Schaaf[1], A. Hogendoorn[1], A. Ahmad[1,2,3] & A. van der Wal[1]
[1]*Evides Water Company, Department of Water Technology and Source Protection, Rotterdam, The Netherlands*
[2]*Wageningen University, Department of Environmental Technology, Wageningen, The Netherlands*
[3]*KWR Watercycle Research Institute, Water Systems and Technology, Nieuwegein, The Netherlands*

ABSTRACT: For drinking water production, the influence of cation exchange and pellet softening on subsequent arsenic (As) removal by iron was investigated. It was found that Ca^{2+}(aq) removal had limited effect on iron floc growth, but the presence of fine carry-over hindered floc growth severely. Compared to a situation without softening, implementing cation exchange resulted in an increase of 25% in the required iron dose for As removal, while it had to be doubled after pellet softening. Jar test do, however, underestimate As adsorption when compared to pilot filters, since they only simulate reactions in the filter supernatant.

1 INTRODUCTION

Evides Water Company investigates the implementation of softening at a treatment plant which uses dune infiltrated water as source. Two common softening techniques are cation exchange (catIEX) and pellet softening (PS), but their implementation can potentially decrease arsenic (As) removal, since:

- Iron floc growth is hindered at low Ca^{2+}(aq) and Mg^{2+}(aq) concentrations (de Ridder *et al.*, 2018), and
- Arsenate adsorption onto iron flocs decreases at higher pH values (Dixit & Hering 2003).

Our aim was to quantify the impact of both catIEX and PS on As adsorption.

2 MATERIALS AND METHODS

2.1 *Process scheme*

Surface water is infiltrated in the dunes and recovered after a residence time of 30 days. The treatment steps following dune filtration are spray aeration, rapid sand filtration (RSF), activated carbon filtration (ACF) and ultrafiltration (UF). For enhanced As removal iron is dosed prior to both RSF and UF.

Pellet softening can be implemented either as first (anaerobic) treatment step, after aeration or after rapid sand filtration. In our current investigation, the latter was selected, and RS filtrate was used in our experiments.

CatIEX benefits from a low particle load, since this reduces the backwash frequency of the resin bed. Disinfection is, however, still required after catIEX. For this reason, catIEX was considered after ACF, and AC filtrate was used in our experiments.

2.2 *Procedure*

For both softening techniques, the goal was to reach 1,4 mmol/L hardness in the final treated water. When water demand is high, separately treated natural groundwater is used as an additional source, and mixed in the clear water reservoir. This water has a high hardness, and as a result, a hardness goal of 0,5 mmol/l is set for the main treatment in order to reach 1,4 mmol/L after mixing in the clear water reservoir.

Softening was carried out in a 23 L batch for both PS and catIEX, with continuous mixing for 30 minutes. For catIEX softening, 230 g Dowex Marathon 1300 Na resin was used, while 2.2 kg washed and broken calcite pellets of 0.4–0.6 mm were used for PS. The PS reaction was initiated by dosing 43 ml of 5% NaOH solution. After 30 minutes, the pH was corrected to its equilibrium value if it had not yet been reached.

Arsenic adsorption experiments were carried out using a jar test apparatus. Each experiment consisted of series of 1.5 L batches with iron concentrations ranging from 0.4–2.0 mg/L. After an initial rapid mixing phase (2 min, 200 RPM) during which the iron solution was dosed, a steady mixing of 60 RPM was maintained for 60 minutes. The concentrations of As and Fe were measured at the start and end of an experiment, both unfiltered and after filtration over a 0.45 μm syringe filter. In addition, Ca, Mg, Si, PO_4 and DOC concentrations were measured.

DOI: 10.1201/9781003317395-169

3 RESULTS AND DISCUSSION

3.1 *Baseline experiments*

Arsenic concentrations were 11 µg/L in raw water, 2.1 µg/L in RS filtrate and 1.7 µg/L in AC filtrate. Due to these differences in initial concentration, a comparison based on removal percentages was not feasible. Instead, As removal was expressed in an adsorption isotherm (Figure 1). A steeper isotherm slope reflects more efficient adsorption. Jar test experiments revealed that As adsorption was most effective in AC filtrate, most likely since it contains the lowest DOC and PO_4 concentrations. Full scale filters showed more effective As adsorption than their jar test counterparts. In sand filters, As oxidation occurs within the filterbed, enhancing the adsorption as compared to the jar test with raw water. Possibly, As adsorption is enhanced in Coagulation-Ultrafiltration (C-UF; fed by AC filtrate) because of shorter diffusion distances as compared to the jar test with the same water matrix.

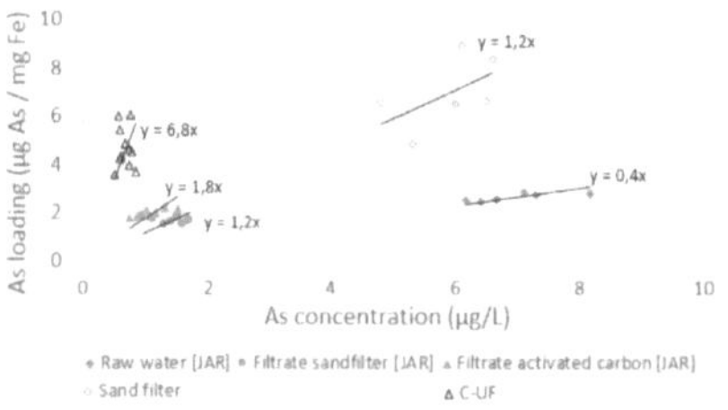

Figure 1. Arsenic adsorption isotherms in raw water, RS filtrate, AC filtrate.

3.2 *Influence softening on iron flocculation*

The influence of catIEX on iron flocculation was limited, with >85% iron removal by a 0.45 µm filter at the end of the jar test. In contrast, only 40–80% iron could be removed after PS at 0.5 mmol/L hardness (Figure 2). Most likely, the fine carry-over particles act as seed for iron flocculation, resulting in the formation of smaller flocs.

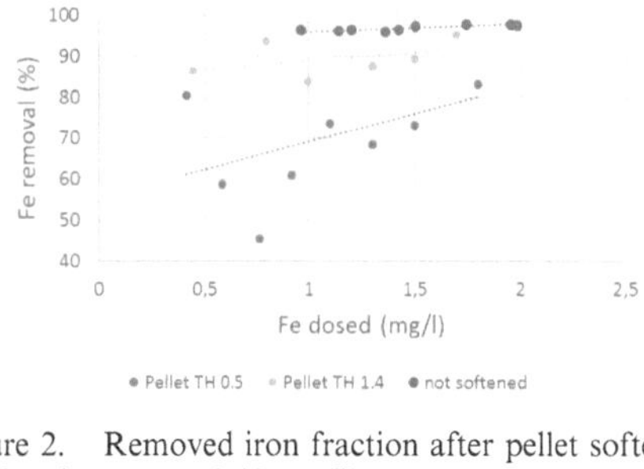

Figure 2. Removed iron fraction after pellet softening by filtration over a 0.45 µm filter.

3.3 *Influence softening on arsenic removal*

Arsenic removal after softening decreased, both after catIEX and PS. Based on the adsorption isotherms, it is possible to calculate the required iron dose for various scenarios. Table 1 shows the required iron doses when removing As from 5 µg/l to 1 µg/l.

Table 1. Required iron doses to remove arsenic from 5 µg/L to 1 µg/L.

AC filtrate (JAR)	
No softening	1.9 mg Fe/l
catIEX 1,4 mmol/L	2.1 mg Fe/l
catIEX 0,5 mmol/L	2.5 mg Fe/l
RS filtrate (JAR)	
No softening	1.5 mg Fe/l
PS 1,4 mmol/L	1.8 mg Fe/l
PS 0,5 mmol/L	2.9 mg Fe/l
C_UF	
No softening	0.6 mg Fe/l

At a hardness of 0,5 mmol, the required iron dose doubles for PS, while it increases only 25% for catIEX. Jar tests may over-estimate the required iron dose, since they only simulate a reaction in supernatant water.

4 CONCLUSIONS

- Iron floc size is reduced after pellet softening, which is likely related to carry over acting as a seed for iron floc growth. This is not observed after catIEX, indicating that Ca removal is not responsible.
- Arsenic removal decreases after softening, and this is more pronounced for pellet softening, since iron floc formation is hindered.
- Jar test experiments are useful to obtain rapid results, but underestimate As removal in pilot filters.

REFERENCES

de Ridder D.J., van de Wetering T.S.C.M., van Dijk T. & van Halem D. 2018. Influence of HPO_4^{2-}, H_2SiO_4, Ca^{2+}, Mg^{2+} on Fe floc growth and As(III) removal in aerated, natural groundwater. *J. Wat. Proc. Eng.* 25: 149

Dixit S. & Hering J.G. 2003. Comparison of arsenic(V) and arsenic (III) sorption onto iron oxide minerals: implications for arsenic mobility. *Environ, Sci. Technol.* 37: 4182

Arsenic in the Environment: Bridging Science to Practice for Sustainable Development – van der Wal, Ahmad, Petrusevski, Weijma, Savic, van der Wens, Beerendonk, Bhattacharya, Bundschuh & Naidu (Eds)
© 2023 the Author(s), ISBN 978-1-032-32928-4

Removal of low concentration of arsenic by coagulation-ultrafiltration

P. Vollaard[1], B. Hofs[1], A. Jeworrek[1], D. de Ridder[1], A. Hogendoorn[1], A. Ahmad[1,2,3] & A. van der Wal[1,2]
[1]*Evides Water Company, Department of Water Technology, Rotterdam, The Netherlands*
[2]*Wageningen University, Department of Environmental Technology, Wageningen, The Netherlands*
[3]*KWR Watercycle Research Institute, Water Systems and Technology, Nieuwegein, The Netherlands*

ABSTRACT: A In order to reduce arsenic concentrations in drinking water to very low levels (below <µgl/L), we investigated the effect of coagulation with iron(III) prior to ultrafiltration on arsenic removal. It was found that arsenic could be removed to below 1 µg/L, as long as only infiltrated dune water is used as a source. When natural groundwater is also abstracted, removal of arsenic by coagulation – ultrafiltration is not sufficient due to resulting higher concentrations of arsenic and various anions in the feed water. At a dose of 0.4 mg Fe/L and a flux of 55 L m-2 h-1, membrane fouling could be controlled by chemically enhanced backwashing with citric acid and sodium hypochlorite.

1 INTRODUCTION

The World Health Organization (WHO) limit for arsenic (As) in drinking water is 10 µg/L. However, the Dutch drinking water sector has voluntarily agreed to reduce arsenic concentrations in drinking water to below 1 µg/L. Here, we investigate the option to use coagulation with ferric chloride (FeCl3) in combination with an ultrafiltration (UF) treatment step at a drinking water treatment plant (DWTP) to reduce arsenic concentrations from 3 µg/L to below 1 µg/L.

2 MATERIALS AND METHODS

2.1 *DWTP Ouddorp*

The main raw water source of DWTP Ouddorp is surface water which is pre-treated and subsequently infiltrated into the dunes. In addition, a limited amount of natural groundwater is used as a source. During dune infiltration arsenic is mobilized.

DWTP Ouddorp consists of aeration – rapid sand filtration (RSF), granular activated carbon filtration (GAC) and ultrafiltration (UF). Prior to the RSF iron chloride ($FeCl_3$) is dosed in order to partly remove As by means of coprecipitation with in situ formed Fe(III) precipitates. The predominant form of As in the raw water is arsenite. The remaining As in the RSF effluent is completely oxidized from arsenite to arsenate. The latter is considered to be easier removable at neutral pH (6-8) (Dixit & Hering 2003).

2.2 *Pilot description*

For the execution of the pilot research one pressure vessel, containing four Norit S-225 35 m^2 polyether-sulfone hollow fibre membranes, was isolated from one of the full scale UF skids. Subsequently, the isolated pressure vessel was attached to a fully automated pilot installation which included an inline Fe(III) dosing at the feed side of the membranes. Iron was dosed in the range of 0.1 – 1.0 mg Fe/L. Periodically, the membranes were cleaned with backwashes and chemical enhanced backwashes (CEB's). The applied CEB's were citric acid (2.000 or 4.000 mg/L) and sodium hypochlorite (100 mg/L). In order to achieve a stable membrane performance, a combination of different operational parameters were varied throughout the study (e.g. flux, backwash interval, CEB regime).

UF feedwater both pre- and post Fe(III) dosing and UF permeate was sampled. The samples were analyzed for arsenic, iron, ortho-P, Si and dissolved organic carbon (DOC). All analyses were executed at Aqualab Zuid, an accredited laboratory. Table 1 provides details of infiltrated dune

Table 1. Average quality parameters ($N \geq 17$) of the raw water sources, RSF effluent and UF feed of DWTP Ouddorp.

Parameter	Unit	Infiltrated dune water	Natural groundwater	RSF effluent	UF Feed
pH (-)	–	7.6 ± 0.2	7.5 ± 0.0	7.7 ± 0.1	7.6 ± 0.1
As Dissolved	µg/L	10 ± 2.5	<0.5	3.0 ± 1.1	3.0 ± 1.1
Fe, Dissolved	mg/L	0.78 ± 0.17	0.50 ± 0.40	0.01 ± 0.01	<0.005
Si	mg/L	4.6 ± 2.9	24 ± 1	4.4 ± 2.7	4.4 ± 2.7
Ortho-P	mg/L	n.b.	0.36 ± 0.1	<0.02	<0.02
DOC	mg/L	3.8 ± 0.7	6.7 ± 1.1	2.8 ± 0.5	2.2 ± 0.5

DOI: 10.1201/9781003317395-170

water, natural groundwater, RSF effluent and UF feed water for the aforementioned parameters.

3 RESULTS AND DISCUSSION

3.1 *Arsenic removal*

In Figure 1, the concentration of As in the coagulation-ultrafiltration (C-UF) permeate at different applied Fe(III) dosages is displayed. Independently of the operational settings an As concentration of lower than 1 μg/L could be reached with a Fe(III) dosage of $0.3 \geq Fe \leq 0.5$ mg/L, but only in the absence of natural groundwater (red triangles; September, October, December). It is evident that during admixture of natural groundwater in July, August and November, As could not be lowered below 1 μg/L by C-UF. The worsened As removal can be attributed either to a higher concentration of arsenic in the feed water of the C-UF, or changes in water quality of the raw water, for example higher concentrations of silica, phosphate and/or DOC (Ahmad *et al.*, 2019).

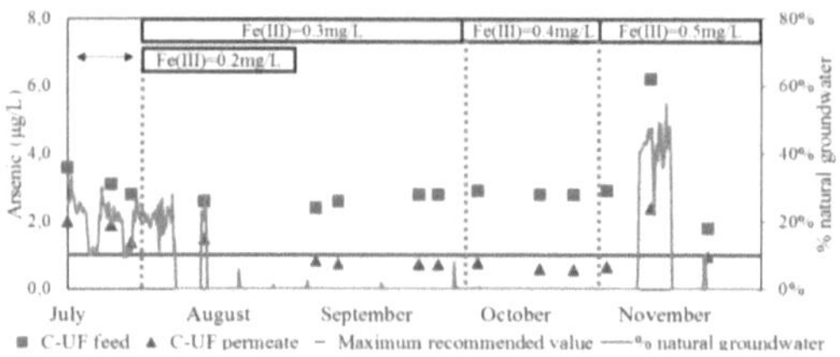

Figure 1. Effect of coagulation dose and percentage natural groundwater on arsenic removal by C-UF.

3.2 *Fouling control*

During coagulation, the membrane resistance (R) increases due to buildup of a layer of iron precipitates. Figure 2 shows the increase in R over ten filtration cycles, including one CEB per cycle, with a dose of 0.4 mg Fe/L and a flux of 55 L/m²/h. The CEB is shown to be effective as R does not increase in time. R^2 is low and this shows that irreversible fouling is either absent or small using the above-mentioned settings. At higher fluxes or lower citric acid soaking times, irreversible fouling occurs (Table 2).

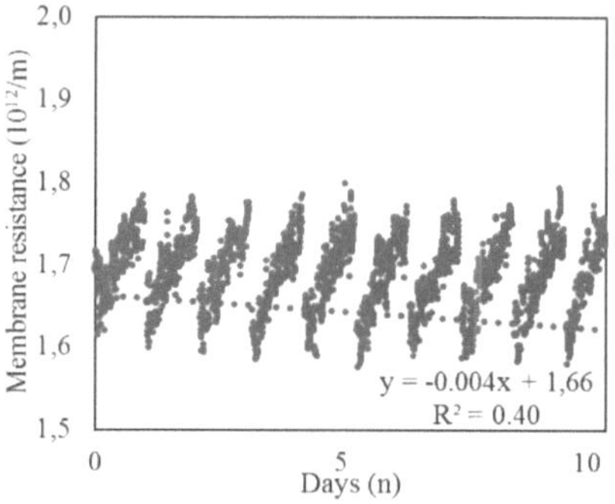

Figure 2. Changes in membrane resistance at 0.4 mg Fe/L, 55 L/m²/h and citric acid soaking time of 60 min. Ten filtration cycles including the effect of a CEB are shown.

Table 2. Slopes and R^2 of membrane resistance development of C-UF in time with an iron dose of 0.4 mg Fe/L.

Flux (L/m²/h)	Soaking time citric acid (min)	Slope	R^2
55	60	−0.004x	0.40
55	30	0.011x	0.94
120	60	0.084x	0.96

4 CONCLUSIONS AND RECOMMENDATIONS

C-UF with a dose of 0.4 mg Fe/L and flux 55 L/m²/h can be successfully applied to remove arsenic concentrations from 3 to below 1 μg/L. In these circumstances, fouling of the UF modules can be controlled if a daily CEB with citric acid followed by sodium hypochlorite is applied.

REFERENCES

Ahmad A., Rutten S., Eikelboom M., de Waal L., Bruning H., Bhattacharya P. & van der Wal A. 2019. Impact of phosphate, silicate, and natural organic matter on the size of Fe(III) precipitates and arsenate co-precipitation efficiency in calcium containing water. *J. Sep. Pur. Technol.* 235: 116117.

Dixit S. & Hering J.G. 2003. Comparison of arsenic(V) and arsenic (III) sorption onto iron oxide minerals: implications for arsenic mobility. *Environ. Sci. Technol.* 37: 4182

Arsenic in the Environment: Bridging Science to Practice for Sustainable Development –
van der Wal, Ahmad, Petrusevski, Weijma, Savic, van der Wens, Beerendonk, Bhattacharya, Bundschuh & Naidu (Eds)
© 2023 the Author(s), ISBN 978-1-032-32928-4

Implementation of co-precipitation combined with membrane filtration for arsenic removal: The effect of the water matrix on ferric hydroxide formation and membrane performance

S.B. Rutten
Department of Environmental Technology, Wageningen University and Research, Wageningen, The Netherlands
KWR Water Research Institute, Nieuwegein, The Netherlands

ABSTRACT: In recent years the health impacts of prolonged exposure to low concentrations of arsenic (As) have been established. Dutch drinking water companies, with a proactive vision towards potential changes in legislation of As concentrations in drinking water, actively try to reduce As concentrations under 1 μg/L. In this research the implementation of ferric chloride dosing combined with membrane filtration at the terminal end of a drinking treatment plant as As removal technique has been studied. The study aimed to assess the effect of phosphate, silica and dissolved organic matter on ferric hydroxide formation, growth and As adsorption in both synthetic and natural groundwater. Moreover, two partial treated drinking waters were used to assess the effectiveness of the combined co-precipitation/membrane filtration method. As shown in the results, DOC significantly affected particle formation and growth which subsequently reduced As adsorption. Arsenic removal was influenced Si and P while particle growth was not significantly affected. The study concluded that $FeCl_3$ dosing combined with membrane filtration suffices as a treatment method to reduce As concentrations below 1 μg/L in drinking water.

1 INTRODUCTION

Groundwater is widely used as a source of drinking water throughout the world. While using groundwater has proven to be more beneficial than surface waters, potentially polluted with anthropogenic sources, arsenic (As) contamination poses a significant threat to human health (Ahmad *et al.*, 2017a; Nriagu *et al.*, 2007). Dependent on the redox conditions, As exists in two species in groundwater sources. In anoxic conditions, As occurs as its trivalent species, arsenite (As(III)), while in oxic states, pentavalent As, arsenite (As (V)) is most prevalent. The health effects of As exposure at high concentrations have been known for multiple centuries, but the adverse impact on human health due to prolonged exposure with low concentrations has only been known for a couple of decades (Ahmad & Bhattacharya 2019; Havelaar & Melse 2003). This prolonged exposure has been linked to various types of cancer, reproductive disorders, and diabetes mellitus (Islam *et al.*, 2012; Kapaj *et al.*, 2006). Given these findings, the World Health organisation established a recommended As concentration lower than 10 μg/L in drinking water (WHO 2017).

While As is only found as a trace element in Dutch groundwater sources, with concentrations ranging between 0.1 – 70 μg/L, Dutch drinking water companies, with a proactive mentality towards potential alterations in the maximum As concentration limit, aim to lower As concentrations under 1 μg/L (Ahmad *et al.*, 2017b, 2020; van der Wens *et al.*, 2016). A potential method researched in this study to reach this concentration limit is implementing co-precipitation of arsenate with ferric chloride ($FeCl_3$) and membrane filtration at the terminal end of the drinking water treatment plant (Ahmad *et al.*, 2017b, 2020). Previous studies have established that the major species of As in drinking water treatment plants, after rapid sand filtration, consists of arsenate, which readily adsorbs to ferric hydroxides ($Fe(OH_3)$) formed during ferric chloride dosing. The study aimed to assess the effectiveness of As removal in this combined system of co-precipitation and direct dead-end filtration.

2 RESULTS AND DISCUSSION

2.1 *Ferric hydroxide growth*

During the study, particle growth of ferric hydroxides in the presence of phosphate, silica, and DOC in synthetic and real drinking water sources was studied. The concentration of phosphate, silica, and DOC in the artificial drinking waters was varied within the range found in Dutch groundwater sources. These experiments were performed using a batch reactor (Applikon Bioreactors, Delft, NL) connected to a Mastersizer 2000 (Malvern Panalytical, Almelo, NL) for particle sizing.

2.2 *Lab-scale UF membrane filtration experiments*

The membrane filtration experiments were performed using two partially treated drinking water sources provided by Evides (Rotterdam, NL) and Dunea (Katwijk, NL). Optimal ferric chloride doses for As removal were determined by jar-test experiments. The results of these jar-test experiments are not included in this abstract. The experiments were performed using a custom build lab-scale membrane filtration set-up. UFC M5 (Pentair, Enschede, NL) hollow fibres, extracted from a full-scale module were repotted to reach a surface area of ~20 cm^2.

Each experiment consisted of six dead-end filtration cycles of an hour with subsequent backwashes. The transmembrane pressure, water flux, and As concentration in the feed, permeate and backwash reject were monitored throughout the cycles. Based on these results, the contribution of ferric chloride dosing towards filtration resistance and the As removal efficiency were assessed.

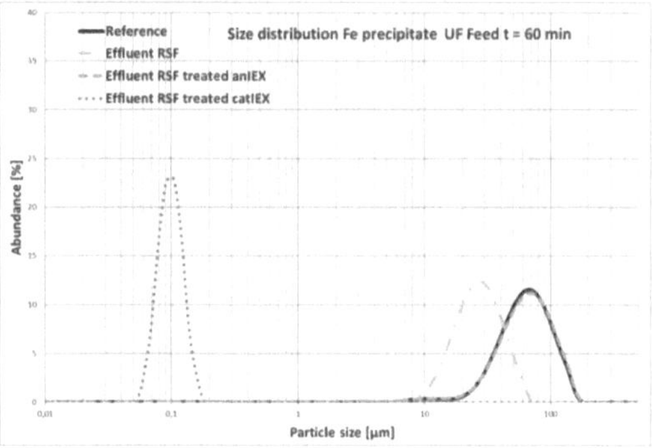

Figure 2. Particle size of ferric hydroxides in untreated and anion and cation treated membrane filtration feed water.

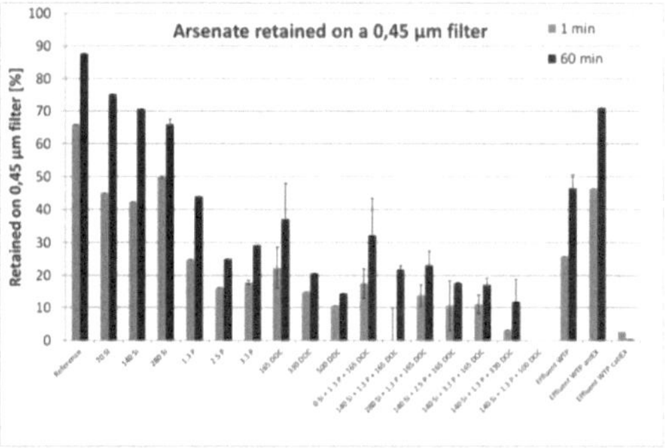

Figure 3. Arsenic removal by co-precipitation and 0.45 µm syringe filtration.

3 RESULTS AND DISCUSSION

As shown in Figure 1, DOC affected particle growth most significantly, which subsequentially reduced As adsorption and removal most. Results with an actual groundwater source, shown in Figure 2, show the effect of pre-treatment by anion or cation exchange on particle growth. As observed, the removal of cations significantly hampers particle growth and particle size due to the loss in stabilizing ions such as calcium and magnesium. The removal of anions reduces the effect of particle growth inhibition, which is visible in an increased particle size compared to the untreated groundwater sample.

As shown in Figure 3, adsorption competition between arsenate, silica, and phosphate significantly reduces the overall As removal. Reduced As removal in the presence of DOC is hypothesized to occur due to particle growth inhibition, leading to limited adsorption sites for arsenate. Similarly, cation exchange pre-treatment significantly reduces arsenate removal since it limits particle growth and formation. In comparison, anion exchange treatment improved As removal. It is assumed that the removal of competition species, silica and phosphate, by anion exchange improved the overall As removal.

Due to ferric chloride dosing, the build-up of a reversible, fouling layer contributed most significantly to the increased filtration resistance, while irreversible fouling only minorly occurred (Figure 4). While this increase in filtration resistance is undesirable, optimised fouling control and cleaning methods can limit its effect on the overall performance of an ultrafiltration treatment skid.

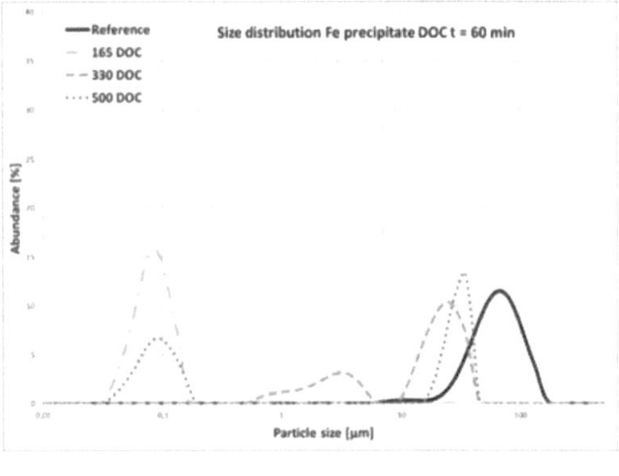

Figure 1. Particle size distribution of Ferric hydroxide as affected by dissolved organic matter. Values in legend correspond to the µmol concentration reported in Dutch groundwaters.

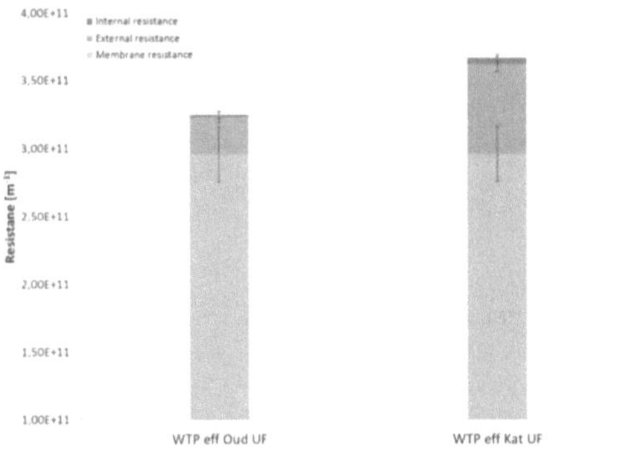

Figure 4. Contribution of filtration resistance per type (membrane, reversible fouling, irreversible fouling).

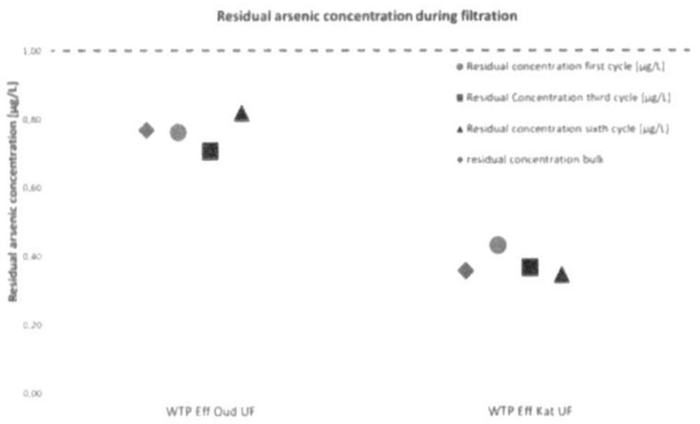

Figure 5. Residual arsenic concentration during experimental runs (1st, 3rd, 6th cycle, and bulk).

Irrespective of the filtration resistance build-up and filtration cycle, As concentrations were lowered below 1 µg/L (Figure 5), showing potential in full-scale drinking water treatment plants.

4 CONCLUSIONS AND RECOMMENDATIONS

Combined co-precipitation and membrane filtration is a feasible method to reduce As below 1 µg/L for drinking water purposes. Anion exchange treatment could promote overall As removal, while cation exchange treatment will significantly hamper precipitate formation and As removal. Further, long-term experiments are recommended to study the proposed method's efficacy at larger time scales.

ACKNOWLEDGEMENTS

This research was performed as part of the master program Environmental Technology at the Wageningen University and Research (WUR). We acknowledge the WUR and KWR Water Research Institute for supporting this research and providing their expertise and equipment.

REFERENCES

Ahmad A. & Bhattacharya P. 2019. Arsenic in drinking water: is 10 µg/L a safe limit? *Curr. Pollut. Rep.* 5(1): 1–3

Ahmad A., Richards L.A. & Bhattacharya P. 2017a. Arsenic remediation of drinking water: an overview. (chapter 7). In: P. Bhattacharya, D.A. Polya & D. Jovanovic (eds.) *Best Practice Guide on the Control of Arsenic in Drinking Water.* IWA Publishing, UK. pp. 79–89.

Ahmad A., Van Dijk T., Gude J., van de Wetering S., Groenendijk M. & Bhattacharya P. 2017b. Remediation case study: drinking water treatment by aocf to target <1 µg l–1 effluent arsenic concentration. In: P. Bhattacharya, D.A. Polya & D. Jovanovic (eds.) *Best Practice Guide on the Control of Arsenic in Drinking Water. Metals and Related Substances in Drinking Water Series,* IWA Publishing, UK, pp. 219–225

Ahmad A., van der Wens P., Baken K, de Waal L., Bhattacharya P. & Stuyfzand P. 2020. Arsenic reduction to <1 µg/L in dutch drinking water. *Environ. Int.* 134:105253.

Havelaar A.H. & Melse J.M. 2003. *Quantifying Public Health Risk in the WHO Guidelines for Drinking Water Quality: A Burden of Disease Approach.* RIVM Report. 734301022/2003, RIVM, The Netherlands.

Islam M.R., Khan I., Hassan S.M.N., McEvoy M., Deste C., Attia J., Peel. R., Sultana M., Akter S. & Milton A.H. 2012. Association between type 2 diabetes and chronic arsenic exposure in drinking water: a cross sectional study in Bangladesh. *Environ. Health: Global Access Sci. Source,* 11 (1)

Kapaj S., Peterson H., Liber K. & Bhattacharya P. 2006. Human health effects from chronic arsenic poisoning – a review. *J. Environ. Sci. Health A* 41(10): 2399–242

Nriagu J.O., Bhattacharya P., Mukherjee A.B., Bundschuh J., Zevenhoven R. & Loeppert R.H. 2007. Arsenic in soil and groundwater: an overview. In: P. Bhattacharya, A.B. Mukherjee, J. Bundschuh, R. Zevenhoven & R.H. Loeppert (eds.) *Arsenic in Soil and Groundwater Environment: Biogeochemical Interactions, Health Effects and Remediation.* Trace Metals and Other Contaminants in the Environment, Volume 9, Elsevier B.V. Amsterdam, The Netherlands. pp. 3–60.

Van der Wens P., Baken K. & Schriks M. 2016. Arsenic at low concentrations in dutch drinking water: assessment of removal costs and health benefits. In: P. Bhattacharya, M. Vahter *et al.* (eds.) *Arsenic Research and Global Sustainability (As2016),* CRC Press, pp. 563–564.

WHO. 2017. *Guidelines for Drinking-Water Quality: Fourth Edition Incorporating the First Addendum.* (ISBN 978-92-4-154995-0) World Health Organization, Geneva.

Arsenic in the Environment: Bridging Science to Practice for Sustainable Development – van der Wal, Ahmad, Petrusevski, Weijma, Savic, van der Wens, Beerendonk, Bhattacharya, Bundschuh & Naidu (Eds)
© 2023 the Author(s), ISBN 978-1-032-32928-4

Controlling factors for arsenic removal by Iron amended bio-sand filters: A case example of Kanchan Arsenic Filters in Nepal

S. Padhi[1], M. Sakamoto[2], T. Tokunaga[1], J. Otomo[3] & R. Ogata[4]
[1]*Geosphere Environment Systems, Department of Environment Systems, GSFS, The University of Tokyo, Japan*
[2]*Department of International Studies, GSFS, The University of Tokyo, Japan*
[3]*Environmental Chemical Energy Engineering, Department of Environment Systems, GSFS, The University of Tokyo, Japan*
[4]*Japan International Cooperation Agency (JICA), Amman, Jordan and The University of Tokyo, Japan*

ABSTRACT: Kanchan Arsenic Filters (KAF) are widely used to provide arsenic (as) free drinking water in Nepal and many other countries. The performance of these filters for As removal and the controlling factors were evaluated. It was found that changing the filter media increased As removal capacity, whereas cleaning and using the same media with As free water or changing the Iron (Fe) nails did not improve As removal performance. Field experiments are in progress by introducing Fe in raw water to improve As removal capacity of these filters.

1 INTRODUCTION

To provide Arsenic (As) free drinking water, there are two viable options: First, finding and preparing alternative water sources that are free of arsenic, or, second, treating arsenic contaminated water with arsenic removal technology. In Nepal, iron amended bio-sand filters, known as Kanchan Arsenic Filters (KAF) were widely used. KAFs were also used in Cambodia and India (Ogata *et al.*, 2014) and a field based trial was conducted in China (Smith *et al.*, 2017). However, different studies have reported different performance levels in As removal by KAFs (Chiew *et al.*, 2009; Ngai *et al.*, 2007; Ogata *et al.*, 2019; Singh *et al.*, 2014; Smith *et al.*, 2017). In addition, there is no clear explanation on the controlling factors for As removal by these filters. The overall objectives of this study are (1) to understand the efficiency of As removal by KAFs and the mechanism of As removal under different groundwater compositions, and (2) to maximize the performance of KAF to provide safe and sustainable drinking water to the local community in developing countries.

2 CURRENT STATUS OF UNDERSTANDING THE MECHANISM OF ARSENIC REMOVAL BY KAF AND ITS PERFORMANCE

The KAF (Figure 1) was designed to remove As from contaminated groundwater by adsorption of As to Fe (III) oxides formed by rusting Fe nails. In Nepal, As(III) is the dominant inorganic As species in groundwater (Ngai *et al.*, 2002). Fe in groundwater ranges from 0–10 mg/l and P ranges from 0–1.0 mg/L. Oxidation of As (III) to As (V) is a desired step in As removal technologies with the existence of Fe for better performance as As(V) sorbs more efficiently on Fe (oxy)hydroxides than that of As (III). The mechanism of parallel As (III) and Fe (II) oxidation in the presence of dissolved oxygen (DO) has been proposed with reactive intermediates such as Fe(IV) (Roberts *et al.*, 2004) and Fe(IV) at low pH, and Fe(VI) at circumneutral pH, respectively (Katsoyiannis *et al.*, 2008).

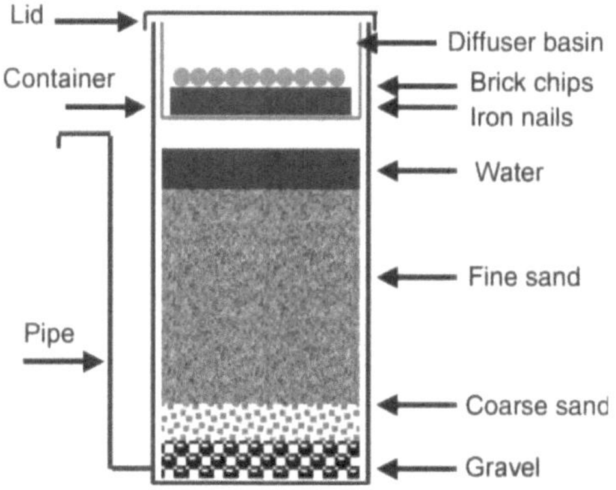

Figure 1. Components of the Kanchane Arsenic Filter (Ngai *et al.*, 2007).

Reported performance of KAF for As removal in Nepal varies, e.g., in Nawalparsi, Ogata *et al.* (2019) reported average arsenic removal rate of 75% for filters in use for 0–10 years, whereas that reported by

DOI: 10.1201/9781003317395-172

Singh *et al.* (2014) was 54%. Chiew *et al.* (2009) found KAF to be ineffective to remove As in groundwater to below guideline level in Cambodia and attributed it to high P (>0.5 mg/L) and low Fe (<5 mg/L) concentrations and insufficient contact time between influent and Iron nails. Smith *et al.* (2017) also reported KAF performance to decrease to 50% after six months of operation and reasoned the same as that of Chiew *et al.* (2009), and demonstrated that putting Fe nails below the sand layer improved the results by increasing the contact time between the inflowing water and Fe nails.

Studies showed that introducing Fe (II) in batches, rather than Fe (III) to the raw water performed better in removing As in presence of P and silicate (Roberts *et al.*, 2004). Gude *et al.* (2018) found addition of Fe (III) to the raw water increased As removal efficiency.

All these observations indicate that As removal by Fe based sand filters should be studied carefully in order to make these filters sustainable.

3 FIELD EXPERIMENTS, RESULTS AND PLANNED EXPERIMENTS

We carried out detailed analysis of some of the installed KAF in Nawalparsi, Nepal, in terms of filter performance, chemistry of influent water, supernatant water and filtered water, effect of standing water depth, and filter media. The following experiments were conducted and the performance of KAFs were evaluated: (1) by introducing new filter material, (2) by washing the filter media and using the media again, and (3) by changing the iron nails.

Our results showed that when the filter medium was changed, the performance of As removal was 96% in the same day and 70% after 20 days, with respect to raw water As, which was only 14% with the existing filter material. Washing the media with As free water increased As removal efficiency of the filter in the same day, however, more As in the effluent than that of the raw water was detected thereafter. Changing the iron nails does not change filter performance significantly.

Supernatant water (standing water), in all of the filters, had lower As and higher Fe than that of raw water in contrast to that found by Chiew *et al.* (2009), implying that Fe nails provide some dissolved Fe, however, not sufficient to remove all the As from the raw water, as the dissolved Fe in the raw water itself is not enough high.

SEM-EDX analysis of the filter media suggested that As was mostly accumulated in the top 5 cm, implying that the As sorbed Fe (oxy)hydroxide was accumulated in the top layer. A detailed analysis of the iron nails, the used filter media is still under investigation. Our experiment adding Fe to the raw water under the prevailing groundwater conditions is also ongoing, which will be discussed later.

REFERENCES

Chiew H., Sampson M.L., Huch S., Ken S. & Bostick B. C. 2009. Effect of groundwater iron and phosphate on the efficacy of arsenic removal by iron-amended bio-sand filters. *Environ. Sci. Technol.* 43: 6295–6300.

Gude J.C.J., Joris K., Huysman K., Rietveld L.C. & van Halem D. 2018. Effect of supernatant water level on as removal in biological rapid sand filters. *Water Res. X* 1: 100013.

Katsoyiannis I.A., Ruettimann T. & Hug S.J. 2008. pH dependence of fenton reagent generation and As(III) oxidation and removal by corrosion of zero valent iron in aerated water. *Environ. Sci. Technol.* 42: 7424–7430.

Ngai T.K.K., Murcott S., Shrestha R.R., Dangol B. & Maharjan M.J. 2007. Design for sustainable development: household drinking water filter for arsenic and pathogen treatment in Nepal. *Environ. Sci. Health* 42 (12): 1879–1888.

Ogata R., Dangol B. & Sakamoto M. 2019. Sustainability assessment of long-term, widely used household Kanchan arsenic filters in Nepal. *J. Environ. Sci. Health A* 55(5): 517–527.

Roberts L.C., Hug S.J., Ruettimann T., Billah M., Khan A. W. & Rahman M.T. 2004. Arsenic removal with Iron(II) and Iron(III) waters with high silicate and phosphate concentrations. *Environ. Sci. Technol.* 38: 307–315.

Singh A., Smith L. S., Shrestha S. & Maden N. 2014. Efficacy of arsenic filtration by kanchan arsenic filter in Nepal. *J. Water Health* 12: 596–599.

Smith K., Li Z., Chen B., Liang H., Zhang X., Xu R., Li Z., Dai H., Wei C. & Liu S. 2017. Comparison of sand-based water filters for point-of-use arsenic removal in China. *Chemosphere* 168: 155–162.

Arsenic in the Environment: Bridging Science to Practice for Sustainable Development –
van der Wal, Ahmad, Petrusevski, Weijma, Savic, van der Wens, Beerendonk, Bhattacharya, Bundschuh & Naidu (Eds)
© 2023 the Author(s), ISBN 978-1-032-32928-4

Arsenic removal by iron coagulation/flocculation and double filtration

L.G. Romero-Esquivel[1], A. Lazo[2], A. Araya[3], L. Gomez[4] & R.G. Fernández[5]
[1]*Environmental Protection Research Center (CIPA), School of Chemistry, Instituto Tecnológico de Costa Rica, Cartago, Costa Rica*
[2]*Applied Research, Research & Development Department, Costa Rican Institute for Drinking Water and Sanitation (AyA), San José, Costa Rica*
[3]*School of Construction, Instituto Tecnológico de Costa Rica, Cartago, Costa Rica*
[4]*PIENSA, Engineering National University of Nicaragua (UNI), Managua, Nicaragua*
[5]*Sanitary Engineering Center (CIS), National University of Rosario, Rosario, Argentina*

ABSTRACT: Iron coagulation/flocculation for arsenic (As) removal was evaluated in a double filtration system based in the CIS-UNR® process. Two pilot systems, each one consisting in an up-flow gravel filter (UGF) followed by a rapid sand filter (RSF) were evaluated in two wells with As polluted water. One well water presented variable As (10 to 40 μg As/L) and iron (1.2 ± 1.0 mg Fe/L) concentration, the latter worked as coagulant. The other one presented around 46 ± 5 μgAs/L and 0.1 mgFe/L, therefore iron was dosed (4 ± 2 mg Fe/L,11.6 ± 5.8 mg $FeCl_3$/L). In the first case 20 to 40% removal was reached by the UGF and further removal (42–83%) after the RSF. The second case showed (76 ± 15%) removal after the UGF and up to 82 ± 16% in the rapid filter water. Both reached less than 10 μg As/L. The CIS-UNR® process, originally designed for working with polyaluminum chloride (PACl), performed well at low iron coagulant concentration and the need of the RSF as a polishing step was demonstrated.

1 INTRODUCTION

Effectiveness of iron coagulation/flocculation for arsenic (As) removal has been reported widely (Fernandez *et al.*, 2009a; Hering *et al.*, 1996; Laky & Licsko 2011). Arsenic is removed by sorption into hydrous ferric oxide formed by iron hydrolysis of ions originally present in the water or by added ones (as $FeCl_3$). The flocs containing As can be removed by sedimentation/filtration or direct filtration. Similarly, aluminum coagulation/flocculation can be applied for As removal. Fernández *et al.* (2009b) and Ingallinella & Fernandez (2011) reported the effective As removal using polyaluminum chloride (PACl) as coagulant followed by a double filtration system, CIS-UNR® process. The CIS-UNR® consists in the separation of the flocs formed by a gravel up flow prefiltration unit followed by a down flow rapid sand filter (Figure 1). The up-flow roughing filtration stage is a process with a high efficiency in the removal of light flocs (Ingallinella *et al.*, 1998) and permits longer filtrations runs in the rapid filter. The system is capable to reduce As and fluoride concentrations below the Argentinian regulation (50 μg As/L and 0.7 mg F/L to 1.7 mg F/L depending on water temperature (CAA 2019). However, As removal below the WHO and local regulation (10 μg As/L) is difficult to be achieved. Contrary, iron coagulation/flocculation can reach such values, therefore, the aim of this study was to evaluate the use of the CIS-UNR® double filtration system in combination with iron coagulation/flocculation.

2 MATERIALS AND METHODS

The CIS-UNR® system was evaluated in two different wells under the administration of Costa Rican Institute for Drinking Water and Sanitation (AyA). One well, in La Uruca, San José, Costa Rica, used natural iron originally present in the raw water at 1.2 ± 1.0 mg Fe/L, variable As concentration (10 to 40 μg As/L) and pH of 6.1 ± 0.1. The other one, located in Bagaces, Guanacaste, Costa Rica, presented low iron level (<0.1 mg Fe/L), 46 ± 5 μg/L of As and pH of 6.3 ± 0.1. The latter needed the addition of commercial iron chloride (4 ± 2 mg Fe/L,11.6 ± 5.8 mg/L $FeCl_3$) previously determined by jar test, in both cases no pH adjustment was used. The average groundwater temperature in La Uruca and Bagaces sites were close to 27°C and 30°C respectively.

2.1 *General description of the pilot system*

Both systems were identical and constructed in PVC plastic (Figure 1). The up-flow gravel filter (UGF) had 30 cm diameter, 50 cm of bed support (25–50 mm gravel) and one meter of filter bed (25–9.5 mm). The rapid sand filter (RSF) had

 DOI: 10.1201/9781003317395-173

10 cm diameter, 5 layers of bed support (0.85–25 mm), and 80 cm of sand (0.6–0.85 mm) for filtration. The UGF one was operated at 1 m/h and the RSF at 8 m/h, treating around 70 L/h.

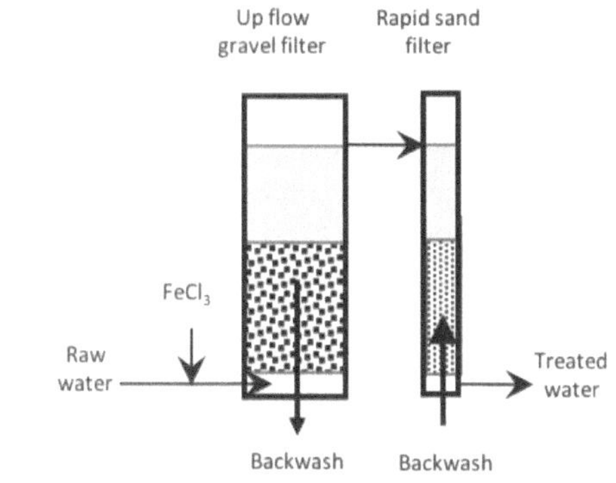

Figure 1. Pilot system design.

2.2 *Water sampling and laboratory analysis*

The system in La Uruca was operated from mid of August to mid of September 2016. The one in Bagaces, functioned during two periods in April 2017 and in February 2018. Water sampling was carried out during 5 days in La Uruca and for a total of 20 days in Bagaces.

Water quality was monitored on samples of raw water, outlet of UGF and outlet of RSF. Arsenic, iron and silicate as silicon were analyzed by atomic absorption using an AAnalyst 800 Perkin Elmer, USA. Tubidity with a 2100Q (Hach) and pH with a HQ30d (Hach) device. Silicon was only analyzed in raw and RSF water at the Bagaces system. Sampling, preservation and analysis were performed following Standard Methods (APHA, AWWA & WPCF 2005) or equipement manufacture recommendations.

3 RESULTS AND DISCUSSION

3.1 *Arsenic removal efficiency*

The As concentration in La Uruca raw water varied during the monitoring period, probably due to variations in the well pumping regime. However, it can be seen in Figure 2 that in most of the cases the As removal started in the UGF (20–45%), further total removal (42–83%) was achieved after sand filtration. As shown in Figure 2, the As concentration at the exit of the RSF reached values below 10 μgAs/L when the concentration in the raw water was lower.

As shown in Figure 3, contrary to La Uruca case, in Bagaces, the well water As concentration was stable, and the addition of iron chloride was needed in (at) very low doses (4 ± 2 mgFe/L,11.6 ± 5.8 mg/L $FeCl_3$). Similarly, the UGF showed important removal, reducing the As concentration to 11 ± 8 μgAs/L (76 ± 15% removal) with 9 of 20 samples below the WHO guidelines. After passing the RSF the treated water presented 9 ± 8 μgAs/L (82 ± 16%) with most of the samples below the WHO limit (14 of 20).

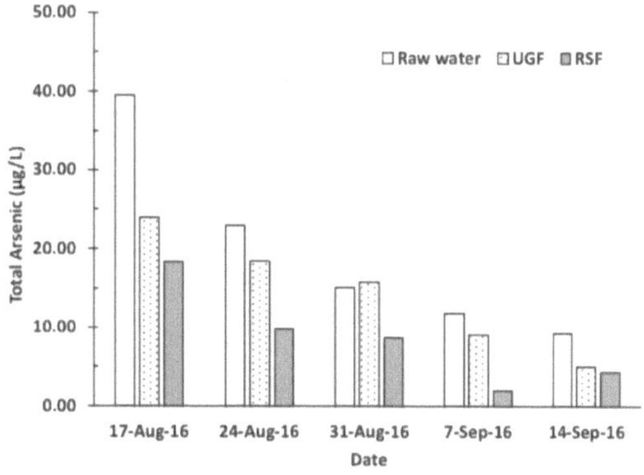

Figure 2. Arsenic removal in La Uruca system.

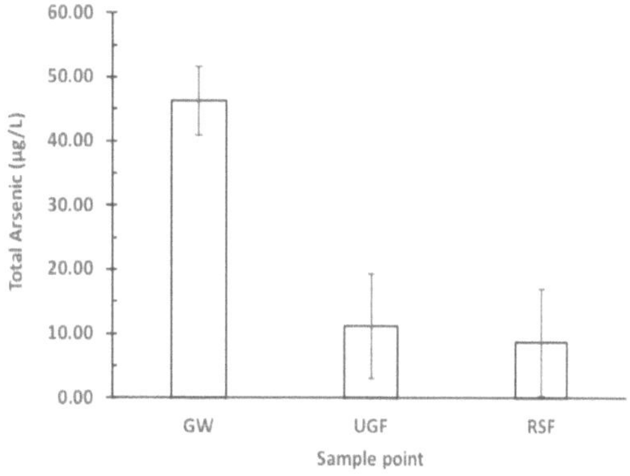

Figure 3. Arsenic removal in Bagaces system.

3.2 *Behavior of other water quality parameters*

Considering the formation of hydrous ferric oxide flocs formation special attention must be given to turbidity, pH and iron. As shown in Table 1, the pH is appropriate for iron hydrolysis in the two sides. Regarding turbidity and iron, La Uruca raw water presented high values of both parameters. After the UGF, they were reduced and the RSF water presented values well below the local regulation. In the case of Bagaces, as mentioned before, no iron was present, and the turbidity was very low. As expected, the flocs formed in the UGF were not fully retained inside it, and a slight increase in turbidity and iron is observed, again the RSF works as a polisher and both parameters were reduced to pass the local regulation. Besides, in both cases, an increase in the water quality was

Table 1. Others water quality parameters.

Sample point	pH	Turbidity (NTU)	Total iron (mgFe/L)
GW-La Uruca	6.21 ± 0.09	18.14 ± 13.20	1.20 ± 1.04
UGF-La Uruca	6.11 ± 0.07	12.99 ± 9.28	0.63 ± 0.50
RSF-La Uruca	6.54 ± 0.26	2.19 ± 2.64	< 0.01
GW-Bagaces	6.23 ± 0.19	0.19 ± 0.10	0.06 ± 0.04
UGF-Bagaces	6.17 ± 0.16	0.43 ± 0.26	0.56 ± 0.33
RSF-Bagaces	6.57 ± 0.24	0.19 ± 0.08	0.13 ± 0.14
Local regulation (CMH 2019)	6.0–8.0	5.0	0.3

observed as the experiment proceeds, suggesting the effect of ripening of the filters.

Finally, silicate has been reported as an interferent in As removal by iron coagulation/flocculation, e.g. Laky & Licsko (2011) reported the need to increase the coagulant those 2.5–3.5 in the presence of 30–50 mg/L SiO_2 to achieve the target removal. Therefore, raw water that presented 128 ± 12 mg/L SiO_2 was monitored in Bagaces. The silicate remains practically the same during the treatment presenting 131 ± 14 mg/L SiO_2 at the RSF exit. Thus, further research is needed to determine the effect of silicate in the system.

4 CONCLUSIONS AND RECOMMENDATIONS

Data from both sites showed an important contribution of the UGF for the As removal. However, a second RSF stage is necessary for polishing and reaching concentrations below the WHO guidelines.

Flocs formed during the process increase turbidity and iron content in the water, however, the UGF and the RSF were enough to maintain those parameters under the local regulation.

The adapted CIS-UNR® double filtration system performed well for the removal of As in groundwater using low levels of naturally occurring iron or little iron doses. Besides, the system permitted to achieve WHO guidelines without the need of pH adjustment. Future research is recommended for the As polluted sludge management and reduction.

ACKNOWLEDGEMENTS

This research was founded by the Research and Extension Council of ITCR, Costa Rica project no. 1460066. We would like to acknowledge to the Pan American Health Organization (PAHO) for the master scholarship awarded to Lidia Gomez.

REFERENCES

APHA, AWWA & WPCF. 2005. *Standard Methods for the Examination of Water and Wastewater*. 21st edn. American Public Health Association, Washington, DC, USA.

CAA. 2019. *Código Alimentario Argentino*, Rev. November 2019. Article 982: Drinking Water.

CMH. 2019. Executive Decree No. 41499-S. *Costa Rican Ministry of Health*, Official Gacette. 01.22.2019, pp. 2–16.

Fernández R., Petrusevski B., Schippers J. & Sharma S. 2009a. Arsenic removal from groundwater using ferric chloride and direct filtration (chapter 60). In: J. Bundschuh, M.A. Armienta, P. Birkle, P. Bhattacharya, J. Matschullat & A.B. Mukherjee (eds.) *Natural Arsenic in Groundwater of Latin America — Occurrence, Health Impact and Remediation*. CRC Press/Balkema, Leiden, The Netherlands, pp. 555–564.

Fernández R., Ingallinella A. & Stecca L. 2009b. Arsenic removal from groundwater by coagulation with polyaluminum chloride and double filtration. (chapter 64). In: J. Bundschuh, M.A. Armienta, P. Birkle, P. Bhattacharya, J. Matschullat & A.B. Mukherjee (eds.) *Natural Arsenic in Groundwater of Latin America — Occurrence, Health Impact and Remediation*. CRC Press/Balkema, Leiden, The Netherlands, pp. 589–594.

Hering J.G., Chen P.Y., Wilkie J.A., Elimelech M. & Liang S. 1996. Arsenic removal by ferric chloride. *J. Am. Water Works Assoc.* 88: 155–167.

Ingallinella A., Stecca L. & Wegelin M. 1998. Up-flow roughing filtration: rehabilitation of a water treatment plant in tarata, bolivia. *Water Sci. Technol.* 37(9): 105–112.

Ingallinella A. & Fernández R. 2011. Experiencia Argentina En la Remoción de Arsénico Por Diversas Tecnologías. (Chapter 9). In: M. Litter, A.M. Sancha & A.M. Ingallinella (eds.) *Tecnologías Económicas para el Abatimiento de Arsénico en Aguas*. (ISBN 978-84-96023-74-1), Argentina.

Laky D. & Licsko I. 2011. Arsenic removal by ferric-chloride coagulation – effect of phosphate, bicarbonate and silicate, *Water Sci. Technol.* 64: 1046–1055.

Arsenic in the Environment: Bridging Science to Practice for Sustainable Development – van der Wal, Ahmad, Petrusevski, Weijma, Savic, van der Wens, Beerendonk, Bhattacharya, Bundschuh & Naidu (Eds)
© 2023 the Author(s), ISBN 978-1-032-32928-4

Arsenic attenuation by alluvial soils of Sutlej River deposits, Rupnagar District, Punjab, India

N. Kaur & S. Paikaray
Department of Geology, Panjab University, Chandigarh, India

ABSTRACT: The mobility, toxicity and bioavailability of arsenic (As) in environment is mainly dependent on retention efficiency of different soil types that it interacts in aquifer systems. Attenuation potential of a representative soil for As was evaluated in this study in order to understand the control of aquifer soil on As bioavailability around a severely contaminated site (up to ~128 mg/kg) of Rupnagar District, Punjab, India. The soil belongs to an agricultural field with minor As content (~1.2 mg/kg) and is the dominant soil type in this region composed mainly of sand, silt and clay. Commonly encountered geochemical variables like PO_4^{3-}, NO_3^- and pH were evaluated. The results indicate that As attenuation by the soil was maximum at pH 6 (1 g/kg) that lowered to ~43% at pH 10 possibly due to soil surface repulsive force. The positive and negative correlation of released PO_4^{3-} and SO_4^{2-} with residual As, respectively at different pH is suggestive of $HAsO_4^{2-}$-PO_4^{3-} aqueous complexation and likely $HAsO_4^{2-}$-SO_4^{2-} ion exchange governing the overall attenuation mechanisms. Increasing aqueous PO_4^{3-} and NO_3^- concentrations as controls exhibited a slight inverse relation with residual As partly supporting As-PO_4 complexation, while NO_3^- seems non-responsive. The study suggests that the aquifer soils possess considerable retentive efficiency for As, especially under near-neural pH conditions in presence of dissolved PO_4^{3-} through both ion-exchange and aqueous complexation processes.

1 INTRODUCTION

Arsenic (As), the oldest pollutant, is raising severe concerns throughout the world owing to its toxicity and negative impacts on human health, prompting research into how to effectively remove As from groundwater. Due to the chemistry of As, the inherent attenuation capabilities of soil can immobilize it, and soluble or mobile As can be transported into the particulate phases by applying the appropriate adsorbent (Lin & Puls 2003; Sailo & Mahanta 2016; Sengupta *et al.*, 2018). Natural attenuation is a cost-effective and ecologically friendly in situ remediation technology. Adsorption, desorption, reduction, oxidation, co-precipitation, and dissolution all contribute to natural attenuation of arsenic in the subsurface (Lin & Puls 2003). Among these processes, the adsorption–desorption reaction of As species on minerals plays a crucial role. The immobilization through sorption to solid phases such as (hydro)oxides of iron (Fe), aluminum (Al), and manganese (Mn), organic matter, and clay minerals are the major mechanisms that cause As to be attenuated naturally (Wang & Mulligan 2006). Many factors influence attenuation capacity, however, the investigation of As adsorption on sediments has been primarily tied to environmental parameters such as As contents, pH, other solutes concentrations that compete with As for adsorption sites and the sediments' adsorption capacity (Sailo & Mahanta 2016; Stollenwerk *et al.*, 2007; Wang & Mulligan 2006).

The present study area is a severely As contaminated site of Rupnagar District, Punjab, India with As contents up to 128 mg/kg in soil and 1446 µg/L in groundwater (Kaur & Paikaray 2021). The major objective of the research was to evaluate the natural attenuation potential of a representative soil for As under normal atmospheric conditions (which represents exactly the study area) in order to understand the control of aquifer soil on As bioavailability. Batch experiments were performed to evaluate the effects of As concentration, varying pH, and competing anions on the adsorption of As by the sediments.

2 MATERIALS AND METHODS

2.1 *Study area*

The sampling area (31°1′51.17″N & 76°31′48.4″E) is located in the Rupnagar District of Eastern Punjab, India. The soil belongs to an agricultural field with minor As content (~1.2 mg/kg) and the dominant soil type in this region is mainly composed of sand, silt and clay. The area has hazardous waste deposits of the cement plant (CP) and the thermal power plant (TPP), major rivers flowing through it, i.e., River Sutlej and its tributary i.e., River Sirsa.

2.2 *Batch experiments*

In the laboratory batch experiments, 10 g soil was added to 1 L deionized water and the sediment solution was interacted with variable As content (10 mg/L) for 1, 5, 10, 24, 72 and 120 h. After the kinetic study, equilibrium time period was obtained and the samples were centrifuged, filtered and analysed for As

DOI: 10.1201/9781003317395-174

contents. The attenuation potential of soil as natural sinks of As was evaluated using 10 g soil in 1 L solution containing 10 mg/L of As under different geochemical conditions i.e., PO_4^{3-} (10, 20 and 30 mg/L), NO_3^- (10, 30 and 50 mg/L) and pH (6, 8 and 10) for the obtained equilibrium time period. The salts of KH_2PO_4, KNO_3 and $Na_2(HAsO_4).7H_2O$ were used for making standards for PO_4^{3-}, NO_3^- and As while pH buffers were used for setting the pH 6, 8 and 10,

2.3 *Analysis*

Portable pH meter (HANNA instruments) was used for measuring the pH. The analysis of NO_3^-, PO_4^{3-} and As was conducted by UV-VIS spectrophotometer (Lasany Double beam LI-2004, Japan).

3 RESULTS AND DISCUSSION

3.1 *Kinetic study*

The results of the kinetic batch experiment of As vs. time by the soil is shown in Figure 1a. The results inferred that adsorption of As by soil was fast in the initial period that followed by an equilibrium point in almost 24 h by removing 100% As that suggests that the transfer of As to adsorption sites of minerals and its aggregates can be a rate limiting step.

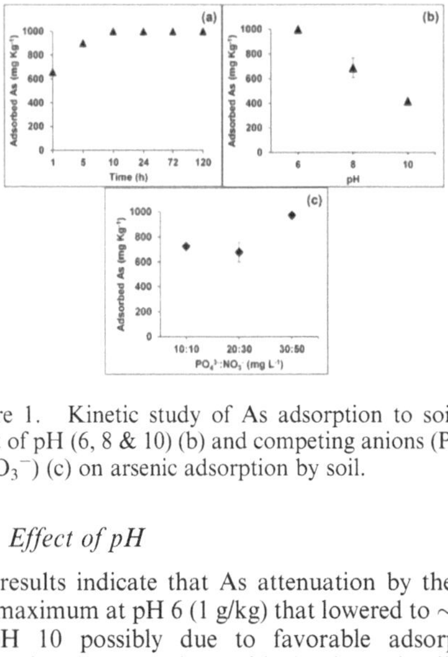

Figure 1. Kinetic study of As adsorption to soil (a); effect of pH (6, 8 & 10) (b) and competing anions (PO_4^{3-} & NO_3^-) (c) on arsenic adsorption by soil.

3.2 *Effect of pH*

The results indicate that As attenuation by the soil was maximum at pH 6 (1 g/kg) that lowered to ~43% at pH 10 possibly due to favorable adsorption mechanism between the positively charged soil surface and the predominant As(V) species i.e., $H_2AsO_4^-$ at low pH while as pH increases, soil surface repulsive forces emerge owing to the formation of OH^- ions, which produce reduced adsorption via electrostatic repulsion with the oxyanions (Figure 1b).

3.3 *Effect of competing anions*

Due to their high charge density, phosphate and nitrate compete with As for adsorption sites. Although previous studies reported very less uptake of As in presence of competing anions and stronger for PO_4^{3-} due to its high charge density as compared to NO_3^- (Guo *et al.*, 2012). But in the present study, investigated soil sample attenuated As best at 30 mg/L of PO_4^{3-} and 50 mg/L of NO_3^- (0.97 g/kg) (Figure 1c). A slight inverse relation of residual As with increasing aqueous PO_4^{3-} and NO_3^- concentrations partly supports As-PO_4^{3-} complexation, while NO_3^- seems non-responsive.

4 CONCLUSIONS

The attenuation investigation reveals that the study area's aquifer soils have high As retentive efficacy, notably at near-neutral pH and in the presence of dissolved PO_4^{3-}, due to ion-exchange and aqueous complexation processes. The use of a proper adsorbent in the study area is also recommended that can transfer the mobile or soluble As into particulate phases.

ACKNOWLEDGEMENTS

NK acknowledges Department of Science and Technology (DST), Government of India for supporting this research through DST-INSPIRE Fellowship and Panjab University, Chandigarh, India for providing all the technical and infrastructural facilities for the laboratory and field work.

REFERENCES

Guo Y., Zhu Z., Qiu Y. & Zhao J. 2012. Adsorption of arsenate on Cu/Mg/Fe/La layered double hydroxide from aqueous solutions. *J. Hazard. Mater.* 239: 279–288.

Kaur N. & Paikaray S. 2021. Arsenic-rich surface and groundwater around eastern parts of rupnagar district, punjab, india. In: P.K. Shit, P.P. Adhikary & D. Sengupta (eds.). *Spatial Modeling and Assessment of Environmental Contaminants.* Springer, Cham. pp. 379–393.

Lin Z. & Puls R.W. 2003. Potential indicators for the assessment of arsenic natural attenuation in the subsurface. *Adv Environ. Res.* 7(4): 825–834.

Sailo L. & Mahanta C. 2016. Natural attenuation processes of arsenic in the groundwater of the Brahmaputra floodplain of Assam, India. *Environ. Sci. Process Impacts.* 18(1): 115–125.

Sengupta S., Sracek O., Jean J.S., Yang H.J., Wang C. H., Kar S., Babek O., Lee C.Y. & Das S. 2018. Difference in attenuation among Mn, As, and Fe in riverbed sediments. *J. Hazard. Mater.* 341: 277–289.

Stollenwerk K.G., Breit G.N., Welch A.H., Yount J.C., Whitney J.W., Foster A.L., Uddin M.N., Majumdar R.K. & Ahmed N. 2007. Arsenic attenuation by oxidized aquifer sediments in Bangladesh. *Sci. Total Environ.* 379(2–3): 133–150.

Wang S. & Mulligan C.N. 2006. Natural attenuation processes for remediation of arsenic contaminated soils and groundwater. *J. Hazard. Mater.* 138(3): 459–470.

Section 5: Sustainable mitigation and management

5.1 Approaches for system strengthening and scaling up drinking water safety in Bangladesh

Arsenic in the Environment: Bridging Science to Practice for Sustainable Development – van der Wal, Ahmad, Petrusevski, Weijma, Savic, van der Wens, Beerendonk, Bhattacharya, Bundschuh & Naidu (Eds)
© 2023 the Author(s), ISBN 978-1-032-32928-4

System strengthening and scaling up drinking water safety and reduction of the risks of arsenic in the communities of Bangladesh

N. Akter[1], D. Johnston[1], M. Niang[1], Z. Jurji[1], E.R. Khan[2], M.S. Rahman[2], T.M.S. Khan[2], M.T. Islam[3], K.M. Ahmed[4] & P. Bhattacharya[3]
[1]*UNICEF Bangladesh, Dhaka, Bangladesh*
[2]*Department of Public Health Engineering, Dhaka, Bangladesh*
[3]*KTH-International Groundwater Arsenic Research Group, Department of Sustainable Development, Environmental Science and Engineering, KTH Royal Institute of Technology, Stockholm, Sweden*
[4]*Department of Geology, University of Dhaka, Dhaka, Bangladesh*

ABSTRACT: Despite 98.5% access to improved water sources, significant challenges remain to achieve the target of SDG 6.1. This includes the availability of safely managed drinking water that is on premises, available when needed, arsenic safe and free from microbial contamination. Currently, 11.8% of the population – 17.5 million people – remain exposed to arsenic contamination at drinking water source above Bangladesh's standard of 50 µg/L. There were many initiatives taken by Government and donors on arsenic mitigation, but the situation did not improved significantly. Various stakeholder consultations have indicated that the slow rate of progress is primarily associated with challenges in enforcement of policies, legislative and regulatory framework, harmonizing approaches, institutional arrangements, capacities and social behavior. UNICEF Bangladesh with partners implemented a new harmonized approach of arsenic mitigation by addressing entire population of a village with arsenic safe water, improved sanitation and hygiene behavior through vulnerability risk assessment, appropriate technologies, evidence based decision making and strong community involvement so that poor and extreme poor people are addressed. This project also aims at triggering the institutionalization of strategic interventions that address the systemic weaknesses of the sector such as (a) poor intra and inter sectoral coordination; (b) insufficient private sector capacity and regulation; (c) low institutionalization of water quality monitoring, surveillance and water safety planning; (d) insufficient use of alternate water sources and evidence for technical decision making; and the (e) insufficient access to reliable data. The project is jointly implemented with the Government partner DPHE, NGOs and Royal Institute of Technology (KTH), Sweden.

1 INTRODUCTION

Bangladesh has made laudable progress towards achieving its goal of ensuring access to safe drinking water for all its citizens and achieved the Millennium Development Goal (MDG). However, in the context of the SDG 6.1 there is a gap between the coverage and safety of the improved water sources. Although 98.5% of the population has access to improved water sources nationally, about 65% of the population lack access to drinking water that is arsenic safe and free from microbial contamination and 17.5 million people are exposed to arsenic from improved water sources that exceed the BDWS (50 µg/L) (BBS & UNICEF 2019).

The Sector Development Plan (2011–2025) identifies the challenges to scale up drinking water safety in Bangladesh as geogenic arsenic, manganese & salinity; inadequate prioritization of arsenic prone areas; private sector capacity; and absence of harmonized sector wide approaches. This paper focuses on the sustainable services delivery model implemented by UNICEF with GO-NGO partnership to address he challenges in the provision of arsenic safe drinking water in Bangladesh.

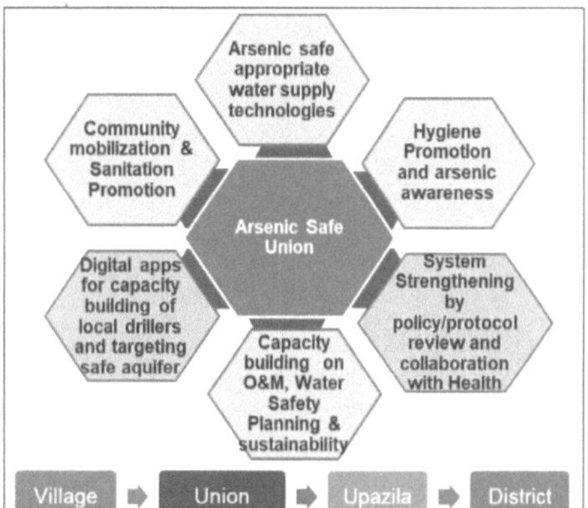

Figure 1. Arsenic safe union model.

UNICEF in collaboration with DPHE and partners developed arsenic safe village concept in 2015 and piloted between 2016 and 2017. The concept has been improved and expanded to arsenic safe union model

DOI: 10.1201/9781003317395-175

with system strengthening and private sector capacity building and implemented between 2018–2021.

2 APPROACH

System strengthening and scaling up drinking water safety needs a specific and systematic approach to create a smooth enabling anvironment. The include: i) Partnership – with GO, NGOs, academic Institutions (KTH consortium) and private sector; ii) Vulnerability assessment – to target vulnerable areas based on arsenic contamination, safe water coverage, hard to reach, arsenicosis patients, climate vulnerabilities; iii) Arsenic Screening – Hotspots identification through blanket arsenic testing for equitable resources allocation; iv) Evidence generation – for local level administration for water point's allocation and equity based sites selection for installation through community situation analysis and arsenic contamination map from screening; v) Feasibility assessment – for targeting context specific technology with safe water source; vi) Private Sector (local drillers) capacity building – for arsenic safe drilling and increasing safe water access; vii) Community capacity building – through mass awareness on water safety, safe sanitation, hygiene and O&M system development for triggering community and local administration for long term sustainability; viii) Verification and declaration – is the final steps for he declaration, verification, and certification of the union as arsenic safe.

3 RESULTS: KEY CONTRIBUTIONS TO THE SECTOR

3.1 *Arsenic mapping to target most unserved*

All drinking water sources (n = 21,540) in the intervention sites were screened for arsenic (As). From this survey it is found that 88% tube wells are shallow (<500 feet) and 12% are deep wells. 45% shallow wells are contaminated with arsenic, whereas it is only 1%in deep wells. Another important finding is that 83% wells are privately drilled by local drillers hired by houseowner without any guidance or knowledge of arsenic.

3.2 *Adoption of equity based site selection -no one is left behind*

Community WSAH situation were mapped based on step by step systematic approaches and using participatory methods; e.g. social map, wealth being ranking, WASH related diseases mapping, safe / unsafe water point mapping, hygienic latrine mapping, action plan, etc. Around 2000 equity-based sites for poor and unserved locations were selected based on this social maps, arsenic screening maps and DPHE UNICEF site selection criteria. With these evidence based, negotiation with LGIs carried out for approval of unserved and pro-poor locations for installation of safe water points. The approach is highly appreciated by the government due to remarkable change from the past when allocation was not based on need or degree of contamination.

3.3 *Scalable DPHE-UNICEF Arsenic Safe Union model*

The Project reached 203,053 people with safe drinking water services and 323,000 people with safe sanitation and increased awareness about arsenic and hygiene between 2018 to 2021. As a result, ten unions (the smallest rural administrative and local government unit in Bangladesh) has been declared as arsenic safe where all people are getting arsenic safe water with improved sanitation and hygiene. Development partners and Government agreed to scale up this programme for a sustainable arsenic mitigation.

3.4 *System strengthening and private sector capacity development*

Under policy support and system strengthening a policy hub has been established for policy gap analysis and upgradation/ new policy protocol development. National Implementation Plan on Arsenic Mitigation (IPAM) 2018, Multiple Indicator Cluster Survey (MICS) 2018–2019 are two major outcomes of this project. Collaboration with KTH for Enhancing Private Sector Capacity resulted in 150 local drillers and technocrats capacitated with arsenic-safe drilling and groundwater monitoring through 61 monitoring wells installation.

4 CONCLUSION AND NEXT STEPS

The Arsenic Safe Union model with system strengthening is a scalabele model agreed by donor, sector partners and government. As per GoB's request in 2021, UNICEF is planning to provide technical assistance to DPHE on establishing WQ dbase for planning & monitoring, engagement of private sector for safe drilling and sustainable O&M, equity based site selection and community awareness following the arsenic safe union model. UNICEF is working with KTH Royal Institute of Technology in Sweden to integrate science with indigenous knowledge to improve safe water access. In 2022, Government of Bangladesh has adopted the model in Government's $240 million arsenic mitigation project in 31 arsenic prone districts.

ACKNOWLEDGEMENTS

The financial contribution of the Embassy of Sweden (Contribution 52170040) is appreciated.

REFERENCES

BBS & UNICEF. 2019. *Bangladesh Multiple Indicator Cluster Survey 2018–2019, Progotir Pathey: Final Report*. Dhaka, Bangladesh.: Bangladesh Bureau of Statistics and UNICEF Bangladesh.

Flanagan S.V., Johnston R.B. & Zheng Y. 2012. Arsenic in tube well water in Bangladesh: health and economic impacts and implications for arsenic mitigation. *Bull. World Heal. Organ.* 90(11): 839–846.

General Economics Division (GED). 2016. *Millennium Development Goals (MDGS): End-Period Stocktaking and Final Evaluation (2000–2015)*. Dhaka, Bangladesh: General Economics Division (GED), Bangladesh Planning Commission, Government of the People's Republic of Bangladesh.

Arsenic in the Environment: Bridging Science to Practice for Sustainable Development –
van der Wal, Ahmad, Petrusevski, Weijma, Savic, van der Wens, Beerendonk, Bhattacharya, Bundschuh & Naidu (Eds)

Screening of water points as first step for effective arsenic risk reduction in Bangladesh

E.R. Khan[1], M.S. Rahman[1], T.M.S. Khan[1], B.C. Dey[1], N. Akter[2], M. Niang[2], D. Johnston[2], M.T. Islam[3], K.M. Ahmed[4], M.J. Alam[4] & P. Bhattacharya[3]

[1]*Department of Public Health Engineering, Dhaka, Bangladesh*
[2]*UNICEF Bangladesh, Dhaka, Bangladesh*
[3]*KTH-International Groundwater Arsenic Research Group, Department of Sustainable Development, Environmental Science and Engineering, KTH Royal Institute of Technology, Stockholm, Sweden*
[4]*Department of Geology, University of Dhaka, Dhaka, Bangladesh*

ABSTRACT: Arsenic screening or blanket survey of water points to assess the arsenic risk at community is a critical step for an effective arsenic mitigation programme. The mapping of existing safe and unsafe water options based on screening results are important aspects in allocation of new water points and optimization of sites for installation of water points. DPHE-UNICEF Arsenic Safe Union Project (2017–2021) targeted 22 unions under three severely arsenic affected upazilas under Cumilla, Sylhet and Satkhira districts for arsenic screening and thereafter installation of Arsenic (As) safe drinking water points as required to declare the whole union arsenic safe. One of the most important outcomes of arsenic screening results was to plot the location of water points in each union with a clear and visible symbol representing Arsenic safe and unsafe water wells using GIS. The distribution pattern along with the summary result helped technical committee for appropriate planning and decision making on allocation/optimization of the locations for new water point installations to bring all people under one union in the coverage of arsenic safe water. Moreover, the screening process helped to identify the unserved population and remote pocket areas those are overlooked by regular water supply interventions. At the same time, it provides required testimony for selection of context specific appropriate technology and raise awareness on arsenic contamination within the local people as well as their capacity for arsenic testing.

1 INTRODUCTION

Over the past one decade, Bangladesh has made significant progress towards achieving its goal of universal access to improved water supply (Akter *et al.*, 2023; Hossain 2006; Onabolu *et al.*, 2019). However, the country faces considerable challenges in terms of the safety and sustainability of water supply. According to the Multiple Indicator Cluster Survey (MICS 2012–2013) data, Bangladesh has 97.9% improved water source coverage; however, only about 35% of the population has access to water that is arsenic (As) and bacteriologically safe (MICS 2019–2021).

According to the Multiple Indicator Cluster Survey (MICS) 2012–2013 data, Bangladesh has 97.9% improved water source coverage; however, only about 35% of the population has access to water that is arsenic (As) and bacteriologically safe. The proportion of the population that are using arsenic-contaminated water above the recommended 50 µg/L, has been reduced from 26.6% in 2001 to 12.4% in 2013 and to 10.6% in 2019 (Figure 1). The scenario varies widely because of the geographic and socio-economic disparities.

Figure 1. Trend of the changing proportion of population using arsenic contaminated water sources.

Due to poorly managed surface water, around 97% of the population of Bangladesh are dependent on groundwater. However, this has become a matter of concern since Arsenic (As) was first detected in groundwater in 1993 by DPHE. According to Bangladesh Water Quality Standard (BWQS), drinking water is arsenic safe if the concentration is within 50 µg/L (ECR 1997); although WHO guideline recommendation is 10 µg/L.

Despite many mitigation measures over the last three decades (Figure 1), Bangladesh is facing great problems with poisonous arsenic in groundwater and 17.5 million people are still exposed to arsenic contaminated water (MICS 2019).

DOI: 10.1201/9781003317395-176

The Government of Bangladesh is presently undertaking the 'Arsenic Risk Reduction Project for Water Supply (ARRP-WS)'with an overall objective of improving the health and wellbeing of people living in most arsenic affected areas in Bangladesh. The specific objectives are: i) to assess the present situation of arsenic contamination in the water supply through screening; ii) to provide arsenic safe water to very high to medium priority areas (more than 60% arsenic contamination and water supply coverage of less than 60%); and iii) to ensure the use of safe water by increasing knowledge and capacity of LGIs and water user groups.

Accordingly, there is a strong need to screen the arsenic concentration in all water sources, and map all tested water points to be identified as *safe* or *unsafe*. This screening eventually determines vulnerability with detailed variation of arsenic exposure within the targeted project area. Mapping of the results can be used for assessment, optimization of sites and choice of technology and allocation of resources for the installation of new safe water points to bring all people under the coverage of safe drinking water.

2 THE APPROACH

In order to generate an evidence-based arsenic mitigation program, arsenic screening is being carried in 3,200 unions with known incidence of high arsenic in the drinking water wells (Figure 2). This includes: i) training master trainers for conducting training on arsenic screening and reporting; ii) conduct training for local level testers (19,200 testers) on arsenic screening and reporting; iii) arsenic screening of an estimated 8 million water points with 10% supervision; iv) analysis of data, reporting and visualization.

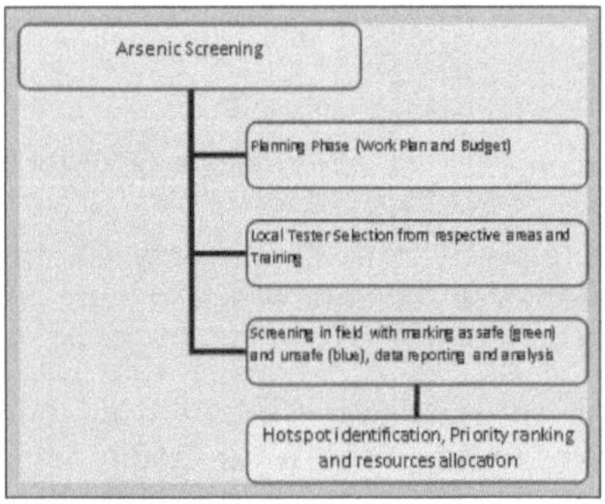

Figure 2. Approaches of arsenic screening.

Following this approach, blanket survey was planned to screen the water points for arsenic and some other important parameters i.e. salinity in SW part of Bangladesh. Aligned with this activity, a screening program was conducted in three different geological areas with limited safe water coverage (20%–60%), located in Daudkandi, Cumilla district, Gowainghat, Sylhet district, and Assasuni, Satkhira district (Figure 3) where more than 70% arsenic contaminated wells have been installed at various depths at shallow, intermediate-deep and deep levels.

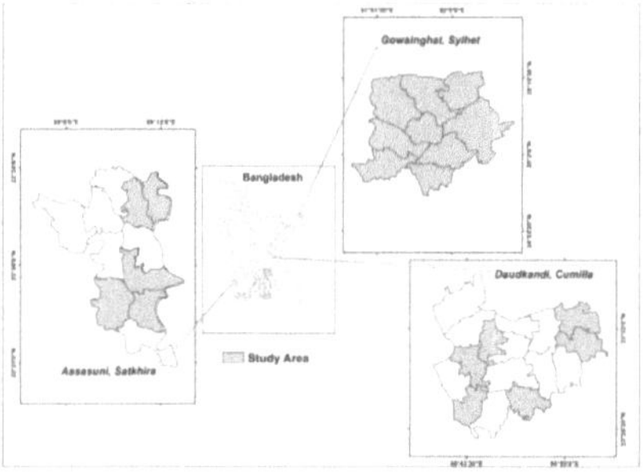

Figure 3. GoB-UNICEF project intervention areas.

3 RESULTS AND DISCUSSION

3.1 *Database development*

Data can be collected by using approved mobile application ASMITAS (Sharma *et al.*, 2019), where data preservation for future references in hard copy is highly recommended. A structured database can be exported as excel and portable document format for further analysis as well as visualization on a digital dashboard (Figure 4).

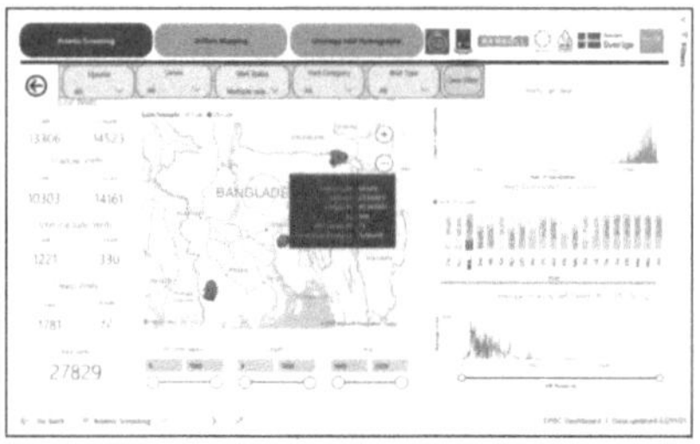

Figure 4. Interactive ASMITAS – digital dashboard for capturing, analysis, management and integration of arsenic screening data.

3.2 *Statistical analysis*

All water points using by local level community people for drinking and housing purpose were covered under this survey. The survey data all information's were grouped into three different segments based on depth such as the shallow, intermediate deep and deep wells (Hossain *et al.*, 2017, 2023). Different maps can be produced for different aquifers to understand the spatial distribution pattern with respect to arsenic and other water quality parameters (Figures 5 & 6).

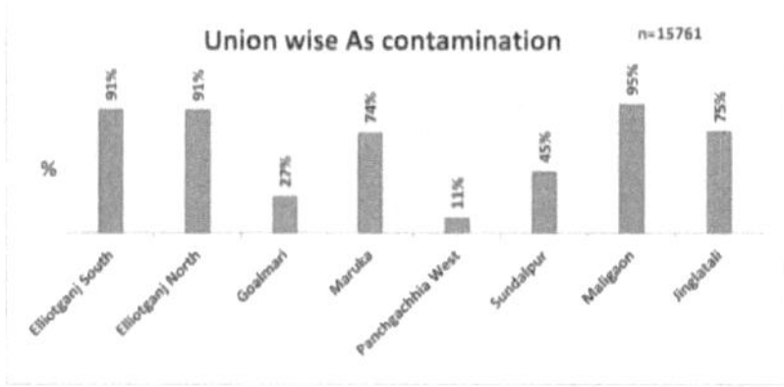

Figure 5. Union wise arsenic contamination based on well screening.

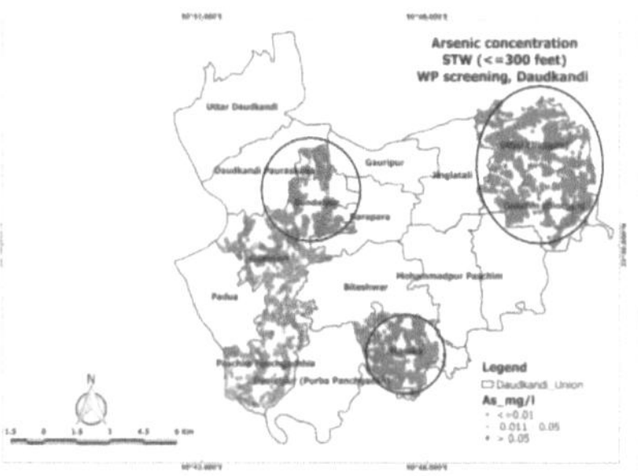

Figure 6. Spatial distribution of arsenic contamination for targeting area of interest.

3.3 *Spatial analysis*

Spatial analysis to identify hot spots where more interventions is necessary

From this map priority of intervention's area can be identified. Big circle areas have higher arsenic contamination rate (94%–97%) where more allocations is necessary to mitigate arsenic exposure.

3.4 *Resources allocation of new safe water points*

A technical review committee (TRC) formed by DPHE and UNICEF will assess above mentioned analysis. They will consider safe water coverage and other climatic aspects. Finally TRC will come up with some numbers for arsenic safe water options as a part of resources allocations. This approach will also minimize the possibility of elite capture to a great extent and will prioritize the most unserved areas.

4 CONCLUSIONS

This approach of enhancing safe drinking water access to the people in a systematic manner through area wide testing, developing digital database, analyzing results, mapping of water points, identifying the most demanding areas for interventions and eventually allocation of new safe water options has been found a suitable strategy for arsenic mitigation. For further scaling up, replication of this strategy in other arsenic hotspots in near future could be way forward to bring people exposed to arsenic contamination under safe water coverage.

REFERENCES

Ahmed K.M., van Geen A. & Bhattacharya P. 2019. Implementation of arsenic mitigation: insights from araihazar and matlab – two extensively studied areas in Bangladesh. In: Y.G. Zhu *et al.* (eds.) *Environmental Arsenic in a Changing World (As2018)*. CRC Press/Taylor and Francis (ISBN 978-1-138-48609-6), pp. 565–566.

Akter N., Johnston D., Niang M., Jurji Z., Khan E.R., Rahman M.S. Khan T.M.S., Islam M.T., Ahmed K.M. & Bhattacharya P. 2023. System strengthening and scaling up drinking water safety for and reduction of the risks of arsenic in the communities of bangladesh. In: In: A. van der Wal *et al.* (eds.) *Arsenic in the Environment: Bridging Science to Practice for Sustainable Development (As2021)*. CRC Press/Taylor and Francis (ISBN 978-1-032-32928-4) (This volume).

E.C.R. 1997. *The Environment Conservation Rules. Government of Bangladesh.*

Hossain M., Bhattacharya P., Jacks G., von Brömssen M., Ahmed K.M., Hasan M.A. & Frape S.K. 2017. Sustainable arsenic mitigation – from field trials to implementation for control of arsenic in drinking water supplies in bangladesh. In: P. Bhattacharya D. A. Polya & D. Jovanovic (eds.) *Best Practice Guide on the Control of Arsenic in Drinking Water*. Metals and Related Substances in Drinking Water Series, IWA Publishing, UK, pp. 99–116. https://doi.org/10.2166/9781780404929_099

Hossain M., Bhattacharya P., Shaun K. Frape S.K, Ahmed K.M., Jacks G., Hasan M.A., von Brömssen M., Shahiruzzaman M. & Mörth C.-M. 2023. A potential source of low-manganese, arsenic-safe drinking water from intermediate deep aquifers (IDA), Bangladesh. *Groundw. Sustain. Dev.* 21: 100906.

Hossain M.F. 2006. Arsenic contamination in Bangladesh—an overview. *Agric. Ecosyst. Environ.* 113(1–4): 1–16.

MICS. 2019. *Multiple Indicator Cluster Survey, 2019.* Government of the Peoples Republic of Bangladesh, Bangladesh Bureau of Statistics & United Nations Children's Fund.

Onabolu B., Khan E., Chowdhury J., Akter N., Ghosh S., Rahman S., Bhattacharya P., Johnston D., Alam K., Ahmed F., Amin R., Uddin S., Khanam S., Hassan M., von Brömssen M. & Ahmed K.M. 2019. Integrating policy, system strengthening, research and harmonized services delivery for scaling up drinking water safety in Bangladesh. In: Y.G. Zhu *et al.* (eds.) *Environmental Arsenic in a Changing World (As2018)*. CRC Press/Taylor and Francis (ISBN 978-1-138-48609-6), pp. 535–537.

Sharma S., Bhattacharya P., Kumar D., Perugupalli P., von Brömssen M., Islam M.T. & Jakariya M. 2019. Asmitas – a novel application for digitalizing the sasmit sediment color tool to identify arsenic safe aquifers for drinking water supplies. In: Y.G. Zhu *et al.* (eds.) *Environmental Arsenic in a Changing World (As2018)*. CRC Press/Taylor and Francis (ISBN 978-1-138-48609-6), pp. 629–632.

Arsenic in the Environment: Bridging Science to Practice for Sustainable Development –
van der Wal, Ahmad, Petrusevski, Weijma, Savic, van der Wens, Beerendonk, Bhattacharya, Bundschuh & Naidu (Eds)
© 2023 the Author(s), ISBN 978-1-032-32928-4

Preliminary investigations to assess potentialities of applying the SASMIT protocol in three contrasting hydrogeological environments in Bangladesh

K.M. Ahmed[1], M.J. Alam[1], A.S.M. Woobaidllah[1], M.M.A. Akif[1], M.M. Bishal[1], I. Rahman[1], P. Bhattacharya[2], M.T. Islam[2], M. von Brömssen[3], S. Sharma[4], E.R. Khan[5] & N. Akter[6]
[1]*Department of Geology, University of Dhaka, Dhaka, Bangladesh*
[2]*KTH-International Groundwater Arsenic Research Group, Department of Sustainable Development, Environmental Science and Engineering, KTH Royal Institute of Technology, Stockholm, Sweden*
[3]*Ramböll Sweden AB, Stockholm, Sweden*
[4]*ExcelDots AB, Stockholm, Sweden*
[5]*Department of Public Health Engineering, Kakrail, Dhaka, Bangladesh*
[6]*WASH Section, UNICEF Bangladesh, Dhaka, Bangladesh*

ABSTRACT: The SASMIT protocol, based on the linkage between dissolved arsenic (As) concentrations and color of sediments in the screened zone, has been developed during an As study in Matlab in southeast Bangladesh during 2005–2009. The current study aims at applying the same protocol in three diverse geological environments in Bangladesh, viz. Gowainghat in the northeast piedmont plain; Daudkandi in the Meghna flood plain and Assassuni in the Ganges deltaic plain. Detail sediment characterization from targeted depths at different drilling locations and creation of a digital platform has been assessed to unravel the potentials for application of the SASMIT protocol i.e. the wells in oxidized layers and intermediated depth aquifers with grey to light grey sediments are showing low arsenic concentrations. The digital platform was implemented during this intervention period and acted as a decision making tool as well as data capturing and preservation platform. The drillers and implementing agencies in installing public water supply wells at safe aquifers; this would also support decision makers to propose alternative water supply options other than tubewells in regions with high As in shallow and deep aquifer. Implementations of the approach combined with ASU of UNICEF-DPHE can improve safe water access in Bangladesh significantly.

1 INTRODUCTION

High levels of arsenic (As) has been reported in groundwater in Bangladesh since 1993, and investigations have been carried out to understand the occurrence and distribution patterns for designing mitigation plans. BGS & DPHE (2001) study conducted the national hydrochemical survey and reported that 27% and 46% of tested 3534 wells exceeded Bangladesh national drinking water standard (50 µg/L) and WHO provisional guideline value (10 µg/L) respectively. Multi Cluster Indicator Survey (MICS) conducted by UNICEF in collaboration with the Bangladesh Bureau of Statistics provides data on gradual changes in As exposure since 2013. The latest survey conducted in 2019 (BBS 2019) reports that 18.6% of the tested drinking sources contained As in access of 10 µg/L, compared to 25.5% in 2013, and 11.8% in excess of 50 µg/L, compared to 12.5% in 2013 (Figure 1a).

Although the proportions of population drinking high As water have declined significantly, still a large number are exposed to unacceptable levels of As. Despite efforts by Government of Bangladesh and other agencies, As remains a major public health concern in the country. At the same time occurrence of microbiological contaminations is another major public health concern. Arsenic and microbiological contamination combinedly reduce safe water coverage remarkable (Figure 1b).

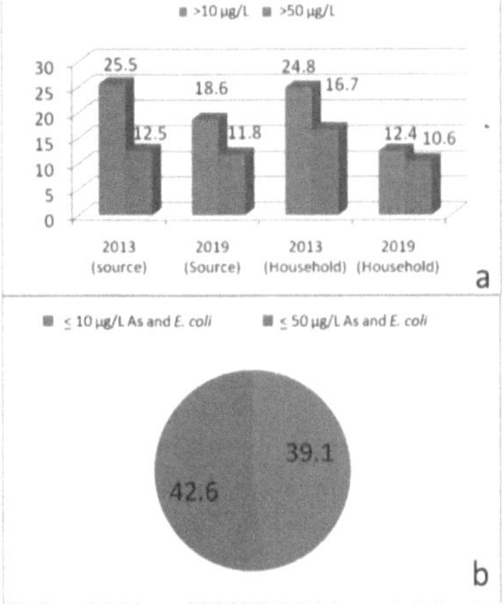

Figure 1. a) Arsenic in sources and household drinking water in Bangladesh. b) Access to safe after adjusting for As and E. coli in drinking water in Bangladesh (BBS 2019).

DOI: 10.1201/9781003317395-177

The present study aims to investigate the potentialities of applying the SASMIT protocol, developed earlier (SASMIT 2014), in three contrasting geological environments in Bangladesh to improve their existing mitigation plans. The study also aims to digitize the previously developed color tool (Hossain *et al.*, 2014) in order to build capacity of local drillers and communities in finding safe aquifers.

2 MATERIALS AND METHODS

2.1 *Study area*

SASMIT concept was developed in Matlab area of the Meghna Deltaic plain in central Bangladesh. The present study has been extended over three geological provinces, i.e. the Northeast Piedmont Plains (Gowainghat of Sylhet District), Meghna Deltaic Plain (Daudkandi of Comilla District) and Ganges Deltaic Plain (Assassuni of Satkhira District) as shown in Figure 2 along with variations in As concentrations patterns.

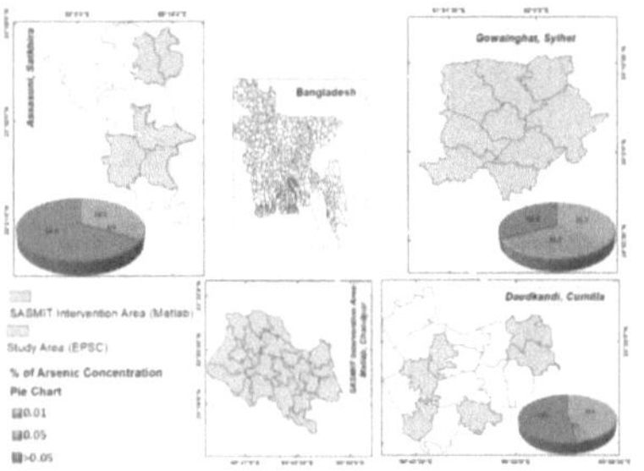

Figure 2. Location map of the study areas in Bangladesh in relation to Matlab. The pie charts show distribution of As concentrations in the existing wells.

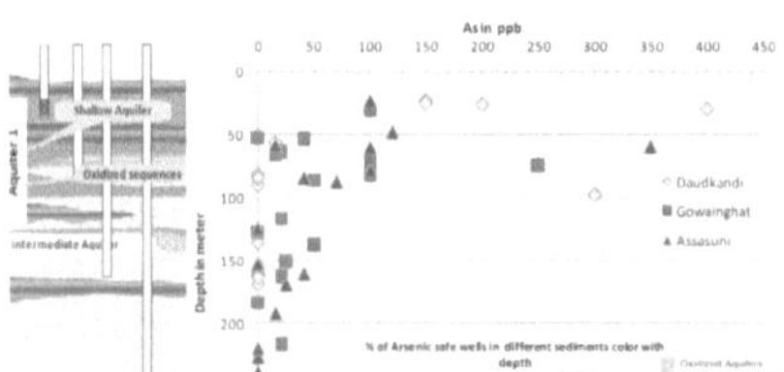

Figure 3. Distribution of As in groundwater three study upazilas showing varying proportions of safe and unsafe wells as well as different depth distribution patterns.

2.2 *Hydrogeological mapping for sediment characterization*

Detailed hydrogeological mapping has been carried out in the three study areas for characterization of sediments in relation to spatial and vertical variations of dissolved As. Sites for detailed investigatios were selected primarliy based on existing well As data and constrained by surface geophysical resistivity survey data. Drilling were conducted in selcted locations and sediment samples were collected during the installation of multilevel piezometers for monitoring temporal variations and groundwater levels and water quality. ASMITAS (Arsenic Mitigation at Source) digital platform (Sharma *et al.*, 2019) was used to record the sediment colours using a Nix Sensor. ASMITAS was used to record data of 61 new wells for data capturing and preservation in the digital platform which helped to enhance decision making baseed on the Digital Sediment Color Tool instead of manual colour matching.

3 RESULTS AND DISCUSSION

The three aquifer systems, viz. Shallow, Intermediate and Deep, have been identified in all areas with characteristic As concentrations. Oxidized aquifer has also been identified at certain locations (Figure 3).

4 CONCLUSIONS AND RECOMMENDATIONS

The integration of different mapping techniques along with use of the digital platform significantly enhanced the decision making capabilities for arsenic mitigation at all three areas. This approach combined with social criteria developed under the Arsenic Safe Union concept of UNICEF-DPHE can help in improving water access for high arsenic areas of Bangladesh.

ACKNOWLEDGEMENTS

We acknowledge Sida for supporting the SASMIT and EPSC projects through UNICEF-Bangladesh. We also acknowledge the supports of AAN, EPRC, VERC) and the Practical Action.

REFERENCES

Bangladesh Bureau of Statistics 2019. *Progotir Pathey, Bangladesh Multiple Indicator Cluster Survey 2019, Key Findings*. Bangladesh Bureau of Statistics Dhaka, Bangladesh.

BGS & DPHE 2001. *Arsenic Contamination of Groundwater in Bangladesh*. Vol. 2. BGS Technical Report WC/00/19.

Hossain M., Bhattacharya P., Frape S.K., Jacks G., Islam M.M., Rahman M.M., von Brömssen M., Hasan M.A., Ahmed K.M. 2014. Sediment color tool for targeting arsenic-safe aquifers for the installation of shallow drinking water tubewells. *Sci. Total Environ.* 493: 615–625. 277.

SASMIT 2014. Sustainable Arsenic Mitigation. Community Driven Initiatives to Target Arsenic Safe Groundwater as Sustainable Mitigation Strategy. *Sida Contribution*: 73000854

Sharma S., Bhattacharya P., Kumar D., Perugupalli P., von Brömssen M., Islam M.T. & Jakariya M. 2019. ASMITAS–a novel application for digitalizing the SASMIT sediment color tool to identify arsenic safe aquifers for drinking water supplies. In: Y.G. Zhu *et al.* (eds.) *Environmental Arsenic in a Changing World (As2018)*. CRC Press/Taylor and Francis (ISBN 978-1-138-48609-6), pp. 629–632.

*Arsenic in the Environment: Bridging Science to Practice for Sustainable Development –
van der Wal, Ahmad, Petrusevski, Weijma, Savic, van der Wens, Beerendonk, Bhattacharya, Bundschuh & Naidu (Eds)*
 ISBN 978-1-032-32928-4

Drillers mapping – as supportive tool to assess local hydrogeologic setting to enhance private sector capacity for scaling up safe water access

M.J. Alam[1], K.M. Ahmed[1], M.M.A. Akif[1], M.M. Bishal[1], I. Rahman[1], P. Bhattacharya[2], M.T. Islam[2], M. von Brömssen[3], S. Sharma[4], E.R. Khan[5] & N. Akter[6]
[1]*Department of Geology, University of Dhaka, Dhaka, Bangladesh*
[2]*KTH-International Groundwater Arsenic Research Group, Department of Sustainable Development, Environmental Science and Engineering, KTH Royal Institute of Technology, Stockholm, Sweden*
[3]*Ramböll Sweden AB, Stockholm, Sweden*
[4]*ExcelDots AB, Stockholm, Sweden*
[5]*Department of Public Health Engineering, Kakrail, Dhaka, Bangladesh*
[6]*WASH Section, UNICEF Bangladesh, Dhaka, Bangladesh*

ABSTRACT: Water supply in Bangladesh is mainly dependent on GW and there are some specialized/technical processes to bring that water from underground to surface and drillers are the most important and first tier of resources in this context. In 2020 an extensive survey and analysis have been done in three different parts in Bangladesh which includes Gowainghat, Daudkandi, and Assasuni with the support of DPHE-UNICEF and found that approximately > 85% tubewell are installed by local drillers individually which can extensively influence safe water coverage. But there is gap between their authorization process and knowledge. Among those surveyed 126 drillers about 33% do not have any knowledge about As and 30% driller consider clear water for a successful well instead of As. In these areas about 38180 private well are installed by local drillers and approx. 48% wells are still As contaminated. It's not practical to have geological and Hydrogeological information from all over the country. So these add-ons can help to do some detail study. In Daudkandi, Gowainghat and Assasuni general ideas of oxidized sediments distribution in shallow aquifers were identified and expanded from driller's mapping. From their long experiences aquifers and aquitard distribution can be assumed and used in extensive analysis. So they can play an important role for scaling up safe water coverage in Bangladesh.

1 INTRODUCTION

High arsenic (As) concentration in groundwater controls safe water coverage in Bangladesh from 1993 when it was first explored. Many options developed but hand tubewell is the most cost effective and can be maintained by users easily. Currently in many areas of Bangladesh people have been switched from any other alternative options to tubewell (Hossain *et al.*, 2015). These hand tubewell are installing by local drillers (local entrepreneur). In Bangladesh 90% tubewell out of 10 millions are installed by local drillers (Hossain *et al.*, 2014). In this study the number is about 92% (Figure 1). They are the first knowledge hub from where anyone can get initial idea about local Hydrogeological condition i.e. GWQ and current practice for As mitigation (local) if

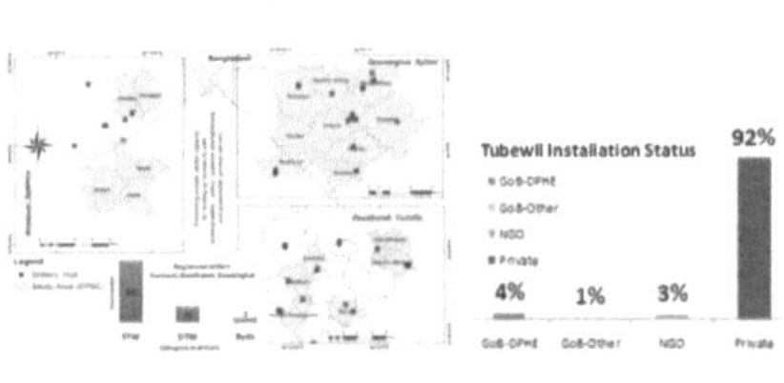

Figure 1. Drillers mapping intervention areas with registered driller number (left) and their contribution in total number of water points (right).

there are no other relevant information. But there is gap between local drillers and authorized organization. They are doing their drilling activity independently. They do not have any binding and the authority does not have any control to them and their activities. If GoB, NGO, INGO and LGIs can control drillers activities and mapped out their knowledge and experience then these large number (92%) can be improved.

2 MATERIALS AND METHODS

2.1 *Study area*

Driller's survey was first initiated in Matlab under SASMIT program in 2010. Now under EPSC it was expanded more scientifically to national level to three different part of Bangladesh i.e. Gowainghat of Sylhet District, Daudkandi of Cumilla District and Assasuni of Satkhira District with numbers of surveyed drillers (Figure 1).

2.2 *Drillers mapping*

2.2.1 *Hub identification and dialogue with local drillers*

Driller's information were collected from local hardware shop known as "driller's hub" (Figure 1). A

 DOI: 10.1201/9781003317395-178

comprehensive questionnaire (Google forms) was finalized which includes general and Technical section. About 126 drillers were surveyed and lots of information's were collected.

2.2.2 *Knowledge and capacity assessment*
Based on their feedback a detail analysis was completed to assess their knowledge about local GWQ, GWL, geology and also local practice for As mitigation according to their perception and their capacity. During their capacity assessment it was explored that about. After assessing their knowledge gaps were identified where their capacity need to be enhanced.

2.3 *Validation with scientific tool*

Result from drillers survey and mapping were validated with some scientific study includes piezometer drilling under EPSC program.

2.3.1 *Water Quality with sediment color*
Groundwater quality with sediments color has been analyzed and found that off-white or red color sediments have low As concentration while black color sediments have high (von Brömssen *et al.*, 2007). Local drillers are targeting these off-white or red sediments which they sometimes called "ghiya" from their long experiences with no scientific knowledge.

2.3.2 *GWL for targeting sustainable options*
GWL can be assumed from this mapping from their long experiences which will be validated with EPSC nest data.

2.3.3 *Subsurface geological information*
Some draft ideas of sand and clay layers distribution can be explored from technical session and will be incorporated with extensive geological analysis after validation with EPSC piezometer data.

3 RESULTS AND DISCUSSION

3.1 *Knowledge and capacity assessment*

It was observed that about drillers about 33% do not have any knowledge about As and 30% driller consider clear water for a successful well instead of As

3.2 *Spatial distribution of oxidized layers*

Oxidized layers are identified and validated with EPSC nest data (Figure 2).

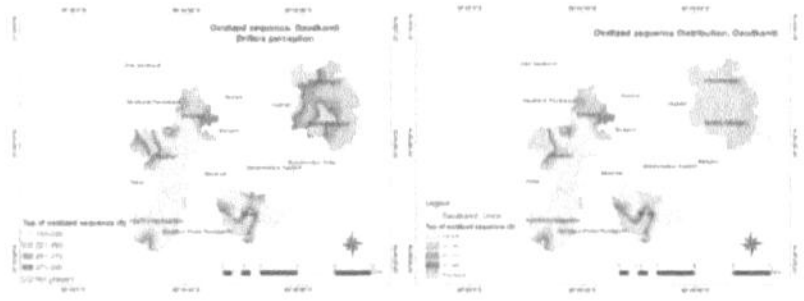

Figure 2. Distribution of oxidized layers in different part of Daudkandi after drillers mapping (left) and with validation through EPSC nest (right).

3.3 *Groundwater arsenic concentration in wells installed with driller perceived sediments color*

Drillers have their own perception about groundwater quality with sediments color and they assumed that brown/red color sediments are suitable for As and iron free water which was validated by water quality (field kit) data from the EPSC nest (Figure 3). Also black color sand has high As and iron. White sand with less black color minerals mixtures in deeper and sometimes shallow depth may have low As and iron.

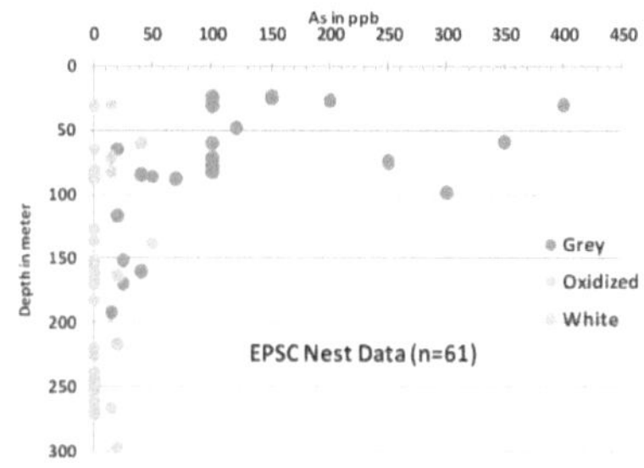

Figure 3. Arsenic distribution in groundwater with sediments colour based on driller's perception.

4 CONCLUSIONS AND RECOMMENDATIONS

Drillers mapping can be an important tool to do some initial assessment for any geographical areas having not much geological/Hydrogeological information's. They can be mapped out through some scientific approach to have some helpful information about their knowledge, capacity, current practice for As mitigation (local), local geology and GWQ. This information can be a supportive tool for further detail analysis. This mapping can be a helpful tool to fill the gaps between them and authorized organizations as well as their current knowledge which can be enhanced through proper training.

ACKNOWLEDGEMENTS

We acknowledge the Swedish International Development Cooperation Agency (Sida) for supporting the SASMIT project (Contribution No. 73000854) and EPSC project through Unicef-Bangladesh (BCO/PCA/2018/015-2018/001, Contribution: 52170040).

REFERENCES

Hossain M., Bhattacharya P., Frape S.K. *et al.* 2014. Sediment color tool for targeting arsenic-safe aquifers for the installation of shallow drinking water tubewells. *Sci. Total Environ.* 493: 615–625.
Hossain M., Rahman S.N., Bhattacharya P. *et al.* 2015. Sustainability of arsenic mitigation interventions-an evaluation of different alternative safe drinking water options provided in Matlab, an arsenic hot spot in Bangladesh. *Front. Environ. Sci.* 3: 1–15.
von Brömssen M., Jakariya M., Bhattacharya P., Ahmed K. M. *et al.* 2007. Targeting low-arsenic aquifers in Matlab Upazila, Southeastern Bangladesh. *Sci. Total Environ.* 379: 121–132.

Arsenic in the Environment: Bridging Science to Practice for Sustainable Development –
van der Wal, Ahmad, Petrusevski, Weijma, Savic, van der Wens, Beerendonk, Bhattacharya, Bundschuh & Naidu (Eds)
 ISBN 978-1-032-32928-4

Method for systematic and automatized data capturing of hydrogeological information, for provision of safe drinking water and sustainable groundwater management

M. Boberg[1], M. Selander[1], C. Haugwitz[2], M. von Brömssen[1,3], P. Bhattacharya[1], S. Sharma[4], K.M. Ahmed[5], M.T. Islam[1,6], M.J. Alam[5] & N. Akter[7]

[1]*KTH-International Groundwater Arsenic Research Group, Department of Sustainable Development, Environmental Science and Engineering, KTH Royal Institute of Technology, Stockholm, Sweden*
[2]*Ramboll Denmark A/S, Copenhagen, Denmark*
[3]*Ramboll Sweden AB, Stockholm, Sweden*
[4]*ExcelDots AB, Stockholm, Sweden*
[5]*Department of Geology, University of Dhaka, Dhaka, Bangladesh*
[6]*WaterAid-Bangladesh, Dhaka, Bangladesh*
[7]*WASH Section, UNICEF Bangladesh, Dhaka, Bangladesh*

ABSTRACT: The SASMIT protocol, based on the relation between arsenic (As) in groundwater and color of sediments, has previously been developed in Matlab, SE Bangladesh. The current study aimed to investigate the potentialities of a digital platform for collecting hydrogeological data with a mobile app/software called ASMITAS, automatically transferring data to the GeoGIS 2020 system – used for managing and analyzing hydrogeological data for the study area Daudkandi, Bangladesh. Detailed sediment characteristics, groundwater pressure heads and As concentrations, according to the SASMIT protocol, was collected from 5 piezometer nests (23 piezometer) for the aquifer delineation analysis. Three main aquifer systems i.e. Shallow, Intermediate and Deep, were identified with characteristic As concentrations. Oxidized brown sediments were identified and safe aquifers for the installation of drinking water tube wells could be outlined. The integration of the digital platform for data collection, and it's management including the post-processing can enhance the decision-making capabilities for arsenic mitigation at national and local level as well as for producing sustainable groundwater management strategies. Furthermore, providing information transparent at local as well as national level will ensure accountability of decision makers at all levels and may be an efficient tool for combating elite capturing of safe drinking wells.

1 INTRODUCTION

Although the proportions of population drinking high Arsenic (As) water in Bangladesh have declined significantly in recent years, still a large population are exposed to unacceptable levels of As and it remains a major public health challenge to combat. At the same time occurrence of microbiological contaminants as well as elevated saline water in aquifers are other major public health- and water management concerns. Arsenic and microbiological contamination combinedly reduce safe water coverage remarkable in Bangladesh (BBS 2019).

In regions with increased water stress or elevated levels of As, microbial contaminants and saline groundwater, capturing large amount of hydrogeological field data can lay the foundation for a platform for well informed decision making regarding sustainable water management as well as provision of safe drinking water.

Major challenges for collecting hydrogeological data in rural Bangladesh includes the lack of a common protocol and a tool for collecting and transferring larger data sets to a common database. A common tool for quality assurance of hydrogeological quality assurance, post-processing of the data as well as a possibility to forward analyzed data to national-as well as local stakeholders.

Therefore, the present study aimed to investigate the potentialities of applying a digital platform for collecting and managing hydrogeological data called ASMITAS, in combination with GeoGIS 2020 – an advanced software for managing geological, hydrogeological and groundwater chemical databases in Daudkandi, Bangladesh (Figure 1).

Data used for this study were collected under ongoing Sida-DPHE-Unicef-KTH supported 'Systems strengthening and scaling up Drinking Water Safety in

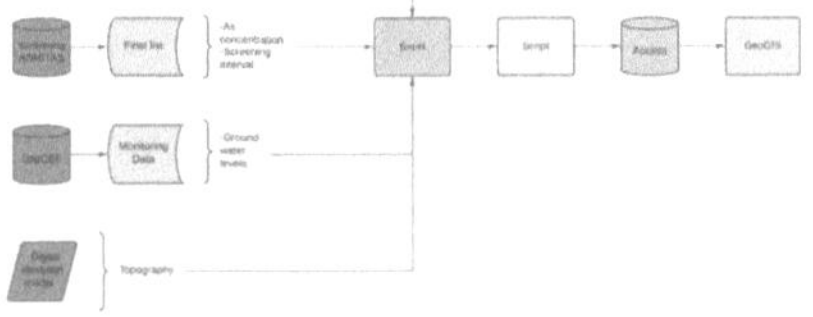

Figure 1. Data management using ASMITAS and GeoGIS.

 DOI: 10.1201/9781003317395-179

Bangladesh' project aiming rural water safety where a harmonized Sustainable Arsenic Mitigation (SASMIT 2014) protocol and Arsenic Safe Union (ASU) approach have been instigated. The SASMIT protocol aim at building capacity of local drillers and communities in targeting safe aquifers in regions with high As groundwater using straightforward analysis of lithological profiles and groundwater piezometric levels from piezometer nests (Hossain *et al.*, 2014; von Brömssen *et al.*, 2007).

2 MATERIALS AND METHODS

2.1 *Study area*

The SASMIT concept was originally developed in the Matlab area of the Meghna Deltaic plain in central Bangladesh. Based on that research, this study was conducted in a nearby area in the Daudkandi upazila (Comilla district) as shown in Figure 2. Daudkandi is situated in the southeast of Bangladesh and located near the Meghna river, one of the three rivers that forms the Ganges delta.

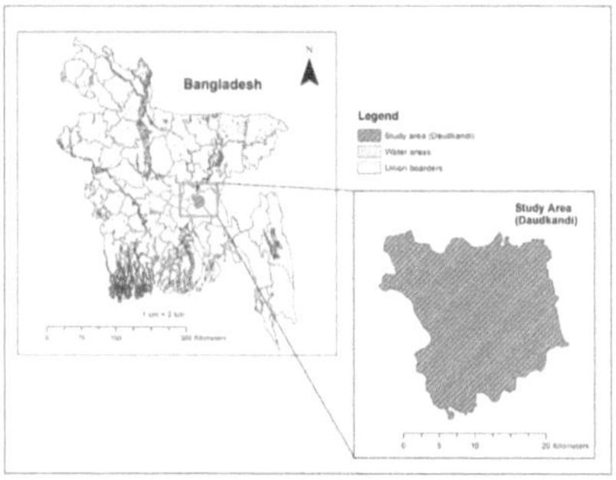

Figure 2. Location map of the study area (Daudkandi) in Bangladesh.

2.2 *Automatic transferring field data to GeoGIS*

Visualizations of hydrogeological data in the Daudkandi area have been created based on field data recorded by the digital solution ASMITAS (Arsenic Mitigation at Source, Sharma *et al.*, 2019) and digital survey by Dhaka University, DPHE, and implementing partners of UNICEF. Aquifer delineation and borehole logs were created in the software GeoGIS, used as a common platform for data from different sources. Essential data was structured systematically by identifiers that connect geological sample data and chemical data to the right borehole from 23 installed piezometers in the area. Thereafter data was transferred automatically (imported) to GeoGIS using a Python-script and finally incorporated into GeoGIS.

3 RESULTS AND DISCUSSION

Combining the power of ASMITAS and GeoGIS system automatization of data capturing and management process as well increasing the capability of post-processing hydrogeological data has proved to be an efficient way of performing hydrogeological analysis. Examples of essential data visualized in the models consisted of the piezometers monitored As levels, strata and variation in groundwater levels. Furthermore, three main aquifer systems i.e. Shallow, Intermediate and Deep, have been identified in the study area with characteristic of As concentrations. Oxidized brown sediments were also identified and safe aquifers for the installation of drinking water tube wells were outlined.

4 CONCLUSIONS

The integration and ascending use of digital platform for data collection, and it's management including the post-processing may significantly enhance the decision-making capabilities for arsenic mitigation at national and local level as well as for producing sustainable groundwater management strategies. This harmonized approach combined with social criteria developed under the ASU approach along with SASMIT protocol can help in improving water access to high arsenic areas of Bangladesh. Furthermore, providing information transparent at local as well as national level will ensure accountability of decision makers at all levels and may be an efficient tool for combating elite capturing of safe drinking wells provided by local government institutions.

ACKNOWLEDGEMENTS

We acknowledge the projects Sida-SASMIT (Contribution: 73000854) and Sida-UNICEF project–Systems Strengthening and Scaling Up Drinking Water Safety in Bangladesh (Contribution 52170040) for financial support and AAN, EPRC, VERC for support for field implementation.

REFERENCES

Bangladesh Bureau of Statistics 2019. *Progotir Pathey, Bangladesh Multiple Indicator Cluster Survey. Key Findings.* Bangladesh Bureau of Statistics Dhaka, Bangladesh.

BGS & DPHE 2001. *Arsenic Contamination of Groundwater in Bangladesh*, Vol. 2. BGS Technical Report WC/00/19.

Hossain M., Bhattacharya P., Frape S.K., Jacks G., Islam M.M., Rahman M.M., von Brömssen M., Hasan M.A., Ahmed K.M. 2014. Sediment color tool for targeting arsenic-safe aquifers for the installation of shallow drinking water tubewells. *Sci. Total Environ.* 493: 615–625.

SASMIT 2014. Sustainable Arsenic Mitigation. Community Driven Initiatives to Target Arsenic Safe Groundwater as Sustainable Mitigation Strategy. *Sida Contribution: 73000854*.

Sharma S., Bhattacharya P., Kumar D., Perugupalli P., von Brömssen M., Islam M.T. & Jakariya M. 2019. ASMITAS–a novel application for digitalizing the SASMIT sediment color tool to identify arsenic safe aquifers for drinking water supplies. In: *Environmental Arsenic in a Changing World As2018*. CRC Press/Taylor and Francis (ISBN 978-1-138-48609-6), pp. 629–632.

von Brömssen M., Jakariya M., Bhattacharya P., Ahmed K. M., Hasan M.A., Sracek O., Jonsson L., LundellL., Jacks G. 2007. Targeting low arsenic in Matlab Upazila, Southeastern Bangladesh. *Sci. Total Environ.* 379(2–3): 121–132.

Arsenic in the Environment: Bridging Science to Practice for Sustainable Development –
van der Wal, Ahmad, Petrusevski, Weijma, Savic, van der Wens, Beerendonk, Bhattacharya, Bundschuh & Naidu (Eds)
 ISBN 978-1-032-32928-4

Developing machine learning approach for predicting groundwater quality based on arsenic and other major contaminants data to accelerate the SDG 6.0 agenda on drinking water safety

S. Basak[1], S. Sharma[1], P. Bhattacharya[2], M.T. Islam[2], A. Ahmad[2,3], M. von Brömssen[4], K.M. Ahmed[5], M.J. Alam[5] & N. Akter[6]
[1]*ExcelDots AB, Bromma, Sweden*
[2]*KTH-International Groundwater Arsenic Research Group, Department of Sustainable Development, Environmental Sciences and Engineering, KTH Royal Institute of Technology, Stockholm, Sweden*
[3]*Sibelco Ankerpoort NV, Maastricht, The Netherlands*
[4]*Ramböll Water AB, Stockholm, Sweden*
[5]*Department of Geology, Dhaka Universiy, Dhaka, Bangladesh*
[6]*WASH Section, UNICEF Bangladesh, Dhaka, Bangladesh*

ABSTRACT: Alike several other countries, Bangladesh has large dependency on groundwater for drinking, agriculture and industrial usages. This study aimed at developing a new methodology based on supervised machine learning to predict Groundwater Quality Index (GWQI) which helps to know about Groundwater Quality Class (GWQC) in groundwater samples during pre-monsoon and post-monsoon seasons. The proposed methodology contains five major inputs i.e. well depth (WD), arsenic (As), manganese (Mn), iron (Fe) and calcium (Ca) to determine GWQI and GWQC. We successfully implemented the popular supervised ML models and concluded that the role of MLR is best in both pre-monsoon and post-monsoon to predict GWQI with MAE closes to 0.03 and 0.02, respectively. We also applied various classifiers where Naïve Bayes plays a significant role with 85% accuracy in pre-monsoon and 94% accuracy in post-monsoon respectively.

1 INTRODUCTION

Groundwater is one of the important natural resource of safely livelihood. Unfortunately, geogenic and emerging health contaminants in groundwater poses serious health issues globally. From agriculture to industry application to drinking purpose groundwater is used by 97% of peoples in Bangladesh with very high levels of arsenic (As) in groundwater (Hossain *et al.*, 2014; SASMIT 2014). Over the past couple of decades, Government of Bangladesh has been instrumental in developing and implementing policies to tackle the challenge of As, however around 15 million of Bangladesh population still on verge of the risk of As exposure from drinking water sources. Besides As, high level of iron (Fe), manganese (Mn), calcium (Ca), sodium (Na) are have been reported by various agencies.

Machine learning (ML), Artificial Intelligence (AI) and IOT have been extensively embedded in water sector to build smart systems for monitoring of management of environmental resources (Zulkafli *et al.*, 2017). Sharma *et al.* (2019) developed ASMITAS digital platform, which uses AI for As risk in groundwater. In recent years several ML methods have been developed to predict WQI, such as the Feedforward Artificial Neural Networks (FENN, Ahmad *et al.*, 2017) and Artificial Neural Network (ANN) for prediction of water quality parameters. Recently Mukherjee *et al.* (2021) proposed random forest model as a preferred ML model for predicting As variability. In the present study, we use the datasets obtained from SASMIT studies (Hossain *et al.*, 2014) to determine GWQI and GWQC as per WHO recommended guideline. After thorough analysis, our study considered five input parameters i.e., well depth (WD), As, Mn, Fe and Ca and their overall impact on the overall GWQI. Several Machine Learning models were used for the regression and classification analyses.

2 MATERIALS AND METHODS

2.1 *Study area*

The study area is located in Matlab Upazila in Chandpur district Bangladesh, situated 55 km southeast from Dhaka with area 410 km^2 with high levels of As in groundwater. Fifteen different piezometer nests were installed at various places in Matlab, Bangladesh to monitor water quality at various depth levels (Figure 1). The groundwater samples were colected from 81 different wells and analyzed various water quality parameters.

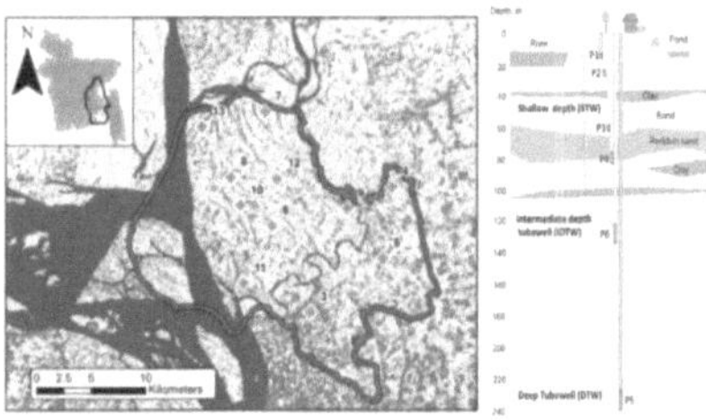

Figure 1. The study area in Matlab, Bangladesh with the location of the piezometer nests (SASMIT 2014).

 DOI: 10.1201/9781003317395-180

3 RESULTS & DISCUSSION

The summary of the overall water quality parameters are presented in Tables 1 and 2. World Health Organization guideline values (WHO 2019) were used to determine the GWQI and GWQC.

Table 1. Variation of water quality parameters during pre-monsoon season.

Parameters	Minimum	Maximum	Mean	Median	SD	Variance
EC	248.0	6170	1164	957.0	928.4	861943
pH	6.30	7.70	6.95	6.90	0.23	0.05
As	5.60	739.8	158.5	43.97	197.1	38850
Mn	0.01	3.90	0.63	0.29	0.81	0.66
Fe	0.08	20.6	4.99	4.00	3.91	15.28
Cl	3.90	2085	299.4	194.1	376.3	141570
SO_4	0.04	325.2	5.73	1.21	27.2	740.3
Na	0.15	1527	215.3	145.4	265.9	70704
Ca	2.69	230.4	67.91	60.8	44.5	1981
Mg	1.62	158.6	32.32	27.5	25.9	670.7
K	1.96	50.4	8.74	6.73	8.13	66.0

Table 2. Variation of water quality parameters during post-monsoon season.

Parameters	Minimum	Maximum	Mean	Median	SD	Variance
EC	260.0	6250	1180	1042	854.0	729260
pH	6.10	7.40	6.80	6.80	0.27	0.07
As	5.60	725.9	160.7	57.9	193.5	37445.9
Mn	0.01	4.16	0.65	0.27	0.86	0.75
Fe	0.11	30.9	5.54	4.12	4.87	23.75
Cl	0.86	2111	295.5	191.1	373.3	139356.0
SO_4	0.00	81.8	3.36	0.53	11.47	131.52
Na	8.15	1599	219.8	151.6	277.6	77064.7
Ca	3.03	236.0	68.9	59.8	44.5	1976.5
Mg	1.70	159.1	31.2	25.9	23.9	572.2
K	1.18	48.8	7.70	5.89	7.05	49.7

Several commonly used supervised machine learning algorithms were used to assess GWQI in pre-monsoon and post-monsoon water samples (Table 3). Multiple Linear Regression (MLR) performed efficiently in both seasons with RSE 1 to predict GWQI. In case of classifying GWQC, Naïve Bayes classifier perform best with high accuracy for both pre-monsoon and post-monsoon seasons (Figure 2). We noticed seasonal impact on GWQI due to the mobility of As and other contaminants in shallow aquifers. The intermediate deep tube-wells have shown reasonably good quality groundwater.

In order to determine the significant parameters that can effect the water quality, a correlation matrix was derived that suggests As, Fe, Mn, and Ca along with well depth (WD) effect the GWQI and GWQC significantly.

Table 3. GWQI and GWQC of wells in pre- and post-monsoon seasons.

		Pre-monsoon			Post-monsoon		
GWQC	GWQI	2009	2010	2011	2009	2010	2011
<50	Excellent	1	4	6	1	4	5
50 – 100	Good	2	9	8	3	8	7
100 – 200	Poor	11	20	14	18	20	12
200 – 300	Very Poor	9	11	7	3	13	9
>300	Not Sustainable	17	28	16	18	28	17

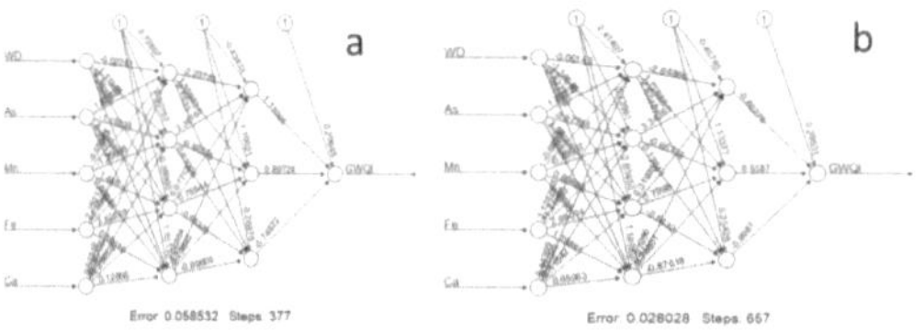

Figure 2. Results of the ANN regressions for the data sets on measures groundwater quality parameters and the ML outpots for a) pre-monsoon period and b) for post-monsoon period.

4 CONCLUSIONS AND RECOMMENDATIONS

Groundwater quality and its acceptance were mainly dependent on present of toxic contaminants (As, Mn, F etc), salinity (Na, K, Mg etc.), odor (due to presence of Fe), hardness (due to presence of Ca) are important key performance indicator (KPIs) for rationalising the GWQI and GWQC from the WQ data. Among the various supervised ML algorithms used to assess GWQI, Multiple Linear Regression (MLR) performs efficiently for both seasons with RSE 1 to predict GWQI. In case of classifying GWQC, Naïve Bayes classifier perform best with high accuracy in both pre-monsoon and post-monsoon data sets. This study will be relevant for decision makers and researcher to develop strategies for sustainable drinking water supplies and optimise the treatment strategy based on GWQI and GWQC. The most suitable ML models can be integrated in the process automation for optimised treatment strategy.

REFERENCES

Ahmad Z., Rahim N.A. & Bahadori A. 2017. Improving water quality index prediction in Perak river basin Malaysia through a combination of multiple neural networks. *Int. J. River Basin Manag.* 15(1): 79–87.

Hossain M., Bhattacharya P., Frape S.K., Jacks G., Islam M. M., Rahman M.M., Hasan M.A. & Ahmed K.M. 2014. Sediment color tool for targeting arsenic-safe aquifers for the installation of shallow drinking water tubewells. *Sci. Total Environ.* 493: 615–625. 277.

Mukherjee A., Sarkar S., Chakraborty M., Duttagupta S., Bhattacharya A., Saha D., Bhattacharya P., Mitra A. & Gupta S. 2021. Occurrence, predictors and hazards of elevated groundwater arsenic across India through field observations and regional-scale ai-based modeling. *Sci. Total Environ.* 759: 143511.

SASMIT 2014. Sustainable Arsenic Mitigation. Community Driven Initiatives to Target Arsenic Safe Groundwater as Sustainable Mitigation Strategy. *Sida Contribution:* 73000854. http://dx.doi.org/10.13140/2.1.1290.3200

Sharma S., Bhattacharya P., Kumar D., Perugupalli P. *et al.* 2019. ASMITAS–a novel application for digitalizing the sasmit sediment color tool to identify arsenic safe aquifers for drinking water supplies. In: *Environmental Arsenic in a Changing World As2018*. CRC Press/Taylor and Francis: 629–632.

Zulkafli Z., Perez K., Vitolo C., Buytaert W., Karpouzoglou T. *et al.* 2017. User-driven design of decision support systems for polycentric environmental resources management. *Environ. Model. Softw.* 88: 58–73.

5.2 Arsenic in drinking water sources: citizen science and community based interventions

Arsenic in the Environment: Bridging Science to Practice for Sustainable Development –
van der Wal, Ahmad, Petrusevski, Weijma, Savic, van der Wens, Beerendonk, Bhattacharya, Bundschuh & Naidu (Eds)

Building resilience for drinking water safety in arsenic prone rural communities in the global south

T. van der Voorn[1], P. Bhattacharya[2], A. Ahmad[2,3] & M.T. Islam[2,4]
[1]*Institute of Environmental Systems Research, University of Osnabrück, Osnabrück, Germany*
[2]*KTH-International Groundwater Arsenic Research Group, Department of Sustainable Development, Environmental Science and Engineering, KTH Royal Institute of Technology, Stockholm, Sweden*
[3]*SIBELCO Ankerpoort NV, Maastricht, The Netherlands*
[4]*WaterAid-Bangladesh, Dhaka, Bangladesh*

ABSTRACT: Rationale for developing global resilience towards the arsenic risk mitigation involves the mitigation or elimination of ongoing vulnerabilities, including exposure to constant environmental stresses that damage the well-being and health of the population. The occurrence of geogenic arsenic (As) in groundwater is a severe challenge for global drinking water quality with extreme severity in several countries in the global south especially South Asia, Africa and Latin America. Drinking water practice has changed from surface water to groundwater and millions of groundwater-based drinking water access points have been installed during last three decades. Unfortunately, the discovery of As and its widespread occurrence in shallow aquifers drastically reduced the safe water access and huge population is now exposed to high levels of As from drinking water sources. This imposes the need for building resilience for drinking water safety in arsenic prone rural communities in the global south based on the best practices that can be scaled up for developing the resilience capacities for drinking water safety for such communities.

1 INTRODUCTION

The geogenic occurrence of arsenic (As) in groundwater is a severe challenge on global drinking water quality (Bundschuh *et al.*, 2017; van der Voorn 2008) with extreme severity in several countries in South Asia and Latin America. Drinking water practice has changed from surface water to groundwater and millions of tubewells (TW) has been installed during the last three decades. Unfortunately, the presence of As and its widespread occurrence in shallow aquifers drastically reduced the safe water access and huge population is now exposed to high levels of As from drinking water sources (Kapaj *et al.*, 2006). The socioeconomic infrastructure, population density, lack of awareness and the failure of involving the local initiatives resulted into limited mitigation attempts so far (Hossain *et al.*, 2015, 2017). The concentration of As exceeded the World Health Organization (WHO) drinking water guideline value of 10 μg/L and Bangladesh Drinking Water Standard (BDWS) of 50 μg/L (Ahmed *et al.*, 2004; Ahmad & Bhattacharya 2018). Arsenic concentration in TW water varies widely, with respect to depth and even within short distances, and this is mainly because of the local variation of geology (Ahmed *et al.*, 2004; Ligate *et al.*, 2021; Litter *et al.*, 2019; Quino Lima *et al.*, 2019, 2021; von Brömssen *et al.*, 2007). This makes it difficult to predict a generalised condition and thereby to adapt to a specific mitigation approach. In this context, a quick screening of the TWs based on visible features may help to identify the respective TWs and thereby to delineate the contaminated domains within a given area. It is difficult to identify the contaminated aquifers by testing a small number of TWs through conventional methods; e.g. field kits and laboratory analysis.

The challenge for developing global resilience towards the As risk mitigation involves the primary task involving the identification of the contaminated drinking water sources, other ongoing vulnerabilities, including WASH challenges and other exposure to constant environmental stresses that damage the well-being and health of the population (Lake & Fenner 2019). The relationship between health and resilience comprises both local and global processes. It is at play both in the long and the short term and embraces ecological, social and economic systems. Health can be seen as an important integrative index that reflects the status – and in the long term, the sustainability – of our environment and our socio-economic climate (Figure 1).

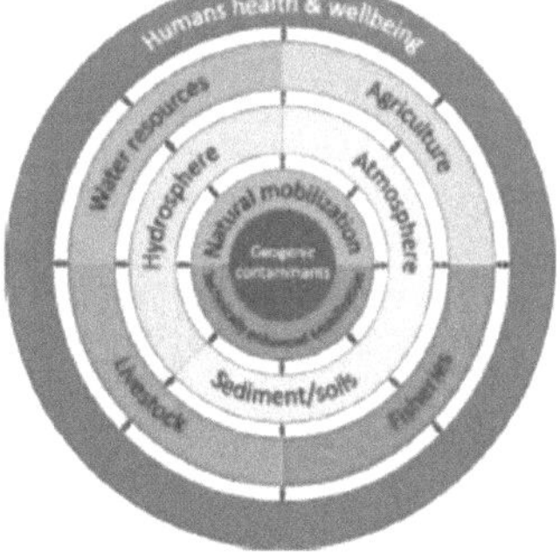

Figure 1. System assessment of geogenic contaminants (Bundschuh *et al.*, 2017).

DOI: 10.1201/9781003317395-181

2 A HARMONIZED APPROACH FOR BUILDING RESILIENCE

The resilience of a drinking water supply system is reflected in its ability to ensure the provision of the system functions in the face of increasingly complex and accumulating economic, social, environmental and institutional shocks and stresses, through capacities of robustness, adaptability and transformability, which are rooted in literature on adaptive cycles and adaptive governance (Holling 1996; Meuwissen *et al.*, 2019). This resilience perspective differs from much of the social-ecological resilience literature in its focus on output and in considering a socially determined flexibility in this output, namely the set of desired functions e.g., drinking water supply (Ge *et al.*, 2016), but can be helpful to develop plans for building resilience for drinking water safety in As prone communities based on the specific learnings from the various global interventions (Figure 2).

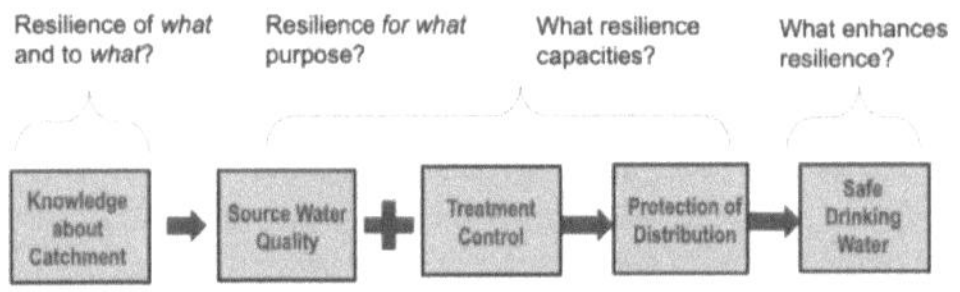

Figure 2. A stepwise approach for building resilience.

Water Safety Plans (WSPs) entail the safety of a drinking-water supply through the use of a comprehensive risk assessment and risk management approach that encompasses all steps in water supply from catchment to consumer. WSPs include three key components: system assessment, effective operational monitoring and management. *System assessment* determines whether the drinking-water supply chain (up to the point of consumption) as a whole can deliver water of a quality that meets health-based targets. *Monitoring* to control the identified risks and ensure compliance with health-based targets. *Management* prescribes actions to be taken during normal operation or incident conditions and documenting the system assessment (including upgrade and improvement). However, the effectiveness of WSPs is often compromised due to a lack of adequate monitoring, communication and surveillance, which impedes meeting health-based targets. This calls for a harmonized approach for building resilience capacities to enable effective and sustainable WSP implementation; integrate health into water safety planning for better preparedness and response to health risks; increase awareness of WSP benefits in securing public health, water security and resilience to public health crises.

ACKNOWLEDGEMENTS

We acknowledge Sida for supporting the SASMIT and EPSC projects through UNICEF-Bangladesh, the Sida-Bolivia and Sida-Tanzania cooperation on research capacity building in Bolivia and Tanzania.

REFERENCES

Ahmad A. & Bhattacharya P. 2019. Arsenic in drinking water: is 10 µg/L a safe fimit?. *Curr. Pollut. Rep.* 5(1): 1–3.

Ahmed K.M., Bhattacharya P., Hasan M.A. *et al.* 2004. Arsenic contamination in groundwater of alluvial aquifers in Bangladesh: an overview. *Appl. Geochem.* 19(2): 181–200.

Bundschuh J., Maity J.P., Mushtaq S., Vithanage M. *et al.* 2017. Medical geology in the framework of the sustainable development goals. *Sci. Total Environ.* 581–582: 87–104.

Ge L., Anten N.P.R. *et al.* 2016. Why we need resilience thinking to meet societal challenges in bio-based production systems. *Curr. Opin. Environ. Sustain.* 23: 17–27.

Holling C.S. & Meffe G.K. 1996. Command and control and the pathology of natural resource management. *Conserv. Biol.* 10: 328–337.

Hossain M., Bhattacharya P. *et al.* 2014. Sediment color tool for targeting arsenic-safe aquifers for the installation of shallow drinking water tubewells. *Sci. Total Environ.* 493: 615–625.

Hossain M., Bhattacharya P. *et al.* 2017. Sustainable arsenic mitigation – from field trials to implementation for control of arsenic in drinking water supplies in Bangladesh. In: P. Bhattacharya *et al.* (eds.) *Best Practice Guide on the Control of Arsenic in Drinking Water.* IWA Publishing, UK, pp. 99–116.

Kapaj S., Peterson H., Liber K. & Bhattacharya P. 2006. Human health effects from chronic arsenic poisoning – a review. *J. Environ. Sci. Health A.* 41(10): 2399–2428.

Lake P. & Fenner R. 2019. The influence of underlying stresses from environmental hazards on resilience in Bangladesh: a system view. *Int. J. Disaster Risk Sci.* 10(4): 511–528.

Ligate F., Ijumulana J., Ahmad A., Kimambo V. *et al.* 2021. Groundwater resources in the East African Rift Valley: understanding the geogenic contamination and water quality challenges in Tanzania. *Sci. Afr.* 13: e00831.

Litter M.I., Ingallinella A.M., Olmos V. *et al.* 2019. Arsenic in Argentina: occurrence, human health, legislation and determination. *Sci. Total Environ.* 676: 756–766.

Meuwissen M.P.M., Feindt P.H. *et al.* 2019. A framework to assess the resilience of farming systems. *Agric. Syst.* 176: 102656.

Quino Lima I., Ormachea Muñoz M., Ramos Ramos O. E., Bhattacharya P. *et al.* (2019) Hydrochemical assessment with respect to arsenic and other trace elements in the lower Katari Basin, Bolivian Altiplano. *Groundw. Sustain. Dev.* 8: 281–293

Quino Lima I., Ormachea Muñoz M. *et al.* 2021. Hydrogeochemical contrasts in the shallow aquifer systems of the lower Katari Basin and Southern Poopó Basin, Bolivian Altiplano. *J. South Am. Earth Sci.* 105: 102914.

van der Voorn T. 2008. The hidden language of rural water supply programmes. In: P. Bhattacharya *et al.* (eds) *Groundwater for Sustainable Development Problems, Perspectives and Challenges.* Taylor & Francis Group, London, pp. 423–433.

von Brömssen M., Jakariya M., Bhattacharya P., Ahmed K.M. *et al.* 2007. Targeting low-arsenic aquifers in groundwater of Matlab Upazila, Southeastern Bangladesh. *Sci. Total Environ.* 379: 121–132.

Arsenic in the Environment: Bridging Science to Practice for Sustainable Development – van der Wal, Ahmad, Petrusevski, Weijma, Savic, van der Wens, Beerendonk, Bhattacharya, Bundschuh & Naidu (Eds)

Can community deep tubewells provide safe drinking water? Evidence from a randomized experiment in rural Bangladesh

S. Cocciolo[1], S. Ghisolfi[2], A. Habib[3,4], S.M.A. Rashid[3], A. Tompsett[2,5] & P. Bhattacharya[3]
[1]*World Bank, USA*
[2]*Institute for International Economic Studies, Stockholm University, Stockholm, Sweden*
[3]*KTH-International Groundwater Arsenic Research Group, Department of Sustainable Development, Environmental Science and Engineering, KTH Royal Institute of Technology, Stockholm, Sweden*
[4]*NGO Forum for Public Health, Dhaka, Bangladesh*
[5]*Beijer Institute for Ecological Economics, Royal Academy of Sciences, Stockholm, Sweden*

ABSTRACT: Meeting Sustainable Development Goal (SDG) 6 requires countries to ensure that their citizens have access to sources of drinking water that are free from both fecal and priority chemical pollutants. This presents a major policy challenge for many developing countries, including Bangladesh. Using a randomized experiment, this study evaluates the impact of a program to provide deep tubewells on access to safe drinking water in rural Bangladesh. The program reduces exposure to arsenic, a major natural pollutant, but not fecal contamination. Improvements in source water quality with respect to fecal contamination are unexpectedly modest. These source water quality improvements are further offset by these source water quality improvements are further offset by increases in travel time and possibly by changes in storage behavior, which increase recontamination. Our results suggest that while deep tubewell construction programs may be effective in reducing exposure to arsenic contamination in Bangladesh, they may have a limited effect on exposure to fecal contamination. Meeting SDG 6 in Bangladesh may necessitate complementary or alternative approaches.

1 INTRODUCTION

Sustainable Development Goal (SDG) 6 has reframed the challenge of providing safe drinking water in the developing world. While Millennium Development Goal 7 could be met by ensuring access to any improved source, SDG 6 can only be met by ensuring access to sources that are free of both fecal and priority chemical contamination. Bangladesh faces a particularly stark challenge: while 98% of the Bangladeshi population were using improved drinking water sources by 2015, only 52% were using sources that were free from both fecal contamination and arsenic, a natural pollutant that is common in shallow groundwater in Bangladesh (BBS & UNICEF 2015).

Most attempts to improve safe water provision in rural Bangladesh have focused on deep tubewell construction. Deep tubewells draw water from aquifers that are free from both fecal contamination and arsenic contamination. Deep tubewells provide a viable technical solution to the problem of arsenic contamination, because arsenic contamination only occurs at source. However, it is less clear whether or not deep tubewells can solve the problem of fecal contamination in drinking water. This is because drinking water may be contaminated with fecal pathogens during transport and storage, as well as at source (Wright *et al.*, 2004). Using a randomized control trial, we evaluate the impact of deep tubewell construction program in rural Bangladesh on household water quality. Prior evidence is mixed as to the relative importance of source water quality, and transport and storage behaviour in determining household drinking water quality (e.g. Clasen *et al.*, 2015; Fewtrell *et al.*, 2005). Disentangling the effects of source water quality, transport, and storage empirically is difficult, because households that live near safe water sources likely differ from those who live further away in other respects that also affect their drinking water quality, such as income or education. We also estimate effects on source water quality, transport time, and storage, and we describe how changes in source water quality, transport, and storage relate to changes in household drinking water quality.

2 MATERIALS AND METHODS

The program comprises a package of subsidies and technical advice to install deep tubewells. We implement the program in 129 communities and compare outcomes in these communities to outcomes in 42 control villages, which receive no intervention. All study communities are located in north-western Bangladesh. The primary inclusion criteria is arsenic contamination in 25% of community water sources. We evaluate the impact of the program using a georeferenced household survey in more than 6,000 households. We test water quality in both samples of household drinking water and samples collected directly from all water sources

DOI: 10.1201/9781003317395-182

used by study households, before and after the intervention. We measure both arsenic contamination and contamination with fecal bacteria in all samples.

3 RESULTS

The program successfully installs a total of 107 new deep tubewells in the 129 treated communities. The wells installed provide arsenic safe drinking water. However, 34% of installed tubewells unexpectedly test positive for fecal contamination, compared to 46% of other tubewells in the same communities. Since the aquifers from which the deep tubewells draw water are isolated from fecal contamination, fecal contamination must occur either through shallow groundwater entering the tubewell system or through the pump body, potentially via the tubewell mouth (ICDDRB & UNICEF 2018).

The program reduces arsenic contamination in both source and household drinking water. Each tubewell installed effectively eliminates arsenic contamination at the WHO level for around five households. However, the program does not reduce fecal contamination in household drinking water. Each tubewell installed introduces fecal contamination into drinking water for around two households, although we cannot reject a small reduc- tion or no effect on fecal contamination in household drinking water. Fecal contamination in source water reduces modestly, but these reductions appear to be offset by increased recontamination via transport and storage. Our best estimates suggest that walking an extra minute to collect drinking water increases the risk of fecal contamination by around 1.7% while storing drinking water in the house increases the risk of fecal contamination by around 7%. The negative consequences of recontamination via increased transport and storage are modest, however, because few households walk more than a minute to collect drinking water, and the majority of households do not change their storage behavior as a result of the intervention.

4 DISCUSSION

Our results have important policy implications in Bangladesh. Recent studies raise the concern that arsenic mitigation programs could have inadvertently increased exposure to fecal contamination, either because households switch to sources that are arsenic safe but fecally contaminated or because households increase transport and storage times when they adopt more distant, safe sources (Buchmann *et al.*, 2019; Wu *et al.*, 2011). Our results suggest that if such inadvertent effects occur, they are more likely to arise through source-switching than through increased transport or storage. Our results may allay fears that providing safer but more distant sources could result in large increases in exposure to pathogens via contamination in transport or storage. However, our results also suggest that construction of deep tubewells alone may be insufficient to reduce exposure to fecal contamination in drinking water in rural Bangladesh.

Our results support previous findings (e.g. Kremer *et al.*, 2011) which suggest that household drinking water contamination may persist even if fecal contamination in source water is eliminated. Further, we find that sources that are designed to be free from fecal contamination appear in practice to become contaminated during collective use. Eliminating exposure to fecal pathogens through drinking water may require strategies that go beyond improved community water sources. One caveat is that our measure of fecal contamination is coarse. Future research is needed to better quantify the extent of fecal contamination in deep tubewells and to better understand the ways in which deep tubewells become contaminated.

ACKNOWLEDGEMENTS

We acknowledge the International Initiative for Impact Evaluation (3ie), the Swedish Research Council Formas, and the International Growth Centre for the necessary financial support.

REFERENCES

BBS & UNICEF 2015. *Bangladesh Multiple Indicator Cluster Survey 2012–2013: Progotir Pathey*. Bangladesh Bureau of Statistics and United Nations Children's Fund.

Buchmann N., Field E.M. *et al.* 2019. Throwing the Baby Out With the Drinking Water: Unintended Consequences of Arsenic Mitigation Efforts in Bangladesh. *NBER Working Paper* 25729.

Clasen T., Alexander K., Sinclair D. *et al.* 2015. Interventions to improve water quality for preventing diarrhoea. *Cochrane Database Syst. Rev.* 2015(10): CD004794.

Fewtrell L., Kaufmann R.B. *et al.* 2005. Water, sanitation, and hygiene interventions to reduce diarrhoea in less developed countries: a systematic review and meta-analysis. *Lancet Infect. Dis.* 5(1): 42–52.

ICDDRB & UNICEF 2018. A bacteriological water quality issue: Rohingya "forcefully displaced Myanmar citizen" camp in Cox's Bazar, Bangladesh. *Fact Sheet 8*.

Kremer M., Leino J. *et al.* 2011. Spring cleaning: rural water impacts, valuation, and property rights institutions. *Q. J. Econ.* 126: 145–205.

United Nations 2016. *Sustainable Development Goals*. Goal 6: Facts and Figures. Report.

Wright J., Gundry S. & Conroy R. 2004. Household drinking water in developing countries: a systematic review of microbiological contamination between source and point-of-use. *Trop. Med. Int. Health* 9(1): 106–117.

Wu J., van Geen A., Ahmed K.M., Alam Y.A.J. *et al.* (2011). Increase in diarrheal disease associated with arsenic mitigation in Bangladesh. *PLOS One* 6(12): e29593.

Arsenic in the Environment: Bridging Science to Practice for Sustainable Development –
van der Wal, Ahmad, Petrusevski, Weijma, Savic, van der Wens, Beerendonk, Bhattacharya, Bundschuh & Naidu (Eds)
 ISBN 978-1-032-32928-4

Scope for integrated arsenic mitigation, water supply and response to COVID-19 pandemic: A case from rural coastal Bangladesh

B.A. Hoque[1], S. Khanam[1], N. Akter[2], M.N. Mahmud[2], M.A. Zahid[1], E.R. Khan[3] & A.N.M.K. Zillany[4]
[1]*Environment and Population Research Centre (EPRC), Dhaka, Bangladesh*
[2]*UNICEF Bangladesh, Dhaka, Bangladesh*
[3]*Department of Public Health Engineering (DPHE), Dhaka, Bangladesh*
[4]*CARE Bangladesh, Dhaka, Bangladesh*

ABSTRACT: Safely managed water, sanitation, and hygiene (WASH) services have been universally recommended for preventing and protecting human health during infectious disease outbreaks, including the current COVID-19 pandemic. We have discussed our experiences gained in integrating the comprehensive educational activity to promote prevention of COVID-19 into water, sanitation, and hygiene project. We have observed the integration in real time to benefit thousands of populations without additional costs to the project.

1 INTRODUCTION

Corona virus pandemic is the longest unprecedented global challenge that the present generation have faced in the world. Combination of wearing masks, cleaning-hands, and physical distancing have been universally recommended measures to prevent COVID-19. All of the this basic preventive measures will require risk communications mentioned in the global and national strategies. Cleaning-hands (hand washing) are linked to access to water, which might have special perspectives in water stressed areas. Frequent and proper hand hygiene will need access to water. The volume of water needed for hand hygiene over a day is few times higher than that of drinking water by a family.

The proportion of population drinking arsenic-affected water in Bangladesh dropped from 26.6% in 2000 to 11.8% in 2019 (UNICEF 2021). The country is still experiencing the largest proportion of people exposed to arsenic contamination in the world.

Approximately 35 million people, representing the 29% of population, live in the coastal zone where arsenic, salinity, iron, bacteriological, and other contamination in water reported.

Experiences gained during integrated arsenic mitigation water supply and COVID-19 risk communication intervention has been presented in real time in rural coastal Satkhira district.

2 MATERIALS AND METHODS

The study is a part of the project entitled 'System Strengthening for Scaling up Drinking Water Safety in Bangladesh at Khulna Division'. The project is being conducted in Assasuni sub-district that lies in between 22°21' and 22°40' N latitudes and in between 89°03' and 89°17' E longitudes (Figure 1). The project period is 2019 February-May 2021 with extension requested under consideration.

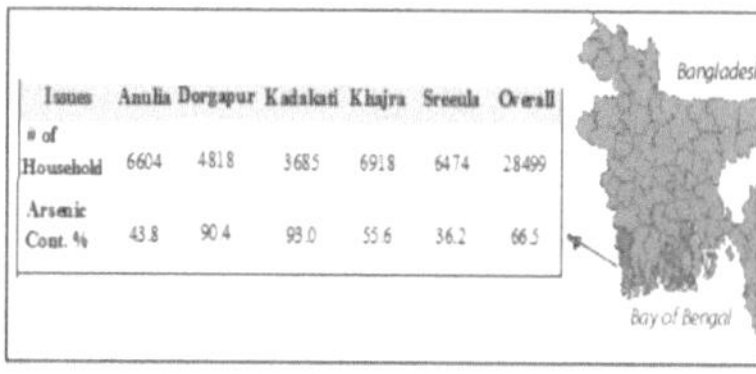

Issues	Anulia	Dorgapur	Kadakati	Khajra	Sreeula	Overall
# of Household	6604	4818	3685	6918	6474	28499
Arsenic Cont. %	43.8	90.4	93.0	55.6	36.2	66.5

Figure 1. Map of the study area with a summary of the households and extent of arsenic contaminated water.

The main objectives of the project are to: (i) achieve a quantitative and measurable reduction in arsenic exposure through building up areas of contiguous arsenic drinking water safety in the sequence: village to union to upazila to district to scale up arsenic mitigation in drinking water systematically, and (ii) promote zero open defecation, improved sanitation facilities, safe hygiene behavior. The project is strengthening system for scale up drinking water safety in 5 unions, namely Anulia, Durgapur, Kadakati, Khajra, and Sreeula of rural Assasuni sub-district in Satkhira.

The main activities under safe water supply includes: arsenic screening, community situation analyses, baseline survey, formation of community based organizations (CBO), and their capacity building, planned creation of awareness about WASH related risks and its appropriate solutions through CBOs and project staff facilitation, feasibility assessments for safe water technology, local government institution engagement at all levels (WATSAN), Hygiene training of trainers among representatives from all stakeholders (schools, health, NGOs, etc.), new water point /sites, following the selection criteria, etc.

The decision and plan about integrating Arsenic Mitigation Water Supply and COVID-19 Risk Communicationwere made by the partners in 2020 April. The communications have been carried out by incorporating the message delivery and

DOI: 10.1201/9781003317395-183

demonstrations with its regular awareness creation and educational activities. The messages were drawn based on UNICEF guidelines/materials. Hand-washing messages were supported by demonstrating construction of 'do-it-yourself' facilities for hand washing.

3 RESULTS AND DISCUSSION

The summary of selected results from intervention to promote safe drinking water, with integrated COVID-19 risk communication as the pandemic hit the area is presented in Table 1.

Approximately 67% of the 28,500 households in the project area exposed to risks for drinking arsenic contaminated tube well water. Although salinity level was claimed to be high in STW and many DTW, STW was the main source of drinking water during the baseline survey. The safe water intervention for increasing the access to safe drinking water in accordance with the SDG 6 requirement for 'basic improved source' within 30 minutes round trip of collection time planned to reach 71% from 31% based on ongoing installation and preliminary feasibility data. Washing of hands with safe water is associated with reduced contamination of hands (Hoque 2003). The reduction of number of users per improved facility has been found helpful to promote adoption of social distancing, particularly during the peak hours of water collection.

Table 1. Various interventions to promote safe drinking water integrated with COVID-19 risk communication in the pandemic hit the area.

	In 5 Unions		
Issues	Min.	Max	Overall
Access to arsenic safe water in May 2019 (B.L)			
i) Within 30 minutes %	15.0	56.3	30.8
ii) Avg. user Hhs size per technology			
– Deep tube well (DTW)	80	85	83
– Shallow tube well (STW) As free	22	59	63.5
– Pond Sand Filter (PSF)	304	590	436
– Rain water Harvesting (RWH)	1	22	6.5
– Direct Pond	50	496	323
– Reverse osmosis (RO)	230	767	585
Access to arsenic safe water planned by end-line (E.L)			
i) Within 30 minutes %	51.3	91.0	71.0
ii) Avg. user Hhs size per technology			
– DTW	20	25	23
– STW As free	1	5	3
– PSF	30	65	47
– RWH	1	1	1
– Pipe water	250	400	130
COVID-19 awareness, % of house-holds reached:			
– Started April 2020	1.8	4.1	2.5
– April 2021	47.5	83.5	58.0
% hand wash facilities present			
i) May 2019	3.5	22.2	8.9
ii) April 2021	23.7	54.6	34.7

Information about COVID-19 prevention, protection and other public health issues (hand-washing, other WASH, mask wearing and social distancing) were communicated and discussed one or more representatives of about 60% households (from > 0.12 million people) about 6 times among in one year. Representatives from health, educational, local government, administration, NGOs, and other institutions/sectors were also trained. The number of handwash facility, with water and soap, increased from about 9 to 35% during the year. The 2019 MICs survey showed presence of facility to be 71.4% in rural areas. A recent study on hand washing has found level of presence of hand washing facility significantly lower than the 71.4% (Hoque *et al.*, 2023).

The project shown innovative way to help, empower, and build trust for integrated WASH and COVID-19 prevention-protection improvements among the vulnerable rural coastal populations and stakeholders has programmatic and policy implications towards sustainable development goals.

4 CONCLUSION AND RECOMMENDATIONS

COVID-19 prevention has been integrated into existing arsenic mitigation water supply project in real time and through innovative way to benefit a large population without additional costs. We recommend research and development for sound, timely, and cost effective integrated WASH and COVID-19 prevention program in different contexts towards sustainable development with policy advocacy.

ACKNOWLEDGEMENTS

This study was supported by the Arsenic Safe Union project funded by Sida-UNICEF Bangladesh and technical support by DPHE. We acknowledge the community people of Assasuni Upazila for participation & support and thank to CARE-B for collaboration.

REFERENCES

Hoque B.A. 2003. Hand washing practices and challenges in Bangladesh. *Int. J. Environ. Health Res.* 13(Suppl. 1): S81–S87.

Hoque B.A., Khanam S., Shipon Z., Akter N., Haque S. *et al.* 2023. Insights and monitoring challenges regarding hand-washing: a rural Bangladesh experience. *Groundw. Sustain. Dev.* 20: 100854.

Ravenscroft P., Kabir A. *et al.* 2014. Effectiveness of public rural water points in bangladesh with special reference to arsenic mitigation. *J. Water Sanit. Hyg. Dev.* 4(4): 545–562.

UNICEF 2021. (https://www.unicef.org/bangladesh/en/better-access-safe-drinking-water)

World Health Organization. https://www.who.int/news-room/fact-sheets/detail/arsenic [accessed May 19 2021].

Arsenic in the Environment: Bridging Science to Practice for Sustainable Development – van der Wal, Ahmad, Petrusevski, Weijma, Savic, van der Wens, Beerendonk, Bhattacharya, Bundschuh & Naidu (Eds)
 ISBN 978-1-032-32928-4

Utilizing citizen science to develop knowledge exchange and sampling-analysis schemes on geogenic arsenic in groundwater in Patna, India

S.T. Addison, D.A. Polya & L.A. Richards
Department of Earth and Environmental Sciences and Williamson Research Centre for Molecular Environmental Science, University of Manchester, Manchester, UK

ABSTRACT: In this study we have investigated the use of citizen science in schools and colleges to develop knowledge exchange and sampling-analysis schemes to learn more about arsenic contamination of drinking water in areas of Northern India. Citizen science is most often used because it provides an efficient strategy for large-scale data collection whilst also serving as an effective and engaging platform for education and raising awareness. The study has indicated that co-designing a sampling-analysis scheme with local coordinators can allow citizen science to efficiently provide a tailored knowledge exchange platform to fit the needs of the volunteers, whilst maintaining the collection of large data sets for research purposes.

1 INTRODUCTION

Citizen science is the undertaking of research using the collaboration of scientists and volunteers and has made significant contributions to scientific discoveries (Harnik & Ross 2003). Citizen science has proved to be an efficient strategy for larger-scale data collection; allowing for the collection of data on a scale difficult to attain by individual researchers or research teams (Bonney *et al.*, 2014). Importantly, it has also shown to engage and inspire participants involved, thus leading to a deep learning experience (Ryan *et al.*, 2018). Therefore, we endeavor to assess the feasibility of citizen science platforms to increase public awareness of the groundwater quality (notably arsenic) in Patna (Bihar, India), whilst also being able to collect larger datasets for research purposes. Patna a city in Bihar is part of the Middle Ganga plain, one of the most thickly populated areas of India (Chakraborti *et al.*, 2003). In India 'groundwater is the backbone of drinking water security' (Sukumaran *et al.*, 2015), thus the arsenic contamination of groundwater in this area is of much concernr.

2 OBJECTIVES

To develop and assess the feasibility of different citizen science approaches for knowledge exchange to raise awareness whilst also maintaining the ability to collect large datasets for research purposes.

3 METHODS

- Contact institutionoften building upon partnerships that have been previously created through the activities of associated projects and people.
- Co-design a project and sampling-analytical scheme with institution staff that facilitates the needs of the institution.
- Introductory lecture on arsenic and the project to potential volunteers.
- Volunteers collect water samples and additional supplementary information. Depending on the approach co-designed with the institution, this could be up to a few to hundreds of volunteers
- Samples are analyzed in the laboratories of The University of Manchester.
- A report is provided with every sample analyzed. The report provides the arsenic and fluoride concentration of the sample. Resources for further information are also suggested, the resources produced by national, local and regional authorities and international organizations.
- Depending on the co-design, volunteers may undertake projects utilizing the data and findings.

Utilizing a citizen science approach serves an efficient mechanism to gather data on the arsenic contamination of groundwater in tested areas. However, ensuring that this leads to benefits for those involved is crucial. Therefore, by co-designing the samplinganalytical scheme with each institution, this ensures that volunteers at every institution gain

DOI: 10.1201/9781003317395-184

a tailored experience to maximize the knowledge exchange suitable for the volunteers.

We provide 1) an introduction on groundwater/arsenic to contextualize the project, 2. accurate and up to date water analysis data, and 3) links to further information. These 3 sets of information allow us to tackle a key barrier to the arsenic crisis, the barrier of "a lack of awareness" (Polya & Richards 2017). Depending on the co-designed methodology at the various institutions, this is the limit to the knowledge exchange, however, some co-designed schemes will lead to projects being developed by volunteers using the data collected from the sampling-analytical scheme. In either scenario, we ensured that through the citizen science activities, volunteers received accurate, current and readily available information by being provided with a standardized framework for results feedback. Following on from the knowledge exchange, the actions and motives of volunteers towards arsenic contamination of groundwater was not analyzed, as this was outside of the aims of the project. However, this could be the work of future research to ensure that the knowledge exchange is acted upon.

Hobbs & White (2012) have shown that citizen science participation is reduced by common barriers, thus a key element of our methodology was to en-sure simplicity of procedures and low costs for the potential volunteers. Each volunteer involved col-lected a groundwater sample from a drinking water source, calculated the GPS (using smartphones) and completed a sample collection information sheet. The process was made to take a few minutes and was guided through a step-by-step guide. Prelabelled sample vials were provided and the location of sam-ple collection was chosen by the volunteers them-selves.

A key element of the project was to maintain adapt-ability. Adaptability allows for the project to be highly scalable should a suitable framework be de-veloped; similar studies (e.g. Tyson 2015) have demonstrated the potential possibilities. We aim to spatially expand and efforts as so far encouraging; suggesting expansion from a local project to a re-gional project is easily possible. Furthermore, the study is temporally scalable, allowing for two key issues to be tackled. Firstly, as Tyson (2015) shows, engagement between the public and scientists can be challenging yet important to create a lasting aware-ness of arsenic-contamination issues. Secondly, var-ious studies (e.g. Malakar *et al.*, 2016) suggest a po-tential temporal variability of arsenic. With 1 school already we have co-designed a self-sustained student driven system which will maintain student involve-ment within the sampling-analytical scheme. This is encouraging and, in the future, we would like to in-crease the number of schools that are engaged in this style of project.

4 RESULTS

Whilst the project is in its early stages, we already have gone through the sampling-analytical cycle with two colleges. The two colleges were visited within 1 day and we were able to return and collect 58 samples which shows the ease of data collection for ourselves, whilst taking little of the volunteer's time. Our preliminary results suggest that the two colleges allowed us to gain knowledge of the areas sampled in an efficient way. The 58 samples had an average concentration of 1μg/L and a maximum of 6 μg/L of arsenic. Spatially the results provided a clustered sampling spread around each college. The level of knowledge exchange with the two colleges, however, was relatively low. Currently we are undergoing the co-design of a self-sustained project with a school which will focus on increased knowledge exchange. We have learned that a co-design with the institution of the sampling-analytical scheme allows for us to vary the style of knowledge exchange for volunteers involved whilst still allowing for a large amount of data collection.

5 CONCLUSIONS

Crucially our study has piloted the use of citizen science frameworks to assess the feasibility of such methodologies in developing partnerships with schools and colleges in northern India. Our study suggests that rather than using one model for application to all institutions, a co-design between both parties of the citizen science scheme should be used. This allows for knowledge exchange to be tailored for the needs of the institution whilst maintaining an efficient mode of sample collection. Our study provides a simple framework which can easily be applied for different uses, such as the development of hazard maps, which we would have the current data and capabilities of undertaking.

ACKNOWLEDGEMENTS

This research has been supported by the DST-Newton Bhabha-NERC-EPSRC Indo-UK Water Quality Project (NE/R003386/1 & DST/TM/INDO- UK/2K17/55(C) & 55(G) 2018 – 2021) FAR-GANGA; an Engineering and Physical Sciences Research Council (EPSRC IAA award via University of Manchester to DP *et al.*), a Dame Kathleen Ollerenshaw Fellowship (to LR) and facilitated through SA's MPhil project (University of Manchester).

REFERENCES

Bonney R., Shirk J.L., Phillips T.B., Wiggins A., Ballard H.L., Miller-Rushing A.J. & Parrish J.K. 2014. Next steps for citizen science. *Science* 343(6178): 1436–1437.

Chakraborti D., Mukherjee S.C., Pati S., Sengupta M. K., Rahman M.M., Chowdhury U.K., Lodh D., Chanda C.R., Chakraborti A.K. & Basu G.K. 2003. Arsenic groundwater contamination in middle ganga plain, bihar, india: a future danger?. *Environ. Health Perspect.* 111(9):1194–1201.

Harnik P.G. & Ross R.M. 2003. Developing effective k-16 geoscience research partnerships. *J. Geosci.* 51 (1): 5–8.

Hobbs S. J. & White P.C.L. (2012) Motivations and barriers in relation to community participation in biodiversity recording. *J. Nat. Conserv.* 20: 364–73.

Malakar A., Islam S., Ali M.A. & Ray S. 2016. Rapid decedal evolution in the groundwater arsenic content of Kolkata, India and its correlation with the fractices of her dwellers. *Environ. Monit. Assess.* 188(10): 584.

Polya D.A. & Richards L.A. 2017. Arsenic and the provision of safe and sustainable drinking water: aspects of innovation and knowledge transfer. Asia-Pacific technical monitor, UNESCAP (United Nations Economic and Social Commission for Asia and the Pacific) APCTT (Asian and Pacific Centre For Transfer of Technology) July–September 2017 Silvertown, J., 2009. A New Dawn for Citizen Science. *Trends in Ecology & Evolution*, 24(9), pp.467–471.

Sukumaran D., Saha R. & Saxena R.C. 2015. Groundwater quality index of Patna, the capital city of Bihar, India. *American Journal of Water Resources*, 3(1): 17–21.

Tyson J. 2015. Integrating arsenic-related environmental topics into the education of the next generation of citizens for arsenic-hit communities: awareness and mobilization. *Int. J. Environmental Monitoring and Analysis* 3(1): 50–55.

Arsenic in the Environment: Bridging Science to Practice for Sustainable Development –
van der Wal, Ahmad, Petrusevski, Weijma, Savic, van der Wens, Beerendonk, Bhattacharya, Bundschuh & Naidu (Eds)
© 2023 the Author(s), ISBN 978-1-032-32928-4

Science and policy in action: A localized intervention for arsenic risk mitigation in the Philippines

C. Faulmino[1], A. Rola[2], K. Solis[1], R. Macasieb[1] & A. Resurreccion[1]
[1]*Institute of Civil Engineering, College of Engineering, University of the Philippines, Diliman, Quezon City, Philippines*
[2]*Institute of Governance and Rural Development, College of Public Affairs and Development, University of the Philippines, Los Baños, College, Laguna, Philippines*

ABSTRACT: Toxic levels of arsenic (As) in drinking water sources is an emerging concern in the Philippines. An investigation of groundwater quality and epidemiological symptoms of arsenicosis in 2014 confirmed the need for intervention in affected municipalities in Central Luzon. This study presents how scientific and policy analyses were undertaken by a multi-disciplinary team from the academe in partnership with a local government to respond to the impending crisis of As poisoning in their municipality. Groundwater samples were collected monthly from wells and analyzed for As concentrations for one year. Laboratory results determined the extent and location of barangays at risk and aided in selecting the site for a pilot As removal water treatment plant. Moreover, policy mapping with stakeholders were undertaken to determine courses of action to mitigate As exposure in all affected barangays. Adopting a transdisciplinary approach is thus recommended as a means to strengthen intervention sustainability.

1 INTRODUCTION

Hazardous levels of arsenic (As) in Philippines groundwaters were of little concern until an upsurge of cases of As poisoning in Pampanga and nearby provinces occurred in 2014. A report by Sy *et al.* (2017) confirms at least 81 cases of chronic As poisoning among examined residents of a community in Central Luzon.

Addressing the problem has so far been a challenge for local governments who are at the forefront of enforcing water quality regulations and implementing strategies for water safety. Moreover, for issues as As in drinking water, devising appropriate response strategies must be backed up by scientific data.

With this, the University of the Philippines (UP) performed its duty as the country's national university by partnering with the affected municipality. The multi-disciplinary team from the technical sciences of engineering and chemistry with governance and sociology teamed up with the local government to mitigate As risks through science, technology, and policy. This study presents the processes and outcomes so far of the undertaking, which can guide future efforts much needed to combat the rising As crisis in the Philippines.

2 METHODS

2 1 *Water sampling and laboratory analysis*

Sampling site profile was obtained using Garmin eTrex 20x GPS and through interviews with locals. Groundwater samples from Central Luzon were collected in acid-washed 250 mL HDPE bottles and transported to the Institute of Civil Engineering, UP Diliman. The samples were preserved using concentrated HNO_3, filtered with a 0.4 μm syringe filter, and stored in 15-mL polypropylene centrifuge tubes for analysis. The As levels of the water samples were measured using the Teledyne-Leeman Prodigy7 hydride-generation ICP-OES which runs on Salsa software version 5.0 Approximately 10 mL of filtered water samples were acidified with conc. HCl in a 1:10 v/v ratio and 250 mL of fresh 0.13 M $NaBH_4$ solution is prepared before ICP analysis for As hydride generation. Initial calibration method and continuous calibration method was performed throughout the analysis while QA check 50 μg/L ± 10% is performed every 15 samples.

2.2 *Policy analysis*

Problem structuring, policy reviews, and local stakeholder consultations were undertaken by the

DOI: 10.1201/9781003317395-185

governance team members from the academe to determine possible courses of action to avert the emerging As crisis in the municipality. The problem structuring exercise explored the substantive and formal dimensions of the problem. The problem structure was then delineated through a review of municipal government mandates, which aided in identifying problem focus and establishing jurisdictional authority of the municipal government. From the review of pertinent national laws and policies and consultations with various stakeholders, alternative and supplementary courses of action to implement in the municipality were crafted.

3 RESULTS AND DISCUSSION

3.1 *Barangays with arsenic contamination*

Of the 31 barangays (the smallest administrative division in the Philippines) in the municipality, at least 18 (58%) barangays have handpumps drawing groundwaters with As concentrations beyond the maximum acceptable level (MAL) of 10 μg/L set by the Philippine National Standards of Drinking Water (PNSDW). A map presenting the location and extent of As concentration is shown in Figure 1.

Figure 1. Arsenic values from various sampling points (N = 97) in the study site. Large, orange circles indicate [As] >50 μg/L.

3.2 *Site determination for a pilot-scale arsenic removal water treatment plant*

The water quality data shortlisted the barangays to be prioritized for the As removal plant called electrochemical As remediation (ECAR) to be piloted in the municipality. Albeit promising, the limited volume capacity of the pilot plant can only serve a small portion of the at-risk populations, thus the necessity for more encompassing policy actions to mitigate health risks of the broader public.

3.3 *Policy actions for arsenic risk reduction*

The Sanitation Code of the Philippines (Presidential Decree No. 856), National Policy on Water Safety Plan (Department of Health Administrative Order 2014-27), and Philippine National Standards of Drinking Water (Department of Health Administrative Order 2017-0010) were among the key policies reviewed in consonance with stakeholder consultations with national government agencies and the partner local government. Aligning the focus with the problem structure yieled the following courses of action to be undertaken by the local government for the short-, medium- and long-terms:

- Regulation of wells through signages indicating "Water unfit for drinking" based on water quality monitoring results;
- Conduct of educational outreach and promotion of self-protection policies; and
- Constitution or re-activation of the multi-sectoral Local Drinking Water Quality Monitoring Committee (LDWQMC) which will be in charge of water quality surveillance beyond the jurisdiction of other local water service providers.

4 CONCLUSIONS AND RECOMMENDATIONS

Scientific data is critical for designing appropriate interventions and sound decision making amid complex human-environment problems. This study demonstrates how science and policy can be linked to come up with actionable steps to reduce As risks that compromise drinking water safety. It is thus recommended that a transdisciplinary approach as adopted in this study be employed in similar endeavors to strengthen intervention sustainability.

ACKNOWLEDGEMENTS

This study is an output of the project "Philippine electrochemical As remediation: implementation of a novel technology to affordably and effectively remove As from contaminated Philippine groundwaters" funded by the Commission on Higher Education (CHED) of the Philippines through the Philippine-California Advanced Research Institutes (PCARI).

REFERENCE

Sy S.M.T., Salud-Gnilo C., Yap-Silva C. & Tababa E.J. L. 2017. A retrospective review of the dermatologic manifestations of chronic arsenic poisoning in the Philippines. *Int. J. Dermatol.* 56(7): 721–725.

5.3 Arsenic in drinking water: technology and management challenges

Arsenic in the Environment: Bridging Science to Practice for Sustainable Development –
van der Wal, Ahmad, Petrusevski, Weijma, Savic, van der Wens, Beerendonk, Bhattacharya, Bundschuh & Naidu (Eds)
© 2023 the Author(s), ISBN 978-1-032-32928-4

Arsenic concentrations "as low as reasonably possible": Leaders, laggards and possibilities for a guideline <10 μg/L

J. Schullehner[1,2], L. Ramsay[3], M.M. Petersen[4,5], B. Hansen[2], P. van der Wens[6], D. Voutchkova[2] & S.M. Kristiansen[5]
[1]*Department of Public Health, Aarhus University, Denmark*
[2]*Geological Survey of Denmark and Greenland (GEUS), Aarhus, Denmark*
[3]*VIA University College, Horsens, Denmark*
[4]*COWI, Aarhus, Denmark*
[5]*Department of Geoscience, Aarhus University, Denmark*
[6]*Brabant Water, Breda, The Netherlands*

ABSTRACT: Worldwide exposure to naturally occurring arsenic in drinking water has received much attention due to adverse irreversible effects on human health. In 1993, the WHO suggested a provisional arsenic guideline value of 10 μg/L, but added that "every effort should be made to keep concentrations as low as reasonably possible", as the mechanism by which arsenic causes cancer remains to be determined. Until 2003, Denmark had a 50 μg/L guideline value in drinking water which then de facto was lowered to 5 μg As/L for all public waterworks. In this paper, we show changes in As concentrations at the Danish waterworks violating the new guideline and, based on this, discuss the compliance at the waterworks including technical possibilities and leaders and laggards motivations. The possibilities for lowering the guideline even below 5 μg/L in high-income, low-dose countries are also discussed.

1 INTRODUCTION

Even in low concentrations, arsenic in drinking water can potentially have adverse effects on public health worldwide, as long-time exposure to this metalloid may give acute and chronic effects (Nordstrom 2002). During the last decades, WHO has accordingly recommended a guideline value of 10 μg As/L. However, WHO also made it clear that "*in view of the practical difficulties in removing arsenic from drinking water and the practical quantification limit for arsenic, the guideline value of 10 μg/L is retained as a goal and designated as provisional.*" From a health perspective, the concentration of arsenic in drinking water below which no adverse health effects can be observed remains to be determined. There is an urgent need for identification of the most sensitive health-related end-point, and of the mechanism by which arsenic causes cancer (WHO 2017). Furthermore, the practical removal of arsenic to concentrations below 10 μg/L is difficult in many circumstances. The current WHO paradigm is hence that "*every effort should be made to keep concentrations as low as reasonably possible*".

Since 2003, Denmark has had a de facto 5 μg/L guideline value for drinking water as all public waterworks had to distribute water with this maximum concentration. More recently the Dutch drinking water companies set an internal guideline at the level of 1 μg/L (Ahmad *et al.*, 2020; Van der Wens *et al.*, 2016). The health and socio-economic effects of lower guideline value remain to be studied relative to the increased technical costs at the waterworks, as well as how and if such "*as low as possible*" guideline values can be implemented in high-income countries.

Drinking water for the 5.6 million consumers affiliated with public water works in Denmark is entirely derived from groundwater and without any complex treatment as chlorination or similar, the consumption of bottled water is among the lowest in the EU, and data are available in the public database Jupiter (Voutchkova *et al.*, 2018; Wodschow *et al.*, 2019). Denmark is hence very well suited for follow-up studies of the health and socio-economic effects on drinking water arsenic concentrations. One example is to study how both authorities and waterworks made efforts to keep concentrations as low as possible after the guideline value was lowered from 50 to 5 μg/L in 2003.

The aim of this paper is hence to discuss the compliance of the waterworks and the drivers for some being leaders and some which continued to be laggards in a high-income, low-dose country.

DOI: 10.1201/9781003317395-186

2 METHODS

Data are based on drinking water analyses from the publicly accessible Jupiter database from the period 2002–2016, including analyses from all Danish public waterworks (n = 3125 during this period). In this study, a total of 47,646 arsenic analyses are used to follow the development in drinking water arsenic since the lowering of the national criterion from 50 to 5 µg/L in December 2003. More details on the methodology are provided in Ramsay *et al.* (2021).

3 RESULTS AND DISCUSSION

In Figure 1, we show that during the first 13 years after the Danish guideline was lowered from 50 to 5 µg/L, more than 100 initially non-compliant water works succeeded in distributing water with <5 µg/L, although some remained non-compliant.

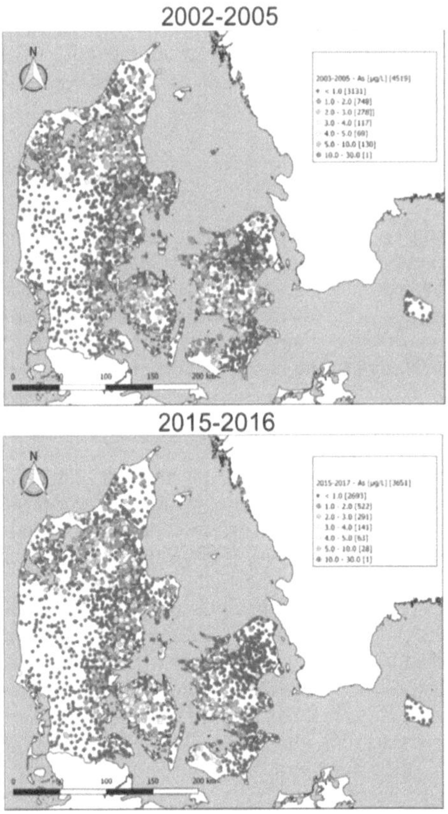

Figure 1. Map showing arsenic in drinking water in Danish public waterworks from the first period before the new 5 µg/L criterion was put in action in 2003, and in the period 2015–2016, respectively. Please notice that a few waterworks may be incorrectly labelled as discussed in Ramsay *et al.* (2021).

In Ramsay *et al.* (2021) we show that the arsenic concentrations at 2087 public Danish waterworks in the period 2015–16 had the following distributions: 1% had >5 µg/L, 3% had 5-4 µg/L, 6% had 4-3 µg/L, 12% had 3-2 µg/L, while 28% had 2-1 µg/L. The geographical distribution of these waterworks is shown in Figure 2 where the clustering in some parts of the country is very evident. Similar observations are made in the US where some states have higher As levels in public drinking and private well water (Flanagan *et al.*, 2018; Foster *et al.*, 2019). These findings are discussed in more details in Ramsay *et al.* (2021).

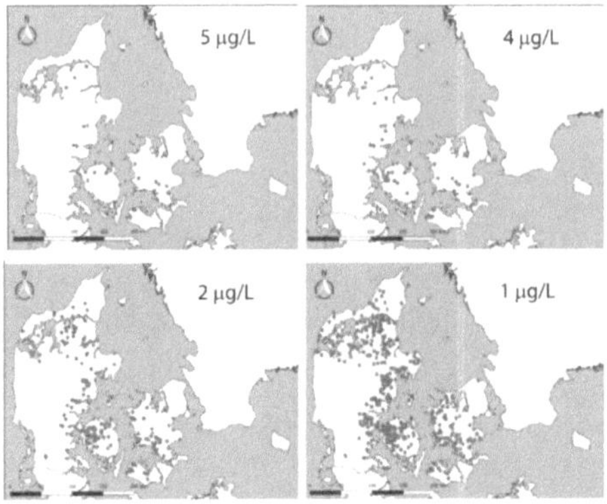

Figure 2. Map of Denmark showing (blue dots) the waterworks that in 2015–16, based on public database Jupiter, were non-compliant with a guideline values for 5 µg As/L, and if the guideline value was lowered to 4, 2 or 1 µg As/L respectively. Please notice that a few waterworks may be incorrectly labelled as discussed in Ramsay *et al.* (2021).

Measures implemented at the originally non-compliant waterworks to improve the drinking water quality were determined by interviewing the utilities. This revealed several drivers which prompt corrective action, and barriers which hinder corrective action. However, further studies are needed to determine the relative importance of especially the barriers as some waterworks remained noncompliant even 13 years after the guideline was lowered.

4 CONCLUSIONS

Based on the findings presented above, and in Ramsay *et al.* (2021), we find that compliance at the individual waterworks, and the motivations for respectively the leaders and laggards among them, need to be discussed further. Additionally, the possibilities for lowering the guideline even below 5 µg/L is possible for high-income, low-dose countries. However, knowledge gaps about the potential health benefits at very low As concentrations need to be filled and a cost-benefit analysis weighing treatment costs against consumers' health benefits is warranted.

REFERENCES

Ahmad A., van der Wens P., Baken K., de Waal L., Bhattacharya P. & Stuyfzand P. 2020. Arsenic reduction to <1 mg/L in dutch drinking water. *Environ. Int.* 134: 105253.

Flanagan S.V. & Zheng Y. 2018. Comparative case study of legislative attempts to require private well testing in New Jersey and Maine. *Environ. Sci. Policy* 85, 40–46.

Foster S.A., Pennino M.J., Compton J.E., Leibowitz S. G. & Kile M.L. 2019. Arsenic drinking water violations decreased across the United States following revision of the maximum contaminant level. *Environ. Sci. Technol.* 53: 11478–11485.

Nordstrom D.K. 2002. Worldwide occurrences of arsenic in groundwater. *Science* 296: 2143–2145.

Ramsay L., van der Wens P., Petersen M., Hansen B., Schullehner J. & Kristiansen S.M. 2021. Drinking water criteria for arsenic in high-income, low-dose countries: the impact of legislation on public health. *Environ. Sci. Technol.* 55: 3483–3493.

van der Wens P., Baken K. & Schriks M., 2016. Arsenic at low concentrations in dutch drinking water: assessment of removal costs and health benefits. In: P. Bhattacharya *et al.* (eds.) *Arsenic Research and Global Sustainability (As2016)*: Taylor & Francis Balkema The Netherlands, pp. 563–564.

Voutchkova D. D., B. Hansen V. Ernstsen & Kristiansen S.M. 2018. Nationwide drinking water sampling campaign for exposure assessments in Denmark. *Int. J. Environ. Res. Public Health* 15: 467.

WHO. 2017. *Guidelines for Drinking-Water Quality. 4th Edition, Incorporating the 1st Addendum.* World Health Organization Geneva.

Wodschow K., B. Hansen J. Schullehner & A. K. Ersbøll (2018). Stability of major geogenic cations in drinking water—an issue of public health importance: a Danish study, 1980–2017. *Int. J. Environ. Res. Public Health* 15: 1212.

Arsenic in the Environment: Bridging Science to Practice for Sustainable Development –
van der Wal, Ahmad, Petrusevski, Weijma, Savic, van der Wens, Beerendonk, Bhattacharya, Bundschuh & Naidu (Eds)
© 2023 the Author(s), ISBN 978-1-032-32928-4

Arsenic and arsenicosis threat to achieve the sustainable development goals

M.M.R. Sarker[1], M.M. Ahmad[2] & U. Deb[3]
[1]*Department of Agricultural Statistics, Sher-e-Bangla Agricultural University, Dhaka, Bangladesh*
[2]*Regional and Rural Development Planning, Department of Development and Sustainability, School of Environment, Resources and Development, Asian Institute of Technology, Thailand*
[3]*Department of Aquaculture and Fisheries, University of Arkansas, Pine Bluff, AR, USA*

ABSTRACT: Arsenic (As) pollution in the drinking water creates a great threat for public health in South Asia. Waters show that several sources of drinking water have elevated concentrations above national and international standard in Bangladesh. The SDGs are a set of 17 interdependent thematic goals which themselves are broken down into 169 specific targets and 232 monitoring indicators. The present study determines the link between As related health hazared and SDGs and deveplop the model for As, arsenicosis and the SDGs. The sixth goal, SDG 6 focus on water and sanitation and third goal SDG 3 concentrates specifically on good health and well being related issues. In line with this interdependence between SDGs, As and health related targets are also either explicitly or indirectly linked to many other SDGs. For example, the SDGs on water, health and wellbeing contain targets that are directly contingent on mitigating the As problem. Almost all the SDGs are directly related to As and arsenicosis or will contribute to indirectly. Arsenic is direct threat for SDG 6 and arsenicosis is for SDG 3 however, almost all other goals are related to or contribute to health and well-being and water issues. More than 50 SDG indicators, across more than 12 goals, have been identified which are directly or indirectly linked to As and arsenicosis crisis.

1 INTRODUCTION

Arsenic (As) is a systemic poison, the continued ingestion of which can lead to a wide range of diseases and premature death. Arsenic contamination in soil and water is a growing problem worldwide, and millions of people face the risk of cancer and poisoning due to As in their drinking water and food supplies. Arsenic contamination of groundwater used for domestic water supplies has been reported from over 70 countries, affecting the health of an estimated 150 million people (Ravenscroft *et al.*, 2009). Arsenic contamination of groundwater poses a threat to the availability of clean drinking water in various countries in the South Asia, especially in Bangladesh, India, Cambodia, and Vietnam. Arsenic contamination of groundwater in many South and Southeast Asian countries, the As hot spots in the world, and irrigation with As-contaminated groundwater in rice cultivation has resulted in high As accumulation in rice grains. In addition, cooking rice with As-contaminated water increases its burden in cooked rice. Therefore, rice is considered to be a health hazard for the population of the rice dominant region and other regions who consume imported rice from As-contaminated countries. Agenda 2030 provides a scope of reference for global development up to 2030. For the water and heath wellbeing sectors, the SDG target of achieving universal access by 2030 is particularly ambitious in South Asia and they will therefore need to call upon all their resources. The SDGs should be implemented progressively and the goals and targets need to be adapted and adopted at both the national and local levels, to ensure that the specific features of each environment are taken into account. The SDG 6 focuses on water-related issues. It consists of eight specific targets including six on water related outcomes (targets 6.1 to 6.6), and two on implementing the outcome targets (targets 6.a and 6.b). One goal (SDG3) specifically sets out to "*Ensure healthy lives and promote well-being for all at all ages*" which is directly related to arsenicoosis. The integrated approach to the 2030 agenda recognizes that most aspects of society, development, sustainable growth and the environment are symbiotic. An integrated approach has important implications. It means that progress towards SDG 6 and SDG 3 can enable and drive progress in most other SDGs; equally, the success of SDG 3 and 6 will also depend on most other SDGs meeting their targets. The source, distribution and mobilization of As in aqueous environment differs by country and within same country differs by location and are associated with either geogenic or anthropogenic processes. The literacy rate, daily wage rate of agricultural labor, As status, and percentage of red mark tube well usage in districts were found to contribute positively and significantly to households' knowledge about arsenicosis (Sarker 2012). Lower As concentration to <1 μg/L in the Netherlands is shown to result in an annual benefit of 7.2–14 M€ (Ahmad *et al.*, 2020). Therefore, the present study aims to investigate the effect of As and arsenicosis on SDGs and identify the probable links.

DOI: 10.1201/9781003317395-187

2 MATERIALS AND METHODS

2.1 *Study location*

Unfortunately, Bangladesh has been hit by environmental catastrophe in last few decades. A large volume of its ground water, the major source of drinking water in the country, has been severely contaminated by As. The presence of high As in the groundwater in South Asia has become a priority environmental and health issue. The distributions of As in wells less than 150 m deep in Bangladesh are also of major concern.

2.2 *Source of data and information*

Secondary data were used for the study from the different sources; such as Bangladesh Multiple Indicator Cluster Survey (MICS) (BBS & UNICEF 2014), Statistical Yearbook of Bangladesh (BBS 2019), Sustainable Development Goals: Bangladesh Progress Report 2018 (BPC 2018) etc.

2.3 *Breakdown into goals, targets and indicators*

Each SDG is broken down into specific targets. In turn, each target is linked to one or several monitoring indicators with a standardized methodology to assess the present and future situation and progress made towards achieving the targets.

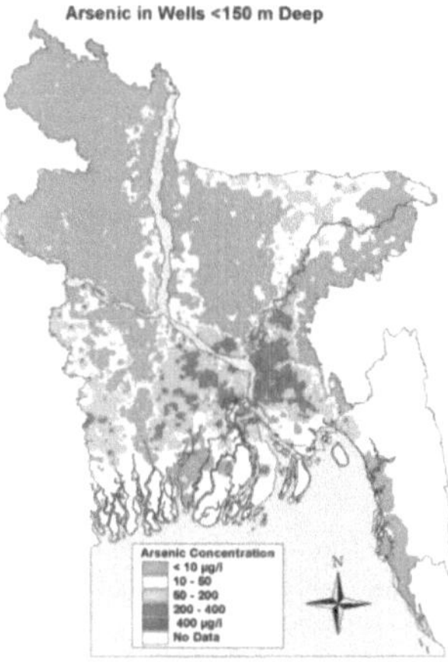

Figure 1. Distribution of arsenic in wells less than 150m deep in Bangladesh (Ravenscroft *et al.*, 2009).

3 RESULTS AND DISCUSSION

3.1 *Interdependence of targets*

The thematic scope of the SDGs and their targets overlap and so As and arsenicosis are therefore addressed through a number of interdependent targets. Arsenic patients are often subject to gross discrimination: children may be excluded from education, women may be excluded from marriage or forced to divorce and may be prevented from working. While As contamination of the drinking water is the major focus, there are other significant routes of As exposure. Exposure to As from food poses a major risk and food can take up As through the soil, irrigation water or cooking water.

3.2 *Link between the SDGs and arsenic and arsenicosis*

The SDGs are a set of 17 interdependent thematic goals. The SDG 6 and SDG 3 focus specifically on water and health related issues, respectively. In line with this interdependence between SDGs, As and its health related goals are also either explicitly or indirectly linked to many other SDGs. For example, the SDGs on water and health contain targets that are directly contingent on As and its health hazard. Figure 2 shows the probable map of link among the SDGs and As and health hazard from As.

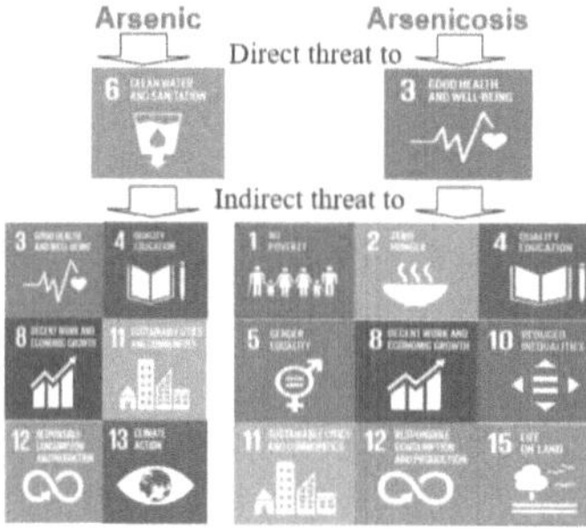

Figure 2. Link among the SDGs and arsenic and related health hazards.

4 CONCLUSIONS AND RECOMMENDATIONS

The challenge of meeting future demands of safe water and achieving sustainable water and sound health and well being for all is significant. Strong political will and commitment are required. However, there is no standard approach for sequencing the policies and institutional developments and investments required for effective management of pure water and provision of services that will be valid for less develop countries and under all circumstances. Finding sustainable development pathways will be challenging for water-insecure countries, especially for Africa and South Asian countries.

REFERENCES

Ahmad A., Wens P.V.D., Baken K., de Waal L.D., Bhattacharya P. & Stuyfzand P. 2020. Arsenic reduction to <1 µg/L in dutch drinking water. *Environ. Int.* 134(1): 105253.

Ravenscroft P., Brammer H. & Richards K.S. 2009. *Arsenic Pollution: A Global Synthesis*. Wiley-Blackwell. http://onlinelibrary.wiley.com/book/10.1002/9781444308785

Sarker M.M.R. 2012. Spatial modelling of households' knowledge about arsenic pollution in Bangladesh. *Soc. Sci. Med.* 74 (8): 1232–1239.

Arsenic in the Environment: Bridging Science to Practice for Sustainable Development –
van der Wal, Ahmad, Petrusevski, Weijma, Savic, van der Wens, Beerendonk, Bhattacharya, Bundschuh & Naidu (Eds)
© 2023 the Author(s), ISBN 978-1-032-32928-4

Assessment and management of risks in supply of safe drinking water through alternative water supply options in arsenic affected area of Bangladesh

K. Sufia, B.A. Hoque, M.A. Zahid, S. Ahmed, M.S. Huque & M.A.I. Khan
Environment and Population Research Centre, New DOHS, Mohakhali, Dhaka, Bangladesh

ABSTRACT: Water is a vital natural resource which is a pre-requisite for the proper functioning of every life form as well as the ecosystem. Due to different natural and anthropogenic activities both the quality and quantity of water degrading day by day at present. Water safety plan (WSP) is one of the most effective means of consistently ensuring the safety of a drinking-water supply. This study analyzed the effectiveness of WSP interventions in rural areas of two Arsenic vulnerable sub-districts of the Narail District in Bangladesh in terms of sanitary scores and microbiological contamination. After WSP intervention, source water fecal contamination percentage improved slightly (3%) while in storage water it shows a considerable improvement (8%). On the other hand, in case of sanitary scoring, low-risk scores improved to 81% from 56% after intervention at the source level. The study shows that the chances of contamination were high during water handling for storage of water, only 21% and 32% were in no risk category in baseline and post intervention. The sanitary risk score method available and used was found useful and easy to apply at the field level. But when it showed 'low' score stating 'no action' under the group, microbiological pollution was observed in a significant proportion of options in that group. This means that there were other risk factors not included in the method and chances for leaving gaps in the holistic approach.

1 INTRODUCTION

The target 6.1 of Sustainable Development Goals (SDG) aims at achieving universal and equitable access to safe and affordable drinking water for all by 2030. The achievement of this target is measured using indicator 6.1.1, which emphasizes the proportion of the population using safely managed drinking water services. However, in recent years, mainly due to human activities, a drastic change has occurred both in the quality and quantity of drinking water sources. Environmental pollution has led to a shortage of clean microbial safe water for drinking purposes. In Bangladesh, the presence of a high concentration of Arsenic (11.8%) and E. coli (82%) at drinking water at source and households consumption point has been reported in the Multiple Indicator Cluster Survey, 2019. Moreover, various studies (Ahmed *et al.*, 2005) noted that water became heavily microbiologically contaminated after supply/collection and storage. Different actors in the chain of water management contribute to microbial contamination between the supply and consumption point. The most effective means of consistently ensuring the safety of a drinking-water supply is through the use of a comprehensive risk assessment and risk management approach that encompasses all steps in water supply from catchments to consumers. In the WHO Guidelines, such approaches are termed as water safety plans (WSP) (WHO 2005 & 2008).

In Bangladesh, concerns about the microbiological quality of water have been reported in the recent & ongoing efforts for safe water supply through alternative options in arsenic affected areas (Hoque *et al.*, 2006). The Government and its development partners are undertaking efforts to ensure microbiologically safe drinking water practices through the promotion of WSP. In this study, the scopes of WSP for drinking water risk management in an arsenic vulnerable area of Bangladesh in SDG perspectives, particularly under microbial safe drinking water and Water Safety Framework in Bangladesh has been analyzed.

2 MATERIALS AND METHODS

This study considers the effectiveness of Water Safety Plans (WSP) in two sub-districts (Kalia and Lohagara) of Narail District located in the south-western part of Bangladesh. 546 water supply points at baseline and 265 points at post-intervention were selected for this study of which

DOI: 10.1201/9781003317395-188

most points are either DTW or STW. Sanitary inspections, interviews, water quality tests, and focus group discussions were conducted during every survey. The educational interventions mainly included the WSPs focusing sanitary/hygienic management of the water at the source as well as through collection, transportation, and storage by the caretakers and users.

Sanitary inspection indicators were selected from WHO and Water Safety Framework (GoB-WHO 2011) in Bangladesh. For the microbiological quality of water, the presence of fecal coliform (FC) bacteria was tested at the EPRC Dhaka laboratory following standard method (APHA 1997).

3 RESULTS AND DISCUSSION

The result shows the quality of drinking water based on sanitary inspection variables and microbiological concentrations.

3.1 *Sanitary condition at source or supply point*

Sanitary scores based on inspection results were low in most of the options. It shows that the rates of compared sanitary risk high score changed from 11% to 3% in post-intervention at the supply points in Figure 1(A). Sanitary scores were 'zero' of 13% in baseline and of 25% in the post-intervention survey. About half of the water points have had one or more of the important sanitary inspection problems. The rates of 'presence of 1 or more latrine within 30 feet' reduced to about 14% in post-intervention from about 41% in the baseline survey, 'ok platform' improved to 88% from 67% and 'clean platform' improved to 71% from 52 %. The mean value of sanitary inspection risk score improved to 2.0 from 3.5. The main indicators for sanitary problems were associated with the presence of latrines, dirty water, drainage problems, poultry/animal farms and priming near the water source which improved considerably after the WSP intervention.

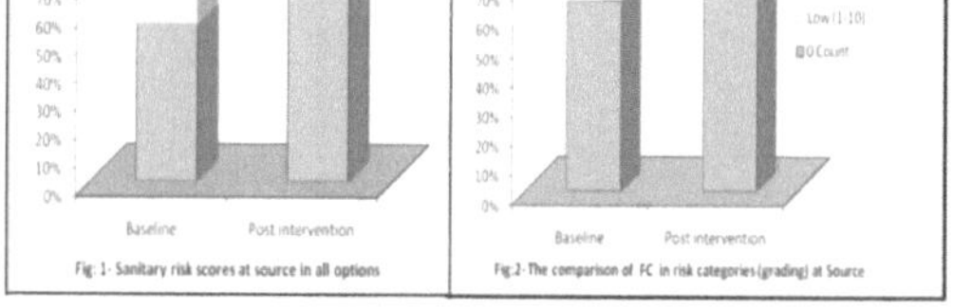

Figure 1. Sanitary risk scores at source (A) and comparison of FC in risk categories at source (B).

3.2 *Microbiological quality of water at source or supply points*

From Figure 1(B), the rates of FC bacteria colony forming unit (cfu/100 mL) in source water conformed to BD standard (nil) changed negligibly between post-intervention and baseline from 65 % to 68 % while in storage water it improved to 32% from 21% only. The results show that the mean FC count was reduced from 21.15 to 1.73 cfu/100 mL and range between 0–5000 to 0–26 cfu/100 mL during baseline to post-intervention respectively. The rate of association between sanitary risk indicators (based on the assigned inspection variables) and microbiological quality (FC) was low. It may be due to the absence of the right indicators in the sanitary inspection list.

3.3 *Water handling during transportation & storage*

Water collection, transportation, and storage are important to assess the risk of water contamination. Water handling practices about placing the containers at high and clean places improved from 39% to 73%. The percentage of containers covered with lid rose to 93% from 64%.

3.4 *Microbiological quality of storage water*

The quality of storage water is important as it refers to the quality of water consumed. Storage water quality improved slightly after WSP from 54% to 64% conformity and low risk in FC categories. The median value of Fecal Coliform (FC) reduced to 11.3 (cfu/100 mL) at post intervention from 68.7 (cfu/100 ml) at baseline. Besides, the microbiological quality of water improved slightly at the source point (3%) and considerably at the storage point (8%) at post intervention. The summary of the results is shown in Table 1.

Table 1. Summary of FC concentration of drinking water.

FC (cfu/100 mL) Statistics function	Source point Baseline	Source point Post-Int.	Storage point Baseline	Storage point Post-Int.
Number of samples	546	265	546	265
Median	21.2	1.73	68.7	11.3
Maximum	5000	26.0	5000	93.0
% met BD standard	65	68	12	20

4 CONCLUSIONS AND RECOMMENDATIONS

The study show that the implementation of WSP is feasible for safe water in arsenic affected areas. The sanitary risk score method available and used was found useful and easy to apply at field level. But when it showed 'low' score stating 'no action' under the group, microbiological pollution was observed in significant proportion of options in that group. This means that there were other risk factors not included in the method and chances for leaving gaps in the holistic approach. The sanitary risk variables should be further studied in local contexts to identify the other variables. Change of practices of certain behaviors from collection to consumption much has been achieved yet wide scale adoption of WSPs remains a challenge. It may require multi faceted communication through different channels and mode over a longer period.

ACKNOWLEDGMENTS

The study was sponsored by EKN, the Netherlands and UNICEF Bangladesh. We appreciate the leaders and all stakeholders in the study areas who showed patience, trust and support.

REFERENCES

Ahmed M.F., Shamsuddin S.A.J., Mahmud S.G., Rashid H., Deere D. & Howard G. 2005. Risk Assessment of Arsenic Mitigation Options (RAAMO). PSU.

APHA. 1997. Standard methods for the examination of water and wastewater, 1995. *Search PubMed, p.20.*

GoB-WHO. 2011. *Water Safety Framework (WSF) in Bangladesh. Govt. of Bangladesh*

Hoque B.A., Yamaura S., Sakai A., Khanam S. *et al.* 2006. Arsenic mitigation for water supply in Bangladesh: appropriate technological and policy perspectives. *Water Qual. Res. J. Canada.* 41(2): 226–234.

WHO. 2005 & 2008. *Guidelines for Drinking Water Quality*, World Health Organization, Geneva.

Arsenic in the Environment: Bridging Science to Practice for Sustainable Development – van der Wal, Ahmad, Petrusevski, Weijma, Savic, van der Wens, Beerendonk, Bhattacharya, Bundschuh & Naidu (Eds)
© 2023 the Author(s), ISBN 978-1-032-32928-4

Poor economics and arsenic: the key role of end-user and science informed co-designed policy and action, stakeholder guidance, transparency and the proactively enhanced role of women for better interventions in India

D.A. Polya[1], B. Chakraborti[2], A. Mukherjee[3], A.K. Ghosh[4], D. Saha[5], H. Joshi[6], D. Mondal[7], D. Gooddy[8], S. Krause[9] & L.A. Richards[1]

[1]*Department of Earth and Environmental Sciences, & Williamson Research Centre for Molecular Environmental Science, University of Manchester, Manchester, UK*
[2]*National Institute of Hydrology, Patna, India*
[3]*Department of Geology and Geophysics, IIT Kharagpur, West Bengal, India*
[4]*Mahavir Cancer Sansthan and Research Centre, Patna, Bihar, India*
[5]*Partners in Progress, India*
[6]*Department of Hydrology, IIT Roorkee, Uttarakhand, India*
[7]*School of Environment and Life Sciences, University of Salford, Manchester, UK*
[8]*Groundwater Processes, British Geological Survey, Wallingford, UK*
[9]*School of Geography, Earth and Environmental Sciences, University of Birmingham, Birmingham, UK*

ABSTRACT: Groundwater derived arsenic in drinking water supplies in India is overwhelmingly, though far from exclusively, a problem for the rural poor. In the year that Banerjee, Duflo and Kremer have been awarded the Nobel Prize for Economics "for their experimental approach to alleviating global poverty", it seems timely to consider how their work might speak to efforts to improve the lives of the rural poor in India through drinking water arsenic remediation and mitigation. Five major themes are highlighted: (i) end-user involvement in priorities and co-design; (ii) science informed policy: randomized controlled trials represent an approach that would enable better more robustly informed unbiased decision-making about the "best" remediation/mitigation approaches than hitherto often opportunistic implementations; (iii) transparency (and consequent improvement in governance and reduction in asymmetric business relationships) including through appropriate (iv) stakeholder guidance is also required to optimize decision-making; and (v) a more equitable role for women is indicated in both better decision-making and implementation processes.

1 INTRODUCTION

Increasing population, increasing affluence and climate change are driving an increased dependence in India on groundwater (Water Resources Group 2009). Arsenic in Indian aquifers is widespread, particularly in northern major river basins, controlled on a large scale by tectonic (Mukherjee *et al.*, 2014) related processes and on smaller scales by sedimentary (Donselaar *et al.*, 2017; Saha & Saha 2015) and indigenous microbial processes (Islam *et al.*, 2004). There is still considerable debate (Harvey *et al.*, 2002; Lawson *et al.*, 2016) over whether or not massive groundwater abstraction for irrigation accelerates arsenic mobilisation. Usage of arsenic bearing groundwaters (Bhattacharya *et al.*, 2017; UNICEF 2017) in India (Mukherjee *et al.*, 2018); gives rise to material arsenic exposure to at least tens of millions of people (Chakraborti *et al.*, 2017), resulting, directly and indirectly [through groundwater irrigated food crops] (Banerjee *et al.*, 2013; Mondal & Polya 2008), in at least tens of thousands of avoidable premature deaths every year as well substantial morbidity in addition to individual & societal economic losses (Polya *et al.*, 2020).

Many remediation strategies, including in the context of water safety plans, have been implemented (Polya & Richards 2017; Richards 2017; UNICEF 2017), but the effectiveness of such strategies remains variably sub-optimal – reasons for this include: importance of other challenges to life, health & economic prosperity, poverty, lack of will, lack of knowledge, asymmetrical relationships between end-users and remediation technology suppliers, poor maintenance and monitoring as well as issues with governance and transparency (Polya *et al.*, 2020).

In the year that Banerjee, Duflo and Kremer have been awarded the Nobel Prize for Economics "for their experimental approach to alleviating global poverty" (Nobel Media AB 2020), we explore the relevance of their of the work, notably Banerjee and Duflo's (2011) "Poor Economics" to efforts to improve the lot of particularly the rural poor in India through the selection and implementation of remediation/mitigation interventions for drinking water containing hazardous concentrations of arsenic. The implications for the principles of further developing stakeholder guidance are considered.

DOI: 10.1201/9781003317395-189

2 END-USER INFORMED CO-DESIGNED POLICY AND ACTION

Given the devastating impacts of chronic arsenic poisoning the lack of willingness to pay (WTP) for remediation of moderate to high arsenic groundwaters utilized as drinking water may seem at first glance to many to be a surprising result (Gibson *et al.*, 2016). This may, in part, reflect a lack of effective knowledge transfer to impacted communities, probability neglect (Sunstein & Zeckhauser 2011) as well as the "hidden" nature of both dissolved arsenic and its health impacts (e.g. cardiovascular diseases and immune system suppression not obviously linked to chronic arsenic exposure).

The widespread low WTP for arsenic remediation is, however, mirrored in observations recorded by Banerjee & Duflo (2011) in regard to expenditure more broadly on protection of health, viz. (i) quite reasonable (given life and social circumstances) preferences for expenditure on other good and services which make life more tolerable and interesting; and (ii) a tendency, when expenditure is incurred to not select optimal interventions, for example (op cit., p. 81) "why do the poor sometimes reject inexpensive effective sanitation – the cheap and easy way to dramatically improve people's health – in favour of spending a lot of money on things that don't help"?

The broader implications of Banerjee and Duflo's (2011) observations is that effective interventions, whether they relate to waterborne pathogens or waterborne chemical hazards such as arsenic, necessarily should take into account the reasonable preferences of the end-users and the social contexts that make equally (theoretically) technological interventions of highly variable effectiveness.

3 SCIENCE INFORMED CO-DESIGNED POLICY AND ACTION

3.1 *Best practice and emerging best practice*

There is a substantive volume of best practice published in relation to groundwater arsenic mitigation, much of it summarized, for example, in Bhattacharya *et al.* (2017), and UNICEF (2017) and the references therein. This best practice is based upon the theoretical and in-field experiences a considerable number of professionals. Notwithstanding this two general observations might be made: (i) "best practice" is continually evolving meaning that "best practice" now may not be the "best practice" is 10 or even 5 years time – e.g. the increasing evidence (Polya *et al.*, 2018) that chronic exposures to what hitherto have been considered "low" concentrations of arsenic in drinking water are leading to detrimental health outcomes of concern is driving the development of technologies to reduce arsenic concentrations in drinking water from 10 µg/L to 1 µg/L levels (Ahmad *et al.*, 2020); novel technologies involving graphene may be too expensive to deploy widely now but are a promising future technology once that are developed to a higher Technology Readiness Level (Kumar *et al.*, 2014); (ii) "best practice" is not always demonstrated robustly to be best practice, but rather denoted as such based upon selective experience or for opportunistic reasons, in part because of the variety and complexity of human-physical environments in which technologies and mitigations are sought to be implemented.

In relation to the selection of optimal interventions, we speculate that, in general, there might actually be no general mathematically solvable optimal solution. Irrespective of this (i) "Best" solution may be strongly dependent on groundwater & stakeholder characteristics; (ii) "best" begs the question – "best for whom" as there are, in general, multiple stakeholders, with not necessarily comprehensively overlapping interests; and, in part as result of this, (iii) near-optimal solutions may the "best" practice.

3.2 *Randomized controlled trials*

Randomized controlled trials (RCTs) are described by UNICEF (2014) as "a way of doing impact evaluation in which the population receiving the programme or policy intervention is chosen at random from the eligible population, and a control group is also chosen at random from the same eligible population. It tests the extent to which specific, planned impacts are being achieved." The principal advantages of RCTs are that serve to reduce bias inherrent in other methods to assess the effectiveness of interventions under consideration. Disadvantages include that relatively large sample sizes are often required in order to detect statistically significant effects and consequently they may be relatively complex and expensive to implement.

Although RCTs are widely used in the pharmaceutical and medical sectors and their use in assessing the effective of social and other interventions is highly recommended (Banerjee & Duflo 2011; Hernandez *et al.*, 2019), there are comparatively few examples of their application to arsenic remediation/mitigation and those have been largely implemented countries other than India, notably Bangladesh (Bennear *et al.*, 2012; George *et al.*, 2012; 2013; 2016; Inauen *et al.*, 2014; Inauen & Mosler 2014; Johnston *et al.*, 2014; Milton *et al.*, 2007; Smits *et al.*, 2019). These examples provide evidence for the relative success of certain types of intervention, notably those involving effective education of end-users. The utility of RCTs to inform arsenic mitigation approaches in India is therefore indicated provided that (Whittington *et al.*, 2020) results are objectively interpreted and acted upon.

4 TRANSPARENCY

Banerjee & Duflo (2011) identify lack of transparency and knowledge transfer as a major impediment to the poor improving their lives – it impacts on the ability to make well-informed sound decisions, may lead to disillusionment about prospects and mistrust as to motivations and effectiveness of various agencies, and makes them vulnerable to be at the receiving end of asymmetric business transactions. Equally (op. cit.) they suggest that lack of transparency can contribute to a lack of effective oversight of the operations of various agencies and which, in turn, creates a driver for less than optimal performance at both an individual and hence institutional level. Transparency and effective knowledge transfer are self-evidently likely to be indicated to improve both governance and the effectiveness of groundwater arsenic interventions.

5 STAKEHOLDER GUIDANCE

Stakeholder guidance needs to be both of a generic nature but also differentiated to address the specific requirements and language of different types of stakeholders – e.g. regulators, government and non-government water supply agencies, technology developers, technology suppliers, monitors, providers of finance and/or insurance, and different types (e.g. urban, rural) end-users. Figure 1 shows recommended questions for end-users to ask suppliers before selecting groundwater arsenic technologies and mitigations.

What are your credentials?
Do you have any conflicts of interest?
How much arsenic will the unit remove?
Is it suitable for my water supply?
Does the unit treat other contaminants?
How much does it cost?
What are the running costs?
What is the anticipated lifetime of the unit?
How complicated is it to run? Is training provided?
What are the consequences of poor maintenance?
Will you check the unit long-term?
Will you/others provide long-term support?
Where can I go to for impartial advice?
How does your unit compare to that of other operators? Is it really the "best"? And, if not, how sub-optimal is the unit?

Figure 1. Questions for end-users to ask arsenic removal technology providers.

6 ROLE OF WOMEN

In addition to the variable differences in the exposure and susceptibility of women and men to groundwater arsenic (Vahter *et al.*, 2007), and in addition to the fundamental equality of rights of men and women under the United Nations charter, the generally more dominant role of women in many rural Indian settings in (i) raising children (with their generally much greater susceptibility than adults to detrimental health outcomes arising from chronic arsenic exposure); and in (ii) procuring water (Sultana 2009), means that their voices are particularly important in determining what interventions are, in practice, workable or more workable. Anecdotally, the experience of several of us is that the stated views of men and of women in arsenic impacted rural villages may be very different, especially in relation to who is actually likely to have access to installed low arsenic water units, irrespective of the technical characteristics of the unit. Women also have a critical role to play in education/knowledge transfer, including to the next generation (Sen 1999), in balancing of attitudes to risk and relative perceived importance of potential interventions (e.g. health/water vs transportation links) (Banerjee & Duflo (2011) and in the nature of decision-making processes.

7 CONCLUSIONS AND RECOMMENDATIONS

The judicious use of randomized controlled trials; greater transparency/knowledge transfer to stakeholders at all levels; and the proactive and systematic incorporation of the views and experience of end-users, proactively including women, in selecting and implementing appropriate interventions are all indicated to improve reductions in exposure of the rural poor in India and other countries to arsenic from drinking water.

ACKNOWLEDGEMENTS

We acknowledge funding from the NERC (UK) (NE/R003386/1) and DST (India) (DST/TM/INDO-UK/2K17/55(C) & 55(G)) under the 2017 Newton Fund India-UK Water Quality Call for the Indo-UK FAR-GANGA project (www.farganga.org/). We thank Narayan Ghosh (formerly National Institute of Hydrology, Roorkee) for this unique and prominent role in bringing that partnership together. LR acknowledges receipt of a Leverhulme Trust Early Career Fellowship (ECF-2015-657) and a Dame Kathleen Ollerenshaw Fellowship. The introduction to this study has been largely taken from (most of) the authors' abstract and presentation (Polya *et al.*, 2020) at the NIT Durgapur hosted WEES2020 meeting – we thank Kalyan Adhikari for the opportunity to present that work there. Lastly we thank the numerous householders and other stakeholders in India and beyond whose advice and experience have helped us refine this work.

REFERENCES

Ahmad A. van der Wens P., Baken K., de Waal L. Bhattacharya P. & Stuyfzand P. 2020. Arsenic reduction to <1 μg/L in Dutch drinking water. *Environ. Int.* 134: 105253.

Banerjee A. & Duflo E. 2011. *Poor Economics: A Radical Rethinking of the the Way to Fight Global Poverty.* Public Affairs.

Banerjee M., Banerjee N., Bhattacharjee P., Mondal D. *et al.* 2013. High arsenic rice is associated with elevated genotoxic effects in humans. *Sci. Rep.* 3: 2195.

Bennear L., Tarozzi A., Pfaff A., Balasubramanya S. *et al.* 2012. Impact of a randomized controlled trial in arsenic risk communication on household

water-source choices in Bangladesh. *J. Environ. Econ. Manage.* 65(2): 225–240.
Chakraborti D. Rahman M.M., Das B. *et al.* 2017. Groundwater arsenic contamination and its health effects in India. *Hydrogeol. J.* 25: 1165–1181.
Donselaar M.E., Bhatt A.G & Ghosh A. 2017. On the relation between fluvio-deltaic flood basin geomorphology and the wide-spread occurrence of arsenic pollution in shallow aquifers. *Sci. Total Environ.* 574: 901–913.
George C., van Geen A., Slavkovich V., Singha A. *et al.* 2012. A cluster-based randomized controlled trial promoting community participation in arsenic mitigation efforts in Singair, Bangladesh. *Environ. Health* 11: 41.
George C. Inauen J., Rahman S.M. & Zheng Y. 2013. The effectiveness of educational interventions to enhance the adoption of fee-based arsenic testing in bangladesh: a cluster randomized controlled trial. *Am. J. Trop. Med. Hyg.* 89(1): 138–144.
George C., Inauen J., Perin J., Tighe J. *et al.* 2017. Behavioural determinants of switching to arsenic-safe water wells: an analysis of a randomized controlled trial of health education interventions coupled with water arsenic testing. *Health Educ. Behav.* 44(1): 92–102.
Gibson J., Rigby D., PolyaD.A. & Russell N. 2016. Discrete choice experiments in developing countries: willingness to pay vs willingness to work. *Environ. Resour. Econ. (Dordr)* 65: 697–721.
Harvey C.F., Swartz C.H., Badruzzaman A.B.M. *et al.* 2002. Arsenic mobility and groundwater extraction in Bangladesh. *Science* 298: 1602–1606
Hernandez D., Boden K., Paul P., Bandaru S. *et al.* 2019. Strategies for successful field deployment in a resource-poor region: arsenic remediation technology for drinking water. *Dev. Eng.* 4: 1000045.
Inauen J. & Mosler H.J. 2014. Developing and testing theory-based and evidence-based interventions to promote switching to arsenic-safe wells in Bangladesh. *J. Health Psychol.* 19(12): 1483–1498.
Inauen J., Tobias R. & Mosler H.-J. 2014. The role of commitment strength in enhancing safe water consumjption: mediation analysais of a cluster-randomized trial. *Br. J. Health Psychol.* 19: 701–719.
Islam F.S., Gault A.G., Boothman C., Polya D.A. *et al.* 2004. Role of metal reducing bacteria in arsenic release in Bengal delta sediments. *Nature* 430: 68–71.
Johnston R., Hug S.J., Inauen J., Khan N.I. *et al.* 2014. Enhancing arsenic mitigation in Bangladesh: findings from institutional, psychological, and technical investigations. *Sci. Total Environ.* 488–499: 477–483.
Kumar S., Nair R.R., Pillai P.B., Gupta S.N., Iyengar M.A.R. & Sood A.K. 2014. Graphene oxide–MnFe2O4 magnetic nanohybrids for efficient removal of lead and arsenic from water. *ACS Appl. Mater. Interfaces* 6(20): 17426–17436.
Lawson M., Polya D.A., Boyce A.J., Bryant C. & Ballentine C.J. 2016. Tracing organic matter composition and distribution and its role on arsenic release in shallow cambodian groundwaters. *Geochim. Cosmochim. Acta* 178: 160–177.
Milton A.H., Smith W., Dear K., Ng J. *et al.* 2007. A randomised intervention trial to assess two arsenic mitigation options in Bangaldesh. *J. Environ. Sci. Health A* 42: 1897–1908.
Mondal D. & Polya D.A. 2008. Rice is a major exposure route for arsenic in Chakdha block, West Bengal: a probabilistic risk assessment. *Appl. Geochem.* 23: 2986–2997
Mukherjee A., Verma S., Gupta S., Henke K.R. & Bhattacharya P. 2014. Influence of tectonics, sedimentation and aqueous flow cycles on the origin of global groundwater arsenic: paradigms from three continents. *J. Hydrol.* 518: 284–299.
Mukherjee A. (ed.) 2018. *Groundwater of South Asia*, Springer, ISBN 978-981-10-3889-1.
Nobel Media AB. 2020. The Prize in Economic Sciences 2019. https://www.nobelprize.org/prizes/economic-sciences/2019/summary/. Last accessed 24/2/2020.
Polya D.A., Charkaborti B., Mukherjee A. *et al.* 2020. Guidance for Stakeholders in the Usage, Supply, Remediation, Regulation and Monitoring of Water Supplied From Arsenic Prone Aquifers in India. *Abstract & Programme Volume WEES 2020*, NIT Durgapur, January 2020.
Polya D.A. & Richards L.A. 2017. Arsenic and the Provision of Safe and Sustainable Drinking Water: Aspects of Innovation and Knowledge Transfer. *Asia-Pacific Technical Monitor*, UNESCAP (United Nations Economic and Social Commission for Asia and the Pacific) APCTT (Asian and Pacific Centre For Transfer of Technology) July–Sept 2017. https://tinyurl.com/arsenic-innovation/
Polya D.A. Xu L., Launder J., Gooddy D. & Ascott M. 2018. Distribution of arsenic hazard in public water supplies in the united kingdom – methods, implications for health risks & recommendations. In: Y.G. Zhu, H.M. Guo, P. Bhattacharya *et al.* (eds.) *Environmental Arsenic in a Changing World (As2018)*. CRC Press (ISBN 978-1-138-48609-6), pp. 22–25.
Sahu S. & Saha D. 2015. Role of shallow alluvial stratigraphy and holocene geomorphology on groundwater arsenic contamination in the middle Ganga plain, India. *Environ. Earth Sci.* 73: 3523–3536.
Sen A. (1999) *Development as Freedom*. Oxford Univ. Press.
Smits, J.E. *et al.* 2019. Food as medicine: selenium enriched lentils offer relief against chronic arsenic poisoning in Bangladesh. *Environmental Research* 176, 108561.
Sultana F. 2009. Fluid lives: subjectivities, gender and water in rural Bangladesh. *Gender Place and Culture*, 16, 427–44.
Sunstein C.R. & Zeckhauser R. 2011. Overreaction to fearsome risks. *Environ. Resour. Econ. (Dordr)* 48: 434–449.
UNICEF (2017) Arsenic Primer (updated);
Vahter M., Åkesson A., Liden. C., Ceccatelli S. & Berglund M. 2007. Gender differences in the disposition and otxicity of metals. *Environ. Res.* 104: 85–95.
Water Resources Group. 2009. *Charting Our Water Future*. http://www.oecd.org/env/resources/44864100.pdf
White, H., Sabarwal, S. & de Hoop, T. 2014. Randomized Controlled Trials (RCTs). *UNICEF Methodological Briefs*, Impact Evaluation No. 7.
Whittington D., Radin M. & Jeulund M. 2020. Evidence-based policy analysis? the strange case of the randomized controlled trials of community-led total sanitation. *Oxf. Rev. Econ. Policy* 36(1): 191–221.

Arsenic in the Environment: Bridging Science to Practice for Sustainable Development –
van der Wal, Ahmad, Petrusevski, Weijma, Savic, van der Wens, Beerendonk, Bhattacharya, Bundschuh & Naidu (Eds)
 ISBN 978-1-032-32928-4

Is the colour of the sediment a pointer to predict arsenic safe water? Exploration from the field in Bihar, India

B.K. Thakur[1], M. Jakariya[2] & P. Bhattacharya[3]
[1]*Department of Economics, FLAME University, Pune, India*
[2]*Department of Environment Science and Management, North South University, Dhaka, Bangladesh*
[3]*KTH-International Groundwater Arsenic Research Group, Department of Sustainable Development, Environmental Science and Engineering, KTH Royal Institute of Technology, Stockholm, Sweden*

ABSTRACT: Many regions of the Bihar is facing high concentration levels of arsenic (As) in drinking water which causes several health problems besides socioeconomic and other issues. The major source of drinking water is through the tubewell extracted from the groundwater. We investigate the sediment colours at various depths and associated As and Fe in drinking water. A sample of 10 shallow and deep aquifers tubewells were tested for As, Mn, Fluoride and Fe and As concentration was found from less than 1 μg/L to a maximum of 300 μg/L.We observed the sediment at different depths through the installation process by local drillers for 2 shallow aquifer tubewells and 1 deep aquifer tubewells. We suggest sediment colour concept can also be applied in Bihar to target safe water through depth despite having differences in terms of soil characteristics and other geological conditions.

1 INTRODUCTION

Bihar along with many other states in India is facing a widespread occurrence of geogenic arsenic in groundwater. Due to the occurrences of high concentrations of arsenic (As) and iron (Fe) in the alluvial shallow aquifers, safe sources of drinking water are becoming scarce. This situation is leading not only to serious health threats for the millions of people of Bihar but also other vulnerabilities in the arsenic enriched Gangetic plains of Bihar (Saha *et al.*, 2011; Thakur & Gupta 2019; Thakur *et al.*, 2020). In a recent study through the extensive field observation in India, the occurrence, predictions and observations of elevated groundwater As through the AI-based modeling predicted 82% to 84% of accuracy and impressive 0.88 and 0.89 for validation and test data, respectively (Mukherjee *et al.*, 2021) This study investigates the sediment colour of the soil at various depths and safe sources of arsenic (As) and iron (Fe) in the drinking water.

2 METHODS

A sediment colour concept based study is being considered in Bangladesh and West Bengal, India for providing safe water (Annaduzzaman *et al.*, 2018; Biswas *et al.*, 2012; Jakariya *et al.*, 2007; Hossain *et al.*, 2014). Based on this concept the present study aimed to investigate the sediment color of sediment in Bihar at various depths and safe sources of drinking water without higher concentrations of As and Fe.

A sample of 10 shallow and deep aquifers were se-lected to test the water sample. Water samples were analysed for As, Fe, Mn and Fluoride using field test kit, and later verified by using an atomic absorption spectroscopy (AAS) technique in the laboratory. Three tubewells (2 shallow aquifers and 1 deep aquifers) were verified at the construction of those tubewells and the sediments colours were taken for sampling for various depths, and the discrepancies were insignificant. The shallow aquifers weredefined as < 60 m while > 60 m we defined as deep aquifers tubewells. Figure 1 presents the sample study area.

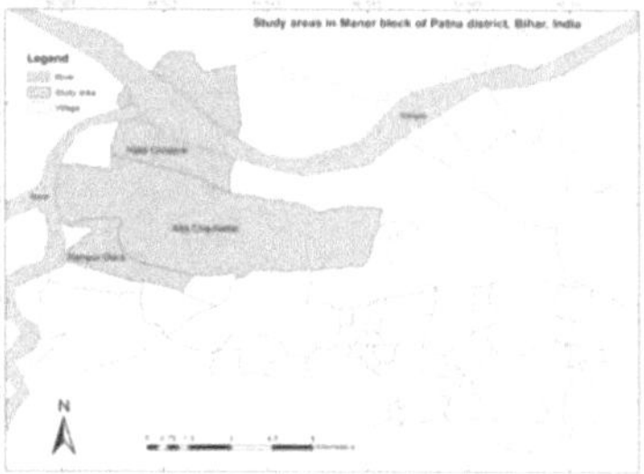

Figure 1. Map of the study area.

3 RESULTS AND DISCUSSION

It was found from the local drillers that deeper tubewells are being preferred at present in Bihar than the earlier days due to lesser concentrations of As and Fe at deeper depths. After thorough discussions and information by the local drillers, 10 shallow and deep aquifers were selected to test water samples. Initially, the water samples were tested using field test kit for four parameters: As, Fe, fluoride, and manganese (Mn). Later, all the water samples were retested using atomic absorption spectroscopy (AAS) technique in the lab and was found consistent with the field kit

DOI: 10.1201/9781003317395-190

results. The black color sediments are not used for drinking water and usually had higher As and Fe concentrations. According to the test results, the range of As concentration was found from less than 1 μg/L to a maximum of 300 μg/L. However, in the deeper tubewells (> 60 meters depth) Mn was present while the concentration of As, Fe, and Fluoride was found within the WHO permissible limit of 10 μg/L. While drilling tubewells, local drillers identified five different types of sediment colours: black; white; off-white (buff); brown; and red (Thakur *et al.*, 2019). Now a days, brown, off-white or red sediment layers are targeted to install tubewells instead of using white and off-white sediments, which was the case in the past. In order to test the applicability of the sediment colour concept in Bihar, three tubewell (2 shallow tubewells and 1 deep tubewell) installation process by local drillers was observed and sediments from different depths (12 meters; 24 meters; 33–36.5 meters; 42 meters; 45 meters; 60 meters; 70 meters; and 76 meters) were collected for comparison with that of the Bangladesh sediment colours (Hossain *et al.*, 2014; von Brömssen *et al.*, 2007; van Geen *et al.*, 2014). The sediments colours at different depths of shallow aquifers and deep aquifers is presented in Figure 2.

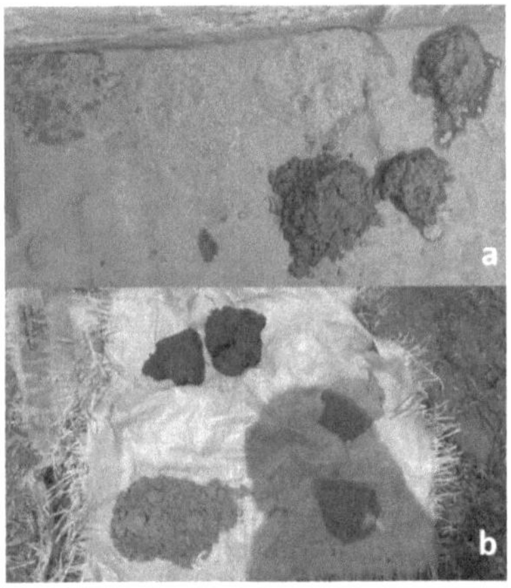

Figure 2. Colors of aquifer sediments identified by local drillers during tubewell drilling. Five different types of sediment colours were recovered at depths of: a) 24 m, 33.5 m, 36.5 m, 41 m, and 44 m); and b) 12 m, 33.5 m, 70 m, and 76 m.

4 CONCLUSIONS

Based on the perception of the local drillers from both Bangladesh and West Bengal, India and the present study, it can be concluded that sediment colour concept can also be applied in Bihar to target safe water depth despite having differences in terms of sediment characteristics (Hossain *et al.*, 2014; Saha *et al.*, 2018; von Brömssen *et al.*, 2007; van Geen *et al.*, 2014). However, the process requires further investigation both in terms of quality and quantity of the available water at targeted depths. Therefore, this study may be used as an indicator to target the safe aquifers while installing the tubewells by local drillers in the Gangetic belt.

REFERENCES

Annaduzzaman M., Bhattacharya P., Biswas A., Hossain M., Ahmed K.M. & van Halem D. 2018. Arsenic and manganese in shallow tubewells: validation of platform color as a screening tool in Bangladesh. *Groundw. Sust. Dev.* 6: 181–188.

Biswas A., Nath B., Bhattacharya P., Halder D., Kundu A.K., Mandal U., Mukherjee A., Chatterjee D., Mörth C.-M. & Jacks G. 2012. Hydrogeochemical contrast between brown and grey sand aquifers in shallow depth of Bengal Basin: consequences for sustainable drinking water supply. *Sci. Total Environ.* 431: 402–412.

Hossain M., Bhattacharya P., Frape S.K., Jacks G., Islam M.M., Rahman M.M., von Brömssen M., Hasan M.A., Ahmed K.M. 2014. Sediment color tool for targeting arsenic safe aquifers for the installation of shallow drinking water tubewells. *Sci. Total Environ.* 493: 615–625.

Jakariya M., Vahter M., Rahman M., Wahed M.A., Hore S.K., Bhattacharya P., Jacks G. & Persson L.A. 2007. Screening of arsenic in tubewells water with field test kits: evaluation of the method from public health perspective. *Sci. Total Environ.* 379: 167–175.

Mukherjee A., Sarkar S., Chakraborty M., Duttagupta S., Bhattacharya A., Saha D., Bhattacharya P., Mitra A. & Gupta S. 2021. Occurrence, predictors and hazards of elevated groundwater arsenic across India through field observations and regional-scale AI-based modeling. *Sci. Total Environ.* 759: 143511.

Saha D., Sinha U.K. & Dwivedi S.N. 2011. Characterization of recharge processes in shallow and deeper aquifers using isotopic signatures and geochemical behavior of groundwater in an arsenic-enriched part of the Ganga plain. *Appl. Geochem.* 26(4): 432–443.

Saha R., Dey N.C., Rahman S., Galagedera L. & Bhattacharya P. 2018. Exploring suitable sites for installing safe drinking water wells in coastal Bangladesh. *Groundw. Sust. Dev.* 7: 91–100.

Thakur B.K. & Gupta V. 2019. Valuing health damages due to groundwater arsenic contamination in Bihar, India. *Econ. Hum. Biol.* 35: 123–132.

Thakur B.K., Bhattacharya P. & Jakariya M. 2019. Sediment colour based tubewell installation process in Bihar, India. *Geol. Soc. Amer., Abstracts with Programs* 51(5): doi: 10.1130/abs/2019AM-335557.

Thakur B. K., Gupta V., Bhattacharya P., Jakariya M. & Islam M.T. 2020. Arsenic in drinking water sources in the middle gangetic plains in Bihar: an assess-ment of the depth of wells to ensure safe water supply. *Groundw. Sust. Dev.* 12: 100504.

van Geen A., Ahmed E.B., Pitcher L., Mey J.L., Ahsan H., Graziano J.H. & Ahmed K.M. 2014. Comparison of two blanket surveys of arsenic in tubewells conducted 12 years apart in a 25 km^2 area of Bangladesh. *Sci. Total Environ.* 488–489: 484–492.

von Brömssen M., Jakariya M., Bhattacharya P., Ahmed K.M., Hasan M.A., Sracek O., Jonsson L., Lundell L. & Jacks G. 2007. Targeting low-arsenic aquifers in Matlab Upazila, Southeastern Bangladesh. *Sci. Total Environ.* 379: 121–132.

 ISBN 978-1-032-32928-4

Phytoremediation strategies of arsenic-contaminated surface water near former mining pit 'Kolong' using arsenic hyperaccumulator *Pteris vittata* as mitigation measurement in Bangka Island, Indonesia

G. Widyastuti & G. Surya
Department of Environmental Health, Faculty of Public Health, Universitas Indonesia, Depok, Indonesia

ABSTRACT: Bangka Island, Indonesia has been acknowledged as the second world's largest tin producer. Some mining pits are still untreated by the stakeholders and it increases the risk of arsenic contamination to the society through clean and drinking water. In addition, the increasing contamination risk is related to the usage of former open mining pit pond and contaminated rivers as raw water by Indonesia's state-owned water utility company (PDAM). In this study, descriptive method has been used by holding literature review and secondary data analysis. The result indicates that *Pteris vittata* along riverside for arsenic phytoremediation can cut down any residual or hazardous chemical remain that is likely to enhance the risk of adverse health effects. To maximize the impact to the society, the cooperation between several parties are needed. This phytoremediation strategies are expected to be a sustainable measure in mitigating damages caused by arsenic.

1 INTRODUCTION

Indonesia is endowed with rich and abundant natural resources. This makes Indonesia the world's second-largest tin producer after China and the world's largest tin exporter. Located in Bangka Belitung Province, it is account for 90 percentage of the country's annual tin export (ICDX 2020).

Tin mining is thought to have negative impacts to 15 rivers in Bangka Belitung. Most of the people still rely on these rivers and former mining pit as their clean water sources. Indonesia's state-owned water utility company (PDAM) in Bangka Belitung which supposedly provides clean tap water supplies, still used former open mining pit pond as their raw water. Numbers of untreated mining pits give rise to the risk of environmental damage and heavy metal contamination which will be harmful to the environment and public health.

Arsenic (As) is one of heavy metals that possibly present in mine water. Chronic exposures to As causes various adverse health effects (Hajalilou *et al.*, 2011). The main objective of this study was to give mitigation measure by implementing phytoremediation to prevent As contamination of water.

2 MATERIALS AND METHODS

This paper uses a descriptive method by holding literature review and secondary data analysis. The secondary data is taken from national and international journals, government publication, and documents related to the studied case.

3 RESULTS AND DISCUSSION

3.1 *Health and environmental effect of arsenic*

Arsenic is known with its multiple effects in many aspects, especially on health and environmental aspect. Chronic exposure of As can cause many diseases in human. Through inhalation, As can generate oxidative stress, inflammation, and functional damage of lungs. Through ingestion, As can cause skin lesion, cancer, non alcoholic fatty liver disease (NAFLD), and cardiovascular diseases. Dermatitis also can be generated by As exposure through dermal absorption. Meanwhile, acute exposure of As will cause gastrointestinal effects, such as nausea and diarrhea, and neurological effects in central and peripheral nervous system (Lesmana *et al.*, 2017). In the other hand, As is also able to contaminate soil and water (ground water) that will potentially decrease the quality of water and soil. Furthermore, As that is contained in soil can be absorbed by plants and accumulated in it. Plants that contained the accumulated As are potentially consumed by human and increasing the health risk caused by As.

3.2 *Chinese brake fern (Pteris vittata)*

Pteris vittata, also known as Chinese brake or ladder fern, is a large fern genus belong to the family *Pteridaceae*. It originates from Asia is a quick-growing fern that is widely naturalized throughout the tropics and subtropics (Zhu *et al.*, 2014). *Pteris vittata* grows in alkaline habitat especially those that contaminated by As and copper (Ma & Zheng in Mumpuni 2016). It is also adapted to any variety of soils that makes it able to extract heavy metals

DOI: 10.1201/9781003317395-191

contained in soils or known with phytoremediation (Pritchard 2020). *Pteris vittata* is able to live in around 0–2.000 m above sea level and survive in extreme environment (Saputra *et al.*, 2011). It is usually found in roadsides, calcareous substrates, savanna ecosystem, mixed tropical forest ecosystem, and in mountains up to 1.800 m above sea level (Mumpuni 2016). In its native habitat it favors open sites on limestone, but in urban environment it grows well on walls and in concrete. These facts support that *Pteris vittata* is feasible to be used as a phytoremediation in Indonesia due to its environmental condition.

3.3 *Phytoremediation strategies*

Phytoremediation can be defined as the use of plants and associated microorganisms in order to remove, degrade, or isolate toxic substances from the environment. Phytoremediation has various techniques, depending on the chemical nature, properties of the contaminant, and the characteristic of the plants. Generally, phytoremediation comprises of six strategies (phytodegradation, phytostabilization, phytovolatilization, phytoextraction, phytofiltration, and rhizodegradation). Nevertheless, it is very likely to be used by the plant with more than one strategy simultaneously (Favas 2014). Phyextraction is considered as the most effective and suitable for As contaminated soils (Gonzaga 2006).

3.4 *Arsenic hyperaccumulator Pteris vittata implementation*

Specifically, the using of *Pteris vittata* as a phytoremediation is implemented in rivers close to former mining pit (locally called as '*kolong*'). In this case, the implementation is concentrated in Kurau River, Baturusa River, and Selindung River in Bangka Island. Those rivers are the closest to the biggest *kolong* and also close to the residential areas. Regarding to the feasibility, *Pteris vittata* were planted in riverside parts close to the residential area (at least 10 m). *Pteris v.* in the area have to be replanted every 5 years maximum, due to its decreasing capability to accumulate As within that period of time (Sussmilch *et al.*, 2017). In its implementation, the cooperation between government, mining enterprises, environmentalists, and the society needs to be done.

3.5 *Impact expected*

By using *Pteris vittata,* As level in soil and water is decreased, especially in the environment around mining area. Besides, by implementing this biological control, there are no any residual or hazardous chemical remain left to the environment that can potentially give any side effects. Besides, phytoremediation using *Pteris vittata* is a sustainable measure in mitigating damages caused by As.

4 CONCLUSIONS AND RECOMMENDATIONS

Bangka Island, Indonesia has been acknowledged as the second world's largest tin producer. Some mining pits are still untreated by the stakeholders and it increases the risk of As contamination to the society through clean and drinking water. In addition, the increasing contamination risk is related to the usage of former open mining pit pond and contaminated rivers as raw water by Indonesia's state-owned water utility company (PDAM). By utilizing *Pteris vittata* along riverside for As phytoremediation can cut down any residual or hazardous chemical remain that is likely to enhance the risk of adverse health effects. To maximize the impact, the cooperation between several parties are needed.

REFERENCES

Favas P.J.C., Pratas J., Varun M., D'Souza R. & Paul M.S. 2014. Phytoremediation of soils contaminated with metals and metalloids at mining areas: potential of native flora. In: M.C. Hernandez-Soriano (ed.) *Environmental Risk Assessment of Soil Contamination.* InTech. doi:10.5772/57469.

Gonzaga M. 2006. *Effects of Soil and Plant on Arsenic Accumulation by Arsenic Hyperaccumulator Pteris vittata.* University of Florida.

Hajalilou B., Mosaferi M., Khaleghi F., Jadidi S., Vosugh. B, & Fatehifar E. 2011. Effects of abandoned arsenic mine on water resources pollution in north west of Iran. *Health Promot. Perspect.* 1(1): 62–70.

ICDX. 2020. Tin [Online]. Cited: Feb 28, 2020. Available at: https://www.icdx.co.id/products-indonesia-commodity-and-derivatives-exchange/tin

Lesmana S.O., Kusnoputranto H. & Wulandari R.A. 2017. Hubungan kadar metaloid arsen (As) pada air minum dengan kejadian lesi kulit di pulau obi propinsi Maluku Utara. *Jurnal Kesmas Jambi* 1(1): 61–69.

Mumpuni Mugi. (2016). Variasi morfologi *Pteris vittata* L. (pteridaceae; pteridophyta) dan Korelasinya dengan Ketinggian Lokasi tempat Tumbuhnya di Jawa. *Biolink* 3 (1): 100–110.

Pritchard K. 2020. Plant 168: *Pteris vittata L. (Pteridaceae)* Chinese Brake. BRAHMS Online [Available online] https://herbaria.plants.ox.ac.uk/bol/plants400/Profiles/OP/Pteris. [Accessed on Feb 25, 2020].

Puspitasari D.S., Chikmawati T. & Praptosuwiryo T.N. 2015. Gametophyte morphology and development of six species of pteris (pteridaceae) from Java Island Indonesia. *J. Trop. Life Sci.* 5(2): 98–104.

Saputra Qotrunnada. (2011). Study of pteridophytes diversity and vegetation analysis in Jatikerep, Legonlele And Nyamplung, Karimunjawa Island, Central Java. *J. Bio. Ind.* 7(2): 207–212.

Sussmilch F.C., Atallah N.M., Brodribb T.J., Banks J.A. & McAdam S. 2017. Abscisic acid (ABA) and key proteins in its perception and signaling pathways are ancient, but their roles have changed through time. *Plant Signal. Behav.* 12(9): e1365210.

Zhu L.J., Guan D.-X., Luo J., Ratinsabapathi B. & Ma L.Q. 2014. Characterization of arsenic-resistant endophytic bacteria from hyperaccumulators Pteris vittata and Pteris multifida. *Chemosphere* 113: 9–16.

Arsenic in the Environment: Bridging Science to Practice for Sustainable Development – van der Wal, Ahmad, Petrusevski, Weijma, Savic, van der Wens, Beerendonk, Bhattacharya, Bundschuh & Naidu (Eds)
© 2023 the Editor(s), ISBN 978-1-032-32928-4

Author index